A Textbook of BUILDING CONSTRUCTION

(For students of B.E./B.Tech. & M.E./M.Tech. Civil Engineering)

Dr. S.K. Sharma
B.Tech. (*Hons.*) (*Civil Engg.*)
M.Sc. (*Highway Engg.*),
Ph.D. (*Civil Engg. F.I.E.*)
Former Professor and Head of Civil Engineering,
PEC University of Technology,
Chandigarh

S Chand And Company Limited
(ISO 9001 Certified Company)

S Chand And Company Limited

(ISO 9001 Certified Company)

Head Office: Block B-1, House No. D-1, Ground Floor, Mohan Co-operative Industrial Estate, New Delhi – 110 044 | Phone: 011-66672000

Registered Office: A-27, 2nd Floor, Mohan Co-operative Industrial Estate, New Delhi – 110 044 Phone: 011-49731800

www.**schandpublishing.com**; e-mail: **info@schandpublishing.com**

Branches

Chennai	:	Ph: 23632120; chennai@schandpublishing.com
Guwahati	:	Ph: 2738811, 2735640; guwahati@schandpublishing.com
Hyderabad	:	Ph: 40186018; hyderabad@schandpublishing.com
Jalandhar	:	Ph: 4645630; jalandhar@schandpublishing.com
Kolkata	:	Ph: 23357458, 23353914; kolkata@schandpublishing.com
Lucknow	:	Ph: 4003633; lucknow@schandpublishing.com
Mumbai	:	Ph: 25000297; mumbai@schandpublishing.com
Patna	:	Ph: 2260011; patna@schandpublishing.com

First Edition 1960
Subsequent Editions and Reprints 1964, 1965, 1966, 1969, 1976 (Twice), 1980, 1981, 82, 83, 84, 85, 86, 1988, 1990
Revised and Enlarged Edition 2013
Reprints 2014 (Twice), 2016, 2017, 2018, 2019
Reprint 2020

ISBN : 978-81-219-0479-7 **Product Code :** H3BUI46CIVE10ENZX13O

PRINTED IN INDIA

By Vikas Publishing House Private Limited, Plot 20/4, Site-IV, Industrial Area Sahibabad, Ghaziabad – 201 010 and Published by S Chand And Company Limited, A-27, 2nd Floor, Mohan Co-operative Industrial Estate, New Delhi – 110 044.

PREFACE TO THE REVISED EDITION

The author is very much beholden to his readers for the constant appreciation of the utility of the book, critical review and constructive suggestions for the improvement of the text.

The present edition has been thoroughly revised and enlarged to include the latest information covered in various Codes of practice in vogue and the developments in the field of building construction. The provisions of the National Building Code of India 2005, brought out by the Bureau of Indian Standards, New Delhi have been incorporated. An introductory chapter dealing with the classification of buildings, their orientation, parts of buildings and special requirements of buildings has been added to acquaint the reader with the various aspects of building construction. To aid in the structural design of buildings, a chapter on desion loads has been included. A chapter dealing with lifts and escalators, considered to be necessary adjuncts for high rise buildings, has also been incorporated. Precast and prefabricated construction is becoming very popular these days for accelerated construction. This aspect has been covered in chapter 20. Information on special treatments for buildings with regard to thermal insulation, ventilation, air-conditioning, acoustics, sound insulation and noise control, fire and earthquake protection has been updated and dealt with in separate chapters.

To imporve further the scope of the book from examination point of view, numerical problems on design of foundation, design of concrete mixes and lay-out of staircases have been solved.

We extend our grateful thanks to the entire management and editorial staff of S. Chand and Co. Pvt. Ltd., New Delhi who took great pains to get the book published in such a nice form.

It is fervently hoped that the book, in its present form, will prove to be more useful to both the students and the field engineers. The author shall look forward for suggestions to improve the text further.

Dr. S.K. Sharma

PREFACE TO THE FIRST EDITION

The auther claims no originality in computing this text-book on Building Construction except for the treatment of the subject matter. It was felt that there existed not a single book on this subject which dealt comprehensively with the Syllabi of all Indian universities, technical boards and competitive examinations. This book has been written keeping in view all these courses.

One of the special features of this book is the incorporation of the latest information based on research work conducted in various countries on allied topics. The outcome of research organisations in India, namely, the National Building Organisation, New Delhi, and the Central Building Research Institute, Roorkee and that of other countires, such as research papers published by Her Majesty's Stationery Office, London, etc., has been included in this book.

In the field of Building Construction, concrete and steel constructions are gaining popularity these days. Most of the existing books have laid very little emphasis on these topics. In this text care has been taken to deal with the fundamental aspects of these constructions, their design criteria and construction details in the easiest possible manner. Chapters on Air-conditioning and Ventilation, Thermal Insulation, Fire-protection, Acoustics and Sound Insulation, etc., have also been incorporated to bring this publication up-to-date.

The book is illustrated with numerous sketches all of which have been very carefully drawn to make them self-explanatory. While making these sketches, building construction practices followed in India have been kept in view.

The book not only contains syllabi of different examinations but also deals with the constructional procedures and difficulties with which engineers in general are faced. It is, therefore, hoped that the book will prove very useful to the students and the field engineers in particular.

I am grateful to the various authors whose works have been referred to in this text. I am also indebted to my colleagues for their kind assistance and collaboration. My sincere thanks are due to Messrs. S. Chand & Co. who have taken great pains in bringing out this book in such a nice form.

Suggestions for improvement will be highly appreciated.

Chandigarh *S. K. Sharma*
September 4, 1960

CONTENTS

DETAILED CONTENTS

Introduction

CLASSIFICATION OF BUILDINGS

Buildings are classified based on the character of occupancy or according to the use and type of construction.

1.1 OCCUPANCY CLASSIFICATION

All buildings can be classified in one of the following groups:

Group A : Residential
Group B : Educational
Group C : Institutional
Group D : Assembly
Group E : Business
Group F : Mercantile (includes both retail and wholesale stores)
Group G : Industrial (includes low, moderate and high fire hazards)
Group H : Storage
Group I : Hazardous

Buildings under various groups are further subdivided as given in Table 1.1. A brief description of the various classes of buildings follows:

1.1.1 Group A : Residential Buildings

These include any building in which sleeping accommodation is provided for normal residential purposes, with or without cooking or dining or both facilities, except any building classified under Group C.

(*a*) ***Subdivision A-1. Lodging or rooming houses.*** These include any building or group of buildings under the same management, in which separate sleeping accommodation for a total of not more than 40 persons (beds) on transient or permanent basis, with or without dining facilities, but without cooking facilities for individuals, is provided. This includes inns, clubs, motels and guest houses.

A lodging or rooming house shall be classified as a dwelling in Subdivision A-2 if no room in any of its private dwelling units is rented to more than three persons.

Table 1.1: Occupancy classification: Buildings

Group A	Group B	Group C	Group D	Group E	Group F	Group G	Group H	Group I
Residential	*Educational*	*Institutional*	*Assembly*	*Business*	*Mercantile*	*Industrial*	*Storage*	*Hazardous*
A-1 : Lodging or rooming houses	B-1 : Schools upto secondary level	C-1 : Hospitals and sanatoria	D-1 : With a theatrical or motion picture or any other stage and fixed seats for more than 1000 persons	E-1 : Offices, banks, professional establishments like offices of architects, doctors, lawyers, and police stations	F-1 : Shops, stores, departmental stores, markets with area up to 500 m^2	G-1 : Buildings for low hazard industries		
A-2 : One or two family private dwellings	B-2 : All others / training institutions	C-2 : Custodial institutions	D-2 : With theatrical or motion picture or any other stage and fixed seats upto 1000 persons	E-2 : Laboratories, research establishment, libraries and test houses	F-2 : Shops, stores, departmental stores, markets with area more than 500 m^2	G-2 : Buildings for moderate hazard industries		
A-3 : Dormitories		C-3 : Penal and mental institutions	D-3 : Without a permanent stage for 300 or more persons but no permanent seating arrangements	E-3 : Computer installation	F-3 : Underground shopping centres, storage and service facilities incidental to the sale of merchandise and located in the same building	G-3 : Buildings for high hazard industries		

(Contd.)

Group A	Group B	Group C	Group D	Group E	Group F	Group G	Group H	Group I
Residential	*Educational*	*Institutional*	*Assembly*	*Business*	*Mercantile*	*Industrial*	*Storage*	*Hazardous*
A-4 : Apartment houses (flats)			D-4 : Without a permanent stage for less than 300 persons	E-4 : Telephone exchange but no permanent seating arrangements				
A-5 : Hotels			D-5 : All other structures including temporary structures D-1 to D-4	E-5 : Broadcasting stations and T.V. stations.				
A-6 : Hotels (starred)			D-6 : Building having mixed occupancies such as shopping, cinema theatres and restaurants					
			D-7 : All other elevated or underground structures not covered under D1 to D6					

(*b*) ***Subdivision A-2. One-or two-family private dwellings.*** These include any private dwelling which is occupied by members of one or two families and has a total sleeping accommodation for not more than 20 persons.

If rooms in a private dwelling are rented to outsiders, these shall be for accommodating not more than three persons per room.

If sleeping accommodation for more than 20 persons is provided in any one residential building, it shall be classified as a building in Subdivision A-1, A-3 or A-4 as the case may be.

(*c*) ***Subdivision A-3. Dormitories.*** These include any building in which group sleeping accommodation is provided, with or without dining facilities, for persons who are not members of the same family, in one room or a series of closely associated rooms under joint occupancy and single management, for example, school and college dormitories, student's hostels and military barracks.

(*d*) ***Subdivision A-4. Apartment houses (flats).*** These include any building or structure in which living quarters are provided for three or more families, living independently of each other and with independent cooking facilities, for example, apartment houses, mansions and *chawls*.

(*e*) ***Subdivision A-5. Hotels.*** These include any building or group of buildings under single management, in which sleeping accommodation is provided with or without dining facilities, for hotels classified up to 4 star category.

(*f*) ***Subdivision A-6. Hotels (starred).*** These include the hotels duly approved by the concerned authorities as Five Star and above Hotels.

1.1.2 Group B : Educational Buildings

These include any building used for school, college other training institutions for day-care purposes involving assembly for instruction, education or recreation for not less than 20 students.

(*a*) ***Subdivision B-1. (Schools upto Senior Secondary level).*** This includes any building or group of buildings under single management for students not less than 20 in number.

(*b*) ***Subdivision B-2. (All others/training institutions).*** This includes any building or group of buildings under single management used for students not less than 100 in number.

1.1.3 Group C : Institutional Buildings

These include any building or part thereof, which is used for purposes, such as medical or other treatment or care of persons suffering from physical or mental illness, disease or infirmity; care of infants, convalescents or aged persons and for penal or correctional detention in which the liberty of the inmates is restricted. Institutional buildings ordinarily provide sleeping accommodation for the occupants.

(*a*) ***Subdivision C-1. (Hospitals and sanatoria).*** This subdivision includes any building or a group of buildings under single management, which is used for housing persons suffering from physical limitation because of health or age, for example, hospitals, infirmaries, sanatoria and nursing homes.

(*b*) ***Subdivision C-2. (Custodial institutions).*** This subdivision includes any building or a group of buildings under single management, which is used for the custody and care of persons, such as children, convalescents and the aged, for example, homes for the aged and infirm, convalescent homes and orphanages.

(*c*) ***Subdivision C-3. (Penal and mental institutions).*** This subdivision includes any building or a group of buildings under single management, which is used for housing persons under restraint, or who are detained for penal or corrective purposes, in which the liberty of the inmates is restricted, for example, jails, prisons, mental hospitals, mental sanatoria and reformatories.

1.1.4 Group D: Assembly Buildings

These include any building or part of a building, where groups of people congregate or gather for amusement, recreation, social, religious, patriotic, civil, travel and similar purposes, for example, theatres, motion picture houses, assembly halls, auditoria, exhibition halls, museums, skating rinks, gymnasiums, restaurants, places of worship, dance halls, club rooms, passenger stations and terminals of air, surface and marine public transportation services, recreation piers and stadia, etc.

(*a*) ***Subdivision D-1.*** This subdivision includes any building primarily meant for theatrical or operatic performances and exhibitions and which has a raised stage, proscenium curtain, fixed or portable scenery or scenery loft, lights, motion picture booth, mechanical appliances or other theatrical accessories and equipment and which is provided with fixed seats for over 1000 persons.

(*b*) ***Subdivision D-2.*** This subdivision includes any building primarily meant for use as described for Subdivision D-1, but with fixed seats up to 1000 persons.

(*c*) ***Subdivision D-3.*** This subdivision includes any building, its lobbies, rooms and other spaces connected thereto, primarily intended for assembly of people, but which has no theatrical stage or theatrical and/or cinematographic accessories and has accommodation for 300 persons or more for example, dance halls, night clubs, halls for incidental picture shows, dramatic, theatrical or educational presentation, lectures or other similar purposes, having no theatrical stage except a raised platform and used without permanent seating arrangement; art galleries, exhibition halls, community halls, marriage halls, places of worship, museums, lecture halls, passenger terminals; and Heritage and Archeological Monuments.

(*d*) ***Subdivision D-4.*** This subdivision includes any building, primarily intended for use as described in Subdivision D-3, but with accommodation for less than 300 persons with no permanent seating arrangement.

(*e*) ***Subdivision D-5.*** This subdivision includes any building or structure permanent or temporary meant for assembly of people not covered by Subdivisions D-1 to D-4, for example, grandstands, stadia amusement park structures, reviewing stands and circus tents.

(*f*) ***Subdivision D-6.*** This includes any building for assembly of people provided with multiple services/facilities like shopping, cinema theatres and restaurants, for example, multiplexes.

(*g*) ***Subdivision D-7.*** Any building or structure permanent or temporary meant for assembly of people not covered by D-1 to D-6, for example, underground or elevated railways.

1.1.5 Group E : Business Buildings

These include any building or part of a building which is used for transaction of business (other than covered by Group F) for keeping of accounts and records and similar purposes, professional establishments, service facilities, etc. City halls, town halls, court houses and libraries are classified in this group so far as the principal function of these is transaction of public business and keeping of books and records.

Business Building are further subdivided as follows:

(*a*) ***Subdivision E-1.*** Offices, banks, professional establishments, like offices of architects, engineers, doctors, lawyers and police stations.

(*b*) ***Subdivision E-2.*** Laboratories, research establishments, libraries and test houses.

(*c*) ***Subdivision E-3.*** Computer installations.

(*d*) ***Subdivision E-4.*** Telephone exchanges.

(*e*) ***Subdivision E-5.*** Broadcasting stations and T.V. stations.

1.1.6 Group F : Mercantile Buildings

These include any building or part of a building, which is used as shops, stores, market, for display and sale of merchandise, either wholesale or retail.

Mercantile buildings shall be further subclassified as follows:

Subdivision F-1 : Shops, stores, departmental stores, markets with area up to 500 m^2.

Subdivision F-2 : Shops, centres, departmental stores, markets with area more than 500 m^2.

Subdivision F-3 : Underground shopping centres.

Storage and service facilities incidental to the sale of merchandise and located in the same building shall be included under this group.

1.1.7 Group G : Industrial Buildings

These include any building or part of a building or structure, in which products or materials of all kinds and properties are fabricated, assembled, manufactured or processed, for example, assembly plants, laboratories, dry cleaning plants, power plants, pumping stations, smoke houses, laundries, gas plants, refineries, dairies and saw-mills, etc.

The hazard of occupancy shall be the relative danger of the start and spread of fire, the danger of smoke or gases generated, the danger of explosion or other occurrence potentially endangering the lives and safety of the occupants of the buildings.

(*a*) ***Subdivision G-1. (Buildings used for low hazard industries)***: This subdivision includes any building in which the contents are of such low combustibility and the industrial processes or operations conducted therein are of such a nature that there are no possibilities for any self-propagating fire to occur and the only consequent danger to life and property may arise from panic, fumes or smoke, or fire from some external source.

(*b*) ***Subdivision G-2. (Buildings used for moderate hazard industries).***: This subdivision includes any building in which the contents or industrial processes of operations conducted therein are liable to give rise to a fire which will burn with moderate rapidity and give off a considerable volume of smoke, but from which neither toxic fumes nor explosions are to be feared in the event of a fire.

(*c*) ***Subdivision G-3. (Buildings used for high hazard industries)***: This subdivision includes any building in which the contents or industrial processes or operations conducted therein are liable to give rise to a fire which will burn with extreme rapidity or from which poisonous fumes or explosions are to be feared in the event of a fire.

1.1.8 Group H : Storage Buildings

These include any building or part of a building used primarily for the storage or sheltering (including servicing, processing or repairs incidental to storage) of goods, wares or merchandise (except those that involve highly combustible or explosive products or materials), vehicles or animals, for example, warehouses, cold storages, freight depots, transit sheds, storehouses, truck and marine terminals, garages, hangars (other than aircraft repair hangars), grain elevators, barns and stables.

Storage properties are characterized by the presence of relatively small number of persons in proportion to the area. Any new use which increases the number of occupants to a figure comparable with other classes of occupancy shall change the classification of the building to that of the new use, for example, hangars used for assembly purposes, warehouses used for office purposes, garage buildings used for manufacturing.

1.1.9 Group I : Hazardous Buildings

These include any building or part of a building which is used for the storage, handling, manufacture or processing of highly combustible or explosive materials or products which are liable to burn with extreme rapidity and/or which may produce poisonous fumes or explosions; for storage, handling, manufacturing or processing which involve highly corrosive, toxic or noxious alkalies, acids or other liquids or chemicals producing flame, fumes and explosive, poisonous, irritant or corrosive gases; and for the storage, handling or processing of any material producing explosive mixtures of dust which result in the division of matter into fine particles subject to spontaneous ignition. Examples of buildings of this class are those buildings which are used for:

(*a*) Storage, under pressure of more than 0.1 N/mm^2 and in quantities exceeding 70 m^3, of acetylene, hydrogen, illuminating and natural gases, ammonia, chlorine, phosgene, sulphur dioxide, carbon dioxide, methyloxide and all gases subject to explosion, fume or toxic hazard, cryogenic gases, etc.;

(*b*) Storage and handling of hazardous and highly flammable liquids, liquefiable gases like LPG, rocket propellants, etc;

(*c*) Storage and handling of hazardous and highly flammable of explosive materials, (other than liquids); and

(*d*) Manufacture of artificial flowers, synthetic leather, ammunition, explosives and fireworks.

1.2 CLASSIFICATION BASED ON TYPE OF CONSTRUCTION

This classification is based on the fire resistance offered by a building. Buildings are classified into four categories, as given below:

(*a*) Type 1

(*b*) Type 2

(*c*) Type 3

(*d*) Type 4

For details refer to chapter 29.

The main problem of the hot and arid zone is to keep out the day time heat or the solar radiation. The most effective method to tackle the problem is the correct orientation. Another method is shading the openings with *Chajjas*. Past methods of providing thick walls with small openings, more height and deep verandahs will prove ineffective and uneconomical in the present day circumstances.

The walls and roofs should have adequate thermal capacity and time-lag. Sunlit surfaces should be reflective and/or be heavily shaded. Air movement should be restricted to the minimum required for ventilation. But fresh air is very useful during the evening to counteract the heat gained by the structure.

Summer nights are very uncomfortable in hot and arid zone. Interiors are unbearable as surrounding walls and roof radiate thermal heat stored by them during the day. Terraces, open verandahs, staggered sleeping out balconies are necessary for comfortable sleep at night.

The main problem of hot/warm and humid zone is to provide maximum of ventilation, to counteract discomfort of humidity. The best solution is to make maximum use of prevailing winds.

Screens and *Jallies* are more effective as they not only prevent glare but also allow breeze to pass through them.

An effort should be made to maintain indoor temperatures, as near as possible, to the ambient or shade temperature. Walls and roofs should be of light weight construction having low thermal capacity. The external surfaces should absorb as little solar radiation as possible and the internal surface should emit minimum of long wave radiation. Air movement should be speedy up to 50 to 100 cm/s, which may have to be obtained artificially if prevailing winds do not help to do so.

The problems of the cold zone are that during summer if the sun's radiation is excessive, there should be some shading during day but at night both in summer and winter cold air should be shut out. During day, in winter, sunshine should be welcome.

In the cold zone, especially in hills, walls and roofs should be heavy weight with high thermal capacity for absorbing and storing heat during day and using the stored heat to warm the inside of the building during the night. If heavy weight roofs are not feasible, they should be insulated against loss of heat of the interior during night. Air movement and ventilation should be restricted to the minimum. Regular flues or devices should be provided to avoid concentration of carbon monoxide, when open coal fires are used for heating in enclosed rooms, in winter, especially in hill town.

ORIENTATION OF BUILDINGS

Orientation of buildings refers to the direction of the normal to the long axis. For example, if the length of the building is east-west, its orientation is north-south.

The chief aim of orientation of buildings is to provide physically and psychologically comfortable living inside the building by creating conditions which suitably and successfully ward off the undesirable effects of severe weather to a considerable extent by judicious use of the recommendations and knowledge of climatic factors.

1.3 BASIC ZONES AND THEIR PROBLEMS

For the purpose of orientation, it will be convenient to divide the country into three broad climatic zones:

(*a*) hot and arid,

(*b*) hot/warm and humid, and

(*c*) cold.

It is to be remembered that there may not be uniform climatic factors in a particular zone. They might even vary during day and night in the same zonal region. Each zone, all the same, poses certain basic problems, which are to be solved.

1.4 FACTORS AFFECTING ORIENTATION OF BUILDINGS

1.4.1 Climatic Factors

From the point of view of lighting and ventilation, the following climatic factors influence the optimum orientation of the building:

(*a*) Solar radiation and temperature,

(*b*) Clouds,

(*c*) Relative humidity, and

(*d*) Prevailing winds.

1.4.2 Solar Radiation and Temperature

The best orientation from solar point of view requires that the building as a whole should receive the maximum solar radiation in winter and the minimum in summer. For practical evaluation, it is necessary to know the duration of sunshine, and hourly solar intensity on the various external surfaces on representative days of the seasons. The total direct diurnal solar loads per unit area on vertical surface facing different directions are given in Table 1.2 for two days in the year, that is, 16 May and 22 December, representative of summer and winter for latitudes corresponding to some important cities all over India. From Table 1.2, the total heat intake can be calculated for all possible orientation of the building for these extreme days of summer and winter.

Wherever possible, suitable sunbreakers have to be provided to cut off the incursion of direct sunlight to prevent heat radiation and to avoid glare.

In order to ascertain good and bad aspects and to decide whether or not to take advantage of the sun's rays, day temperatures of the region, for which orientation is to be decided, should be studied in relation to the following broad classification of temperature ranges:

Below 15°C	Cold	Sun's rays advantageous
15 to 20°C	Cool	
20 to 30°C	Temperate	
30° to 15°C	Hot	Protection from sun's rays advantageous
Above 35°C	Very hot	

Table 1.2: Daily total direct solar radiation on vertical surfaces in g. cal/cm² day for two representative days

	8°N		13°N		19°N		23°N		29°N	
	16 May	22 Dec	16 May	22 Dec	16 May	22 Dec	16 May	22 Dec	16 May	22 Dec
North	187	—	140	—	83	—	64	—	46	—
North East	228	35	214	27	194	20	188	15	180	9
East	225	187	232	173	240	157	247	146	253	126
South East	100	291	115	294	141	295	158	297	188	281
South	—	358	—	377	—	393	18	398	64	390
South West	100	291	115	294	141	295	158	297	188	281
West	225	187	232	173	240	157	247	146	253	126
North West	228	35	214	27	194	20	188	15	180	9

1.4.3 Clouds

The clouds reduce not only the direct radiation from sun but also make sun protection devices of little advantage. It is, therefore, desirable to take note of cloudy periods of the year and if they are long enough and also coincide with hot periods, then the ideas of sun protection should be given up, although high day temperatures may demand such protection.

1.4.4 Relative Humidity and Prevailing Winds

The discomfort due to high relative humidity in air when temperatures are also high can be counteracted, to a great extent, by circulation of air with electric fans or by ventilation. In the past simultaneously with heavy construction and surrounding verandahs to counter the effect of sun's radiation, there was also an overemphasis on prevailing winds to minimize the adverse effects of high humidity with high temperatures. With the introduction of electric fan to effectively circulate air and owing to taking into account the rise in cost of construction of buildings, it would perhaps be better to shift the emphasis on protection from solar radiation where temperatures are very high. When, however, there is less diurnal variation between morning and mean maximum temperatures along with high humidity, as in coastal areas, the emphasis should be on prevailing winds.

For the purpose of orientation, it is necessary to study the velocity and direction of the wind at each hour and in each month instead of relying on generalization of a month or a period or for the year as a whole. This helps to spot the right winds for a particular period of day or night.

It is generally found that variation up to 30° with respect to the prevalent wind direction does not materially affect indoor ventilation (average indoor air velocity) inside the building.

A comparative study of relative humidity can be made under the following categories:

0—25 Per cent	Very dry
25—50 Per cent	Dry
50—75 Per cent	Humid
75—100 Per cent	Very humid

When relative humidity is of the category of dry and very dry, advantage can be taken of evaporative cooling in summers to cool the air before introducing it into the building. This, however, raises the relative humidity to some extent. But when the atmosphere is already humid or very humid, it is desirable either to regulate the rate of air movement with the aid of electric fans or to take advantage of prevailing winds.

1.4.5 Aspects of Daylighting

Since the clear design sky concept for daylighting takes care of the worst possible situation, orientation is not a major problem for daylighting in multistoreyed buildings, except that direct sunshine and glare should be avoided. However, due allowance should be given to the mutual shading effects of opposite facades.

1.4.6 Planting of Trees

Planting of trees in streets and in open spaces should be done carefully to take advantage of both shades and sunshine without reducing the flow of natural winds. Their advantage in abating glare and in providing cool and/or warm pockets in developed areas should also be taken. Some trees shed leaves in winter while retaining thick foliage in summer. Such trees will be very advantageous, particularly where southern and western exposures are concerned, by allowing maximum sun during winter and effectively blocking in summer.

PARTS OF BUILDINGS AND THEIR REQUIREMENTS

1.5 PLINTH

Plinth is the portion of a structure between the surface of the surrounding ground and surface of the floor, immediately above the ground.

The plinth or any part of a building or outhouse should be so located with respect to the surrounding ground level that adequate drainage of the site is assured. The height of the plinth should not be less than 45 cm from the surrounding ground level.

Every interior courtyard shall be raised at least 150 mm above the level of the nearest street and shall be satisfactorily drained.

1.6 HABITABLE ROOMS

Height: The height of all rooms for human habitation should not be less than 2.75 m measured from the surface of the floor to the lowest point of the ceiling (bottom of slab). In the case of pitched roof, the average height of rooms should not be less than 2.75 m. The minimum clear head room under a beam folded plates or eaves shall be 2.4 m. In the case of air-conditioned rooms, a height of not less than 2.4 m measured from the surface of the floor to the lowest point of airconditioning duct or the false ceiling should be provided.

The above requirements apply to residential, business and mercantile buildings. For educational and industrial buildings, the following minimum requirements apply:

(*a*) Educational buildings: Ceiling height 3.6 m for all regions; in cold regions, 3 m.

(*b*) Industrial buildings: Ceiling height 3.6 m except when air-conditioned, 3 m (Factory Act 1948 and rules therein shall govern such heights, where applicable).

Size: The area of habitable room should not be less than 9.5 m^2, where there is only one room with a minimum width of 2.4 m. Where there are two rooms, one of these should not be less than 9.5 m^2 and the other not less than 7.5 m^2, with a minimum width of 2.1 m.

1.7 KITCHEN

Height: The height of a kitchen measured from the surface of the floor to the lowest point in the ceiling (bottom slab) shall not be less than 2.75 m, except for the portion to accommodate floor trap of the upper floor.

Size: The area of a kitchen, where separate dining area is provided, shall not be less than 5.0 m^2 with a minimum width of 1.8 m. Where there is a separate store, the area of the kitchen may be reduced to 4.5 m^2. A kitchen, which is intended for use as a dining area also, shall have a floor area of not less than 7.5 m^2 with a minimum width of 2.1 m.

Other Requirements: Every room to be used as kitchen shall have:

(*a*) unless separately provided in a pantry, means for the washing of kitchen utensils which shall lead directly or through a sink to a grated and trapped connection to the waste pipe;

(*b*) an impermeable floor;

(*c*) a flue, if found necessary; and

(*d*) a window or ventilator or opening.

1.8 BATHROOMS AND WATER CLOSETS

Height: The height of a bathroom or water closet measured from the surface of the floor to the lowest point in the ceiling (bottom of slab) shall not be less than 2.1 m.

Size: The size of a bathroom shall not be less than 1.8 m^2 with a minimum width of 1.2 m. The floor area of water closet shall be 1.1 m^2 with a minimum width of 0.9 m. If bath and water closet are combined, its floor area shall not be less than 2.8 m^2 with a minimum width of 1.2 m.

Other Requirements: Every bathroom or water closet shall:

(*a*) be so situated that at least one of its walls shall open to external air;

(*b*) not be directly over or under any room other than another water closet, washing place, bath or terrace, unless it has a water-tight floor;

(*c*) have the platform or seat made of water-tight non-absorbent material;

(*d*) be enclosed by walls or partitions and the surface of every such wall or partition shall be finished with a smooth impervious material to a height of not less than 1 m above the floor of such a room;

(*e*) be provided with an impervious floor covering, sloping towards the drain with a suitable grade and not towards verandah or any other room; and

(*f*) have a window or ventilator, opening to a shaft or open space, of area not less than 0.3 m^2 with side not less than 0.3 m.

No room containing water closets shall be used for any purpose except as a lavatory and no such room shall open directly into any kitchen or cooking space by a door, window or other opening. Every room containing water closet shall have a door completely closing the entrance to it.

1.9 LEDGE OR TAND/LOFT

A ledge or tand is a shelf-like projection, supported in any manner, whatsoever, except by means of vertical supports within a room itself but not having projection wider than one metre.

A loft is an intermediate floor between two floors on a residual space in a pitched roof, above normal floor level with a maximum height of 1.5 m and which is constructed or adopted for storage purposes.

Height: It shall have a minimum head-room of 2.2 m.

Size: A ledge or tand in a habitable room shall not cover more than 25 per cent of the area of the floor on which it is constructed and shall not interfere with the ventilation of the room under any circumstances.

Loft: The maximum height of loft shall be 1.5 m. A loft, if provided, on a kitchen, shall not exceed 25 per cent of the area of the kitchen, leaving a minimum headroom of 2.2 m for kitchen under loft. On bathroom, water closet and corridor, the loft can be 100 per cent.

1.10 MEZZANINE FLOOR

It is an intermediate floor, between two floors, above ground level, accessible only from the lower floor.

Height: It shall have a minimum height of 2.2 m.

Size: The minimum size of the mezzanine floor, if it is to be used as a living room, shall not be less than 9.5 m^2. The aggregate area of such mezzanine floor in a building shall in no case exceed one-third the plinth area of the building.

Other Requirements: A mezzanine floor may be permitted over a room or a compartment provided:

(*a*) it conforms to the standards of living rooms as regards lighting and ventilation in case the size of mezzanine floor is 9.5 m^2 or more,

(*b*) it is so constructed as not to interfere under any circumstances with the ventilation of the space over and under it;

(*c*) such mezzanine floor is not subdivided into smaller compartments;

(*d*) such mezzanine floor or any part of it shall not be used as a kitchen; and

(*e*) in no case shall a mezzanine floor be closed so as to make it liable to be converted into unventilated compartments.

1.11 STORE ROOM

Height: The height of a store room shall not be less than 2.2 m.

Size: The size of a store room, where provided in a residential building, shall not be less than 3 m^2.

1.12 GARAGE

Height: The height of a garage shall not be less than 2.4 m.

Size: The size of garages shall be as below:

(*a*) ***Private Garage***: 3 m × 6.0 m minimum: and

(*b*) ***Public Garage***: Based on the number of vehicles parked, etc.

1.13 BASEMENT

Basement or cellar is the lower storey of a building below or partly below ground level.

The basement is not used for residential purposes.

The basement to be constructed within the building envelope and subject to maximum coverage on floor 1 (entrance floor) may be put to only the following uses:

(*a*) Storage of household or other goods of ordinarily non-combustible material;

(*b*) Strong rooms, bank cellars, etc. :

(*c*) Air-conditioning equipment and other machines used for services and utilities of the building; and

(*d*) Parking spaces.

The basement should have the following requirements:

(*a*) Every basement shall be at least 2.4 m in height from the floor to the underside of the roof slab or ceiling;

(*b*) Adequate ventilation shall be provided for the basement. The ventilation requirements shall be the same as required by the particular occupancy according to byelaws. Any deficiency may be met by providing adequate mechanical ventilation in the form of blowers, exhaust fans, air-conditioning systems, etc.;

(*c*) The minimum height of the ceiling of any basement shall be 0.9 m and the maximum 1.2 m above the average surrounding ground level;

(*d*) Adequate arrangements shall be made such that surface drainage does not enter the basement;

(*e*) The walls and floors of the basement shall be watertight and be so designed that the effects of the surrounding soil and moisture, if any, are taken into account in design and adequate damp proofing treatment is given; and

(*f*) The access to the basement shall be separate from the main and alternative staircase providing access and exit from higher floors. Where the staircase is continuous in the case of buildings served by more than one staircase, the same shall be of enclosed type serving as a fire separation from the basement floor and higher floors. Open ramps shall be permitted if they are constructed within the building line subject to the provision of (*d*). Adequate exit requirements in basements should be provided.

1.14 CHIMNEYS

The chimneys shall be built at least 0.9 m above flat roofs, provided the top of the chimney is not below the top of the adjacent parapet wall. In the case of sloping roofs, the chimney top shall not be less than 0.6 m above the ridge of the roof in which the chimney penetrates.

1.15 PARAPET

Parapet walls and handrails provided on the edges of roof-terraces, balcony, verandah, etc. shall not be less than 1.0 m and not more than 1.2 m in height from the finished floor level.

1.16 CABIN

The size of cabins shall not be less than 3.0 m^2. The clear passages within the divided space of any floor shall not be less than 0.75 m and the distance from the farthest space in a cabin to any exit shall not be more than 18.5 m. In case the subdivided cabin does not derive direct lighting and ventilation from any open spaces/mechanical means, the maximum height of the cabin shall be 2.2 m.

1.17 BOUNDARY WALL

The requirements of the boundary wall are given below:

(*a*) Except with the special permission of the Authority, the maximum height of the compound wall shall be 1.5 m above the centre line of the front street. Compound wall up to 2.4 m height may be permitted if the top 0.9 m is of open type construction of an approved design.

(*b*) In the case of corner plot, the height of the boundary wall shall be restricted to 0.75 m for a length of 10 m on the front and side of the intersections and the balance height of 0.75 m if required in accordance with (a) may be made up of open type construction (through railings).

(*c*) However, the provisions of (a) and (b) are not applicable to boundary walls of jails. In industrial buildings, electric substations, transformer stations, institutional buildings like sanatoria, hospitals, industrial buildings like workshops, factories and educational buildings like schools, colleges, including hostels, and other uses of public utility undertakings height up to 2.4 m may be permitted.

1.18 WELLS

Wells, intended to supply water for human consumption or domestic purposes, where provided, shall comply with the following requirements.

1.18.1 Location

The well shall be located:

(*a*) not less than 15 m from any ash pit, refuse pit, earth closet or privy and shall be located on a site upwards from the earth closet or privy;

(*b*) not less than 18 m from any cess pit soakway or borehole latrine and shall be located on a site upwards from the earth closet or privy;

(*c*) that contamination by the movement of sub-soil or other water is unlikely; and

(*d*) not under a tree or otherwise it should have a canopy over it, so that leaves and twigs may not fall into the well and rot.

1.18.2 Requirements

The well shall:

(*a*) have minimum internal diameter of not less than 1 m;

(*b*) be constructed to a height not less than 1 m above the surrounding ground level, to form a parapet or kerb and to prevent surface water from flowing into a well, and shall be surrounded with a paving constructed of impervious material which shall extend for a distance of not less than 1.8 m in every direction from the parapet from the kerb forming the well head and the upper surface of such a paving shall be sloped away from the well;

(*c*) be of sound and permanent construction (*pucca*) throughout. Temporary or exposed (*kutcha*) wells shall be permitted only in fields or gardens for purposes of irrigation; and

(*d*) have the interior surface of the lining or walls of the well rendered impervious for a depth of not less than 1.8 m measured from the level of the ground immediately adjoining the well-head.

1.19 SEPTIC TANKS

Where a septic tank is used for sewage disposal, the location, design and construction of the septic tank shall conform to following requirements:

1.19.1 Location of Septic Tanks and Subsurface Absorption Systems

A subsoil dispersion system shall not be closer than 18 m from any source of drinking water, such as well, to mitigate the possibility of bacterial pollution of water supply. It shall also be as far removed from the nearest habitable building as economically feasible but not closer than 6 m to avoid damage to the structures.

1.19.2 Requirements

(*a*) ***Dimensions of septic tanks.*** Septic tanks shall have a minimum width of 750 mm, a minimum depth of 1 m below the water level and a minimum liquid capacity of 1 m^3. The length of tanks shall be 2 to 4 times the width;

(*b*) Septic tanks may be constructed of brickwork, stone masonry, concrete or other suitable materials;

(*c*) Under no circumstances shall effluent from a septic tank be allowed into an open channel drain or body of water without adequate treatment;

(*d*) The minimum normal diameter of the pipe shall be 100 mm. Further at junctions of pipes in manholes, directions of flow from a branch connection shall not make an angle exceeding 45° with the direction of flow in the main pipe;

(*e*) The gradients of land drains, under-drainage as well as the bottom of dispersion trenches and soakways shall be between 1 : 300 and 1 : 400.

(*f*) Every septic tank shall be provided with ventilating pipe of at least 50 mm diameter. The top of the pipe shall be provided with suitable cage of mosquito-proof wire mesh. The ventilating pipe shall extend to a height which would cause no smell nuisance to any building in the area. Generally, the ventilating pipe may extend to a height of about 2 m when the septic tank is at least 15 m away from the nearest building and to a height of 2 m above the top of the building when it is located closer than 15 m.

(*g*) When the disposal of septic tank effluent is to a seepage pit, the seepage pit may be of any suitable shape with the least cross-sectional dimension of 0.9 m and not less than 1.0 m in depth below the invert level of the inlet pipe. The pit may be lined with stone, brick or concrete blocks with dry open joints which should be backed with at least 75 mm of clean coarse aggregate. The lining above the inlet level should be finished with mortar. In the case of pits of large dimensions, the top portion may be narrowed to reduce the size of the RCC cover slabs. Where no lining is used specially near trees, the entire pit should be filled with loose stones. A masonry ring may be constructed at the top of the pit to prevent damage by flooding of the pit by surface runoff. The inlet pipe may be taken down a depth of 0.9 m from the top as an ant-mosquito measure; and

(*h*) When the disposal of the septic tank effluent is to a dispersion trench, the dispersion trench shall be 0.5 m to 1.0 m deep and 0.3 m to 1.0 m wide excavated to a slight gradient and shall be provided with 150 mm to 250 mm of washed gravel or crushed stones. Open jointed pipes placed inside the trench shall be made of unglazed earthenware clay or concrete and shall have a minimum internal diameter of 75 mm to 100 mm. Each dispersion trench shall not be longer than 30 m and trenches shall not be placed closer than 1.8 m.

1.20 OFFICE-CUM-LETTER BOX ROOM

In the case of multistoreyed multifamily dwelling apartments constructed by existing and proposed Cooperative Housing Societies or Apartment Owners Associations, limited companies and proposed societies, an office-cum-letter box room of dimension 3.6 m × 3 m shall be provided on the ground floor. In case the number of flats is more than 20, the maximum size of the office-cum-letter box room shall be 20 m^2.

Business Buildings: Provision shall be made for letter boxes on the entrance floor as per the requirements of the postal department.

1.21 METER ROOMS

For all buildings above 15 m in height and in special occupancies, like educational, assembly, institutional, industrial, storage, hazardous and mixed occupancies with any of the aforesaid occupancies having area more than 500 m^2 on each floor, provision shall be made for an independent and ventilated meter (service) room, as per requirements of electric (service) supply undertakings on the ground floor with direct access from outside for the purpose of termination of electric supply from the licensee's service and alternative supply cables. The door/doors provided for the service room shall have fire resistance of not less than two hours.

1.22 STAIRCASE/EXIT REQUIREMENTS

The minimum clear width, minimum tread width and maximum riser of staircases for buildings shall be as given below:

Minimum Width: The minimum width of staircase shall be as follows:

(*a*) Residential buildings (dwellings) 1.0 m

NOTE: For row housing with 2 storeys, the minimum width shall be 0.75 m.

(*b*) Residential hotel buildings 1.5 m

(*c*) Assembly buildings like auditoria, theatres and cinemas 2.0 m

(*d*) Educational buildings 1.5 m

(*e*) Institutional buildings 2.0 m

(*f*) All other buildings 1.5 m

Minimum Tread: The minimum width of tread without nosing shall be 250 mm for residential buildings. The minimum width of tread for other buildings shall be 300 mm.

Maximum Riser: The maximum height of riser shall be 190 mm for residential buildings and 150 mm for other buildings and these shall be limited to 12 per flight.

The minimum head-room in a passage under the landing of a staircase shall be 2.2 m. The minimum clear head-room in any staircase shall be 2.2 m.

All aspects of exit requirements for corridors, doors, staircases, ramps, etc. in respect of widths, travel distance shall be as given in chapter 29.

1.23 ROOFS

The roof of a building is so constructed or framed as to permit effectual drainage of the rain-water therefrom by means of sufficient rain-water pipes of adequate size wherever required, so arranged, jointed and fixed as to ensure that the rain-water is carried away from the building without causing dampness in any part of the walls or foundations of the building or those of an adjacent building.

The Authority may require rain-water pipes to be connected to a drain or sewer to a covered channel formed beneath the public footpath to connect the rain-water pipe to the road gutter or in any other approved manner.

Rain-water pipes shall be affixed to the outside of the external walls of the building or in recesses or chases cut or formed in such external walls or in such other manner as may be approved by the Authority.

It is desirable and in certain cases mandatory to conserve rain water using suitable rain water harvesting techniques including by roof water collection.

SPECIAL REQUIREMENTS OF BUILDINGS

1.24 RAT PROOFING AND TERMITE PROOFING OF BUILDINGS

Every building or part thereof that is designed or intended for use as a dwelling or for the handling, storage or sale of foodstuffs shall conform to the following requirements:

Every such building unless supported on posts shall have continuous foundation walls extending from at least 600 mm below ground level to at least 150 mm above ground level or shall have a continuous floor of masonry or reinforced concrete or other equally effective rat-proof materials.

All openings in such foundations or floors, windows and drains, and all junctions between foundation walls and building walls shall be effectively rat-proofed, that is, windows and doors shall be tight fitting and other openings shall be securely covered with rat-proof screening or grillage or shall be tightly closed with metal sheeting, concrete or other equally effective rat-proof material.

Termite control in buildings is very important as the damage likely to be caused by termites to wooden structures of buildings and other household articles like furniture, clothing, stationery, etc. is considerable. Anti-termite measures in buildings shall be taken in accordance with good practice.

1.25 LIGHTING AND VENTILATION

The lighting and ventilation requirements for different uses and occupancies are covered in Chapter 25.

1.25.1 Lighting and Ventilation of Rooms

Rooms shall have, for the admission of light and air, one or more openings, such as windows and ventilators, opening directly to the external air or into an open verandah with a maximum width of 2.4 m.

Notwithstanding the area of openings, the minimum aggregate area of such openings, excluding doors inclusive of frames, should not be less than:

(*a*) one-tenth of the floor area for dry hot climate;
(*b*) one-sixth of the floor area for wet hot climate;
(*c*) one-eighth of the floor area for intermediate climate; and
(*d*) one-twelfth of the floor area for cold climate.

Note 1: If a window is partly fixed, the openable area shall be counted.

Note 2: No portion of a room shall be assumed to be lighted if it is more than 7.5 m away from the opening assumed for lighting that portion.

Note 3: The area of openings as given in (a) to (d) above shall be increased by 25 per cent in the case of a kitchen.

1.26 AIR-CONDITIONING AND HEATING

For information regarding design, construction and installation of air-conditioning, heating and mechanical ventilation systems, reference may be made to Chapter 26.

1.27 ACOUSTICS, SOUND INSULATION AND NOISE CONTROL

For requirements regarding the desired noise levels and sound insulation in different occupancies, reference may be made to Chapter 27.

1.28 HEAT INSULATION

For information on solar radiation on buildings and recommended limits of thermal transmittance of roofs and walls for different parts of the country and heat transmission losses due to different constructions, reference may be made to Chapter 24.

1.29 PROVISION OF LIFTS AND ESCALATORS

Provision for lifts shall be made for buildings 15 m or more in height. Planning, designing and installation etc. of lifts and escalators shall be done in accordance with the provisions given in Chapter 15.

1.30 LIGHTNING PROTECTION OF BUILDINGS

Lightning protective system is required to be installed on all buildings and structures vulnerable to lightning strokes owing to their height or exposed situation, buildings of public or strategic importance, buildings of public resort, buildings housing valuable materials, ancient monuments, observatories, etc. For information regarding details of various factors that affect the risk of the structure being struck, the consequential effects of a stroke and other details, reference may be made to National Building Code of India 1983-Part viii Building Services, Section 2 : Electrical installation.

QUESTIONS

1. How are the buildings classified on the basis of occupancy and the types of construction?
2. Give the sub-classification of the following classes of buildings:
   (*a*) Residential (*b*) Institutional
   (*c*) Assembly (*d*) Business
   (*e*) Mercantile and (*f*) Industrial
3. Give details of occupancy classification.
4. (*a*) What are the three basic zones for the purpose of orientation of buildings? What are the problems posed by them? Suggest suitable remedial measures.
   (*b*) Discuss the factors affecting orientation of buildings.
5. List the various parts of a building and state their requirements.
6. Discuss the requirements of the following components of a building:
   (*a*) kitchen (*b*) bathrooms and W.C.
   (*c*) septic tanks.
7. What are the special requirements of a building?

REFERENCES

1. *Bureau of Indian Standards New Delhi*: National Building Code of India 2005–Part 3.
2. *Bureau of Indian Standards*: National Building Code of India 2005–Part 8 Section 1: Lighting and Ventilation.
3. *IS: 7662 (Part I)–1974*: Recommendations for orientation of buildings.

Design Loads

DESIGN LOADS

The various basic design loads and forces which are taken into consideration for structural design of buildings are:

(1) Dead load
(2) Imposed (Live) load
(3) Wind load
(4) Seismic or earthquake load
(5) Snow load
(6) Special loads
  (*a*) Temperature effects
  (*b*) Hydrostatic and soil pressure
  (*c*) Fatigue
  (*d*) Erection loads
  (*e*) Accidental loads, etc.
(7) Other Loads

A description of these design loads follows.

2.1 DEAD LOAD

2.1.1 Assessment of Dead Load

The dead load in a building comprises the weight of all walls, partitions, floors and roofs and includes the weights of all other permanent constructions in the building. Where positioning of partitions is not decided, a uniformly distributed load of one-third of weight of partition subject to a minimum of 1.0 kN/m^2 in case of office floors may be assumed.

The unit weight of commonly used materials is given in Table 2.1.

Table 2.1: Unit weights of building materials (I.S. 875 (Part 1)—1987)

Material	Unit weight/mass		
	kN	kg/m^2	Per
1. *Aggregate coarse*			
Broken stone ballast:			

(*Contd.*)

	Material	Unit weight/mass		
		kN	kg/m²	Per
	Dry, well shaken	15.70 to 18.35	1600 to 1870	m^3
	Perfectly wet	18.85 to 21.95	1920 to 2240	m^3
	Shingles, 3 mm to 38 mm	14.35	1460	m^3
	Broken bricks:			
	Fine	14.20	1450	m^3
	Coarse	9.90	1010	m^3
	Foam slag, foundry pumice	6.85	700	m^3
2.	*Aggregate, cinder fine*	7.85	800	m^3
	Sand:			
	Dry, clean	15.10 to 15.70	1540 to 1600	m^3
	River	18.05	1840	m^3
	Wet	17.25 to 19.60	1760 to 2000	m^3
	Brick dust (surkhi)	9.90	1010	m^3
3.	*Bitumen*	0.102	10.40	m^3
4.	*Blocks*			
	Lime-based solid blocks	8.65 to 12.55	880 to 1280	m^3
	Hollow blocks	1.41 to 0.94	14.4 to 9.6	m^3
	Solid concrete blocks	17.65	18.00	m^3
5.	*Bricks*			
	Common burnt clay bricks (I.S. 1077-1987)	15.70 to 18.85	1600 to 1920	m^3
	Engineering bricks	21.20	2160	m^3
	Heavy duty bricks (I.S. 2180-1985)	24.50	2500	m^3
	Pressed bricks	17.25 to 18.05	1760 to 1840	m^3
	Refractory bricks	17.25 to 19.60	1760 to 2000	m^3
	Sand cement bricks	18.05	1840	m^3
	Sand lime bricks	20.40	2080	m^3
	Brick dust (surkhi)	9.90	1010	m^3
6.	*Cement*			
	Ordinary and aluminous	14.10	1440	m^3
	Rapid hardening	12.55	1280	m^3
7.	*Cement concrete, plain and reinforced*			
	Cement concrete, plain			
	Aerated	7.45	760	m^3
	No fines, with heavy aggregates	15.70 to 18.80	1600 to 1920	m^3
	No fines, with light aggregates	8.65 to 12.55	880 to 1280	m^3
	with burnt clay aggregates	17.25 to 21.20	1760 to 2160	m^3
	with expanded clay aggregates	9.40 to 16.50	960 to 1680	m^3

(*Contd.*)

	Material	Unit weight/mass		
		kN	kg/m^2	Per
	with clinker aggregates	12.55 to 17.25	1280 to 1760	m^3
	with pumice aggregates	5.50 to 11.00	560 to 1120	m^3
	with sand and gravel or			
	crushed natural stone aggregates	22.50 to 23.50	2240 to 2400	m^3
	Reinforced with			
	1% steel	22.75 to 24.20	2310 to 2470	
	2% steel	23.25 to 24.80	2370 to 2530	
	5% steel	24.80 to 26.50	2530 to 2700	
8.	*Cement concrete, prestressed*	23.50	2400	m^3
9.	*Lime concrete*			
	with burnt clay aggregates	18.80	1920	m^3
10.	*Masonry stone*			
	Cast	22.55	2300	m^3
	Dry rubble	20.40	2080	m^3
	Granite ashlar	25.90	2640	m^3
	Granite rubble	23.55	2400	m^3
	Limestone ashlar	25.10	2560	m^3
	Marble dressed	26.50	2700	m^3
	Sandstone	22.00	2240	m^3
11.	*Mortar/Plaster*			
	Cement	20.40	2080	m^3
	Gypsum	11.80	1200	m^3
	Lime	15.70 to 18.05	1600 to 1840	m^3
12.	*Soils*			
	Alluvial ground, undisturbed	15.70	1600	m^3
	clay fills			
	undisturbed	18.85	1920	m^3
	Undisturbed, gravelly	20.40	2080	m^3
	Dry, lumps	10.20	1040	m^3
	Earth:			
	Dry	13.89 to 18.05	1410 to 1840	m^3
	Moist	15.70 to 19.60	1600 to 2000	m^3
	Mud river, wet	17.25 to 18.85	1760 to 1920	m^3
	Silt, wet	17.25 to 18.85	1760 to 1920	m^3
13.	*Tar, coal*	9.90	1010	m^3
14.	*Steel*			
	Gray cast	68.95 to 69.90	7030 to 7130	m^3
	White cast	74.30 to 75.70	7580 to 7720	m^3

(*Contd.*)

Material	Unit weight/mass		
	kN	kg/m²	Per
Pig	70.60	7200	m^3
Wrought	75.50	7700	m^3
15. *Timber*	5.90 to 7.40	600 to 750	m^3

2.2 IMPOSED (LIVE) LOAD

2.2.1 Imposed (Live) Loads on Floors Due to Use and Occupancy

The imposed loads on floors comprise all loads other than dead loads. Floors shall be investigated for both the uniformly distributed load (UDL) and the corresponding concentrated load specified in Table 2.2 and designed for the most adverse effects but they shall not be considered to act simultaneously. The concentrated load specified in Table 2.2 is assumed to act over an area of 0.3 m × 0.3 m. However the concentrated loads need not be considered where the floors are capable of effective lateral distribution of this load.

Table 2.2: Imposed Floor Loads for Different Occupancies

Sl. No.	*Occupancy Classification*	*Uniformly Distributed Load (UDL)*	*Concentrated Load*
(1)	(2)	(3)	(4)
(*i*)	***Residential Buildings***	kN/m²	kN
	(*a*) Dwelling houses:		
	(1) All rooms and kitchens	2.0	1.8
	(2) Toilets and bathrooms	2.0	
	(3) Corridors, passages, staircases including fire escapes and store rooms	3.0	4.5
	(4) Balconies	3.0	1.5 per metre run concentrated at the outer edge
	(*b*) Dwelling units planned and executed in accordance with [6-1(2)] only:		
	(1) Habitable rooms, kitchens, and toilets and bathrooms	1.5	1.4
	(2) Corridors, passages and staircases including fire escapes	1.5	1.4
	(3) Balconies	3.0	1.5 per metre run concentrated at the outer edge
	(*c*) Hotels, hostels, boarding houses, loading houses, dormitories and residential clubs:		
	(1) Living room, bed rooms and dormitories	2.0	1.8
	(2) Kitchen and laundries	3.0	4.5
	(3) Billiards rooms and public lounges	3.0	2.7

(*Contd.*)

Sl. No. (1)	Occupancy Classification (2)	Uniformly Distributed Load (UDL) (3)	Concentrated Load (4)
	(4) Store rooms	5.0	4.5
	(5) Dining rooms, cafeterias and restaurants	4.0	2.7
	(6) Office rooms	2.5	2.7
	(7) Rooms for indoor games	3.0	1.8
	(8) Baths and toilets	2.0	—
	(9) Corridors, passages staircases including fire escapes and lobbies as per the floor services (excluding stores and the like) but not less than	3.0	4.5
	(10) Balconies	Same as rooms to which they give access but with a minimum of 4.0	1.5 per metre run concentrated at the outer edge
	(*d*) Boiler rooms and plant rooms to be calculated but not less than	5.0	6.7
	(*e*) Garages:		
	(1) Garage floors (including parking area and repair workshops for passenger cars and vehicles not exceeding 2.5 tonnes gross weight, including access ways and ramps—to be calculated but not less than	2.5	9.0
	(2) Garage floors for vehicles not exceeding 4.0 tonnes gross weight (including access ways and ramps)—to be calculated but not less than	5.0	9.0
(*ii*)	***Educational Buildings***		
	(*a*) Class rooms and lecture rooms (not used for assembly purposes)	3.0	2.7
	(*b*) Dining rooms, cafeterias and restaurants	3.0[1]	2.7
	(*c*) Offices, lounges and staff rooms	2.5	2.7
	(*d*) Dormitories	2.0	2.7
	(*e*) Projection rooms	5.0	—
	(*f*) Kitchens	3.0	4.5
	(*g*) Toilets and bathrooms	2.0	—
	(*h*) Store rooms	5.0	4.5
	(*i*) Libraries and archives:		
	(1) Stack room/stack area	6.0 kN/m^2 for a minimum height of 2.2 m + 2.0 kN/m^2 per metre height beyond 2.2 m	4.5
	(2) Reading rooms (without separate storage)	4.0	4.5

(*Contd.*)

Sl. No.	*Occupancy Classification*	*Uniformly Distributed Load (UDL)*	*Concentrated Load*
(1)	**(2)**	**(3)**	**(4)**
	(3) Reading rooms (with separate storage)	3.0	4.5
	(*j*) Boiler rooms and plant rooms—to be calculated but not less than	4.0	4.5
	(*k*) Corridors, passages, lobbies, staircases including fire escapes—as per the floor serviced (without accounting for storage and projection rooms) but not less than	4.0	4.5
	(*l*) Balconies	Same rooms to which they give access but with a minimum of 4.0	1.5 metre run concentrated at the outer edge
(*iii*)	***Institutional Buildings***		
	(*a*) Bed rooms, wards, dressing rooms, dormitories and lounges	2.0	1.8
	(*b*) Kitchens, laundries and laboratories	3.0	4.5
	(*c*) Dining rooms, cafeterias and restaurants	3.0[1]	2.7
	(*d*) Toilets and bathrooms	2.0	—
	(*e*) X-ray rooms, operating rooms and general storage areas—to be calculated but not less than	3.0	4.5
	(*f*) Office rooms and O.P.D. rooms	2.5	2.7
	(*g*) Corridors, passages, lobbies, staircases including fire escapes—as per the floor serviced (without accounting for storage and projection rooms) but not less than	4.0	4.5
	(*h*) Boiler rooms and plant rooms—to be calculated but not less than	5.0	4.5
	(*i*) Balconies	Same as rooms to which they give access but with a minimum of 4.0	1.5 metre run concentrated at the outer edge
(*iv*)	***Assembly Building***		
	(*a*) Assembly areas:		
	(1) With fixed seats[2]	4.0	—
	(2) Without fixed seats	5.0	3.6
	(*b*) Restaurants (subject to assembly), museums and art galleries and gymnasia	4.0	4.5
	(*c*) Projection rooms	5.0	—
	(*d*) Stages	5.0	4.5
	(*e*) Office rooms, kitchens and laundries	3.0	4.5
	(*f*) Dressing rooms	2.0	1.8
	(*g*) Lounges and billiards rooms	2.0	2.7

(*Contd.*)

Sl. No. (1)	Occupancy Classification (2)	Uniformly Distributed Load (UDL) (3)	Concentrated Load (4)
	(h) Toilets and bathrooms	2.0	—
	(i) Corridors, passages and staircases including fire escapes	4.0	4.5
	(j) Balconies	Same as rooms to which they give access but with a minimum of 4.0	1.5 metre run concentrated at the outer edge
	(k) Boiler rooms and plant rooms including weight of machinery	7.5	4.5
	(l) Corridors, passages, subject to loads greater than from crowds, such as wheeled vehicles, trolleys and the like corridors, staircases and passages in grandstands	5.0	4.5
(v)	***Business and Office Buildings***		
	(a) Rooms for general use with separate storage	2.5	2.7
	(b) Rooms without separate storage	4.0	4.5
	(c) Banking halls	3.0	2.7
	(d) Business computing machine rooms (with fixed computers or similar equipment)	3.5	4.5
	(e) Records/files store rooms and storage space	5.0	4.5
	(f) Vaults and strong rooms—to be calculated but not less than	5.0	4.5
	(g) Cafeterias and dinning rooms	3.0[1]	2.7
	(h) Kitchens	3.0	2.7
	(i) Corridors, passages, lobbies, staircases including fire escapes—as per the floor serviced (excluding stores) but not less than	4.0	4.5
	(j) Bath and toilets rooms	2.0	—
	(k) Balconies	Same as rooms to which they give access but with a minimum of 4.0	1.5 metre run concentrated at the outer edge
	(l) Stationary stores	4.0 for each metre of storage height	9.0
	(m) Boiler rooms and plant rooms—to be calculated but not less than	5.0	6.7
	(n) Libraries	See Sl. No.(ii)	
(vi)	***Merchantile Buildings***		
	(a) Retail shops	4.0	3.6
	(b) Wholesale shops—to be calculated but not less than	6.0	4.5
	(c) Office rooms	2.5	2.7

(Contd.)

Sl. No.	*Occupancy Classification*	*Uniformly Distributed Load (UDL)*	*Concentrated Load*
(1)	(2)	(3)	(4)
	(*d*) Dining rooms, restaurants and cafeterias	3.0[(1)]	2.7
	(*e*) Toilets	2.0	—
	(*f*) Kitchens and laundries	3.0	4.5
	(*g*) Boiler rooms and plant rooms—to be calculated but not less than	5.0	6.7
	(*h*) Corridors, passages, staircases including fire escapes and lobbies	4.0	4.5
	(*i*) Corridors, passages, staircases subject to loads greater than from crowds, such as wheeled vehicles, trolleys and the like	5.0	4.5
	(*j*) Balconies	Same as rooms to which they give access but with a minimum of 4.0	1.5 metre run concentrated at the outer edge
(vii)	***Industrial Buildings***[(4)]		
	(*a*) Work areas without machinery/equipment	2.5	4.5
	(*b*) Work areas with machinery/equipment[(3)]		
	(1) Light duty — To be calculated but not less than	5.0	4.5
	(2) Medium duty — To be calculated but not less than	7.0	4.5
	(3) Heavy duty — To be calculated but not less than	10.0	4.5
	(*c*) Boiler rooms and plant rooms—to be calculated but not less than	5.0	6.7
	(*d*) Cafeterias and dinning rooms	3.0[(1)]	2.7
	(*e*) Corridors, passages, staircases including fire escapes	4.0	4.5
	(*f*) Corridors, passages, lobbies, staircases subject to machine loads and wheeled vehicles—to be calculated but not less than	5.0	4.5
	(*g*) Kitchens	3.0	4.5
	(*h*) Toilets and bathrooms	2.0	—
(viii)	***Storage Buildings***[(4)]		
	(*a*) Storage rooms (other than cold storage) and warehouses—to be calculated based on the bulk density of materials stored but not less than	2.4 kN/m^2 per metre of storage height with a minimum of 7.5 kN/m^2	7.0
	(*b*) Cold storage—to be calculated but not less than	5.0 kN/m^2 per metre of storage height with a minimum of 15 kN/m^2	9.0
	(*c*) Corridors, passages, staircases including fire escapes—as per the floor serviced but not less than	4.0	4.5

(*Contd.*)

Sl. No. (1)	Occupancy Classification (2)	Uniformly Distributed Load (UDL) (3)	Concentrated Load (4)
	(d) Corridors, passages subject to loads greater than from crowds, such as wheeled vehicles, trolleys and the like	5.0	4.5
	(e) Boiler rooms and plant rooms	7.5	4.5

(1) Where unrestricted assembly of persons is anticipated, the value of UDL should be increased to 4.0 kN/m^2.

(2) With fixed seats' implies that the removal of the seating and the use of the space for other purposes is improbable. The maximum likely load in this case is, therefore, closely controlled.

(3) The loading in industrial buildings (workshops and factories) varies considerably and so three loadings under the terms 'light', 'medium' and 'heavy' are introduced in order to allow for more economical designs but the terms have no special meaning in themselves other than the imposed load for which the relevant floor is designed. It is, however, important particularly in the case of heavy weight loads, to assess the actual loads to ensure that they are not in excess of 10 kN/m^2; in case where they are in excess, the design shall be based on the actual loadings.

(4) For various mechanical handling equipment which are used to transport goods, as in warehouses, workshops, store rooms, etc., the actual load coming from the use of such equipment shall be ascertained and design should cater to such loads.

2.2.1.1 Load Application

The uniformly distributed loads specified in Table 2.2 shall be applied as static loads over the entire floor area under consideration or a portion of the floor area whichever arrangement produces critical effects on the structural elements.

In the design of floors, the concentrated loads are considered to be applied in the positions which produce the maximum stresses and where deflection is the main criterion in the positions which produce the maximum deflections. Concentrated load, when used for the calculation of bending and shear, are assumed to act at a point. When used for the calculation of local effects, such as, crushing or punching, they are assumed to act over an actual area of application of 0.3 m × 0.3 m.

2.2.1.2 Loads Due to Light Partitions

In office and other buildings, where actual loads due to light partitions cannot be assessed at the time of planning the floors and the supporting structural members shall be designed to carry, in addition to other loads, uniformly distributed loads per square metre of not less than 33.33 per cent of weight per metre run of finished partitions, subject to a minimum of 1 kN/m^2, provided total weight of partition walls per m^2 of the wall area does not exceed 1.5 kN/m^2 and the total weight per metre length is not greater than 4.0 kN.

2.2.2 Reduction in Imposed Loads on Floors

2.2.2.1 For Members Supporting Floors

The following reductions in assumed total imposed loads on the floors may be made in designing columns, load bearing walls, piers, their supports and foundations.

Number of Floors (Including the Roof) to be Carried by Member Under Consideration (1)	*Reduction in Total Distributed Imposed Load on All Floors to be Carried by the Member Under Consideration Per cent* (2)
1	0
2	10
3	20
4	30
5 to 10	40
Over 10	50

(*a*) No reduction is made for any plant or machinery which is specifically allowed for, or for buildings for storage purposes, warehouses and garages. However, for other buildings, where the floor is designed for an imposed floor load of 5.0 kN/m^2 or more, the reductions shown above may be taken provided that the loading assumed is not less than it would have been if all the floors had been designed for 5.0 kN/m^2 with no reductions.

Note: In case if the reduced load in the lower floor is lesser than the reduced load in the upper floor, then the reduced load of the upper floor will be adopted.

2.2.2.2 For Beams in Each Floor Level

Where a single span of beam, girder or truss supports not less than 50 m^2 of floor at one general level, the imposed floor load may be reduced in the design of the beams, girders or trusses by 5 per cent for each 50 m^2 area supported subject to a maximum reduction of 25 per cent. However, no reduction shall be made in any of the following types of loads:

(*a*) any superimposed moving load,

(*b*) any actual load due to machinery or similar concentrated loads,

(*c*) the additional load in respect of partition walls; and

(*d*) any impact or vibration.

Note: The above reduction does not apply to beams, girdes or trusses supporting roof loads.

2.2.3 Imposed (Live) Loads on Roofs

Live loads on various types of roofs: On flat roofs, sloping roofs and curved roofs, allowance for live load shall be as given in Table 2.3.

Roofs of buildings used for promenade or incidental assembly purposes shall be designed for the appropriate imposed floor loads given in Table 2.2 for the occupancy.

Snow load. If a roof is subjected to snow load it should be designed for the actual load due to snow or for the imposed loads specified in Table 2.2 whichever is more severe. Actual load due to snow will depend upon the shape of the roof and its capacity to retain the snow; and each case shall be treated on its own merits. In the absence of any specific information, the loading due to the collection of snow may be assumed to be 0.25 kN/m^2 per centimetre depth of snow. The possibility of total or partial snow load should be considered, that is, one-half of the roof fully loaded with the design snow load and the other half loaded with half the design snow load. In the case of roofs with slopes greater than 50°, snow load may be disregarded, where, however, there are possibilities of formation of snow pockets, these should be taken into account.

Loads due to rain. On surfaces whose positioning, shape and drainage system are such as to make accumulation of rainwater possible, loads due to such accumulation of water and the live loads for the roofs as given in Table 2.2 shall be considered separately and the more critical of the two shall be adopted in the design.

Loads on members directly supporting the roof coverings. Every member of the supporting structure which is directly supporting the roof covering(s) shall be designed to carry the more severe of the two load(s) indicated below:

(*a*) The loads transmitted to the member from the roof covering(s) and

(*b*) An incidental concentrated load of 0.90 kN placed at the most unfavourable position on the member.

Where it is ensured that the roofs would be traversed only with aid of planks and ladders capable of distributing the loads on them to two or more supporting members, the intensity of concentrated load indicated in (*b*) above may be reduced to 0.50 kN.

Load concentrated on roof coverings. To provide for loads incidental to maintenance, unless otherwise specified, all roof coverings (other than glass) shall be capable of carrying an incidental load of 0.90 kN concentrated on an area 125 mm square. The intensity of the concentrated load may be reduced where it is ensured that the roof coverings would not be traversed without suitable aids.

Horizontal loads on Parapets and Balustrades. Where it is desired to design parapets and balustrades against horizontal pressures, the loads given in Table 2.4, expressed as horizontal loads acting at handrail or coping level shall be provided for. The values given are for guidance only and where values for actual loadings are available, they shall be used instead;

Table 2.3: Imposed Loads on Various Types of Roofs

Sl. No.	Type of Roof	Imposed Load Measured on Plan Area	Minimum Imposed Load Measured on Plan
(*i*)	Flat, sloping or curved roof with slopes up to and including 10 degrees		
	(*a*) Access provided	1.5 kN/m^2	3.75 kN uniformly distributed over any span of one metre width of the roof slab and 9 kN uniformly distributed over the span of any beam or truss or wall
	(*b*) Access not provided except for maintenance	0.75 kN/m^2	1.9 kN uniformly distributed over any span of one metre width of the roof slab and 4.5 kN uniformly distributed over the span of any beam or truss or wall
(*ii*)	Sloping roof with slope greater than 10°	For roof membrane sheets or purlins–0.75 kN/m^2 less 0.02 kN/m^2 for every degree increase in slope over 10°	Subject to a minimum of 0.4 kN/m^2

(*Contd.*)

Sl. No.	Type of Roof	Imposed Load Measured on Plan Area	Minimum Imposed Load Measured on Plan
(*iii*)	Curved roof with slope of line obtained by joining springing point to the crown with the horizontal, greater than 10°	$(0.75 - 0.52\ \alpha^2)$ kN/m^2 where $\alpha = h/l$ h = height of the highest point of the structure measured from its springing; and l = chord width of the roof if singly curved and shorter of the two sides if doubly curved. Alternatively, where structural analysis can be carried out for curved roofs of all slopes in a simple manner applying the laws of statistics, the curved roofs shall be divided into minimum 6 equal segments and for each segment imposed load shall be calculated appropriate to the slope of the chord of each segment as given in (*i*) and (*ii*).	Subject to a minimum of 0.4 kN/m^2

Notes:

1. The loads given above do not include loads due to snow, rain, dust collection, etc. The roof shall be designed for imposed loads given above or for snow/rain load, whichever is greater.
2. For special types of roofs with highly permeable and absorbent material, the contingency of roof material increasing in weight due to absorption of moisture shall be provided for.

Table 2.4: Horizontal Loads on Parapets, Parapet Walls and Balustrades

Sl. No.	Usage Area	Intensity of Horizontal Load kN/m Run
(*i*)	Light access stairs, gangways and like not more than 600 mm wide	0.25
(*ii*)	Light access stairs, gangways and like, more than 600 mm wide; stairways, landings, balconies and parapet walls (private and part of dwellings)	0.35
(*iii*)	All other stairways, landings and balconies and all parapets and handrails to roofs [except those subject to overcrowding covered under (iv)]	0.75
(*iv*)	Parapets and balustrades in place of assembly, such as theatres, cinemas, churches, schools, places of entertainment, sports and buildings and buildings likely to be overcrowded.	2.25

2.2.4 Loading Effects due to Impact and Vibrations

For structures carrying imposed (live) loads which induce impact or vibration, as far as possible, calculations shall be made for the increase in the live load due to impact or vibration. In the absence of sufficient data for such calculation, the increase in the live load shall be as follows:

Structure	Impact Allowance Per cent Minimum
(*a*) For frames supporting lifts and hoists	100
(*b*) For foundations, footings and piers supporting lifts and hoisting apparatus	40
(*c*) For light machinery, shaft or motor units	20
(*d*) For reciprocating machinery or power units	50

Concentrated imposed loads with impact and vibrations. Concentrated imposed loads with impact and vibration which may be due to installed machinery shall be considered and provided for in the design. The impact factor shall not be less than 20 per cent which is the amount allowable for light machinery.

Provision shall also be made for carrying any concentrated equipment loads while the equipment is being installed or moved for servicing and repairing.

2.2.4.1 Impact Allowance for Crane Girders

For crane gantry girders and supporting columns, the allowances given in Table 2.5 shall be deemed to cover all forces set up by vibration, shock from slipping of slings, kinetic action of acceleration and retardation, and impact of wheel loads.

Table 2.5: Impact Allowance for Crane Girders

Type of Load	Additional Load
(*a*) Vertical loads for electric overhead cranes	25. per cent of maximum static loads for crane girders for all class of cranes
	25 per cent for columns supporting Class III and Class IV cranes.
	10 per cent for columns supporting Class I and Class II cranes.
	No additional load for design of foundations.
(*b*) Vertical loads for hand operated cranes	10 per cent of maximum wheel loads for crane girders only
(*c*) Horizontal forces transverse to rails:	
(1) For electric overhead cranes with trolley having rigid mast for suspension of lifted weight (such as, soaker crane, stripper crane, etc).	10 per cent of weight of crab and the weight lifted by the cranes, acting on any one crane track rail, acting in either direction and equally distributed amongst all the wheels on one side of rail track

(*Contd.*)

Type of Load	*Additional Load*
	For frame analysis, this force, calculated as above, shall be applied on one side of the frame at a time in either direction.
(2) For all other electric overhead cranes and hand operated cranes	5 per cent of weight of crab and the weight lifted by the cranes, acting on any one crane track rail, acting in either direction and equally distributed amongst the wheels on one side of rail track For the frame analysis, the force, calculated as above, shall be applied on one side of the frame at a time in either direction.
(*d*) Horizontal traction forces along the rails for overhead cranes, either electrically operated or hand operated	5 per cent of all static wheel loads.

2.2.4.2 Overloading Factors in Crane Supporting Structures

For all ladle cranes and charging cranes where there is possibility of overloading from production considerations, an overloading factor of 10 per cent of the maximum wheel loading shall be taken.

2.2.5 Crane Load Combinations

In the absence of any specific indications, the load combinations shall be as indicated below.

2.2.5.1 Vertical Loads

In an aisle, where more than one crane is in operation or has provision for more than one crane in future, the following load combinations shall be taken for vertical loading:

Forces specified in (*c*) and (*d*) shall be considered as acting at the rail level and being appropriately transmitted to the supporting system. Gantry girders and their vertical supports shall be designed on the assumption that either of the horizontal forces in (*c*) and (*d*) may act at the same time as the vertical load.

(*a*) Two adjacent cranes working in tandem with full load and with overloading according to 2.2.4.2; and

(*b*) For long span gantries, where more than one crane can come in the span, the girder shall be designed for one crane fully loaded with overloading according to 2.2.4.2 plus as many loaded cranes as can be accommodated on the span but without taking into account overloading to give the maximum effect.

2.2.5.2. Lateral Surge

For design of columns and foundations, supporting crane girders, the following crane combinations shall be considered:

(*a*) ***For Single Bay Frames***: Effect of one crane in the bay giving the worst effect shall be considered for calculation of surge force; and

(*b*) ***For Multi-Bay Frames***: Effect of two cranes working, one each in any of two bays in the cross-section to give the worst effect shall be considered for calculation of surge force.

2.2.5.3 Tractive Force

(*a*) Where one crane is in operation with no provision for future crane, tractive force from only one crane is taken.

(*b*) Where more than one crane is in operation or there is provision for future crane, tractive force from two cranes giving maximum effect is considered.

Note: Lateral surge force and longitudinal tractive force acting across and along the crane rail respectively shall not be assumed to act simultaneously. However, if there is only one crane in the bay, the lateral and longitudinal forces may act together simultaneously with vertical loads.

2.3 WIND LOAD

2.3.1 General

Wind is air in motion relative to the surface of the earth. The primary cause of wind is traced to differences in solar and terrestrial radiations setting up irregularities in temperature which give rise to convection either upwards or downwards. Gravity is the operative force working in some cases through the agency of pressure difference.

Since vertical components of atmospheric motion are relatively small, especially near the surface of the earth, the term 'winds' denotes almost exclusively the horizontal wind; vertical wind velocities are assessed with the aid of anemometers or anemographs which are installed at meteorological observatories, at heights generally varying from 10 to 30 metres above ground.

Very strong winds (greater than 80 km/h) are generally associated with cyclonic storms, thunderstorms, duststorms or vigorous monsoons. A feature of the cyclonic storms over the Indian area is that they rapidly weaken after crossing the coasts and moves as depressions inland. The influence of a severe storm after crossing the coasts and a severe storm after striking the coast does not, in general, exceed 60 to 70 kilometres, though sometimes, it may extend even up to 120 kilometres.

The wind speeds recorded at any locality are extremely variable and in addition to steady wind at any time, there are effects of gusts which may last for a few seconds. These gusts cause increase in air pressure but their effect on the stability of the building may not be quite important; often gusts affect one part of the building and the increased local pressures may be outbalanced by a momentary reduction in the pressure elsewhere. Because of the inertia of the building, short period gusts may not cause any appreciable increase in stress in the main components of the building, although the wall panels and the roof sheetings may be more seriously affected. Gusts can also be extremely important for the design of structures with high slenderness ratio.

The liability of a building to high wind pressure depends not only upon the geographical locations and proximity of other obstructions to air flow but also upon the characteristics of the structure itself.

The effect of wind on the structure as a whole is determined by combined action of external and internal pressures acting upon it. In all cases, the calculated wind loads act normal to the surface to which they apply. Buildings are designed with due attention to the effects of wind on the comfort of people inside and outside the building.

The stability of the building as a whole is checked considering the combined effect, as well as separate effects of imposed loads and wind loads on vertical surface roofs and other parts of the building above the general roof level.

2.3.2 Basic Wind Speed (V_b)

Figure 2.1 gives basic wind speed map of India, as applicable to 10 m height above mean ground level for different zones of the country. Basic wind speed is based on peak gust velocity averaged over a short time interval of about 3 seconds and corresponds to mean heights above ground level in an open terrain. Basic wind speeds presented in Fig. 1 have been worked out for a 50 year return period.

2.3.3 Design Wind Speed (V_z)

The basic wind speed (V_b) for any site shall be obtained from Fig. 2.1 and shall be modified to include the following effects to get V_z, design wind speed at any height for the chosen structure.

(*a*) risk level;

(*b*) terrain roughness, height and size of structure; and

(*c*) local topography.

It can be mathematically expressed as follows:

$$V_z = V_b\, k_1\, k_2\, k_3$$

where

V_z = design of wind speed at any height z in m/s;

V_b = basic wind speed in m/s (Fig. 2.1);

k_1 = probability factor (risk coefficient)

k_2 = terrain, height and structure size factor; and

k_3 = topography factor

Note: Design wind speed up to 10 m height from mean ground level shall be considered constant. For details of k_1, k_2 and k_3 refer to National Building Code of India, 2005.

2.3.4 Design Wind Pressure

The design wind pressure at any height above mean ground level shall be obtained by the following relationship between wind pressure and wind velocity:

$$p_z = 0.6\, V_z^2$$

where

p_z = design wind pressure in N/m^2 at height Z, and

V_z = design wind velocity in m/s at height Z.

Note: The coefficient 0.6 (in SI units) in the above formula depends on a number of factors, and mainly on the atmospheric pressure and air temperature. The value chosen corresponds to the average appropriate Indian atmospheric conditions.

2.3.5 Offshore Wind Velocity

Cyclonic storms form far way from the sea coast and gradually reduce in speed as they approach the sea coast. Cyclonic storms generally extend up to about 60 km inland after striking the coast. Their effect on land is already reflected in basic wind speeds specified in Fig. 2.1. The influence of wind speed off the coast up to a distance of about 200 km may be taken as 1.15 times the value on the nearest coast in the absence of any definite wind data.

2.3.6. Wind Pressure and Forces on Buildings/Structure

2.3.6.1 General

The wind load on a building is calculated for:

(*a*) the building as a whole;

(*b*) individual structural elements as roofs and walls; and

(*c*) individual cladding units including glazing and their fixings.

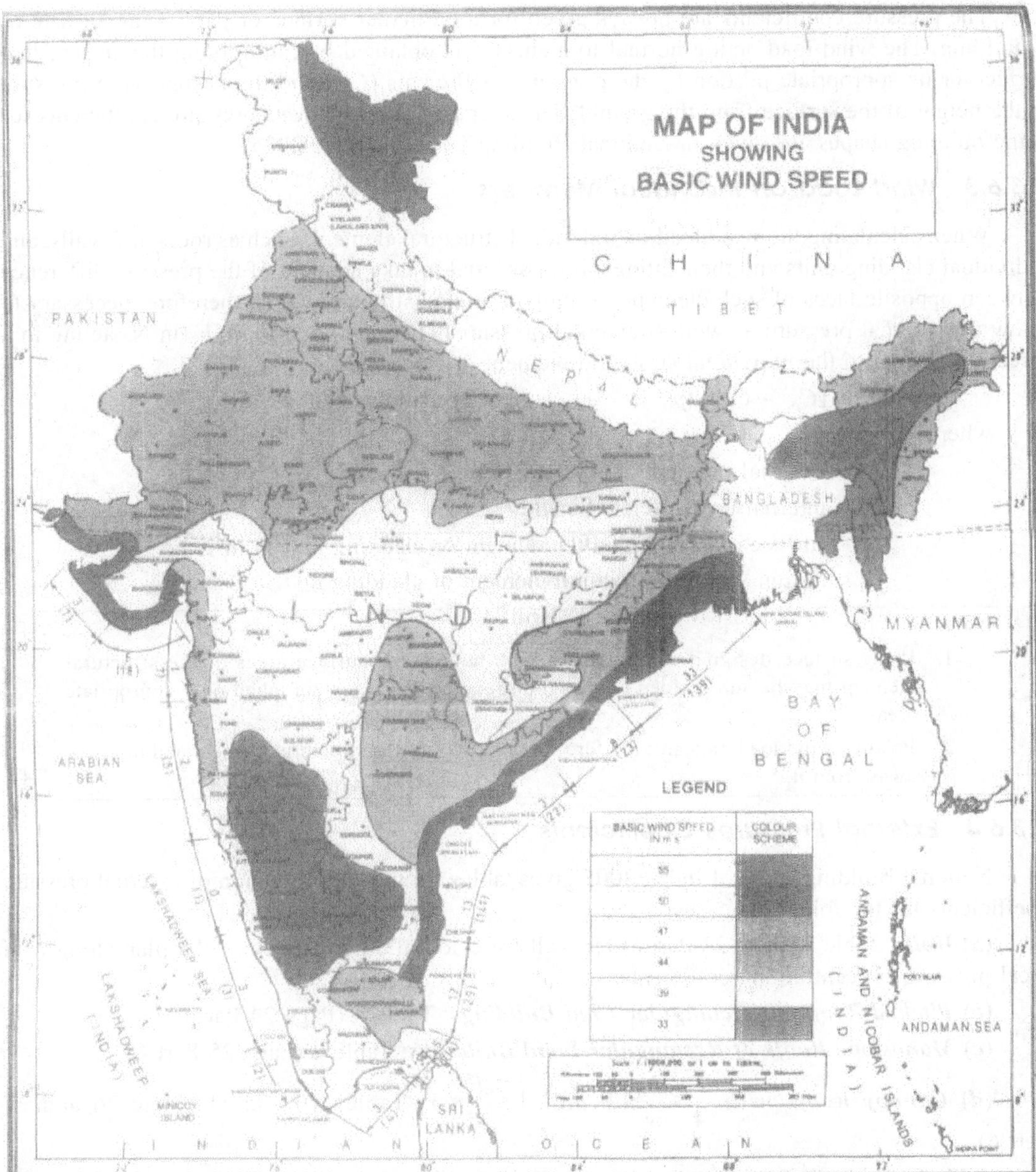

Based upon Survey of India Outline Map printed in 1993.

The territorial waters of India extend into the sea to a distance of twelve nautical miles measured from the appropriate base line.
The boundary of Meghalaya shown on this map is as interpreted from the North-Eastern Areas (Reorganisation) Act, 1971, but has yet to be verified.
Responsibility for correctness of internal details shown on the map rests with the publisher.
The state boundaries between Uttaranchal & Uttar Pradesh, Bihar & Jharkhand and Chhattisgarh & Madhya Pradesh have not been verified by Governments concerned.

Fig. 2.1: Basic wind speeds

2.3.6.2 Pressure Coefficients

The pressure coefficients are always given for a particular surface or part of the surface of a building. The wind load acting normal to a surface is obtained by multiplying the area of that surface or its appropriate portion by the pressure coefficients (C_p); and the design wind pressure at the height of the surface from the ground. The average values of these pressure coefficients for some building shapes are given in National Building Code of India, 2005.

2.3.6.3 Wind Load on Individual Members

When calculating the wind load on individual structural elements such as roofs and walls, and individual cladding units and their fittings, it is essential to take account of the pressure difference between opposite faces of such elements or units. For clad structures, it is, therefore, necessary to know the internal pressure as well as external pressure. Then the wind load, F (in N) acting in a direction normal to the individual structural element or cladding unit is:

$$F = (C_{pe} - C_{pi})\, A p_d$$

where

C_{pe} = external pressure coefficient;
C_{pi} = internal pressure coefficient;
A = surface area of structural element or cladding unit in m^2; and
p_d = design wind pressure in N/m^2

Notes:

1. If the surface design pressure varies with height, the surface areas of the structural element may be sub-divided so that the specified pressures are taken over appropriate areas.
2. Positive wind load indicates the force acting towards the structural element and negative away from it.

2.3.6.4 External Pressures Coefficients

National Building Code of India, 2005 gives tables for the determination of external pressure coefficients for the following cases:

(a) Walls: Table 7 (Page 23 Part 6) for walls of clad buildings of rectangular plan along with local pressure concentration coefficients.

(b) Pitched Roofs of Rectangular Clad Buildings: Table 8 (Page 24 Part 6)

(c) Monoslope Roofs of Rectangular Load Buildings: Table 9 (Page 25 Part 6)

(d) Canopy Roofs with $\frac{1}{4} < \frac{h}{w} < 1$ ***and*** $1 < \frac{l}{w} < 3$: Tables 10 and 11 (Page 26 and 27 Part 6)

Tables provide coefficients separately for monopitch and double pitch canopy roofs such as open-air parking garages, shelter areas, outdoor areas, railway platforms, stadium and theatres. The coefficients take account of the combined effect of the wind exerted on and under the roof for all wind directions. Tables 12 to 17 (Pages 28 to 32 Part 6) deal with some specific cases for which aspect ratios and roof slopes have been specified.

(e) Curved Roofs: Table 18 (Page 33 Part 6)

(f) Pitched and Saw-Tooth Roofs of Multi-span Buildings: Table 19 (Page 34 Part 6)

(g) Roofs and Bottom of Cylindrical Related Structure: The external pressure coefficients for roofs and bottoms of cylindrical elevated structures as given in Table 22 (Page 37 Part 6) (*see* also Fig. 2.2)

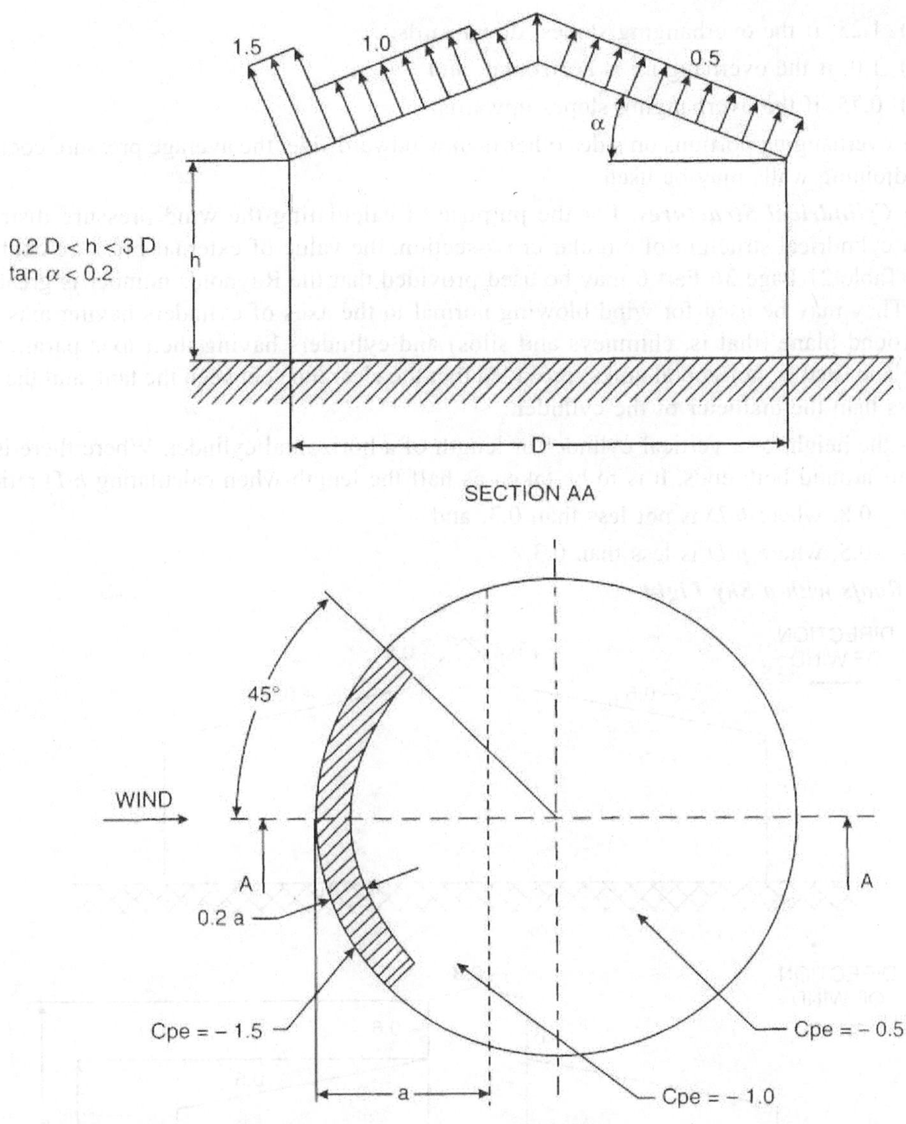

(For force coefficient corresponding to shell portion *see* Table 23 of Code)

Fig. 2.2: External pressure coefficients on the upper roof surface of singular circular standing on the ground

The total resultant load (P) acting on the roof of the structure is given by the following formula:

$$P = 0.785\ D^2\ (C_{pi} - C_{pe})\ p_d$$

The resultant of P for roofs lies at 0.1 D from the centre of the roof on the windward side.

Combined Roofs and Roofs with a Sky Light: The average external pressure coefficients for combined roofs and roofs with a sky light are shown in Table 23. (Page 39 Part 6)

(g) Pressure Coefficients on Overhangs from Roofs: The pressure coefficients on the top overhanging portion of the roofs is taken to be the same as that of the nearest top portion of the roof. Pressure coefficients for the underside surface of the overhanging portions is taken as follows and shall be taken as positive if the overhanging portion is on the windward side:

(1) 1.25, if the overhanging slopes; downwards;

(2) 1.0, if the overhanging is horizontal; and

(3) 0.75, if the overhanging slopes upwards.

For overhanging portions on sides other than windward side, the average pressure coefficients on the adjoining walls may be used.

(h) Cylindrical Structures: For the purpose of calculating the wind pressure distribution around a cylindrical structure of circular cross-section, the value of external pressure coefficients given in Table 21 Page 36 Part 6 may be used provided that the Raynolds number is greater than 10 000. They may be used for wind blowing normal to the axes of cylinders having axis normal to the ground plane (that is, chimneys and silos) and cylinders having their axis parallel to the ground plane (that is, horizontal tanks) provided that the clearance between the tank and the ground is not less than the diameter of the cylinder.

h is the height of a vertical cylinder or length of a horizontal cylinder. Where there is a free flow of air around both ends, h is to be taken as half the length when calculating h/D ratio.

(1) –0.8, where h/D is not less than 0.3; and

(2) –0.5, where h/D is less than 0.3.

(I) Roofs with a Sky Light

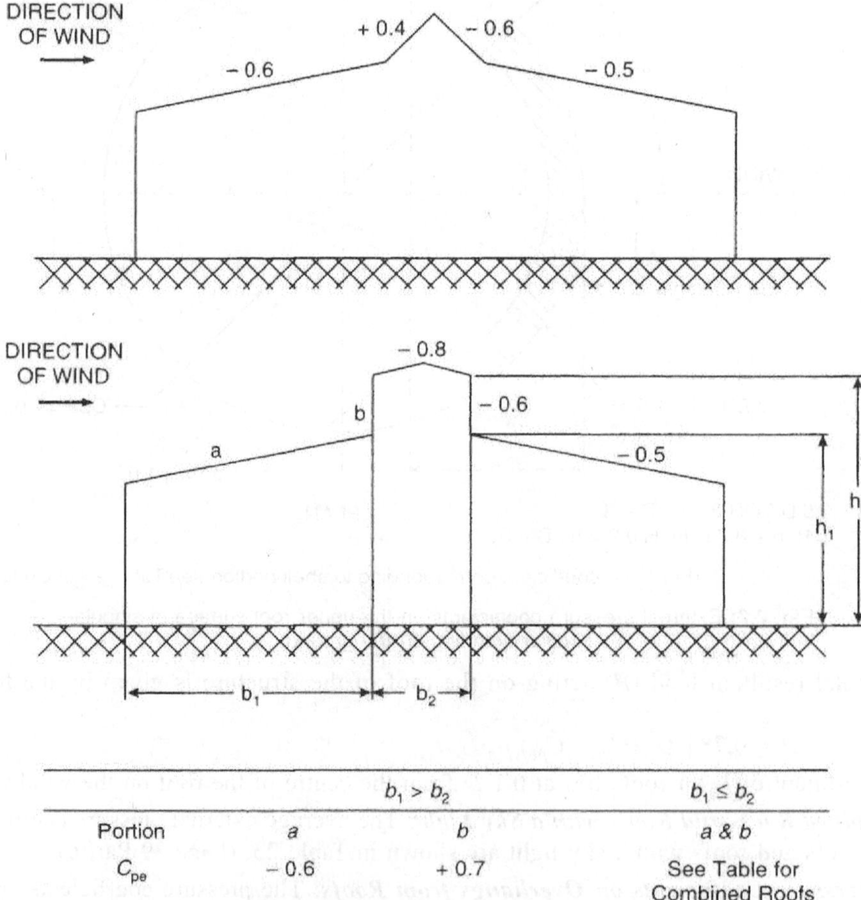

	$b_1 > b_2$		$b_1 \leq b_2$
Portion	*a*	*b*	*a & b*
C_{pe}	– 0.6	+ 0.7	See Table for Combined Roofs

Fig. 2.3: External pressure coefficients for roof with a sky-light

(*j*) ***Grandstands***: The pressure coefficients on the roof (top and bottom) and rear wall of a typical grandstand roof, which is open on three sides, is given in Table 24 (Page 41 Part 6). The pressure coefficients are valid for a particular ratio of dimensions as specified in the Table, but may be used for deviations up to 20 per cent. In general, the maximum wind load occurs, when the wind is blowing into the open front of the stand causing positive pressure under the roof and negative pressure on the roof.

(*k*) ***Upper Surface of Round Silos and Tanks***: The pressure coefficients on the upper surface of round silos and tanks standing on ground shall be as given in Fig. 2.2.

(*l*) ***Spheres***: The external pressure coefficients for spheres shall be as given in Table 25 (Page 42 Part 6).

2.3.6.5 Internal Pressure Coefficients

Internal air pressure in a building depends upon the degree of permeability of the cladding to the flow of air. The internal air pressure may be positive or negative depending on the direction of flow of air in relation to the openings in the buildings.

(*a*) In the case of buildings where the claddings permit the flow of air with openings not more than about 5 per cent of the wall areas but where there are no large openings, it is necessary to consider the possibility of the internal pressure being positive or negative.

Two design conditions shall be examined, one with an internal pressure coefficient of +0.2 and another with an internal pressure coefficient of – 0.2.

(*b*) ***Building with medium and large openings***: Building with medium and large openings may also exhibit either positive or negative internal pressure depending upon the direction of wind. Buildings with medium openings between about 5 to 20 per cent of wall area shall be examined for an internal pressure coefficient of +0.5 and later with an internal pressure coefficient of – 0.5, and the members shall be adopted. Buildings with large openings, that is, openings larger than 20 per cent of the wall area shall be examined once with an internal pressure coefficient of +0.7 and again with an internal pressure coefficient of – 0.7, and the analysis which produces greater distress on the members shall be adopted.

Buildings with one open side or openings exceeding 20 per cent of wall area may be assumed to be subjected to internal positive pressure or suction similar to those for buildings with large openings.

(*c*) In buildings with roofs but no walls, the roofs will be subjected to pressure from both inside and outside.

2.3.6.6 Force Coefficients

The value of force coefficients apply to a building or structure as a whole, and when multiplied by the effective frontal area, A_e of the building or structure and by design wind pressure, p_d give the total wind load on that particular building or structure.

$$F = C_f A_e p_d$$

where F is the force acting in a direction specified in the respective tables and C_f is the force coefficient for the building.

2.3.7 Frictional Drag

In certain buildings of special shape, a force due to frictional drag is taken into account. For rectangular clad buildings, this addition is necessary only where the ratio d/h or d/b is greater than 4. The frictional drag force, F' in the direction or the wind given by the following formulae:

If $h \le b$, $F' = C'_f (d - 4h)\, bp_d + C'_f (d - 4h)\, 2hp_d$

or If $h > b$, $F' = C'_f (d - 4b)\, bp_d + C'_f (d - 4b)\, 2hp_d$

The first term in each case gives the drag on the roof and the second on the walls. The value of C'_f has the following values:

$C'_f = 0.01$ for smooth surfaces without corrugations or ribs across the wind direction;

$C'_f = 0.02$ for surfaces with corrugations or ribs across the wind direction;

$C'_f = 0.04$ for surfaces with ribs across the wind direction.

For other buildings, the frictional drag has been indicated, where necessary, in the tables of pressure coefficients and force coefficients.

2.3.8 Force Coefficients for Clad Buildings

(a) Clad buildings of uniform section: The overall force coefficients for rectangular clad buildings of uniform section with flat roofs in uniform flow shall be as given in Fig. 2.4 and for other clad buildings of uniform section (without projections, except where otherwise shown) shall be as given in Table 2.6.

(b) Buildings of circular shapes: Force coefficients for buildings of circular cross-section shall be as given in Table 2.6 and Fig. 2.5.

(c) Low walls and hoardings: Force coefficients for low walls and hoardings less than 15 m high shall be as given in Table 2.7 provided the height shall be measured from the ground to the top of the walls or hoarding, and provided that for walls or hoardings above the ground the clearance between the wall or hoarding and the ground shall be not less than 0.25 times the vertical dimension of the wall or hoarding.

To allow for oblique winds the design shall also be checked for the net pressure normal to the surface varying linearly from a maximum of 1.7 C_f at the up wind edge to 0.44 C_f at the down wind edge.

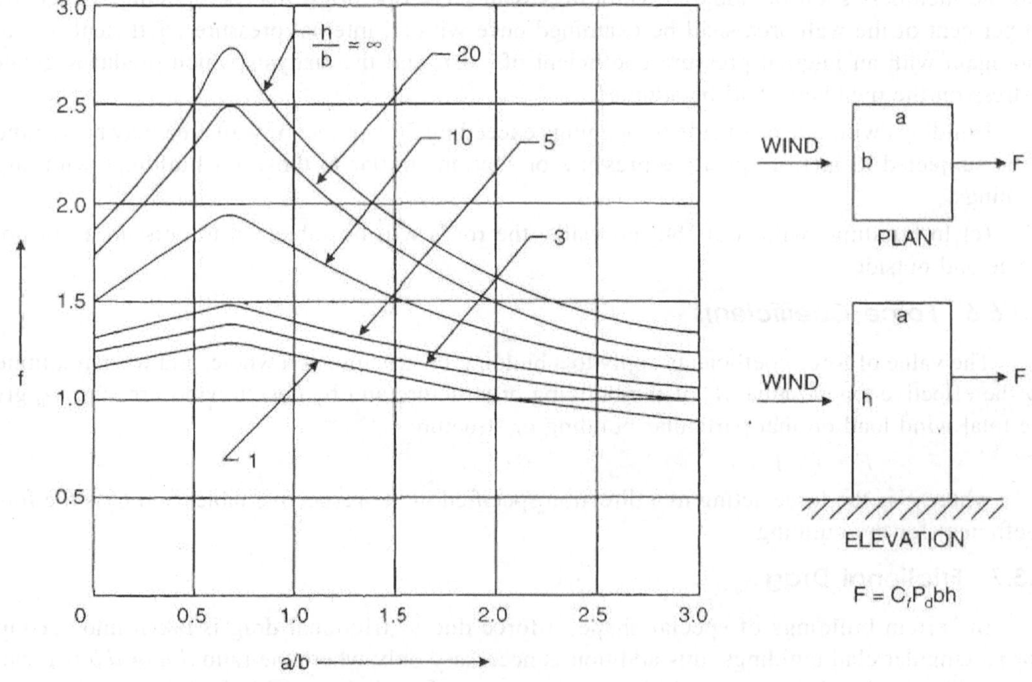

(a) Values of C_f *versus* a / b for $h / b \geq 1$

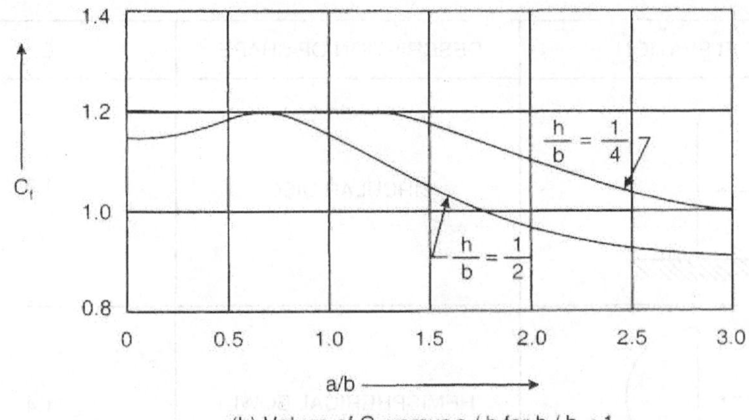

(b) Values of C_f *versus* a / b for h / b < 1

Fig. 2.4: Force coefficient for rectangular clad building in uniform flow

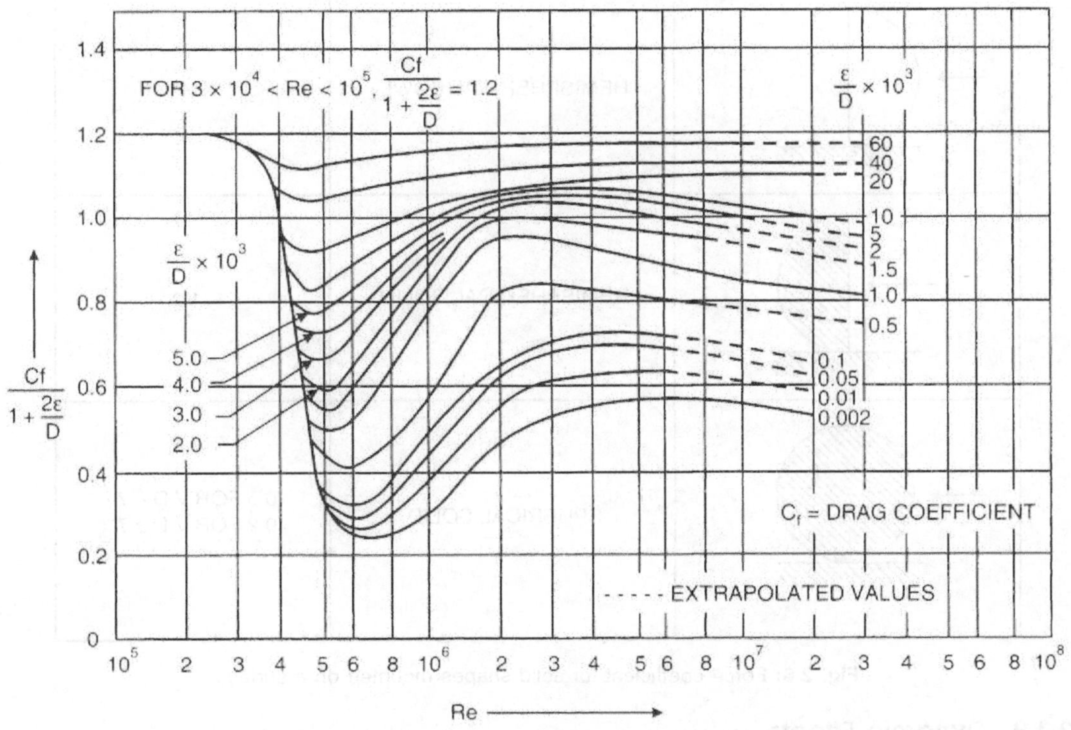

Fig. 2.5: Variation of $\frac{C_f}{1+\frac{2\varepsilon}{D}}$ with $Re < 3 \times 10^4$ for circular sections

The wind load on appurtenances and supports for hoardings shall be accounted for separately by using the appropriate net pressure coefficients. Allowance shall be made for the shielding effects of one element or another.

(d) Solid circular shapes mounted on a surface: The force coefficients for solid circular shapes mounted on a surface shall be as given in Fig. 2.6.

SIDE ELEVATION	DESCRIPTION OF SHAPE	C_f
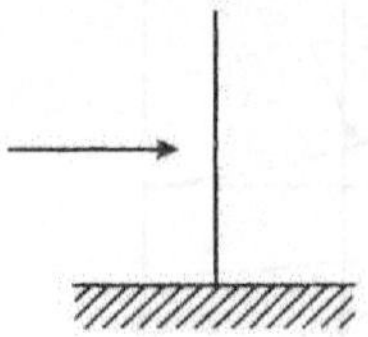	CIRCULAR DISC	1.2
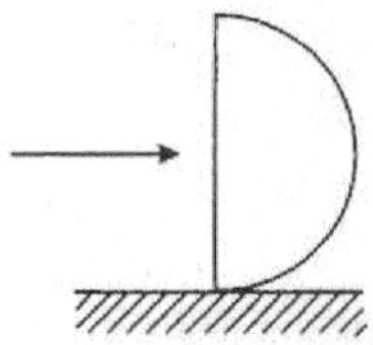	HEMISPHERICAL BOWL	1.4
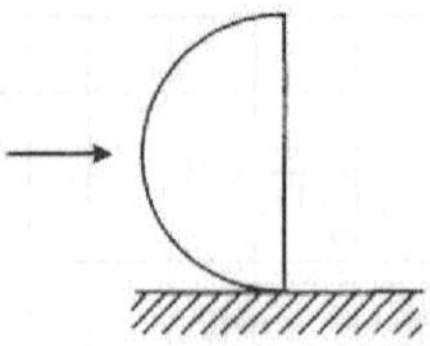	HEMISPHERICAL BOWL	0.4
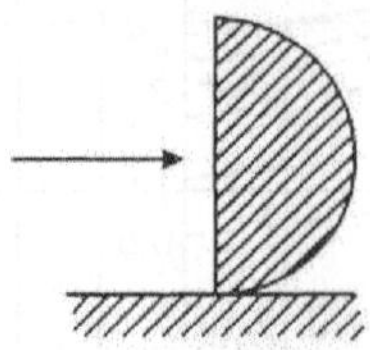	HEMISPHERICAL SOLID	− 1.2
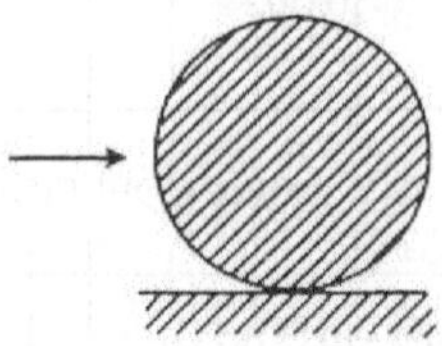	SPHERICAL SOLID	0.5 FOR V D < 7 0.2 FOR V D ≥ 7

Fig. 2.6: Force coefficient for solid shapes mounted on a surface

2.3.9 Dynamic Effects

Flexible slender structures and structural elements shall be investigated to ascertain the importance of wind induced oscillations for excitations along and across the direction of wind.

In general the following guidelines may be used for examining the problems of wind induced oscillations:

(*a*) Buildings and closed structures with a height to minimum lateral dimension ratio of more than about 5.0; or

(*b*) Buildings and structures whose natural frequency in the first mode is less than 1.0 Hz. Any building or structure which satisfies either of the above two criteria shall be examined for dynamic effects of wind.

Table 2.6: Force Coefficients C_f for Clad Buildings of Uniform Section (Acting in the Direction of Wind)

Plan Shape		$V_z b$ m²/s	C_f for Height/Breadth Ratio: Up to 1/2	1	2	5	10	20	∞
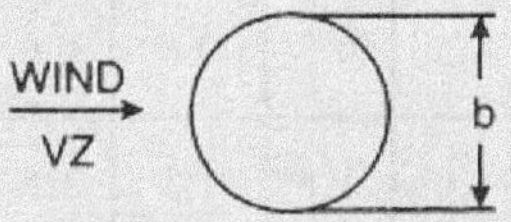 Circle (WIND VZ →, b) (*see also* Annex E)	All surfaces	< 6	0.7	0.7	0.7	0.8	0.9	1.0	1.2
	Rough or with projections	> 6							
	Smooth	≥ 6	0.5	0.5	0.5	0.5	0.5	0.6	0.6
Ellipse $b/d = 1/2$ (d, b)		< 10	0.5	0.5	0.5	0.5	0.6	0.6	0.7
		≥ 10	0.2	0.2	0.2	0.2	0.2	0.2	0.2
Ellipse $b/d = 2$ (d, b)		< 8	0.8	0.8	0.9	1.0	1.1	1.3	1.7
		≥ 8	0.8	0.8	0.9	1.0	1.1	1.3	1.5
$b/d = 1$, $r/b = 1/3$ (r)		< 4	0.6	0.6	0.6	0.7	0.8	0.8	1.0
		≥ 4	0.4	0.4	0.4	0.4	0.5	0.5	0.5
$b/d = 1$, $r/b = 1/6$ (r)		< 10	0.7	0.8	0.8	0.9	1.0	1.0	1.3
		≥ 10	0.5	0.5	0.5	0.5	0.6	0.6	0.6
$b/d = 1/2$, $r/b = 1/2$ (d, b, r)		< 3	0.3	0.3	0.3	0.3	0.3	0.3	0.4
		≥ 3	0.2	0.2	0.2	0.2	0.3	0.3	0.3
$b/d = 1/2$, $r/b = 1/6$ (d, b, r)		All values	0.5	0.5	0.5	0.5	0.6	0.6	0.7
$b/d = 2$, $r/b = 1/12$ (d, b)		All values	0.9	0.9	1.0	1.1	1.2	1.5	1.9

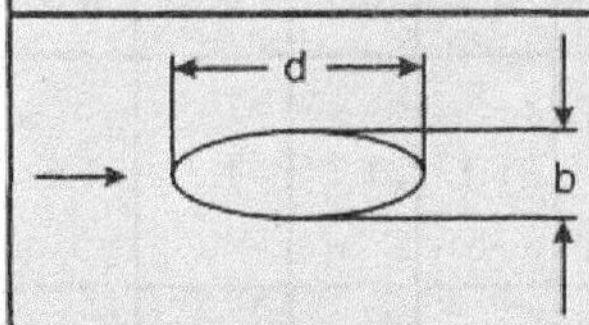

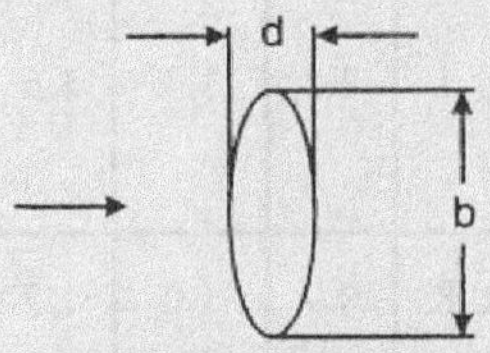

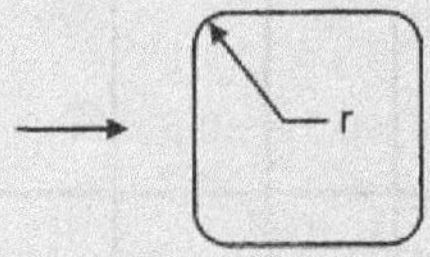

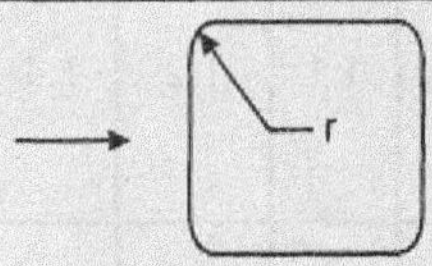

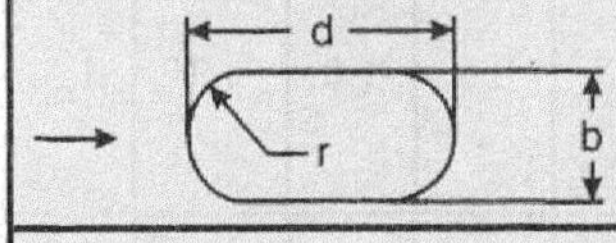

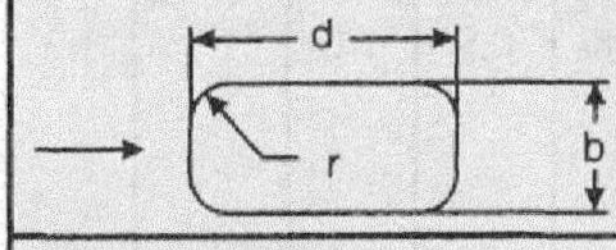

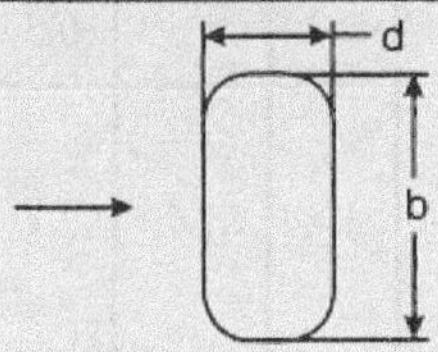

(Contd.)

Plan Shape	$V_z b$	C_f for Height/Breadth Ratio						
	m²/s	Up to 1/2	1	2	5	10	20	∞
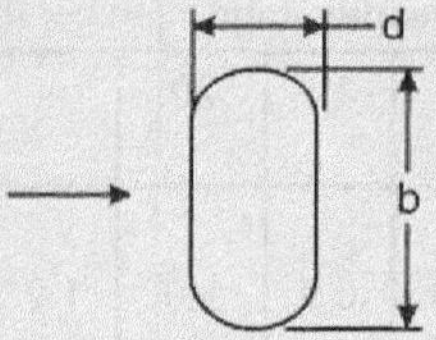 All surfaces b/d = 2 r/b = 1/4	< 6	0.7	0.8	0.8	0.9	1.0	1.2	1.6
	≥ 6	0.5	0.5	0.5	0.5	0.5	0.6	0.6
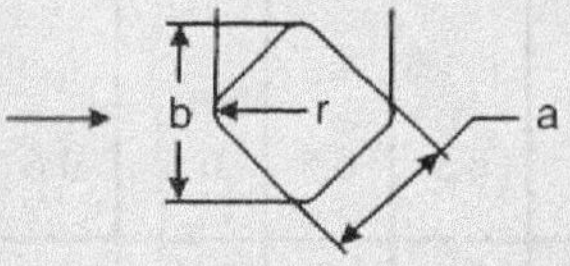 r/a = 1/3	< 10	0.8	0.8	0.9	1.0	1.1	1.3	1.5
	≥ 10	0.5	0.5	0.5	0.5	0.5	0.6	0.6
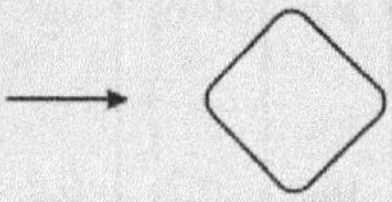 r/a = 1/12	All values	0.9	0.9	0.9	1.1	1.2	1.3	1.6
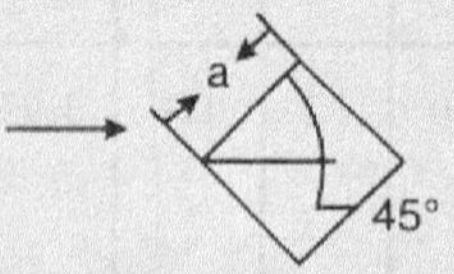 r/a = 1/48	All values	0.9	0.9	0.9	1.1	1.2	1.3	1.6
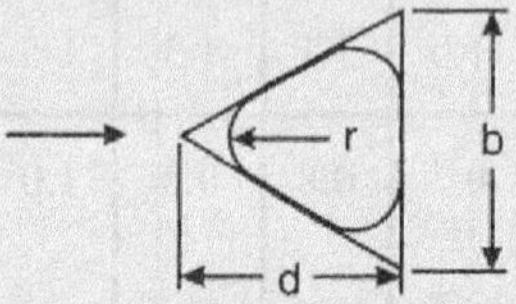 r/b = 1/4	< 11	0.7	0.7	0.7	0.8	0.9	1.0	1.2
	≥ 11	0.4	0.4	0.4	0.4	0.5	0.5	0.5
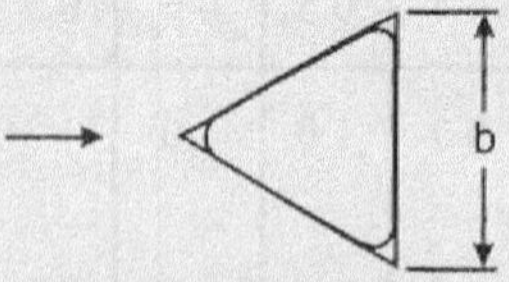 r/b = 1/12	All values	0.8	0.8	0.8	1.0	1.1	1.2	1.4
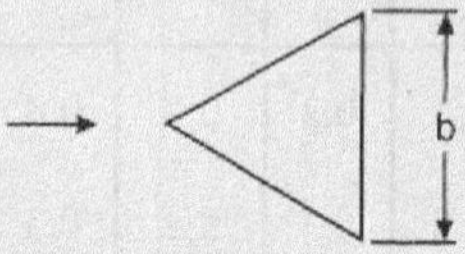 r/b = 1/48	All values	0.7	0.7	0.8	0.9	1.0	1.1	1.3
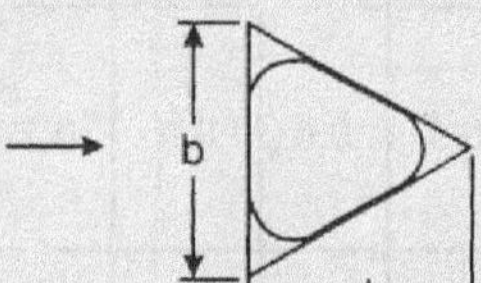 r/b = 1/4	< 8	0.7	0.7	0.8	0.9	1.0	1.1	1.3
	≥ 8	0.4	0.4	0.4	0.4	0.5	0.5	0.5
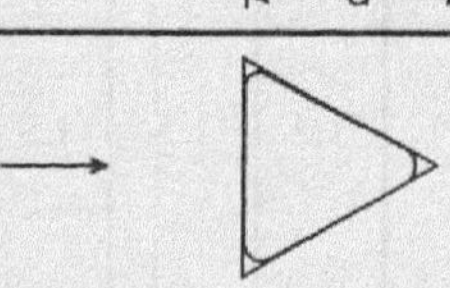 1/48 < r/b < 1/12	All values	1.2	1.2	1.2	1.4	1.6	1.7	2.1

(Contd.)

Plan Shape	$V_z b$	C_f for Height/Breadth Ratio						
	m^2/s	Up to 1/2	1	2	5	10	20	∞
12 sided polygon (→ d)	< 12	0.7	0.7	0.8	0.9	1.0	1.1	1.3
	≥ 12	0.7	0.7	0.7	0.7	0.8	0.9	1.1
Octagon	All values	1.0	1.0	1.1	1.2	1.2	1.3	1.4
Hexagon	All values	1.0	1.1	1.2	1.3	1.4	1.4	1.5

Note: Structures that, because of their size and the design wind velocity, are in the supercritical flow regime may need further calculation to ensure that the greatest loads do not occur at some wind speed below the maximum when the flow will be subcritical.

The coefficients are for buildings without projections, except where otherwise shown. In this table V b is used as an indication of the airflow regime.

Table 2.7: Force Coefficients for Low Walls or Hoardings (< 15 m High)

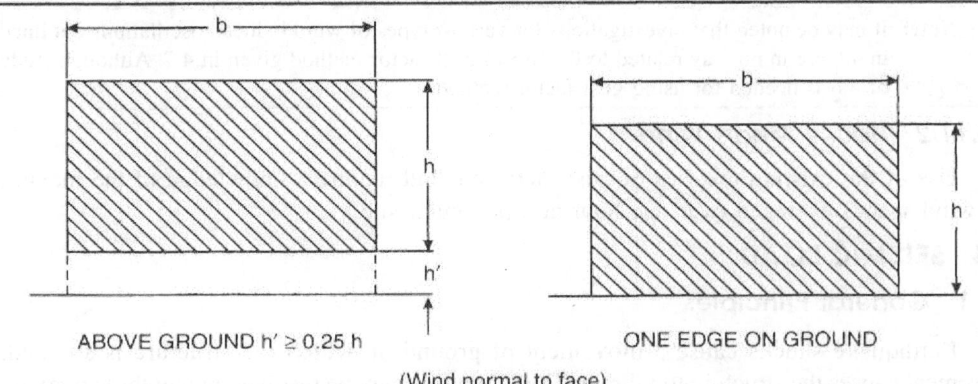

Width to Height Ratio, *b/h*		Drag Coefficient, C_t
Wall above ground	Wall on ground	
From 0.5 to 6	From 1 to 12	1.2
10	20	1.3
16	32	1.4
20	40	1.5
40	80	1.7
60	120	1.8
80 or more	160 more 20	2.0

2.3.10 Motions Due to Vortex Shedding

Slender structures: For a structure, the shedding frequency, η shall be determined by the following formula:

$$\eta = \frac{SV_d}{b}$$

where

S = Strouhal number,

V_d = Design wind velocity, and

b = The breadth of a structure or structural members in the horizontal plane normal to the wind direction.

(*a*) ***Circular Structure***: For structures circular in cross-section:

$S = 0.20$ for bV_z not greater than 7, and

$S = 0.25$ for bV_z greater than 7.

(*b*) ***Rectangular Structures***: For structures of rectangular cross-section:

$S = 0.15$ for all values of bV_z.

2.3.11 Gust Factor (GF) or Gust Effectiveness Factor (GEF) Method

2.3.11.1 Application

Only the method of calculating load along wind or drag load by using gust factor method is given in the section since methods for calculating load across-wind or other components are not fully matured for all types of structures. However, it is permissible for a designer to use gust factor method to calculate all components of load on a structure using any available theory. However, such a theory must take into account the random nature of atmospheric wind speed.

Note: It may be noted that investigations for various types of wind induced oscillations out lined in 4.6 are in no way related to the use of gust factor method given in 4.7. Although study of 4.6 is needed for using gust factor method.

2.3.11.2 Hourly Mean Wind

Use of the existing theories of gust factor method require a knowledge of the maximum of the wind speeds averaged over one hour at a particular site.

2.4 SEISMIC LOAD

2.4.1 General Principles

Earthquake shocks cause a movement of ground on which the structure is situated. This movement causes the structure to vibrate. The vibrations may be resolved in any three perpendicular directions and the design of structures, made safe for the component vibrations in the three directions acting simultaneously, shall be considered safe unless otherwise specifically stated. The predominant direction of vibration is horizontal.

The characteristic (intensity, duration, etc.) of seismic ground vibrations expected at any location depends upon the magnitude of earthquake, the depth of focus, the distance from the epicentre, characteristics of the path through which the seismic waves travel and the soil strata on which the structure stands. The important structures shall be designed for the maximum vibration intensity expected at the place.

The response of the structure to the ground vibration is a function of the nature of foundation soil; materials, form, size and mode of construction of the structure; and the duration and the intensity of ground motion. This section specifies design acceleration for structures standing on soils which will not considerably settle or slide appreciably due to vibration lasting for a few seconds.

In the case of structures designed for horizontal seismic force only it is considered to act in any one direction at a time. Where both horizontal and vertical seismic forces are taken into account, horizontal force in any one direction at a time may be considered simultaneously with the vertical force as specified.

The vertical seismic coefficient shall be considered only in the case of structures in which stability is a criterion of design or for overall stability except as otherwise stated.

Though the basis for the design of different types of structure is covered in this section, it is not implied that structural analysis should be made in every case. There might be cases of less importance and provided small structures for which no analysis need be made, relatively certain simple precautions are taken in the construction. For example, suitably proportioned diagonal bracings in the vertical panels of steel and concrete structures add to the resistance of frames to withstand earthquake forces. Similarly in highly seismic areas, construction of a type which entails heavy debris and consequent loss of life and property, such as masonry, particularly mud masonry and rubble masonry, should be avoided in preference to construction of a type which is known to withstand seismic effects better, such as construction in light weight materials and well braced timber-framed structures.

2.4.2 Assumptions in Earthquake Resistant Design

The following assumptions are made in the earthquake resistant design of structures:

(*a*) Earthquake causes impulsive ground motion which is complex and irregular in character, changing in period and amplitude each lasting for small duration. Therefore, resonance of the type as visualized under steady state sinusoidal excitations will not occur as it would need time to build up such amplitudes; and

(*b*) Earthquake is not likely to occur simultaneously with wind or maximum flood or maximum sea waves.

(*c*) The value of elastic modulus of materials, wherever required, may be taken as for elastic analysis unless a more definite value is available for use in such condition.

2.4.3 Permissible Increase in Stresses

Permissible increase in material stresses. Whenever earthquake forces are considered along with other normal design forces, the permissible stresses in materials, in the elastic method of design, may be increased by one-third. For steels having a definite yield stress, the stress be limited to the yield stress; for steels without a definite yield point, the stress will be limited to 80 per cent of the ultimate strength or 0.2 per cent proof stress, which ever is smaller; and that in prestressed concrete members, the tensile stress in the extreme fibres of the concrete may be permitted so as not to exceed two-thirds of the modulus of rupture of concrete.

Permissible increase in allowable bearing pressure of soils. For permissible increase in allowable bearing pressure of soils, reference may be made to Table 3.2 on page 68.

2.4.4 Design Horizontal Seismic Coefficient for Different Zones

For the purpose of determining the seismic forces, the country is classified into four seismic zones, as shown in Fig. 2.7.

The design horizontal seismic coefficient A_h for a structure shall be determined by the following expression:

$$A_h = \frac{ZI\,S_a}{2\,Rg} \tag{2.1}$$

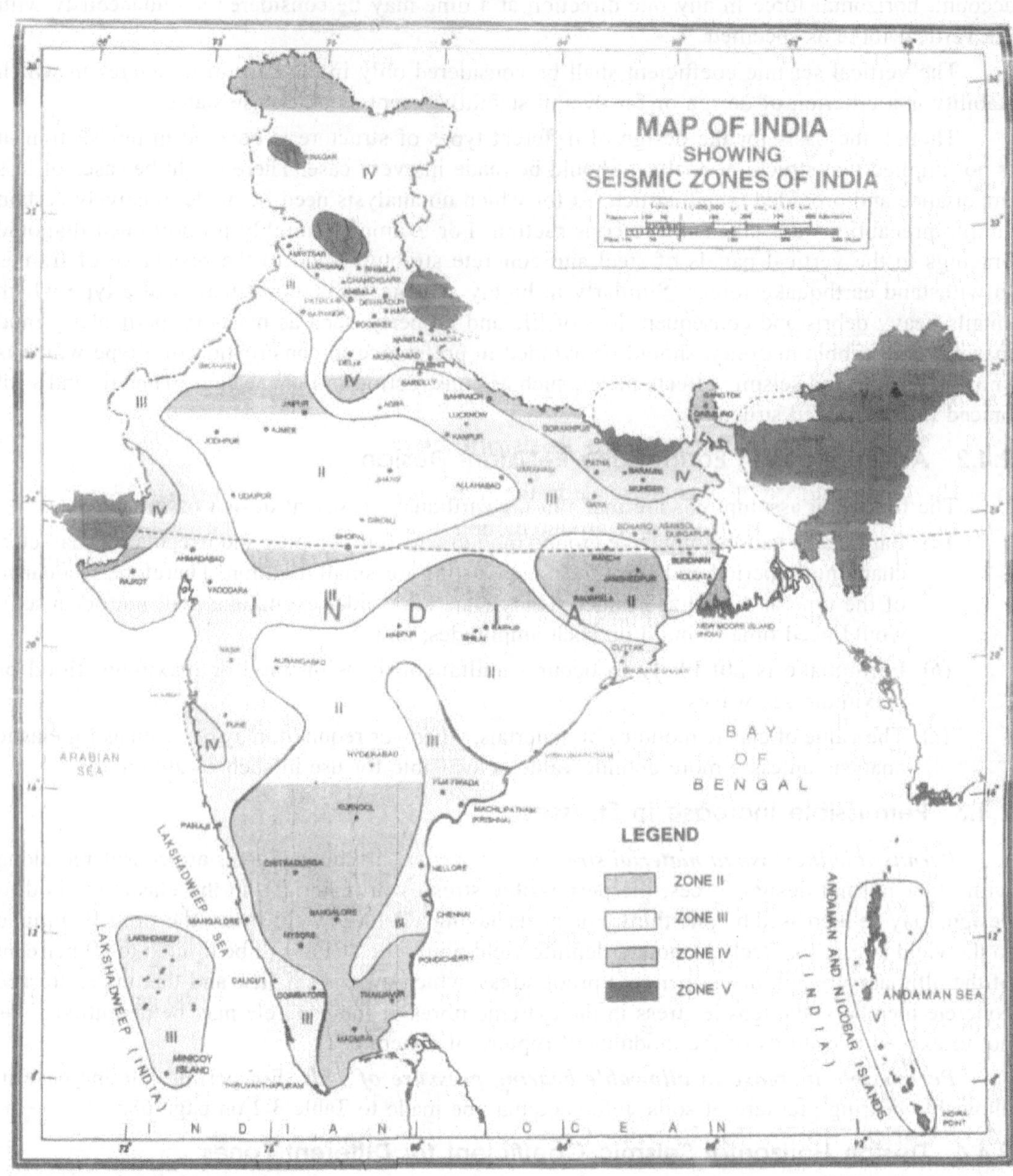

Based upon Survey of India Outline Map printed in 1993

The territorial waters of India extend into the sea to a distance of twelve nautical miles measured from the appropriate base line.
The boundary of Meghalaya shown on this map is as interpreted from the North-Eastern Areas (Reorganisation) Act, 1971, but has yet to be verified.
Responsibility for correctness of internal details shown on the map rests with the publisher.
The state boundaries between Uttaranchal & Uttar Pradesh, Bihar & Jharkhand and Chhatisgarh & Madhya Pradesh have not been verified by Governments concerned.

NOTE — Towns falling at the boundary of zones demarcation line between two zones shall be considered in higher zone.

Fig. 2.7: Seismic zones of India

Provided that for any structure with $T \geq 0.1$ sec, the value of A_h will not be less than $Z/2$ whatever be the value of I/R

where

Z = Zone factor given in Table 2.8, is for the maximum considered earthquake (MCE) and service life of structure in a zone. The factor 2 in the denominator of ≥ is used so as to reduce the maximum considered earthquake (MCE) zone factor to the factor for design basis earthquake (DBE).

I = Importance factor, depending upon the functional use of the structures, characterised by hazardous consequences of its failure, post-earthquake functional needs, historical value, or economic importance (Table 2.9).

R = Response reduction factor, depending on the perceived seismic damage performance of the structure, characterized by ductile or brittle deformations. However, the ratio (I/R) shall not be greater than 1.0. The values of R for buildings are given in Table 2.10.

S_a/g = Average response acceleration coefficient for rock or soil sites as given by Fig. 2.8 and Table 2.11 based on appropriate natural periods and damping of the structure. These curves represent free field ground motion.

Table 2.8: Zone Factor, *Z*

Seismic Zone	**II**	**III**	**IV**	**V**
Seismic Intensity	Low	Moderate	Severe	Very Severe
Z	0.10	0.16	0.24	0.36

Table 2.9: Importance Factors, *I*

Sl. No.	**Structure**	**Importance Factor**
(*i*)	Important service and community buildings, such as hospitals; schools; monumental structures; emergency buildings like telephone exchange, television stations, radio stations, railway stations, fire station buildings; large community halls like cinemas, assembly halls and subway stations, power stations.	1.5
(*ii*)	All other buildings.	1.0

Table 2.10: Response Reduction Factor[1] *R*, for Building Systems

Sl. No.	**Lateral Load Resisting System**	***R***
	Building Frame Systems	
(*i*)	Ordinary RC Moment-Resisting Frame (OMRF)	3.0
(*ii*)	Special RC Moment-Resisting Frame (SMRF)	5.0
(*iii*)	Steel Frame with	
	(*a*) Concentric Braces	4.0
	(*b*) Eccentric Braces	5.0
(*iv*)	Steel Moment Resisting Frame designed as per SP: 6(6)	5.0
	Building with Shear Walls	
(*v*)	Load Bearing Masonry Wall Buildings	
	(*a*) Unreinforced	1.5

(*Contd.*)

Sl. No.	Lateral Load Resisting System	*R*
	(*b*) Reinforced with horizontal RC Bands	2.5
	(*c*) Reinforced with horizontal RC bands and vertical bars at corners of rooms and jambs of openings	3.0
(*vi*)	Ordinary Reinforced Concrete Shear Walls	3.0
(*vii*)	Ductile Shear Walls	4.0
	Buildings with Dual Systems	
(*viii*)	Ordinary Shear Wall with OMRF	3.0
(*ix*)	Ordinary Shear Wall with SMRF	4.0
(*x*)	Ductile Shear Wall with OMRF	4.5
(*xi*)	Ductile Shear Wall with SMRF	5.0

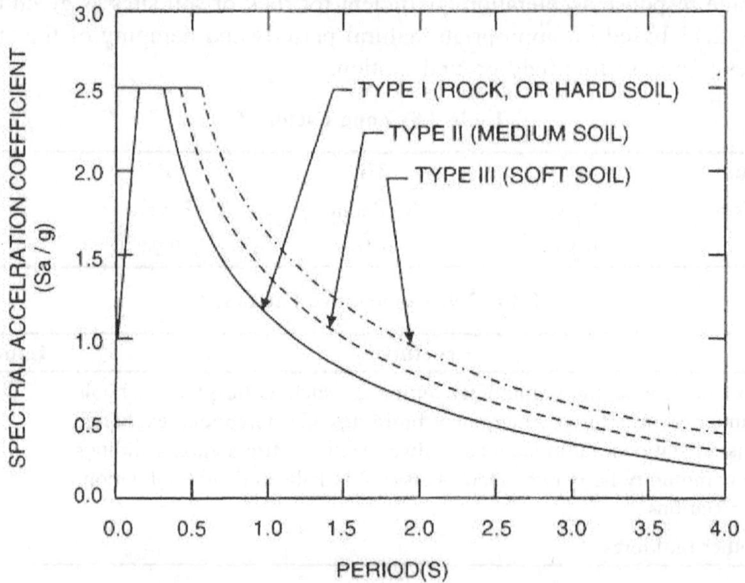

Fig. 2.8: Response spectra for rock and soil sight for 5 per cent damping

Table 2.11: Multiplying Factors for Obtaining Values for Other Damping

Damping per cent	0	2	5	7	10	15	20	25	30
Factors	3.20	1.40	1.00	0.90	0.80	0.70	0.60	0.55	0.50

Where a number of modes are to be considered for dynamic analysis, the value of A_h as defined in above for each mode shall be determined using the natural period of vibration of that mode.

For underground structures and foundations at depths of 30 m or below, the design horizontal acceleration spectrum value shall be taken as half the value obtained from equation 2.1. For structures and foundations placed between the ground level and 30 m depth, the design horizontal acceleration spectrum value shall be linearly interpolated between A_h and 0.5 A_h.

The design acceleration spectrum for vertical motions, when required, may be taken as two-thirds of the design horizontal acceleration spectrum given by equation 2.1.

Figure 2.8 shows the proposed 5 per cent spectra for rocky and soils sites and Table 2.11 gives the multiplying factors for obtaining spectral values for various other dampings.

For Rocky, or hard soil sites

$$\frac{S_a}{g} = \begin{cases} 1+15T; & 0.00 \le T \le 0.10 \\ 2.50 & 0.10 \le T \le 0.40 \\ 1.00/T & 0.40 \le T \le 4.00 \end{cases}$$

For Medium soil sites

$$\frac{S_a}{g} = \begin{cases} 1+15T; & 0.00 \le T \le 0.10 \\ 2.50 & 0.10 \le T \le 0.55 \\ 1.36/T & 0.55 \le T \le 4.00 \end{cases}$$

For Soft soil sites

$$\frac{S_a}{g} = \begin{cases} 1+15T; & 0.00 \le T \le 0.10 \\ 2.50 & 0.10 \le T \le 0.67 \\ 1.67/T & 0.67 \le T \le 4.00 \end{cases}$$

2.4.5 Design imposed Loads for Earthquake

For various loading classes as specified in Table 2.2, the horizontal earthquake force shall be calculated for the full dead load and the percentage of imposed loads as given in Table 2.12.

Table 2.12: Percentage of Imposed Load to be Considered in Seismic Weight Calculation

Imposed Uniformity Distributed Floor Loads (kN/sq.m)	Percentage of Imposed Load
Up to and including	25
Above 3.0	50

Notes 1. The percentage of imposed load given above shall also be used for calculating stresses due to vertical loads for combining with those due to earthquake forces. Under the earthquake condition, the whole frame may be assumed as loaded with live load except the roof.

2. The proportions of the imposed Load indicated above for calculating the horizontal seismic forces are applicable to average conditions. Where the probable loads at the time of an earthquake are more accurately assessed, the designer may alter the proportions indicated or even replace the entire live load proportions by the actual assessed load.

3. If the imposed Load is assessed instead of taking the above proportions for calculating horizontal earthquake force, only that part of the live load shall be considered which possesses mass. Earthquake force shall not be applied on impact effects.

For calculating the earthquake force on roofs, the imposed load may not be considered.

2.5 SNOW LOAD

Roofs should be designed for the actual load due to snow or for the imposed loads specified in 2.2, whichever is more severe.

Mountainous regions in northern part of India (parts of Jammu and Kashmir, Punjab and Himachal Pradesh and Uttarakhand) are subjected to snow fall.

The minimum design snow load on a roof area is obtained by multiplying the snow load on ground, S_o by the shape coefficient μ, as applicable to the particular roof area

$$S = \mu\, S_o$$

where

S = Design snow load in P_a on plan area of roof

μ = Shape coefficient (Refer to National Building Code of India 2005, Page 75, Part 6).

S_o = Ground snow load in P_a ($1\ P_a \simeq 1\ N/m^2$)

Ground snow load at any place depends on the critical combination of the maximum depth of undisturbed snow/fall and its average density.

2.6 SPECIAL LOADS

2.6.1 Temperature Effects

Expansion and contractions due to changes in temperature of the materials of a structure causes stresses. Provision is made either to relieve the stress by either providing expansion/contraction joints or by designing the structure to carry additional stresses due to temperature effects.

The structural analysis must take into account the changes of the mean (through the section) temperature in relation to the initial and the temperature gradient through the section.

(*a*) The changes of mean temperature in relation to the initial are liable to differ as between one structural element and another e.g. between the external walls and the internal elements. The distribution of temperature through single leafed structural elements is assumed to be linear.

(*b*) The effect of mean temperature changes t_1 and t_2 and the temperature gradients v_1 and v_2 in the hot and cold seasons for single leaf structural elements are evaluated on the basis of analytical principles.

2.6.2 Hydrostatic and Soil Pressure

In the design of structures or parts of structures below ground-level, such as basement floors and walls, the pressures exerted by the soil or water or both shall be duly accounted for on the basis of established theories. Due allowance shall be made for possible surcharge from stationary or moving loads.

All foundation slabs and other footings subjected to water pressure are designed to resist a uniformly distributed uplift equal to the full hydrostatic pressure. Checking of overturning of foundation under submerged conditions is done considering buoyant weight of foundation.

While determining the lateral soil pressure on column like structural members, such as pillars which rest in sloping soils, the width of the member shall preferably be taken as follows (*see* Fig. 2.9).

Actual Width of Member	*Ratio of Effective Width to Actual Width*
Less than 0.5 m	3.0
Beyond 0.5 m and upto 1 m	3.0 to 2.0
Beyond 1 m	2.0

The relieving pressure of soil in front of the structural member concerned may generally not be taken into account.

Safeguarding of structures and structural members against overturning and horizontal sliding shall be verified. Imposed loads having favourable effect shall be disregarded for the purpose. Due consideration shall be given to the possibility of soil being permanently or temporarily removed.

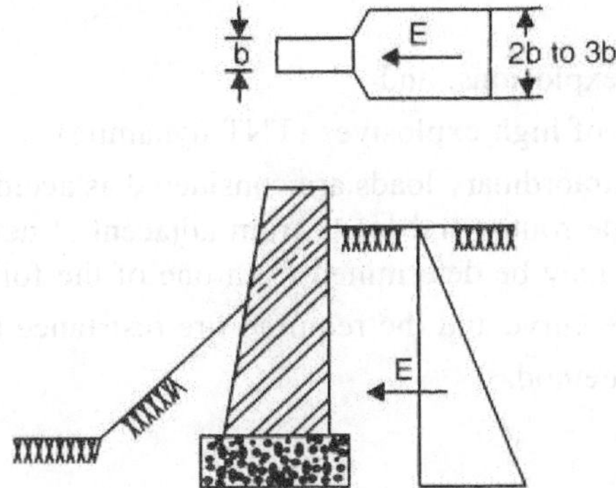

Fig. 2.9: Lateral soil pressure for a pillar resting in sloping soils

2.6.3 Fatigue

Fatigue cracks appear at points of high stress concentration caused by or associated with holes (such as bolt or rivet holes in steel structures), welds and local and general changes in geometry of members. The cracks usually propagate, if loading is continuous.

When necessary permissible stresses shall be reduced to allow for the effects of fatigue. Allowance for fatigue is made for combinations of stresses due to dead load and imposed load. Stresses due to wind and earthquake may be ignored when fatigue is being considered, unless otherwise specified.

2.6.4 Erection Loads

All loads required to be carried by the structure or any part of it due to placing or storage of construction materials and erection equipment including all loads due to operation of such equipment are considered as 'erection loads'. Proper provisions shall be made to take care of all stresses due to such loads.

2.6.5 Accidental Loads

The accidental loads, arising out of human action, include the following:

(*i*) Impact and collisions
(*ii*) Explosions; and
(*iii*) Fire

The causes of accidental loads may be:

(*i*) inadequate safety of equipment (due to poor design or poor safety); and
(*ii*) wrong operation (due to insufficient teaching or training, indisposition, negligence or unfavourable circumstances)

Common sources of *impact* are:

(*a*) Vehicle
(*b*) Dropped objects from cranes, fork lifts, etc.
(*c*) Cranes out of control, crane failures; and
(*d*) Flying fragments

Explosions may cause impulsive loading and are following types:

(*a*) Internal gas explosions
(*b*) Internal dust explosions

(*c*) Boiler failure;

(*d*) External gas cloud explosions; and

(*e*) External explosions of high explosives (TNT dynamite)

During a *fire*, possible extraordinary loads are considered as accidental actions. Examples are loads from people along escape routes and loads from adjacent structure failing because of fire. The thermal effect during fire may be determined from one of the following methods:

(*i*) the time temperature curve and the required fire resistance (minutes) and

(*ii*) an energy balance method.

2.7 OTHER LOADS

Other loads, such as moisture and shrinkage effects, loads due to technical process, etc. should be taken into account where stipulated by building design codes. Adequate provision should be made for the effects of one or more of these phenomena separately or in appropriate combination in accordance with the recommendations of the appropriate design codes.

2.8 LOAD COMBINATIONS

A judicious combination of the working loads specified above keeping in view the probability of their acting together and their disposition in relation to other loads and the severity of stresses or deformations caused by the combination of the various loads is necessary to ensure the required safety and economy in the design of a structure.

The various loads should, therefore, be combined in accordance with the stipulations in the appropriate design codes. In the absence of such recommendations, however, the following load combinations, whichever combination produces the most unfavourable effect in the building, foundation or structural member concerned may be adopted as a general guidance:

(*i*) Dead load

(*ii*) Dead load + Imposed load

(*iii*) Dead load + Wind load

(*iv*) Dead load + Earthquake load

(*v*) Dead load + Temperature load

(*vi*) Dead load + Imposed load + Wind load

(*vii*) Dead load + Imposed load + Earthquake load

(*viii*) Dead load + Imposed load + Temperature load

(*ix*) Dead load + Wind load + Temperature load

(*x*) Dead load + Earthquake load + Temperature load

(*xi*) Dead load + Imposed load + Wind load + Temperature load

(*xii*) Dead load + Imposed load + Earthquake load + Temperature load

While considering the load combinations recommended above, the effects of other forces should be taken into account.

QUESTIONS

1. List the various loads which should be taken into account for the design of a building. How will you account for dead load?
2. What do you understand by the term 'imposed load'? How will you consider the imposed load on floors and roofs?
3. How will you consider the effect of impact and vibration in a building?

4. What is a wind load? What do you understand by basic wind pressures?
5. How wind pressure is taken into account on the following parts:
   (*a*) Walls
   (*b*) Roofs
   (*c*) Structural framework.
6. What are the general principles to account for seismic load? How will you calculate seismic coefficients?
7. Write short notes on:
   (*i*) Soil and hydrostatic pressures
   (*ii*) Erection loads
   (*iii*) Temperature effects, and
   (*iv*) Combination of loads.

REFERENCES

1. Bureau of Indian Standards, New Delhi, SP–7—1983; National Building Code of India 2005, Part 6.
2. I.S. 875 (Part 1)-1987; Code of practice for design loads (other than earthquake) for buildings and structures; Part 1, Dead loads-Unit weights of building material and stored materials (*second revision*).

CHAPTER

Soils and Their Investigation

The materials supporting a building may be broadly classified into two categories, namely, rock and soils. Rock is a natural aggregate of cemented minerals. Soil on the other hand is a natural aggregate of mineral grains which can be separated easily by mechanical means such as sieving or agitation in water.

Rocks are of three classes according to their mode of formation. These are igneous, sedimentary and metamorphic rocks. Igneous rocks are formed from the solidification of molten matter called *magma*. This solidification may occur in the surface of earth or above it. Sedimentary rocks are the result of the accumulation of weathered deposits of igneous rocks. Metamorphic rocks are either igneous or sedimentary rocks whose physical or chemical properties have been altered by the action of intense pressure or heat. Rocks, if encountered as a layer, on which a building is to be constructed, ensure a good support for the building. Rocks can take heavy loads without causing significant settlement or failure. Investigation of rock surfaces is restricted to the determination of any cracks and soft layers underneath. However, even this is necessary only when multi-storeyed buildings have to be constructed. Although rock would provide an ideal surface for constructing a building, yet it is not met with in abundance at building sites.

Rocks are subjected to weathering agencies of wind, water, temperature, frost, waves, etc. These forces of nature weaken the rock considerably and subsequently disintegration takes place. The products of rock weathering located at the place of their origin are termed as residual soils. If they get transported to a different place and deposited there, they are called transported soils. Residual soils are of more uniform character than the alluvial soils. The construction of buildings on alluvial soils needs more detailed investigations to be conducted. Constructional problems also occur in some residual soils, *e.g.*, black cotton soils.

TYPES OF SOILS

Soils may be broadly classified as under:

(1) ***Gravel***. Cohesionless aggregates of angular, rounded or semi-rounded fragments, of more or less unaltered rocks or minerals. This consists of particles of coarser material resulting from the disintegration of rocks. These particles are often transported by water from their original source.

As a result of transportation, these particles are worn out. Fifty per cent or more of the particles in size from 4.75 mm to 80 mm are classified as gravels and those greater than 80 mm are called pebbles. Gravel deposits neither swell when moisture penetrates into them nor they shrink when dried from a moist condition. Any water which collects at their surface is drained off rapidly. They have high strength to bear the loads coming over them. Buildings built on gravel do not show any signs of settlement. Gravel deposits form the second best foundation surface for a building.

(2) *Sands*. Cohesionless aggregates of rounded, subrounded, angular, subangular or flat fragments of more or less unaltered rock or minerals. The particles are visible to the naked eye and have rough surfaces. Clean sand particles will fall apart when collected in a dry state. Their size varies from 0.075 mm to about 4.75 mm. Sand deposits are not affected by the action of frost. They do not swell in size when water penetrates into them. No shrinkage occurs when they are dried from a moist condition. Coarse sands do not allow water to rise through capillary action and are very permeable. Sands in a dense state provide a good foundation surface for buildings and there are no settlement problems associated with them.

(3) *Silt*. This is a fine grained soil with little or no plasticity. The particles are usually equi-dimensional in size, ranging from 0.002 mm to 0.075 mm. Silt particles have a smooth texture. Silt is relatively impervious. It may also exhibit a slight tendency towards swelling or shrinkage. Silt deposits may contain some organic matters. If the silt layers are in lower density due to looseness in structure they can bear stresses of small magnitude. The settlements on account of the loading may also be significant.

(4) *Clays*. An aggregate of microscopic or sub-microscopic particles derived from the chemical decomposition and disintegration of rock constituents. It is plastic within a moderate to wide range of water content. Particles are less than 0.002 mm in size. They have a smooth and flaky appearance. Clays are formed during excessive weathering of the coarser particles of the rock. The elongated shape of clay particles causes a plastic mixture to be formed in wet state. There is very small space between the individual particles of clays and hence water cannot be drained off easily. This keeps the clays in a damp condition for a considerable period after water has penetrated into them. Whenever loads are applied to a clayey soil, settlement takes place gradually and may last even for many years. Buildings constructed on clayey soils are thus liable to a gradual settlement. Clays also show the property of shrinkage if dried. The capacity of wet clays to bear loads is very less. A high capillary rise of moisture is prevalent in clay. It is difficult to construct buildings on a surface of a pure clayey soil.

SOIL INVESTIGATION

The object of soil investigation is to get information regarding:

(*i*) the nature, thickness and variation of soil strata at a place,

(*ii*) physical properties of the soil strata encountered, and

(*iii*) the seasonal variations in ground water table and its effect on the soil strata met with.

The first step in soil investigation is to carry out a preliminary survey of the site. Information can be collected from the existing local knowledge, records of trial pits, bore holes, etc. in the vicinity regarding the type of soils available at different levels and also the location of water table from season to season. A study of any building nearby particularly of a similar nature to the one proposed, will give a fair idea of the behaviour of the soil at the site. In case a good soil is met with, further investigation may not be necessary. However, for high rise buildings and where the character of soil is doubtful, it is essential to carry out a thorough investigation of the soil available at the site.

The methods used for further study are:

(1) Open trial pits.

(2) Borings.

(3) Sub-surface sounding probing.

(4) Geo-physical methods.

A brief description of these follows:

3.1 OPEN TEST PITS

A rectangular pit is excavated at the site and sample of soils collected at different levels. Further idea about the nature of soil layers can be obtained by examining the sides of the pit (*see* Fig. 3.1). This method is, however, limited to depths of 3 metres and cannot be used below ground water table. Moreover, for areas where rock or boulders are encountered after some excavation, this method is unsuitable.

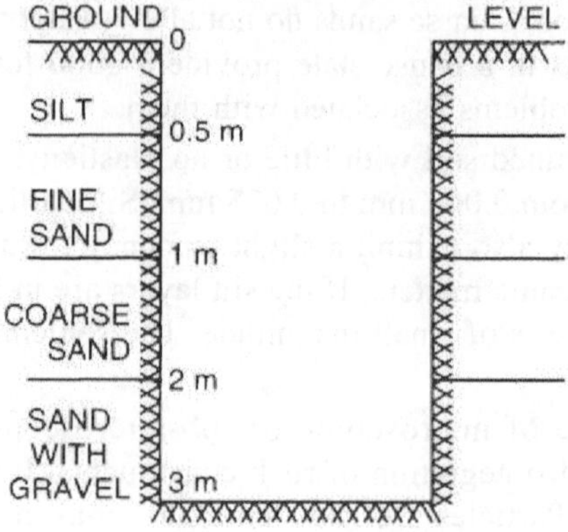

Fig. 3.1: A typical section of a test pit.

The spacing of these test pits depends on the nature of building, the variation in soil met with and the time available for investigations. At least one test pit should be excavated for a 15 m × 15 m area. In case the sides of the test pit do not stand vertically, proper bracing should be provided.

3.2 BORINGS

The types of borings usually adopted are:

(*a*) Thin walled tube sampling

(*b*) Wash borings

(*c*) Auger borings, and

(*d*) Mechanised drilling.

3.2.1 Thin Walled Tube Sampling

This method is used for small jobs. A tube about 2 metres in length with a cutting edge and a flap valve at the lower end is used for taking out the soil sample. It is raised by a rope passing over a pulley which in turn is suspended by a tripod stand. The tube is allowed to fall under gravity inside a casing pipe. The thick mixture formed near the cutting edge enters the tube and is taken out as the tube is raised. The casing is driven as the depth of excavation increases. Water may be added to faciltate the taking out of the slurry. Only a rough idea about the nature of the soil can be had, as the soil coming out is in a mixed state.

3.2.2 Wash Borings

This method is suitable for investigations to deeper depths. The equipment usually consists of a 60 mm dia pipe in 1/2 metre length and is called casing. This serves to support the sides of the hole. Suitable weights are used for driving the casing. A derrick is used for driving casing and handling the weight. A wash pipe 30 mm in dia (usually a thick walled pipe called drill rod) and $1\frac{1}{2}$ to 3 metres in length is used for injecting water and this is connected to a pump through a hose pipe. For boring, the derrick is erected and the casing is driven about 1 metre into the ground. A T-junction is attached to the top of the casing and a short length of pipe connected to it. The wash pipe is introduced into the casing. Water is injected under pressure through the wash pipe and rises through the annular space between the wash pipe and the casing (Fig. 3.2). As the water circulates, the wash pipe is rotated to make the soil loose. The soil comes out in the form of a thin slurry and gives only a rough indication of the soil layers. However, sample tubes can be inserted and samples taken out after certain intervals, *e.g.*, when the colour of the soil slurry changes, the wash pipe is first lifted out from the hole.

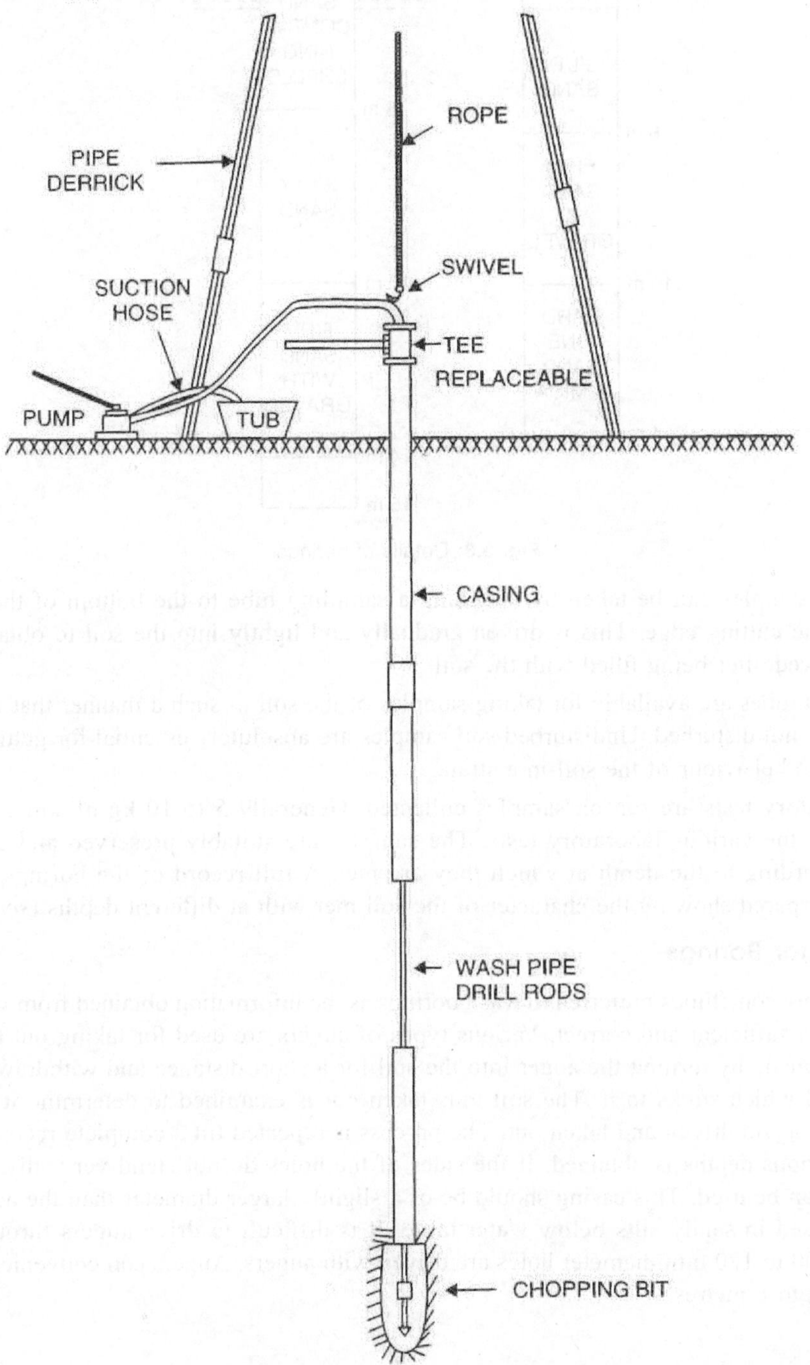

Fig. 3.2: Arrangement for taking wash borings.

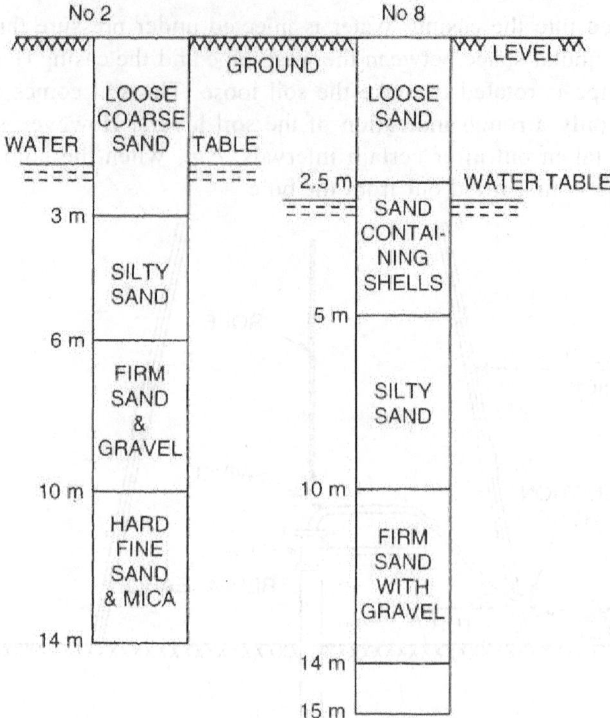

Fig. 3.3: Details of borings.

These samples can be taken by attaching a sampling tube to the bottom of the wash pipe instead of the cutting edge. This is driven gradually and lightly into the soil to obtain a sample and is removed after being filled with the soil.

Special tubes are available for taking samples of the soil in such a manner that the structure of the soil is not disturbed. Undisturbed soil samples are absolutely essential for getting a correct picture of the behaviour of the soil in a strata.

Laboratory tests are run on samples collected. Generally 5 to 10 kg of soil is needed for carrying out the various laboratory tests. The samples are suitably preserved and are arranged serially according to the depth at which they are met. A full record of the borings is kept and a chart is prepared showing the character of the soil met with at different depths (*see* Fig. 3.3).

3.2.3 Auger Borings

These are sometimes preferred to wash borings as the information obtained from wash borings is not always sufficient and correct. Various types of augers are used for taking out the samples. The bore is made by turning the auger into the soil for a short distance and withdrawing it along with the soil which sticks to it. The soil thus taken out is examined to determine its properties. The auger is again driven and taken out. The process is repeated till a complete record of the soil layers at various depths is obtained. If the sides of the holes do not stand vertically, a casing of some pipe can be used. This casing should be of a slightly larger diameter than the auger. Augers cannot be used in sandy silts below water table. It is difficult to drive augers through gravels. Generally 100 to 120 mm diameter holes are driven with augers. Augers can conveniently be used for depths upto 8 metres.

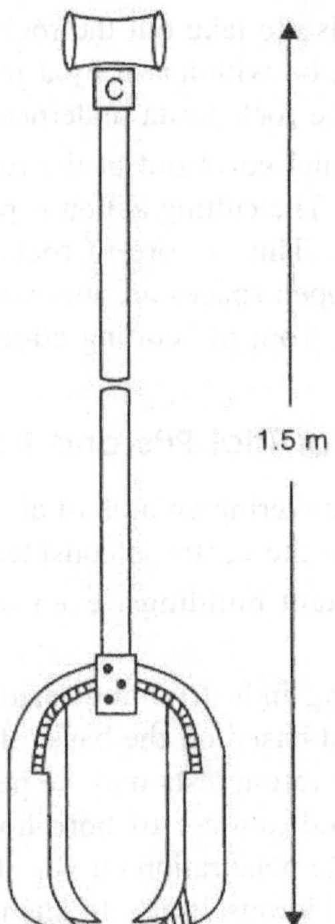

Fig. 3.4: A typical auger. Additional pipes can be fitted for taking samples at deeper depths.

3.2.4 Mechanised Drilling

Soil investigations can be conducted rapidly if machines are used for making bore holes. Several types of machines are available. The rotatory types are preferred. Briefly, the process consists of attaching a cutting tool to the end of a series of drill rods. These drill rods are then coupled with the machine and rotated at a convenient speed and simultaneously pushed into the soil. The bits of soil which get cut from time to time are removed through the bore holes with the help of wash water which is continuously being fed into the soil. As soon as a particular depth is reached, the sample tubes are inserted in the same manner as described earlier and soil samples obtained. When rock is encountered, drilling becomes a necessity. Drilling through rock is done in three ways, *viz*:

(*a*) Churn or percussion drilling,

(*b*) Diamond drilling, and

(*c*) Shot or tooth cutting drilling.

Churn drilling is similar to wash boring except that water, enough only to fill the bottom portion of the hole, is used. The cuttings become mixed with water as the churning proceeds and at intervals, the drill stem is withdrawn. The cuttings and water are removed by means of a pump which is lowered into the hole. This method is used for investigation of boulders and rock. The samples thus obtained give only a rough idea of the soil encountered as the cuttings are mixed with water.

The diamond drill consists of a hollow thick steel tube with cheap diamonds set on its outside and inside edges at the bottom so as to form a cutting bit. The diamonds are set in such a manner that the hole driven is of a slightly greater diameter than that of the tube and also a core is cut which is of a slightly smaller diameter than the inner diameter of the tube. The drill rods are rotated into the rock, water is forced down through the drill rods to the bottom of the hole so as to keep the cutting edge cool and also to take out the rock chippings. After some time, a solid core of rock is formed inside and can be withdrawn by a jerky motion. A continuous record of cores is kept to get a correct idea of the rock strata underneath.

Shot drilling is also used for getting cores out of the rock. A shot-drill consists of a hollow cylinder which is rotated by a drill rod. The cutting action is provided by feeding steel shots which come directly under the rotating edge. Thus a core of rock is formed. Water is fed to take out small chipping. Whenever cavities or open spaces are met with, a cement-water mixture is forced first so that the shots do not get away from the cutting edge. Drilling is resumed only after this mixture has considerably set.

3.2.5 Number and Disposition of Trial Pits and Borings

(*a*) For a compact building site covering an area of about 0.4 hectare, one bore hole or trial pit in each corner and one in the centre is considered adequate.

(*b*) For smaller and less important buildings, even one bore or trial pit in the centre is sufficient.

(*c*) For very large areas covering industrial and residential colonies, the number of bore holes or trial pits are decided based on the basis of the geological nature of the terrain. Dynamic or static cone penetration tests may be performed at every 100 m by dividing the area in a grid pattern and number of bore holes or trial pits may be decided by examining the variation in the penetration curves. The cone penetration test may not be possible at sites having boulderous strata. In such cases, geophysical methods should be resorted to.

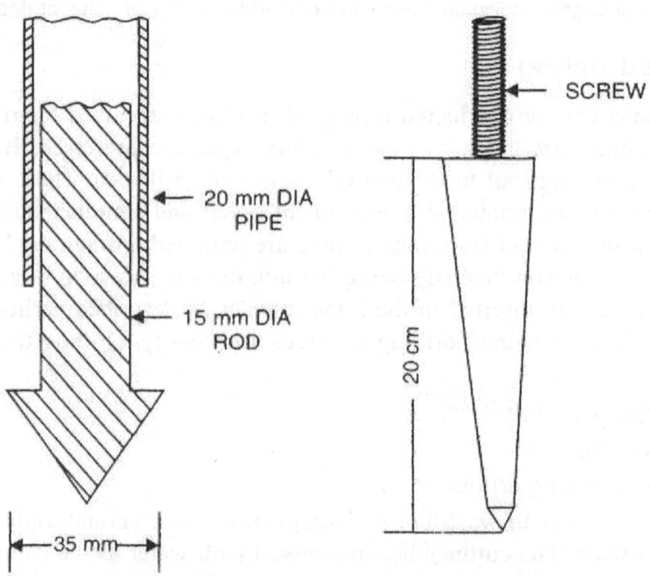

Figs. 3.5 and 3.6: Typical penetrometers.

3.3 SUB-SURFACE SOUNDINGS

These are used for investigation of soil layers which are of a varying nature. They are also used to make sure that the sub-soil does not contain exceptionally soft spots. In olden methods, rails were driven with a hammer weighing about a tonne and falling through a distance of 80 cm. The number of blows needed to drive about 30 cm length of this rail at various elevations below ground would give an idea of the surface met with. Static sounding methods have been developed and these are called penetrometers. They consist of rods of various shapes in standard lengths. After they sink into the soil under their own weight, they are loaded at suitable intervals or else are driven at a steady rate into the soil. The rate of penetration in the first case or the load in the second case will give an idea of the soil type.

The static cone penetrometer is very commonly used now-a-days. This consists of a cone having an area of 10 sq cm. This is driven with the help of a hydraulic jack system using a set of rods. The pressure required to drive the cone is recorded at intervals as the cone penetrates into the soil. The information can also be collected separately for the driven resistance of the cone and the side sleeves used. Typical profile of penetration resistance with depth is shown in Fig. 3.8 and contours of bearing capacity are worked out in Fig. 3.9.

The dynamic method which is known as 'Standard Penetration Test' has become very common and is now an accepted standard procedure. This consists of a thick walled tube having an outer diameter of 5 cm and inner diameter of 3.5 cm and a length of about 60 cm. This thick walled tube can be split into two parts longitudinally so as to facilitate the removal of the soil samples which gets collected inside. The cutting edge is provided at one end and the tube can be screwed to the drill rod at its other end. This tube is driven into the ground with the help of 63 kg weight falling to a distance of 75 cm. The number of blows required to drive this tube through a distance of 60 cm into the ground are recorded and are designated as the Standard Penetration Resistance. Several relationships have been derived between the standard penetration resistance and the load carrying capacity of a soil.

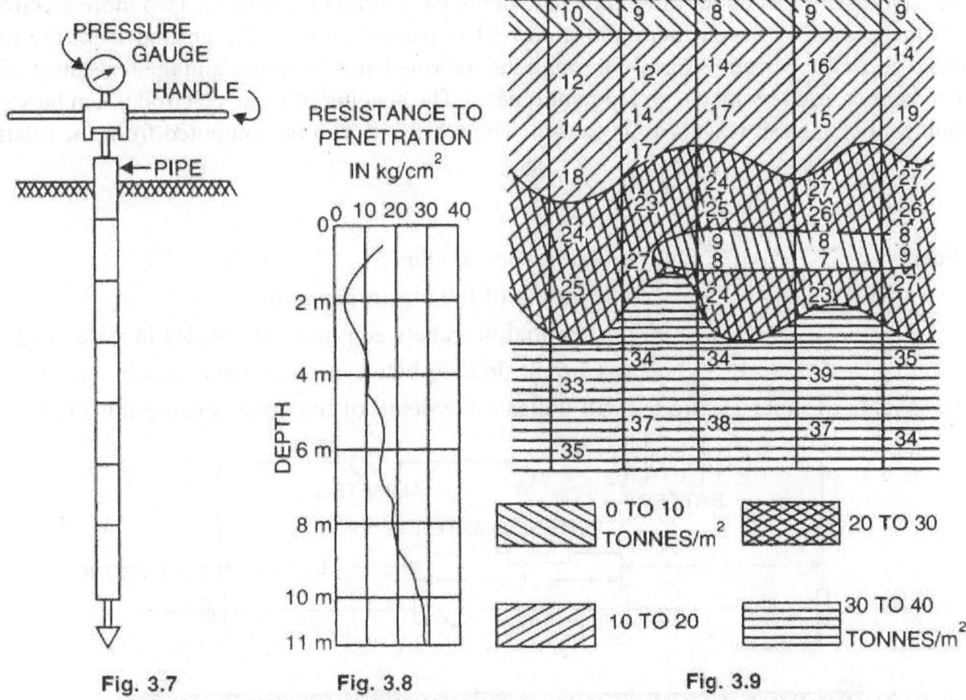

Fig. 3.7 Fig. 3.8 Fig. 3.9

Figs. 3.7 to 3.9: Diagrams showing the set-up for a penetrometer operation with typical results plotted graphically and contours of bearing capacity values worked out.

3.4 GEO-PHYSICAL METHODS

These methods are used where only a rough indication of the soils is needed and also where speed of investigation is of primary importance. Two methods are generally used.

The seismic method (*see* Fig. 3.10) depends upon the fact that sound waves travel faster in rocks than in soils. Similarly sound waves travel faster in solid rocks than in shattered rocks. These waves are created artificially by explosions and get reflected from a stronger layer below. The time of travel can be measured and hence the thickness of strata determined.

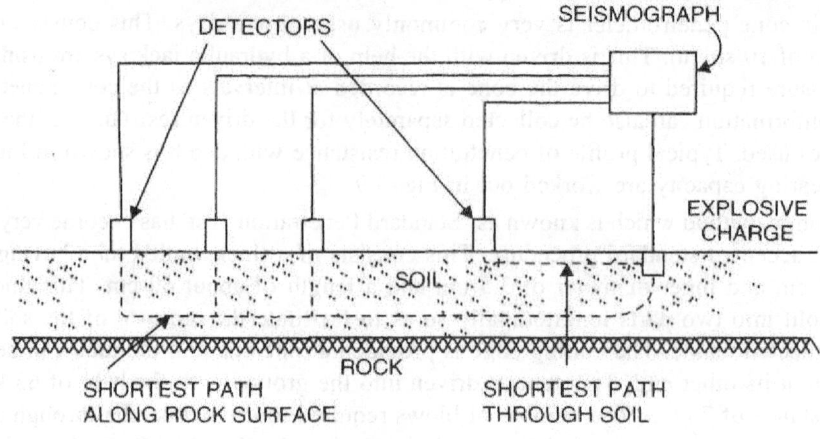

Fig. 3.10: Seismic method.

In the resistivity method (Fig. 3.11) two electrodes are driven into the ground. They are spaced apart a distance of about three times the depth of information needed. Two more electrodes are set in the ground between them and a current is passed through the ground from the outer electrodes. Part of the current is passed through the intermediate electrodes and the resistance of the circuit can be measured by means of a potentiometer. The spacing of these electrodes can be varied till a significant change in resistivity is caused. The resistivity can be computed from the relation:

$$\Omega = 2\pi L \frac{E}{I}$$

where

Ω = Resistivity in Ohm

L = Spacing of the electrodes in m

E = Potential drop between inner electrodes in volts; and

I = Current flowing between outer electrodes in amperes.

The spacing of inner electrodes will indicate the depth of the layer causing this change.

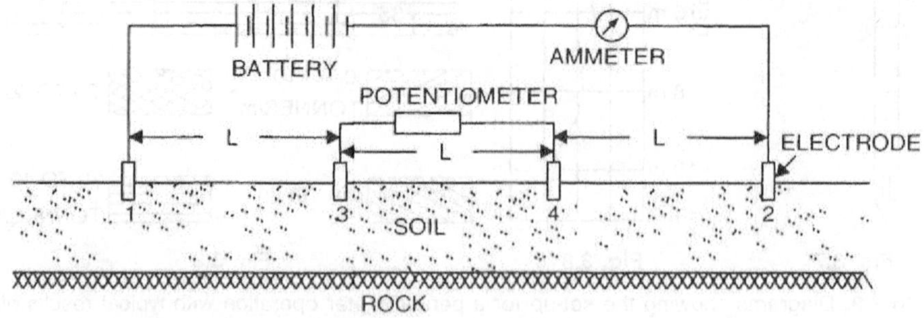

1 & 2 STEEL PIN ELECTRODES
3 & 4 NON-POLARISING ELECTRODES

Fig. 3.11: Resistivity method.

BEARING CAPACITY OF SOILS

The preliminary and the detailed investigations described above will enable the engineer incharge to know the following points about the site:

1. The depth of bed rock, if any.
2. Character of soil, *i.e.*, gravel, sand, silt, clay, etc.
3. Presence of boulders or other obstructions, if any.
4. Elevation of the ground water table.

The next thing is to determine the type of foundation to be used, the depth to which it should be taken and its dimensions so that it can safely transmit the loads from the building to the soil without any failure or significant settlement. For their determination, a knowledge of the safe allowable pressure on the soil is necessary. This is called the bearing capacity of the soil.

The various terms used in connection with the bearing capacity/bearing pressure are defined below:

Bearing capacity, safe. The maximum intensity of loading that the soil will safely carry without risk of shear failure irrespective of any settlement that may occur.

Bearing capacity, ultimate. The intensity of loading at the base of a foundation which would cause shear failure of the supporting soil.

Bearing pressure, allowable (gross or net). The gross allowable bearing pressure is the maximum allowable loading intensity on the ground in any given case (taking cognizance of surcharge), taking into account the maximum safe bearing capacity, the amount and kind of settlement expected and the capability of the structure to take up this settlement. It is, therefore, a combined function of both the site conditions and characteristics of the particular structure.

The net allowable bearing pressure is the gross allowable bearing pressure minus the surcharge intensity.

Bearing pressure, safe. The intensity of the loading that the soil will carry without undergoing settlement more than permissible for the structure.

Factor of safety (with respect to bearing capacity). A factor by which the ultimate bearing capacity (net) must be reducd to arrive at the value of safe bearing capacity (net).

3.5 DETERMINATION OF BEARING CAPACITY OF A SOIL

The various methods of determining the bearing capacity of a soil are:

(1) Local experience in the construction of a similar building.

(2) Bearing capacity by calculation.

(3) Classification of the soil and using a value for the bearing capacity from the standard building code.

(4) Field method for determining allowable bearing pressure.

(5) Laboratory tests.

In the first method information is collected about the values of safe bearing capacities assumed in the design of structures near by and their condition at site is examined. This will lead us to a very approximate value of the bearing capacity and should be used with great caution. This method may be used for small buildings.

The second method is used where the physical characteristics of the soil are available, that is, cohesion, angle of internal friction, density, etc. The bearing capacity is determined from stability considerations of shear, factor of safety of 2.5 is adopted for safe bearing capacity. The effect of interference of different foundations should be taken into account. For details, reference may be made to I.S : 6403-1981 for the determination of allowable bearing pressure on shallow foundations.

The third method consists in identifying and classifying the soils into various groups. After the identification of the soil, standard tables can be referred to determine the value of the safe bearing capacity to be used. These tables are available in the standard building codes of the locality.

National Building Code of India—2005 (SP : 7-1983) gives Table 3.1 for the safe bearing capacity which may be assumed in the absence of test results for the design of foundations of lightly loaded structures (houses, flats and school buildings of not more than two storeys) and for a preliminary design of any structure.

Table 3.1: Safe Bearing Capacity

Sl. No.	Type of Rocks/Soils	Safe Bearing Capacity		Remark
		kN/m^2	t/m^2	
	(a) Rocks			
(1)	Rocks (hard) without lamination and defects, for example, granite, trap and diorite	3240	324	—
(2)	Laminated rocks, for example, sandstone and limestone in sound condition	1620	162	—
(3)	Residual deposits of shattered and broken bed rocks and hard shale, cemented material	880	88	—
(4)	Soft rock	440	44	—
	(b) Non-cohesive soils			
(5)	Gravel, sand gravel, compact and offering high resistance to penetration when excavated by tools	440	44	(See Note 2)
(6)	Coarse sand, compact and dry	440	44	Dry means that the ground water level is at depth not less than the width of foundation below the base of the foundation
(7)	Medium sand, compact and dry	245	24.5	
(8)	Fine sand, silt (dry lumps easily pulverized by the fingers)	150	15	
(9)	Loose gravel or sand-gravel mixture, loose coarse to medium sand, dry	245	24.5	(See Note 2)
(10)	Fine sand, loose and dry	100	10	—
	(c) Cohesive soils			
(11)	Soft shale, hard or stiff clay in deep bed, dry	440	44	This group is susceptible to long term consolidation-settlement
(12)	Medium clay, readily indented with a thumb nail	245	24.5	—
(13)	Moist clay and sand-clay mixture which can be indented with strong thumb pressure	150	15	—

(Contd.)

Sl. No.	Type of Rocks/Soils	Safe Bearing Capacity		Remark
		kN/m^2	t/m^2	
(14)	Soft clay indented with moderate thumb pressure	100	10	—
(15)	Very soft clay which can be penetrated several centimetres with the thumb	50	5	—
(16)	Black cotton soil or other shrinkable or expansive clay in dry condition (50 per cent saturation)	—	—	See Note 3. To be determined after investigation
	(*d*) *Peat*			
(17)	Peat	—	—	See Notes 3 and 4. To be determined after investigation
	(*e*) Made-up ground			
(18)	Fills or made-up ground	—	—	See Notes 2 and 4. To be determined after investigation

Notes 1. Values listed in the table are from shear consideration only.

2. Values are very much rough for the following reasons:

(*a*) Effect of characteristics of foundations (that is, effect of depth, width, shape, roughness, etc.) has not been considered.

(*b*) Effect of range of soil properties (that is, angle of frictional resistance, cohesion, water table, density, etc.) has not been considered.

(*c*) Effect of eccentricity and indication of loads has not been considered.

3. For non-cohesive soils, the values listed in the table shall be reduced by 50 per cent if the water table is above or near the base of footing.

4. Compactness of non-cohesive soils may be determined by driving a cone of 65 mm dia and 60° apex angle by a hammer of 65 kg falling from 75 cm. If corrected number of blows (*N*) for 30 cm penetration is less than 10, the soil is called loose; if *N* lies between 10 and 30, it is medium, and if more than 30, the soil is called dense.

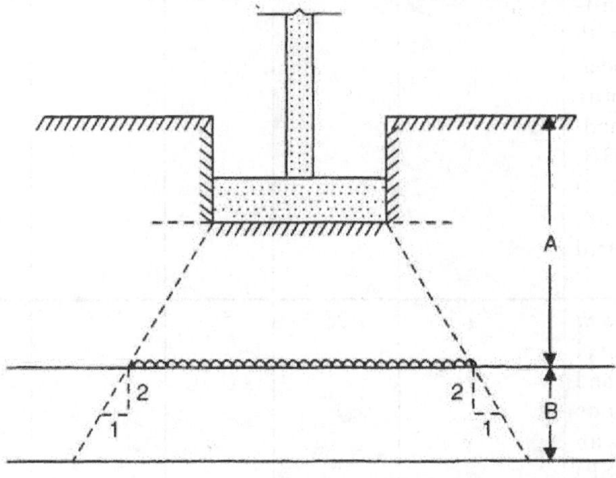

A. STRONG STRATUM
B. WEAKER STRATUM

Fig. 3.12: Practical assumption of pressure distribution through various strata.

Any tabulated values of safe bearing capacity should be used very cautiously. In addition to the soil type, the bearing capacity of a soil depends on many factors such as the degree of compaction, moisture content, size of the foundation below the ground level and the type of foundation used. Some building codes specify the increments in allowable bearing capacities for greater depths below the ground surface. Similarly adjustments for the increased widths and rigidity or flexibility of the foundation are incorporated. Certain building codes are very comprehensive and directions are given to account for a weak strata which underlies a strong strata, *e.g.*, in National Building Code it is specified that the bearing capacity of the weak strata below is to be taken into account and the pressure transmitted to the lower strata is within the prescribed safe limits. It is further stated that when the footings are closely spaced, the pressure transmitted to the underlying soil will overlap. In such cases the pressure in the overlapped zones should be considered. With normal foundations, it is sufficiently accurate to estimate the bearing pressure on the underlying layers by assuming the load to be spread at a slope of 2 (vertical) to 1 (horizontal).

To take into account the effect of wind force, it is provided in the National Building Code that if the bearing pressure due to wind is less than 25 per cent of that due to dead and live loads, it may be neglected in design. Where this exceeds 25 per cent foundations may be so proportioned that the pressure due to combined dead, live and wind loads should not exceed the allowable bearing pressure by more than 25 per cent.

Table 3.2: Permissible increase in allowable bearing pressure or resistance of soils

Sl. No.	Type of Soil mainly constituting the Foundation	Permissible Increase in Allowable Bearing Pressure Per Cent					
		Piles Passing Through Any Soil but Resting on Soil Type 1	Piles not Covered under col (3)	Raft Foundation	Combined or Isolated RCC Footing with Tie Beams	Isolated RCC Footing Without Tie Beams or Unreinforced Strip	Well Foundations
(*i*)	***Type I Rock or Hard Soils***—Well graded gravels and sand-gravel mixtures with or without clay binder, and clayey sands poorly graded or sand-clay mixtures (GB, CW, SB, SW and SC) having *N* above 30, where *N* is the standard penetration value	50	—	50	50	50	50
(*ii*)	***Type II Medium Soils***—All soils with N between 10 and 30 and poorly graded sands or gravelly sands with little or no fines (SP) with N > 15	50	25	50	25	25	25
(*iii*)	***Type III Soft soils***—All soils other than SP with N < 10	50	25	50	25	—	25

When earthquake forces are also to be taken into account, the permissible increase in allowable bearing pressure of soil should correspond to Table 3.2, depending upon the type of foundation of the structure.

Notes 1. The allowable bearing pressure shall be determined in accordance with IS: 6403—1981.

2. If any increase in bearing pressure has already been permitted for forces other than seismic forces, the total increases in allowable bearing pressure when seismic force is also included shall not exceed the limits specified above.
3. In the case of submerged loose sands and soils falling under classification SP with standard penetration values less than the values specified in Note 5 below, the vibrations caused by earthquake may cause liquefication or excessive total and differential settlements. In important projects, this aspect of the problem need be investigated and appropriate methods of compaction or stabilization adopted to achieve suitable *N*. Alternatively, deep pile foundation may be provided and taken to depths well into the layers which are not likely to liquefy.
4. The piles should be designed for lateral loads neglecting lateral resistance of soil layers liable to liquefy.
5. Desirable field values of *N* are as follows:

Zones III, IV and V 15
Zones I and II 10

For multi-storeyed and important buildings it is necessary to carry out a field test or laboratory test on the soil. The simplest type of field test is called 'Plate load Test' and is described below:

3.6 PLATE LOAD TEST

This is a field test used in the case of unusual soil types and for all heavy and important buildings. The plate load test is used to determine the ultimate bearing capacity of the soil in place, assuming that the soil strata is reasonably uniform. The load test can also be used to find modulus of subgrade reaction, which is useful in the design of raft foundations.

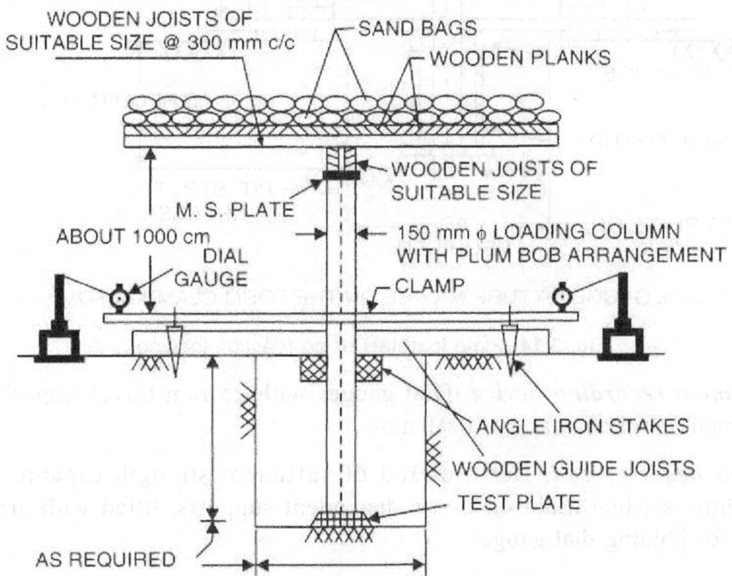

Fig. 3.13: Plate load test using gravity loading.

The method consists essentially in loading a rigid plate at the foundation level, observing the settlement for each increment of load after intervals, plotting the time-settlement curve and then determining the ultimate bearing capacity, which is taken as the load at which the plate starts sinking at a rapid rate.

3.6.1 Apparatus

(*i*) ***Bearing plates***. Circular or square bearing plates of mild steel, varying in size from 300 to 750 mm and thickness not less than 25 mm, with chequered or grooved bottom, provided with handles for convenient setting and centre marked. As an alternative, cast-in-situ or precast concrete blocks may be used with depths not less than two-third the width.

(*ii*) ***Hydraulic jack*** of required capacity, with properly calibrated load measuring device such as pressure gauge, electronic load cell or proving ring.

(*iii*) ***Loading platform truss*** of sufficient size and properly designed members.

The typical set-ups used are shown in Figs. 3.13 to 3.15 for the following cases:

(*a*) Gravity loading,

(*b*) Reaction loading, and

(*c*) Loading truss.

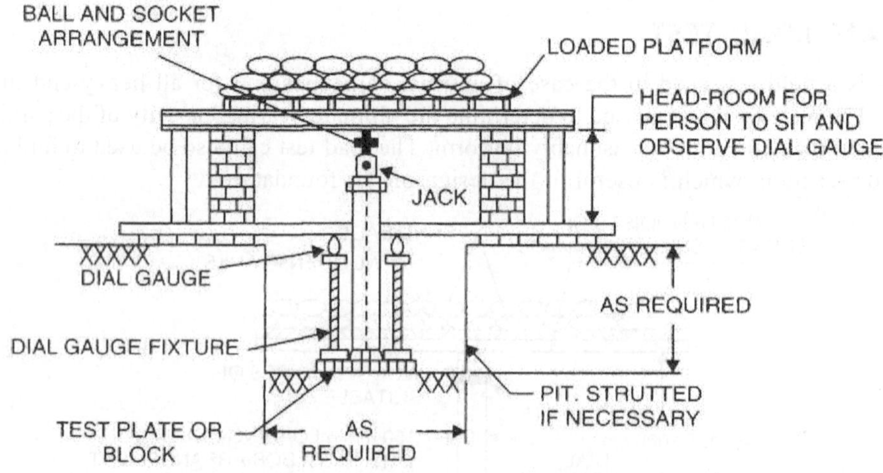

Fig. 3.14: Plate load test using reaction loading.

(*iv*) ***Settlement recording device***. Dial gauges with 25 mm travel, capable of measuring settlement to an accuracy of 0.01 mm.

(*v*) ***Datum beam or rod***. Beam or rod of sufficient strength capable of maintaining straightness when fitted on two independent supports, fitted with arms or magnetic bases for holding dial gauges.

(*vi*) ***Miscellaneous apparatus***. A ball and socket arrangement, loading columns, wooden blocks, collar, steel shims, reaction girder with cradles for independent fitting to the reaction platform.

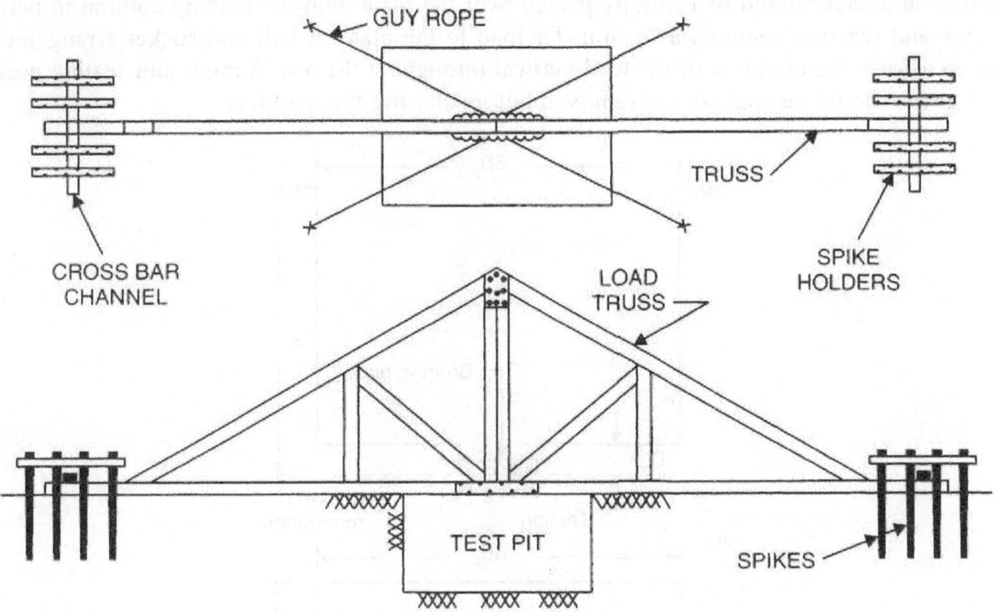

Fig. 3.15: Plate load test using loading truss.

3.6.2 Procedure

A test pit, of width equal to five times the test plate or block, is dug up to the required level, carefully levelled and bottom cleaned and protected against disturbance or changes in natural formation. The pit is usually carried up to the foundation level. In case the water table is within the depth equal to the width of the test plate, the test shall be conducted at water table level.

Next the size and shape of the plate is selected. Except for circular footings, square plates are adopted. For clayey and silty soils and for loose to medium dense sandy soils with the standard penetration resistance value, N (as per IS: 2131-1981) less than 15, a 450 mm square plate or concrete blocks shall be used. In the case of dense sandy or gravelly soils, with N between 15 and 30, three plates of sizes 300 mm to 750 mm shall be used depending upon practical considerations of reaction loading and maximum grain size. The thickness of the plate shall be at least four times the maximum size of the soil particle present at the test location.

The test plate is made to rest in the centre of the pit in a depression of the same area as that of the test plate. The depth of the depression should be such that the ratio of depth, D_p over width B_p, of the loaded area is the same as that for the actual foundation *i.e.* in Fig. 3.16.

$$\frac{D_p}{B_p} = \frac{\text{Depth of foundation}}{\text{Width of foundation}} = \frac{D}{B}$$

The test plate is placed over a fine sand layer of maximum thickness of 5 mm.

A loading platform is then built which may be either of gravity loading or reaction loading or truss loading, as shown in Figs. 3.13 to 3.15. The loading platform should be supported by suitable means at least 2.5 m from the test area with a height of 1 m or more above the bottom of the pit to provide sufficient working space. No support of loading platform should be located within a distance of 3.5 times size of the test plate from its centre.

The centre of the test plate should coincide with the centre of the reaction girder/beam, with the help of a plumb and bob and horizontally levelled by a spirit level to avoid eccentric loading.

The hydraulic jack should be centrally placed over the plate with the loading column in between the jack and reaction beam so as to transfer load to the plate. A ball and socket arrangement is inserted to keep the direction of the load vertical throughout the test. A minimum seating pressure of 70 g/cm^2 should be applied and removed before starting the load test.

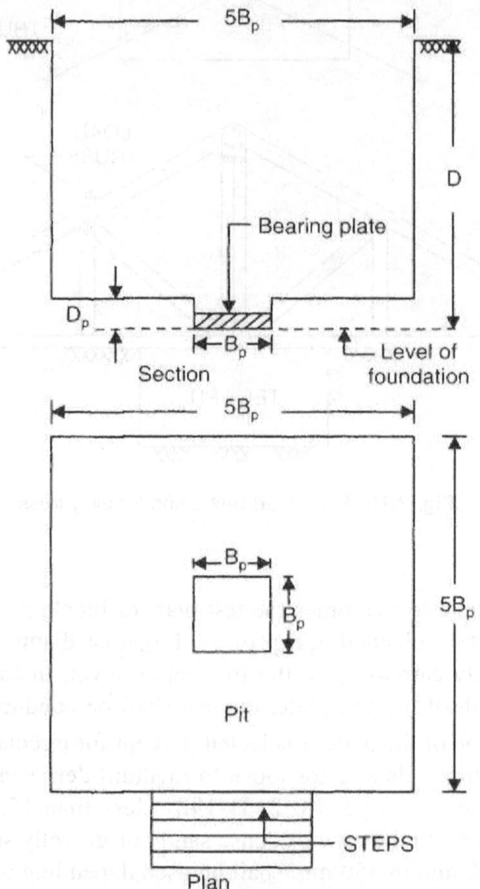

Fig. 3.16: Test pit details.

The two supports of the reference beam or datum rod should be placed over firm ground, fixed with minimum two dial gauges resting at diametrically opposite ends of the plate.

Apply the load to the soil in cumulative equal increments up to 1 kg/cm^2 or one-fifth of the estimated ultimate bearing capacity, whichever is less. The load is applied without impact, fluctuation or eccentricity and in case of hydraulic jack load is measured over the pressure gauge, attached to the pumping unit kept over the pit, away from the test plate through extending pressure pipes.

Settlement is observed for each increment of load after an interval of 1, 2.25, 4, 6.25, 9, 16 and 25 minutes and thereafter at hourly intervals to the nearest 0.02 mm. The test is continued till a settlement of 25 mm under normal circumstances or 50 mm in special cases such as dense gravel, gravel and sand mixture, is obtained or till failure occurs, whichever is earlier. Alternatively where settlement does not reach 25 mm, the test should be continued to at least two times the estimated design pressure.

A load settlement curve is then plotted. From this load-settlement curve, the zero correction, which is given by the intersection of the early straight lines or the nearly straight line part of the curves with zero deadline shall be determined and subtracted from the settlement readings to allow for the perfect seating of the bearing plate and other causes.

Four typical curves are shown in Fig. 3.17. Curve A is typical for loose to medium cohesion less soil, it is a straight line in the earlier stages but flattens out after some time, but there is no clear point of failure. Curve B is for cohesive soil, it may not be quite straight in the early part and leans towards settlement axis as the settlement increases. For partially cohesive soil curve C possessing the characteristics of both the curves A and B is obtained while curve D is purely for dense cohesionless soils.

From the corrected load settlement curves, it may be seen that the failure is well defined in the case of cohesive soils or dense cohesionless soils (curves B and D) and the ultimate bearing capacity can be easily determined by extending the straight line portion of the curves. But in the case of curves A and C, where the yield point is not well defined, settlements should be plotted as abscissa against corresponding load intensities as ordinate, both to logarithmic scales, as shown in Fig. 3.18. This gives two straight lines, the intersection of which gives the yield value of the soil.

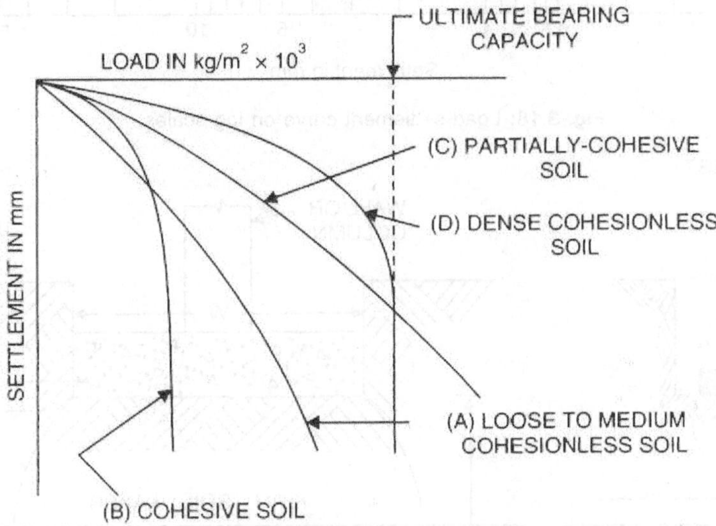

Fig. 3.17: Load settlement curves.

From Fig. 3.17, the safe bearing pressure for medium and dense sands can be read, corresponding to a settlement, S_p, which is calculated as under:

$$S_t = S_p \left[\frac{B\,(B_p + 0.3)}{B_p(B + 0.3)} \right]^2$$

where S_t = permissible settement of footing, in m, (as per IS : 1904-1978)

B = the size of footing in m

B_p = the size of test plate in m

S_p = settlement of test plate in m

From this formula total settlement of footing, S_t, can be calculated, taking S_p as observed total settlement of plate.

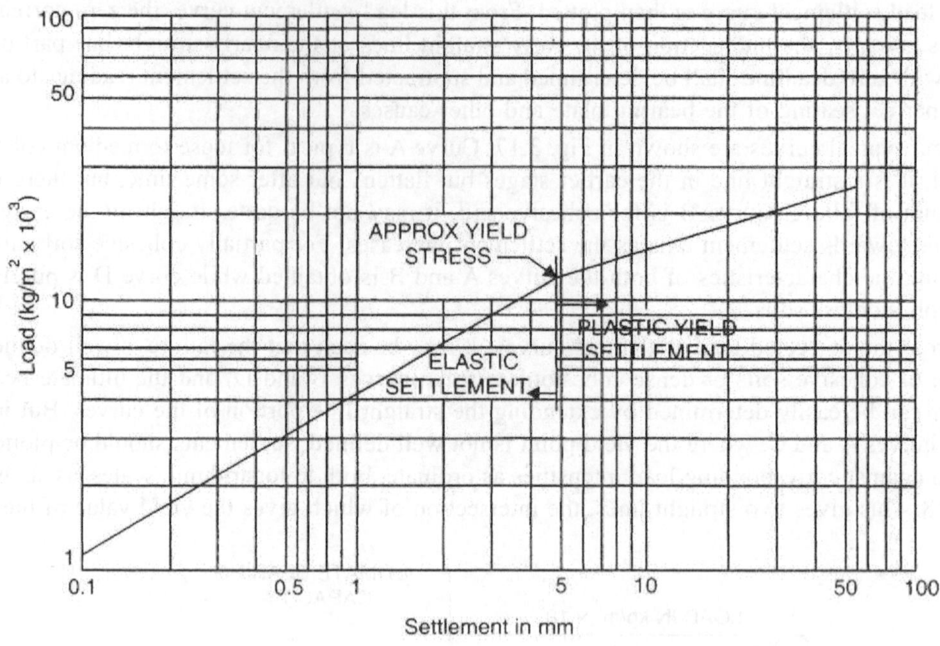

Fig. 3.18: Load settlement curve on log scales.

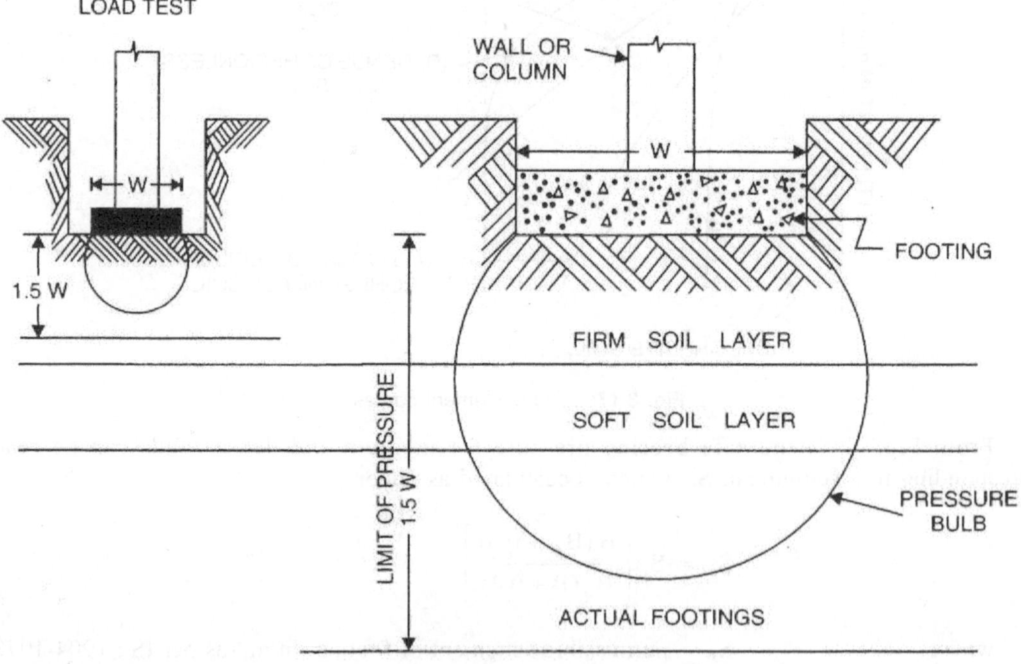

Figs. 3.19 and 3.20: Pressure distribution for a load test and for actual footings.

3.6.3 Limitations

(1) The test results reflect only the character of the soil located within a depth of less than twice the width of the bearing plate. Since the foundations are generally larger than the test plates, the settlement and shear resistance will depend upon the properties of a much thicker stratum; as shown in Figs. 3.19 and 3.20. Thus the results of the test are likely to be misleading, if the character of the soil changes at shallow depths, which is not uncommon. Hence any soft strata lying within the influence of pressure created by the actual building is liable to greater settlement. Hence if any weak strata is encountered below the level of the footing, it should be ensured that the pressure coming on that strata should be less than its allowable safe bearing pressure.

(2) Another limitation is concerning the effect of size of foundation. For clayey soils, the bearing capacity (from shear consideration) for a larger foundation is almost the same as that for the smaller test plate. But in dense sandy soils, the bearing capacity increases with the size of the foundation. Thus tests with smaller size plate tend to give conservative values in dense sandy soils. It may, therefore, be necessary to test with plates of at least three sizes and the bearing capacity results extrapolated for the size of the actual foundation.

3.7 LABORATORY TESTS

Whenever load tests are not feasible, approximate values of bearing capacity can be determined by testing the samples of the soils in the laboratory. Various tests are specified for sandy and clayey soils. The values of bearing capacities can be calculated after determining the physical properties of these soils. Charts are used to help in the calculations. A reference may be made to a standard book on Soil Mechanics for details of laboratory tests.

3.8 METHODS TO IMPROVE THE BEARING CAPACITY OF A SOIL

The various methods generally employed to improve the bearing power of a soil are:

(1) ***Increasing the depth of foundation.*** Usuallys soils have got greater bearing capacity at deeper depths. This is especially true of sandy soils, for which the bearing capacity increases due to the weight of the overlying material. But this method may not prove to be of great advantage as the load of the building foundation will also increase with depth.

(2) ***By draining the soil.*** The bearing capacity of every soil decreases when its water content increases. Submergence of a soil strata decreases the ultimate bearing capacity of a sandy soil by as much as fifty per cent. The bearing capacity of clayey soil also decreases considerably by an increase in water content, may be by a rise in ground water table. Cohesionless soils, *i.e.*, sandy soils and gravels can be drained easily, either by gravity pipe drainage system or by installing shallow tubewells.

(3) ***By compacing the soil.*** Compaction of the soil reduces the voids between the individual particles and hence they are less liable to displacement, thereby increasing the bearing capacity.

The various methods of compacting the soil are:

(*i*) Ramming the soil at the optimum moisture content with the aid of hand rammers or mechanically operated frog rammers or vibratory rollers.

(*ii*) By packing gravel or rubble into the foundation trench and ramming it.

(*iii*) Flooding the loose sands to increase its density.

(*iv*) Vibration with compactors or heavy vibratory rollers. This method is also suitable for granular soils.

(*v*) *Vibroflotation.* This method combines the effects of vibration and jetting. A heavy cylinder, known as fibro-float, is inserted in the ground, water added under pressure and clean sand added into a crater which develops on the surface.

(*vi*) *Compaction by pre-loading.* Pre-loading results in accelerated consolidation. The method is suitable for clayey soils.

(vii) ***Driving sand piles***. Hollow pipes are driven in the ground, thereby compacting soil between adjacent pipes. Pipes are gradually removed filling the hole with sand and ramming it. This results in the formation of sand piles.

(4) ***By confining the soils***. Sheet piles are driven to form an enclosure, thereby confining the movement of the soil. This can help in increasing the bearing power.

(5) ***By grouting***: Cement grout can be injected under pressure in the foundation to seal off any cracks or fissures which otherwise reduce the bearing capacity of the soil. This method is mostly employed for fissured cracks.

(5) ***Chemical treatment***. Chemical solutions like silicates are injected under pressure into the soil. They form a gel and thereby unite to develop a compact mass. This method is very costly and is adopted in exceptional cases.

QUESTIONS

1. Describe the various investigations you would make for the examination of soils for the construction of buildings.
2. Name the type of soils generally encountered in the construction of a building and briefly describe their properties stating clearly their suitability over each other.
3. Describe briefly the various methods of a soil investigation with the aid of borings for ordinary soils.
4. Write short notes on:
   (i) Wash borings
   (ii) Churn drilling
   (iii) Penetrometers
   (iv) Geo-physical method of soil investigation.
5. What is meant by the term 'Bearing capacity' of soil? Describe briefly how the bearing capacity of soil is determined with the help of plate load test.
6. (a) Define the following terms:
   (i) Bearing capacity, ultimate
   (ii) Being capacity, safe
   (iii) Bearing pressure, allowable
   (v) Bearing pressure, safe
   (b) What are the values of the bearing capacity generally allowed for:
   (i) Black cotton soil,
   (ii) Gravel,
   (iii) Moorum,
   (iv) Trap rook,
7. What are the methods normally used to improve the bearng capacity of a soil?

REFERENCES

1. **Taylor, D.W.:** *Fundamentals of Soil Mechanics*
2. **Wadia, D.N.:** *Indian Geology.*
3. **Terzaghi, K. and Peck, R.B:** *Soil Mechanics in Engineering Practices.*
4. **Henry, F.D.C.:** *Design and Construction of Engineering Foundation.*
5. **H.M.S.O.:** *Soil Mechanics for Road Engineers.*
6. **Davis:** *Foundations for Bridges and Buildings.*
7. **National Buildings Organisation, India:** *Foundations.*

8. **Huntington, W.C.:** *Building Construction.*
9. **American Society of Civil Engineers:** *A symposium on Carrying of Load Test for Evaluating Bearing Capacity of Soil.*
10. **IS: 1882—1982:** *Method of Load Test on Soil.*
11. **IS 1498—1970:** *Classification and Identification of Soils for General Engineering Purposes.*

Foundations

Foundation is that part of a building which is in direct contact with the soil and transmits loads to it. Foundations are normally placed below the ground surface. The function of a foundation is to distribute the load over a larger area at a uniform rate so that the pressure on the soil does not exceed its allowable bearing capacity. Foundations increase the stability of a building and, moreover, enable the construction to be carried out systematically.

A foundation whose width is greater than its depth is called shallow foundation. The shearing resistance of the soil above the base level of the foundation is neglected. On the other hand, in the case of deep foundation, its depth is greater than its width.

Foundation can be built in various types of material. Generally bricks, stones, concrete, steel, timber, etc., are used in different forms for constructing the foundation of a building. However, the type of building which is supported on a particular foundation has a great bearing on the selection of the material, *e.g.*, the use of R.C.C. foundation for reinforced concrete building.

4.1 FUNCTIONS OF FOUNDATION

Foundation serves the following functions:

(1) ***Distribution of load***. Foundations help to distribute the loads of the super-structure to a large area of the soil, so that the intensity of load at its base does not exceed the safe bearing capacity of the soil. In the case of deep foundations, the super imposed loads are transmitted either through end bearing or both by side friction and end bearing.

(2) ***Stability against sliding and overturning***. Foundation imparts lateral stability to the super-structure by anchoring it to the ground. It increases the stability against sliding and overturning due to horizontal forces of wind, earthquake, etc.

(3) ***Minimise differential settlement***. Foundations distribute the super-imposed loads evenly on the sub-soil, even in the case of non-uniform loads. This can be achieved by constructing combined footings or raft foundation.

(4) ***Safety against undermining***: Foundations provide safety against scouring or undermining by flood water or burrowing animals.

(5) ***Provide level surface***. It provides level surface over which super-structure can be raised.

(6) ***Minimises distress against soil movement***. Distress or failure due to expansion or contraction of the sub-soil due to moisture variation in clayey and black-cotton soils are minimised by the provision of special type of foundations.

4.2 ESSENTIAL REQUIREMENTS OF FOUNDATION

The essential requirements of a good foundation are:

(1) The foundations should be capable of sustaining the design loads (dead, imposed (live), wind or seismic whichever is critical) without exceeding the permissible stresses of the materials of the foundations.

(2) The design loads should be transmitted by the foundations to the sub-soil so that the bearing pressure of the supporting soil is within the allowable limits or the settlement (total or differential) of the foundation is within permissible limits.

(3) Foundatios should be taken sufficiently deep to control distress or failure caused by shrinkage or swelling of the sub-soil.

4.3 SETTLEMENT OF FOUNDATIONS

The permissible values of differential settlements and tilt (angular distortion) for shallow foundations in soils are given in Table 4.1. The following causes responsible for producing settlement should be thoroughly investigated and taken into account.

4.3.1 Causes of settlement

(1) Elastic compression of the foundation and the underlying soil.

(2) Consolidation including secondary compression.

(3) ***Ground water lowering***. Specially repeated lowering and raising of water level in loose granular soils tend to compact the soil and cause settlement of the footings. Prolonged lowering of water table in fine grained soils may introduce settlement because of extrusion of water from the voids. Pumping water or draining water by tiles or pipes can also cause settlement in granular soils.

(4) Seasonal swelling and shrinkage of expansive clays.

(5) Ground movement on earth slopes *e.g.* surface erosion, slow creep or landslides.

(6) Other causes, such as adjacent excavation, mining, subsidence and underground erosion.

4.3.2 Causes of differential settlements

(1) Geological and physical non-uniformity or anomalies in type, structure, thickness, and density of the soil medium (pockets of sand in clay, clay lenses in sand, wedge like soil strata etc.), an admixture of organic matter, peat, mud;

(2) Non-uniform pressure distribution from foundation to the soil due to non-uniform loading and incomplete loading of the foundations;

(3) Water regime at the construction site;

(4) Overstressing of soil at adjacent site by heavy structures built next to light ones;

(5) Overlap of stress distribution in soil from adjoining structures;

(6) Unequal expansion of the soil due to excavation for footing;

(7) Non-uniform development of extrusion settlement; and

(8) Non-uniform structural disruption or disturbance of soil due to freezing and thawing, swelling and softening and drying of soils.

TYPES OF FOUNDATIONS

The various types of foundations are:

(*a*) ***Shallow Foundations***

(1) Pad or Spread and Strip foundations

Table 4.1: Permissible and Differential Settlements and Tilt (Angular Distortion) for Shallow Foundations in Soils

Sl. No.	Type of Structure	Isolated Foundations						Raft Foundations					
		Sand and Hard Clay			Plastic Clay			Sand and Hard Clay			Plastic Clay		
		Maximum settlement mm	Differential settlement mm	Angular distortion mm	Maximum settlement mm	Differential settlement mm	Angular distortion mm	Maximum settlement mm	Differential settlement mm	Angular distortion mm	Maximum settlement mm	Differential settlement mm	Angular distortion mm
(*i*)	For steel structure	50	0.0033L	1/300	50	0.0033L	1/300	75	0.0033L	1/300	100	0.0033L	1/300
(*ii*)	For reinforced concrete structures	50	0.0015L	1/666	75	0.0015L	1/666	75	0.0021L	1/500	100	0.0020L	1/500
(*iii*)	For multistoreyed buildings												
	(*a*) RC or steel framed buildings with panel walls	60	0.002L	1/500	75	0.002L	1/500	75	0.0025L	1/400	125	0.0033L	1/300
	(*b*) For load bearing walls												
	(1) $L/H = 2$*	60	0.0002L	1/5000	60	0.0002L	1/5000	←Not likely to be encountered→					
	(2) $L/H = 7$*	60	0.0004L	1/2500	60	0.0004L	1/2500						
(*iv*)	For water towers and silos	50	0.0015L	1/666	75	0.0015L	1/666	100	0.0025L	1/400	125	0.0025L	1/400

Note: The values given in the table may be taken only as a guide and the permissible total settlement / differential settlement and tilt (angular distortion) in each case should be decided as per requirements of the designer.

L denotes the length of deflected part of wall/raft or centre-to-centre distance between columns.

H denotes the height of wall from foundation/footing.

* For intermediate ratios of L/H, the values can be interpolated.

(2) Raft foundations

(3) Ring and shell foundations

(*b*) ***Pile Foundations***

(1) Driven cast in-situ concrete piles

(2) Bored cast in-situ concrete piles

(3) Driven precast concrete piles

(4) Bored precast concrete piles

(5) Under-reamed concrete piles; and

(6) Timber piles

(*c*) ***Other foundations***

Pier or well foundations

(a) Shallow Foundations

A foundation which transmits the load to the ground through one or more footings is a shallow foundation. This type of foundation spreads the load over a greater area so that the load transmitted to the soil is less than its allowable bearing capacity, thereby preventing excessive settlement of the foundation. The width of the wall is made wider at the base in a series of steps or footings.

Raft foundation is a sub-structure supporting an arrangement of columns or walls in a row or rows transmitting the load to the soil by means of a continuous slab, with or without depressions or openings.

(b) Pile Foundations

The load is transmitted to the soil by driving long vertical members of either timber or concrete or steel called piles. Loads are supported in two ways, *i.e.*, either by the effect of friction between the pile and the surrounding soil referred to as ***friction pile*** or by resting the pile on a very very hard stratum termed as ***end bearing pile***. The number of piles used will depend on the type of pile and also on the load coming from the building. Group of piles, if used, is connected at the top by a pile layer of concrete which distributes the load uniformly from the walls to the piles.

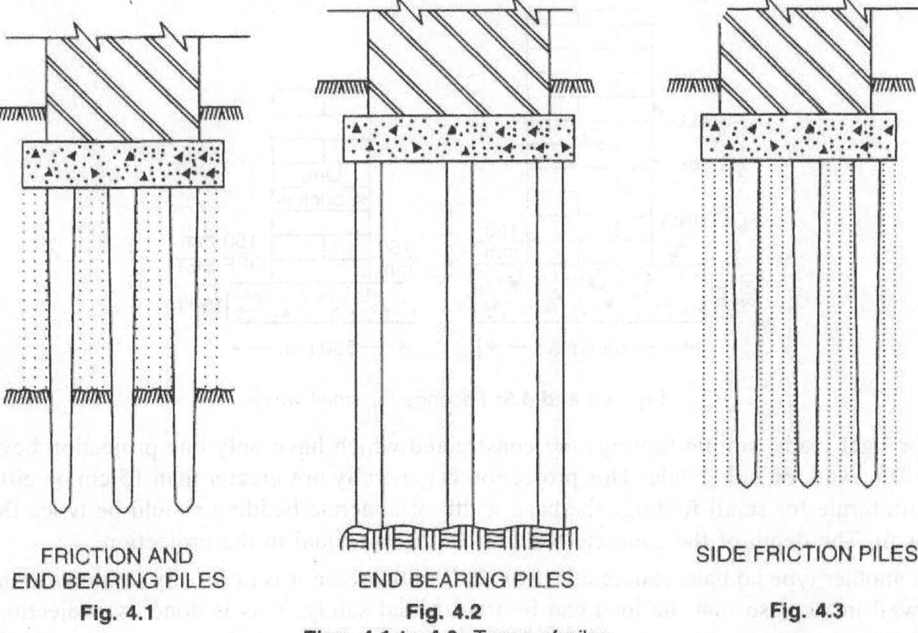

FRICTION AND END BEARING PILES
Fig. 4.1

END BEARING PILES
Fig. 4.2

SIDE FRICTION PILES
Fig. 4.3

Figs. 4.1 to 4.3: Types of piles.

(c) Pier or Well Foundations

These are rarely used for buildings and consist in carrying down through the soil a huge masonry or concrete cylinder which in turn may be supported on a solid rock or by the frictional force between its exterior surface and the soil around it.

4.4 SHALLOW OR SPREAD FOUNDATIONS

These foundations include all those types which are designed to spread the loads. These are provided under walls, columns, piers, fire places and chimneys that bear on soil or rock. The various types are:

(1) Wall footing
(2) Isolated footing
(3) Combined footing
(4) Inverted Arches
(5) Cantilever footing
(6) Continuous footing

A footing is defined as a spread constructed in brickwork, masonry or concrete under the base of a wall or column for the purpose of distributing the load over a larger area. Footings are classified further into groups which are based on the type of construction. These are:

(*i*) ***Simple footing***. These project only a few mm beyond the edges of the walls or other structures supported by them. These foundations are used when only light loads come on the wall, *e.g.*, a wall surrounding a garden has simple footing.

(*ii*) ***Stepped footing***. These are necessary when a wider distribution of the load is to be ensured and also a uniform spreading from the top of the footing to the bottom is desired. The footing consists of layers which increase in widths as the depth of foundation increases.

(*iii*) ***Grillage foundation***. In this type, the footing consists of steel or wooden joists arranged in a stepped manner.

4.4.1 Wall Footing

A footing may have a base course of concrete or may be entirely built up of one material, *e.g.*, bricks or stones. Wall footings can be either simple or stepped.

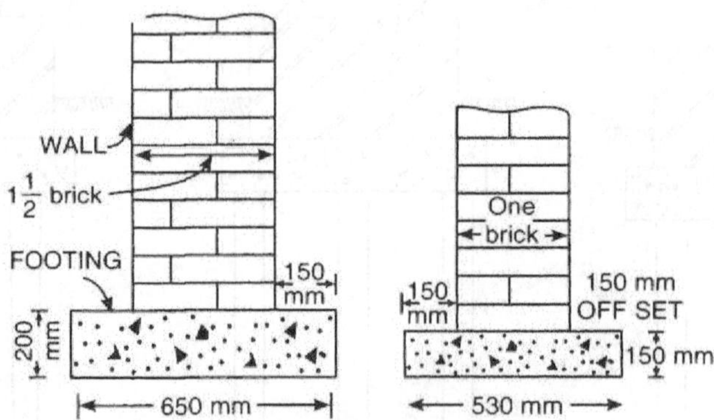

Fig. 4.4 and 4.5: Footings for small walls.

For light loads, simple footings are constructed which have only one projection beyond the width of the wall, on either side. This projection is generally not greater than 15 cm on either side. As a thumb rule for small footings the base width of concrete bedding should be twice the width of the wall. The depth of the concrete bedding is at least equal to the projection.

In another type no base concrete is provided. In that case it is necessary to increase the width of the wall in steps so that the load can be transmitted safely. This is done by projecting bricks regularly to a distance not greater than one-fourth of a brick beyond the edge of the wall. These layers are built in cement mortar and the mix of the mortar should be richer than that used above the foundations. This type of foundation is not used nowadays as the first type is more economical.

When it is necessary to increase the width of the foundation considerably, either the base concrete has to be finished in steps or the bricks are projected and concrete bed is kept level. Projections in concrete are not economical as the construction becomes difficult and hence the bricks are stepped by giving offsets. The concrete projections are at least 150 mm on either side so as to provide a proper working space inside the trench.

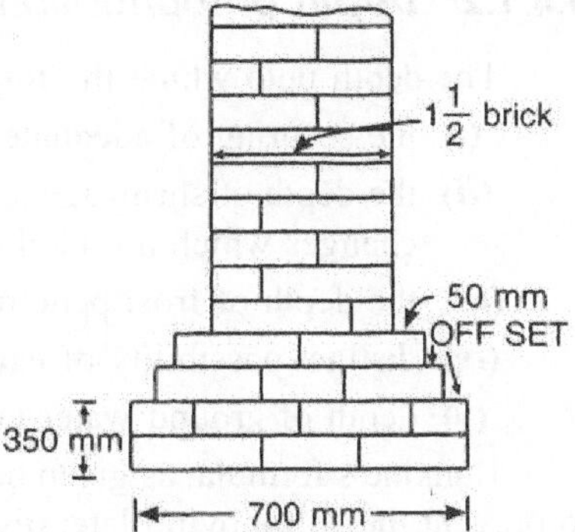

Fig. 4.6: Stepped footings for a $1\frac{1}{2}$ brick thick wall

It should be noted that the number of projections equals the number of half bricks in the thickness of the wall. Further the width of the bottom brick course is equal to twice the thickness of the wall. It is desirable to construct footing courses of as many headers as possible so as to get maximum strength. For breaking the bond at least one fourth brick offsets are essential. The arrangement of these steps can, however, be altered to suit a particular depth of the foundation. A proper design of the steps is essential. The load from the wall is distributed over a greater width of the steps and is further distributed through the concrete bedding. The upward forces of soil try to break the concrete bed as shown in figs. 4.7 to 4.10 and hence the thickness of this bed must be adequate to resist the tendency of cracking.

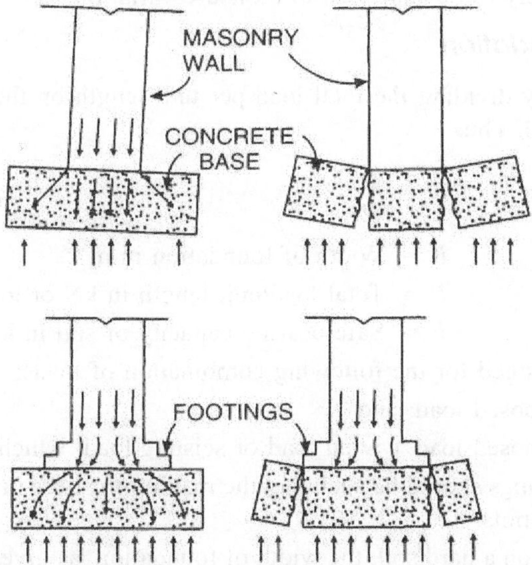

Fig. 4.7 to 4.10: Failures for footings without steps and with steps.

4.4.1.1 *Design of wall footings*

This comprises in determining the following:

(*a*) The depth to which the foundation should be provided below the ground level,

(*b*) Width of foundation; and

(*c*) Thickness of the concrete bed block, if provided.

4.4.1.2 Depth of foundation

The depth upto which the foundations should be carried depends upon:

(*i*) the securing of adequate bearing capacity,

(*ii*) the depth of shrinkage and swelling in the case of clayey soils, due to seasonal weather changes which are likely to cause appreciable movements,

(*iii*) the depth of frost penetration in the case of fine sand and silts,

(*iv*) distinct possibility of excavation close by, and

(*v*) depth of ground water table.

Rankine's formula, as given below, is applied to determine the minimum depth of foundations to prevent the soil moving laterally under pressure.

$$D = \frac{p}{w}\left(\frac{1-\sin\phi}{1+\sin\phi}\right)^2$$

Where D = Depth of foundation below ground level in m.

p = Intensity of load on the base of the footing in kN/m^2 or t/m^2

w = Unit weight of soil in kN/m^3 or t/m^3

ϕ = Angle of repose or angle of internal friction for the soil.

All foundations should extend to a depth of at least 500 mm below natural ground level. A minimum of about 1 m depth should be provided for foundations of load bearing walls. In the case of expansive soils such as clayey soil or black cotton soil, the foundation should be taken to a depth where it is not likely to be subjected to moisture variation.

4.4.1.3 Width of foundation

This is determined by dividing the total load per unit length on the foundation by the safe bearing capacity of the soil. Thus

$$B = \frac{L}{b} \qquad ...(i)$$

where B = Width of foundation in m

L = Total load/unit length in kN or tonnes per m length.

b = Safe bearing capacity of soil in kN/m^2 or t/m^2

The foundation is desiged for the following combination of loads:

(*a*) Dead load + imposed load; and

(*b*) Dead load + imposed load + wind load or seismic load; whichever is critical.

The width of the footings should be such that the maximum stress in the concrete or masonry is within the permissible limits.

In case the wall rests on a hard soil, the width of foundation, as given by the above relations, will be very small. This width, if provided will not be sufficient to prevent tilting of the wall. Thus from practical considerations, National Building Code 2005 (SP-7-2005) has provided that the width of the wall foundation should not be less than that given by:

$$B = 2W + 300 \qquad ...(ii)$$

where B = Width at base in mm; and

W = Width of supported wall in mm.

The final width of foundation adopted is greater of the values given by equations (*i*) and (*ii*).

The angle of spread of the load from the wall base to the outer edge of the ground bearing should not exceed the following values:

Type of construction	*Spread*
Brickwork and stone masonry	1/2 (H) : 1 (V)
Lime concrete	2/3 (H) : 1 (V)
Cement concrete	1 (H) : 1 (V)

4.4.1.4 Thickness of concrete base

The thickness of a concrete bed is given by

$$d_3 = P\sqrt{W/fc}$$

where d_3 = The thickness or depth of concrete in m or cm or mm (see fig. 4.11)
P = Projection or offset of masonry or concrete in m or cm or mm
L = Load coming on the soil in kN/m^2 or kg/cm^2 or t/m^2
fc = Allowable compressive strength of concrete
= 0.03 × ultimate crushing strength of concrete after 28 days in kN/m^2 or kg/cm^2 or t/m^2.

However, it should be ensured that a proper value of fc is used in the design because concrete will gain full strength after about a month and if the footings are built during this period, there is a danger of failure.

National Building Code of India 2005 has recommended that the thickness of different types of footings, if not designed, should be as given in table 4.2.

Table 4.2: Thickness of footings

S. No.	Type of footing	Thickness of footings (minimum)	Remarks
1.	Masonry	(*a*) 250 mm	Select the greater of the two values
		(*b*) Twice the maximum projection from the face of the wall	
2.	Plain concrete		
	For normal structure	(*a*) 200 mm	
		(*b*) Twice the maximum offset in a stepped footing	For footings resting on soil
		(*c*) 300 mm	For footing resting on top of pile
	For lightly loaded structures		
		(*a*) 150 mm	Resting on soil
		(*b*) 200 mm	Resting on pile
3.	Reinforced concrete	(*a*) 150 mm	Resting on soil
		(*b*) 300 mm	Resting on pile

Problem 4.1. *A wall of width W requires a footing width B to transmit the load safely to the foundation. If the angle of spread of the load from the wall base to the outer edge of the bearing is not to exceed the following values:*

In brickwork n_1 *(H) : 1 (V)*

In concrete n_2 *(H) : 1 (V)*

Show that the minimum depth of foundation required is given by:

$$D_{min} = \frac{1}{2n_1}\left[(B-W) - \frac{2P}{n_2}(n_2 - n_1)\right]$$

where L is the offset of the concrete base.

Solution.

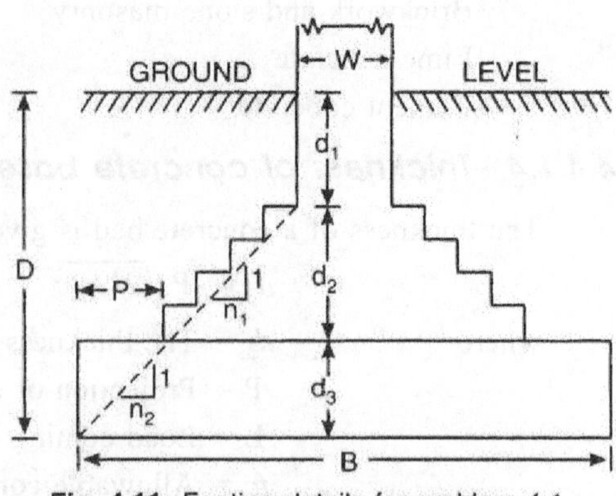

Fig. 4.11: Footing details for problem 4.1.

The total depth D comprise of three depths namely

(*i*) d_1, the depth of wall below G.L. of width, W

(*ii*) d_2, the depth of wall with offsets and having a spread of $n_1 : 1$

(*iii*) d_3, the depth of concrete base, with spread $n_2 : 1$

$$\therefore \quad D = d_1 + d_2 + d_3$$

$$= d_1 + \left(\frac{B}{2} - P - \frac{W}{2}\right)\frac{1}{n_1} + \frac{P}{n_2}$$

The depth will be minimum, when d_1 is zero

$$\therefore \quad D_{min} = \frac{1}{n_1}\left(\frac{B}{2} - P - \frac{W}{2}\right) + \frac{P}{n_2}$$

$$= \frac{1}{2n_1}(B - W) + \left(\frac{P}{n_2} - \frac{P}{n_1}\right)$$

$$= \frac{1}{2n_1}(B - W) + \frac{P}{n_1 n_2}(n_1 - n_2)$$

$$= \frac{1}{2n_1}(B - W) - \frac{2P}{2n_1 n_2}(n_2 - n_1)$$

$$\therefore \quad D_{min} = \frac{1}{2n_1}\left[(B - W) - \frac{2P}{n_2}(n_2 - n_1)\right]$$

If the depth of the concrete bed, d_3 is known, the above expression reduces to:

$$D_{min} = \frac{1}{2n_1}[(B - W) - 2d_3(n_2 - n_1)]$$

Problem 4.2. *Design the foundations of a wall of a single storeyed house resting on sandy soil having bearing capacity as 1.8 kg/cm² and angle of repose of 30°. The unit weight of the soil is 1600 kg/m³. The wall thickness is 30 cm and the total load coming on the wall per metre length is 12,000 kg.*

Solution.

(i) Depth of foundation

Applying Rankine's formula

$$D = \frac{p}{w}\left(\frac{1 - \sin\phi}{1 + \sin\phi}\right)^2$$

$$= \frac{1.8 \times 100 \times 100}{1600}\left(\frac{1 - 0.5}{1 + 0.5}\right)^2$$

$$= \frac{180}{16} \times \frac{1}{9} = 1.25 \text{ m}$$

(ii) Width of foundation

(*a*) By empirical relation

$$B = \frac{L}{b}$$

$$= \frac{12000}{1.8 \times 100 \times 100}$$

$$= 0.67 \text{ m} = 67 \text{ cm}$$

(*b*) From practical considerations

$$B = 2W + 30 = 2 \times 30 + 30 = 90 \text{ cm}$$

Adopt $B = 90$ cm

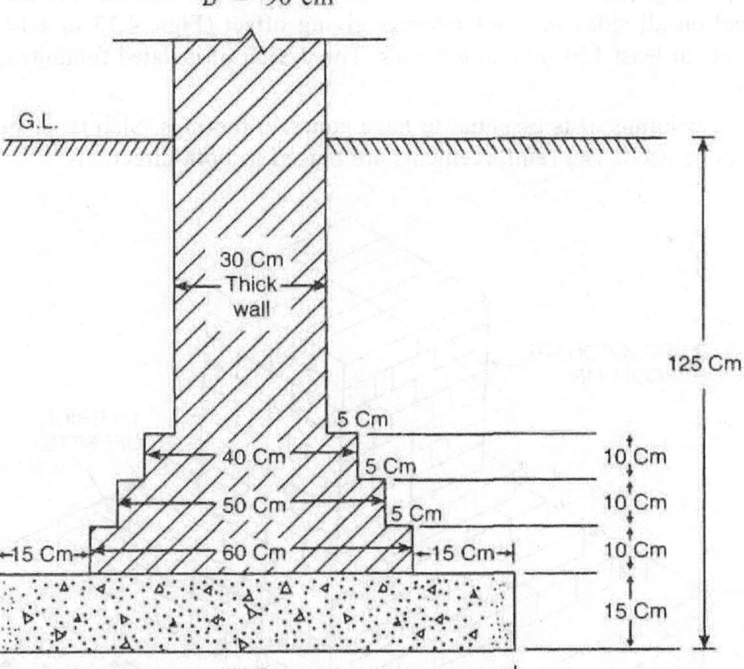

Fig. 4.12: Showing the design features for problem 4.2.

(iii) Thickness of concrete bed

Load on wall/metre length = 12000 kg

Self weight of foundation, assuming it to be 10% of the load on wall

$$= \frac{10 \times 1200}{100} = 120 \text{ kg/m length.}$$

$\therefore$ Total load on the soil = 12000 + 1200 = 13,200 kg/m length

$$B = 90 \text{ cm}$$

$\therefore$ Weight on soil/cm² $= \dfrac{13200}{90 \times 100} = 1.47$ kg/cm².

Using 1 : 4 : 8 plain cement concrete with a strength of M 100

$$fc = 0.03 \times 100 = 3 \text{ kg/cm}^2$$

$$\therefore \quad d_3 = P\sqrt{\frac{W}{fc}}$$

$$= 15\sqrt{\frac{1.47}{3}} = 10.5 \text{ cm.}$$

Since it is a lightly loaded structure, adopt a minmum of 15 cm thickness of concrete 'bed' from table 4.2.

4.4.2 Isolated Footings

Isolated footings are used to support the individual columns. Column footings are either of stepped type or have projections in the base concrete. Since column loads are heavy, reinforcement is provided by placing steel bars at the base of the footings. If columns are made with bricks, these are stepped on all sides in regular layers giving offset (Figs. 4.13 to 4.17). The concrete bedding offsets are at least 150 mm on all sides. The design of isolated footings is similar to that of wall footings.

For concrete columns, it is essential to have concrete footings. Slab type, stepped or sloped footing bases can be used. The reinforcements are placed in both directions.

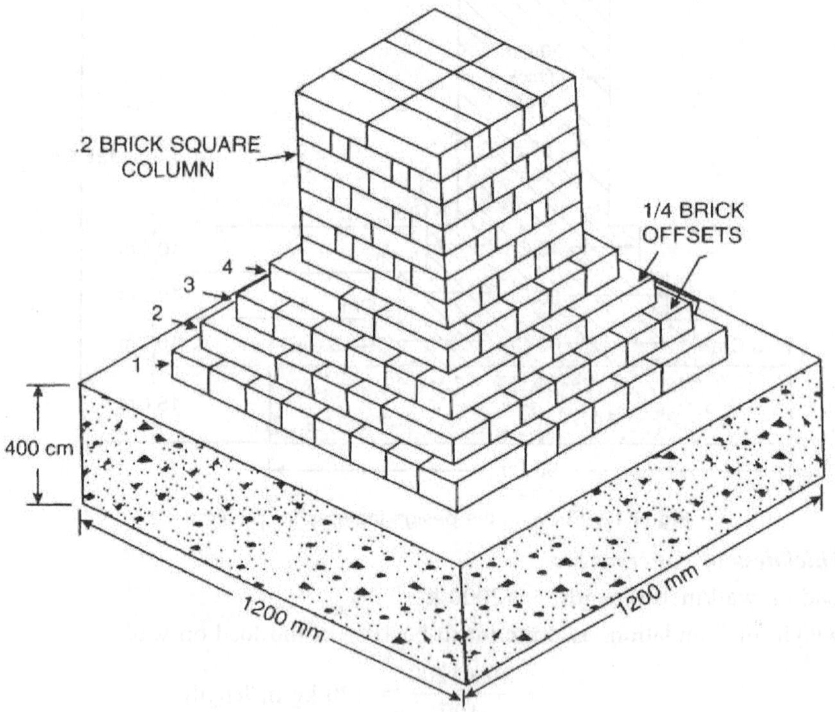

Fig. 4.13.

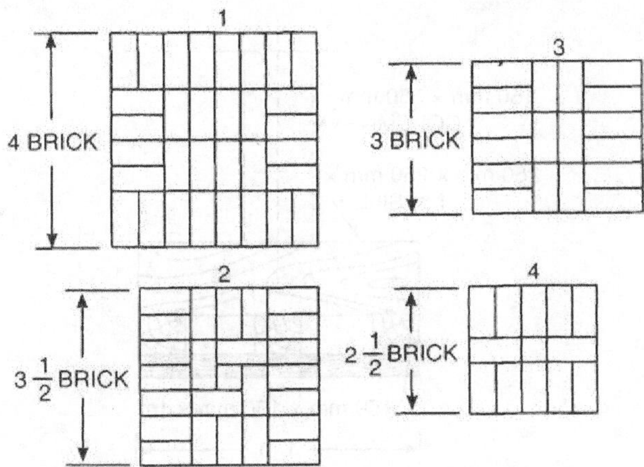

Figs. 4.13 to 4.17: Isometric view and plans of brick courses 1, 2, 3 + 4 of an isolated brick column footing with plans of brick courses.

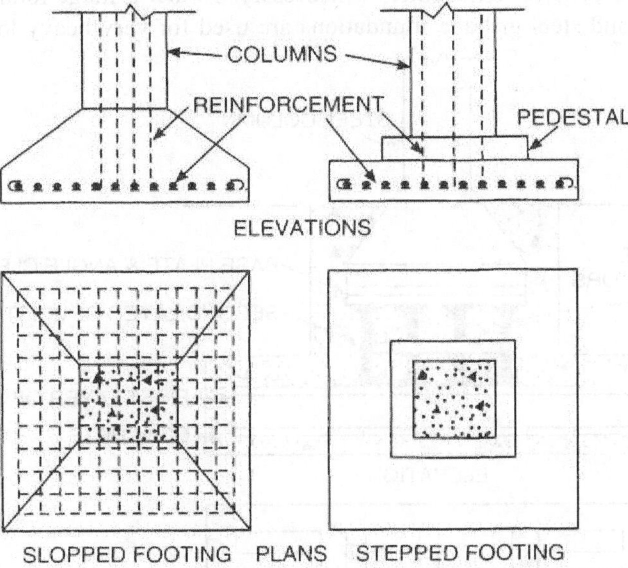

Figs. 4.18 to 4.21: Slopped and stepped footings for isolated R.C.C. columns.

4.4.2.1 Grillage Foundation

Steel columns can have either concrete footings or footings of steel. Generally the latter type is used. This consists of steel beams arranged in layers at right angles to one another and are embedded in concrete. Further, the beams are connected with each other by bolts so as to form a rigid unit. This is known as 'Grillage footing' and shown in Figs. 4.23 to 4.24.

The following points about steel grillage should be kept in view:

(*i*) The pressure from the footing is uniformly distributed over the soil.

(*ii*) The pressure of one tier of beams is uniformly distributed on the tier below,

(*iii*) Each tier acts independently of all other tiers.

(*iv*) The concrete filling in between the beams does not carry any load but only keeps the beams in position and prevents the corrosion of the beams. At least a cover of 100 mm is necessary.

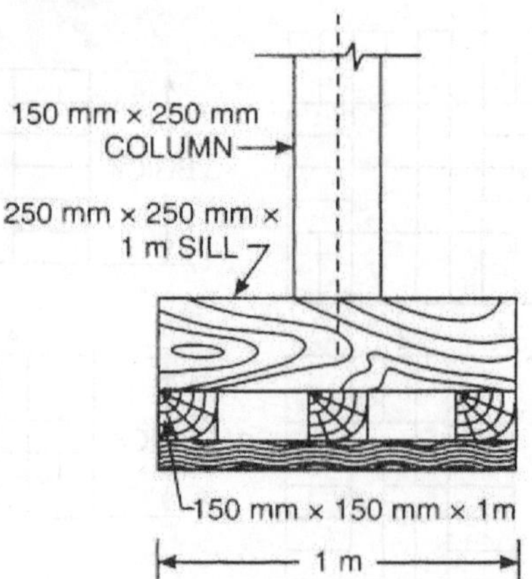

Fig. 4.22: Timber grillage footing.

Timber columns have a grillage foundation built of timber beams in a manner similar to steel grillage (Fig. 4.22). No concrete embedment is necessary. Timber grillage foundations are used for temporary structures and steel grillage foundations are used for very heavy loads.

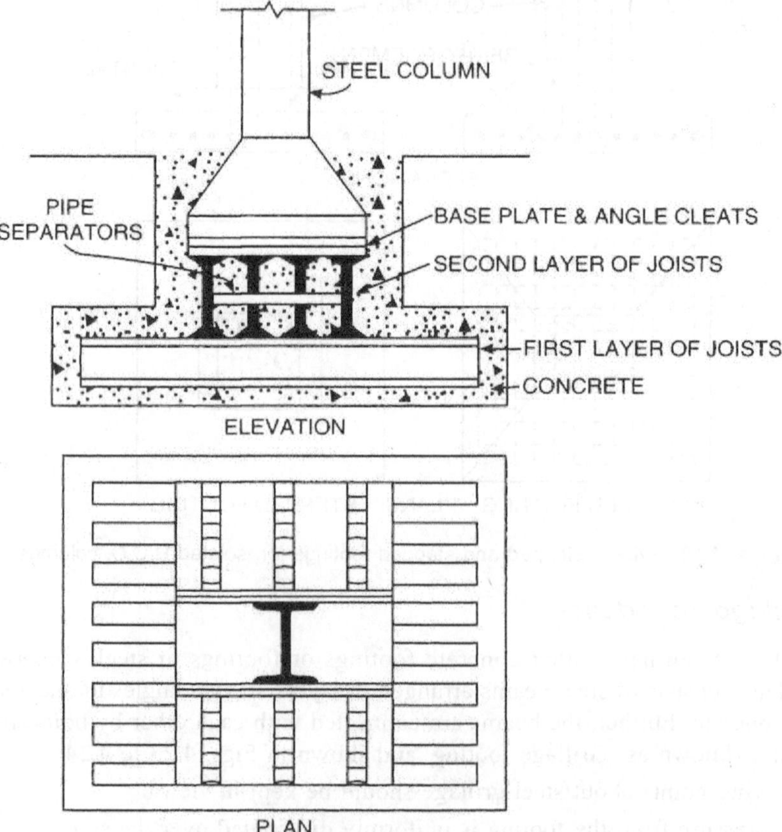

Fig. 4.23 to 4.24: Typical grillage footing for steel columns.

Problem 4.3. *Determine the size of footings and depth of foundation for a brick-pillar 400 mm square. The concentrated load carried by the column at the footing top is 150 kN. The bearing capacity of the soil at the depth of foundation is 135 kN/m². The soil weighs 19.20 kN/m³ and has an angle of repose of 30°. Use 1 : 3 : 6 plain cement concrete (M 10).*

Solution.

Load on brick column upto footing top = 150 kN

Self weight of foundation, assuming it to be 10% of the load on brick column

$$= 150 \times \frac{10}{100} = 15 \text{ kN.}$$

$\therefore$ Total load on soil = 150 + 15 = 165 kN

Base area of concrete block

$$= \frac{165}{135} = 1.22 \text{ sq m}$$

One side of the square concrete block = $\sqrt{1.22}$ = 1.10 m

or = 1100 mm

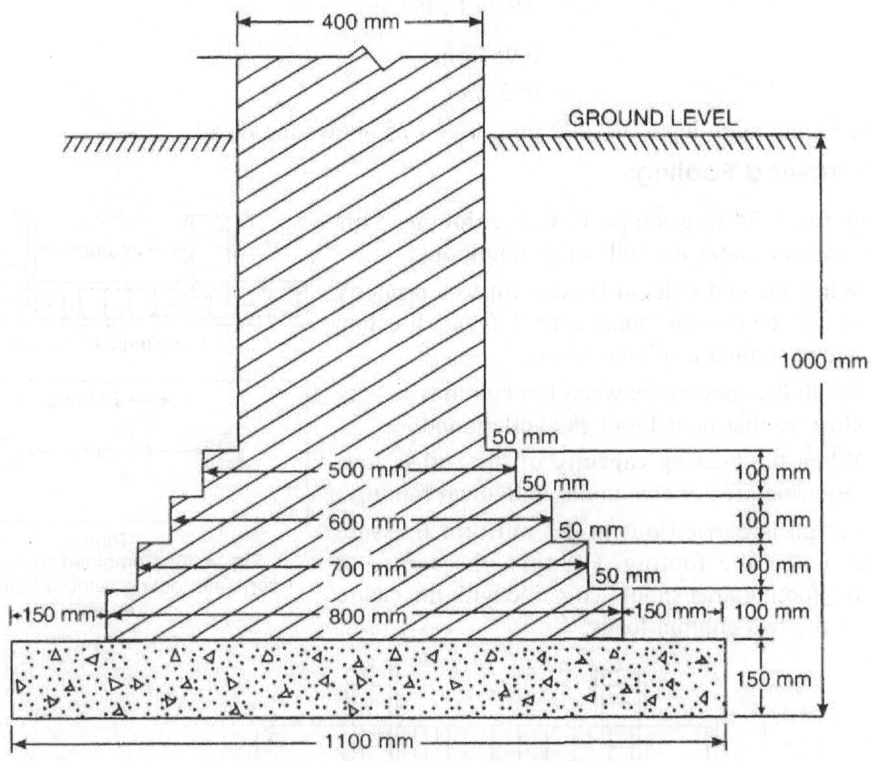

Fig. 4.25: Foundation details for problem 4.3.

Width of bottom-most course of pillar footing

$$= 2W = 2 \times 400 = 800 \text{ mm}$$

Thus offset of concrete, (L) $= \dfrac{1100 - 800}{2} = 150$ mm

$$\text{Load on soil/m}^2 = \frac{165}{1.1 \times 1.1} = 136 \text{ kN/m}^2$$

Depth of concrete block

$$d_3 = P\sqrt{\frac{W}{fc}}$$

Using 1 : 3 : 6 plain cement concrete with a strength of $fc = 0.03 \times 10 = 0.3$ N/mm^2 = 300 kN/m^2

$$\therefore \quad d_3 = 150\sqrt{\frac{136}{300}}$$

$$= 101 \text{ mm}$$

Adopt a minimum thickness of 150 mm

Depth of foundation,

$$D = \frac{p}{w}\left(\frac{1-\sin\phi}{1+\sin\phi}\right)^2$$

$$= \frac{165}{19.2}\left(\frac{1-0.5}{1+0.5}\right)^2$$

$$= 0.9546 \text{ m}$$

$$\simeq 955 \text{ mm}$$

Adopt a minimum depth of 1000 mm. Fig. 4.25 shows the details.

4.4.3 Combined Footings

A combined footing supports two columns. This becomes necessary under the following conditions:

(1) When the end column is adjacent to a property so that the footing cannot spread in that direction due to limited available space.

(2) When the spacing between the columns is very close so that their footings overlap; and

(3) When the bearing capacity of the soil is low, requiring greater area under individual footings.

The design is carried out to get uniform pressure distribution under the footing. For this, the centre of gravity of the footing area should coincide with the centre of gravity of the two column loads.

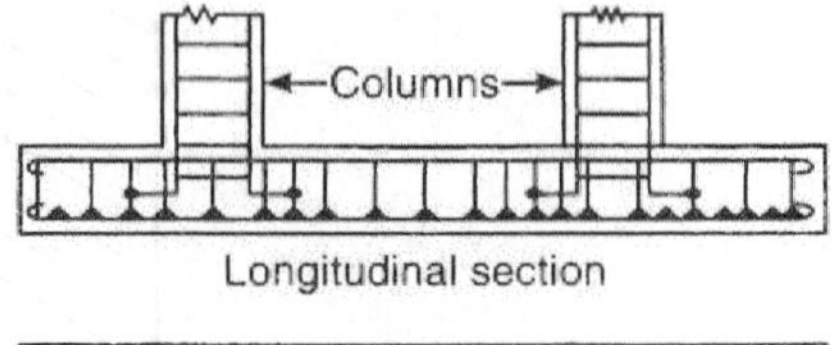

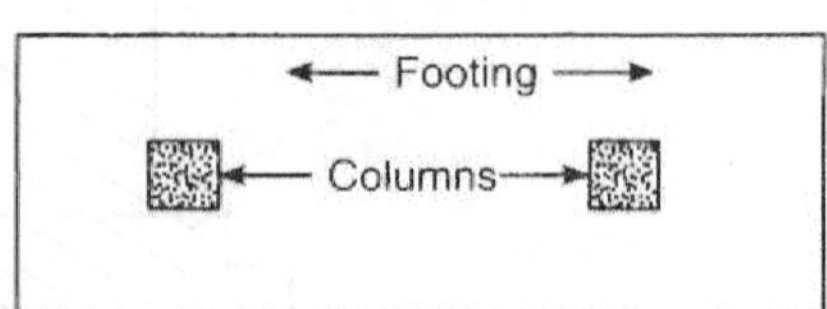

Fig. 4.26: Combined rectangular concrete footing without longitudinal beam.

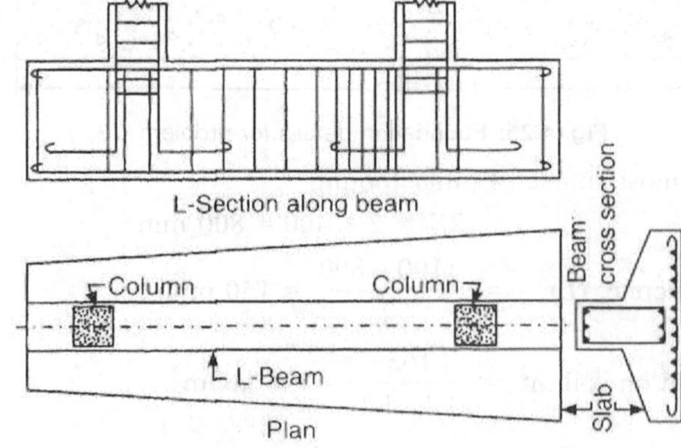

Fig. 4.27: Combined trapezoidal concrete footing with longitudinal beam.

The combined footing may be rectangular or trapezoidal in plan. If the load on the interior column is heavier than the exterior column near the property line, rectangular shape is feasible. If the outer column carries heavier load, trapezoidal shape becomes essential.

The combined footings may be either R.C.C. footing or steel grillage footing. The R.C.C. footings may be without longitudinal beam (see Fig. 4.26) or with longitudinal beam (see Fig. 4.27). Fig. 4.28 and 4.29 show combined trapezoidal steel grillage footings. The bottom tier of joists vary both in length and depth while the top tier of joists is of uniform length and depth.

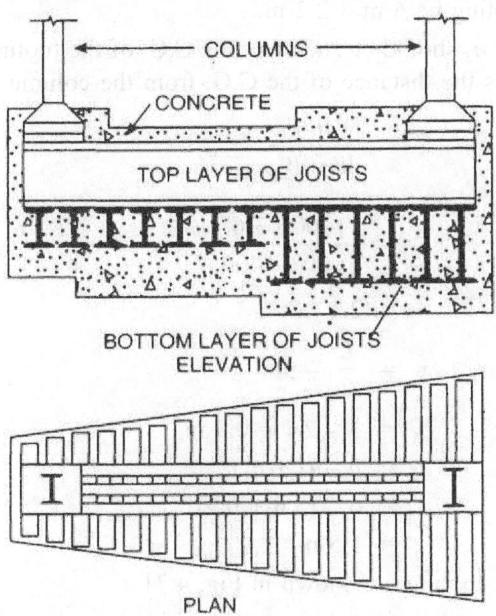

Figs. 4.28 and 4.29: Combined trapezoidal steel grillage footings.

Problem 4.4. *Determine the dimensions of a combined rectangular footing for two columns A and B carrying loads of 1000 kN and 1400 kN respectively. Column A is 500 mm square and column B is 600 mm square. The centre to centre spacing between the columns is 3.6 m. The safe bearing capacity of the soil may be taken as 200 kN/m².*

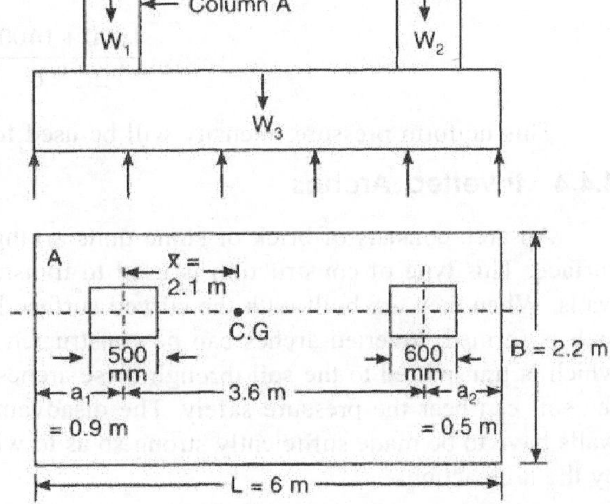

Figs. 4.30 and 4.31: Combined footing for problem 4.4.

Solution.

$$W_1 = 1000 \text{ kN}$$

$$W_2 = 1400 \text{ kN}$$

$$l = 3.6 \text{ m}$$

$$b = 200 \text{ kN/m}^2$$

Assume the weight of footing as 10% of the super-imposed load, $(W_1 + W_2)$.

$\therefore$ Weight of footing $\quad W_3 = \dfrac{10}{100}(1000 + 1400)$

$$= 240 \text{ kN}$$

$$\text{Area of the footing} = \frac{W_1 + W_2 + W_3}{b}$$

$$= \frac{1000 + 1400 + 240}{200}$$

$$= 13.2 \text{ m}^2$$

Let the size of the footing be 6 m × 2.2 m.

The projections a_1 and a_2 should be such that the C.G. of the footing coincides with the C.G. of the column loads. If $\bar{x}$ is the distance of the C.G. from the column A, then

$$\bar{x} = \frac{W_2 l}{W_1 + W_2}$$

$$= \frac{1400 \times 3.6}{1000 + 1400}$$

$$= 2.1 \text{ m}$$

$$\therefore \quad a_1 + \bar{x} = \frac{L}{2} = 3\text{m}$$

or

$$a_1 = 3 - 2.1 = 0.9 \text{ m}$$

Also

$$a_2 = L - (l + a_1).$$

$$= 6 - (3.6 + 0.9)$$

$$= 1.5 \text{ m}$$

The dimensions of the footing are shown in Fig. 4.31.

The net upward pressure P is given by

$$P = \frac{W_1 + W_2}{B \times L}$$

$$= \frac{1000 + 1400}{2.2 \times 6} = 181.82 \text{ kN/m}^2$$

This uniform pressure intensity will be used for the structural design of the footing slab.

4.4.4 Inverted Arches

An arch consists of brick or stone units arranged in such a manner so as to form a curved surface. This type of construction is used to transmit loads above an opening to the supporting walls. When they are built with the curved surface having the concave face upwards, an inverted arch is formed. Inverted arches can be constructed between the two walls at the base. The load which is transmitted to the soil through these arches gets distributed over a wider area and hence the soil can bear the pressure safely. The disadvantage with this type of footing is that the end walls have to be made sufficiently strong so as to withstand the outward pressure which is caused by the arch action.

These are especially suitable when the bearing capacity of the soil is very low and it is desired to distribute the loads over a large area. Further advantage of this type of construction is that the depth of foundation is quite less.

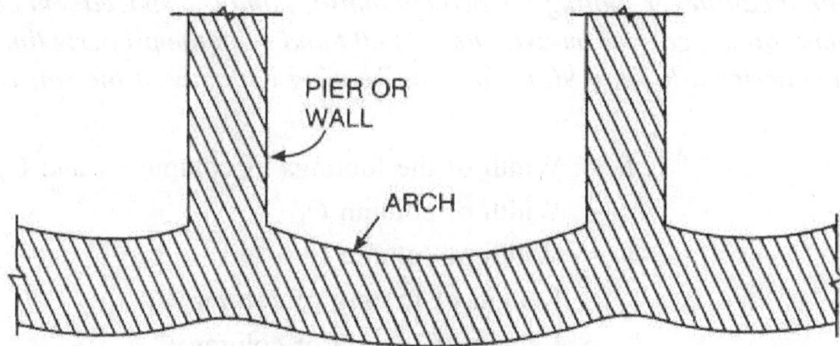

Fig. 4.32: Inverted arch footing.

4.4.5 Cantilever Footings

This type of foundation is useful when it is impossible to place a footing directly beneath a column or other load because of limitations of adjacent buildings or eccentric loading conditions or where footing has to be shallow in depth and where the bearing capacity of the soil is low. A load from the outer column is balanced by the load from the inner column acting about a fulcrum. The product of the load in the interior column multiplied by its corresponding lever arm should be at least 50 per cent greater than the load on the exterior column multiplied by its lever arm under all type of loading conditions. In case the interior column load is not available, a suitable huge concrete block can be built to act as an anchorage. The foundation of the interior column has to be built as strong as possible and must have suitable connecting reinforcement with its base.

In practice, however, point shaped fulcrums are not used and a suitable base is built for the outer column. The connecting beam prevents unequal settlement of the exterior column by holding it in position with respect to the inner column. Cantilever footings can be built of concrete or steel.

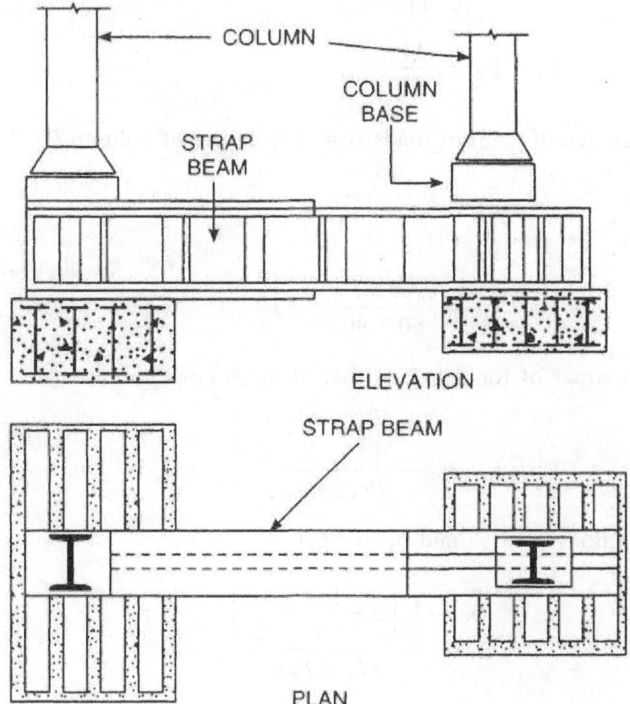

Figs. 4.33 and 4.34: Cantilever grillage footing.

Problem 4.5. *Design a cantilever footing for two columns C_1 and C_2, spaced 5 m centre to centre. Column C_1 is 400 mm square, carries an axial load of 60 t and is on the property line. Column C_2 is 600 mm square and carries a load of 90 t. The safe bearing capacity of the soil is 15 t/m^2.*

Solution

Let

$$B = \text{Width of the footings of column } C_1 \text{ and } C_2$$
$$b_1 = \text{Width of column } C_1$$
$$b_2 = \text{Width of column } C_2$$
$$L_1 = \text{Length of footing of column } C_1$$
$$L_2 = \text{Length of footing of column } C_2$$
$$l = \text{Spacing between the columns}$$
$$W_1 = \text{Load on column } C_1$$
$$W_2 = \text{Load on column } C_2$$
$$W_3 = \text{Weight of footing}$$

Given

$$W_1 = 60 \text{ t}$$
$$W_2 = 90 \text{ t}$$
$$b_1 = 400 \text{ mm} = 0.4 \text{ m}$$
$$b_1 = 600 \text{ mm} = 0.6 \text{ m}$$

Let

$$W_3 = 10\% \text{ of } (W_1 + W_2)$$
$$= \frac{10}{100}(60 + 90) = 15 \text{ t}$$

$$\therefore \quad B(L_1 + L_2) = \frac{60 + 90 + 15}{15} = \frac{165}{15}$$
$$= 11$$

$$\therefore \quad L_1 + L_2 = \frac{11}{B} \qquad \text{...}(i)$$

Let $\bar{x}$ be the distance of C.G. of loads from the centre of column C_2.

$$\therefore \quad \bar{x} = \frac{W_1 l}{W_1 + W_2}$$

$$= \frac{60 \times 5}{60 + 90} = 2 \text{ m}$$

If the C.G. of the areas of footings and that of loads coincide, then

$$\bar{x} = \frac{BL_1\left(l + \frac{b_1}{2} - \frac{L_1}{2}\right)}{B(L_1 + L_2)}$$

Substituting the values of $\bar{x}$, l and b_1, we get

$$2 = \frac{L_1\left(5 + \frac{0.4}{2} - \frac{L_1}{2}\right)}{(L_1 + L_2)}$$

Substituting the value of $(L_1 + L_2)$ from equation (*i*)

$$2 = \frac{BL_1(5 + 0.2 - 0.5L_1)}{11}$$

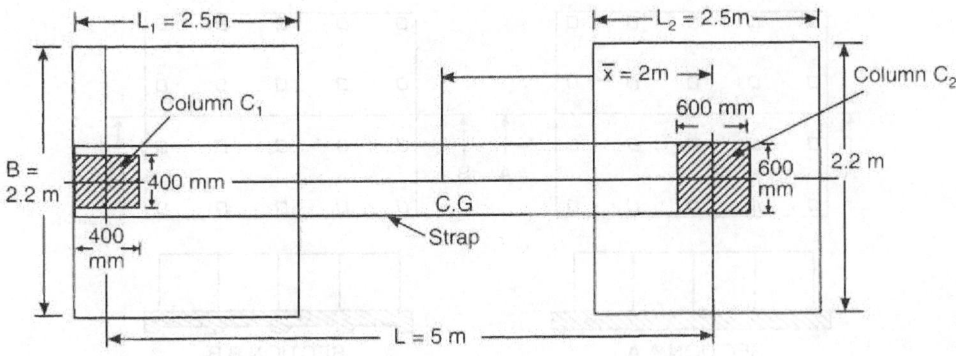

Fig. 4.35: Arrangement of footings for problem 4.5.

Assuming $B = 2.2$ m, we get

$$L_1^2 - 10.4\,L_1 + 20 = 0$$

Solving, we get

$$L_1 = \frac{10.4 \pm \sqrt{(10.4)^2 - 4 \times 20}}{2}$$

$$= 2.55 \text{ m}$$

say $= 2.5$ m

$$\therefore \quad L_2 = \frac{11}{2.2} - 2.5$$

$$= 2.5 \text{ m}$$

The general arrangement of footings is shown in Fig. 4.35.

4.4.6 Continuous Footings

These consist of continuous reinforced concrete slabs extending over three or more columns. A deeper beam can be constructed in between the columns to increase the stability. Suitable steel reinforcement is provided. This type of construction is desired when the differential (uneven) settlement is to be prevented. Further they are suitable wherever safety against earthquake is essential.

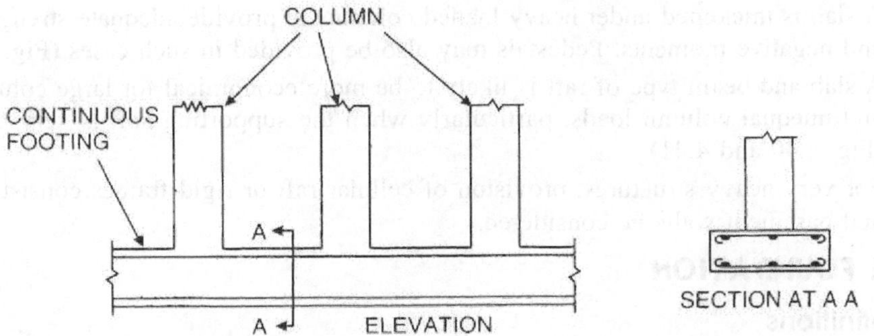

Figs. 4.36 and 4.37: Continuous R.C.C. Footing.

4.4.7 Raft Foundation

These foundations are used when the bearing power of the soil is so low that individual spread footing can not be economically provided.

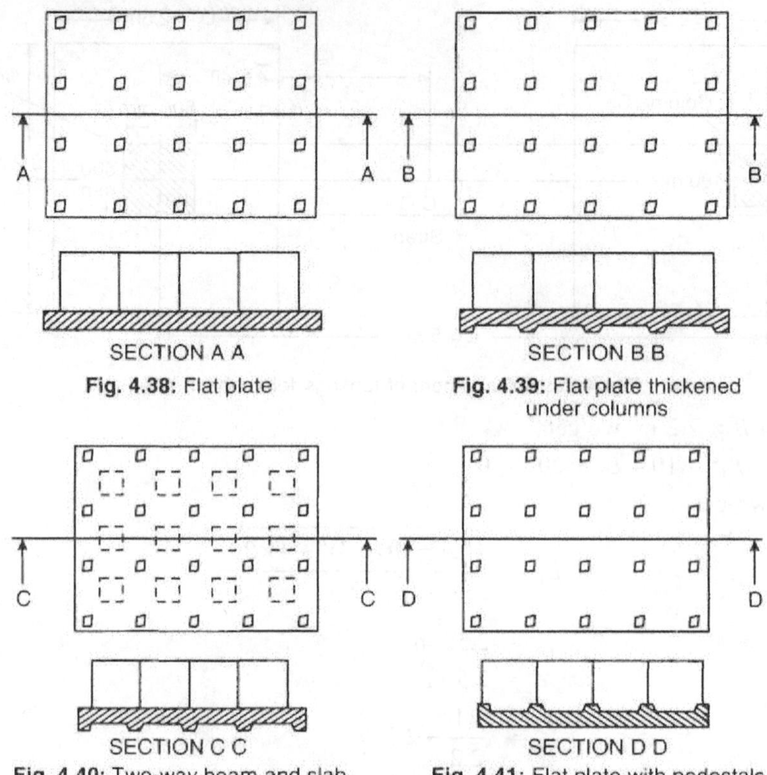

Fig. 4.38: Flat plate

Fig. 4.39: Flat plate thickened under columns

Fig. 4.40: Two-way beam and slab

Fig. 4.41: Flat plate with pedestals

Figs. 4.38 to 4.41: Common types of raft foundations.

Raft foundation is a sub-structure supporting an arrangement of columns or walls in a row or rows, transmitting the loads to the soil by means of a continuous slab covering the entire foundation area, with or without depressions or openings. The common types of raft foundations used are shown in Fig. 4.38 to 4.41. The choice of the raft type may be made as under:

(*i*) For fairly small and uniform column spacing and when the supporting soil is not too compressible, a flat concrete slab having uniform thickness throughout (a true mat) is most suitable (see Fig. 4.38).

(*ii*) A slab is thickened under heavy loaded columns to provide adequate strength for shear and negative moments. Pedestals may also be provided in such cases (Fig. 4.39).

(*iii*) A slab and beam type of raft is likely to be more economical for large column spacing and unequal column loads, particularly when the supporting soil is very compressive (Fig. 4.40 and 4.41).

(*iv*) For very heavy structures, provision of cellular raft or rigid frames consisting of slabs and basement walls be considered.

4.5 PILE FOUNDATION

4.5.1 Definitions

Pile. This is an element of construction placed in the ground either vertically or slightly inclined to increase the load carrying capacity of the soil.

Bearing piles. These transmit the super-imposed load to a stronger strata below.

Friction piles. These are designed to transmit the loads by the frictional force existing between the sides of the pile and the ground.

Battered piles. This is a pile driven at an inclination to the vertical to resist the inclined forces.

Sheet piles. These are thin units of steel or timber driven as piles.

Bored pile. A pile formed with or without casing by excavating or boring a hole in the ground and subsequently filling it with plain or reinforced concrete.

Bored precast pile. A pile constructed in reinforced concrete in a casting yard and subsequently lowered in the pre-bored holes and the space around grouted.

Bored cast-in situ pile. The pile formed within the ground by excavating or boring a hole within it; with or without the use of a temporary casing and subsequently filling it with plain or reinforced concrete. When the liner is left permanently it is termed as cased pile and when the casing is taken out it is called uncased pile.

Bored campaction pile. A bored cast in-situ pile with or without bulb(s) in which the compaction of the surrounding ground and freshly filled concrete in pile bore is simultaneously achieved by a suitable method. If the pile is with bulbs, it is known as underreamed bored compaction pile.

Materials used

Piles can be made of wood, concrete, reinforced cement concrete, steel, prestressed concrete and composite materials.

The various types of piles for foundations are described below:

4.5.2 Wooden Piles

Usually timber trees are used as piles after the bark and the branches are removed. The type of trees which are used in India are Bakul, Owli, Rayani, Jamba, Deodar, Sal, Teak, etc. The circular piles vary in diameter from 30 cm to 40 cm. If square sizes are used, they are obtained from the interior portion of the tree by cutting off the side planks. They also vary in size from 30 to 40 cm square. The lower protion of the piles is tapered to 15 to 18 cm square size. The length of the piles is about 20 times the top width. Longer piles are liable to fail on account of buckling.

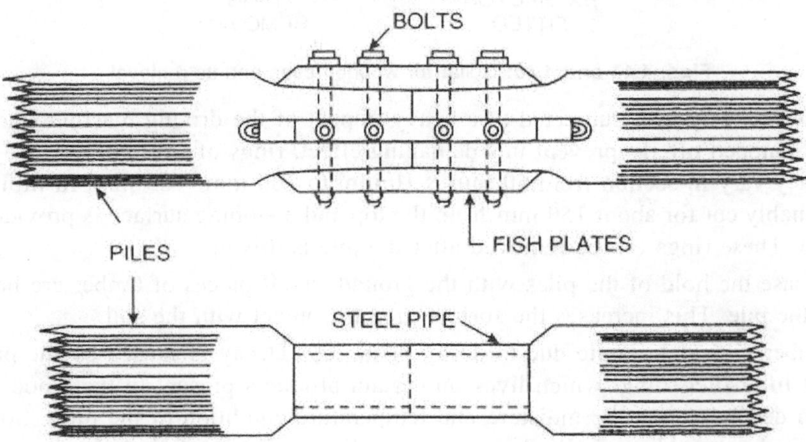

Figs. 4.42 and 4.43: Joining wooden piles.

The foot of a wooden pile should be cut perpendicular to its axis to permit a vertical driving. In soft or silty ground it may not be necessary to sharpen the pile end. If such a pile comes across a harder strata, it will have a wider bearing area over it and hence provide a stronger foundation. The blunt end will also break any obstruction like roots, etc., without tilting the pile. When coarse gravel or boulders are liable to be encountered it is desirable to reduce the area of the pile foot. In stiff clays it is advantageous to make the end pointed. While pointing, the end is formed into a pyramid shape, the edge of which is 80–100 mm square. If this tip is less than the above size, the timber pile end will not be able to resist the action of any hammering. The length of this point can be $1\frac{1}{2}$ to 2 times the diameter of the foot.

The timber point may be protected or replaced by a metalic shoe. Various types of shoes have been designed for use with different types of piles. Shoes are used while driving through soil containing boulders, gravel, hard clays, etc. They may be of cast iron and should have sufficient length of wrought iron straps to connect them with the pile.

Whenever it is necessary to use a longer pile than available in usual lengths, it is customary to splice the piles together with their ends abutting each other. Additional strength can be obtained by the use of steel plates (fish plates) bolted to the sides of the piles. Wooden fish plates can also be used but they increases the section of the pile considerably. As an alternative, a heavy pipe of steel can be put over the pile end and fixed to it.

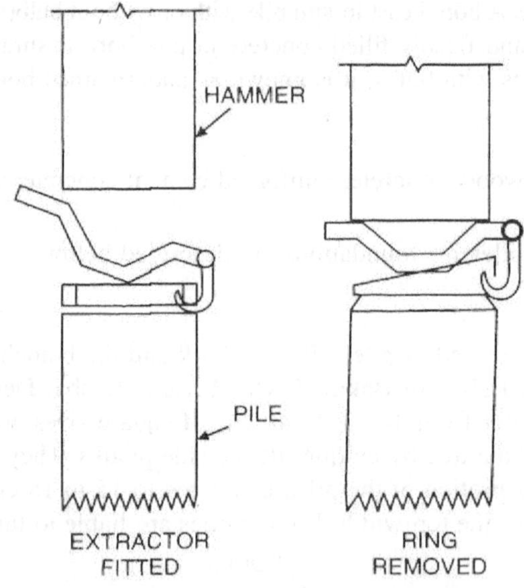

Figs. 4.44 and 4.45: Collar for wooden piles and its removal.

The head of the pile is subjected to a heavy impact of the driving machine and hence gets distorted and chipped off. To prevent this damaging effect, rings of wrought iron are fitted to the pile head. They vary in section from 50 mm × 10 mm to 100 mm × 25 mm. In fitting the rings, the pile is suitably cut for about 150 mm from the top and a sloping surface is provided for a part of this length. These rings can be removed after the pile is driven.

To increase the hold of the piles with the ground, small pieces of timber are bolted around the sides of the pile. This increases the surface area in contact with the soil.

The timber piles deteriorate due to decay or insects. Decay is caused by the presence of a type of plant life called fungi which lives on certain products present in the wood. The rate of disintegration depends upon the moisture and temperature condition of the piles. Some types of these fungi develop quickly under high moisture conditions. Timber fully below water throughout the year does not decay. It is desirable to ensure that the variations in the water level around the timber piles are least.

The termites or insects are of two types, i.e., which grow in wet conditions and those which grow in dry conditions. They live on the cellulose present in the timber. Only few timbers are safe from this attack. Another class of insects which are liable to destroy timber piles are "marine borers" but these are usually found around the coastal areas. The life of timber piles which are not suitably guarded against the action of marine borers is considerably reduced.

Chemical preservation or mechanical protection is commonly provided as a treatment of wooden piles. Treating timber with creosote oil under pressure is usually carried out. The amount of creosote to be a applied depends upon the type of timber used and the degree of protection needed. Several other chemicals like zinc chloride, copper sulphate, mercuric chloride, etc. which are poisonous to the animal life have been used for the preservation of timbers but they are liable to be washed away by the water table fluctuations. The various methods of chemical preservation and their suitability under various conditions are described in detail in books on "Building Materials".

Mechanical protection is afforded by various types of coverings like thin sheets of metal or layer of felt soaked in various chemicals. These treatments are not expensive but are mainly used as a protection against marine borers.

Chemically treated piles should be handled very carefully so as to avoid the breaking of outer fibres which act as protecting coverings.

4.5.2.1 Advantages of wooden piles

(*i*) They are less expensive as most of the local timber available can be used after suitable treatment.

(*ii*) They can be made in longer lengths by joining the individual pieces easily.

(*iii*) Cutting of these piles is very easy.

(*iv*) They can be driven quickly and with lighter machinery.

4.5.2.2 Disadvantages

(*i*) They deteriorate by the action of water or insects.

(*ii*) They have less load bearing capacity.

(*iii*) Whenever long piles are to be driven, it if necessary to join a number of small individual units and this entails lot of joining work and the cost is high.

4.5.3 Concrete Piles

Concrete piles can be broadly classified into two types:

(*i*) ***Precast concrete piles***. These are cast at a suitable place, cured and afterwards driven like a timber pile.

(*ii*) ***Cast-in-situ piles***. These are cast at the place where they have to rest finally. They may have a casing which also remains intact.

Precast Concrete Piles: Precast concrete piles are commonly of square section with chamfered corners. Other shapes, *e.g.*, octagonal types are also available. Octagonal type has a better appearance and steel reinforcement can also be placed in it easily. Whenever these piles are to be driven through hard soils, cast iron or mild steel shoes are used at the end which is driven into the soil. Generally for normal work, 1 : 2 : 4 mix is used whereas for heavy loads and for driving through harder soils, 1 : $1\frac{1}{2}$: 3 mix is employed.

The diameter of these piles varies from 250 to 600 mm. Certain specifications lay down rules for the minimum lateral dimension and least diameter depending upon the length of the pile, *e.g.*, 350 mm, for piles upto 10 metre in length and 400 mm, for piles between 10 and 15 m. Precast concrete piles are available in lengths from 3 to 30 m but generally less than 5 m are not used.

Precast piles are constructed without taper but have pointed lower ends. Whenever tapering is desired, it should not exceed 20 mm per metre length of the pile.

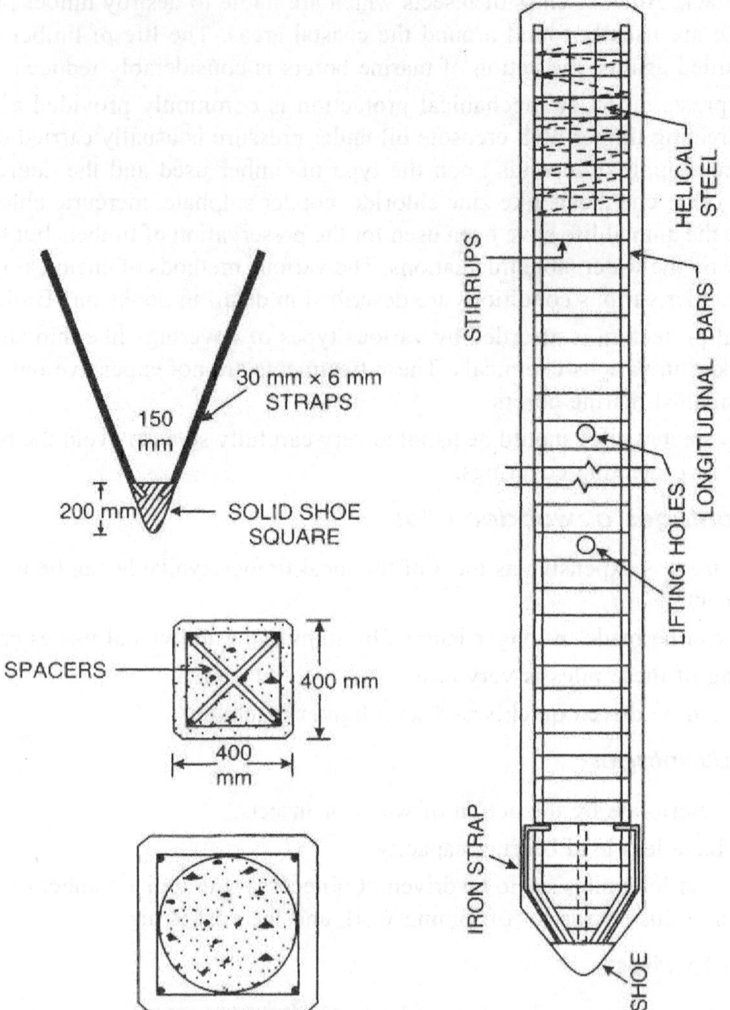

Figs. 4.46 to 4.49: Sections and elevation of R.C.C. piles with details of connectors and shoe.

The reinforcement consists of longitudinal bars with spiral at the top and bottom ends and suitable ties in between. However, in some cases the spiral may extend through the entire length of the pile. The longitudinal reinforcing bars are separated by cast iron or pressed steel forks placed diagonally across the section at 1 to $1\frac{1}{2}$ metre intervals. The ends of these longitudinal bars should bear on the shoe. Sometimes it is desired to project these longitudinal bars beyond the head of the pile and a special hood is built while driving.

These piles are cast in the forms either horizontally or vertically. Generally horizontal casting is preferred. When piles are cast in this position, it should be ensured that an unyielding base is provided for the pile so that concrete may not get disturbed while setting. Use of vibrators is essential. The forms are removed after about 48 hours. The pile is cured in its position for a week. After this, it is removed and placed in stacks for further curing. A period of about 3 weeks is allowed to lapse before they are driven. Whenever a rapid construction is desired, special quick hardening cement may be used so that the pile develops the desired strength earlier.

Special devices are used to permit convenient handling of piles. C.I. bolts screwed into plates are cast in the piles near the centre of cross-section at the points of lifting. The points where the pile is to be lifted are to be suitably calculated.

If a pile is to have a hole in the centre for insertion of a jet pipe for water injection while driving, a pipe is cast along with concrete. Sometimes an oil coated tapered wooden core is left but this has to be constantly rotated to prevent its sticking with concrete.

Adding an extra length to precast piles is done by stripping the head and overlapping the longitudinal bars for a distance not less than 40 times the diameter of the bars. These longitudinal bars may also be welded.

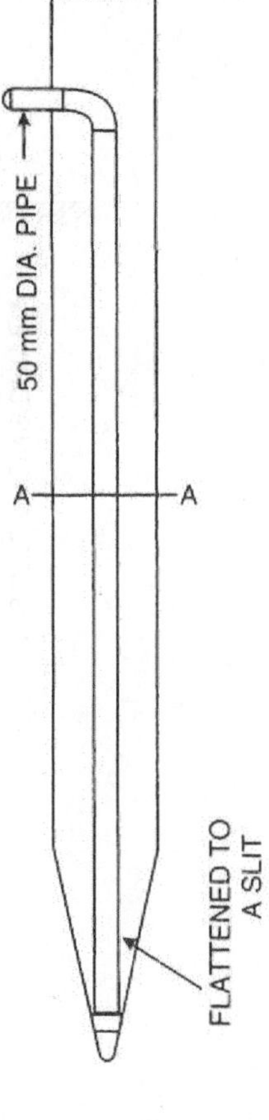

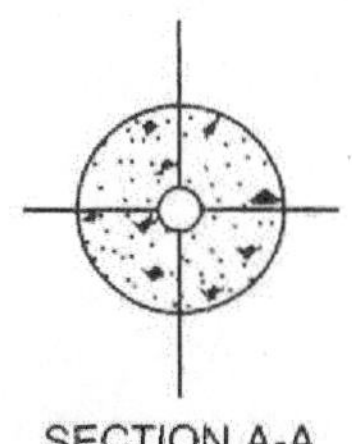

Figs. 4.52 and 4.53: Plan and section of a pile with water jet hole.

Cast-in-situ Concrete Piles. Cast-in-situ pile is a concrete pile built in its permanent location within a hole made for this purpose. This hole may be made be excavating with auger or by driving a suitable casing. The various types of cast-in-situ piles are described below:

(*i*) ***Simplex Pile.*** A hollow cylindrical steel pipe is driven into the ground to the required depth. A cast iron or steel base is placed under the pipe to displace the soil. The reinforcement is placed into the pipe, if needed. Concrete is poured to a depth of about one metre into this pipe. After pouring concrete, the enclosing pipe is withdrawn to some extent. Concrete is again poured into the pipe and the pipe is withdrawn. As the casing is pulled out, the hole gets filled with concrete which acts as a pile.(Figs 4.54 to 4.56)

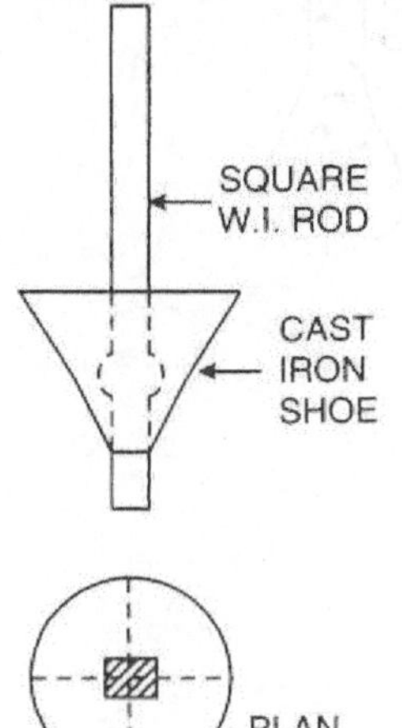

Fig. 4.50 and 4.51: Plan and Elevation of C.I. Shoe.

(*ii*) ***Pedestal Pile.*** The first step in the construction of this type of pile is to drive a casing and a core into the ground. The core is removed and concrete is placed to some depth inside the casing. The core is again placed and casing is pulled up by about one metre while the pressure is being exerted on the concrete with the core. The concrete is rammed so as to form a pedestal. The core is removed and concrete poured into the casing. Finally, as the concrete gets filled, the casing is withdrawn. Hence a concrete pile with a pedestal at the bottom is formed. If reinforcement is necessary, the same is placed after taking out the core and suitable arrangements for ramming concrete are made. (Figs 4.57 to 4.62)

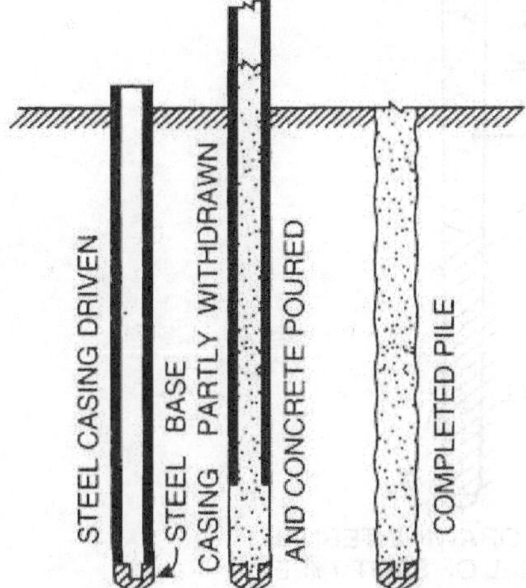

Figs. 4.54 to 4.56: Casting of simplex pile.

(*iii*) ***Raymond Pile.*** A thin sheet steel tapered shell is driven into the soil with a steel mandril inside. The mandril is removed and suitable reinforcement placed, if necessary. The shell is filled with concrete (Fig. 4.63 to 4.65).

(*iv*) ***Mac Arthur Case Pile.*** This is formed by driving into the ground a heavy steel casing in which a core is inserted. The core is removed and a corrugated steel sheet is introduced. The shell is filled with concrete and the casing is withdrawn.

(*v*) ***Vibro Piles.*** Vibro piles are formed by driving a steel tube fitted with a C.I. shoe into the ground. The tube

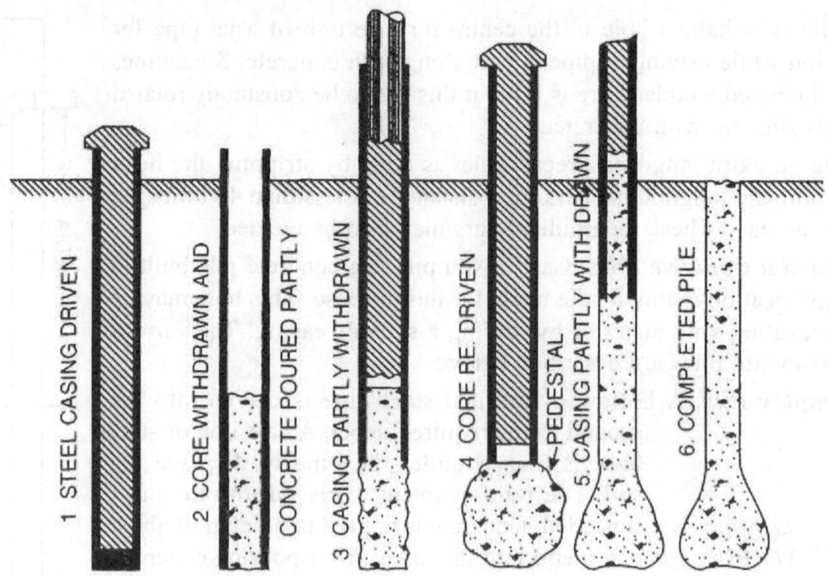

Figs. 4.57 to 4.62: Pedestal piles.

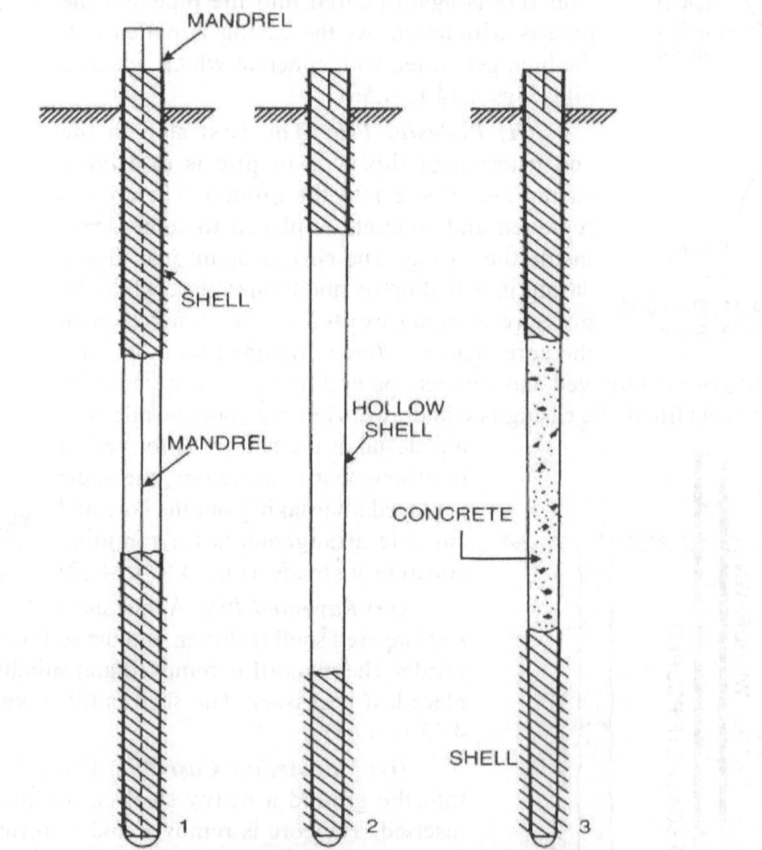

Figs. 4.63 to 4.65: Raymond pile.

is filled with concrete and is extracted by a succession of upward pulling and downward tamping blows. The pile thus gets enlarged every time and fits in the surrounding soil securely. The upward and downward movement is repeated 80 times per minute and a pile of concrete is formed at the rate of about one metre per minute. These piles are cast rapidly and have a uniform dense concrete. (Figs 4.66 to 4.69)

4.5.3.1 *Advantages of Precast Concrete Piles*

(*i*) Best concrete can be prepared by proper workmanship. Any defect can be immediately repaired.
(*ii*) The reinforcement remains in proper position and does not get displaced.
(*iii*) The concrete has only to withstand loads after complete curing has taken place.
(*iv*) They can be cast before hand and a quick driving progress can be ensured.
(*v*) They are more convenient when driven through wet conditions.
(*vi*) They are more suitable when a part of their length is to remain exposed.
(*vii*) They are not affected by any other additional forces which act on them while adjacent piles are driven.

4.5.3.2 *Disadvantages*

(*i*) They are heavy and difficult to transport.
(*ii*) Lapping of additional length means extra cost, labour and energy.
(*iii*) They have to be heavier in section to withstand the handling stresses.
(*iv*) The shocks of driving make them weaker.

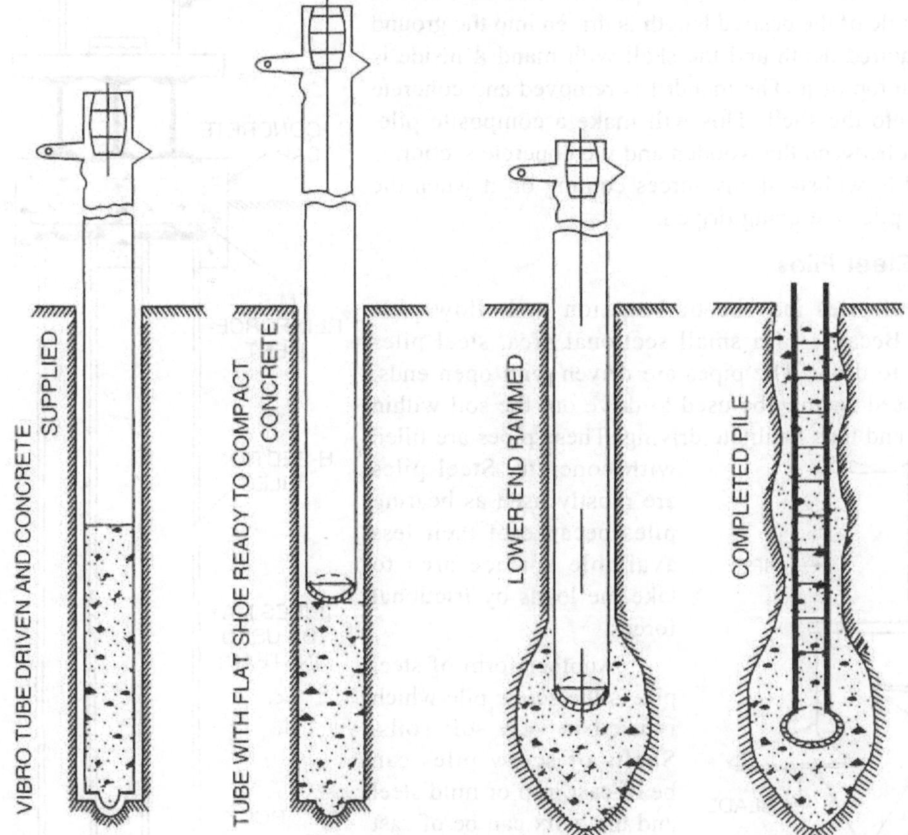

Figs. 4.66 to 4.69: Vibro pile.

4.5.3.3 *Advantages of Cast-in-Situ Piles*

(*i*) There is less wastage of material as exact length of pile is cast.
(*ii*) The time spent on curing etc. is saved.
(*iii*) They can bear heavier loads by improving upon their cross sectional profile, e.g., Pedestal piles.

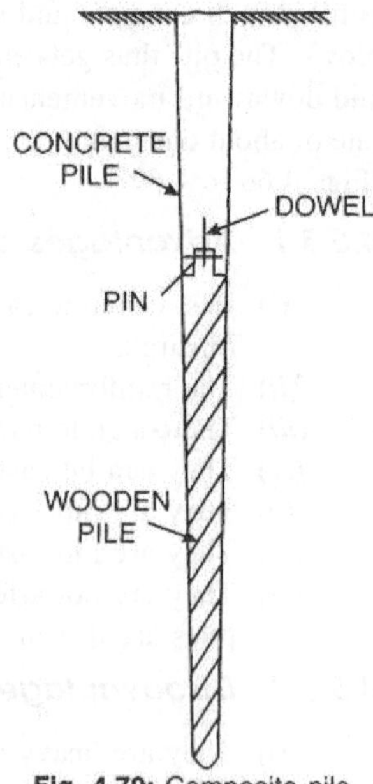

Fig. 4.70: Composite pile.

4.5.3.4 *Disadvantages*

(*i*) Good quality concrete cannot be easily obtained due to unusual height of dumping.
(*ii*) The reinforcements are liable to get displaced.
(*iii*) They cannot be used under water.
(*iv*) The green concrete loses strength after coming in contact with the soil.
(*v*) The shells are affected by casting additional piles adjacent to them.

4.5.4 Composite Piles

Wooden piles can advantageously be used below the ground water level as they are more durable under water. Hence a pile consisting of wood for its lower portion and a concrete section for its upper portion can be used. A wooden pile of the desired length is driven into the ground to its required depth and the shell with mandril inside is placed on top of it. The mandril is removed and concrete poured into the shell. This will make a composite pile. The joint between the wooden and the concrete section is designed to withstand any forces coming on it when the adjacent piles are being driven.

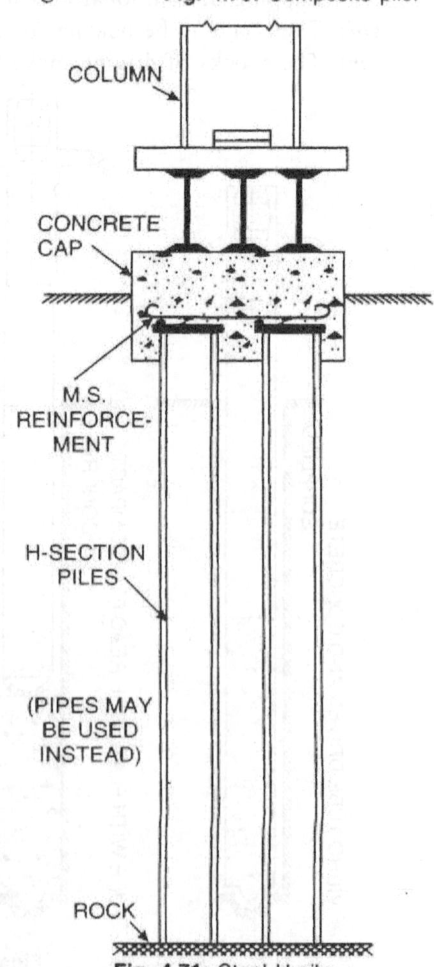

Fig. 4.71: Steel H-pile.

4.5.5 Steel Piles

Steel piles may be of I-section or hollow pipe section. Because of a small sectional area, steel piles are easy to drive. The pipes are driven with open ends. Compressed air may be used to drive out the soil within the pipe and thus facilitate driving. These pipes are filled with concrete. Steel piles are mostly used as bearing piles because of their less available surface area to take the loads by frictional forces.

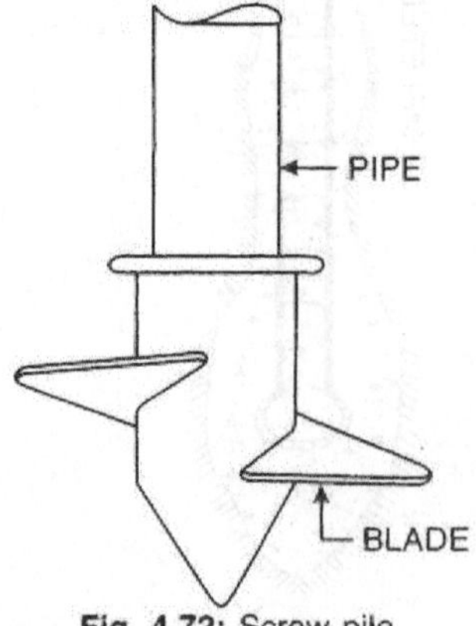

Fig. 4.72: Screw pile.

Another form of steel pile is the screw pile which is used in very soft soils. Shafts of screw piles can be of cast iron or mild steel and the helix can be of cast iron, cast steel or mild steel. The helix should consist of one complete or two half turns and should have sufficient pitch so that stones do not get jammed in between the blades. The diameter of the helix can be as much as 2 metre. They are driven by fixing a capstan with electrical motor to the head of the pile. A framework is built while the pile is being screwed. The spacing of screw piles is about twice the diameter of the helix. Water jets may be used to aid in the driving of screw piles.

A disc pile has a metallic disc attached to its foot so as to enlarge the bearing area. They are not commonly used for building foundation.

4.6 PILE DRIVING

Pile driving is an operation of forcing a pile into the ground without previous excavation. Various methods are used for driving piles. The simplest is to drive with a heavy weight called a drop hammer. This is dropped on the head of the pile and is guided during its fall by suitable staging. The hammer is raised by pulling a rope manually or by a steam engine or electric motor. The weight of this hammer for driving concrete or wooden piles is about 1/2 to 1 tonne and for driving heavier piles it may be as much as 2 to 3 tonnes. The height of drop varies from $1\frac{1}{2}$ to 3 metre. These hammers are made of hard wood or cast iron or cast steel. The pile driving frame is kept in vertical position by suitable guy ropes. Arrangemets for automatic falling of the hammer from desired height are made.

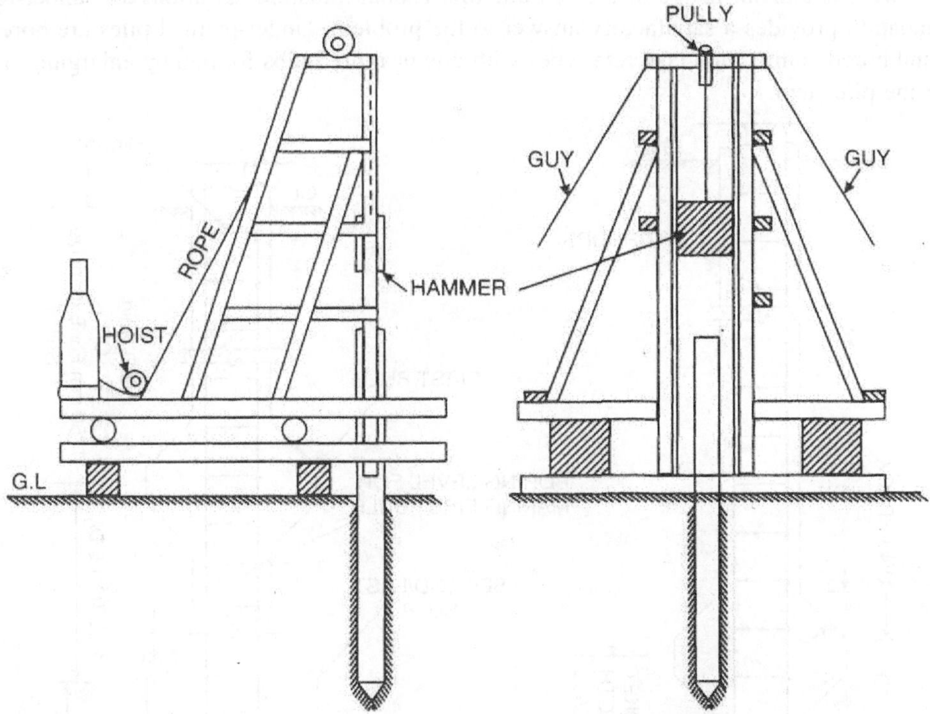

Figs. 4.73 to 4.74: Pile driving with a hammer.

Steam hammer. A heavy hammer is dropped on the pile through a small height but in quick succession. Single acting or double acting steam hammers are available. These hammers are raised with the aid of high pressure steam and are dropped by gravity (single acting) or by steam (double acting). Special devices are used to protect the head of the piles from damage due to excessive blows which they receive. The advantages of steam hammers over the ordinary drop hammer are:

(*i*) The damage to the pile head is less.

(*ii*) The number of piles driven in a given time is more as the pile is practically kept in motion while driving is in progress.

(*iii*) Less vibrations occur in the soil adjacent to the piles and hence the structures in the surrounding area are affected to a lesser extent.

Water jets. These can be used to drive piles easily in hard soils. However they are not much effective in silty or clayey soils. Water jet is applied to the foot of the pile generally by pushing one or two jetting pipes along with the pile as it is being driven. The diameter of the pipe is usually 50 to 80 mm and the nozzle may be about 25 mm in diameter. Whenever it is apparent that difficulty will arise in driving the piles before the process is started it is desirable to have a central hole with a proper nozzle cast along the pile (*see* Figs. 4.73 and 4.74). The jet pressure should be sufficient to overcome the soil pressure easily and hence a minimum of 200 N/mm^2 is desirable. As much as 6,000 litres of water per minute may be needed for harder soils. It is necessary to stop jetting before the final depth is reached. A minimum distance of one metre should be kept as a margin to allow a proper bearing for the pile.

Special Methods. Screw piles and disc piles are driven into the ground with the aid of driving drums and winch crabs. The drum is fixed to the top of the pile which is further rotated gradually by applying a torque from the crab with the help of wire ropes.

4.7 UNDER-REAMED PILES

Structures built on expansive soils often crack due to the differential movement caused by the alternate swelling and shrinkage of the soil due to seasonal moisture variations and under-reamed pile foundation provides a satisfactory answer to the problem. Under-reamed piles are bored cast in-situ and bored compaction concrete types with one or more bulbs formed by enlarging the bore hole for the pile stem.

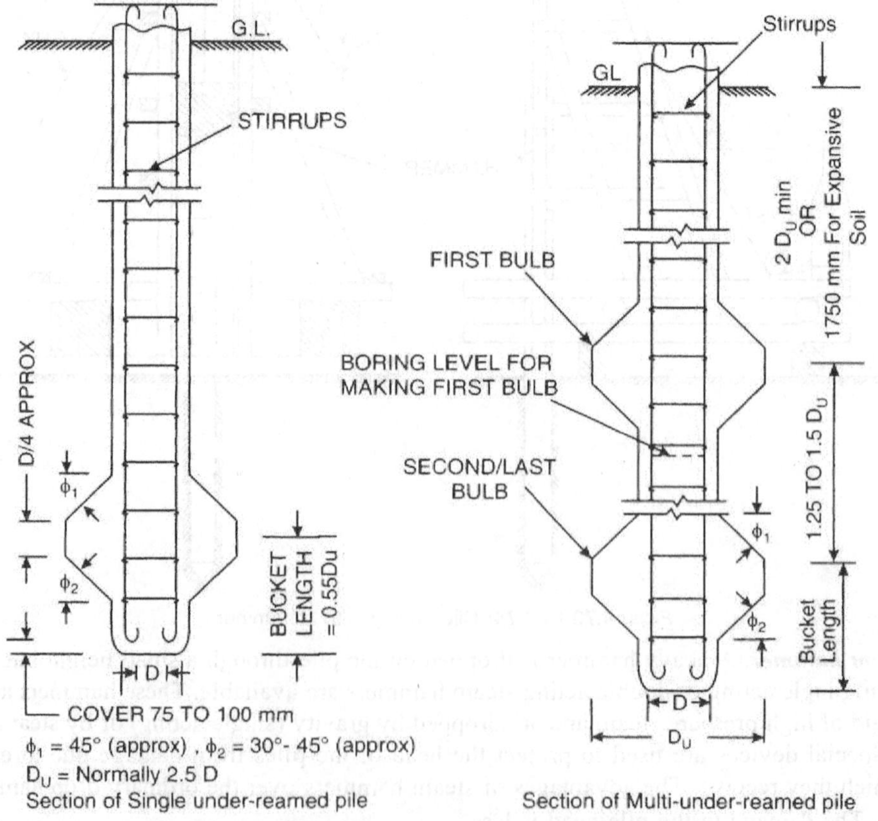

Figs. 4.75 and 4.76: Bored cast-in-situ under-reamed piles.

The principle of this type of foundation is to anchor the structure at a depth where ground movements due to changes in moisture content or consolidation of the poor strata are negligible.

The vertical movement of ground due to seasonal changes in moisture content in deep layers of expansive soils is limited to a depth of 3.5 m. In shallow depths of such soils, the length of piles can be reduced.

Single under-reamed piles may be provided for foundation of lighter structures (up to two storeys) and double under-reamed piles for heavier structures in shallow as well as in deep layers of expansive soils.

Single and double under-reamed piles may also be provided for foundations of structures in poor soils overlying firm soil strata. In such soils for double under-reamed piles, however, both the under-reams rest within the firm soil strata.

Piles of uniform diameter are also provided for foundations of lighter structures in shallow depths (not exceeding 2 m) of expansive soils and in poor soils overlying firm soil strata.

The length of piles will depend on the depth of stable zone and the piles are taken at least 50 cm into it. Uniform diameter piles in expansive soils are however taken down sufficiently deep into the firm soil strata so as to develop the required anchorage against uplift.

Diameter of piles is usually not less than 200 mm.

Diameter of the under-reamed portions are normally two to two and half times the diameter of the shaft and do not exceed three times.

Piles are provided at all the junctions of walls and at intermediate points, where necessary. The spacing of under-reamed piles is normally kept a minimum of twice the under-reamed diameter for normal loading. In double under-reamed piles, the vertical spacing between the two under reamings are also kept equal to one and a half times the under-reamed diameter.

4.8 PILE CAPS

A group of piles is covered and connected together by a cap. Reinforced concrete piles have got R.C.C. piles, the reinforcement of the piles being continued into the caps. The piles are continued into the cap to a distance of at least 8 cm and further the cap projects around the piles by at least 10 cm. The number of piles per cap should not be less than three so as to ensure stability of the group. Whenever less than three piles are used, a general practice is to connect the caps with each other.

4.9 BEARING POWER OF A PILE

The number of piles required to support a given structure was previously found out by dividing the total load coming on the group of piles by the load taken by one pile. This caused excessive settlements in some cases and hence it is desirable to know the settlement produced by the group of piles as a whole. Settlement may be caused by the pile groups yielding as a unit or the soil below may get compacted. The pressure transmitted by the group to the various layers of the soil below can be computed by the aid of formulae given in standard textbooks on **"Foundation Engineering"**.

The bearing power of an individual pile is also computed with the help of empirical formulae or by carrying out load tests. Dynamic formulae are used to compute the bearing power of a pile by noting its behaviour during driving. The energy used in driving and the corresponding penetration produced in the last few blows enables us to compute the bearing power of the pile. Factors like the rate at which pile is driven, its sectional area, etc., are also accounted for. The following formulae are commonly used.

For pile driven with drop hammer

$$P = \frac{160WH}{S + 25}$$

For pile driven with Single acting steam hammer

$$P = \frac{160WH}{S + 2.5}$$

where P = Safe load on the pile, in kN.
W = Weight of the hammer, in kN.
H = Distance through which the hammer falls, in metre.
S = Average penetration for the last three blows, in mm.

Static formulae are derived from the type of soil and the size of pile. These measure the frictional resistance along the surface of the pile and the bearing resistance of the base. Empirical values based on theoretical aspects of '*Soil mechanics*' are used. However, the extent to which pile driving affects the skin friction is unpredictable in deep layers of clayey soils or in soils of varying nature.

Test pile is the most reliable method for calculating the bearing power of a pile. The load on pile is applied with the aid of jacks placed against suitable supports. Whenever uniform soil is available in the vicinity, three test piles are sufficient. The test pile is allowed to recover for some time before additional load is applied. The load is applied in 5 to 10 tonne increments and settlement readings are taken regularly. The design load is kept for at least 24 hours. Maximum load equal to twice the design load is applied. Generally the maximum allowable load is taken as half the load which causes a permanent settlement of 0.25 mm per tonne or 25 mm whichever is less. In the case of piles resting in silty and clayey soils, the maximum load is taken as the load when the settlement increases considerably with very slight increments of load. A suitable factor of safety is used to get the design load.

4.10 LENGTH OF A PILE

The length of a pile can be determined by applying Rankine's theory, according to which

Safe allowable load on pile = Bearing capacity of pile × area of cross section of the pile × coefficient of pressure + frictional resistance of the pile

or

$$P_1 = b \times a \times C_p + \frac{wh^2}{2} \mu \times p \qquad \text{... (1)}$$

where P_1 = Allowable load on the pile, which is obtained by dividing the load on the pile by the factor of safety, varying from 4 to 6.
b = Bearing capacity of the soil, at the foot of the pile
C_p = Coefficient of pressure = $\frac{1 + \sin\phi}{1 - \sin\phi}$, where ϕ is the angle of repose of the soil.
a = Area of cross-section of the pile, depending upon its shape
w = unit weight of soil
h = Length of the pile
μ = Coefficient of friction between the material of the pile and the surrounding soil. Its value varies from 0.3 to 0.35
p = perimeter of the pile.

The values of h can be computed from equation (1) for the known values of p, b, a, c_p, w, μ and p.

Problem 4.6. *Determine the allowable load on a pre-cast R.C.C. pile driven by steam hammer weighting 24 kN. The stroke length of the hammer is 800 mm resulting in average penetration of 3.5 mm in the last three blows. Hence determine the length of the pile, if it is square in cross-section and each side is 200 mm. The angle of internal friction for the soil may be taken as 20° and the unit weight of concrete as 24 kN/m³. Bearing capacity of the soil is 1 kN/m².*

Solution.

Allowable load on the pile is given by

$$P = \frac{160\,WH}{S + 2.5} \qquad \text{(For Steam hammer)}$$

In this problem,

$$W = 24 \text{ kN}$$

$$H = 800 \text{ mm} = 0.8 \text{ m}$$

$$S = 3.5 \text{ mm}$$

$$P = \frac{160 \times 24 \times 0.8}{3.5 + 2.5} = 512 \text{ kN}$$

Assuming a factor of safety of 4

Safe allowable load $\quad P_1 = \frac{P}{4} = \frac{512}{4} = 128 \text{ kN.}$

$$b = 1 \text{ kN/m}^2$$

$$a = 0.20 \times 0.20 = 0.04 \text{ m}^2$$

$$c_p = \frac{1 + \sin 20°}{1 - \sin 20°} = \frac{1 + 0.342}{1 - 0.342} = \frac{1.342}{0.658} = 2.41$$

$$w = 24 \text{ kN/m}^3$$

$$\mu = \tan \phi = 0.364$$

$$p = 4 \times 0.2 = 0.8 \text{ m}$$

Substituting in equation (1)

$$128 = 1 \times 0.04 \times 2.41 + 12\,h^2 \times 0.364 \times 0.8$$

or $$3.494\,h^2 = 128 - 0.0964 = 127.904$$

or $$h^2 = \frac{127.904}{3.494} = 36.61$$

or $$h = \sqrt{36.61} = 6.05 \text{ m}$$

4.11 IMPORTANT POINTS ABOUT PILE FOUNDATIONS

The following points about pile foundations are worth noting:

(*i*) R.C.C. piles are not normally economical for depths less than 4 to 6 metres.

(*ii*) Driving of piles is generally commenced at the centre of the group of piles working outward.

(*iii*) When piles are driven in clay, displacement of the soil occurs which causes heaving in the soil surface. Some type of clays get considerably disturbed and bearing capacity is decreased to a large extent. "H" section piles cause least displacement.

(*iv*) In the case of sandy soils, care should be taken that excessive vibrations are not imparted to the soil otherwise considerable settlement may occur.

(*v*) The minimum spacing of friction pile is 1 to 1.5 metre; closer spacing leading to heaving of the soil.

(*vi*) Not more than 40 mm deviation from the designed alignment of the pile should be accepted, the tolerance on the verticality of the pile being not greater than 1 in 50.

4.12 PIER OR WELL FOUNDATION

Piers are generally cylindrical columns used to support the loads like bearing piles, i.e., the load is transferred through them to a firm strata below or by friction to the surrounding soil. The difference between piers and piles lies in the method of construction. Bearing piles are driven by excavation of the soil which they occupy. Driving of bearing piles had to be discontinued when the pile refuses to be driven in and thus may not reach the desired layer of the firm strata below. In the case of piers the excavation can be carried to the desired depth easily. Piers are preferable when the structure is to be laid on decomposed rock overlying a strata of sound rock as it becomes difficult to drive bearing piles through decomposed rock due to some rock fragments coming under the foot of the pile which prevent it from sinking further. Similarly whenever the top clayey layers offer larger resistance to the driving of a bearing pile, it is desirable to use piers. Piers can also be driven conveniently under water.

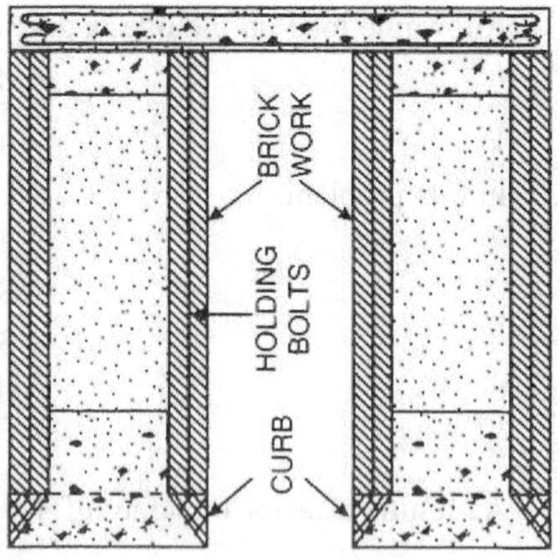

Fig. 4.77: Well foundation.

Piers may be hollow or solid in section. They can be made of brick, plain concrete, R.C.C., steel, cast iron, etc. The hollow space is plugged with concrete so as to form a solid section.

4.12.1 Design Considerations

(a) General

The design of concrete piers shall conform to the requirements for columns. If the bottom of the pier is to be belled so as to increase its load carrying capacity, such bell shall be at least 300 mm thick at its edge. The sides shall slope at an angle of not less than 60° with the horizontal. The least permissible dimensions shall be 600 mm, irrespective of the pier being circular, square or rectangular. Piers of smaller dimensions if permitted shall be designed as piles.

(b) Plain concrete piers

The height of the pier shall not exceed 6 times the least lateral dimension. When the height exceeds 6 times the least lateral dimension, buckling effect shall be taken into account, but in no case shall the height exceed 12 times the least lateral dimension.

When the height exceeds 6 times the least latered dimension, the deduction in allowable stress shall be given by the following formula:

$$f'_c = f_c\left(1.3 - \frac{H}{20\,D}\right)$$

where f'_c = Reduced allowable stress,

f_c = Allowable stress,

H = Height of pier, and

D = Least lateral dimension.

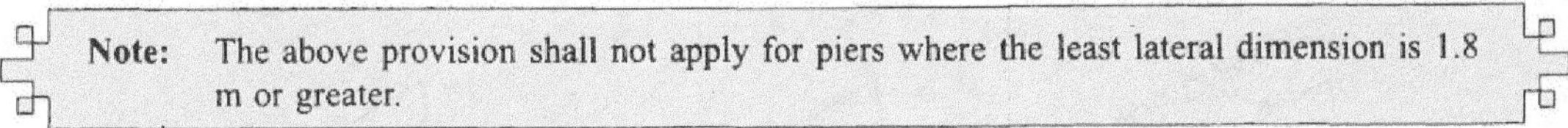
Note: The above provision shall not apply for piers where the least lateral dimension is 1.8 m or greater.

(c) Reinforced concrete piers

When the height of the pier exceeds 18 times its least dimension, the maximum load shall not exceed:

$$P' = P\left(1.5 - \frac{H}{36D}\right)$$

where P' = Permissible load;

P = Permissible load when calculated as axially loaded short column.

H = Height of the pier measured from top of bell, if any, to the level of cut-off of pier; and

D = Least lateral dimension.

There are various methods of driving the piers. These are briefly explained below:

4.12.2 Open Caisson Method

A cassion is an enclosure which is driven into the ground and later on filled up. The sides of this enclosure may be made of masonry or concrete. Typical example of this case is the use of well foundation in India. (See Fig. 4.78). In this type, a special cutting edge and a curb is laid on the ground and masonry built over it. The soil within the hollow inside and underneath the edges is excavated. This well sinks under its own weight or for stiffer soils, it may be loaded. While working under water, special grab dredgers are used for excavating the soil. When the well rests on a sufficiently hard strata (or has enough load carrying capacity on account of side friction) it is plugged by pouring concrete at the top and bottom. Two or three wells may be connected together by covering them with a continuous cap.

4.12.3 Compressed Air Caisson Method

This type is commonly used when the soil is exceedingly soft and rushes in as it is excavated, thereby preventing proper working within the well. The well is sealed at the top with the aid of special airlocks and compressed air pushed into the chamber. This prevents the inflow of soft soil. Men work within this chamber and excavate the soil as described above.

4.12.4 Ordinary Excavation Method

For stiffer soils a sort of well is excavated and the pier is built within this well. This method is used only for smaller depths.

4.12.5 Use of Sheeting and Sheet Piles

Whenever the sides of the excavated pit are liable to fall when the soil within is excavated, it is necessary to support it with the aid of vertical or horizontal wooden planks or steel sheets suitably braced. For still softer soils, thin piles known as sheet piles are driven adjacent to each other to form an enclosure. The soil is excavated within this area and the pier built as usual.

4.13 SETTING OUT FOUNDATION TRENCHES

The first step in excavating the necessary volume of soil to be replaced by a suitable type of footings is to clear the site of all plants, debris, etc. The site is then made somewhat level so as to remove large inequalities.

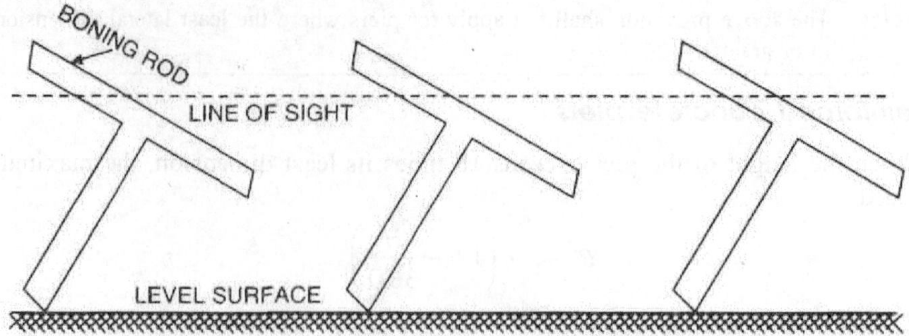

Fig. 4.78: Use of boning rods for levelling

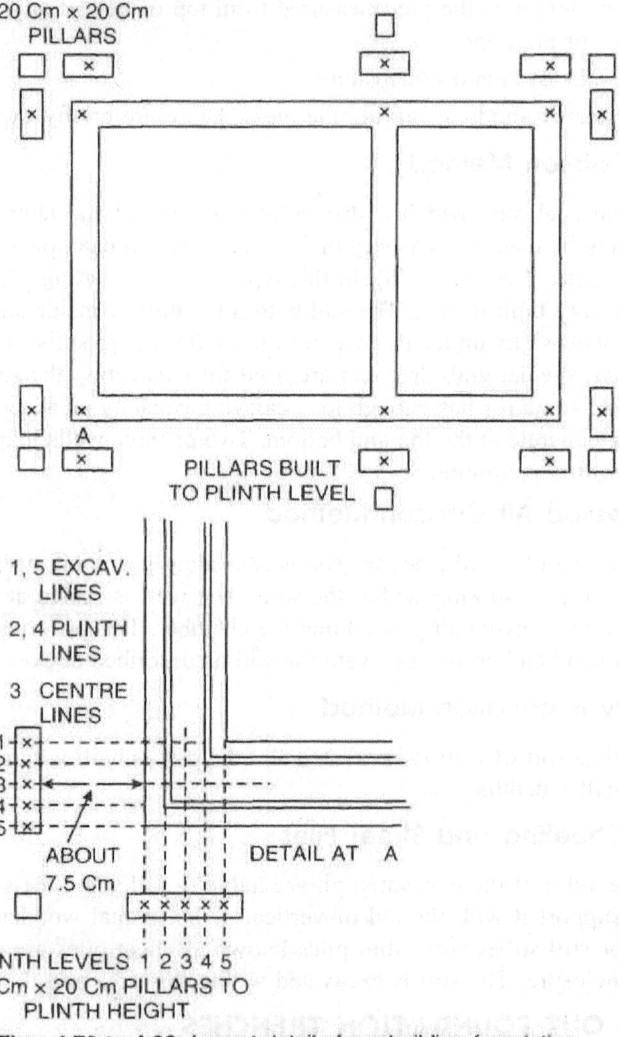

Figs. 4.79 to 4.80: Layout details for a building foundation.

After the foundation design has been finalised, a foundation plan denoting the width of excavation along with other complete dimensions is prepared. The next operation is to mark this plan on the ground surface. This is done with the aid of measuring tapes, wooden squares, pegs, strings, etc. The corners of the building are laid and the sides checked by measuring the diagonals. The entire width of foundation to be excavated is marked on the grourd with the help of lime or by daugh-belling. It is essential to have permanent reference marks which should not get disturbed as the excavation proceeds. Marks denoting the width of excavation, the width of masonry at the plinth, offsets, etc., are made on small brick-pillars built in brick masonry. These pillars are erected at a distance of about one-metre from the edge of the excavation so that they may not get disturbed later on. In larger buildings, help of a theodolite is taken to mark the central lines of the walls and to check the corner angles. The plinth levels are to be accurately set and for this purpose, special brick pillars at the corners are erected and their tops finished at the desired level (*See* Figs. 4.79 and 4.80).

4.14 EXCAVATION OF FOUNDATION

Ordinary excavations are carried out with the aid of tools shown in Figs. 4.81 to 4.90. Most of the work is done with a spade, a kassi (Phaorah), pick-axe and baskets or pans. Whenever rock excavation is to be carried out, chisels, jumpers, wedges, hammers, etc., are used.

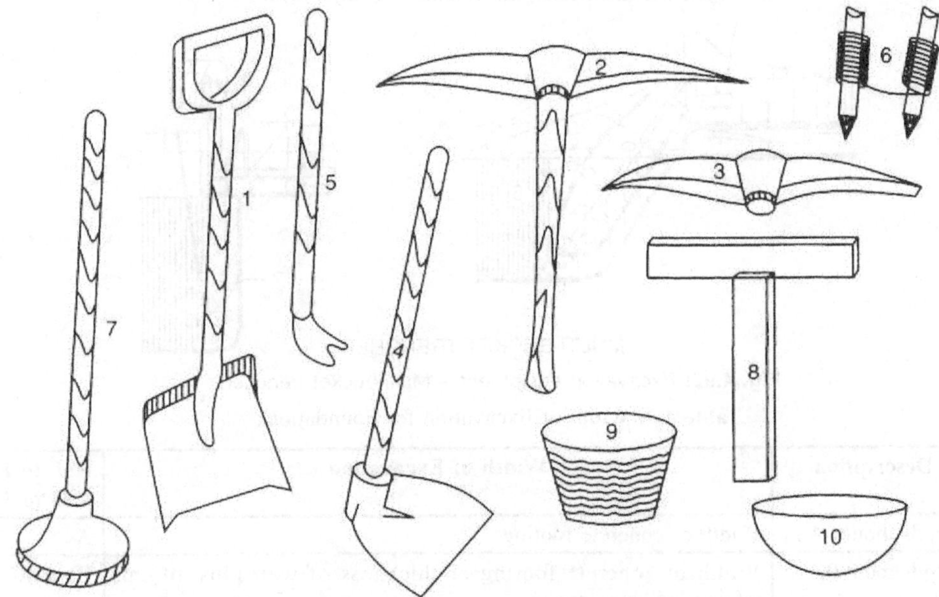

Figs. 4.81 to 4.90: Excavation tools.
(1) Spade, (2) Pickaxe, (3) Pickaxe with flat end, (4) Kassi, (5) Crow bar.
(6) Line and pins, (7) Rammer, (8) Boning rod, (9) Basket, (10) Iron pan

The excavation should not be too wide. At the same time if should be wide enough to ensure safe and efficient working, Table 4.3 gives a guideline of the width of excavation for foundations. The details of width of excavation for different depths of foundations are shown in Figs. 4.93 to 4.99.

Now-a-days mechanical equipment is used on large projects. Drag shovel or multi-bucket trencher is used to excavate the soil. The excavated soil from the trench is carried by buckets and deposited on either side of the trench. The excavation equipment described above are shown in Figs. 4.91 and 4.92.

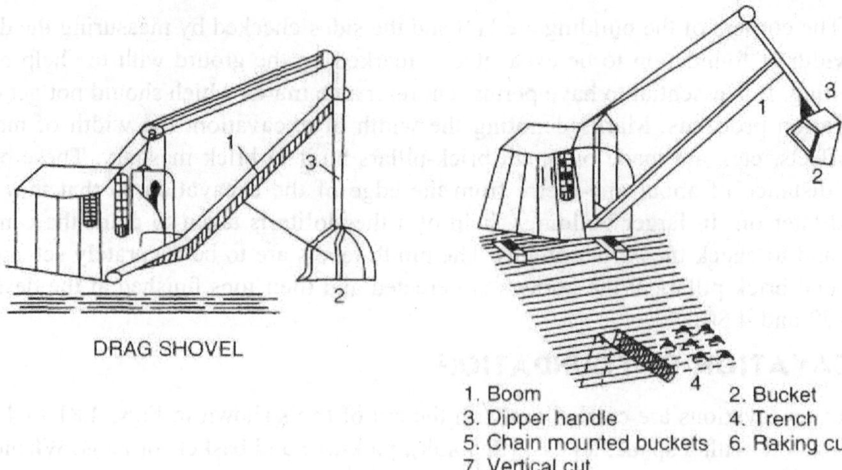

Fig. 4.91: Excavation equipment — Drag shovel.

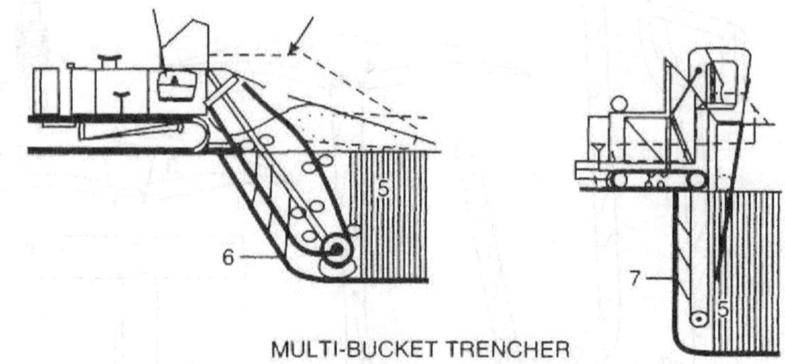

Fig. 4.92: Excavation equipment— Multi-bucket trencher.

Table 4.3: Width of Excavation for Foundations

Sl. No.	Description	Width of Excavation	Ref. to Fig. 4.93 to 4.99
(*i*)	Depth about 0.5 m	Width of concrete footing	A
(*ii*)	Depth about 0.6 m	Width of concrete footing or thickness of wall plus 40 cm, whichever is greater	IB & IC
(*iii*)	Shallow depth, from 0.6 to 0.9 m	10 cm more than that for case (ii) at top, the increased width being obtaind by giving batter on both the sides	ID
(*iv*)	Medium depth, from 0.9 to 1.2 m	Width of concrete footing or thickness of wall plus 65 cm, whichever is greater at bottom and 15 cm more at top, sides being battered	IE
(*v*)	Medium depth, from 1.2 to 1.5 m	Width of concrete footing or thickness of wall plus 80 cm, whichever is greater at bottom and 15 cm more at top, sides being battered	IF
(*vi*)	Depth greater than 1.5 m	Clear working space of 40 cm on either side of wall at bottom, and space of 55 cm at top, sides being battered	IG

Notes 1. Batters are not necessary in trenches in hard soil such as moorum, gravel and in disintegrated rock, if the depth of excavation is not more than 1.2 to 1.5 m.

2. Where shoring of the sides of the trench is found necessary, extra width shall be appropriately provided.

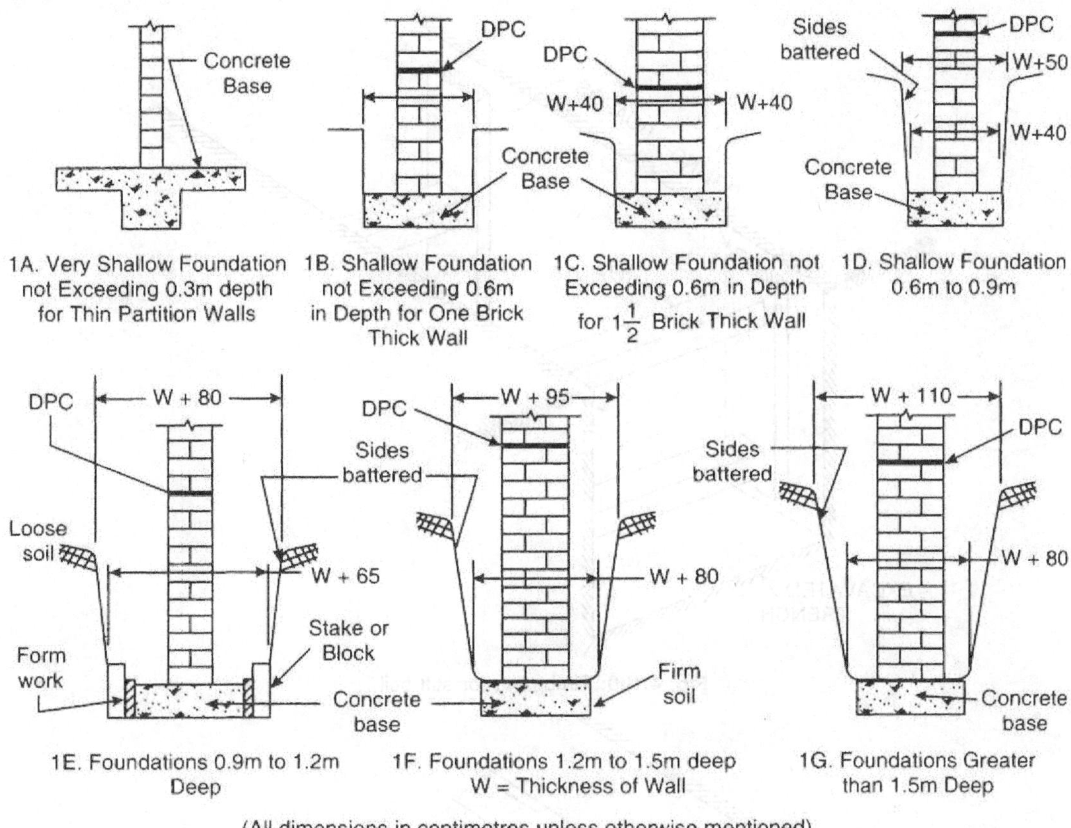

Figs. 4.93 to 4.99: Width of excavation for different depths of foundations.

4.14.1 Excavation with the Aid of Timbering

In the case of hard soils and for small depths' excavation is carried to the desired breadth and depth without the aid of any support for the sides.

For stiff soils and for somewhat deeper excavation, light supports in the form of wooden boards about 4 cm thick and extending to the depth of excavation are used. These supports are spaced at about 2 m centres and are fitted with the help of struts against each other. One strut in the middle of the board or one at top and one at bottom may be used depending upon the firmness of the soil. (*See* Fig. 4.100).

For soils which are moderately firm, two methods are generally used. The first method envisages the use of horizontal boarding called wailing pieces which are strutted as described above. For cheaper work, round wooden struts may be used. The second method is employed for soils which are softer. In this case poling boards, wailing members and struts are used. (*See* Figs. 4.101 and 4.102.)

For soft soils, the poling boards have to be kept side by side otherwise the soil will come out through the intervening space. In addition, wailing members are used to give additional support and these are adequately strutted. Wedges may be inserted at the place shown (Fig. 4.103) to facilitate the tightening and removing of the members.

For deep excavation and in very soft soils, runners, i.e., thick wooden planks pointed at the ends are driven deeper than the desired excavation depth, wailings and struts being used as usual.

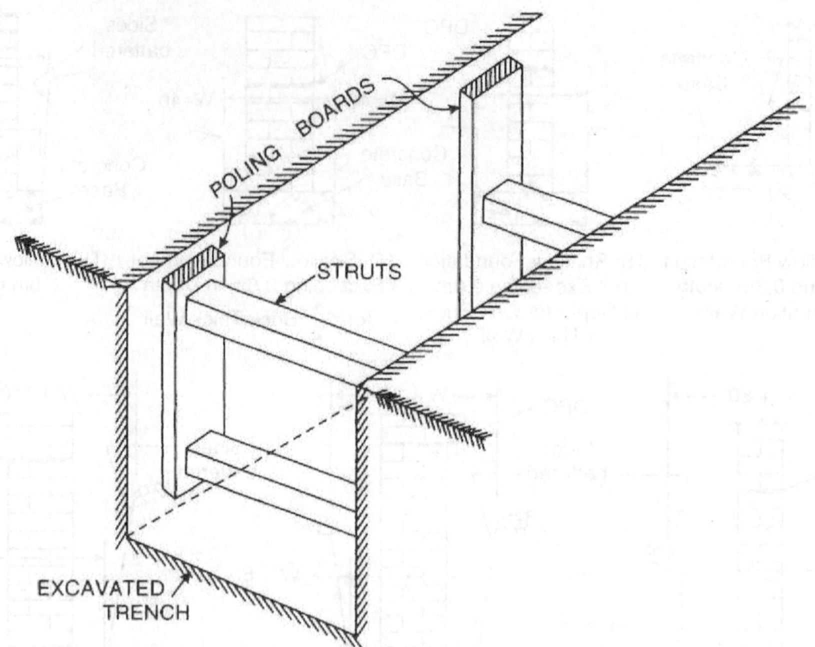

Fig. 4.100: Timbering for stiff soil.

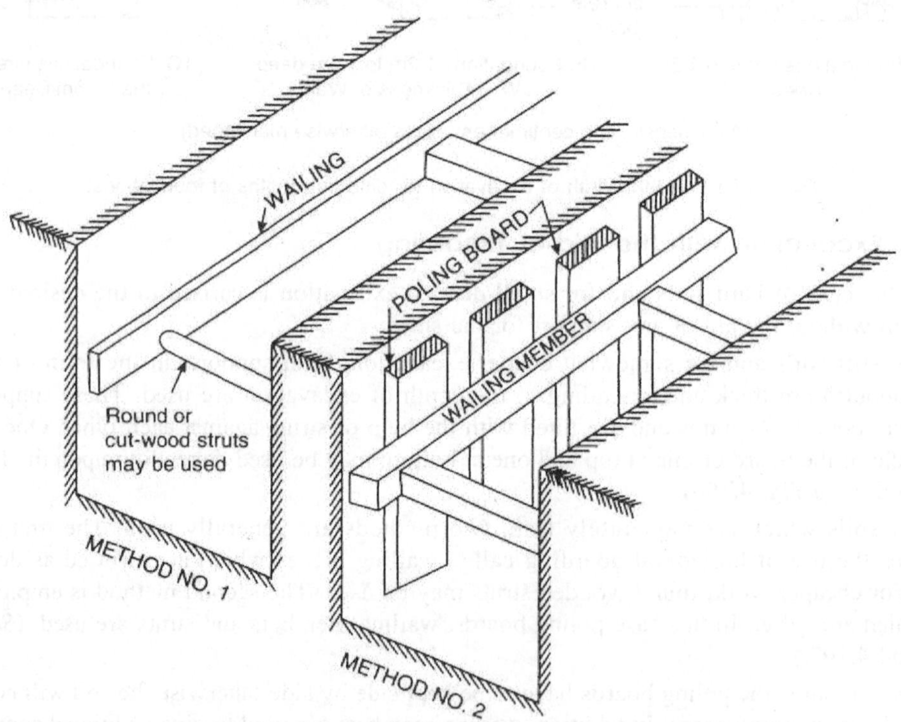

Figs. 4.101 and 4.102: Timbering for moderately firm soil.

For soils which immediately flow as they are excavated, arrangement as shown in Fig. 4.104 is used. The trench is excavated 1 to 2 m in depth and the soil allowed to fall. Sides are then supported by vertical sheets (poling boards), wailed and strutted. A secondary system of runners is used and these are wailed and strutted to support the excavation. Wedges are inserted between runners and wailing pieces. Wedges securing one runner are slightly loosened. The earth from its foot is removed and the runner thus dropped is rewedged. In this manner the whole system of excavation is completed.

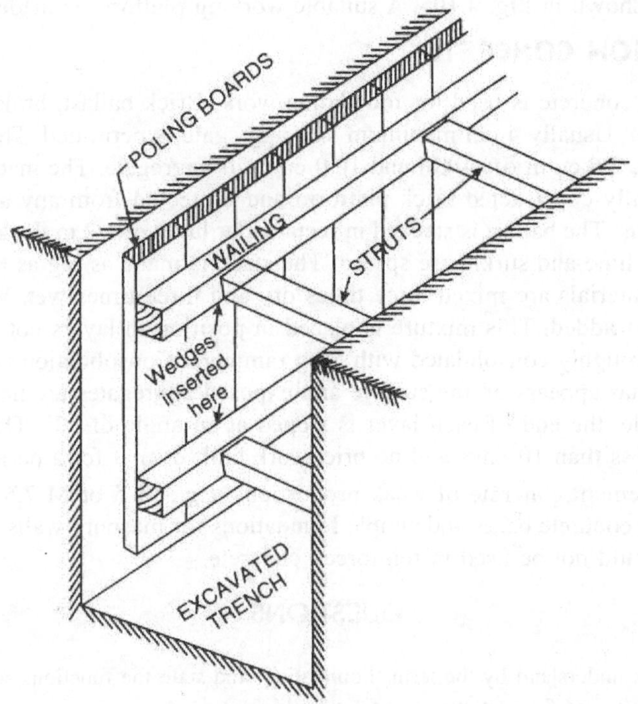

Fig. 4.103: Timbering for soft soil.

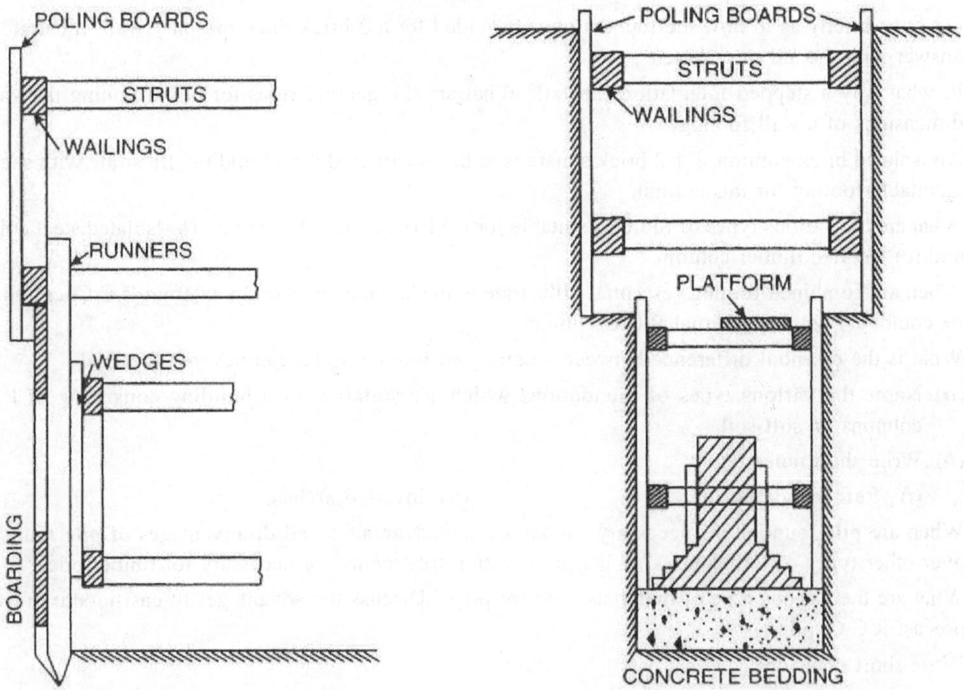

Fig. 4.104: Timbering for running soil.

Fig. 4.105: Diagram showing construction of a stepped footing in a timbered trench.

For waterlogged areas special methods like chemical treatment, freezing method, caisson method, etc. are used.

To build a stepped footing within a timbered trench, the struts are cut and refitted against the footing masonry as shown in Fig. 4.105. A suitable working platform is made.

4.15 FOUNDATION CONCRETE

Generally lime concrete is used for foundation work. Brick ballast, broken stone or shingle are used as aggregate. Usually 4 cm maximum size aggregate is permitted. The usual proportions are 1.3 cu m of lime, 2.6 cu m of surkhi and 10.0 cu m of aggregate. The materials are measured and mixed on specially constructed brick platform and protected from any admixture of dirt or other foreign materials. The ballast is stacked in rectangular layer of 1/2 m thickness and on the top of it properly mixed lime and surkhi are spread. The stack is made as big as to last for one day's concreting. These materials are mixed three times dry and three times wet. Water just sufficient to wet the concrete is added. This mixture is placed in position in layers not exceeding 150 mm in thickness and thoroughly consolidated with 5 kg rammers. Consolidation is taken as complete when a layer of mortar appears on the surface and exposed aggregates are not visible. Wherever joints are unavoidable, the end of each layer is sloped at an angle of 30°. The concrete is cured for a period of not less than 10 days and no brickwork built over it for a period of 7 days.

Alternatively cement concrete of weak proportions, *e.g.*, M 5 or M 7.5 grades of concrete may be used for lean concrete bases and simple foundations for masonry walls. Grades of concrete lower than M 15 should not be used in reinforced concrete.

QUESTIONS

1. (*a*) What do you understand by the term 'Foundation' and state the functions served by them. Name the various types of foundations used for a building.

   (*b*) List the causes of settlement and differential settlement.

2. Describe briefly as to how the foundation is provided for a 2 brick thick masonry wall. Illustrate your answer with the aid of a sketch.

   In what way a stepped foundation can fail? What are the general rules for proportioning the various dimensions of a wall footing?

3. An isolated brick column, 1 1/2 brick square is to be constructed for a building. Illustrate with sketches a suitable footing for this column.

4. What are the various types of footings suitable for (*a*) Isolated R.C.C column; (*b*) Isolated steel column; and (*c*) Isolated timber column.

5. When are combined footings essential? Illustrate with sketch as to how an external R.C.C. column can be combined with an internal R.C.C column.

   What is the essential difference between a combined footing and a cantilevered footing?

6. (*a*) Name the various types of foundations which are suitable for a building consisting of R.C.C. columns on soft soil.

   (*b*) Write short notes on:

   (*i*) Raft foundation. (*ii*) Inverted arches.

7. When are pile foundations necessary? What are the advantages and disadvantages of pile foundation over other types of foundations? What preservation treatments are necessary for timber piles?

8. What are the various type of materials used for piles? Discuss the advantages of cast-in-situ piles over precast R.C.C. piles.

9. Write short notes on:

   (*i*) Precast R.C.C. piles. (*ii*) Pedestal piles.

(*iii*) Composite piles. (*iv*) Screw piles.

(*v*) Pile driving.

10. What are the various methods of installing different types of piles within ground? Briefly describe the water jet method.

11. What are the precautions necessary while driving wooden and concrete piles respectively? What are the various methods of ascertaining the load bearing capacity of an individual pile?

12. What is meant by 'pier foundation'? Briefly describe various methods of constructing a pier foundation. Give the design considerations for plain concrete piles and reinforced concrete piles.

13. What precautions are necessary to be taken while excavating trenches for foundations in very soft soil? Describe the method of timbering trench in soft clayey soil.

14. Design the foundations of a wall of a 2 storeyed house resting on gravel and sand having a safe bearing capacity of 44 kN/m^2 and an angle of internal friction of 30°. The unit weight of the soil is 1.8 kN/m^3. The wall thickness is 40 cm and the total load coming on the wall per m length is 200 kN.

15. A column of width W requires a footing width B to transmit the load safely to the foundation. If the angle of spread of the load from the column base to the outer edge of the bearing is not to exceed *n* (Horiz.): 1 (Vert.), show that the minimum depth of foundation required is given by:

$$D_{min} = \frac{1}{2n}[B - W]$$

16. Determine the size of footing and depth of foundation for a brick pillar 30 cm square. The concentrated load carried by the column at the footing is 10 t. The bearing capacity of the soil at the depth of foundation is 15 t/m^2. The soil weights 1.8 t/m^3 and has an angle of repose of 28°. Use M 15 concrete.

17. Determine the dimensions of a combined footing for two columns A and B carrying loads of 200t and 100t respectively. Column A is 60 cm square and column B is 30 cm square. The centre to centre spacing between the columns is 4 m. The safe bearing capacity of the soil may be taken as 30 t/m^2.

18. Design a cantilever footing for two columns A and B spaced 4 m centre to centre. Column A is 500 mm square, carries an axial load of 800 kN and is on the property line. Column B is 750 mm square and carries a load of 1200 kN. The safe bearing capacity of the soil is 200 kN/m^2.

19. (*a*) State Rankines formula for detemining the length of a pile.

(*b*) Determine the allowable load on a RCC pile driven by steam hammer weighing 25 kN. The stroke length of the hammer is 850 mm, resulting in an average penetration of 4 mm in the last 3 blows. Hence determine the length of the pile having square cross-sectional area of each side 250 mm. The angle of internal friction for the soil may be taken as 25° and the unit weight of concrete 24 kN/m^3. Bearing capacity of the soil is 100 kN/m^2.

REFERENCES

1. **National Building Organisation, India:** *Footings for small buildings.*
2. **Mitchel, G.A.:** *Building construction. Vols. I and II.*
3. **Concrete Publication Ltd.:** *Sheet piling, Cofferdam and Caissons.*
4. **Athavale, J.N.:** *Foundation methods and practice.*
5. **Dunham, C. W.:** *Foundations of structures.*
6. **Salmon, E. H.:** *Materials and structures.*
7. **Henry, F.D.C.:** *Design and construction of Engineering foundations.*
8. **Taschebotariaoff, G.P.:** *Soil Mechanics, Foundations and Earth Structures.*
9. **Peck, Henson, Thournburn:** *Foundation Engineering.*
10. **Hnntington, W.C.:** *Building Construction.*

11. **Terzaghi, K.:** *Theoretical soil mechanics.*
12. **Mackey, W.:** *Building construction, Vol. 1.*
13. **Teng, W.C.:** *Foundation Design, Prentice Hall, 1965.*
14. **IS: 1080—1980:** *Code of practice for design and construction of simple spread foundations.*
15. **Bureau of Indian Standards, New Delhi:** *National Building Code of India 2005- Part-6 Section 2.*
16. **IS: 1892-1979:** *Code of practice for sub-surface investigation for foundations. (first revision)*
17. **IS: 2131-1981:** *Method for standard penetration test for soils (first revision).*
18. **IS: 2911:** *Code of practice for design and construction of pile foundation.*
19. **IS: 2911 (Part 1/Sec. 1) 1979:** *Concrete piles, Section 1, Driven cost in situ concrete piles (first revision)*
20. **IS: 2911 (Part 1, Sec. 2) 1979:** *Concrete piles, Section 2, Bored cast in situ piles (first revision)*
21. **IS: 2911 (Part 3) 1980:** *Under-reamed pile foundation (first revision)*
22. **IS: 2911 (Part 2) 1980:** *Timber piles (first revision)*

Brick Masonry

Bricks arranged systematically and bonded together in mortar to form a homogeneous mass capable of withstanding and transmitting forces, without failure, is called brick masonry.

Ordinary bricks are economical and when laid in good mortar form a durable construction. Bricks are comparatively small in size and, therefore, can be easily handled. Moreover, they can be suitably arranged in various shapes to form different components of a building, *e.g.*, walls, partitions, footings, columns, steps, floors, arches, lintels, etc.

5.1 MATERIALS USED IN BRICK MASONRY

The materials used in brick masonry construction are bricks and mortar.

(1) **Bricks**: The various kinds of bricks are:

***(a) Common bricks*:** This term is applied to bricks which are made of clay and burnt in the usual manner in kilns. Generally, these bricks are graded into three classes:

- *(i)* ***First class bricks*:** They are made from good earth, free from saline deposits and are sand-moulded. They should be thoroughly burnt, of good colour, of regular shape with square edges and parallel faces. These bricks are free from flaws, cracks, chips, stones, nodules of lime, etc. They do not absorb more than one-sixth of their weight of water after being soaked for one hour and should show no signs of afflorescence on drying. They should give ringing sound, when struck together.
- *(ii)* ***Second class*:** These are also fully burnt and give a clear ringing sound when struck. Slight irregularities in shape, size or colour are accepted. They should not absorb more than one-fourth of their dry weight of water after one hour's soaking.
- *(iii)* ***Third class*:** These are not burnt so fully as in the previous two cases but are generally of uniform reddish yellow colour. Defects in uniformity or shape are tolerated but they should not be excessive so as to interfere in the laying of bricks.

Generally first and second class bricks are used for brickwork. They should have a minimum compressive strength of 80 kg per sq cm.

***(b) Face bricks*:** Generally selected first class bricks are used for face work. However, it is essential that face bricks should be uniform in colour, texture, size, etc. Hence special bricks having these qualities are made under controlled supervision. Further, if any markings or surface finishes are needed, special bricks are moulded for face work. These bricks can be used for garden walls, side walks, steps and other exposed work where exceptionally good appearance is desired, regardless of coast.

(c) Pressed bricks: These bricks are moulded under pressure with hand for small jobs and with steam pressure for large works. These bricks have regular smooth faces, sharp edges, square corners and uniform colour. They are used for face work of buildings. Such bricks are specially useful when exact dimensions are required in walls or other structural members.

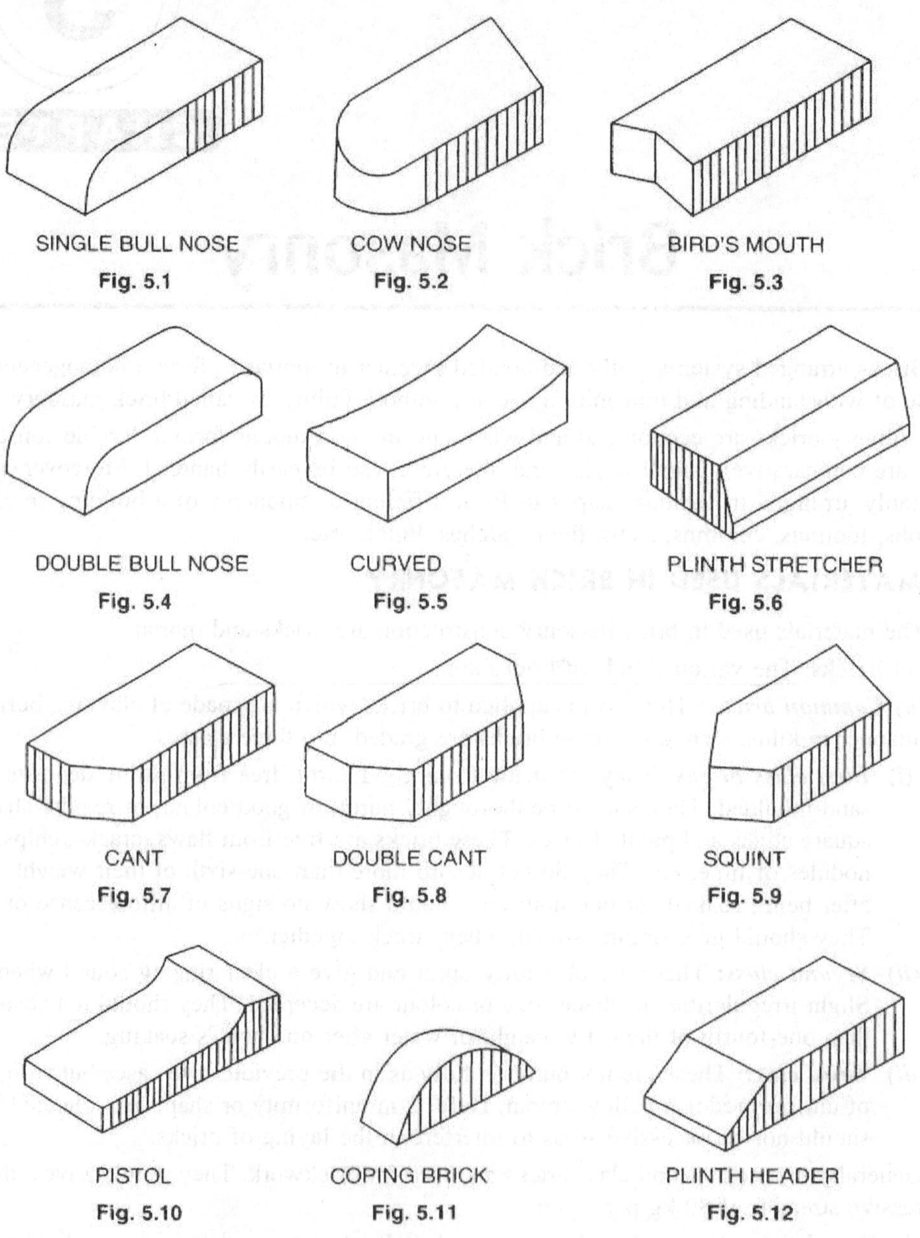

Figs. 5.1 to 5.12: Types of specially shaped bricks.

(d) Fire bricks: These are made of special fire clays. They are used for lining in fire places, furnaces, etc., where high temperatures are prevalent and ordinary bricks get decomposed. They are slightly larger than the ordinary bricks.

(*e*) ***Glazed and coloured bricks***: Coloured bricks are used where special colour treatment is needed or where ordinary bricks of good uniform colour are not available. Glazed bricks have one surface glazed in white or any other colour. These are used for exterior surfaces of walls or partitions, in diaries, hospitals, etc., where cleanliness is important.

(*f*) ***Imitation bricks***: These bricks are similar in size to common bricks but are made of Portland cement and sand. They are unburnt but have the same qualities as that of a good mortar.

(*g*) ***Moulded bricks***: These are special bricks made to suit different requirements, *e.g.*, giving architectural shapes to brickwork. Examples are copings, cornices, string courses, sloping walls, etc. The generally used types are shown in Figs. 5.1-5.12.

(2) **Mortar**: In order to bind the individual bricks to form the whole wall a compact mass, mortar is used. The mortar has also to form a cushion to take up the inequalities in the brick, to distribute the pressure evenly and to fill up the interstices in the bricks. The type of mortar used depends on the total load that the structure has to bear, the weathering agencies, type of finish desired etc. The commonly used mortars are:

(*a*) Lime mortar.

(*b*) Cement mortar.

(*c*) Lime-Cement mortar.

(*d*) Lime-Surkhi mortar.

(*e*) Mud mortar.

5.2 TERMS USED

(1) ***Header***: This denotes the end of a brick as seen in the wall face. (115 mm × 75 mm side).

(2) ***Stretcher***: The side of a brick as seen in elevation in a wall where the brick is laid flat. (230 mm × 75 mm).

(3) ***Bed***: The lower 230 mm × 115 mm surface of a brick when laid flat.

(4) ***Bat***: A portion of a brick cut across the width. These are of different sizes and shapes, *e.g.*, $\frac{1}{2}$ bat, $\frac{3}{4}$ bat and bevelled bat.

(5) ***Closer***: A portion of a brick cut longitudinally with one long face uncut. The types are:

(*a*) ***Queen closer***: This is placed next to the first brick in a header course. This is a half brick cut longitudinally.

(*b*) ***King closer***: This is used for splayed brickwork. It is a special sized brick with its one end having a width of half a brick.

(*c*) ***Bevelled closer***: This has one stretcher face bevelled.

(*d*) ***Mitred closer***: This has the header face splayed.

(6) ***Bull nose***: These are bricks with one edge rounded (single bull-nose) or with two edges rounded (double bull-nose). They are used where slightly curved corner bricks are needed, *e.g.*, copings.

(7) ***Frogs***: These are indentations in the bricks provided to form a key for holding the mortar. The bricks are laid with frogs on top. 75 mm high bricks should have a frog of 10 mm × 4 mm × 1 mm size on one of its flat sides.

(8) ***Course***: This is a complete layer of bricks laid on the same bed. A heading course consists of only headers whereas a stretcher course comprises of stretchers only as seen in elevation. These courses are generally of same thickness.

(9) ***Bond***: This is the method of arranging bricks in courses so that the individual units are tied together. Bonds are distinguished by their elevation or face appearance.

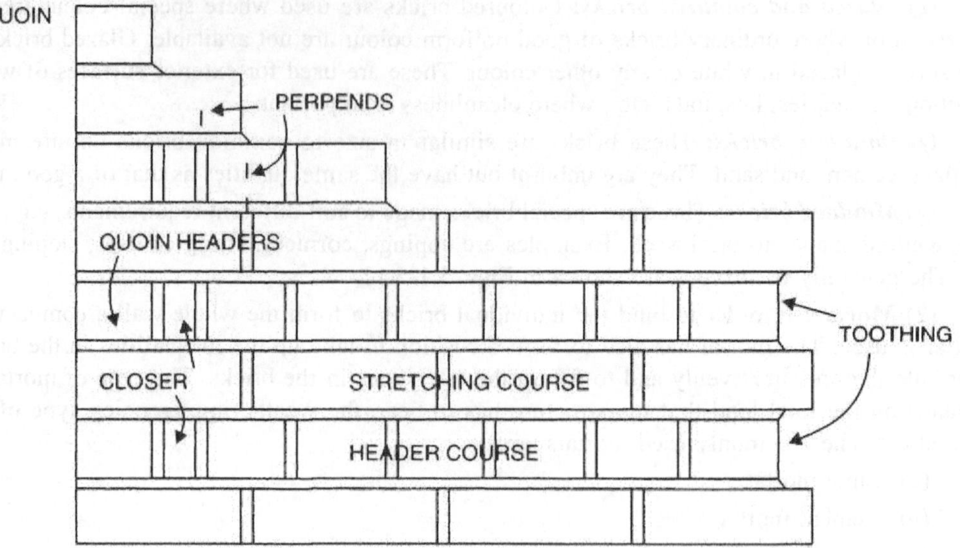

Fig. 5.13 Elevation of a brick-wall

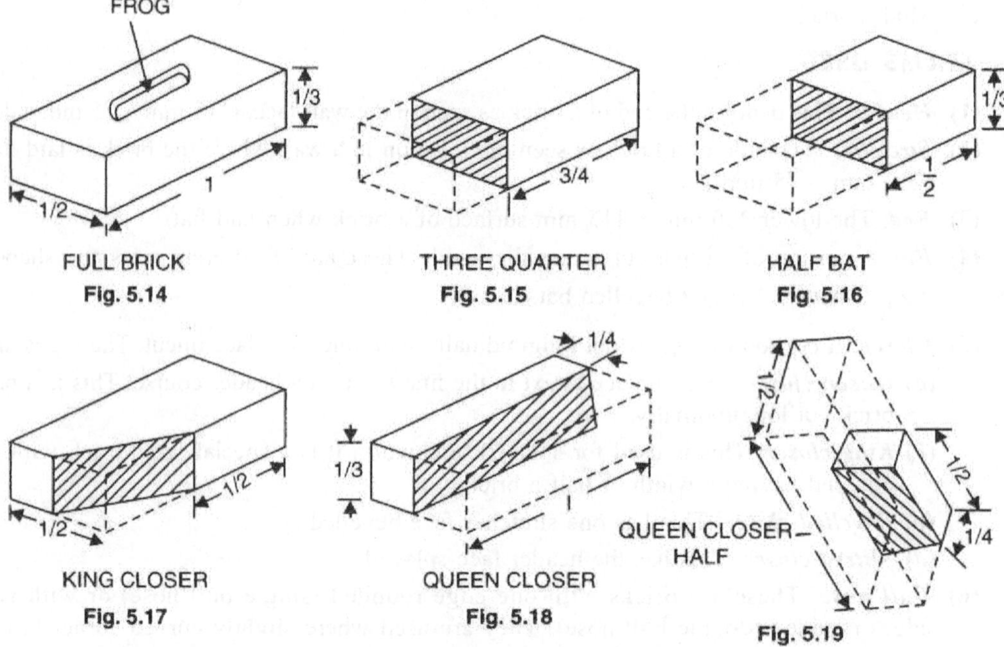

Fig. 5.14

Fig. 5.15

Fig. 5.16

Fig. 5.17

Fig. 5.18

Fig. 5.19

Figs. 5.13 to 5.19: Terms used in brickwork.

(10) ***Bed joints***: Horizontal mortar joints between the two courses.

(11) ***Cross joints***: The short vertical joints between two bricks, also referred to as perpends.

(12) ***Quoin***: A corner or the external angle on the face side of a wall.

(13) ***Facing***: The exposed area of a wall or other structure. The internal surface of the wall is called backing whereas the portion in between the backing and the facing is called hearting.

5.3 CAUSES OF FAILURE OF BRICK MASONRY

The correct approach to the understanding of the basic requirements in any construction is the criterion of failure. Brick masonry may fail due to the following causes:

(1) By crushing, if it is overloaded.

(2) By shearing along any horizontal plane.

(3) By rupture along a vertical joint under vertical loads.

Care can be taken against crushing by providing adequate dimensions. Shearing along any horizontal plane is prevented by providing a strong mortar. Rupture along a vertical plane is avoided by breaking vertical joints in brickwork.

BONDING IN BRICKWORK

It is the process of arranging bricks in courses in order to develop longitudinal and transverse interlocking for individual bricks. The purpose is to achieve a united mass, as strong as practicable, to suit the length, height and thickness of brickwork and the loads and stresses to which it is subjected. The method is to ensure that vertical joints do not come one over the other. A brick wall having a continuous vertical joint will not act as a complete unit but will consist of small portions which act as columns. Between these individual columns, there is no bonding effect except that the mortar causes adhesion between them. In such a case, if the wall is loaded, it is likely that these columns may separate out from each other by the slipping of the column faces from the mortar.

However, if a wall is built so that there are no continuous vertical joints the load will get distributed. This type of construction will be more durable and strong. As bricks are small units which have equal dimensions, the process of bonding is easily performed.

5.4 RULES FOR BONDING

To ensure good bonding, the following rules should be observed:

(*i*) The amount by which the bricks in one course overlap the bricks in the course below should be minimum $\frac{1}{4}$ brick along the length of the wall and $\frac{1}{2}$ brick across the thickness of the wall.

(*ii*) The vertical joints in the alternate courses should fall in plumb (vertical) line from the top of the wall to its base, whether on the face or in the interior of the wall.

(*iii*) Use of bats should be avoided as far as possible.

(*iv*) The bricks should be uniform in size and the proportion of length to breadth be such that the length equals twice the width plus one joint. Otherwise good bond is impossible and the lap would not be uniform.

(*v*) The bricks in the interior thickness of the wall should be laid with their length across the wall, termed headerwise.

(*vi*) It is also recommended that every sixth course on both sides of the wall should be a header course. In walls $1\frac{1}{2}$ brick thick or more, header shall overlap headers to provide continuous tie throughout the wall.

5.5 TYPES OF BONDS

(1) Stretcher Bond

In this type of bonding, all courses are laid as stretchers, *i.e.*, with their lengths in the direction of the length of the wall (*see* Fig. 5.20 to 5.22). This is used for partition walls which are only $\frac{1}{2}$ brick thick. In thick walls, this arrangement is impracticable because there is no bonding across the wall.

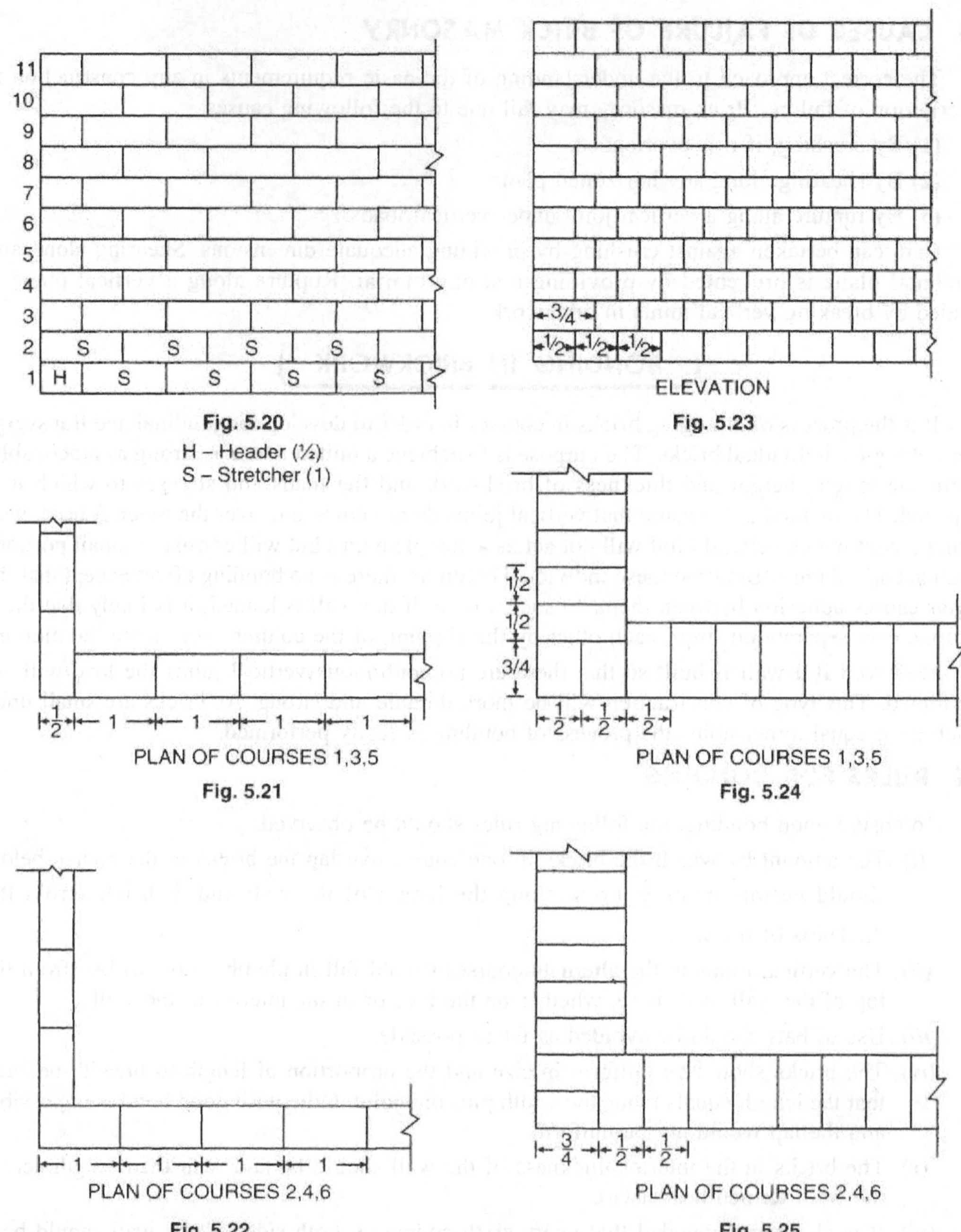

Fig. 5.20 to 5.22: Elevation and plans of alternate courses of stretcher bond.

Fig. 5.23 to 5.25: Elevation and plans of alternate courses of one brick thick wall showing header bond.

(2) Header Bond

This consists of all headers (*see* Figs. 5.23 to 5.25), *i.e.*, bricks are laid with their ends towards the face of the wall. This is used for walls which are one brick thick and is specially useful for curved brickwork where cutting of stretchers would entail lot of inconvenience. This bond does not possess the strength to transmit pressure in the direction of the length of the wall and, consequently, is not employed for walling.

(3) English Bond

This bond is produced by laying alternate courses wholly composed of headers or stretchers. For breaking the joints vertically, it is essential to place queen closer after the header quoin in the heading course. The following points about the English bond should be remembered:

(*a*) A heading course should never start with a queen closer as it would get displaced.

(*b*) Every alternate header comes centrally over the joint between the two stretchers in the course below giving a lap of one-fourth brick.

(*c*) There is no continuous vertical joint except at stopped ends.

(*d*) Walls of an even member of half bricks in thickness present the same appearance on both faces, *i.e.*, a course showing stretchers on the front face will also show stretchers on the back face.

(*e*) Walls of an odd number of half bricks in thickness will show each course comprising of headers on one face and stretchers on another face.

(*f*) The middle portion of each of the thicker walls consists entirely of headers.

(*g*) The number of vertical joints in the header course is twice the number of joints in the stretcher course. Hence the joints in the header course should be made thinner than those in the stretcher course.

In setting out the plan of a course for any width of wall the following steps should be followed:

(*a*) Draw the quoin or the corner brick.

(*b*) Next to the face (which shows header in elevation) place closers to the required thickness of the wall.

(*c*) Set out all the front headers.

(*d*) If the thickness is a even multiple of half bricks, set out the headers in the back face.

Or

If the thickness is an odd multiple of half bricks, set out stretchers in the back face. The in-between space, if any, should be filled with the headers.

The plans of alternate-courses for various thicknesses of walls are shown in Figs. 5.27 to 5.32.

(4) Double Flemish Bond

This is made up of alternate headers and stretchers in the same course. The entire course for facing, backing and hearting or filling is laid in this style. The following points about this bond are worth mentioning:

(*a*) The headers and stretchers appear in the same course alternately on the front and the back faces.

(*b*) A header in any course is in the centre of a stretcher in the course above or below it.

(*c*) Closers are inserted in alternate course next to the quoin header for breaking the vertical joints in successive courses.

(*d*) In walls having a thickness equal to odd number of half bricks, bats are to be used.

The plans of walls built in Double Flemish bond for various thicknesses are shown in Figs. 5.34 to 5.39.

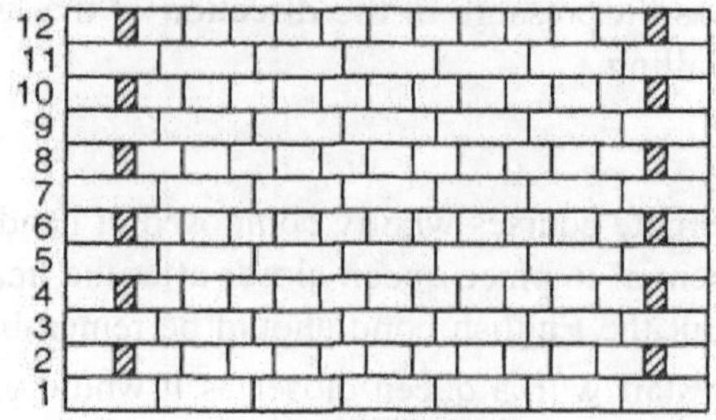

12
11
10
9
8
7
6
5
4
3
2
1
ELEVATION

Fig. 5.26

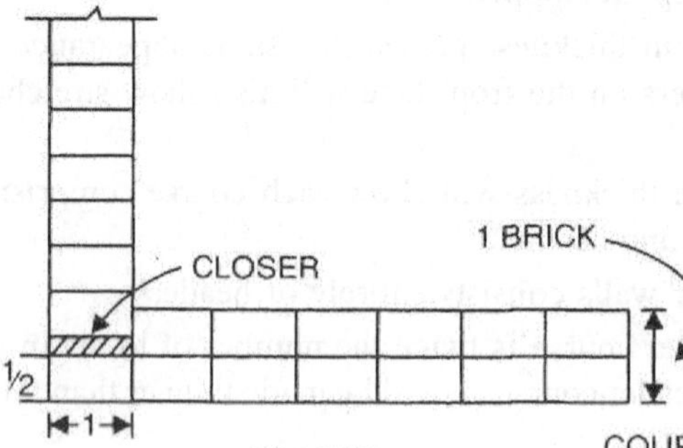

CLOSER
1 BRICK
½
1

Fig. 5.27

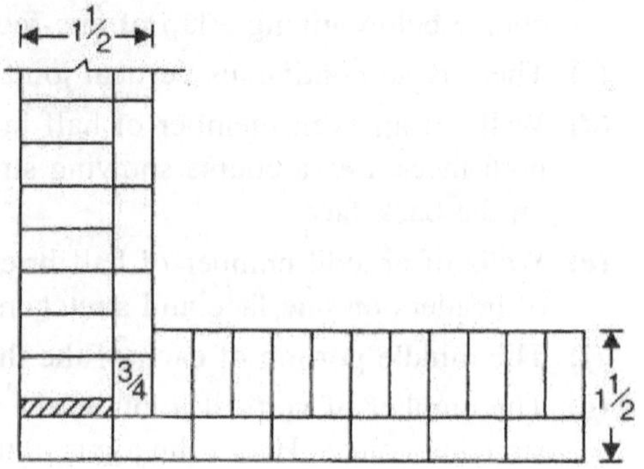

1½
¾
1½

COURSES 1,3,5

Fig. 5.29

1/2
1/4
1/2

Fig. 5.28

STRETCHER
3/4
1½
HEADER

COURSES 2,4,6

Fig. 5.30

Figs. 5.26 to 5.30: Elevation and plans of alternate courses of a one brick and one and a half brick wall corner in English bond. The arrangement of bricks at finished end is also shown.

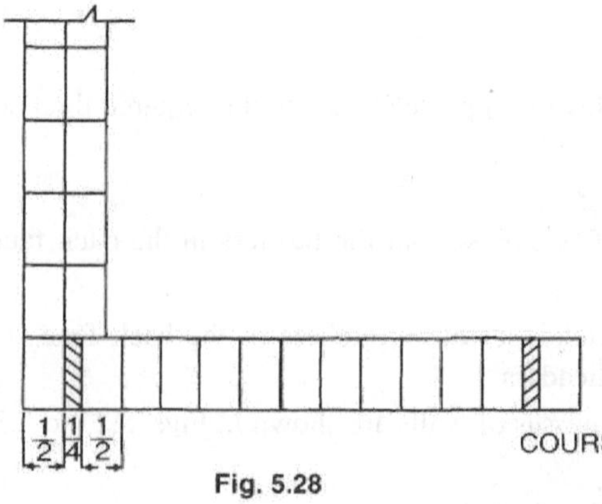

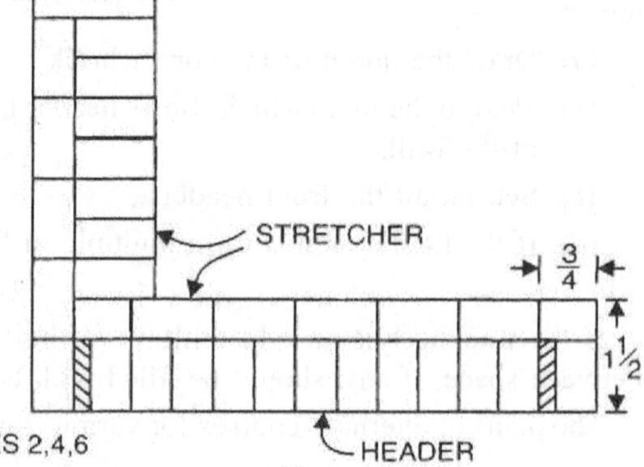

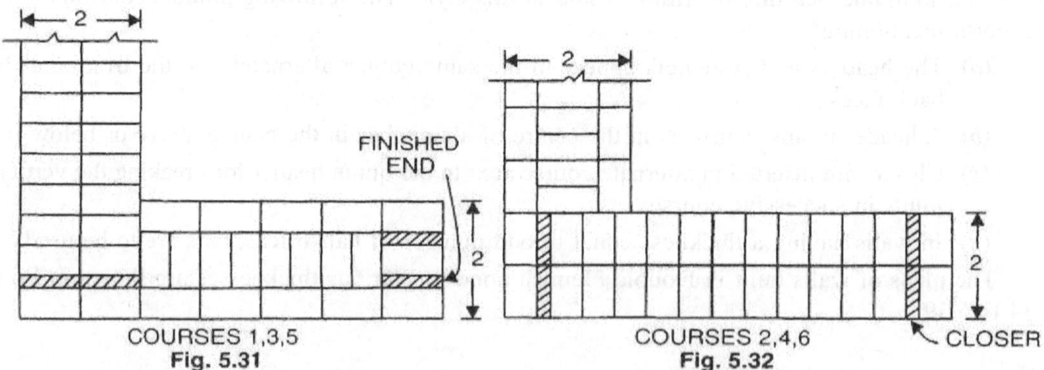

2
FINISHED
END
2

COURSES 1,3,5

Fig. 5.31

2
2
CLOSER

COURSES 2,4,6

Fig. 5.32

Fig. 5.31 and 5.32: Plans of alternate courses of 2 bricks wall in English bond.

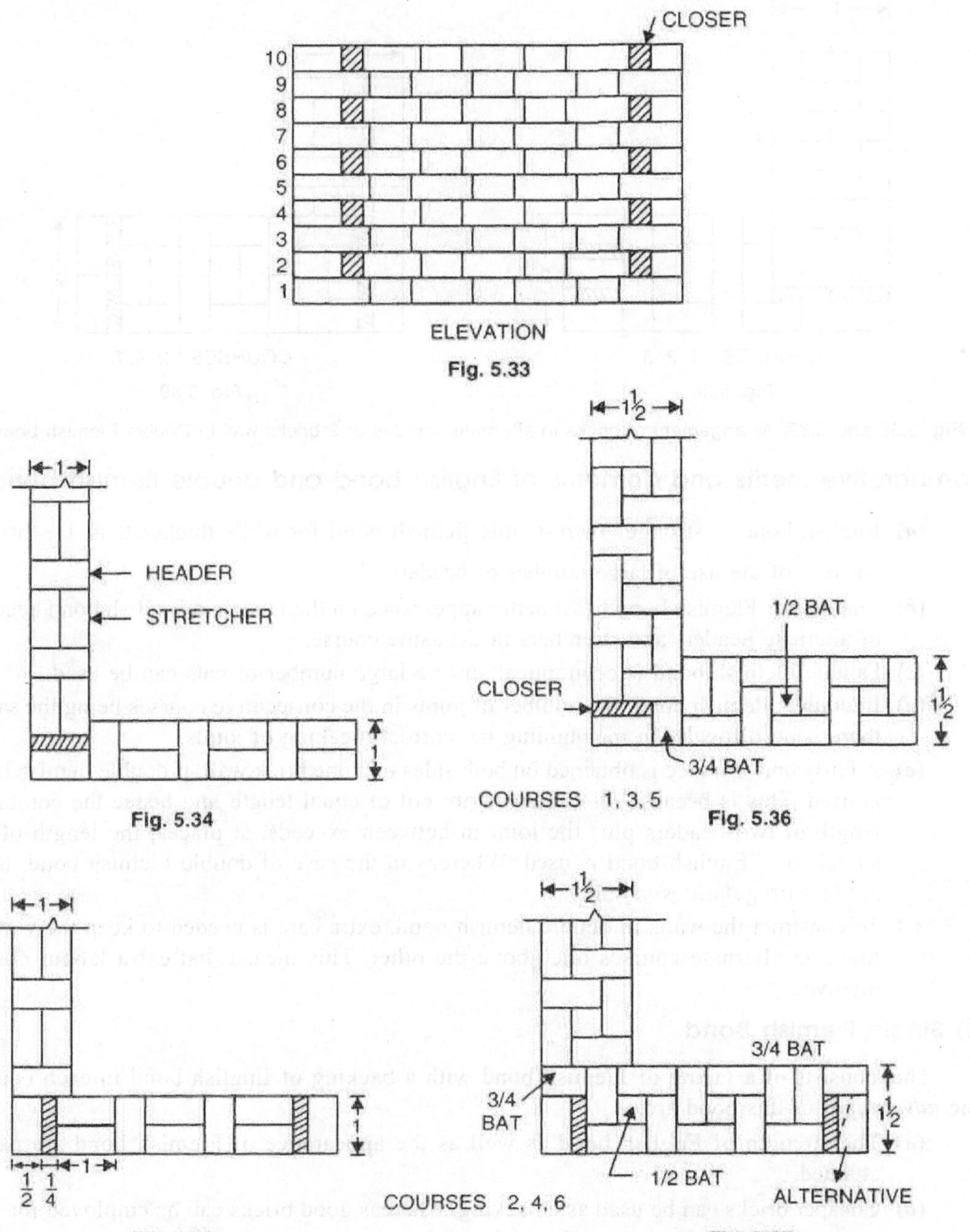

Figs. 5.33 to 5.37: Elevation and plans of alternate courses of one brick and 1 $\frac{1}{2}$ brick walls in Double Flemish bond.

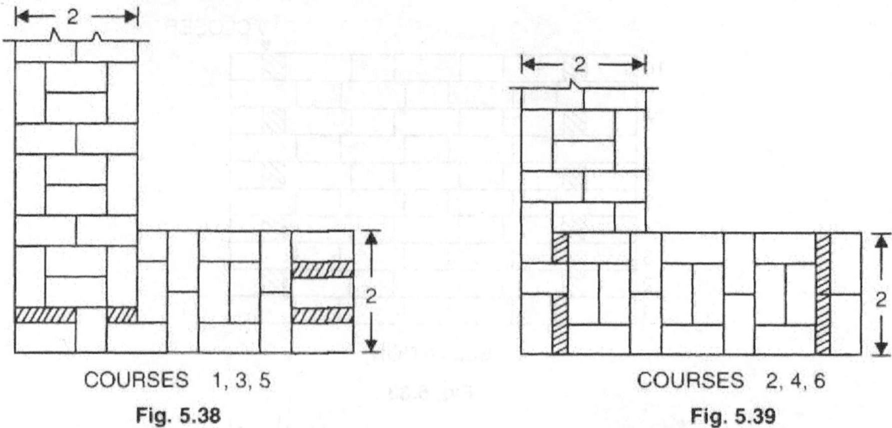

Fig. 5.38

Fig. 5.39

Fig. 5.38 and 5.39: Arrangement of bricks in alternate courses of 2 bricks wall in Double Flemish bond.

Comparative merits and demerits of English bond and double flemish bond

(*a*) English bond is stronger than double flemish bond for walls thicker than $1\frac{1}{2}$ bricks, because of the use of large number of headers.

(*b*) The double Flemish bond has a better appearance on the face than English bond because of alternate headers and stretchers in the same course.

(*c*) Double Flemish bond is economical since a large number of bats can be used.

(*d*) In double Flemish bond, the number of joints in the consecutive courses being the same, there is no difficulty in maintaining the correct breaking of joints.

(*e*) A fairly uniform face is obtained on both sides of a one brick wall, if double flemish bond is used. This is because all stretchers are not of equal length and hence the combined length of two headers plus the joint in between exceeds, at places, the length of the stretchers if English bond is used. Whereas in the case of double Flemish bond, there are less irregularities visible.

(*f*) To construct the walls in double flemish bond, extra care is needed to keep the vertical joints in alternate courses one above the other. This means that extra labour cost is involved.

(5) Single Flemish Bond

This consists of a facing of Flemish bond with a backing of English bond in each course. The ***advantages*** of this bond are:

(*a*) The strength of English bond as well as the appearance of Flemish bond are partly attained.

(*b*) Cheaper bricks can be used as a backing, whereas good bricks can be employed for face work which is to be done in Flemish bond.

The ***disadvantages*** of this bond are:

(*a*) This cannot be used for walls less than $1\frac{1}{2}$ brick in thickness.

(*b*) A long continuous joint occurs in the vertical direction for some portions and thus weakens the wall. (See Figs. 5.40 to 5.45).

(6) Garden Wall Bond

This type of bonding is used for garden compound walls. As already pointed out, it is difficult to make uniform wall of one brick thick in English bond due to the variation in the lengths of the stretcher faces. Stretcher and header bonds can be constructed but they are not very strong. Two types of garden wall bonds are commonly used.

(a) English garden wall bond. This consists of one course of headers after three or five courses of stretchers. A queen closer is placed next to the quoin header. The middle course of stretchers is started with a header to give the necessary lapping.

(b) Flemish garden wall bond. This comprises of one header after three or five stretchers in each course. A $\frac{3}{4}$ bat is introduced next to the quoin header in every alternate course and a header is placed over the middle of each stretcher.

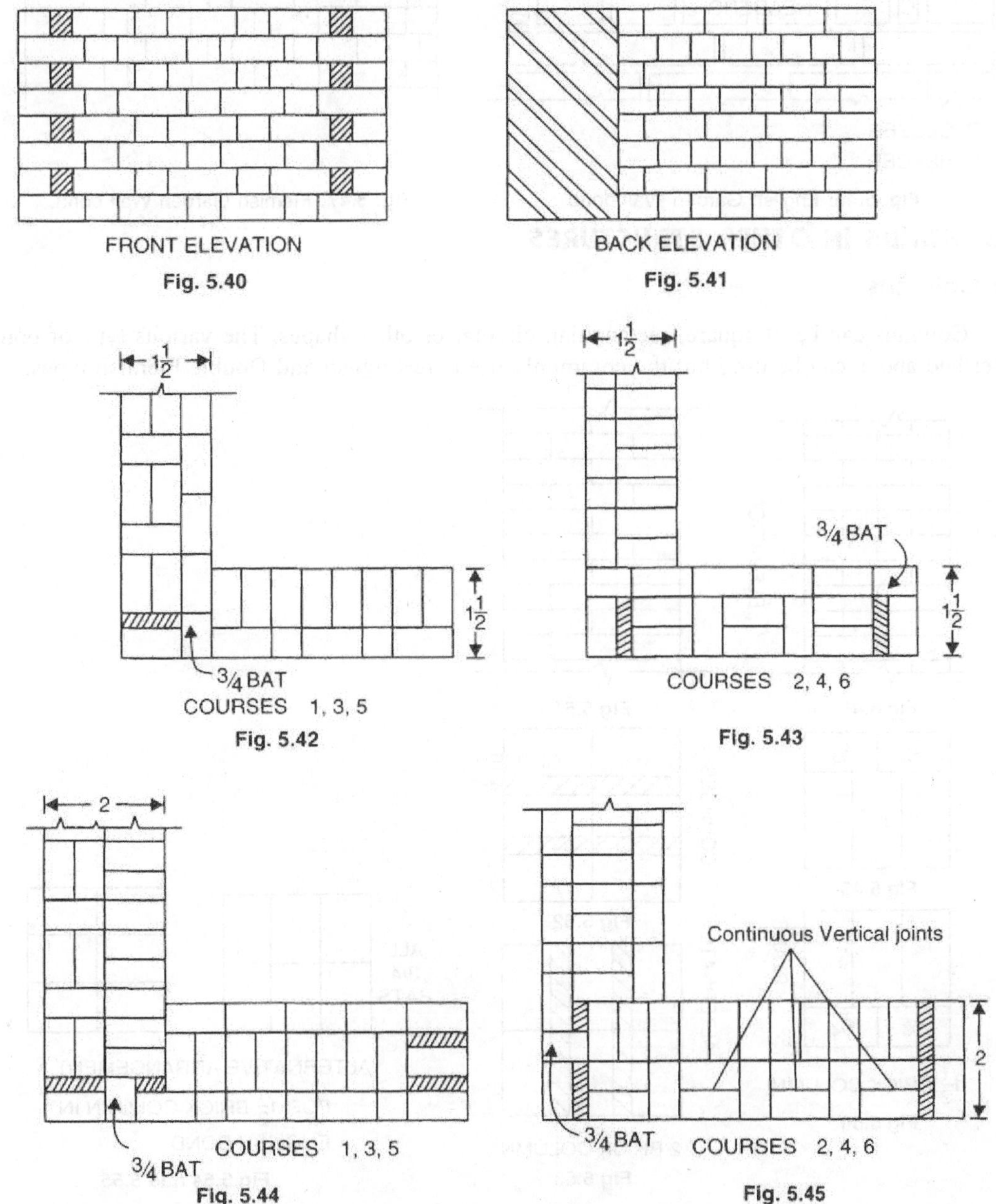

Fig. 5.40 to 5.45: Front and back elevations and plans of alternate courses of 1 $\frac{1}{2}$ and 2 brick wall corners in single Flemish bond.

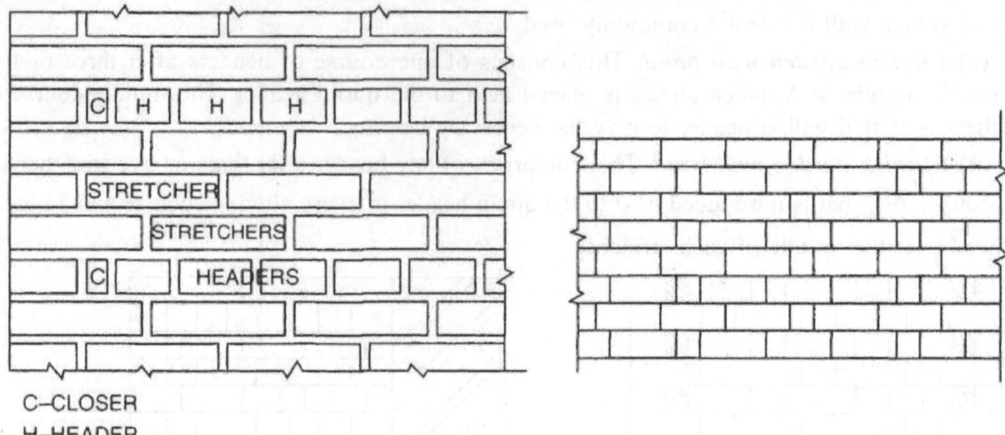

Fig. 5.46: English Garden Wall bond.

Fig. 5.47: Flemish Garden Wall bond.

5.6 BONDS IN OTHER STRUCTURES

(1) Columns

Columns can be of square, rectangular, circular or other shapes. The various type of bonds described above can be used but the commonly used are English and Double Flemish types.

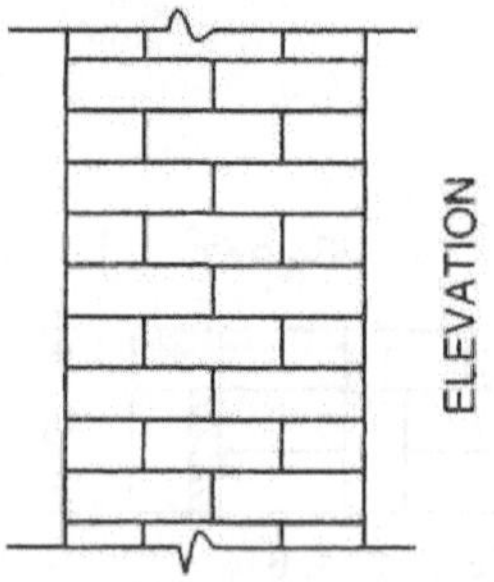

ELEVATION

Fig 5.48

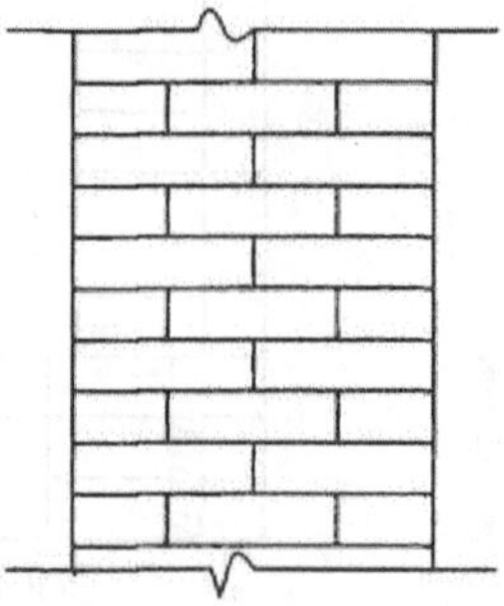

Fig 5.51

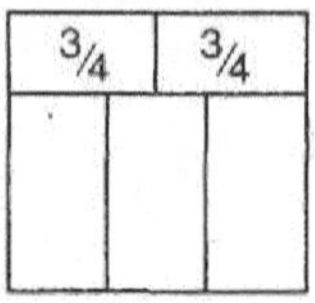

3/4 3/4

Fig 5.49

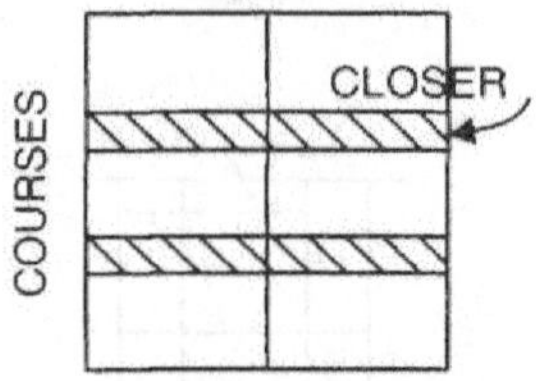

COURSES

CLOSER

Fig 5.52

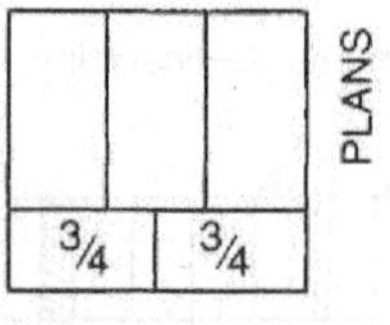

PLANS

ALTERNATE

3/4 3/4

$1\frac{1}{2}$ BRICK COLUMN

Fig 5.50

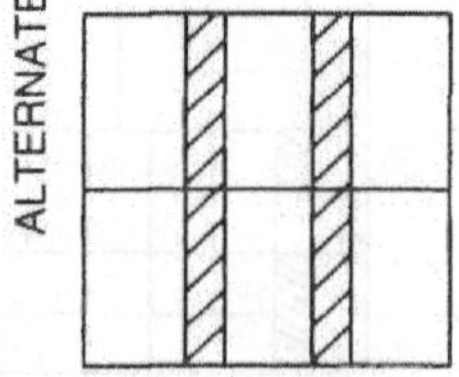

2 BRICK COLUMN

Fig 5.53

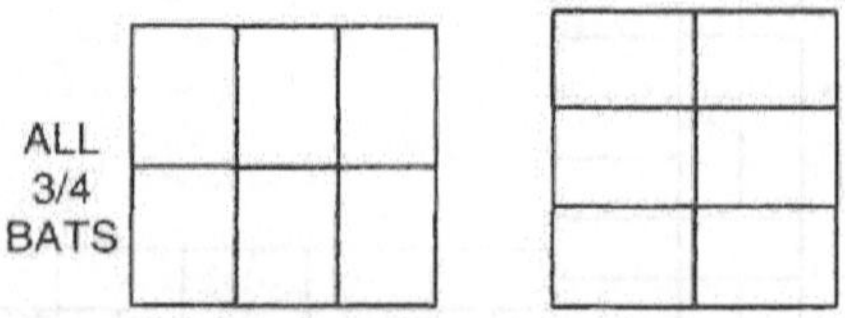

ALL 3/4 BATS

ALTERNATIVE ARRANGEMENT FOR $1\frac{1}{2}$ BRICK COLUMN IN ENGLISH BOND

Fig 5.54 and 5.55

Figs. 5.48 to 5.55: Elevation and Plans of alternate courses of one brick, one and a half brick and two brick columns in English bond.

(a) English Bond. A one brick square column is made by laying two bricks together and the next course is laid at right angles to this.

$1\frac{1}{2}$ brick square column is built in two types. In one type only $\frac{3}{4}$ bats are used. They are placed as shown in Figs. 5.54 and 5.55. In the second type two $\frac{3}{4}$ bats and three full bricks are used in each course (See Figs 5.49 and 5.50). However, there are continuous vertical joints in the second type whereas in the first type the bricks have to be cut to size.

In 2 brick square column, closers of $\frac{1}{2}$ bricks have to be used parallel to the stretcher face. The arrangement as shown in the figure for first course is also used for the second course except that the bricks are turned through one right angle.

In larger sized columns bricks are arranged in a similar fashion and suitable closers are added wherever necessary.

(b) Double flemish bond. The smallest size of the column which can be built in this type of bond is $1\frac{1}{2}$ brick square. A square bat is to be used in every course in the centre of the column and this creates continuous vertical joints which weakens the column. However, the headers do not come over the centre of the stretchers as in the case with the true Flemish bond (*see* Figs. 5.56 to 5.58).

The plans of alternate courses of 2 brick square column in Flemish bond are shown in Figs. 5.60-5.61. Continuous vertical joints occur at certain portions of the column and create zone of weakness.

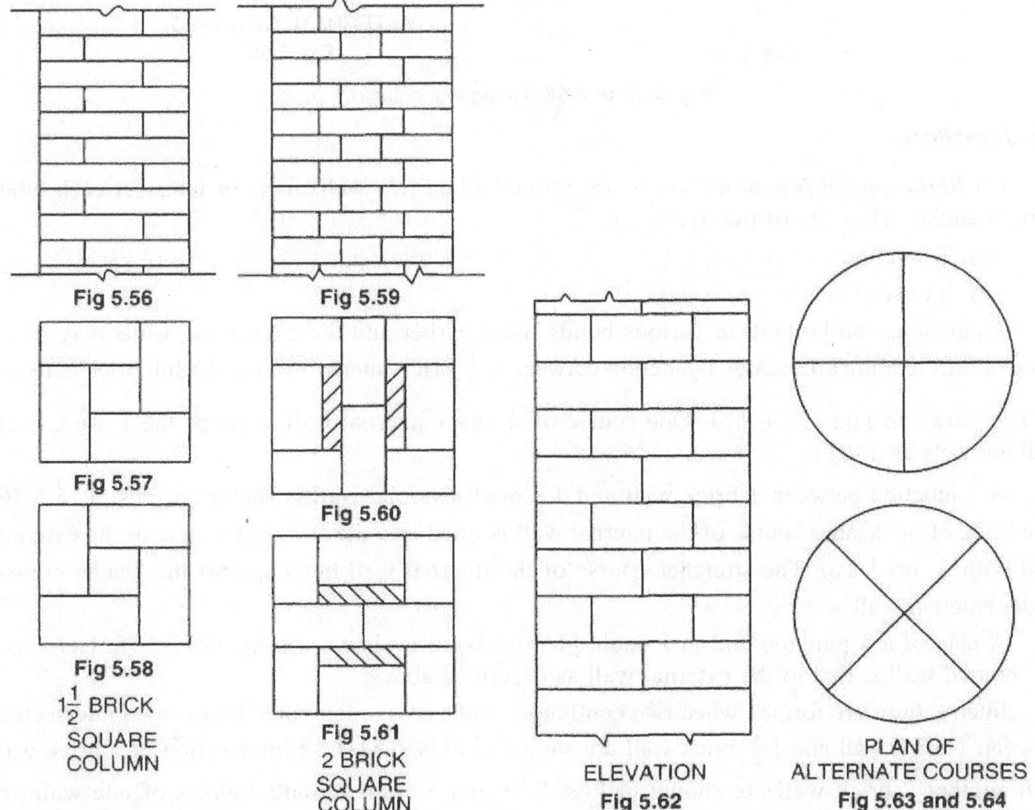

Figs. 5.56 to 5.61: Plans of alternate courses and elevation of $1\frac{1}{2}$ and 2 brick square column in Flemish bond.

Figs. 5.62 to 5.64: Circular column made of special bricks.

Circular and Octagonal Columns. These may be constructed of cut or moulded bricks to suit the size. They are generally made in English bond. At places, bull-nosed brick may be used if fully circular section is not desired. These are shown in Figs. 5.62 to 5.66.

(2) Piers attached to walls

These are used for increasing the stability of the walls and for enhancing the architectural beauty of straight walls. They are projections from the wall acting as columns and can be constructed in English or Double Flemish bond. Plans of courses for some of these piers are shown in Figs. 5.67 to 5.72.

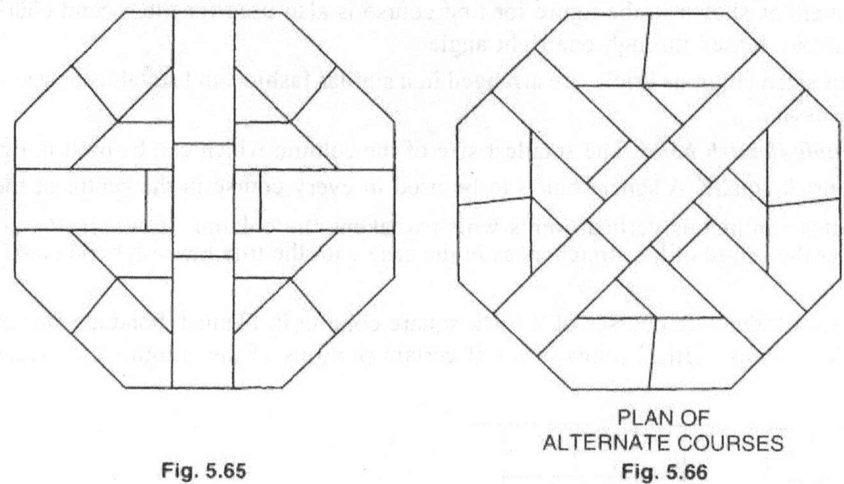

Fig. 5.65

Fig. 5.66

Fig. 5.65 to 5.66: Octagonal column.

(3) *Junctions*

(a) Right-angled junctions. These are formed when two walls meet or intersect each other at right angles. They are of two types, *i.e.*,

(*i*) T-junction.

(*ii*) Intersections.

T-junctions can be built in various bonds listed earlier and the connecting walls may be of same or different thicknesses. A T-junction between a $\frac{1}{2}$ brick internal wall and a full brick external wall is shown in Figs. 5.73-5.74. One course of $\frac{1}{2}$ brick internal wall overlaps the 1 brick main wall and acts as a tie.

A T-junction between 1 brick wall and $1\frac{1}{2}$ brick external wall is shown in Figs. 5.75-5.76. One brick of the header course of the internal wall is fitted into the stretcher course of the external wall with $\frac{1}{4}$ brick lap. The stretcher course of the internal wall butts against the header course of the external wall.

A plan of a T-junction laid in Double Flemish bond is shown in Figs. 5.77-5.78. Here also the internal wall is tied to the external wall as described above.

Intersections are formed when two continuous walls cross each other. Plans of an intersection between 1 brick wall and $1\frac{1}{2}$ brick wall are shown in Figs. 5.81-5.82 Intersection of 2 brick wall with another 2 brick wall are shown in Figs 5.83 and 5.84. Alternate courses of one wall are continuous whereas the bricks in these courses of the other wall about with the bricks of the continuous wall. In the courses which are discontinuous, key headers are fixed into the wall as shown in Figs. 5.81-5.84.

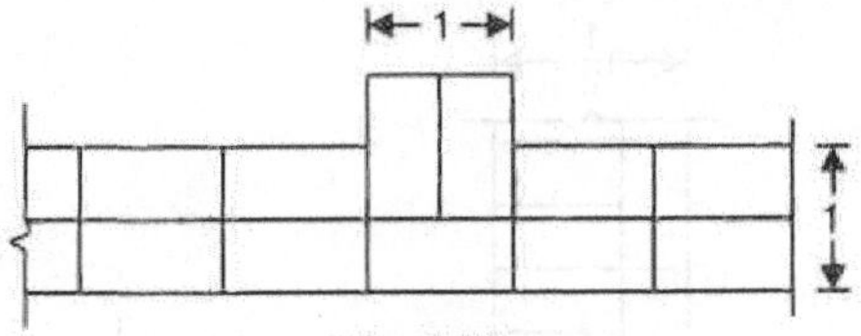

Fig. 5.67

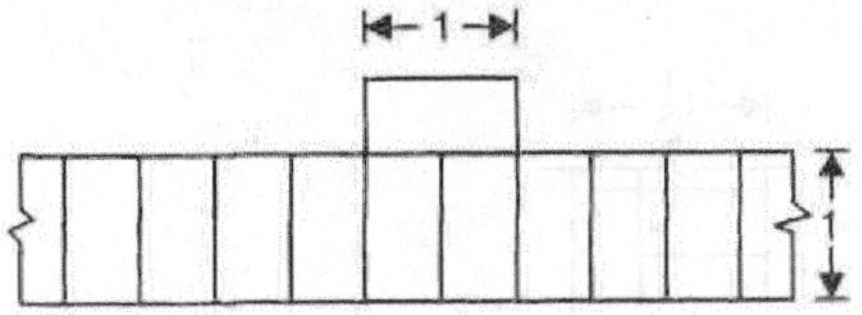

Fig. 5.68

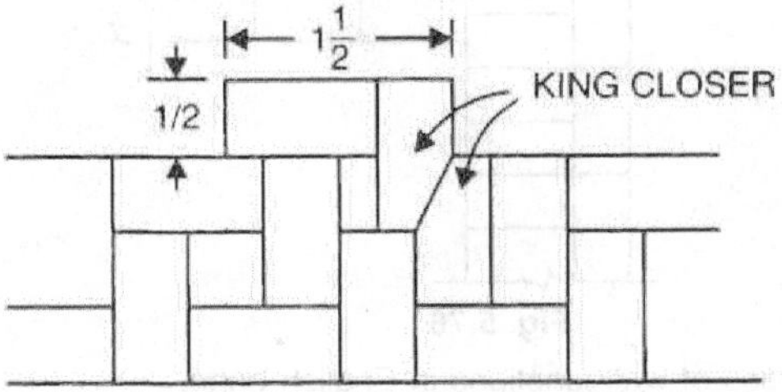

KING CLOSER

Fig. 5.69

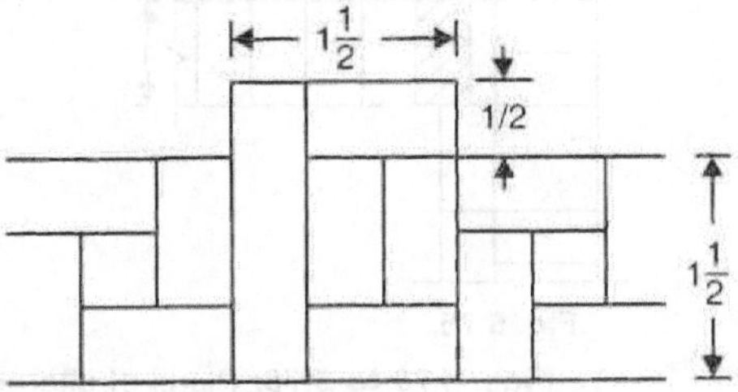

Fig. 5.70

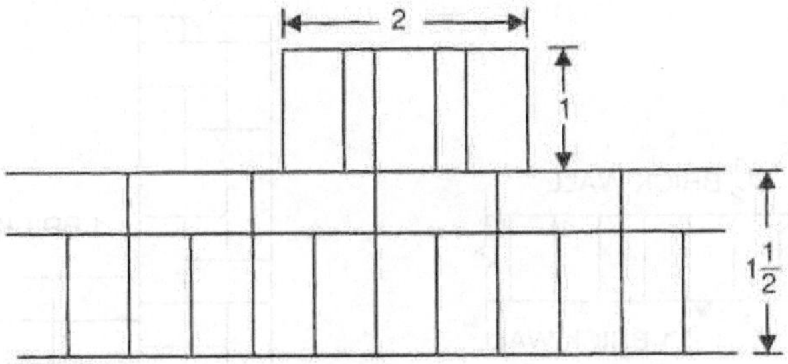

Fig. 5.71

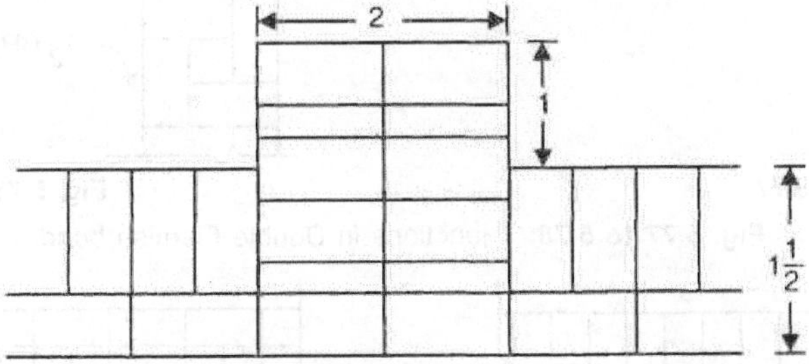

Fig. 5.72

Figs. 5.67 to 5.72: Alternate courses of attached piers in English and Double Flemish bond.

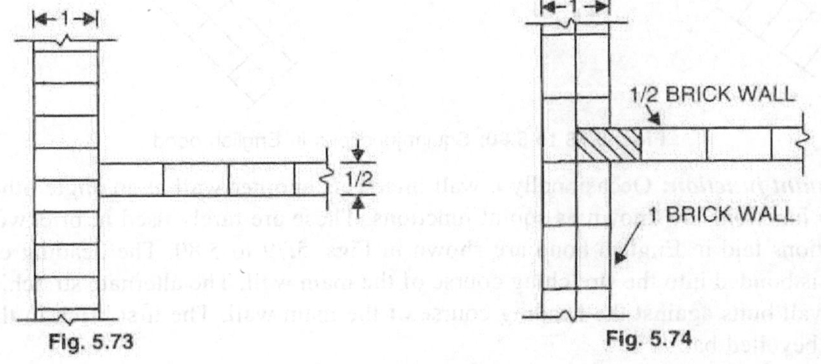

Fig. 5.73

Fig. 5.74

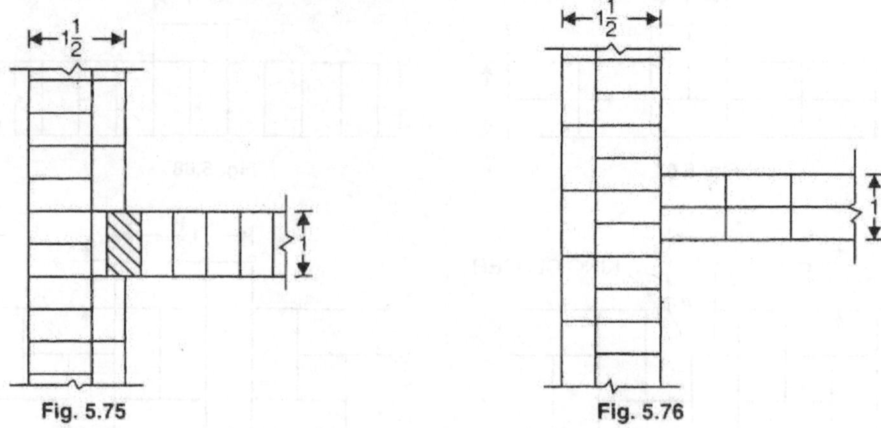

Fig. 5.75 Fig. 5.76

Figs. 5.73 to 5.76: Plans of alternate courses of wall junctions in English bond.

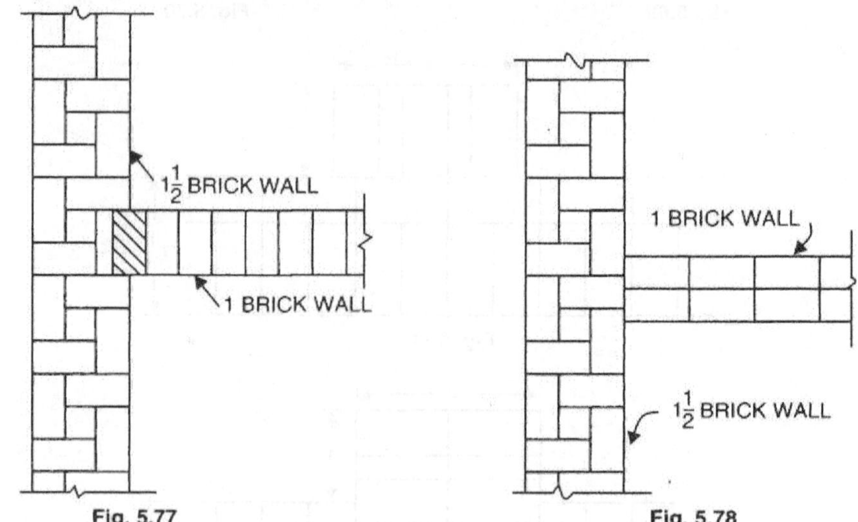

Fig. 5.77 Fig. 5.78

Fig. 5.77 to 5.78: T-junctions in Double Flemish bond.

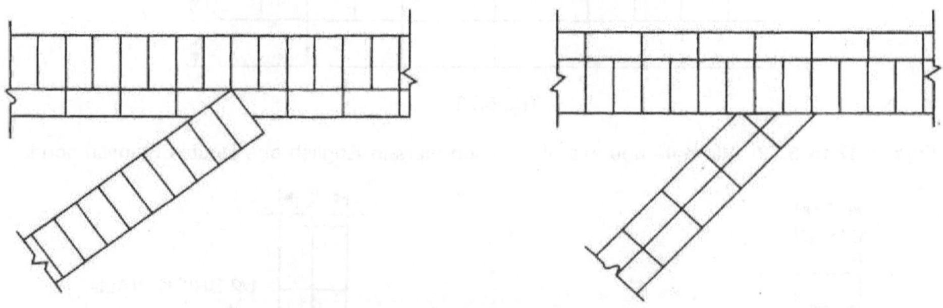

Figs. 5.79 to 5.80: Squint junctions in English bond.

***(b) Squint junction*:** Occasionally a wall intersects another wall at an angle other than right angle. Such junctions are known as squint junctions. These are rarely used in brickwork. Plans of squint junctions laid in English bond are shown in Figs. 5.79 to 5.80. The heading course of the squint wall is bonded into the stretching course of the main wall. The alternate stretching course of the squint wall butts against the heading course of the main wall. The first brick in the stretching course is a bevelled bat.

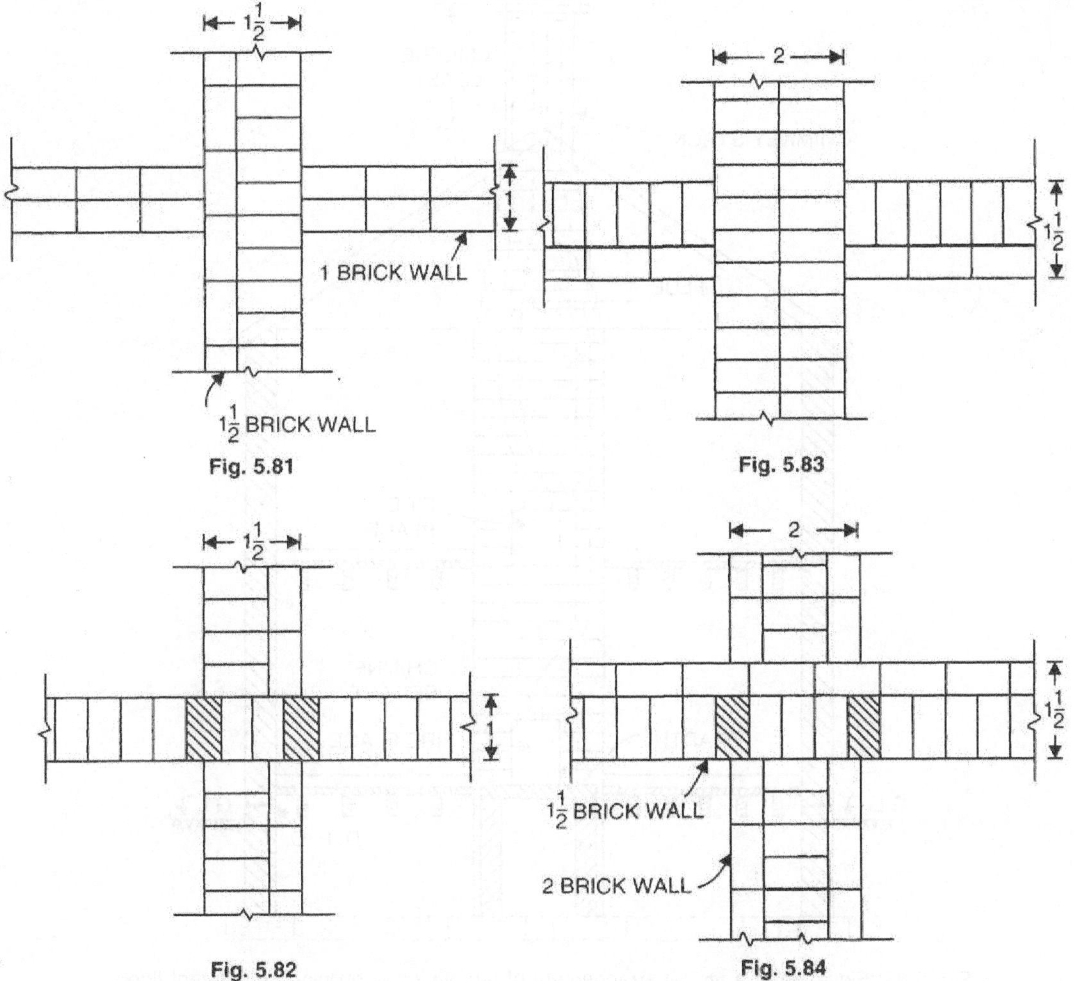

Figs. 5.81 to 5.84: Intersections of walls in English bond.

FIRE-PLACES AND CHIMNEYS

Even today, fire-places are provided in the kitchens and living rooms of a house for fires and heating purposes respectively. They also serve as a means of ventilation. Sometimes boilers are incorporated with the fireplaces and a cheap supply of hot water is obtained. For driving out the smoke and the products of combustion, fireplaces need vertical shafts which are termed as chimneys. These produce a draught which keeps the fire burning constantly.

The following terms are used in connection with the chimneys and fire-places:

(a) Chimney Flue: This is a shaft which carries gases from a fire-place through the building to the atmosphere. This flue is not less than 400 sq cm in sectional area and is carried in brickwork or stone masonry.

(b) Chimney Back: This term applies to the back of a fire-place opening.

(c) Chimney Jambs: These are the vertical sides of a fire-place opening.

(d) Chimney bar: This is a metallic bar which ties the jambs of chimney together.

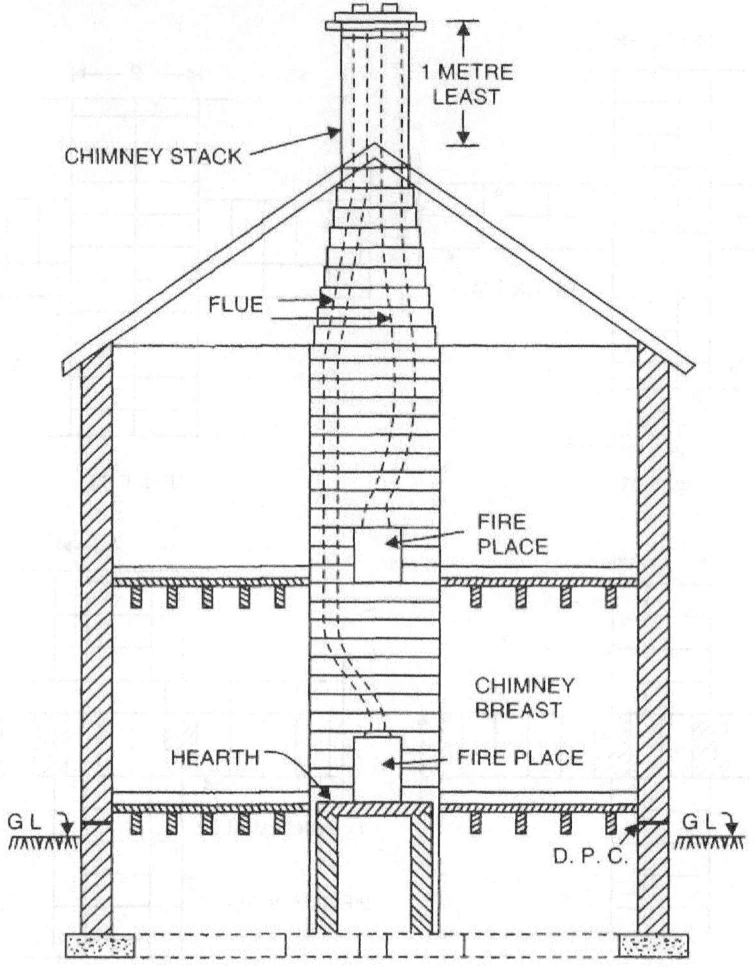

Fig. 5.85: Elevation of a typical arrangement of fire-places of building on different floors.

(*e*) ***Chimney breast***: If the chimney or the fire-place opening projects outside the wall face the portion so projecting is called chimney breast.

(*f*) ***Chimney cowl***: This is a cap or fitting to a chimney and promotes an upward draught in the chimney.

(*g*) ***Chimney gutter***: This is suitably shaped piece of metal which collects and diverts water which otherwise would penetrate the roof at the back of the chimney stack.

(*h*) ***Chimney lining***: This is the rendering on the inner side of a flue; may be of special fire-proof material.

(*i*) ***Chimney piece or mantel piece***: This is an ornamental surrounding to a fire-place.

(*j*) ***Chimney pot***: This is a terra-cotta unit at the top of a chimney stack. This increases the height and prevents down draught.

Chimneys and fire-places are built according to the prevalent building codes. The various parts of a chimney are described below:

(*a*) ***Foundations***: The foundations of a chimney are as deep as that of the adjacent walls. The chimney is bonded properly to the wall.

(b) Chimney breast: A fire-place generally needs greater depth than the thickness of a wall. Hence the chimney is accommodated in a projection called chimney breast. This projection can be avoided if chimney is provided in external walls where the projection can be provided on the outer side.

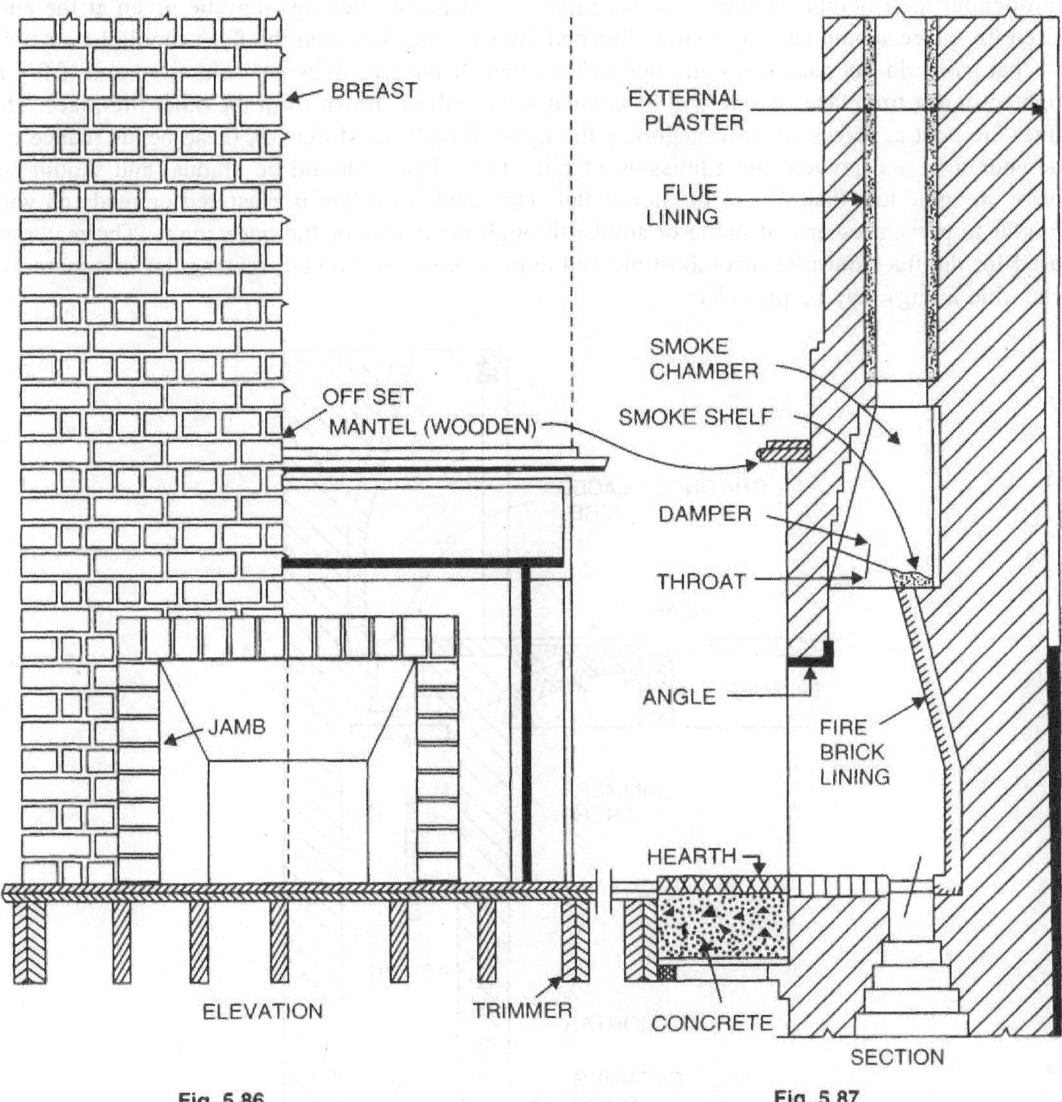

Fig. 5.86 **Fig. 5.87**

Fig. 5.86 and 5.87: Elevation and section of a fire-place.

The jambs of a chimney must not be less than 20 cm in width. The fire-place opening will depend on the size of the room or the type of the fire-place. The smallest room may need a fire-place opening of $\frac{1}{2}$ m width and $\frac{1}{2}$ m height. The chimney breast can be widened and finished suitably so that it may look in proportion to the wall from which it projects.

The head of a fire-place has a masonry arch or a concrete lintel. If the projection of the breast beyond the wall is greater than 10 cm, the arches have to be built on iron bars for support.

The back wall of a chimney should be at least $\frac{1}{2}$ brick thick, if the two fire-places are laid adjacent to each other othrwise a minimum of one brick thickness is needed.

(c) ***Flues***: These are at least 1 brick × 1 brick in size and should be uniform in section throughout their height. However, a throating (*i.e.*, reduced opening) may be given at the end. Each fire-place should have a separate flue and the partitions between the flues should be air-tight so that gases do not pass from one flue to the other. If the two flues get connected and if fire is lighted in one fire-place, smoke will be created as air will be drawn from the other fire-place. The flues are bent at places so as to negotiate the upper fire-place. Moreover, these bends reduce the downdraught and prevent the admission of rain. These bends should be gradual and should not be at an angle less than 45° to the horizontal. The inside of a flue is plastered or rendered with mortar to prevent escape of flame or smoke through the cracks or the open joints. The materials used for the flues must be incombustible and durable. Special circular, rectangular or square fire clay flue linings may be provided.

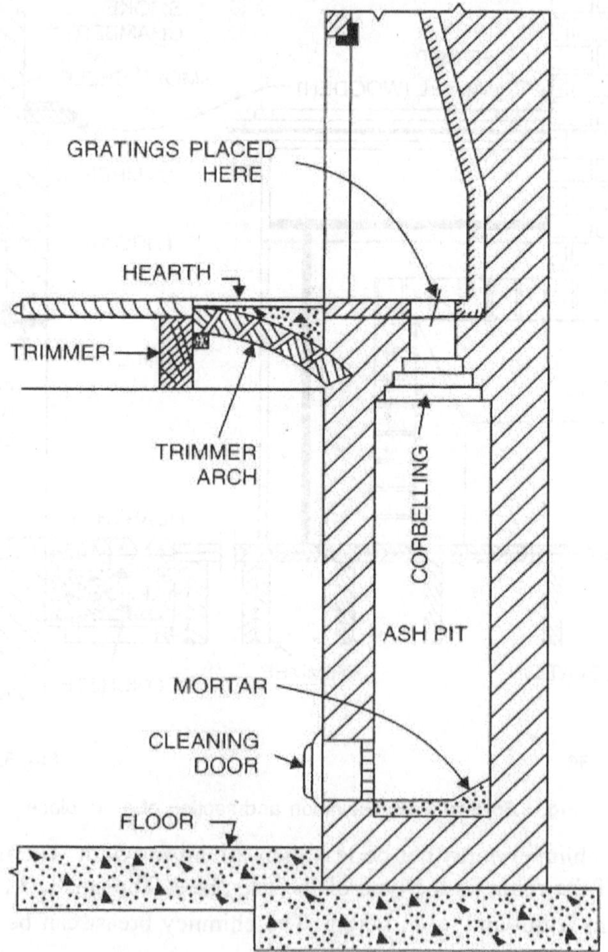

Fig. 5.88: Section of a fire-place ash pit placed in a basement.

(d) ***Chimney stacks***: The chimney breasts are reduced in width when they penetrate above the roofs. According to the building codes, the chimney stacks must be carried upto a height of at least one metre above the highest point of the adjoining roof. To prevent down draught, the chimney stack is generally taken at least one metre above the ridge. The maximum height of the stack is taken as six times the least width. The thickness of the walls is at least $\frac{1}{2}$ brick. Special consideration should be given to prevent the rain water coming down the walls. This can be prevented by constructing the chimney stacks in water-proof cement mortar to a height $\frac{1}{2}$ metre below the lowest point of intersection.

The top of the flue is covered with a chimney pot. This pot is supported on the brickwork and held in a manner so as to throw off the water. Tall chimney pots are not needed unless the flues are short or are liable to produce down-draught.

The front of the hearth must have a thickness of at least 15 cm and should extend on the sides at least 15 cm beyond the inside clear opening of the fire-place. A projection of at least $\frac{1}{2}$ metre is needed outside the face of the breast. This hearth is made of incombustible material and securely supported. All woodwork should be kept away from the fire-place hearth.

The fire-place interior is made of fire clay blocks, metal gratings, metal frame and may be surrounded by glazed tiles. The fire clay block is bedded on the mortar and backed solidly with concrete or brick in cement. The upper surface of this backing is sloped to prevent soot from accumulating.

5.7 JOINTS IN BRICKWORK

Joints are the weakest parts of a masonry structure and unless special care is paid to them, the brickwork cannot be of good strength. Appearance of brickwork also depends on the proper laying and finishing of joints. Most of the defects in brick masonry joints occur on account of improper application of mortar. Some of the important points regarding the joints in the various courses of brickwork are:

(1) ***Bed joints in stretcher courses.*** Mortar for bed joints should be spread evenly and must be quite thick. Generally mortar is not spread ahead for a distance greater than five or six brick lengths. This avoids the drying of mortar and also keeps the mortar plastic so that it can set well after the bricks have been laid. The bricks should be pressed on the mortar and tamped with a trowel.

(2) ***Head joints in stretcher courses.*** These must also be completely filled with mortar. Mortar should be thrown on the end of each brick to be placed. The bricks are placed on the mortar bed and pushed into place till the excess mortar gets squeezed out. An alternate method is to apply mortar to the brick already laid and then push in the next brick. The mortar which squeezes out is removed with a trowel.

(3) ***Cross-joints in header courses.*** Cross joints should be carefully made and should be full of mortar. Before each header is laid, the edge of the brick should be covered with mortar that will stick to it. The header should be pushed into place so that mortar oozes out above the cross joint as well as at the bed joint. The excess mortar is scrapped off with a trowel.

(4) ***Closer joints in stretcher courses.*** The last brick to be placed in a wall along stretchers should be laid in such a way that the head joints are completely filled with mortar. With the bed joints already made, mortar is applied to the ends of the bricks which have previously been laid. Mortar is also placed on both ends of the closer brick. The closer is then pushed inside without disturbing the bricks already laid. The excess mortar is scraped off. A similar procedure is followed in laying the closer bricks in header courses.

5.7.1 Finishing of Joints or Jointing and Pointing in Brickwork

The purpose of finishing the joints is to improve the appearance of brickwork and to make it more waterproof. Merely drawing one edge of trowel along the joint is detrimental for brickwork as it may disturb the adhesion between the mortar and the brickwork. The finishing of joints as the brickwork proceeds is termed as *jointing* whereas finishing of joints after the brickwork has been completed is called ***pointing.***

The edges of the joints in brickwork will be deficient in hardness and density. The expansion and contraction of masonry is liable to cause the mortar in the joint crack or loosen the brick from the mortar. The mortar near the exposed face is, therefore, prone to dislodgement. Unless the wall is to be plastered, it is necessary to refill very compactly the surface joints for a depth of about 2 to 3 cm with a neat mix of cement mortar of proportion 1 : 1. This process is known as pointing. Normally, while laying the mortar in joints at the surface, an offset is left to a depth of about 2 to 3 cm. The joint is subsequently cleaned by scrapping and brushing out all loose material with a wire brush. The rich mortar is then applied, pressed and finished smooth. Pointing should not be allowed to dry rapidly. When cracking due to shrinkage is expected, either high alumina or sulphate resistant cements should be used or the mortar proportions increased from 1 : 1 to 1 : 2 or 1 : 3.

Different kinds of pointing finishes are given below:

(1) *Flush or flat joint.* This is the simplest type of joint and when rubbed gives a good finish. The mortar is pressed into the joint as the work goes on and depressions are filled. A trowel is kept parallel to the face of the wall and the joint is finished. The joint may also be rubbed by a piece of stone held against the wall to improve appearance. This type of joint is needed at places where depressions in brickwork are not desirable, *e.g.*, factories, cellar, garages, etc.

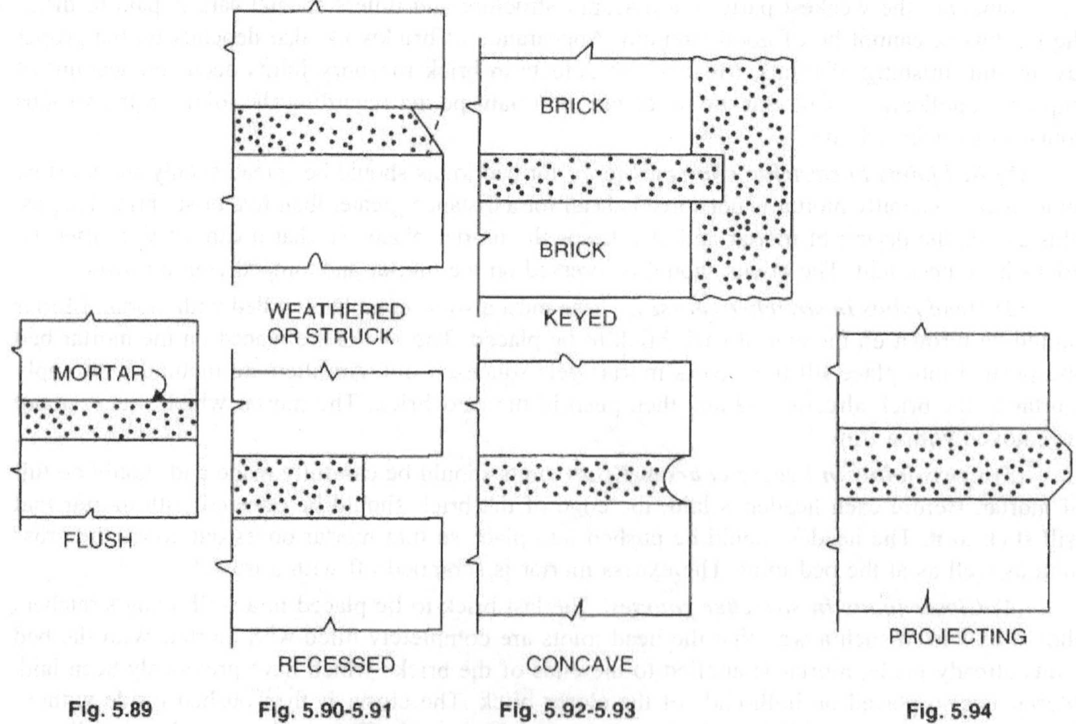

Figs. 5.89 to 5.94: Various types of pointing in brickwork.

(2) *Weathered or struck joint.* This joint permits water to drop off from the face of the brickwork. The joint is made by holding the handle of the trowel below the bed joint and smoothening the mortar in one direction to an angle of about 60°. The vertical joints are first struck followed by bed joints. The appearance of this joint is not satisfactory if ordinary bricks are used because unevenness in the edges of the bricks is visible.

(3) ***Recessed joint.*** This joint is made with a jointing tool, the thickness of the rubber attached in front of the tool being equal to that of the joint. Rubber attachment is used as it adjusts itself to any irregularity of the brick edge. This tool is used after any projecting mortar has been removed. The brick should be carefully selected and should have uniform thickness. This joint is at least 1 cm in thickness. This is satisfactory in face-work for good textured bricks and good quality of mortar.

(4) ***Concave joint.*** This may be formed by a rounded jointer. The vertical joints are formed first and then the bed joints. The jointer is used by two masons in conjunction with pointing rule. The rule is held with the bevelled edge on the same level and parallel to the lower edge of the joint. The jointer resting on this rule is moved to and fro while it presses the mortar to the required depth. This type of joint gives a very attractive appearance to the brickwork.

(5) ***V-joint.*** This is made in a manner similar to the one described above, but a steel jointer with a bevelled edge is used. This is used to give an appearance of narrow joints. The finishing should be done before the mortar sets.

(6) ***Projecting joint.*** Mortar is left projecting from the joints. This type of joint affords a rough surface which can be helpful in keying the plaster which is applied later on. Bricks of various colours, if used with mortars of suitable colours, will provide a beautiful appearance.

(7) ***Stripped joint.*** This is made by placing wooden strips as the masonry is built up. These strips ensure a joint of uniform thickness and may be removed after the mortar has set slightly.

A concave joint ensures good water tightness. Recessed and stripped joints are expensive and are not fully watertight. Joints in the exterior faces of walls should be entirely filled with mortar to reduce water penetration. It is essential that the joints are made smooth and dense by exerting considerable pressure on the tool with which they are made.

REINFORCED BRICKWORK

Plain brick masonry cannot take tensile stresses as the bricks get pulled apart at the mortar joints. Reinforced cement concrete is a structural material which can be put to all type of uses, but owing to its high cost, it is not economical to use it for members which carry a small load and yet cannot be made of plain brickwork. Examples are the construction of small lintels, slabs and beams. To increase the load carrying capacity of plain brickwork, steel reinforcement is introduced in between the mortar joints. This type of brickwork can withstand tensile and shear stresses if the loads are not unduly great. Reinforced brickwork is adopted in the following cases:

(1) When the brickwork is required to carry excessive compressive loads.
(2) When the brickwork has to withstand tensile and shear stresses.
(3) When the brickwork rests on soil subjected to differential settlement.
(4) When the brickwork is provided over opening of doors or windows.
(5) When the brickwork has to resist lateral loads *e.g.*, retaining walls, etc.
(6) When the brickwork is to be constructed in earthquake prone areas.
(7) When it is desired to strengthen the longitudinal bond.

However, the following points about reinforced brickwork construction should be noted:

(1) All the bricks must be of uniform size and structurally sound.
(2) Cement mortar in the proportion of 1 : 3 should be used for bonding the brickwork.
(3) The joints have to be built in such a way that the reinforcement gets sufficient cover and is also prevented from corrosion.

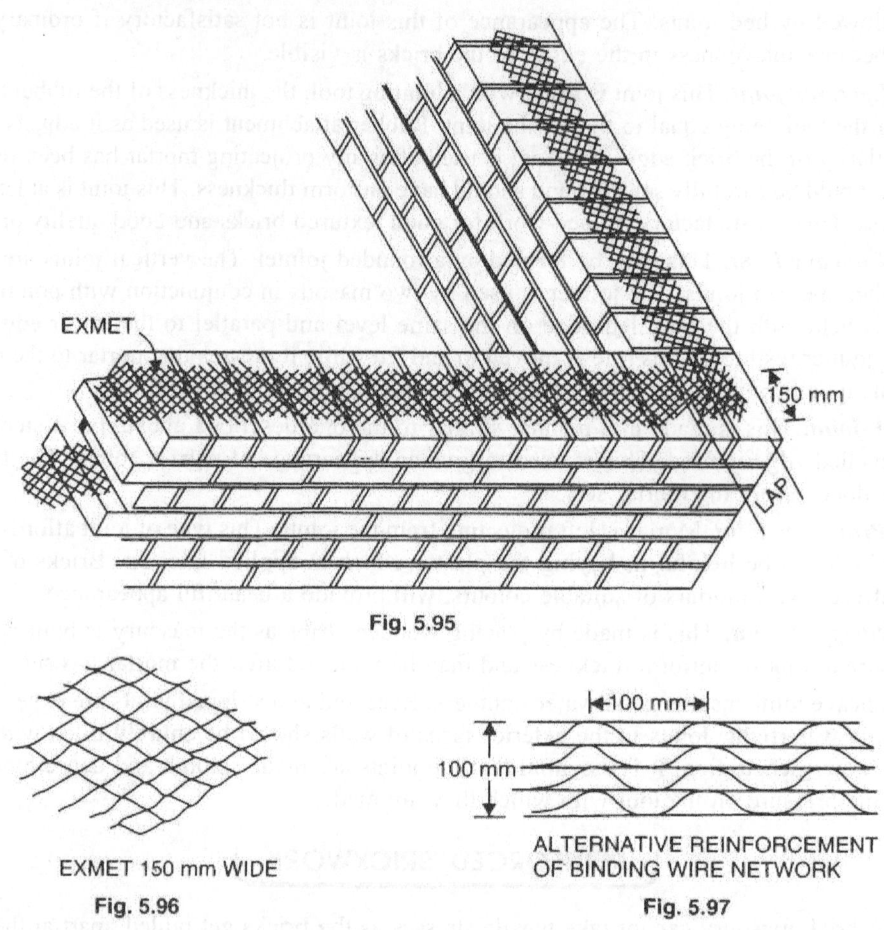

Fig. 5.95

Fig. 5.96

Fig. 5.97

Figs. 5.94 to 5.96: Reinforced brick wall details using mesh reinforcement.

Advantages of reinforced brickwork

(1) The simplicity of work or ease of construction.

(2) Fire-proof construction.

(3) Better bonded structures.

(4) Cheap, sound and permanent construction.

Types of reinforced brickwork

(1) ***Walls***: Reinforcement may consist of iron bars or expanded metal mesh. Standard patented expanded metal meshes are available in different widths and different gauges. This metal mesh is available in coils which can be spread while the brickwork is being laid. Mortar is evenly spread on the brickwork and the expanding metal (XPM) is spread on it. The next course of bricks is laid over it.

Another type of reinforcement which is used for walls is hoop iron. These are steel flats about 25 to 30 mm in width and are from 1.5 to 2.5 mm in thickness. These flats are dipped in molten tar to increase their resistance against rusting and are immediately sanded so as to increase the grip with the mortar. One strip is provided for half brick thickness of wall and every sixth course is reinforced. The joints at the corners are hooked as shown in Figs. 5.98 and 5.99.

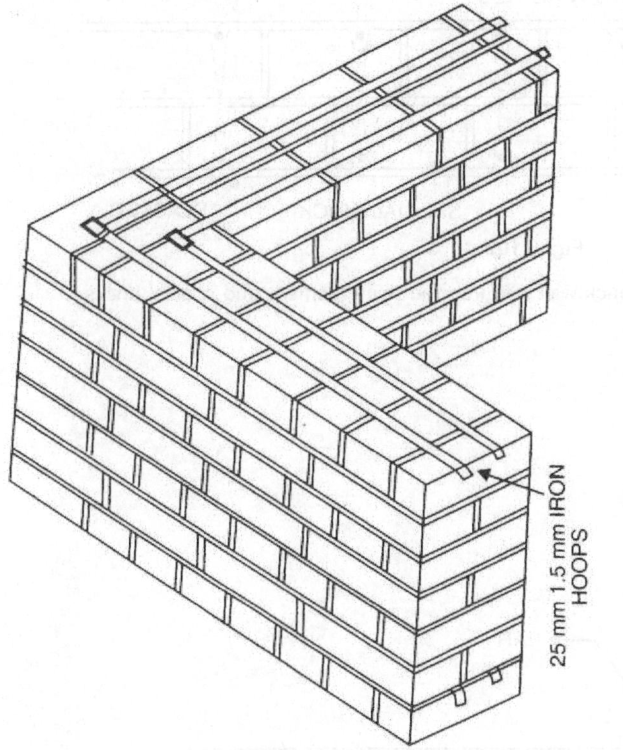

25 mm 1.5 mm IRON
HOOPS

Fig.5.98

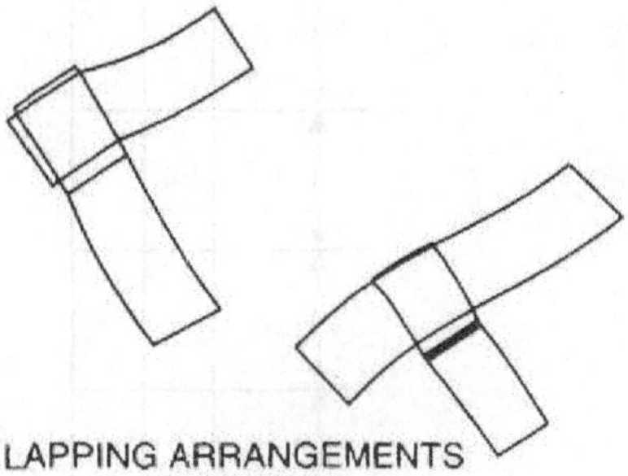

LAPPING ARRANGEMENTS
OF HOOPS

Fig.5.99

Figs. 5.98 and 5.99: Reinforced brick wall using hoop iron reinforcement.

SPECIAL BRICK

Fig. 5.100

ISOMETRIC VIEW

Fig. 5.101

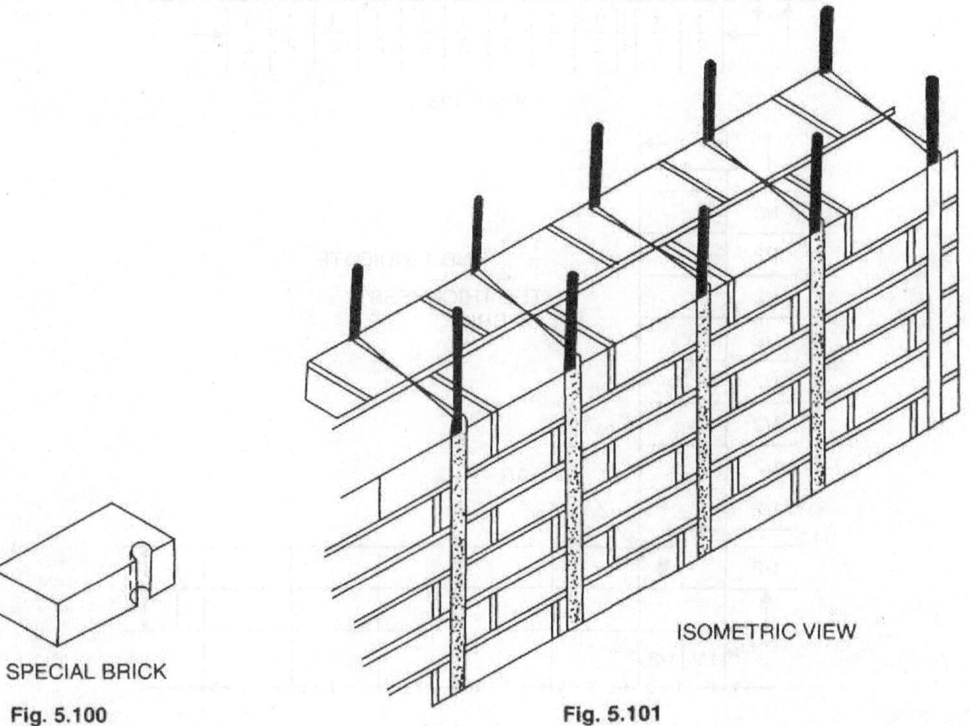

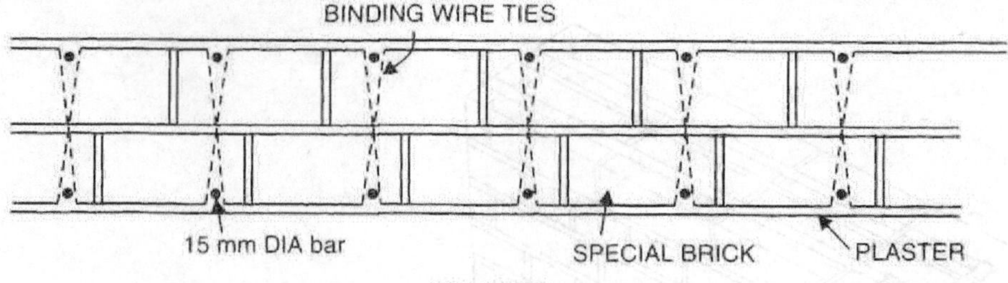

Fig. 5.102

Figs. 5.100 to 5.102: Reinforced brick wall with iron rod reinforcement and special bricks.

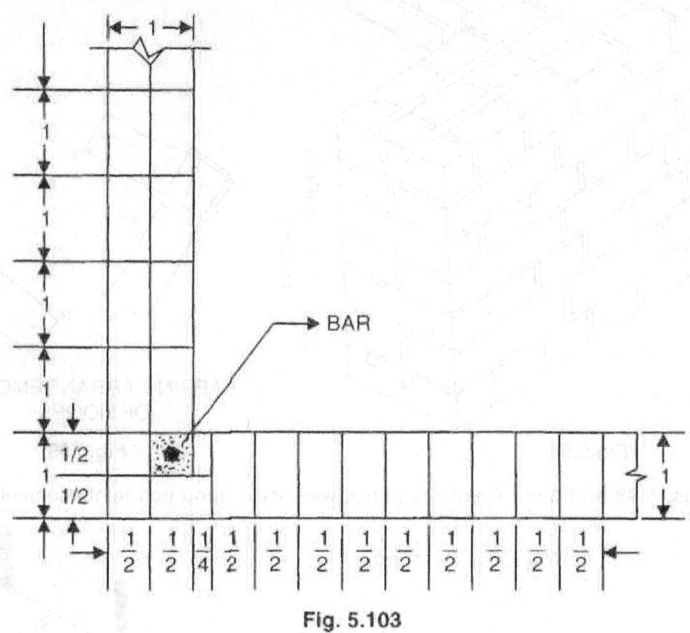

Fig. 5.103

$\frac{1}{4}$, $\frac{1}{2}$ AND 1 INDICATE THE THICKNESS OF BRICK

Fig. 5.104

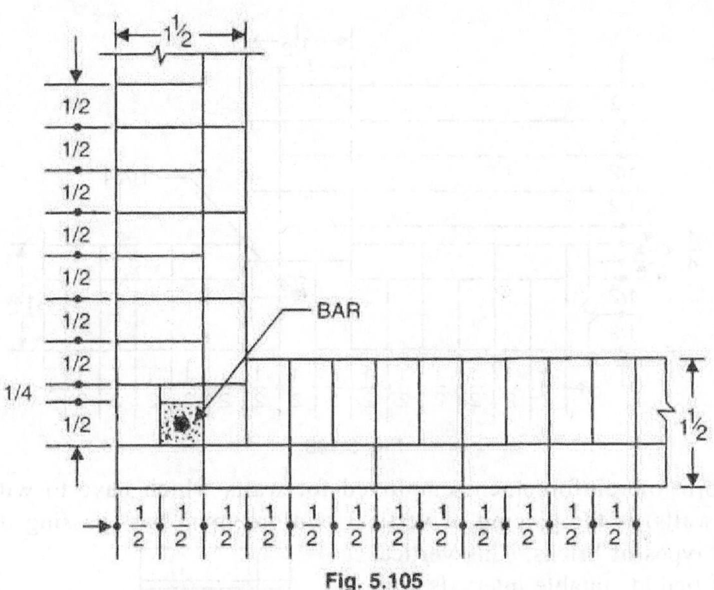

Fig. 5.105

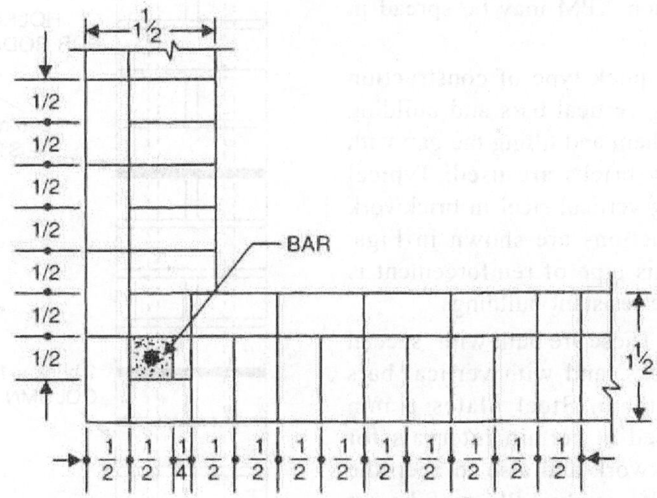

Fig. 5.106

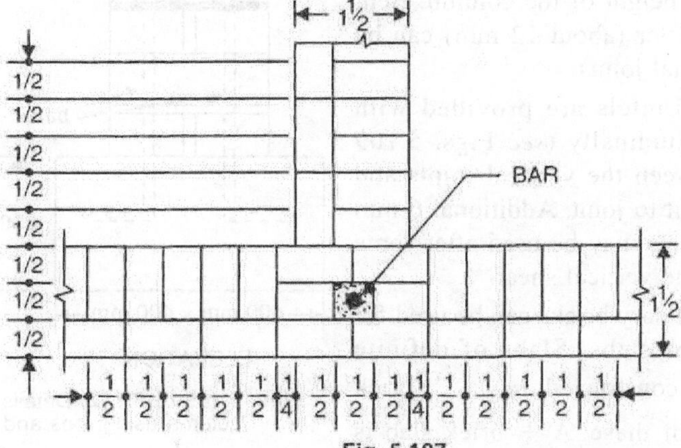

Fig. 5.107

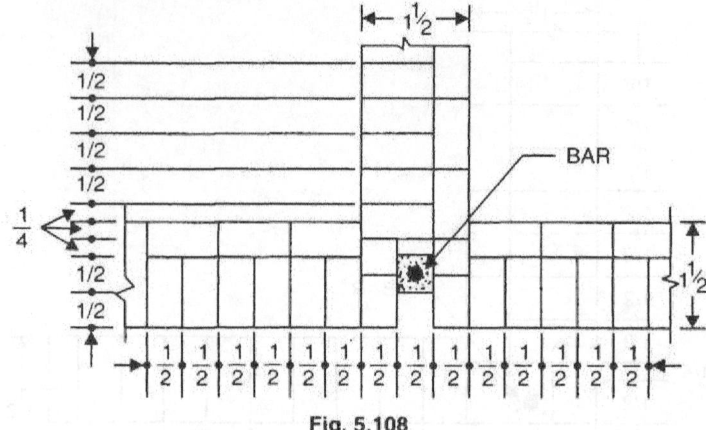

Fig. 5.108

Another form of reinforcement employed for walls which have to withstand pressure (*e.g.*, retaining walls) is the placing of vertical reinforcement bars passing through openings made in special types of bricks. This vertical reinforcement is tied at suitable intervals with fine wire. In addition XPM may be spread in certain courses.

A cheap and quick type of construction is done by erecting vertical bars and building brickwork around them and filling the gap with concrete. Ordinary bricks are used. Typical details of providing vertical steel in brickwork at corners and junctions are shown in Figs. 5.103 to 5.108. This type of reinforcement is used for earthquake resistant buildings.

(2) *Columns*: These are built with 'special purpose made bricks' and with vertical bars running through them. Steel plates 6 mm thick are introduced at certain intervals for strengthening brickwork and also to keep the bars in position. Alternately XPM may be cut to the shape of the pier and laid at suitable intervals along the height of the column. Bent bars of small diameter (about 12 mm) can be used in the horizontal joints.

(3) *Lintels*: Lintels are provided with bars running longitudinally (*see* Figs. 5.109 and 5.112) in between the vertical joints and extending from joint to joint. Additional 6 mm diameter steel stirrups may be used after some intervals to resist the vertical shear.

(4) *Slabs*: Ordinary bricks can be used for the construction of slabs. Slabs of definite depths can only be constructed, *i.e.*, $\frac{1}{2}$, 1 brick or a combination of these. A $\frac{1}{2}$ brick slab is

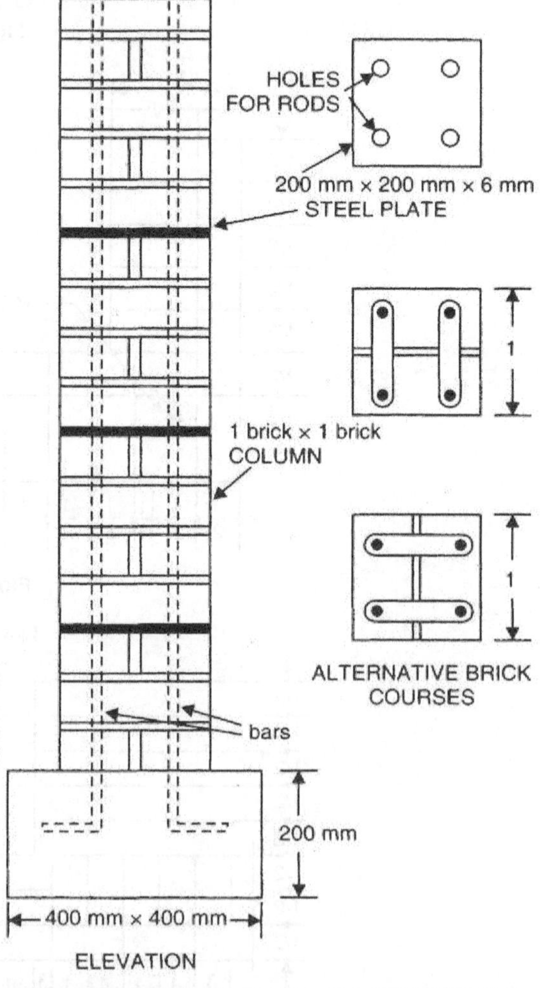

Figs. 5.109 to 5.112: Details of a reinforced brick column using rods and special bricks.

made by laying bricks flat or on an edge and 1 brick by laying two bricks one above the other on edge or flat. However, slabs of a thickness slightly greater than these are formed due to the thickness of the joints being greater than that of the ordinary masonry joints but while measuring the slab, it is usual to give thickness in terms of brick dimensions neglecting the thickness of the joint mortar. Similarly, the reinforcement is placed at intervals which form multiples of brick units.

For laying reinforced brickwork (R.B.) slabs, a centring consisting of supported platform is needed. This is covered with earth, well beaten and finished with sand. The centring is not removed till five to twenty-eight days have passed after the slab is laid, the exact period depends on the loads coming on the slab and the weather condition. In floors and roof slabs, reinforcement not greater than 12 mm in diameter is used. The ends of all rods are bent into semicircular hooks. Wherever lapping of reinforcement is necessary due to short lengths of steel available, a lap of at least 40 times the diameter of the bar should be given.

(5) *Beams*: The construction of R.B. beams is similar to that of R.B. Slabs. However, thick rods up to 20 mm diameter can be used in reinforcement.

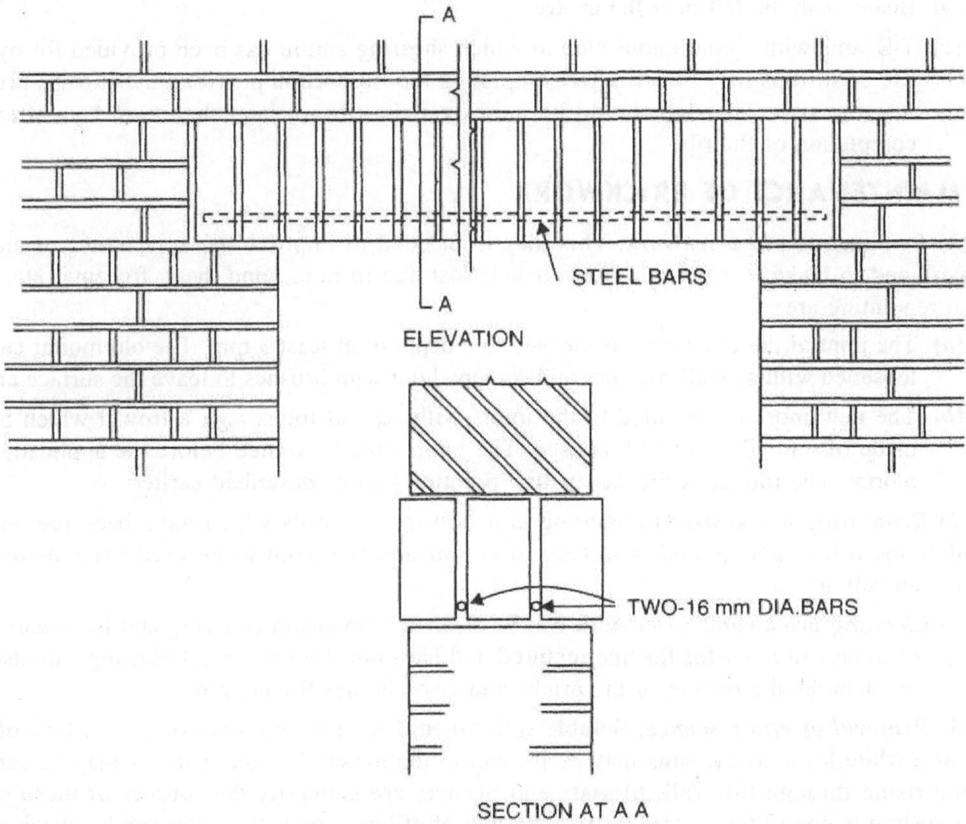

Fig. 5.113-114

Figs. 5.113 to 5.114: Typical reinforced brick lintel details.

Some important points regarding R.B. work are given below:

(1) Breaking Stresses:

Safe compressive stress in bricks	3 N/mm^2
Safe tensile stress in steel	150 N/mm^2

Ratio of the modulus of elasticity for steel to that of brick	40 : 1
Bond stress between steel and mortar	0.6 to 0.7 N/mm^2
Shear in brickwork	0.5 N/mm^2

(2) Before starting work, all rods should be bent to correct shapes and cut to proper sizes.

(3) All joints should be thoroughly filled with mortar and all reinforcement should be well surrounded by mortar. Care should be taken that the bottom rods in the slab have adequate cover of mortar and do not touch the centring.

(4) All work shall be kept moist by means of wet straw or wet sand for the first two days. It should then be profusely watered and kept wet till the centring is removed.

(5) As far as possible the structure should be finished in one day. However, if this is not possible, then the following rules may be kept as a guide:

(*a*) Ordinary slabs supported on two sides may be left after finishing any layer of reinforcement.

(*b*) Beams may be left near the centre.

(*c*) T-Beams with a continuous slab in which shearing action has been provided for by the use of stirrups may be left after completing the rib portion provided the stirrups project into the slab. The slab should be built over the rib not later than two days after the completion of the rib.

5.8 MAINTENANCE OF BRICKWORK

(1) ***Re-pointing old brickwork.*** This may be needed to improve the appearance of an old brickwork and to make it watertight. Mortar gets lost due to rain, wind, heat, freezing, etc. The steps in repointing are:

(*a*) The joint of the old mortar is cleaned to a depth of at least 3 mm. The old mortar can be loosened with a small hammer and scrapped out with brushes to leave the surface clean.

(*b*) The new mortar is applied to the joints with special tools, *e.g.*, a trowel which has a blade of 6 to 12 mm width is used. The joints may be wetted before the application of mortar. The mortar is finished with a pointing tool as described earlier.

(2) ***Repainting brickwork.*** Repainting is necessary for walls which have been previously painted as the bricks get spalled. The type of colour and the paint to be used depends on the treatment already given.

(3) ***Cleaning brickwork.*** Brickwork can be cleaned with steam or steam and hot water jets. This type of treatment is useful for fine textured and hard burnt bricks. Sand blasting can also be done but this disturbs the texture of the bricks and also abrades the pointing.

(4) ***Removal of efflorescence.*** Soluble salts when dried get deposited on the surface of the bricks as a white layer. These salts may be present in the material of the walls or may be carried by damp rising through the wall. Mortars and plasters are generally the sources of these salts. Since moisture is one of the causes for the creation of efflorescence, the latter can be checked by improving on the damp prevention of the building.

Efflorescence can be removed by scrubbing the wall with water and a stiff brush. If this is not successful, a 10% solution of murlatic acid may be used instead of plain water. However, the wall should be rinsed clean with water or with a weak ammonia solution immediately after this treatment.

5.9 JOINT BETWEEN OLD AND NEW MASONRY

While constructing new brickwork along with an old one, it is imperative that the new brick masonry is going to settle much more than the old brick masonry. Hence care must be taken to prevent the cracking and separation of the new masonry from the old one. This is accomplished by leaving bricks projecting from the old work and also in the new work. This is called toothing. Each alternate course projects into the other wall with sufficient lap and also maintains the proper bond. Sometimes these projections are formed of three or four courses with a similar distance in between. Further care can be taken by constructing the new wall in small heights so as not to allow undue settlements.

5.10 ORNAMENTAL BRICKWORK

The decorative effect in brickwork can be produced by the use of special types of bricks, mortar joints and different arrangements of bricks.

Bricks of different thicknesses can be used to give architectural treatment. Machine made bricks with sharp and angular faces lend a beautiful appearance to a wall. A thinner brick appears better in certain locations. Bricks which are coloured can be suitably arranged to produce a good pattern. The texture of the bricks is also important. Smooth faced bricks are preferred at places where dust-storms are prevalent whereas sandy textured bricks give a pleasing appearance.

The texture of joints and the colour of mortar has also to be adjusted in respect to the types of bricks and the treatment desired. Joints smoothened with a trowel should not be used for rough textured bricks for which flush or recessed joints are good. Black mortar should be avoided for face work. Generally mortar which does not present contrast with respect to the bricks should be used, *e.g.*, white mortar should not be used with red faced bricks. The horizontal and vertical joints can be made with different coloured mortars to get decorative effect. Flat joints give a plain surface whereas with recess joints deep shadows are produced which add to the ornamental effect at certain times. The thickness of the joint is also of importance. Thinner joints in certain bricks may be better than thick joints.

A combination of bricks, tiles and stones can produce a good appearance. Sandstone blocks with rough texture and creamy colour present a good appearance when fixed within brickwork at corners or ends of walls. Smooth polished stones cut into thin slabs can be used as a covering for garden walls, parapets, etc.

The arrangement of bricks will affect the appearance considerably. Bricks can be arranged into different designs and also can be left projecting wherever needed. A different texture for the bricks, which are left projecting, can be used.

Brick walls having a large area in elevation can be suitably decorated by the provision of decorative panels. Small bricks can be laid diagonally, vertically or projecting so as to give good patterns (*see* figures 5.115 and 5.116).

Quoins can be built of bricks which have a different colour or which are left projecting. The quoins are generally made darker than the remaining wall or are built of rusticated bricks. A similar treatment can be given to door and window openings.

Parapets are finished at top with bricks laid as headers on edge or two layers of bricks one of which is laid on edge and the other on end. Small tiles can be used to give a decorative appearance. As stated earlier, thin stone slabs can be used as a covering to the brick parapets.

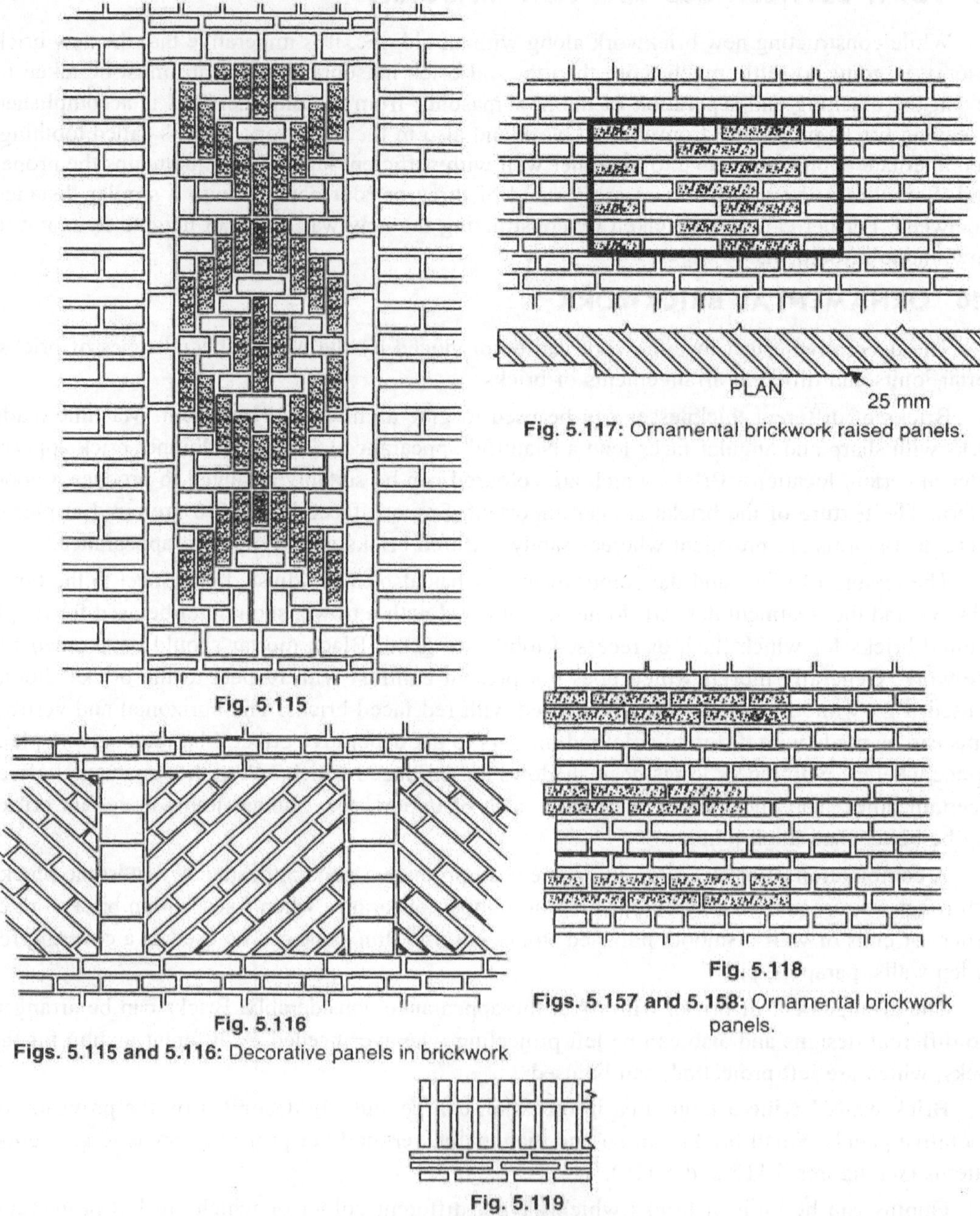

Fig. 5.115

Fig. 5.116

Figs. 5.115 and 5.116: Decorative panels in brickwork

Fig. 5.117: Ornamental brickwork raised panels.

Fig. 5.118

Figs. 5.157 and 5.158: Ornamental brickwork panels.

Fig. 5.119

Fig. 5.120

Figs. 5.119 and 5.120: Ornamental brickwork parapets.

5.11 BRICK KNOGGING

Brick knogging consists of a framework of timber in which bricks are laid and both sides are plastered. The upright posts are 100 mm × 100 mm size and are placed at 1 m centres. The horizontal members are 100 mm × 100 mm and 1.0 m vertically apart. All surfaces of timber coming in contact with the masonry are given a coating of hot coal tar. The bricks are laid in the openings of the framework and are placed in such a way that equal projections of timber are left

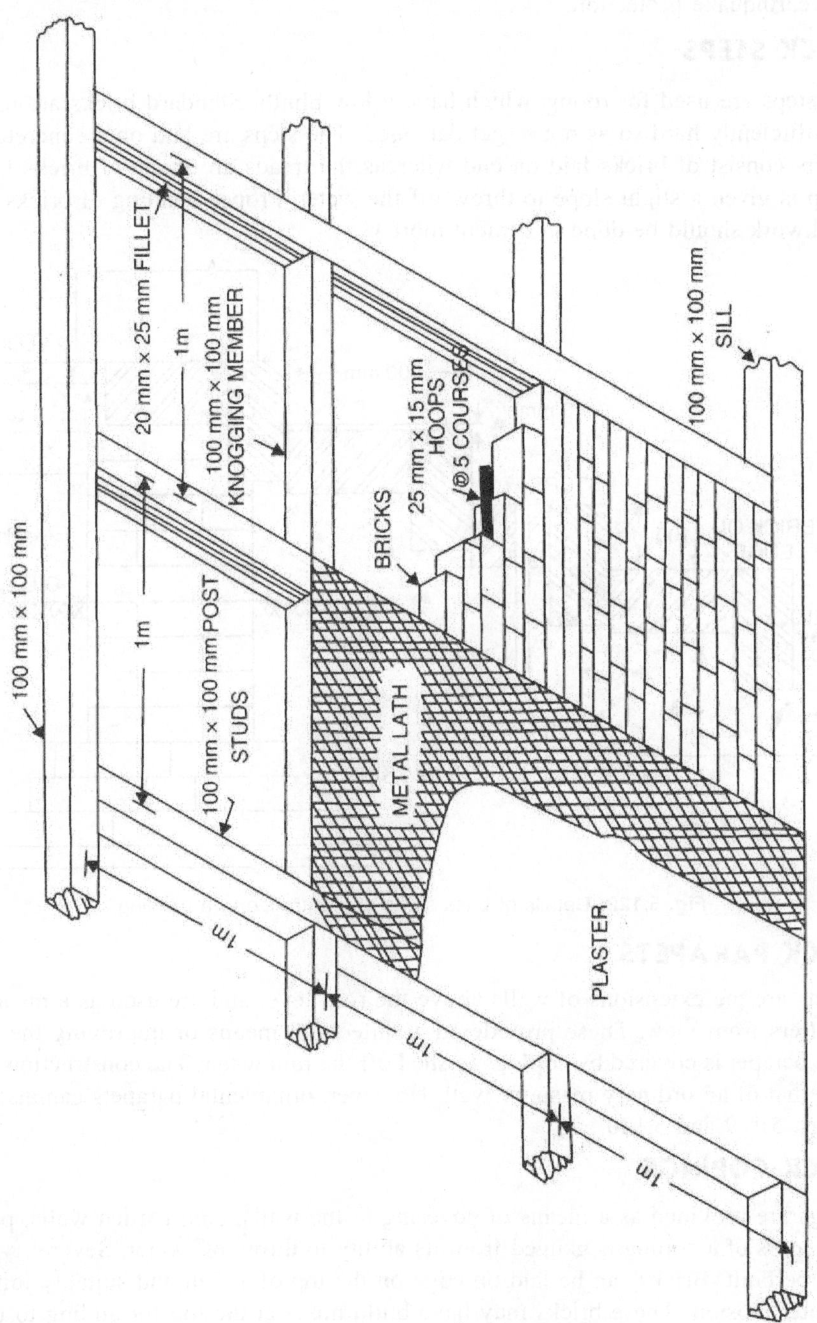

Fig. 5.121: Details of Brick-knogged partition.

on both sides. Reinforcement consisting of 25 mm × 15 mm hoop iron is provided after 4 to 5 courses. Both the sides of bricks are plastered so as to have the surface in flush with the timber. If wooden members are of shorter width and the entire area is plastered, the plaster peels off from the wooden members and falls down. In such a case a metal lath is fixed to both the sides of the knogging and the entire area is then plastered. To provide a proper hold for bricks with the vertical members of wood, small wooden sections of 20 mm × 10 mm size are fitted to the vertical members and the bricks are suitably notched to fit into them. This type of construction is adopted for earthquake protection.

5.12 BRICK STEPS

Brick steps are used for rooms which have a low plinth. Standard bricks are used but they should be sufficiently hard so as not to get damaged. The steps are laid on a concrete foundation and the risers consist of bricks laid on end whereas the treads are made of bricks laid on edge. The top step is given a slight slope to throw off the water. Proper bonding of bricks is essential. All the brickwork should be done in cement mortar.

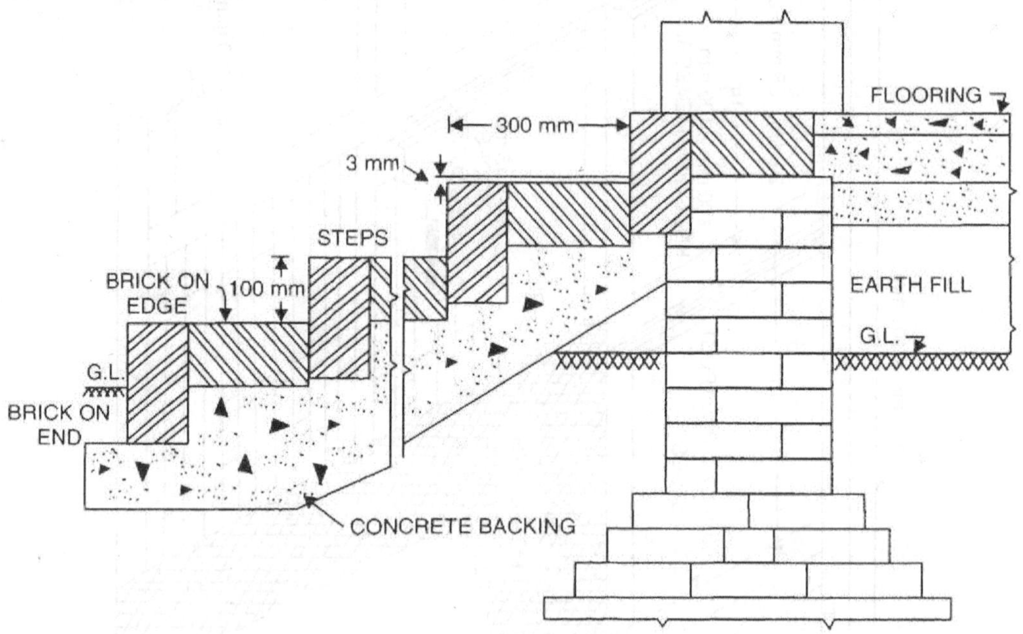

Fig. 5.122: Details of brick steps and entrance to a building.

5.13 BRICK PARAPETS

Parapets are the extensions of walls above the roof level and are used as a means of hiding the back gutters from view. These provide an architectural means of improving the appearance. The top of a parapet is covered by copings to shed off the rain water. The construction of a parapet is similar to that of an ordinary masonry wall. However, ornamental parapets can also be built as shown in figs. 5.119 and 5.120.

5.14 BRICK COPINGS

Copings are provided as a means of covering to the walls, *e.g.*, garden walls, parapets, etc. The effectiveness of a coping is gauged from its ability to throw off water. Several types of brick copings can be built. Bricks can be laid on edge on the top of a wall and suitably joined so as to stand the water erosion. These bricks may have build noses at the top for adding to their beauty. Specially made semi-circular bricks can be laid as coping on the bricks which are built slightly projecting from the wall. This type of coping can shed off the water easily. Tiles can be introduced in between the bricks on edge so as to drip the water easily. A saddle back brick can also be used as a coping.

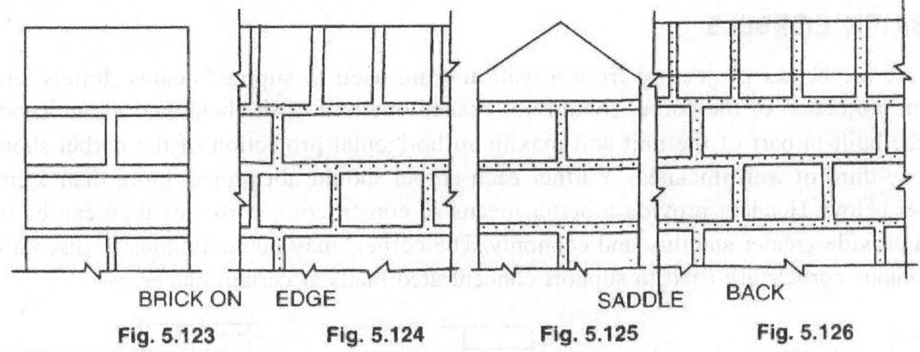

Fig. 5.123 Fig. 5.124 Fig. 5.125 Fig. 5.126

Figs. 5.123 to 5.126: Typical brick copings.

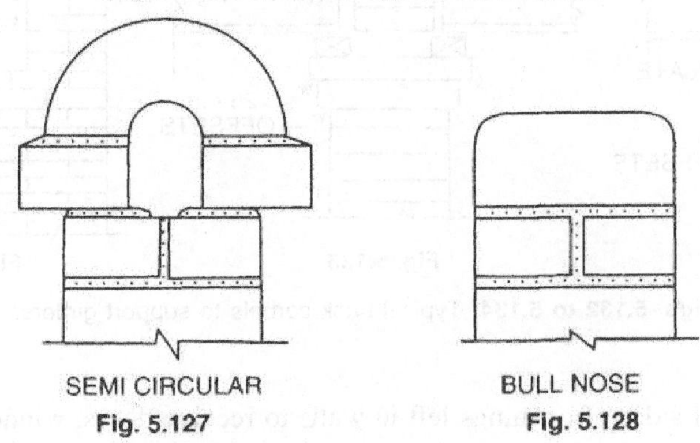

Fig. 5.127 Fig. 5.128

Figs. 5.127 and 5.128: Brick copings.

5.15 BRICK SILLS

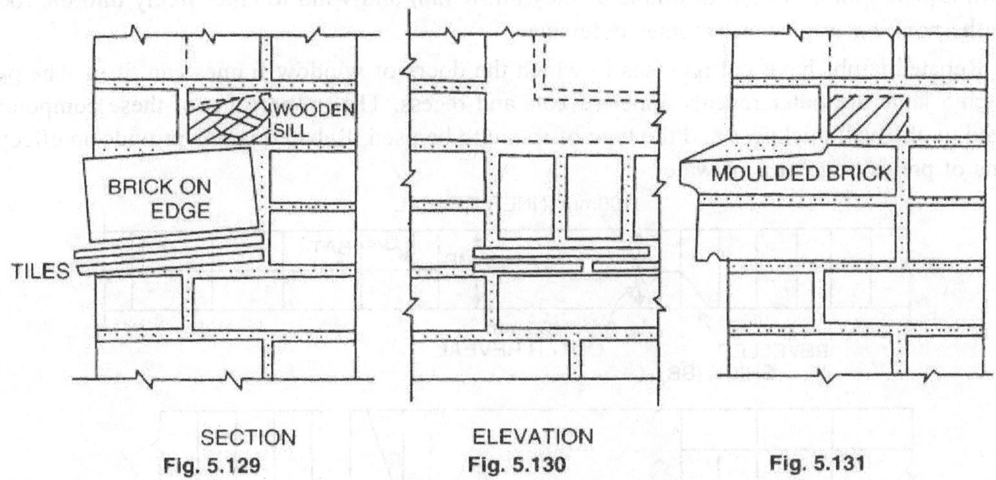

Fig. 5.129 Fig. 5.130 Fig. 5.131

Figs. 5.129 to 5.131: Typical brick sills.

The objectives of providing a window sill are to give a better appearance to the window opening and to afford protection to the wall below. Slight slope should be given to the top of the sill so as to shed off the water from the window. Bricks can be used for making window sills.

They are placed on edge and slightly tilted. Tiles may be introduced below them to improve the appearance. These tiles are left slightly projecting. As an alternative special purpose moulded bricks with suitable throatings can be used for brick sills.

5.16 BRICK CORBELS

These are bricks projecting from a wall and are used to support beams, lintels, etc. The maximum projection of the corbel should not exceed one-half of the height of the unit, nor one-half of the built-in-part of the unit and maximum horizontal projection of the corbel should not exceed one-third of wall thickness. Further each corbel should not project more than 5 cm from the corbel below. Headers provide a better means of constructing corbel as they can be bonded well and provide greater stability and economy. The corbels may be continuous or discontinuous. Discontinuous corbels are used to support concentrated loads at certain places.

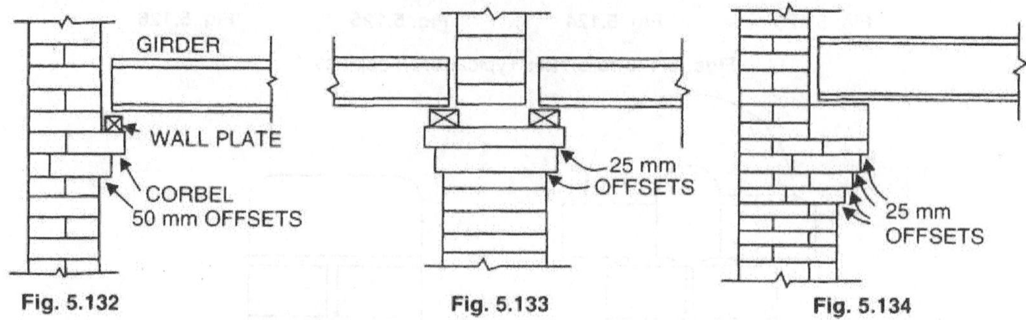

Figs. 5.132 to 5.134: Typical brick corbels to support girders.

5.17 JAMBS

These are vertical sides of openings left in walls to receive doors, windows, fire-places, etc.

The simplest type is to have a square jamb. This means that the wall is finished there as a dead end. The bricks have to be bonded suitably as in the case of dead ends (*see* Figs. 5.135 to 5.136). Square jambs are not desirable as they allow rain and wind to enter freely into the room, once the pointing or plaster becomes defective.

Rebated jambs have got recesses in which the doors or window frames can fit in. The parts of such a jamb are outer reveals, inner reveals and recess. The proportions of these components depend on the wall thickness and the type of frame to be used. Rebated jambs provide an effective means of preventing entry of water.

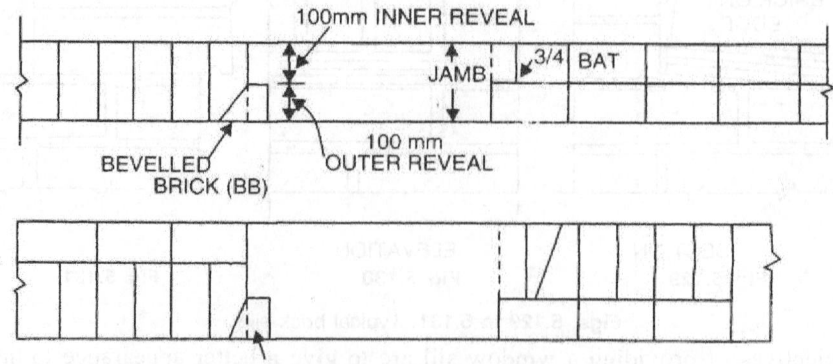

Figs. 5.135 and 5.136: Plans for alternate courses at jambs.

For better lighting of a room or to enhance the architectural beauty, splayed jambs with reveals are used. These are generally made in thick walls. The bonding can be of English or Flem-ish type and should be strong to withstand the forces of arches or lintels which span from jamb to jamb. The cut bricks should be avoided as far as possible. Recesses may be 50 mm or 100 mm in depth.

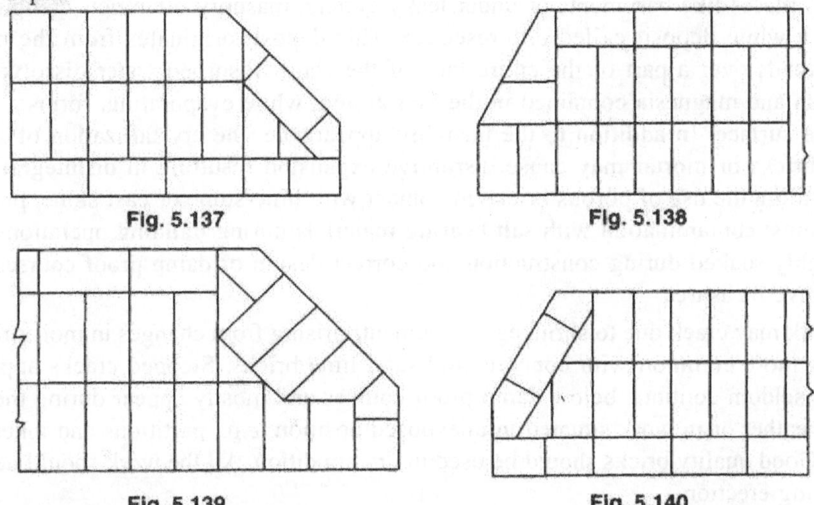

Fig. 5.137

Fig. 5.138

Fig. 5.139

Fig. 5.140

Figs. 5.137 to 5.140: Bonding details of alternate courses for splayed jambs.

5.18 DEFECTS IN BRICKWORK

Common defects occurring in brickwork are:

(*i*) Sulphate attack on mortars.

(*ii*) Use of unsound materials.

(*iii*) Frost action.

(*iv*) Corrosion of iron and steel.

(*v*) Crystallisation of salts.

(*vi*) Linear changes resulting from variation in moisture content.

Defects on account of sulphate attack lead to expansion of mortar thereby causing cracking of brickwork, spalling of brick edges, deterioration of mortar, wide horizontal and vertical cracks in the plaster and the falling off the plastered surface. The cause of this attack is the chemical action between the sulphate salts present in bricks and the aluminium constituents of Portland cement. This action is rapid in the presence of water and hence wherever moisture penetrates or excessive dampness occurs, this type of defect is prevalent. Prevention of moisture penetration will avoid this defect to a large extent. Bricks of low sulphate content and sulphate resistant cement in mortar should be used.

Unsound materials cause the formation of small pits with nodules of friable materials at the mortar joints. General expansion and cracking of brickwork is visible. Unsoundness in lime is caused by the presence of unslaked particles, which may be present in the mortar unless care is taken to effect complete slaking. Similarly, unslaked lime particles may be present in the bricks also.

Defects due to frost action would cause cracking in the brickwork. Prevention of water accumulation would go a long way in preventing this defect.

Brickwork may get opened or cracked and stained due to corrosion of metals lying adjacent to it. Unprotected iron and steel are liable to get corroded when acted upon by moisture and they increase in bulk thereby causing cracks in masonry. Protecting the metal surface with cement mortar, at least 10 mm to 20 mm thick is essential to prevent corrosion. Partially embedded steel or iron members should be surrounded with bituminous compound for portions not embedded in mortar.

Crystallization of salts or efflorescence is a prominent defect in brick masonry. In moist climate, damp places like basements or under leaky gutters, masonry often gets disfigured by the formation of a white deposit called efflorescence. This deposit originates from the mortar and frequently spreads over a part or the entire face of the wall. Absorbed water dissolves the salts of soda, potash and magnesia contained in the mortar and, while evaporating, forms a crystalline deposit on the surface. In addition to the unsighty appearance, the crystallization of salts in the pores of the bricks or mortar may cause disruptive expansion resulting in disintegration due to cracking. Avoiding the use of porous bricks in contact with lime-stone or cast-stone, protection of brickwork against contamination with salt bearing materials during building operations and from being thoroughly soaked during construction and correct design of damp proof courses are some of the preventive measures.

Brickwork may crack due to shrinkage movements arising from changes in moisture contents. This defect is more common with concrete and sand lime bricks. Stepped cracks appear on the surface. They seldom continue below damp proof courses and mostly appear during the first long spell or dry weather or in work situated in unexposed position, *e.g.*, partitions and inner leaves of cavity walls. Good quality bricks should be used in dry condition. All the work should be protected from rain during erection.

QUESTIONS

1. Briefly describe the various types of bricks used in brickwork. What are the different types of mortars used for brick masonry work?
2. Illustrate with sketches the meaning conveyed by the following terms when referred to brickwork:
   (*i*) Headers; (*ii*) Stretchers;
   (*iii*) Closer; (*iv*) Bond;
   (*v*) Perpends; (*vi*) Toothing.
3. What is the significance of bonding in brickwork? Explain by sketches the difference between English bond, Double Flemish bond and Single Flemish bond.
4. What are the advantages and disadvantages of English bond over Flemish bond?
   Draw sketches to illustrate the layout of bricks in alternate courses of a right angled corner of 2 brick walls in English Bond.
5. Write short notes on:
   (*i*) Stretcher bond; (*ii*) Header bond;
   (*iii*) Single Flemish bond; (*iv*) English Garden wall bond;
6. Draw to a suitable scale the arrangement of various bricks in alternate courses of:
   (*a*) A right angled corner of 2 brick wall with $1\frac{1}{2}$ brick wall in English bond.
   (*b*) A right angled corner of a $1\frac{1}{2}$ brick wall with another $1\frac{1}{2}$ brick wall in Double Flemish bond.
7. Briefly describe with sketches the construction of a $1\frac{1}{2}$ brick square and 2 brick square column in English and Flemish bonds.
8. Write short notes on:
   (*i*) T-junctions; (*ii*) Intersections;
   (*iii*) Jambs; (*iv*) A 60° Squint junction.

Draw typical layout for bonding arrangements using one and a half brick wall thicknesses in all the above cases in English bond.

9. What are the characteristics of an efficient fire-place? Draw a typical elevation, section and plan of a fire-place in a living room of a building. Indicate also as to how the flues of two-three fire-places at different floors can be accommodated in one chimney breast.
10. Write a short note on the various types of joints used in brick work. Illustrate with sketches.
11. What are the advantages of reinforced brick work construction? Write short notes on:
    (*i*) Reinforced brick walls; (*ii*) Reinforced brick lintels.
12. Write short notes on:
    (*i*) Brick knogged partitions; (*ii*) Brick copings;
    (*iii*) Jambs; (*iv*) Brick steps.
13. What are the commonly observed defects in brickwork? What preventive measures can be taken in each case?

REFERENCES

1. **Braley Lindsay, E.**: *Brickwork.*
2. **Kay, N. W.**: *Modern Building Encyclopaedia.*
3. **Dalzeil and Tonnsed**: *Masonry Simplified,* Vol. I.
4. **National Buildings Organisation**: *Brick and Brickwork.*
5. **Capital Project Chandigarh**: *Building Byelaws.*
6. **Specification Bombay P.W.D.** Vol. 1.
7. **Specifications Punjab P.W.D.**
8. **Marbs**; *Chimneys and Flues.*
9. **Mitchell, S.A.**: *Building Construction,* Vol. I.
10. **Mackay, W.B.**: *Building Construction,* Vol. I.
11. **National Buildings Organisation**: *Masonry Mortars.*
12. **Mulligm, J.A.**: *Handbook of Brick Masonry Construction.*
13. **Mackay, W.B.**: *Brickwork.*
14. **Dorman, E.G. and Elomes, E.J.**: *Brickwork.*
15. **Lane Publishing Co.** *How to Plan and Build Your Fire-places.*
16. **National Building Code of India 2005; Part VI Section 4**: *Masonry.*
17. **IS: 1077–1992**; *Specification for common burnt clay building bricks (fifth revision).*
18. **IS: 2212–1991**; *Code of practice for brickwork (first revision).*
19. **IS: 2180–1988**; *Specification for heavy duty burnt clay building bricks (third revision).*
20. **IS: 10440:1983**; *Code of practice for construction of RB and RBC floors and roofs.*

Stone Masonry

Stone masonry is used for the construction of walls, columns, lintels, arches, footings, beams, etc., of a building. Stones are abundantly available in nature and when cut and dressed to proper shapes, they provide an economical material for the construction of various parts of a building. However, stones are not available in all parts of India and are costly to work into different shapes. It is also difficult to handle stones because of their heavy weight. Generally, wherever stones are available, bricks cannot be manufactured economically and *vice versa*. Hence stone masonry is an integral feature of buildings located in hilly areas. Stone masonry also provides a good architectural look. Stones are also used for the construction of monumental structures.

6.1 MATERIALS USED FOR STONE MASONRY

The materials used for stone masonry are stones and mortar.

The common types of stones available in India for stone masonry are:

(1) ***Granites***: These are the hardest types of stones and are difficult to work with. They are available in various colours ranging from white to green. These are used for the construction of steps, walls, sills and as facing over other masonry. They are found in Dharwar, Tamil Nadu and Kashmir.

(2) ***Sandstones***: They are made up of quartz cemented by a matrix of silica. They also contain mica, felspar and oxides of iron. The colour of sandstones is due to the presence of other minerals in them. They can be worked easily to take any ornamental shape. Their texture being coarse, they give a good appearance when used along with brick masonry. Coloured sandstones are used in the facework of buildings to give architectural treatment. They are used for walls, columns, facings, steps, flooring, etc. They are mostly found in Maharashtra State, Rajasthan and Andhra Pradesh.

(3) ***Limestones***: These are calcarious rocks and consist of carbonate of lime. They are available in various colours and are easy to work with. They are used for walls, floors, steps, etc. However, they are liable to get disfigured in acidic atmosphere.

(4) ***Marbles***: These are very useful material for flooring and monumental structures. Marbles are available in various colours and can take very good polish. These are available in Rajasthan and Maharashtra.

(5) ***Slates***: These are available in hilly areas and are metamorphic rocks. Generally they have a black colour. Slates can be split in thin sheets along their bedding planes. They are mostly used for roofing work.

The properties of stones which are important for stone masonry are strength and durability. Economy and appearance are additional requirements. The main considerations for durability are the lasting qualities of the stone itself and the locality where it is to be used. Porous stones are unsuitable for areas prone to heavy rainfall and frost. Stones, *e.g.*, marbles having low porosity and low coefficients of expansion and contraction should be used in areas subjected to large variations in rainfall and temperatures.

Generally lime and cement mortars are used for stone-masonry. Their function is to initially provide a workable matrix and ultimately a hard binding material, which renders masonry into a monolithic unit. Hydraulic lime and Portland cement are used as cementing materials and natural or river sand as fine aggregate. Sand serves a twofold purpose, firstly, it is a filler material so economy is obtained and secondly, it adds to density thereby adding strength and also reducing shrinkage.

6.2 DEFECTS IN STONES

Before using a stone in a building, the following defects should be checked:

(*a*) ***Vents***: These are small fissures or hollows in the stone which render it liable to decomposition.

(*b*) ***Shakes***: These are minute cracks containing calcite which form hard veins. As the stone weathers, these veins being stronger, project beyond the surface of the stone and render it a bad appearance.

(*c*) ***Sand holes and clay holes***: These are cracks or holes filled with sandy or clayey matter. They are liable to decompose when subjected to weathering action.

(*d*) ***Mottle***: Stones have spottled appearance due to the presence of chalky substances. Such stones are not fit for use as building material.

(*e*) ***Shelly bars***: These are small fossils or shells and when found in excess, render the stone liable to decay.

6.3 TERMS USED IN STONE MASONRY

(1) ***Bed surface***: The surface of a stone perpendicular to the line of pressure.

(2) ***Bedding planes***: The planes which separate the different layers of stone and are at right angles to the pressure which acts on the stone mass during its formation in nature.

(3) ***Ashlar masonry***: Masonry composed of rectangular units usually larger in size than a brick. The stones are properly bonded, sawn, dressed and have proper thin mortar joints.

(4) ***Rubble masonry***: Any stone work which is not highly finished.

(5) ***String course***: This is a continuous horizontal course of masonry projecting from the face of the wall and is intended to throw-off water.

(6) ***Corbel***: This is a piece of stone projecting from a wall to support a structural member.

(7) ***Cornice***: This is a moulded course of masonry having large projections. It is used for throwing off water.

(8) ***Block in-course***: This is a course of stones erected on the cornice. The object is to hold down the cornice and also to give emphasis to this part of the structure.

(9) ***Cramp***: This is a slate or metal connection used in stone work.

(10) ***Diaper-work***: Surfaces and panels in stone work formed by the use of square stones in an ornamental manner is called diaper work.

(11) ***Drip-stone***: A projected stone moulding having a throated under surface to throw water off the walls.

(12) ***Grouting***: Spreading and working of mortar over the stones to fill up any interstices between them.

(13) ***Mouldings***: These are ornamental features given to constructional members to enhance their aesthetic effect.

(14) ***Templates***: Pieces of stones placed under the end of a beam to distribute the load over a greater area.

(15) ***Throating***: These are grooves cut on the under-surfaces of sills, copings, string courses, etc., to prevent the water from trickling down the walls.

(16) ***Through-stone***: These are stone extending throughout the thickness of the wall. They act as bond stones and increase the stability of the wall.

6.4 CUTTING AND DRESSING OF STONES

Stones, found in nature, have to be quarried from their thick beds. After quarrying large pieces of stones, it is essential to break them into smaller sizes so that they can be used in a building. They are also dressed into suitable shapes and polished to give a smooth surface, if desired. The stones are used in different types of masonry described subsequently and have, therefore, to be cut and shaped to fit in the type of work needed. Various type of finishes and the methods of dressing and cutting the stones to get the desired surfaces are described below:

(1) ***Scabbling***: Irregular edges of the stones are broken off and the stone is shaped somewhat. This work is generally done in a quarry and the edges are broken with a scabbling hammer.

(2) ***Hammer dressed***: Large raised portions of the stones are cut and the stone is made somewhat flat but rough due to hammer marks. These stone blocks are squared somewhat and the bed and vertical sides are dressed to a distance of 80 to 100 mm from the face. This is done to enable the stone to have proper joints. This work is done by the use of Waller's hammer. The types of stones so obtained are termed hammer faced, quarry faced or rustic faced (*see* Figs. 6.1 to 6.6).

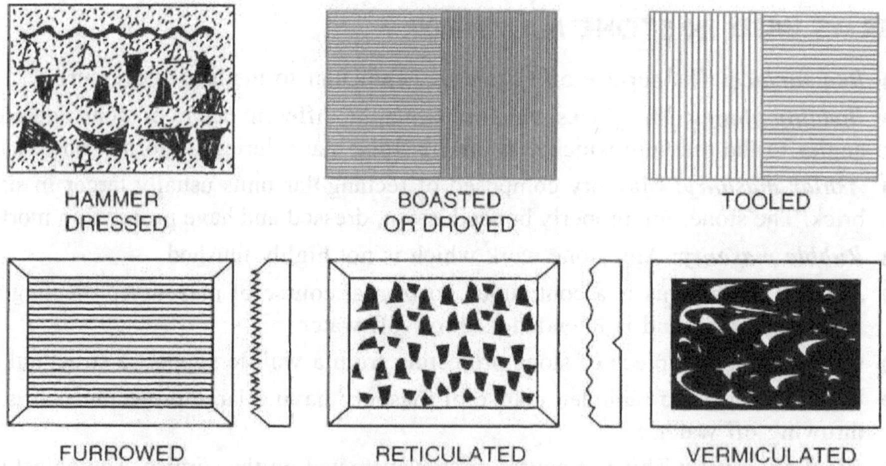

Figs. 6.1 to 6.6: Finishes for stonework

(3) ***Boasted or droved finish***: The stone is cut to a little level face and is finished by means of a boaster (chisel with an edge of 60 mm width). These marks may be horizontal or at an angle. The chisel marks are not continuous across the whole width of the stone.

(4) ***Tooled finish***: In this case the chisel marks are continuous and parallel throughout the width of the stone. These are also deeper than the ones described above.

(5) ***Furrowed finish***: In this case, about 10 mm vertical or horizontal grooves are sunk with a chisel having its end shaped as a hollow semi-circle (gauge). The sides of the rock are kept either chamfered or sunk.

(6) ***Reticulated finish***: In this type of work, irregular shaped sinkings are made within the central portion of the stone having a 20 mm wide margin on its sides. These sinkings are about 6 mm deep. The margin around the sinkings is of constant width. The sunk surfaces may have pinched marks to give a better appearance.

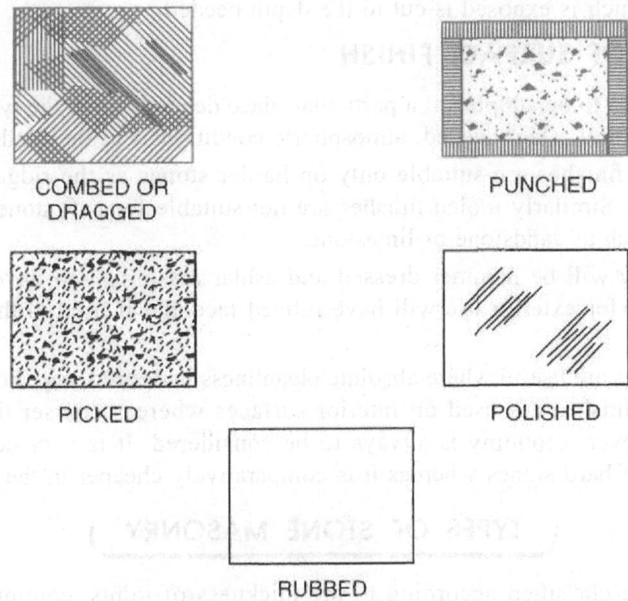

Figs. 6.7 to 6.11: Finishes for stonework

(7) ***Vermiculated finish***: The sinkings are of the reticulated type except that they are more curved and give a worm eaten type of appearance. These types are not very common as they need lot of labour in construction.

(8) ***Combed or dragged finish***: This type of finish is done on soft stones. A comb is driven over the surface of this stone to remove all elevating portions.

(9) ***Punched finish***: Depressions are formed on the rough surface with a punch. A series of ridges and hollows are produced.

(10) ***Picked finish***: This type of finish is obtained by dressing stones with a point and the depressions are smaller than the above type.

(11) ***Chisel drafted margins***: They give a better appearance and help in getting uniform joints. The margins can be chamfered, squared or pitched. They are given a boasted or a tooled finish. Each quoin stone must have drafts at the corners so as to help in putting the stones in plumb.

(12) ***Moulded finish***: Mouldings of various types can be worked on the stones to improve their appearance. These are either hand made or machine made. Various types of mouldings used are described later on.

(13) ***Rubbed finish***: The surfaces of the stones are rubbed to get a smoother finish. One piece of stone is rubbed against the other. Water and sand are added to aid the operation. Machine rubbed stones are also available.

(14) ***Polished surfaces***: Stones which can take polish, *e.g.*, granites, marbles, lime-stones are first rubbed to a smooth surface and then polished by using rubber and pad, sand and water, pumice, and putty powder. However, machines can also be used for polishing.

(15) ***Sand Blasting***: This is done to imprint letterings and designs on the surface of granites. The polished surface is coated with a molten rubber like compound which solidifies on cooling. The desired design is cut on this coating with a sharp tool thereby exposing the stone surface which is to be cut. A blast of sand is then blown with compressed air, the part which is exposed is cut to the depth needed.

6.5 SELECTION OF SURFACE FINISH

The type of finish to be adopted at a particular place depends upon the type of masonry, type of stone, the architectural effect needed, atmospheric conditions and the funds available.

The hammered finishes are suitable only on harder stones as the ridges will not stand, if made of a soft stone. Similarly tooled finishes are not suitable for soft stones. Polishing cannot be done on stones such as sandstone or limestone.

Rubble masonry will be hammer dressed and ashlar masonry will have smooth or rubbed faces. Marble if used for exterior use will have rubbed face and if used in the interior will have a polished surface.

Polished surfaces are useful where absolute cleanliness is desired, *e.g.*, hospitals, bath rooms, etc. Similarly finer finishes are used on interior surfaces whereas coarser finishes are used on external faces. However, economy is always to be considered. It is very costly to get smooth finishes in the case of hard stones whereas it is comparatively cheaper in the case of soft stones.

TYPES OF STONE MASONRY

Masonry can be classified according to the thickness of joints, continuity of courses and finish of face.

Broadly speaking there are two types of stone masonry, namely:

(1) ***Rubble Masonry***: This consists of blocks of stones either undressed or roughly dressed and having wider joints.

(2) ***Ashlar Masonry***: This is built of stones carefully dressed and has narrow joints.

The various sub-divisions of these two types are:

Rubble Masonry
- (*a*) Random Rubble:
  - (*i*) Uncoursed
  - (*ii*) Coursed
- (*b*) Squared Rubble:
  - (*i*) Uncoursed
  - (*ii*) Coursed
  - (*iii*) Built to regular courses
- (*c*) Miscellaneous:
  - (*i*) Polygonal
  - (*ii*) Flint

Ashlar Masonry
- (*a*) Ashlar fine
- (*b*) Ashlar rough tooled
- (*c*) Ashlar rock or quarry faced
- (*d*) Ashlar chamfered
- (*e*) Ashlar facing
- (*f*) Ashlar block-in-course

6.6 RUBBLE MASONRY

The stones in this type of masonry are pyramidal to some extent. They are not of uniform size and shape and are not finely finished. Proper bonding vertically as well as transversely is to be attained. Through stones have to be laid from back to face of wall or for at least more than half the width of the wall. Following points may be noted for all types of rubble masonry.

(1) All stones should be wetted before being laid in masonry.

(2) The stones on the face should have full joints for a specified distance from the face.

(3) Stones from opposite faces should bond with each other, as far as possible.

(4) The backing should have sufficient bond with the facing. Building of backing and facing separately and the filling of intermediate space with small stones is objectionable.

(5) There should be sufficient headers in each course and they should be embedded into the walls for sufficient depth.

(6) The filling may not necessarily be brought to the level of each course but slightly thicker stones than the height of the courses may be allowed.

(7) The height of the stone should never exceed its least horizontal dimension.

(8) Stones should be placed on their widest side so that they may not act as wedges and force out the adjacent stone.

(9) Feather edged stones with insufficient tails should not be used.

(10) The use of chips should not be allowed in bed joints for setting the stones.

(11) The width of the face stone should not be less than the height of the course.

(12) Corner or quoin stone in coursed rubble work should be of the same height as the course. In order that the vertical joint may break, one side of the corner stone is made longer than the other by about 150 mm and the corner stones in alternate courses are placed with their longer sides on the same exposed face. The beds of corner stones must be dressed to a depth of 100 mm and each corner stone should be of such dimensions so as to have the maximum bearing into the wall.

6.6.1 Random Rubble Masonry

Uncoursed: Uncoursed rubble masonry is built practically without any dressing. The stones used vary very much in shape and size. The mason selects the stones at random from a heap and lays them to form a strong bond. Prior to laying, all projecting corners are slightly knocked off with a hammer, if necessary. Larger stones are laid on flat beds and in exceptional cases may be supported on small stones but not on chips. Vertical joints are not formed to plumbness. The joints are filled and finished flush. They vary in width from stone to stone. Large stones are used at corners and jambs to increase strength. Through stones are also provided at intervals to further increase the strength. One through stone is used for every square metre of face work. They are at least 40000 sq mm in area and run to a depth of at least half metre or to the full depth if it is less than half metre. This type of masonry affords a very rough appearance.

If the wall is greater than half metre in thickness, a line of headers overlapping each other by at least 150 mm should be laid through the wall. The backing stones should be of large size and 33% of them should exceed 1/40th cubic metre in content. The quions are of selected stones, hammer dressed to the required angle and laid header and stretcher alternately. No quoin should be less than 1/10th of a cu. m. in content.

Coursed: In this type, the stones are laid to somewhat level courses. In each course, headers of full course height and consisting of hammer dressed stones are placed at certain intervals. Each header has a width not less than its height and projects into the wall at least three times its height. Between the headers, each course is built with smaller stones which are not less than

50 mm in thickness. They may be placed two or three in number, one above the other, in each course. These stones need not be dressed but should have flat beds. The side joints need not be vertical but should also not form an angle greater than 60° with the bed joints. It should be aimed to break the joints in different courses. Joints may be about 15 mm in thickness. In wall of thicknesses up to half metre, headers should run from face to face and act as through stones. In thicker walls, headers from the face and the back should overlap each other by at least 150 mm. The headers of successive courses should not be placed one above the other. The quoins are to be of the same height as that of the course and are about half metre long. They are laid as stretchers and headers alternately. All the stones are to be set in mortar. This type of work is used for small buildings where the walls are not very high.

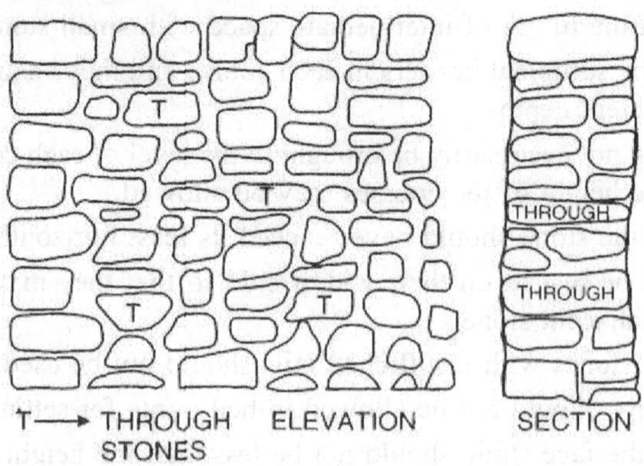

Fig. 6.12 and 6.13: Random rubble uncoursed

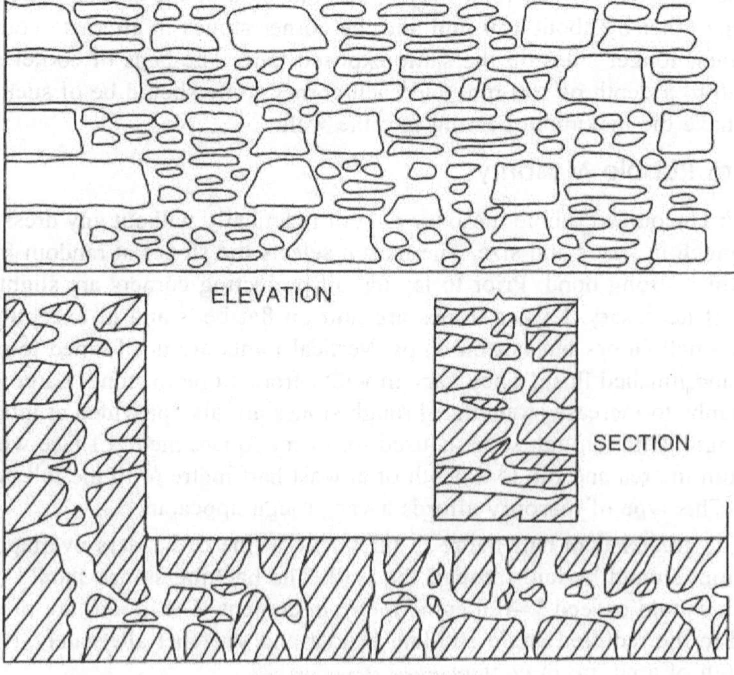

Figs. 6.14 to 6.16: Random rubble coursed. (Elevation, plan of a course and a section)

6.6.2 Squared Rubble

These stones are usually squared and given a hammer dressed finish. They can also have tooled or dragged surface finish.

Uncoursed: The stones are laid in irregular patterns. They are squared on all joints and beds. The beds should be hammer or chisel dressed, true and square for at least 100 mm from the face whereas the joints are finished similarly for at least 50 mm. The faces of the stones should not project more than 500 mm from edges. A good appearance is produced by arranging the stones in series, *i.e.*, a large stone (riser), two thin stones (levellers) and one small stone (check). Most of the joints should break vertically. Bigger sized stones should be used as quoins. Chips should not be used for bedding the stones.

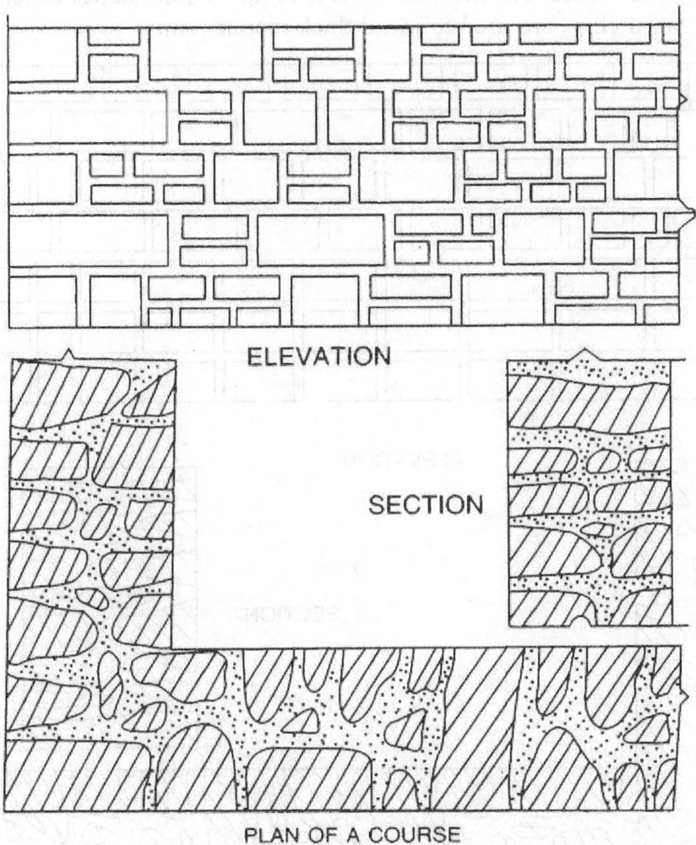

Figs. 6.17 to 6.19: Squared rubble coursed

Coursed: This type of masonry is made up of a facing comprising of hammer squared stones with a backing of rubble masonry and bonded to the former. The stones in each course need not be of the same height. Not more than two stones, one above the other are used in each course. All the courses should be of the same height. However, the lower courses should not be thinner than the upper courses. The stones should break joints with those above them. The hearting and backing should be brought up simultaneously. The hearting is made up of flat bedded stones laid carefully on their beds. Small chips may be used whenever the joint is very thick. However no dry hollow spaces should be left. The quoins should be of stones equal in depth to the height of the course and should at least be half metre long. They are laid alternately as headers and stretchers.

Built to regular courses: The stones are laid in horizontal layers not less than 150 mm in height. Stones are of equal height in a course. No course is thicker than the course below it. The faces of the stones are squared on all joints and beds. The beds are hammered or chisel dressed to a depth of at least 100 mm from the face and the joints to at least 50 mm from the face. The projections do not exceed 50 mm. Stones are not lesser in breadth than height. They fit in the wall to a depth not less than the height and at least 1/3rd of the stones tail into the wall to at least twice their height. Through stones are used at 2 metres apart in every course and extend from one face to the other face of the walls if they are not greater than half metre in thickness. For thicker walls headers should run from back to face with an overlap of at least 150 mm. A header should have a length of at least 3 times the height. The quoins are of the same height as the courses and are at least half metre in length. They are laid as headers and stretchers. Stones are laid in such a manner so as to break the joints vertically. At least half the height of the course is kept as the overlap between the two stones. The hearting is made of flat bedded stones which are set carefully in mortar. Smaller sized chips are used to avoid thick mortar joints.

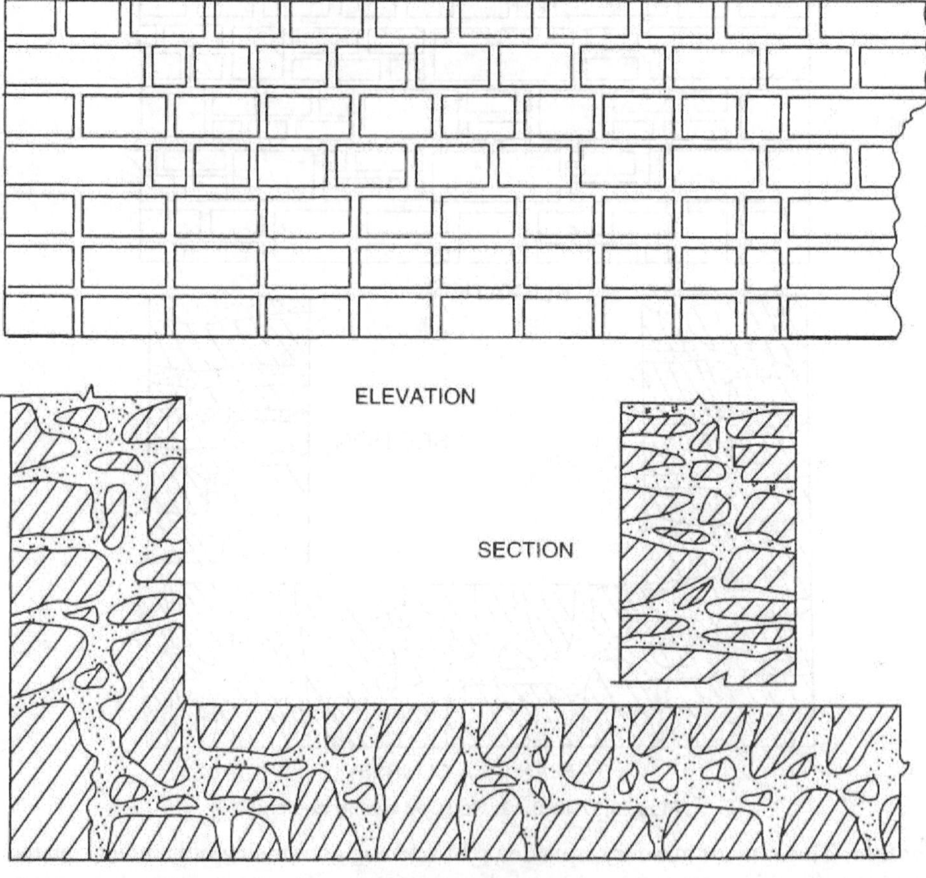

Figs. 6.20 to 6.22: Squared rubble (built to regular courses)

6.6.3 Miscellaneous Types

Polygonal Walling: The stones are roughly dressed to an irregular polygonal shape. Each stone is dressed carefully to form neat and uniform joints. The stones are laid carefully in mortar and tail backed to bond well with the backing. They are not of greater length than the breadth of the face or the length at right angles to the face. One header or through stone is inserted to run through the wall if the wall is less than half metre in thickness or else headers are laid from the face and the back overlapping each other by at least 150 mm. The stones are arranged to break joints as much as possible and long vertical joints should be avoided in face work. Small chips should not be used to support the stones on the face. If the stones are not finished at their edges properly, joints of greater thicknesses are seen. This type of work is slightly inferior.

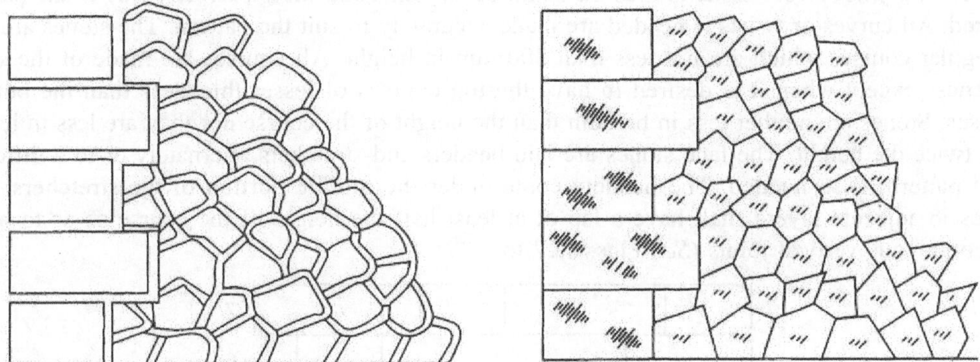

Fig. 6.23 and 6.24: Polygonal walling (Inferior and Superior types)

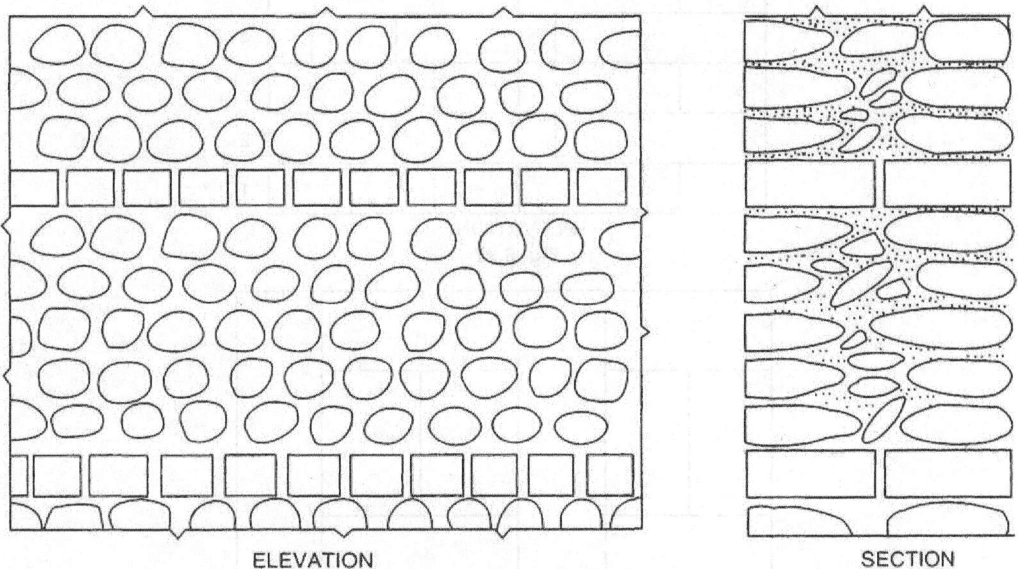

Fig. 6.25 and 6.26: Flint masonry

Flint Walling: The stones used in this type of work are small in size, varying in thickness from 80 mm to 150 mm and in length from 150 mm to 300 mm. Buildings near coasts are constructed of walls with rounded flints procured from the beaches. The walls are about half metre in thickness and may be built with a facing of cut flints and a backing of undressed flints. A second method is to shape the flints into square sizes and use them as facings. A cheap type of work is formed by using undressed flints. Uncoursed or coursed type of work can be made. When the flints are undressed throughout, the external and the internal flints are laid as headers and the hearting of headers and stretchers tightly packed. The mortar joints are slightly raked back with a pointed stick to improve the appearance.

6.7 ASHLAR MASONRY

This class of masonry consists of stone blocks accurately dressed, with very fine joints of the order of 3 mm thick. The backing of such walls may also be built of ashlar masonry or rubble masonry. It should be ensured that the sizes of individual stones are in conformity with the general proportions of the wall in which they are placed. This is the highest grade of masonry and is very costly. Different types of patterns can be formed by suitably arranging the blocks. The various types of ashlar masonry are described below:

Ashlar fine: Every stone is dressed on all beds, joints and faces perfectly true to the pattern desired. All curves or twists if needed are made accurately to suit the pattern. The stones are laid in regular courses which are not less than 300 mm in height. All courses are made of the same thickness except when it is desired to have the top courses of lesser thickness than the bottom courses. Stones are neither less in breadth than the height of the course nor they are less in length than twice the height. The face stones are laid headers and stretchers alternately or to a different bond pattern, if so needed. The headers come under the middle portion of the stretchers. The stones in adjacent layers shall have a lap of at least half the height of the course so as to break the continuous vertical joints (*See* Figs. 6.27 to 6.29).

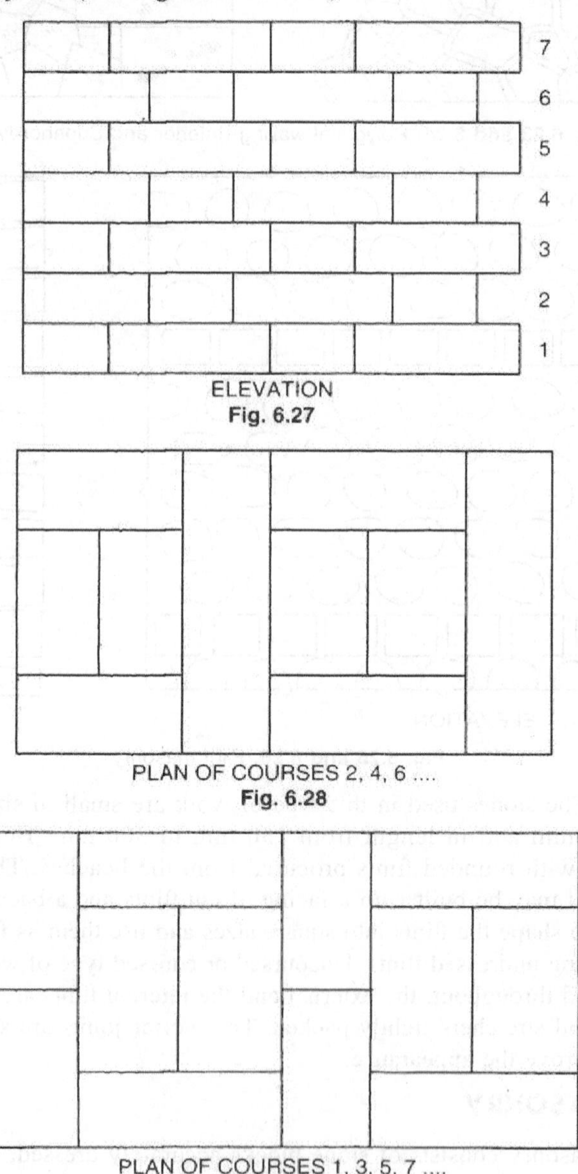

ELEVATION
Fig. 6.27

PLAN OF COURSES 2, 4, 6 ....
Fig. 6.28

PLAN OF COURSES 1, 3, 5, 7 ....
Fig. 6.29

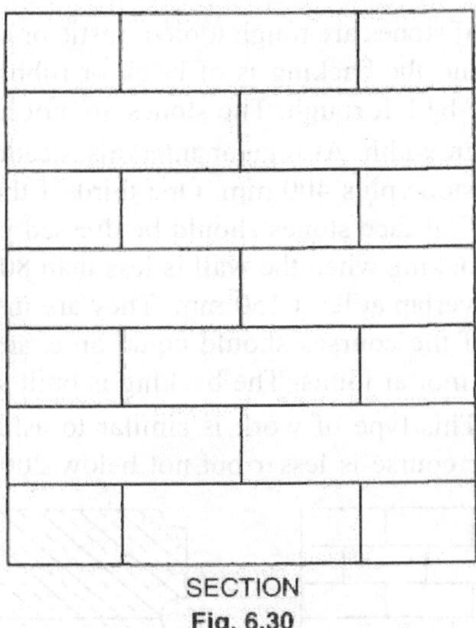

Fig. 6.30

Figs. 6.27 to 6.30: Ashlar fine

Through stones are put in walls which are less than 800 mm in thickness. These are in the form of headers which run from one face to the other. All the joints are either horizontal or vertical. They are made of fine mortar with a thickness not exceeding 3 mm. No broken edges of the stones should be visible at the joints. Stones are struck with a mallet so as to have solid bearing on the beds and the side joints.

Ashlar rough tooled: In this type of masonry, the exposed faces should have a fine dressed chisel drafting. This may be about 25 mm in width. The portion in between the drafts is roughly tooled. All bed joints and vertical joints are similarly drafted. The size of stones, bonds, etc., are as for the Ashlar fine type. However, joints as thick as 6 mm can be allowed in this case.

Ashlar rock or quarry faced: The exposed face between the drafts is not tooled, as described above, but is left unfinished. However, the projections (bushings) should not exceed 100 mm from the planes of drafts. The size of the stones and the bond, etc., is similar to the one described above.

Ashlar chamfered: Stones have the exposed edges bevelled for a depth of about 25 mm. The type of joints, bonds and size of stones is the same as described above.

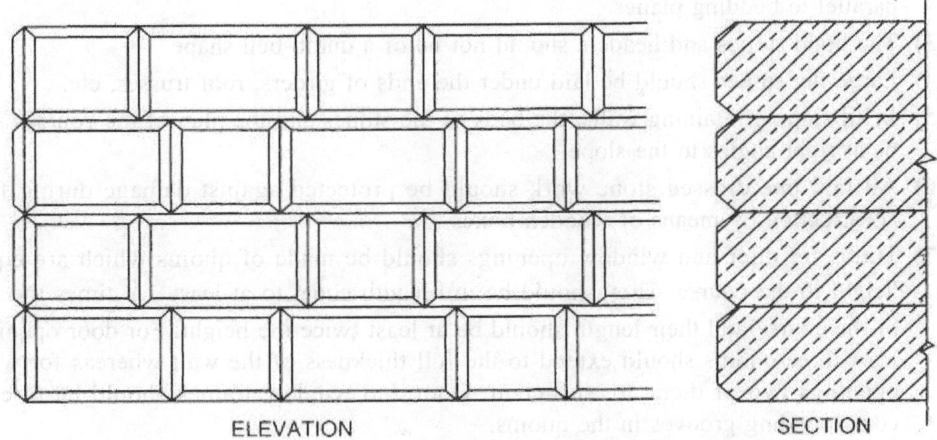

Figs. 6.31 and 6.32: Chamfered Ashlar masonry wall

Ashlar facing: The faces of stones are rough tooled, rustic or chamfered. In this type, the face work is made of these stones and the backing is of brick or rubble masonry or concrete. Hence the back face of the stones may be left rough. The stones are not less than 200 mm in height and are about $1\frac{1}{2}$ times the height in width. At regular intervals, headers are put which are in length greater than the breadth of the stone plus 400 mm. One third of the length of each course should be of headers. The bed joints of all face stones should be dressed perfectly true and square. Bond stones should run through the backing when the wall is less than 800 mm in thickness. For thicker walls, the bond stones should overlap at least 150 mm. They are inserted at every 2 metre interval along the course. The height of the courses should equal an exact number of rubble courses or brick courses plus intermediate mortar joints. The backing is built simultaneously with the facing.

Ashlar block in-course: This type of work is similar to ashlar rough tooled with the only difference that the height of the course is lesser but not below 200 mm.

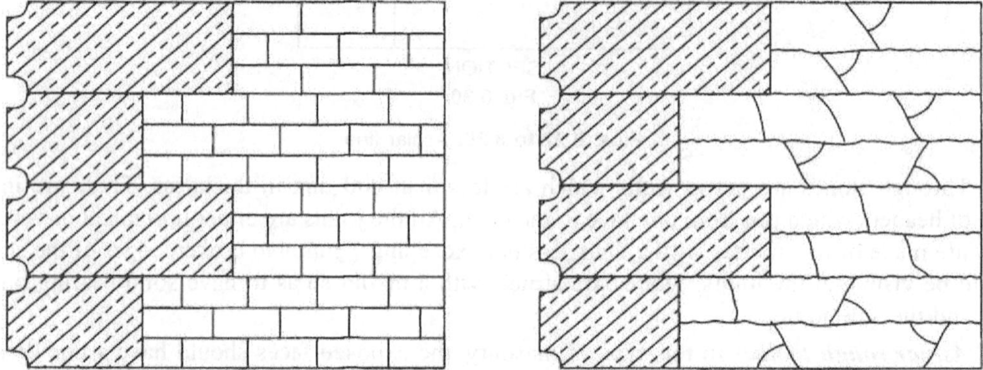

Figs. 6.33 and 6.34: Ashlar facing (*i*) with brick backing (*ii*) with random rubble backing

6.8 GENERAL PRINCIPLES TO BE FOLLOWED IN THE CONSTRUCTION OF STONE MASONRY

(1) The stones used shall be hard, durable and tough. All stones should be laid on its natural bed.

(2) The pressure acting on the stones should not act parallel to the bedding planes. This will try to split the stones. Sometimes stones used in corbels are laid with pressure acting parallel to bedding planes.

(3) The bond stones and headers should not be of a dumb-bell shape.

(4) Large flat stones should be laid under the ends of girders, roof trusses, etc.

(5) In all sloping retaining walls, the beds of the stones and the plan of the courses should be at right angles to the slope.

(6) All laid fine dressed stone work should be protected against damage during further construction by means of wooden boxes.

(7) Jambs for door and window openings should be made of quoins which are equal in height to the course. They should be in breadth equal to at least $1\frac{1}{2}$ times the height of the course and their length should be at least twice the height. For door openings at least three quoins should extend to the full thickness of the wall whereas for window openings two of them are sufficient. Door and window frames should be fitted into corresponding grooves in the quoins.

(8) All the surfaces should be kept wet while the work is in progress and also till the mortar has set.

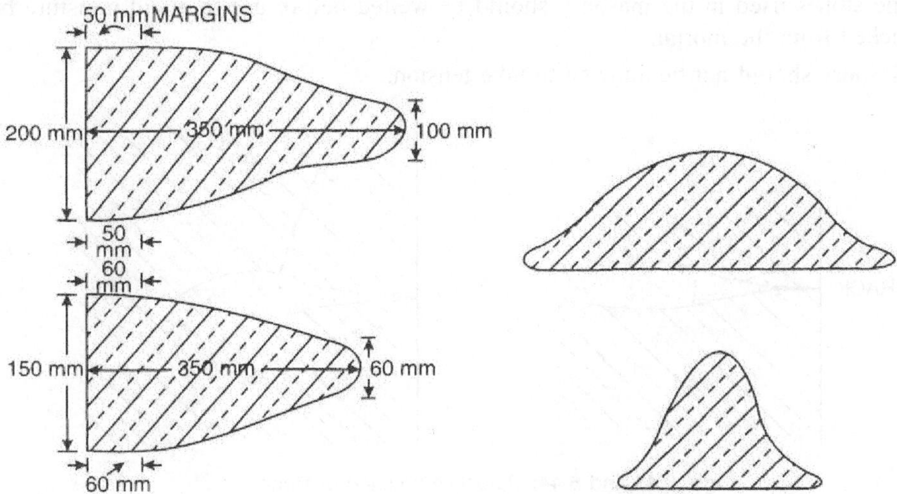

Figs. 6.35 and 6.36: Plans of good headers

Figs. 6.37 and 6.38: Bad stretchers with thin edges

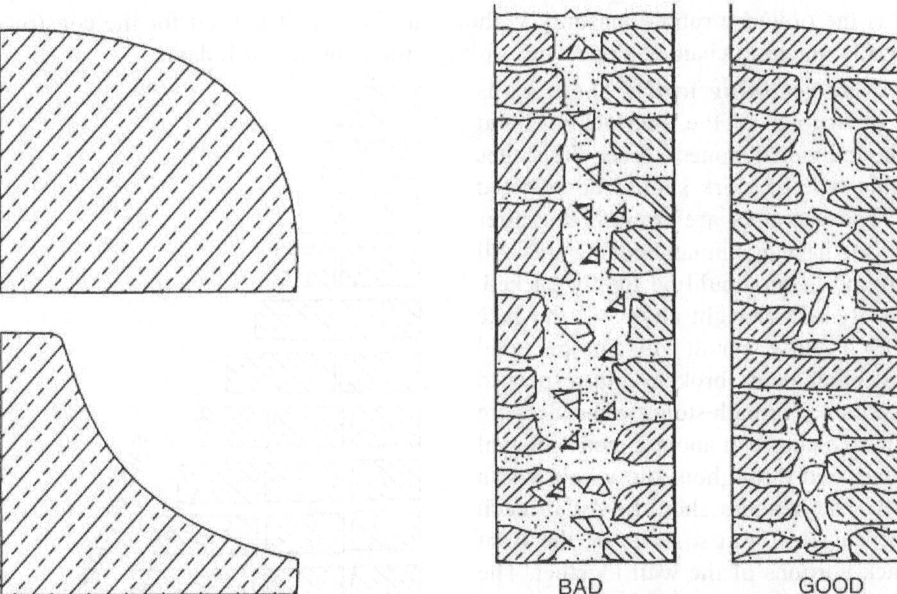

Figs. 6.39 and 6.40: Good and bad quoins

Figs. 6.41 and 6.42: Section through a good and bad masonry wall

(9) Double scaffolding will be used wherever it is difficult to fit in the stones later on.

(10) All the portions of the masonry should be raised uniformly. Wherever this is not possible, the stone work built earlier should be raked (stepped) so that the new work can be bonded well with the old.

(11) Sufficient through stones should be used and they should form $\frac{1}{4}$th to $\frac{1}{8}$th of the area in elevation.

(12) The hearting of the masonry should be properly packed with mortar and chips, if necessary, to avoid any hollows or very thick mortar joints.

(13) Vertical faces of the masonry walls should be checked with a plumb rule and the battered faces should be tested with wooden template corresponding to the batter and a plumb rule to ensure a constant batter.

(14) The stones used in the masonry should be wetted before use to avoid moisture being sucked from the mortar.

(15) Masonry should not be allowed to take tension.

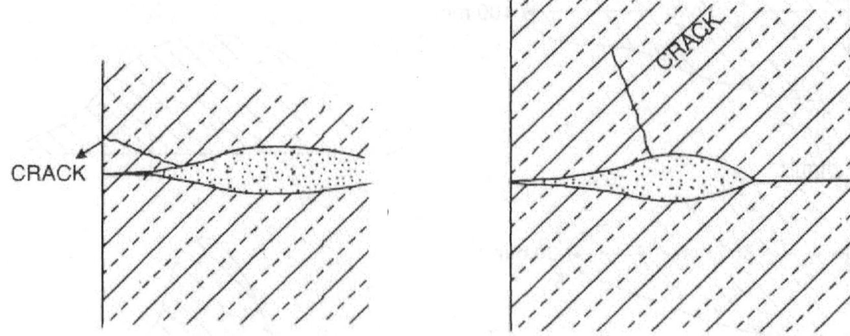

Figs. 6.43 and 6.44: Defective bedding of stones

6.9 DRY STONE MASONRY

This is the ordinary rubble masonry without any mortar. It is used for the construction of retaining walls, pitching channel slopes and rip-rap protection in earth dams.

***Dry stone retaining walls*:** These walls are stable on account of the interlocking bond between the individual stones. Hence the stones used for this type of work should be selected carefully. These stones are roughly hammer dressed so as to have maximum bedding area. All hollows around them should be tightly packed. The stones are laid at right angles to the face batter and each course is built with a proper bond and vertical joints being broken with respect to the other courses. Through stones or headers are provided in each course at about 2 metre interval and are continued throughout the whole width of the wall. The headers should overlap each other by at least 150 mm so as to tie the front and the back portions of the wall together. The filling behind the wall should be made up of stone chips. Earth should not be used for this purpose. Holes running through the width of the walls should be left to allow for the drainage of the back filling. These are called weep holes. The dry stone walls should have at least $\frac{1}{2}$ metre top width and a batter of 1 in 4 for heights upto 5 metres. For greater heights the batter may be decreased or the portions below 5 metres from top should be made up in mortar. Long lengths of dry stone walls should be divided into panels by short lengths of walls (1 metre or so) built in mortar at intervals of 10 metres so as to restrict the damage, if any, within these lengths.

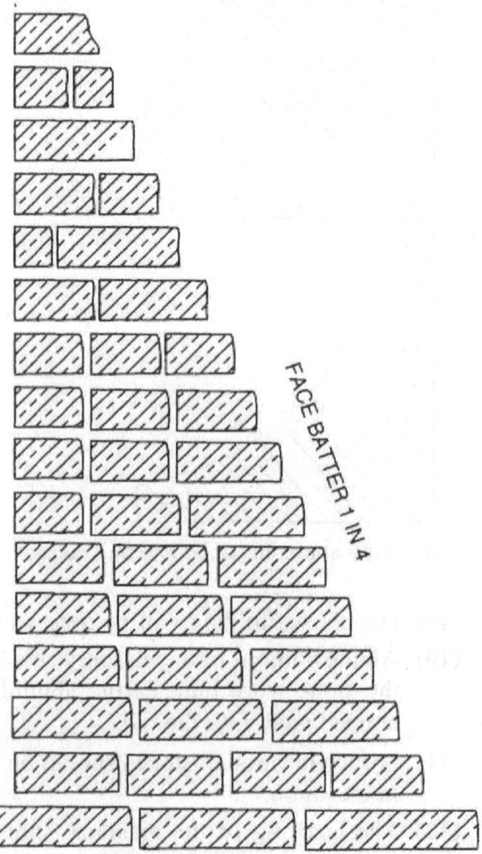

Fig. 6.45: Dry stone masonry wall

Dry stone pitching: When the slope of an earthen cutting cannot stand the scouring action of water, it is protected with a layer of stones laid dry which is called pitching or revetment. The stones used for such type of work should be sound, regular in shape and in length equal to the thickness of the desired pitching. The pitching is generally given a backing of smaller stones packed well or of moorum. This is carried simultaneously along with the pitching and is well watered and rammed; if needed. The stones should be laid tightly with each other in such a manner that their lengths are at right angles to the face of the slope. Small stones may be inserted in between the larger ones to make the bond perfect. For stability at the toe of the embankment, a masonry wall is made to secure the pitching in position or else a dry stone wall of desired thickness is built.

6.10 JOINTS IN STONE MASONRY

The joints in stone masonry may be classified under the following three heads:

(*a*) Strengthening joints

(*b*) Lengthening joints, and

(*c*) Mortar joints.

A brief description of these follows.

6.10.1 Strengthening Joints

There are certain structural locations in a building where the weight of the superstructure and the adhesion of mortar are insufficient to keep the stones in position. This may lead to a possibility of failure along joint by either sliding or separations. Such failure can be checked, if the joint is strengthened by some special means such as dowels, cramps and joggles. Examples where such methods are applicable are coping stones, inclined walls, cornices and corbels; which cannot be kept in position by the surrounding masonry.

Strengthening joints, in stone masonry, are of two types, namely (*a*) joggle joints and dowels and (*b*) ramp joints. The former type is suitable for taking care of sliding of the abutting surfaces, whereas the latter type is suitable for prevention of separation or opening out of the joints.

Joggle joints can be in the form of tongue and groove, table and slate or metal or cement joggles. These joints are suitable to prevent sliding between vertical butt joints (*e.g.* landing of stone-stair), horizontal sliding of stones, lateral movement of stones in copings respectively. Dowels are also used for resisting lateral movements.

Cramps may be metallic or of stone such as slate. Coverings of copings, cornices and projecting string courses are the situations where cramps can be used to resist the parting of stones. Lead plugs can also be used for this purpose.

Details of strengthening joints are given below:

(1) ***Table or bed joggle joint***: The projection of one stone fits into the depression of the adjacent stone. This prevents the movement of the stones along the joining plane.

(2) ***Tongued and grooved***: In this also the projection of one stone fits in the corresponding groove in the adjacent stone. This is also called joggled joint.

(3) ***Slate or metal joggled joint***: In this, a piece of metal or slate is placed in between the corresponding groove of the adjacent stone.

(4) ***Dowel joint***: Wherever stones are liable to be displaced, then a dowel is introduced to connect them. These dowels may be of slate, gun metal, brass, bronze, etc. They are set in cement mortar.

(5) ***Cramped joint***: The cramp is a piece of metal about 20 to 50 mm wide, 6 to 12 mm thick and 200 to 400 mm long with ends turned down. They may be set in depressions in the top of the stone or set beneath the stones with their ends upward. They should be protected by pouring molten lead around them. These cramps can also be made of slate or other stone. They have a dovetailed shape on both ends and are set in cement. However, they are not as effective as the metal cramps.

(6) ***Plug joint***: This is an alternative to the use of cramps and consists of a depression below the top surface dovetailed in plan. The stones are jointed in the usual way and this depression is filled with cement or lead.

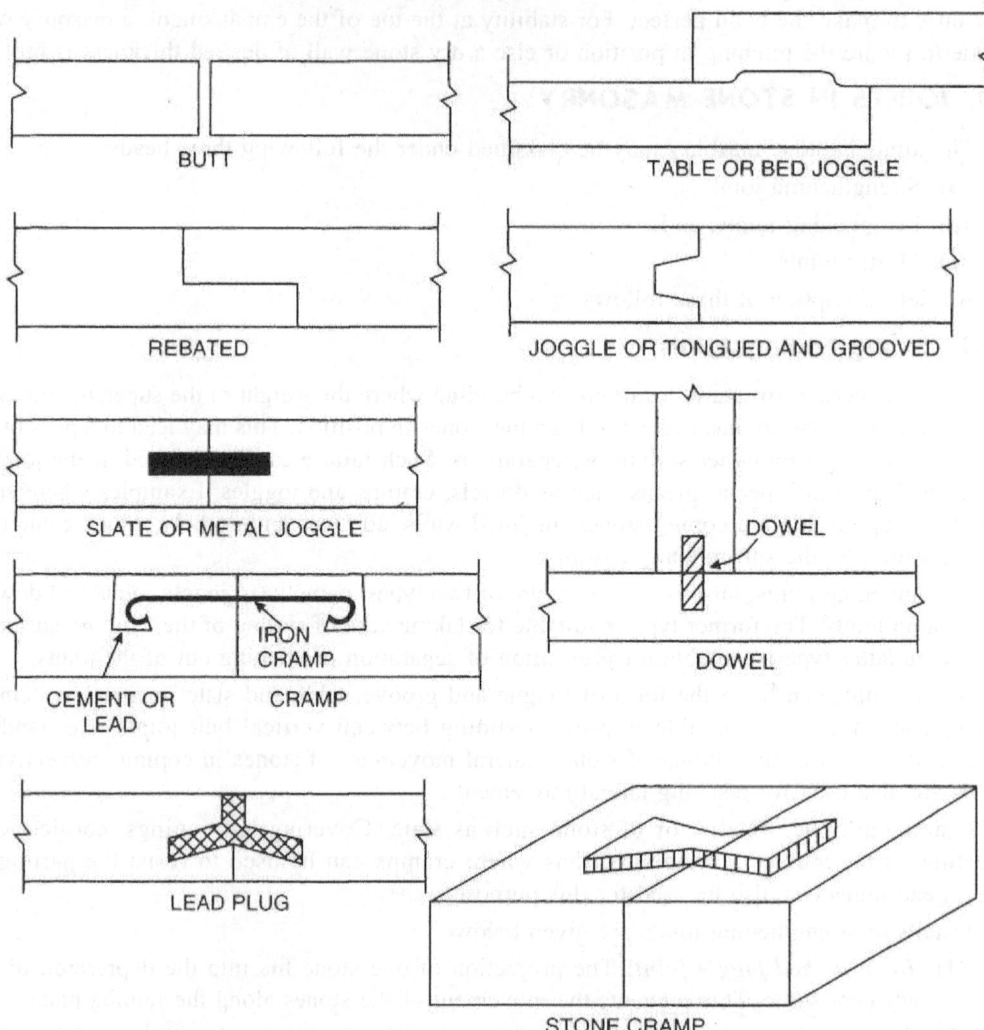

Figs. 6.46 to 6.54: Various types of joints between individual stones

6.10.2 Lengthening Joints

Different kinds of joints used for placing the individual stones together lengthwise are:

(1) ***Butt joints***: The two stones are placed side by side with their square faces abutting each other.

(2) ***Rebated or lapped joints***: In this type of joint, the rebate in one stone laps over the projection of the other stone.

6.10.3 Mortar Joints

Joints used for rubble work are not consistently uniform in thickness. In case of the random rubble or polygonal walling, small pieces of stones or chips should be fitted into the joints to prevent excessive thickness or mortar being held in between the two joints.

In the squared rubble masonry, the joints are more or less horizontal and vertical and are uniform in thickness. Often the joints in this type of work are made flush with the surface and a bead of mortar (may be of a different colour) is provided to give the effect of a narrow joint. However, projecting joints may be used for thick joints. The common types of these joints are shown in figures 6.55 to 6.57.

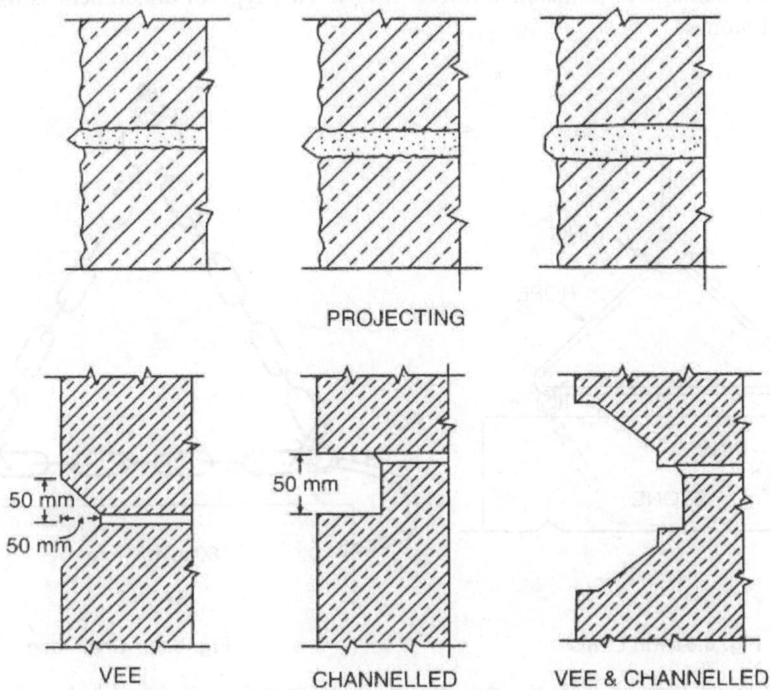

Figs. 6.55 to 6.60: Various types of joints for stone masonry

Ashlar masonry, as described earlier, is accurately dressed so as to have very fine joints. They are mostly flush joints. However, when joints thicker than about 6 mm are used, keyed or V-joints may be used. Other types of joints which are suitable for this type of masonry are the channelled joints or Vee and channelled joints.

6.11 MORTARS USED

Stone masonry mortars have to be selected depending upon the colour of the stone so as to have a good combination of colours on the face work. Generally stone masonry is built in cement mortar of 1 : 3 ratio. To increase the workability of mortar, about 15% of the cement may be replaced by lime. For limestones which get stained by the ordinary cement sand mortar, a non-staining white Portland cement mortar may be used. Generally one part of non-staining cement, one part of hydrated lime and six parts of clean sand are used for stones which are liable to be stained by cement mortar alone. Special non-staining water proof cements may be used instead. For pointing work very rich mortar should not be used.

6.12 LIFTING APPLIANCES

Lifting appliances are necessary for proper handling of heavy stone blocks which cannot otherwise be raised or lowered to the desired level and set properly. Various appliances are used for lifting and placing stones in position. Heavier stones are lifted with a derrick and carried to a place where it is to be positioned in masonry. Gantries or cranes can also be used to move the stones from one place to another. However, certain attachments are necessary for holding the stones while the derrick lifts them. The different types of such attachments are:

(1) ***Grab Hooks***: The pointed ends of these hooks get fitted tightly into the corresponding notches cut in a stone. Small wooden blocks may be used on the sides of the stone to prevent damage to polished surfaces, if any. This type of attachment is used for lifting small stones.

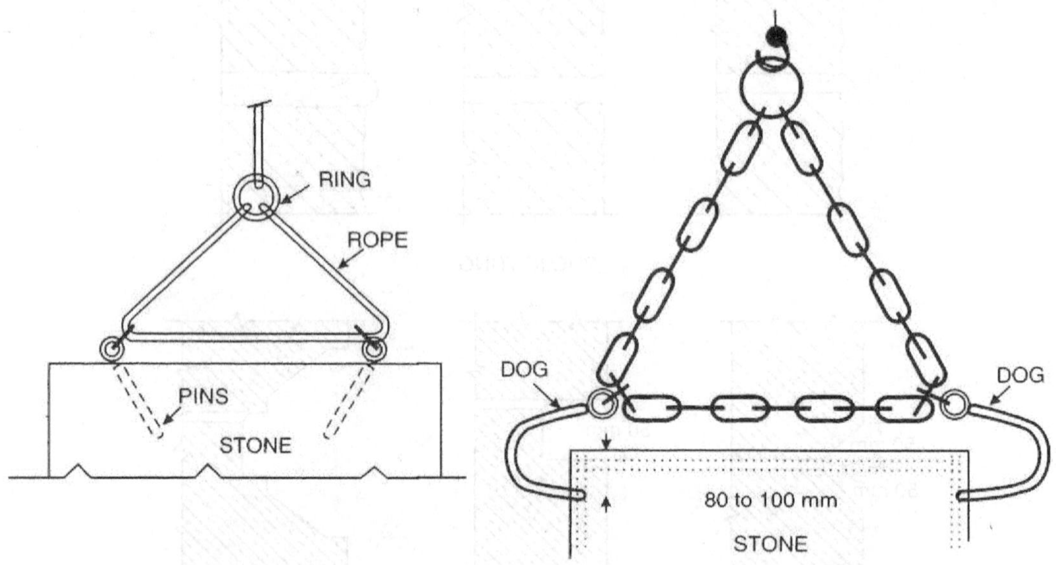

Fig. 6.61 Pin Lewis

Fig. 6.62: Chain dog.

(2) ***Pin Lewis***: Pins of iron are put into the corresponding inclined holes in the stone and attached to a chain or rope bent into a triangular shape as shown in Fig. 6.61. The tightening or pulling of the vertical rope prevents the pins from coming out of the stone and permits the stone to be raised.

(3) ***Chain Dog***: In this type of attachment, the chain used is arranged as described above; but instead of pins at the ends of the triangular rope work, hooks called dogs are fitted. These hooks get fitted into the corresponding depressions about 20 mm deep in the ends of the stones. When the chain is lifted, and dogs get tightened and permit the lifting of the stone. Enough depth (at least 80 to 100 mm) of stone should be available above the end depression so that while lifting the stones, their corners might not give way and they may fall down. This method of lifting is employed for heavy stones with long narrow beds.

(4) ***Chain Lewis***: This is made up of three rings and two curved steel legs which are attached to each other as shown in Fig. 6.63. A hole is formed slightly dovetailed in the top of the stone which is to be lifted. The lewis is placed into the hole with one leg at a time. A wedge piece may be inserted, if necessary. When the crane lifts the top ring, the smaller side rings get pulled apart and cause the lower tips of the legs to grip the

stone. The hole should not be excessively dovetailed as otherwise there is likelihood of the edges being broken and the stone falling down.

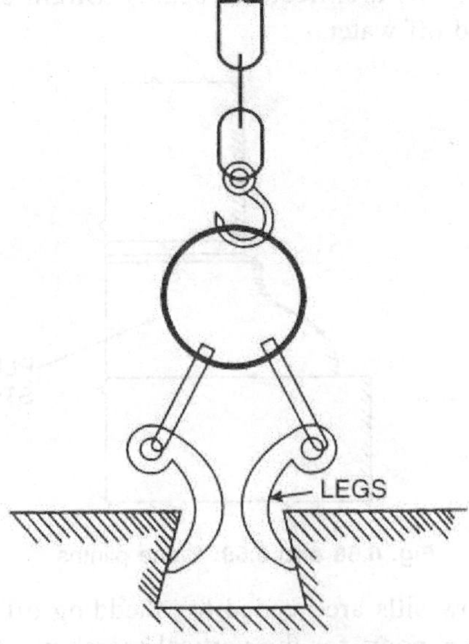

Fig. 6.63 Chain Lewis

(5) ***Three legged common Lewis***: These consist of two dovetailed steel pieces, a parallel piece of steel in between, a shackle at the top and a round steel pin which connects all of these together. A dovetailed hole is made in the top of the stone corresponding to the shape of the legs. First the two end dovetailed pieces are inserted and the central piece is driven next. The pin and the shackle are fitted tightly. The crane hook is passed around the shackle to lift the stone.

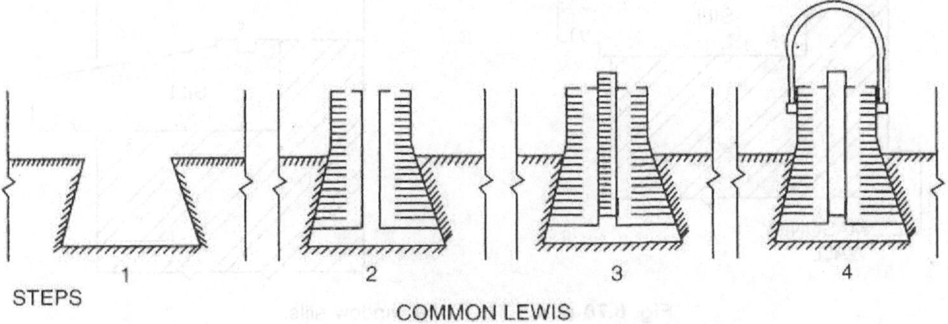

Fig. 6.64 to 6.67: Assembling details of a common lewis (3-legged)

6.13 STRUCTURAL MEMBERS BUILT OF STONE

(1) ***Steps***: Steps are used for entrance and stairs of a building. Stone steps are used to a limited extent these days. They can be given various architectural shapes. Stones used for steps should be hard and durable. They should also take polish. Various types of stone steps and their methods of construction are described in the chapter on Stairs.

(2) ***Plinth and plinth courses***: Plinth is a term applied to the horizontal projecting course at the base of superstructure, *i.e.*, between the ground level and the floor level. The height of the plinth is dependent upon the architectural treatment needed for a building, the

flooding to which the area is subjected, etc. The topmost course of the plinth masonry is called plinth course. Plinth course can be made of stone and may be given special mouldings to enhance its architectural beauty. Slight slope is necessary for the top position so as to shed off water.

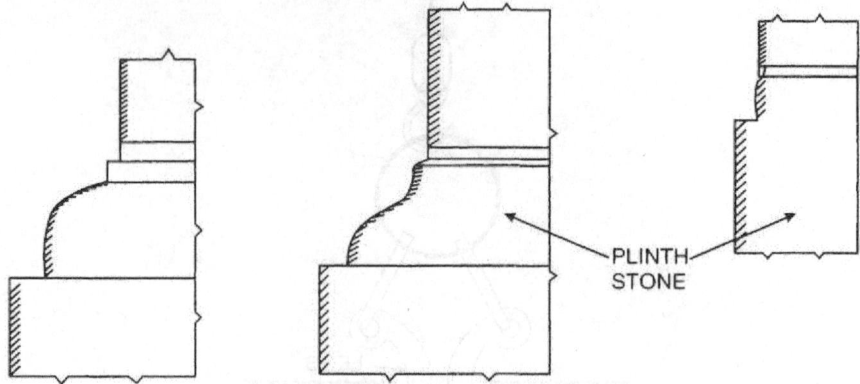

Fig. 6.68 and 6.69: Stone plinths.

(3) ***Window sills***: Window sills are needed for shedding off water falling on the window. They can also act as support for the vertical members of the window. They are made of bricks or stones. Special mouldings have to be given to prevent the water falling on the wall.

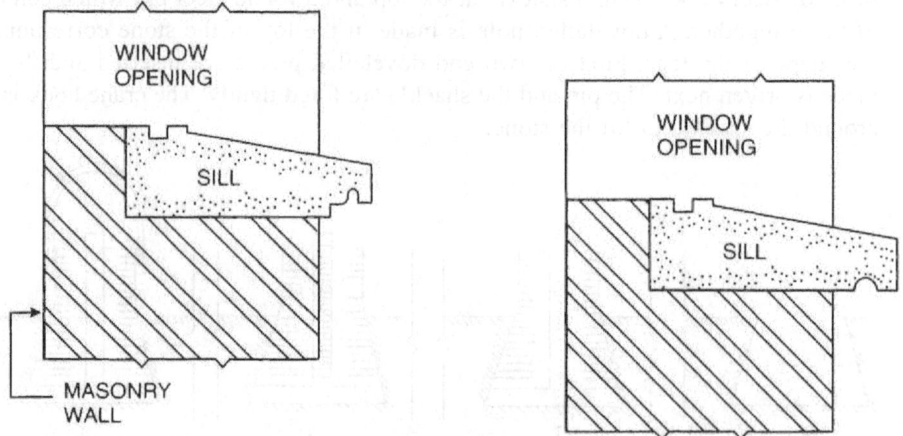

Fig. 6.70 and 6.71: Stone window sills.

(4) ***String courses***: This is a horizontal projection of the masonry throughout the length of the wall and is meant as an architectural feature. Special mouldings can be given to these courses.

(5) ***Frieze***: This is a course below the cornice. This may be in flush with the wall below or may be projecting if there is no projecting member below it.

(6) ***Cornice***: This is a large projecting horizontal course provided on the top of the wall. This adds to the archtitectural beauty of the building and also protects the wall below it from the water trickling down its surface. Special mouldings may be given to the cornices for creating an architectural treatment.

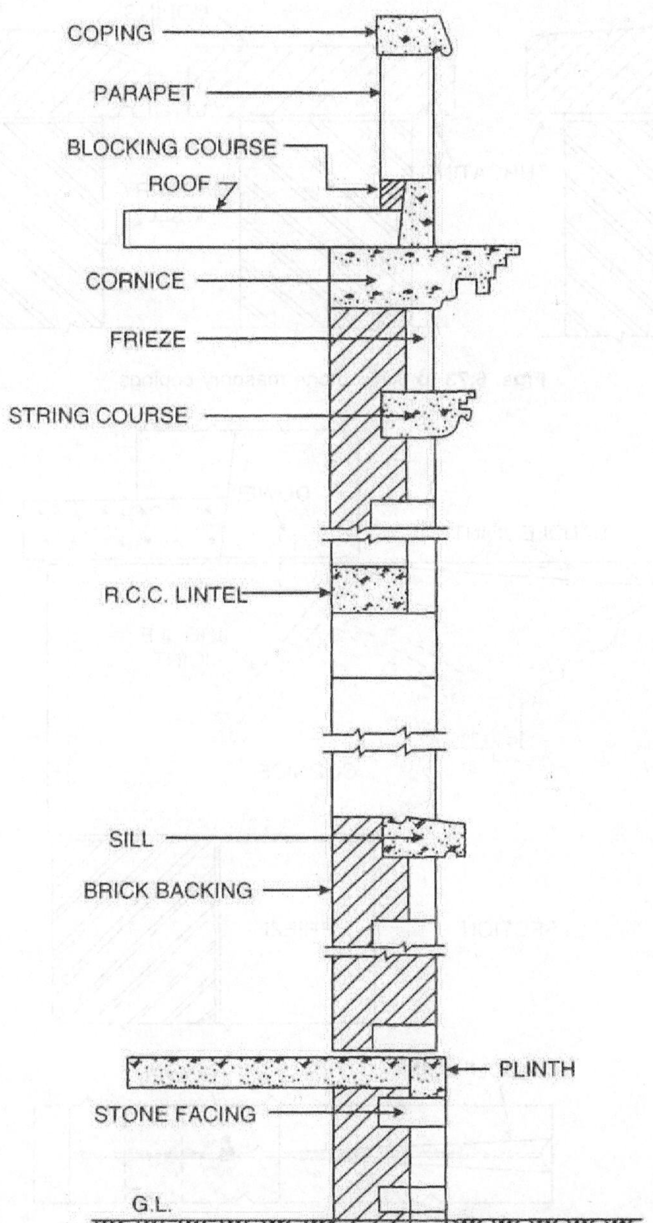

Fig. 6.72: Section through an exterior wall of a building showing different elements of construction in stone masonry

(7) ***Parapets and copings*:** These may also be made of stones. Rubble masonry or ashlar work can be used in their construction depending upon the general type of masonry used in the walls to which the parapets surmount.

Copings can be made of stones cut to different shapes. Mouldings can be given to the face ends of the copings to enhance their beauty. Throatings have to be incorporated to prevent the water from trickling down the face of the wall.

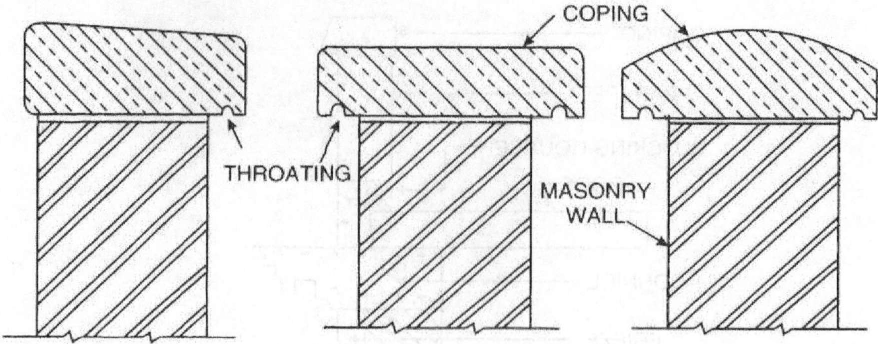

Figs. 6.73 to 6.75: Stone masonry copings

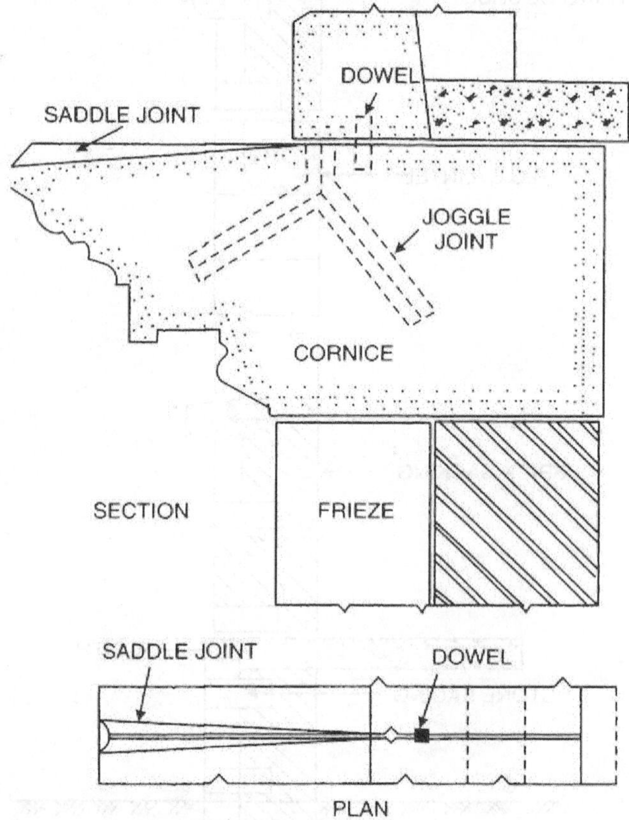

Figs. 6.76 and 6.77: Section and plan of a typical cornice

6.14 MOULDINGS

To give aesthetic effect to the various members of a building, special curved surfaces are carved on them. The moulding to be used depends upon the position of the member in the building. Crowning mouldings are not expected to carry any structure above them. Hence they are used on the top members of a building. Various types of mouldings are used and a few of them are shown in Figs. 6.78 to 6.86.

The supporting mouldings are used in members which have to support loads. They do not have hollow depressions near their upper edges.

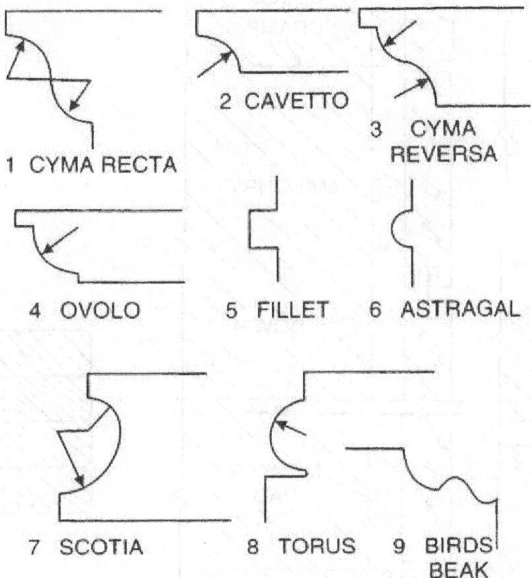

Figs. 6.78 to 6.86: 1-3. Supporting mouldings. 4-6. Connecting mouldings. 7-9 Base mouldings

Connecting mouldings are narrow projections used to divide a group of mouldings on various parts of a structural member.

Base mouldings are constructed to broaden the base of a loaded member and thereby distribute the loads over a large area.

6.15 STONE FACING

Costly stones, *e.g.*, marble are not economical for the construction of walls of large thicknesses. Hence only a face work built of thin slabs of these stones can be used. These facings are not bonded so as to act as load bearing components of walls but have only to be fixed with the backing which takes the entire load of the member. These are also termed *veneered walls*. The stones are cut into thin sheets about 25 mm in thickness and fixed so as to appear as ashlar work or may be suitably cut so as to form panels. Marble facing may be kept about 25 mm away from the face of the backing and can be made to rest on pads of Plaster of Paris. It is kept in position with brass dowels about 6 mm in diameter fixed along the lower edges. The top edges are kept in position by brass cramps which are turned into the slab and fixed in the backing of the wall. The angle joints at corners can be made by the use of screws covered with marble studs or cramps and dowels.

External walls may be faced by thin slabs of stones polished suitably. These stone facing should not be greater than 8 times their thickness in heights. Suitable bond stones may be used for strengthening thick facings whereas for thin facings, cramps have to be used for anchorage with the back of the wall.

A wall in which the facing of stone and the backing of brick work or concrete are bonded together to ensure common action under load is called *faced wall*. This is shown in Figs. 6.87 to 6.93.

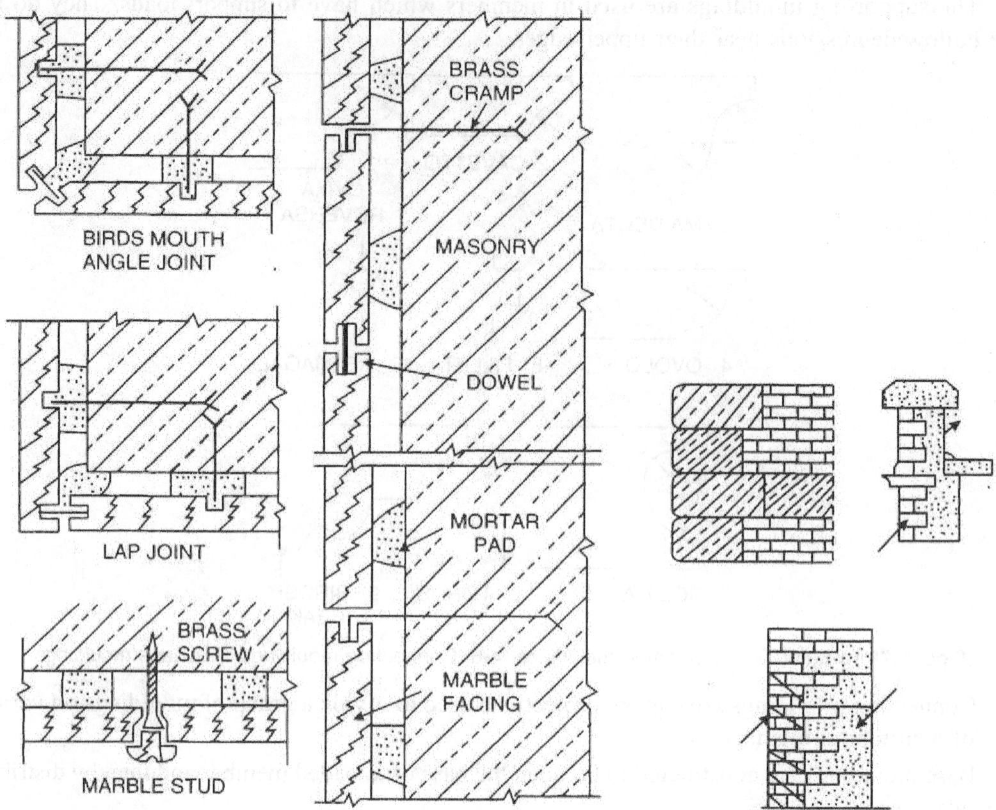

Figs: 6.87 to 6.90: Details of marble linings to masonry walls.

Figs. 6.91 to 6.93: Stone faced walls

6.16 MAINTENANCE OF STONE WORK

(1) ***Efflorescence*:** This is common with certain types of mortars and soluble salts, when dried, get deposited on the surface of stones. This can be prevented by proper drainage of the building. Efflorescence can be removed by scrubbing with water and a stiff brush. If this is not successful, a 10% solution of muriatic acid may be used instead of plain water. The stones should be kept saturated with water so that it may not get discoloured on account of acid action.

(2) ***Stains*:** Stains can be removed easily if they are not old and their origin is known. Iron stains have a rusty appearance and can be removed by rinsing the area with a solution of $\frac{1}{2}$ kg oxalic acid in 5 litres of water. The surface should then be scrubbed with brushes and clean water after 2 to 3 hours. Darker stains may be removed with a mixture of 6 parts of water, one part of sodium citrate, one part of glycerine made into a paste with sufficient whiting. This paste is applied to the surface and left for a few days. For very deep and dark stains, one part of sodium citrate is dissolved in 6 parts of water and the surface of stains is moistened with this solution. The surface is then covered with a thin layer of sodium hydro-sulphite crystals. This is removed after an hour. This treatment may be repeated, if necessary. The final surface is washed.

Copper and bronze stains are removed by applying a mixture of one part of ammonium chloride and 4 parts of powdered talc with ammonia water to form a paste. This paste is covered on the surface and removed after some time.

Smoke and fire stains can be removed by rubbing with powdered pumice or grit. This treatment is applied several times till a clean surface appears.

Oil stains can be removed by scrubbing with benzene or petrol. However, a mixture of acetone and amyl acetate can remove deeper stains.

Tobacco stains can be removed with a weak solution of trisodium phosphate. Chlorinated lime paste is added to make the solution perfect bleaching agent.

Ink stains can be removed by applying a paste of whiting in a strong solution of sodium carbonate and water. Chlorinated lime or ammonia water can also be used as an alternative.

(3) ***Repair of cracks***: Cracks should be repaired only after the settlement causing the crack has ceased. This period may be taken as about one year. Small cracks should be cleaned with a wire brush or with a thin blade. A thick paste of cement mix is forced into the opening. Large cracks should be raked out to get a firm key for the mortar. An inverted V groove of at least 10 mm depth is necessary. A cement mortar of 1 : 2 ratio with less water is suitable. This mortar should not be applied for about an hour after mixing and should be remixed without additional water, immediately before using. This mixture (grout) should be forced into the crack to get a good bond. Expansive agents, *e.g.*, aluminium powder can be added so as to make a tight fit for this grout.

(4) ***Water proofing***: Colourless water proofing materials can keep the stone masonry free from damp, efflorescence, frost action etc. Water proofing materials obtained from heavy petroleum distillates, fatty oils or insoluble soaps are best. Stones having a very close texture are difficult to be water proofed. Water proofing substances containing resins are not suitable. They are applied as a washing coat. The water proofing causes some discoloration of the stones but that goes off in course of time.

6.17 ARTIFICIAL STONE

Wherever stone is not available but still it is desired to give a face treatment of stone work, artificial stone or cast stone is used. These types of stones are nothing but cement concrete blocks with special surface treatment. Aggregates of crushed stone are used to produce a rough surface. Coloured sands may be used with coloured aggregates. Best types of aggregate are essential so as to withstand the weathering effect. Coloured cements or pigments may be used wherever coloured treatment is necessary. Cement aggregate ratio of 1 : 3 is maintained. An excessive quantity of aggregate will produce a weak product whereas a concrete containing less quantity of aggregates is going to have chipped off surfaces. Facing and backing portions of the blocks may be made of different sized aggregates, the smaller sized being used on the face. Special architectural details and mouldings can be cast economically and with ease. A polished surface can also be developed from the cast stone. The cast stone has got some advantages over the natural stone, *e.g.*, no bedding planes are present and hence the precautions regarding the placing of stones with regard to their bedding planes are not necessary.

QUESTIONS

1. Briefly describe the different types of stones used in the construction of stone masonry.
2. What are the factors on which a particular type of finish is selected in stone masonry work? What are the usual types of finishes used?
3. Name the various classes of rubble masonry. What are the important points to be kept in view while carrying out rubble work.
4. Describe briefly the following types of masonry works:

(a) Random Rubble coursed.
(b) Squared Rubble uncoursed.
(c) Polygonal walling.

5. Name the various types of Ashlar masonry. Briefly describe the construction of walls with Ashlar fine masonry.
6. Write short notes on:
(i) Dry stone retaining walls.
(ii) Ashlar facing.
(iii) Ashlar rough tooled.
7. What are the various types of joints used in stone masonry for placing stones together? Illustrate with neat sketches.
8. Write short notes on:
(i) Chain dog
(ii) Common Lewis
(iii) Plinth courses
(iv) Cornice
(v) Frieze
(vi) String courses
(vii) Artificial stones
(viii) Maintenance of stone work.

REFERENCES

1. **Bombay, P.W.D.**: *Specifications Vol. I.*
2. **Huntington, W.C.**: *Building Construction.*
3. **McKay, W.B.**: *Building Construction, Vol. I.*
4. **Mitchell, G.A.**: *Building Construction and Drawing, Vol. I.*
5. **Greenholgh, R.**: *Modern Building Construction.*
6. **Riley, J.W.**: *Building Construction for Beginners*
7. **Tyrrell**: *Principles of Petrology.*
8. **F.G.M. Blyth**: *Geology for Engineers.*
9. **Maccy, F.W.**: *Specifications in Detail.*
10. **Bureau of Indian Standards**; *National Building Code of India*, 2005, *Part* 6, *Section* 4, *Masonry.*

CHAPTER

Hollow Block Masonry

HOLLOW CONCRETE MASONRY

The use of cement concrete blocks for masonry construction has developed rapidly due to the various advantages which they possess over traditional building materials like bricks and stones. The main advantages of concrete blocks are their large size, uniformity in design, easy handling and placing, attractive appearance and strength. Cheapness and the ease with which blocks of any size or form can be manufactured and laid are the two distinct qualities of concrete blocks as a valuable building material. New materials and methods of manufacture for these blocks are being developed and adopted regularly. The use of machine with large manufacturing capacity enables quick production. The method of steam curing enables these blocks to be laid within hours after casting.

There are several kinds of concrete masonry units depending on the shapes and sizes (Figs. 7.1 to 7.13 in which they are manufactured. However, two distinct types of concrete masonry units are:

(1) ***Regular concrete blocks*****:** These are precast cement concrete blocks made from dense aggregates and intended for load bearing walls.

(2) ***Light weight concrete units*****:** Light weight aggregates are used in the construction of these units and although they are not as strong as the heavy concrete blocks, yet they are extensively used for both load bearing and non-load bearing walls. The situations where they are used for load bearing walls are external walls rendered or otherwise protected, the inner leaf of cavity walls, backing for brick work and stone masonry, internal load bearing walls and partitions and infilling panels for framed buildings. As non-load bearing units, they are used in partitions and internal panels for framed buildings.

Concrete block masonry can be built in various sizes depending upon the job requirement. There is also no standard size of concrete blocks. However, as recommended by the Concrete Association of India, the face thickness should not be less than 50 mm and the net area should be at least 55 to 60 per cent of the gross area. Strengthening of the block in the middle, where cracking is liable to take place, is necessary. It is recommended that the cores should be at least two in number and they should preferably be oval-shaped. The common sizes for building blocks are given in the table below:

Type	Actual dimensions in mm		
	Length	Breadth	Height
Size A	390	300	190
Size B	390	200	190
Size C	390	100	190
Tolerance	± 3 mm	± 1.5 mm	± 1.5 mm

Various types of surface finishes are used. Some of them are described below:

(1) *Common finished surface.* Fine to coarse texture can be obtained by varying the mix proportions and by using suitable aggregates. Coloured aggregates can also be used. The aggregates can be exposed by treating the surface of blocks with a dilute solution of acids or by scrubbing it while the concrete has not set completely.

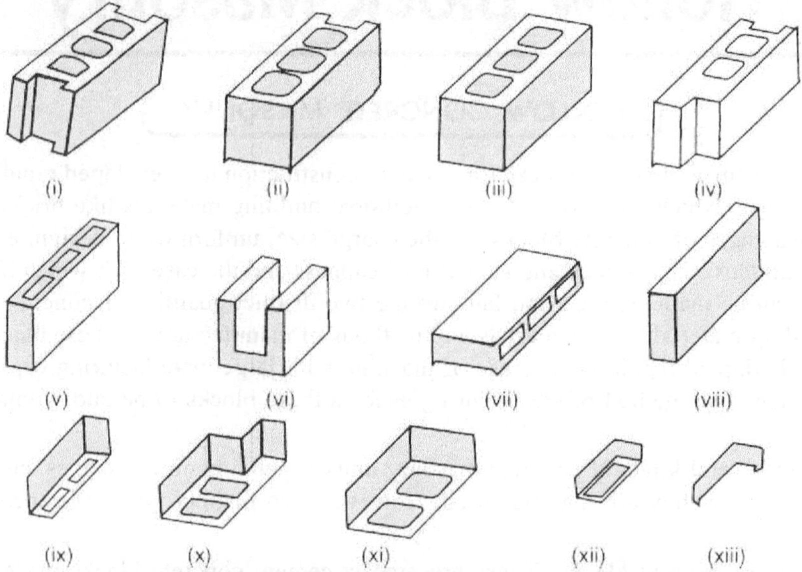

Figs. 7.1 to 7.13: Typical concrete masonry units

(*i*) Stretcher block, (*ii*) Corner block, (*iii*) Double corner or pillar block, (*iv*) Jamb block, (*v*) Partition block, (*vi*) Beam or lintel block, (*vii*) Floor block, (*viii*) Solid block used as facer in framing floor joints, (*ix*) Solid brick block, (*x*) Frogged brick block, (*xi*) Another Stretcher block, (*xii*) Another Jamb block (*xiii*) Another partition block.

(2) *Glazed finish*: Concrete blocks can be finished with a glazed surface in any number of desired shades. This is done in a manner similar to glazing of tiles. These types of blocks are suitable for decorative work and have sufficient water resistant properties.

(3) *Slumped finish*: A concrete of desired slump is prepared so that when the forms are removed, the blocks settle somewhat. This produces a rough surface on the exterior and blocks of various dimensions are used in the face work.

(4) *Specially faced block*: Finishing materials other than concrete, *e.g.*, marble, can be incorporated on the facing side of the blocks. The smooth finish thus obtained enables the blocks to be used in various places, *e.g.*, hospital, cinema, etc.

(5) *Split blocks*: They are used for some decorative finishes in buildings and are produced by dividing the blocks into two parts lengthwise. The surface thus formed is rough and the aggregates are exposed. Hence proper selection of aggregates and their blending has to be done.

(6) ***Coloured finishes***: Various types of pigments can be incorporated in the concrete mix so as to produce a variety of colours from block to block. Similarly special paints can be applied to the surface of the blocks to enhance their beauty.

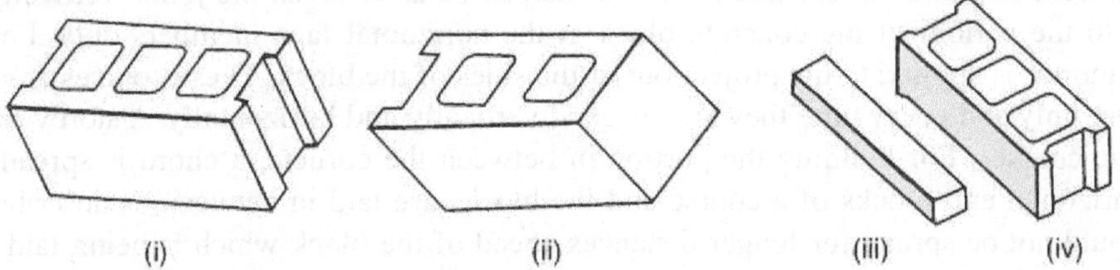

Figs. 7.14 to 7.17: (*i*) Standard block, (*ii*) Scored block, (*iii*) Slump block, (*iv*) Split block.

7.1 MANUFACTURE OF CONCRETE BLOCKS

The manufacture of concrete blocks is similar to that of ordinary concrete masonry. However, the following points regarding the manufacture of these blocks should be noted:

(1) The aggregates should be correctly graded. The mixture of fine aggregates (60%) and coarse aggregates 6 to 12 mm size (40%) is used. The fineness modulus of the mixture should vary between 2.9 to 3.6. Use of fine sand should be avoided.

(2) The concrete mix should be at least of 1 : 6 cement to combined aggregate ratio.

(3) When hand moulding is resorted to, the hollows should be vertical to get better compaction. The concrete should not have a very lean consistency. The concrete block should be taken out from the mould just at the time when it has sufficiently set.

(4) The block should not be disturbed on the moulding platform for at least 24 hours.

(5) It is preferable to use machines for casting the blocks so that a better finish and strength can be obtained.

(6) The blocks should be cured after keeping them under shade for at least 24 hours so that they are protected from the action of wind and sun during this period. These units should then be immersed in a tank of water for at least one week. Blocks should not be stacked for at least two days after casting to eliminate any damage to the corners and edges. After curing, the blocks should be dried for about four weeks before use. Stacking during this period should be done with cells horizontal.

(7) Improper compaction should be avoided.

(8) The compressive strength of the blocks should be at least 3 N/mm^2 at the age of 28 days. However, greater strengths on leaner mixes are possible if good care is taken to see that the blocks are manufactured properly.

7.2 LAYING OF CONCRETE MASONRY

7.2.1 Walls

It is essential to check whether any cutting of blocks is necessary or not because concrete blocks are available in definite sizes. This is done by laying a course of blocks without any mortar and fitting the same within the desired length of the wall. The blocks should be laid with proper clearance to allow for mortar joints.

Next a mortar bed is spread on the foundation concrete and is levelled to have sufficient but uniform thickness of mortar everywhere. The corner block is laid first and placed in position accurately. Mortar is applied to the ends of the other blocks and are placed and pushed against the corner block in turn. After a few blocks are laid, the level of the course is checked. It may be necessary to tap some of the blocks or to put extra mortar underneath them but it should be ensured that every block has at least about 20 mm mortar below it. The first course is also to be carefully put in plumb. The level and verticality of the first course is checked carefully so that the upper courses are brought up regularly in line and level with it.

The succeeding courses are laid in such a manner so as to break the joints vertically. Mortar is applied to the bottom of the concrete block at the horizontal face members only. For vertical joints, the mortar is applied to the projections at the sides of the block. These courses are built first at the corner only and every time they are checked vertically and horizontally. A storey rod is used to guide the courses. For building the portion in between the corners, a chord is spread between the two horizontal end blocks of a course and the blocks are laid in between as described above. Mortar should not be spread for longer distances ahead of the block which is being laid so that it may not get dried when the block at that particular position is fitted. Excess mortar is trowelled off from the face of the joint. The final closing block is fitted carefully. If it is to be cut, the same should be done accurately and the cut edges should be true and square. All the four vertical edges of the block and the edges of the opening are covered with mortar. This block is then pushed into the opening and fitted into place.

The face of the masonry may be pointed by running a tool along the joints after the mortar is stiff but before it has set. The surface of the joint may be concave or V-shaped. Raked, flush and strucked joints are not recommended. The types of joints which can be used are weathered, V-shaped or concave as they shed off water easily. Pointing may also be done.

The following points should be noted for the construction of walls and corners in addition to what has been described above.

(*i*) Only well dried blocks should be used in the construction. The blocks should not be drenched in water before laying in the walls. Only sides of the joints should be slightly wetted.

(*ii*) Blocks for this work should have less than 10 per cent absorption for external walls and less than 15 per cent for internal walls. When blocks with higher water absorption are to be used, they should be painted with water proof materials.

(*iii*) The mortar used shall not be stronger than the concrete mix used for the manufacture of blocks. Cement, lime and sand mortar of 1 : 1 : 10 mix is preferable.

(*iv*) The joints should be about 5 mm to 1 cm in thickness.

(*v*) When the difference in the height of two adjacent walls is more, *e.g.*, when a main wall meets a compound wall, the two walls should be separated by a joint.

(*vi*) The corners of hollow concrete block walls crack due to thermal expansion. In the case of long walls there will be cracks in the centre of the wall. Solid concrete blocks or hollow blocks filled with concrete are used at the junction of walls to avoid cracking.

Controlled joints with metal stirrups may also be used as an alternative.

(*vii*) At the junction of load bearing and non-loading bearing walls, corrugated metal ties are used. The joints between two walls are not inter-bonded.

7.2.2 Columns

Whenever large bearing surfaces are needed, columns are used. They can form an integral unit with the wall or can be built separately. They are made of standard stretcher and corner blocks or other special shapes are used. The hollows within the blocks may be filled with concrete. (See Fig. 7.18 and 7.19).

7.2.3 Window and Door Openings

The blocks used in jambs should have one hollow near the opening which should be filled with concrete. (See Fig. 7.20). The door and window frames are screwed to wooden plugs left in the masonry and fixed in the lintel with small dowels of mild steel. Under the windows and doors, a course of solid concrete block masonry is laid which extends into the adjacent walls to a distance of at least 300 mm on either side. Alternatively a precast concrete sill of solid section may be used. Lintels are made of hollow channel shaped blocks filled with concrete and having steel reinforcement at their bottom.

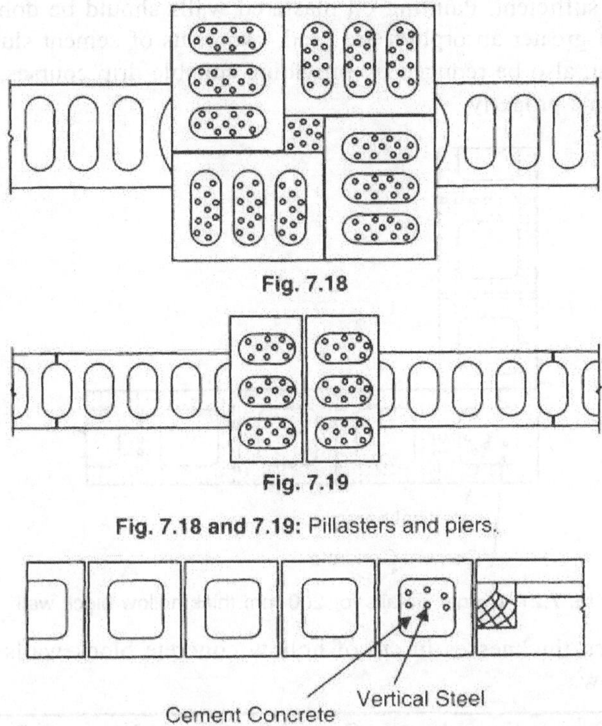

Fig. 7.18

Fig. 7.19

Fig. 7.18 and 7.19: Pillasters and piers.

Fig. 7.20: Jamb details for 200 mm thick hollow block wall.

7.3 EXPANSION JOINTS

The masonry is liable to crack due to unequal stresses created on account of the expansion and contraction of the block. Hence controlled joints are introduced at definite intervals (5 to 10 metre). These arrest the movement of the concrete masonry units. They can be placed at the junction of load bearing and non-load bearing walls or at the junction of walls and columns or at other suitable places. There are many types of controlled joints used. At each joint, an elastic compound consisting of a mixture of chalk, sand and linseed oil is used. This is covered with cement to match the colour. The thickness of the joint is 5 mm to 10 mm. Two G.I. straps 150 mm × 50 mm in size should be provided in every alternate joint for bond with the main walls.

7.4 REINFORCED WALLS

Vertical reinforcement may be provided in the hollow of the corner block and filled with concrete (see Fig. 7.21).

Reinforcement is provided at the horizontal joints for developing higher strength in walls. Further expansion cracks which occur on account of moisture and temperature changes are reduced. Two horizontal bars of 6 mm diameter are placed one each on the face of the wall. This reinforcement may be restricted to top and bottom courses of window openings, door openings, etc. Welded steel mesh can also be used.

7.5 WATER PROOFING INTERIOR WALLS

Walls built of sound concrete blocks are able to resist free leakage of water but leakage may occur through the joints. Two coats of neat cement slurry over 10 mm thick plaster of 1 : 4 cement mortar is helpful in water proofing the masonry. For blocks cast on a vibrating machine, painting only will be sufficient. Painting on plastered walls should be done on completely dry walls. When blocks of greater absorption are used, two coats of cement slurry should invariably be used. Dampness can also be reduced by providing suitable drip courses, proper window sills and by making the joints properly.

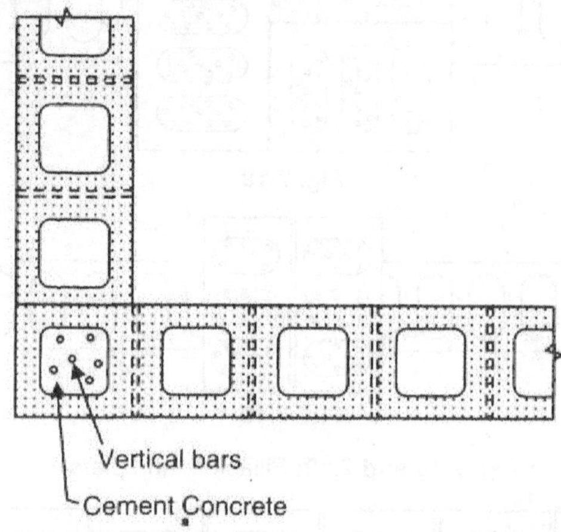

Fig. 7.21: Corner details for 200 mm thick hollow block wall.

The recommended thicknesses in cm of hollow concrete block walls for various floors is given in the table below:

No. of floors	Foundation and Basement	Ground floor	Ist	2nd	3rd
—	300 or 200	200	—	—	—
1	300	200	200	—	—
2	300	300	300	200	—
3	400	400	300	300	200

Hollow concrete block masonry is used in domestic buildings, schools, churches and other public buildings. It is specially suitable for low cost houses.

STRUCTURAL CLAY TILE MASONRY

Structural tile is light in weight, fire proof, resistant to rats and termites and is free from decay caused by the contact with water or chemicals. The use of these tiles reduces the maintenance and constructional costs. Structural clay tile is a term referred to the various sizes and kinds of hollow and practically solid structural units. They are moulded from clay or diatomaceous earth. The latter gives very light tiles which can be easily cut but possess less crushing strength. These tile units are used to build foundations, walls, partitions, floors and other structural members.

The various types of structural clay units are shown in figures 7.22 to 7.33.

It should be noted that these units can be used while being laid on end or on side and are designed accordingly. Load bearing tiles are made in various grades, *e.g.*, load bearing (L.B.) and extra load bearing (L.B.X.). Specifications have been framed in various countries to account for the various physical properties of the tiles. Generally absorption is not greater than 15 per cent for good quality tiles. The compressive strength based on gross area in N/mm^2 is about 10 for tiles laid vertically on ends and about 5 for tiles laid on sides. For non-load bearing walls, these limits are relaxed.

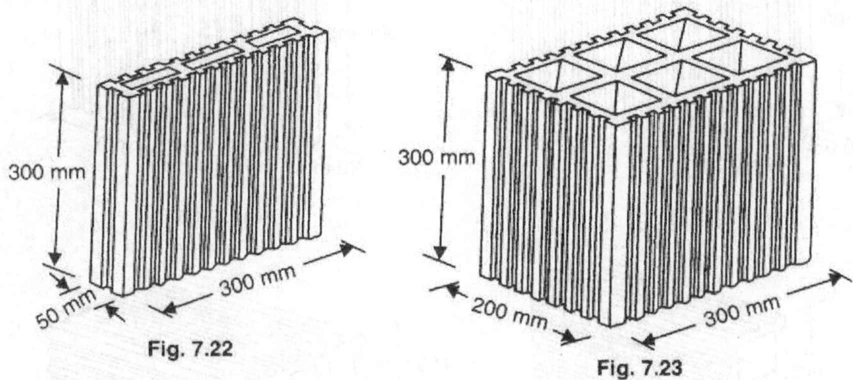

Figs. 7.22 and 7.23: Different types of structural clay units.

The shell of a tile constitutes the four sides surrounding the hollow interior and the webs serve as partitions between the cells. The overall average thickness of the shells should not be less than 20 mm and of the web not less than 10 mm for end construction tiles. The shell for side construction tile should not be less than 10 mm. The width of any cell in the side construction tile, measured in the direction of wall thickness, should not exceed $4\frac{1}{2}$ times the average overall thickness of either the upper or lower bearing shells. All tiles must be free from laminations, cracks, blisters, surface roughness or other defects which would interfere in the construction.

If plaster is to be applied directly to the sides of a tile, the surfaces should be grooved to provide keys for holding the plaster in place. Common and face tile units may have grooves on one or two faces and both ends grooved. Groove should not be less than 3 mm and not more than 250 mm in depth. They should not be more than 25 mm in width. The area covered by grooves should not exceed 50% of the area of the scored faces. For walls and partitions where no plastering is needed, smooth tiles may be used.

Glazed surfaces can be obtained for the tiles in various colours. Other finishes resembling rocks, bricks, etc. can also be obtained.

7.6 LAYING OF CLAY TILE UNITS

This is similar to the laying of concrete hollow block masonry. However, the following points should be noted in this connection:

(*i*) All the units should be dipped in water before being laid in mortar.

(*ii*) The corner tile is laid on end.

(*iii*) Mortar is applied to the inner and the outer faces in the horizontal joints and to ends.

(*iv*) Special closer units can be used for the corners.

(*v*) Columns or pillasters are built of hollow tile units arranged suitably and special blocks are cast. The courses are bonded by providing units which connect the wall with the column in alternate layers.

(*vi*) The jambs of doors and windows are made of special blocks. Lintels are used for bigger openings. They are made up of units which are commonly used for the corners. The concrete is filled into the cores and reinforcement is passed through the lower cells.

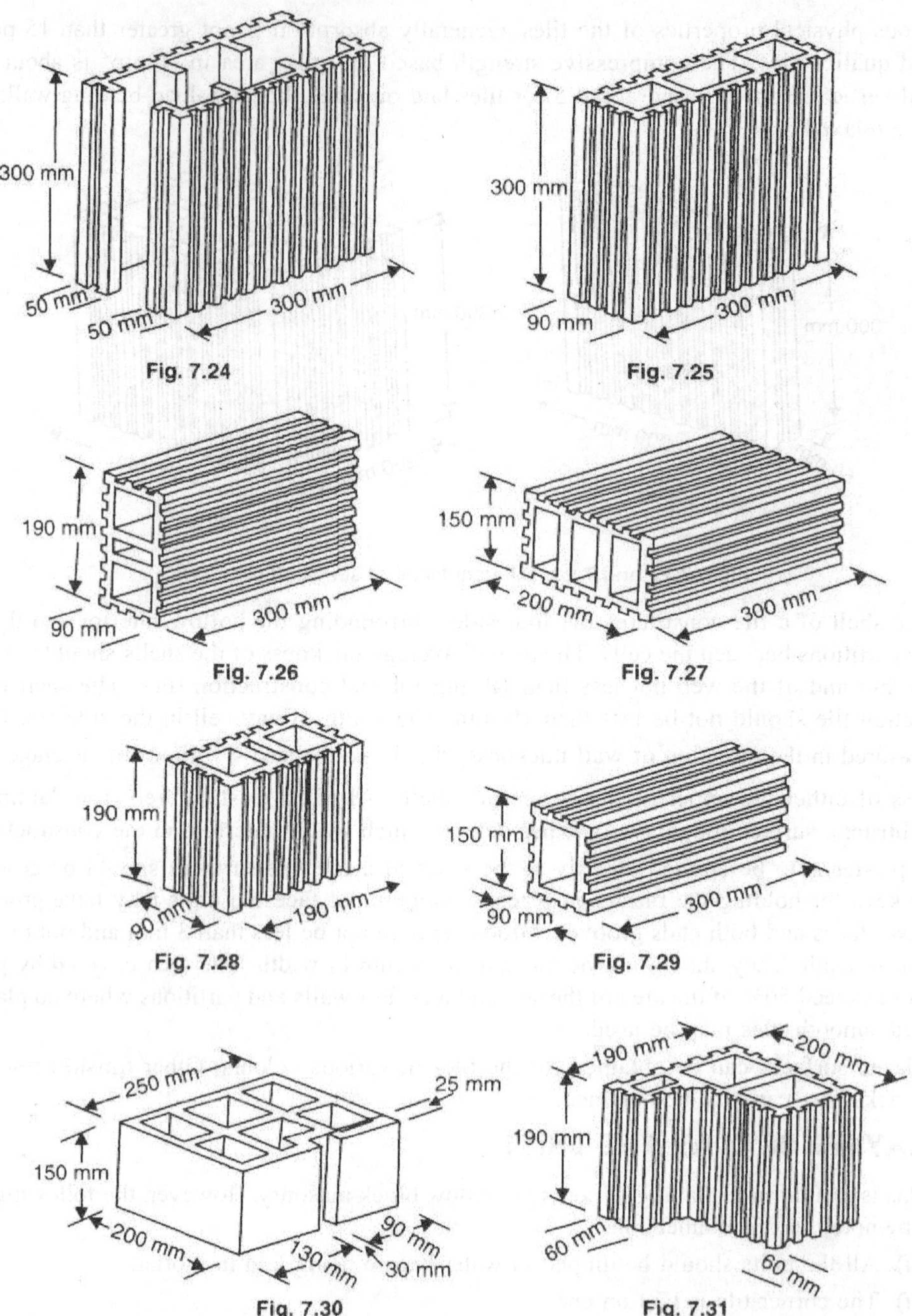

Figs. 7.24 to 7.31: Typical shapes and sizes of structural clay units.

Walls of structural clay are built of units which are of load bearing or non-load bearing type having one or many cells. These units may be glazed or natural, rectangular, square or curved. Tile walls are used for residences, silos, poultry houses, etc. These walls are easy to construct and can bear heat and moisture effectively. Partitions are made of smaller units and are frequently used in construction. Similarly hollow tiles can be used for floors, roofs and foundations.

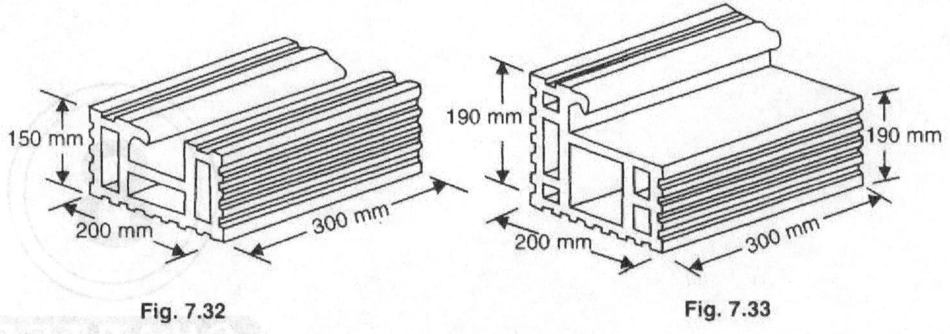

Fig. 7.32

Fig. 7.33

Figs. 7.32 to 7.33: Clay units.

QUESTIONS

1. What are the various types of concrete block masonry and the finishes which can be adopted?
2. How are the concrete blocks manufactured and laid in the walls?
3. What are the precautions to be kept in view while constructing concrete block masonry?
4. What are the advantages of hollow tile block construction?
5. What are the special precautions to be taken for this type of masonry work?

REFERENCES

1. **Dalzell, J.R. and Townsend G.**: *Masonry simplified.*
2. **Concrete Association of India**: Hollow block construction.
3. **N.B.O.**: *Symposium on building materials and construction methods.*

Partition and Cavity Walls

PARTITION WALLS

Partition walls are interior walls one storey or part storey in height, whose main function is to divide the space within a building into rooms or areas. These can be load-bearing or non-load bearing but with the sole purpose of division with strength and stability, sound insulation and fire resistance. When a partition wall is required to partially support girders over it, it is called a load bearing partition wall. A portion of the floor load, which the girders carry, is transferred to the partition wall.

Conventional partition walls are of the non load-bearing type and they are used to economise in space, weight and reduce cost. They do not have to support any other load than their own dead weight.

8.1 TYPES OF NON-BEARING PARTITIONS

The various types of such partition walls are:

(1) Brick-partitions.

(2) Hollow block partitions.

(3) Glass block partitions.

(4) Concrete partitions.

(5) Metal Lath partitions.

(6) Solid plaster partitions.

(7) Corrugated sheet partitions.

(8) Wooden partitions.

These are described in the following articles:

8.1.1 Brick Partitions

They are of the following types:

(*a*) ***Plain Brick Type*:** These partitions are $\frac{1}{2}$ brick thick and are constructed by laying bricks as stretcher courses. Sufficient care should be taken in laying the bricks properly so as to develop full strength. These partitions cannot take a heavy load and hence their height is restricted. Partition walls of $\frac{1}{3}$ brick thickness are also used to enclose rooms of small dimensions and where the partitions are not subjected to lateral forces, *e.g.*, load due to coal or grain. Openings in brick partitions for doors, etc., must have lintels over them.

(*b*) ***Reinforced brick partitions***: These are stronger than the ordinary plain brick partitions. The bricks are reinforced with iron strap about 25 mm wide and 1.5 mm thick which is laid horizontally every fourth or fifth course. If lime mortar is used in the construction of the partitions, the course containing reinforcement must be laid in cement mortar. Instead of hoop iron, two mild steel rods about 6 mm diameter can also be used. Sometimes a wire mesh is used as reinforcement for these partitions.

(*c*) ***Brick knogging***: This type of partition wall consists of brickwork built up within a framework of wooden members. This is described in Chapter 5.

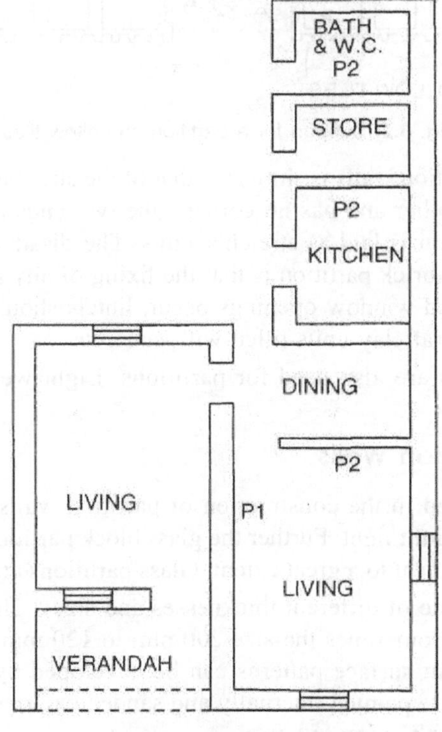

Fig. 8.1

8.1.2 Hollow Block Partition Walls

Structural clay tile units or hollow concrete blocks described in chapter 7 can be used for the construction of partition walls. Tile partition walls are similar to the tile walls except that their thickness is less and varies from 50 mm to 100 mm depending upon the type of partition, attachment of any fixtures and sound insulation desired. Because of large amount of air within the cells of the clay tiles, they are good insulators against heat. Tile partitions are used when they are supported by steel or concrete beams or concrete floors. They should never be used when they are supported on wooden floors or wooden frames. When a partition is supported by a tile floor, there should be a solid concrete foundation below it. This is needed because the tile, if used as a foundation, will not bear the weight coming over it. (*See* Fig. 8.2).

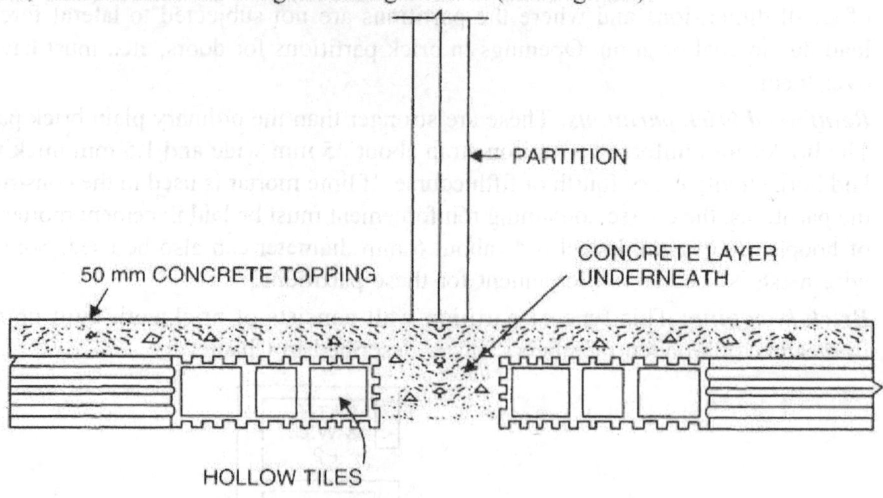

Fig. 8.2: Support for a partition on hollow tiles.

The laying of these partition walls is similar to that of the structural clay walls. If tile partition is laid from one wall to the other and has no corners, the two ends are built up for three or four courses and in between portion is laid as stretcher units. The disadvantage of structural clay tile partition with respect to solid brick partition is that the fixing of any attachment with the partition is difficult. Wherever door and window openings occur, lintels should be provided. These lintels should be made up of structural clay units filled with concrete.

Hollow concrete blocks are also used for partitions. Light weight concrete blocks can be very suitable for insulation.

8.1.3 Glass Block Partition Walls

Glass blocks can be used in the construction of partition walls because they provide good architectural effect and also admit light. Further the glass block partitions act as insulated walls and prevent a passage of sound or heat to a great extent. Glass partitions are easy to clean and maintain.

Glass blocks can be made of different thicknesses and sizes. They are usually made square, of side 140 mm or 190 mm. Sometimes the size 200 mm to 120 mm is also made. The thickness is normally 100 mm. Different surface patterns can be developed by fluting the inner and outer surfaces. The jointing edges are painted internally and sanded externally to form a key for mortar.

They can be of glass brick type which act as individual units and are interconnected with joggled joints. Another type is to have a glass block with hollow inside. A third type is to have glass blocks with hollow spaces, some of which can be filled with concrete. Horizontal and vertical steel rods are placed in between to act as reinforcement.

The glass brick type blocks having joggles are fitted to each other as shown in Fig. 8.3. They are connected to the wall by fixing a fillet to it and fixing the glass blocks in such a manner that their end grooves fit with the fillet.

Other types of glass blocks are connected together with concrete and steel bars or with cement lime mortar. In the latter case fine sand is used and care should be taken that all the joints are filled completely. If blocks are greater than 250 mm in height, every horizontal joint should be reinforced with iron stirrups or expanded metal stirrups. For blocks about 150 mm in height, this reinforcement can be placed after every third or fourth course. The joints can be suitably tooled. Provision for expansion can be suitably made along the jambs and at the head of each panel.

Expansion joint strips about 1 cm thick made of cork, fibre glass or other suitable material should be placed in the expansion joint. Panels which are greater than 16 square metre in area should be provided with structural stiffening members.

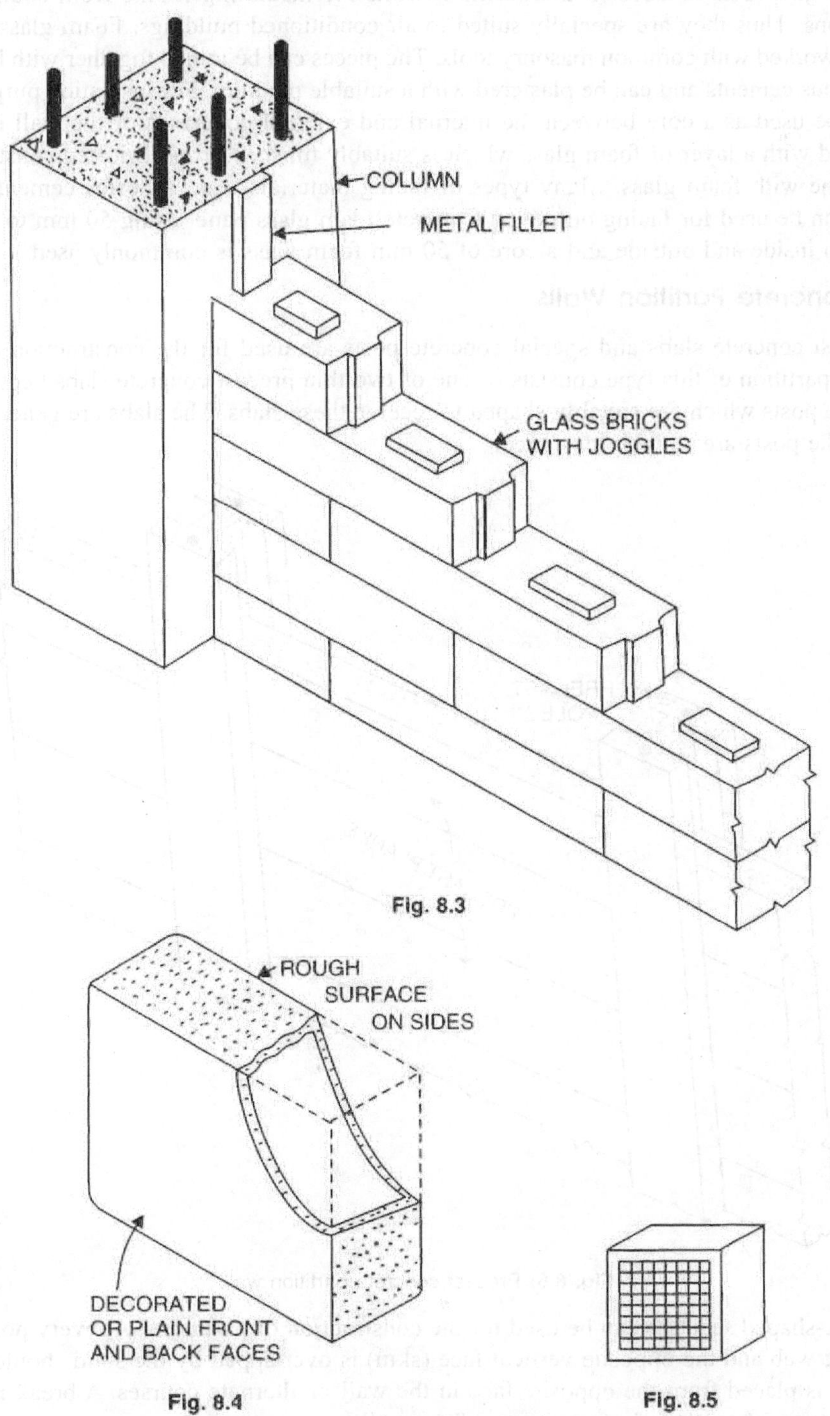

Fig. 8.3

Fig. 8.4

Fig. 8.5

Figs. 8.3 to 8.5: Glass block partitions and typical glass blocks.

Loads other than the self weight of the partitions should not be transmitted to the partitions. The joints may not be broken vertically or horizontally.

Foam glass can be used in the partition walls for insulating rooms from sound and heat transmissions. Thus they are specially suited to air conditioned buildings. Foam glass can be cut easily and worked with common masonry tools. The pieces can be joined together with hot bitumen or bituminous cements and can be plastered with a suitable medium. For insulation purposes, foam glass can be used as a core between the internal and external coverings or the wall is built first and covered with a layer of foam glass which is suitably finished. Pre-fabricated construction can also be done with foam glass. Many types of facing materials, *e.g.*, asbestos cement board and plywood can be used for facing purposes. Concrete foam glass panel using 50 mm to 100 mm of concrete on inside and outside and a core of 50 mm foam glass is commonly used.

8.1.4 Concrete Partition Walls

Precast concrete slabs and special concrete posts are used for the construction of partition walls. The partition of this type consists of one or two thin precast concrete slabs kept in position by concrete posts which are suitably shaped to receive these slabs. The slabs are generally 40 mm thick and the posts are suitably designed.

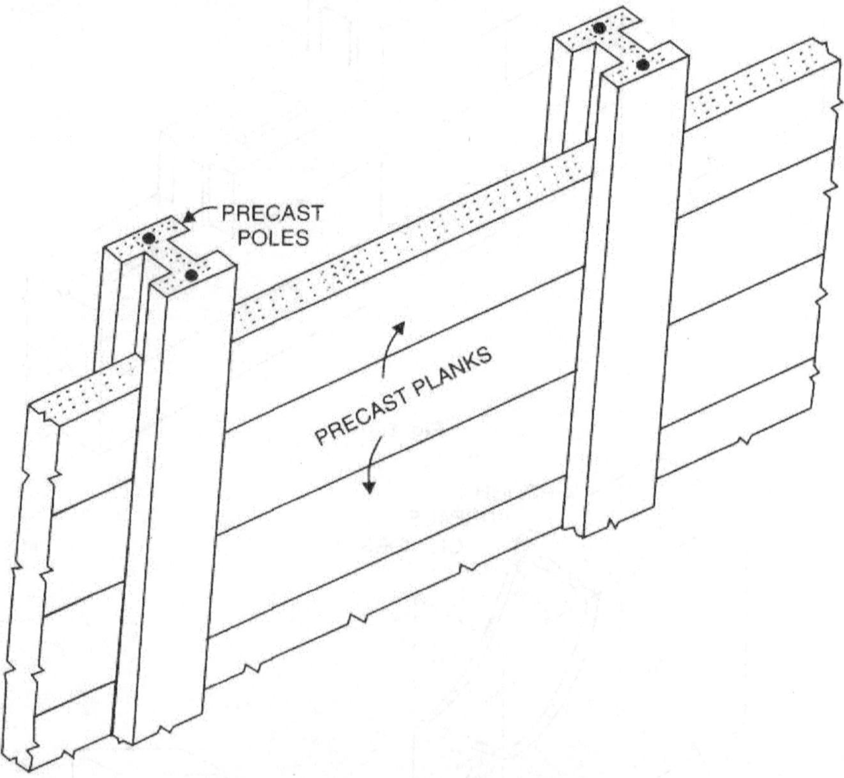

Fig. 8.6: Precast concrete partition wall.

T or L-shaped sections can be used for the construction of partitions. At every point the joint between the web and the opposite vertical face (skin) is overlapped by the solid shoulder of the L or T which is placed from the opposite face in the wall in alternate courses. A break in bond and inter-connection of outer and inner facing of the wall is necessary.

Foamed concrete blocks can be used for the construction of partitions. The thickness used varies from 80 mm to 100 mm. The blocks are made in density of 6 kg/cub metre. These blocks can be easily sawn with ordinary saw to suit the various positions. The mortar used in the construction is lime cement mortar or weak cement mortar. Rich cement mortar, if used, will give cracks along the joints. The mortar may be of 1 : 6 cement sand mix or 1 : 1 : 6 lime, cement, sand mix. Joints in the walls with foamed concrete blocks should be as thin as possible and never greater than 10 mm. The partition walls made of these blocks can be bonded with external walls by driving 15 mm long square nails at every 300 mm spacing in the external walls. For partitions exceeding 8 metres longitudinally, it is desirable to lay two 6 mm dia bars longitudinally at half the height of the wall. No plastering is necessary for internal walls. The surface can be given a coat of coloured cement. If plastering is desired, the same can be done in two layers, the first coat being of weak plaster with coarser sand. Doors and windows can be anchored in the block masonry by means of 15 mm long square nails at every 500 mm along the vertical frame.

Reinforced concrete partitions are not used to a great extent but can be conveniently built and form one of the strongest partitions. For thin walls up to 100 mm thickness the reinforcement is placed centrally in the wall. It consists of horizontal and vertical rods or a woven mesh. They are used where greater resistance of the partition is necessary.

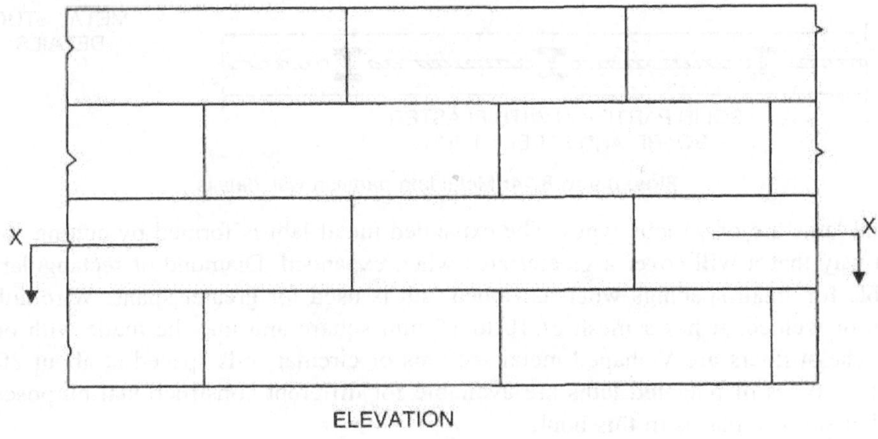

Fig. 8.7

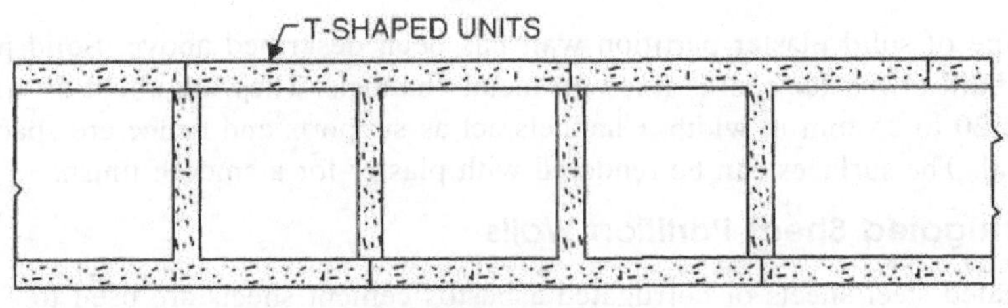

Fig. 8.8

Figs. 8.7 and 8.8: Precast concrete partition using T-shaped units.

8.1.5 Metal Lath Partition Walls

Solid partitions are constructed by placing 20 mm or 25 mm channels vertically and fixing metal lath to it. The metal lath may be fixed on one side and the plaster applied on both the sides (*see* Fig. 8.10). The lath is placed with its greatest dimension horizontally. The sheets are lapped at least 25 mm on the ends and are fixed with the wire to the channel at 150 mm interval. The channels are placed by driving holes into the floors and the roofs. For wooden floors and roofs, these channels can be bent and fixed by nailing.

Hollow partitions can also be made by fixing the metal lath to these steel channels on both sides. These channels are spaced at 300 to 400 mm apart. Special cold formed channels, made up of sheet steel and bent to the required shape are used. They are 50 to 100 mm deep and may have projections to which the lath is fixed. Built up channels can also be used. They consist of two small channels braced by flat iron.

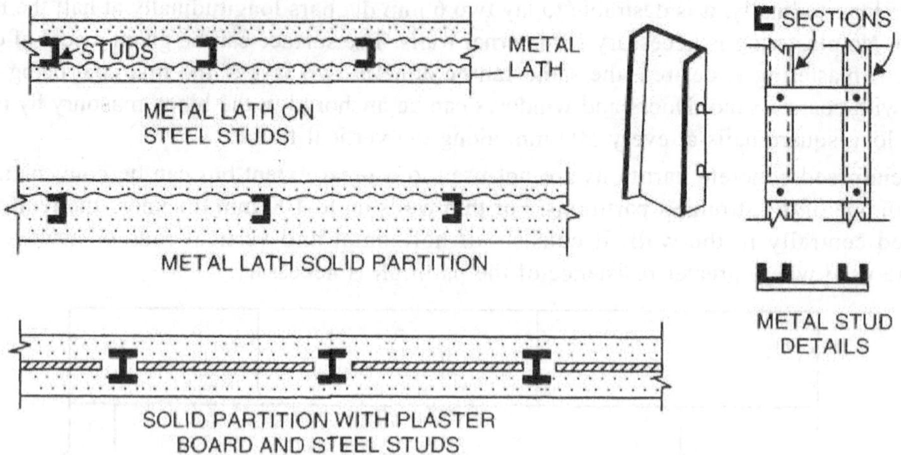

Figs. 8.9 to 8.14: Metal lath partition wall details

Metal laths are of various types. The expanded metal lath is formed by cutting sheet metal in such a way that it will cover a greater area when expanded. Diamond or rectangular meshing are suitable for small spacings whereas ribbed lath is used for greater spans. Wire lathing may be woven or welded. It has a mesh of 10 to 15 mm square and may be made with or without stiffners. The stiffners are V-shaped metal sections or circular rods spaced at about 200 to 300 mm. Various types of patented laths are available for different constructional purposes and are described at suitable places in this book.

8.1.6 Solid Plaster Partition Walls

One type of solid plaster partition wall has been described above. Solid plaster partitions can also be built of plaster boards fixed on metal channels. They are about 50 mm thick and the channels are 20 to 25 mm in width. Channels act as supports and hence are spaced at about half metre interval. The surfaces can be rendered with plaster for a smooth finish.

8.1.7 Corrugated Sheet Partition Walls

Corrugated steel sheets or corrugated asbestos cement sheets are used to a large extent for the construction of partitions. They are specially suitable for cheap type of construction and in factories. These sheets are fixed to a framework of wood or iron. The sheets can be fixed with different types of attachments to the horizontal and the vertical members of the framework. The various types of fixing arrangements to the horizontal members of iron are shown in figures 8.15 to 8.17.

For fixing to wooden members, special screws and washers are used. Openings for doors, etc., can be easily left in these partitions. If it is necessary to have a smooth surface on one side, plywood boards can be fixed to the framework.

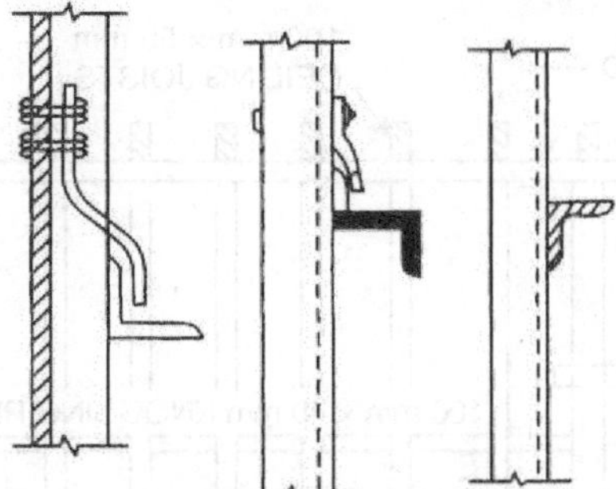

Figs. 8.15 to 8.17: Supporting methods for corrugated partitions.

Patented asbestos cement slabs consisting of two sheets 10 mm thick attached to an inner corrugated sheet about 5 mm thick are used wherever flat surface is desired. These sheets are jointed in cement mortar and can be painted or distempered. They afford durable, fire resistant, insulated and light weight partitions. Asbestos cement sheets having various textures and finishes are also available.

8.1.8 Wooden Partition Walls

These may be divided into two types:

(*a*) Common partition walls.

(*b*) Trussed partition walls.

Common partition walls consist of a framework of head, sill and upright members. The uprights are tenoned into the head and sill. Solid and uniform support for the sill members must be ensured, hence they should be placed just above the girders of a floor or at right angles to the joint of flooring underneath. The struts are usually 100 to 80 mm × 80 to 50 mm in size. They are placed at suitable intervals depending upon the type of facing to be used. The struts are stiffened by horizontal pieces of timbers called knoggings. These are of the same width as the struts and are about 50 mm thick. The width of head and sill is also equal to that of the strut. The head is fixed to the ceiling joint and the sill is fixed to the flooring. Sometimes these members are fitted into the walls which is not necessary if they are securely fixed as described above. If any opening is to be left, it should be ensured that the vertical and horizontal posts for the doors are sufficiently strong and are joined securely. The partition frame thus built can be covered by a layer of plaster on lath on one or both sides or by fixing boarding, plywood sheet, etc. These partitions are very light and need not have any supporting wall below. However, they cannot support any load or cannot bear any big fixture.

Trussed partition walls are required to be constructed where there is no other means of supporting the partitions except at their ends. They are designed to support their own weight and in addition any floor loads coming on them. The main object of trussing is to arrange the members of the frame in such a manner that the loads are transmitted through the members directly on to the supports and a rigid framework is ensured. Hence triangular frameworks are desirable. A simple type of such partition is shown in Fig. 8.26. It consists of a sill, a head, inclined braces, vertical studs, knogging pieces and posts for the door opening. The joints may be held together with iron straps or iron bolts; instead of knogging pieces, knogging strips may be used for strengthening the partition. The ends of the sill and the head rest on suitable bedding stones on the walls. The joint at the head of the brace and the door post and also the joint between the brace and the horizontal member is shown in the figure.

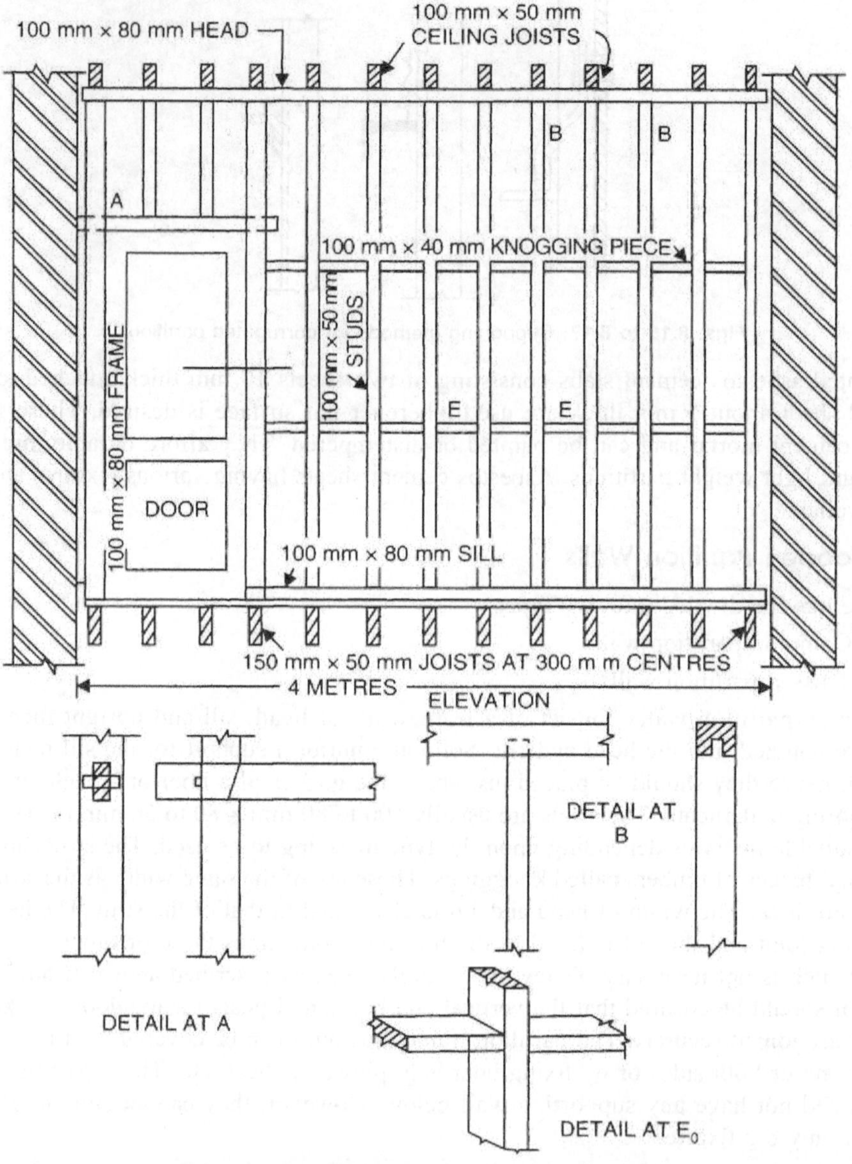

100 mm × 80 mm HEAD
100 mm × 50 mm
CEILING JOISTS
B
B
A
100 mm × 40 mm KNOGGING PIECE
100 mm × 50 mm
STUDS
100 mm × 80 mm FRAME
E
E
DOOR
100 mm × 80 mm SILL
150 mm × 50 mm JOISTS AT 300 m m CENTRES
4 METRES
ELEVATION
DETAIL AT
B
DETAIL AT A
DETAIL AT E0

Figs. 8.18 to 8.23: Common wooden partition.

KNOGGING STRIP
STUDS
STUDS
ELEVATION

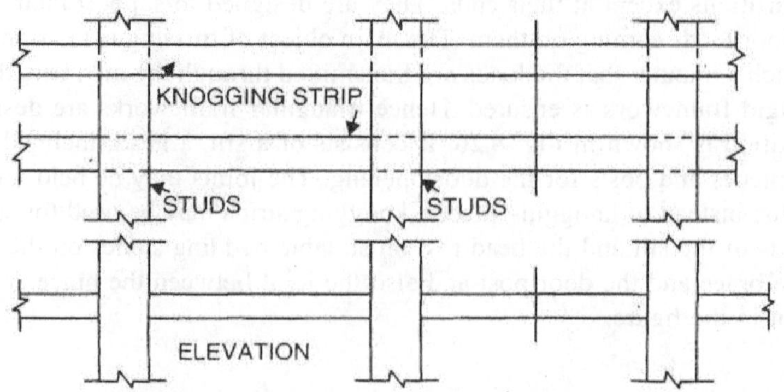

Figs. 8.24 and 8.25: Details of struts and knogging strips.

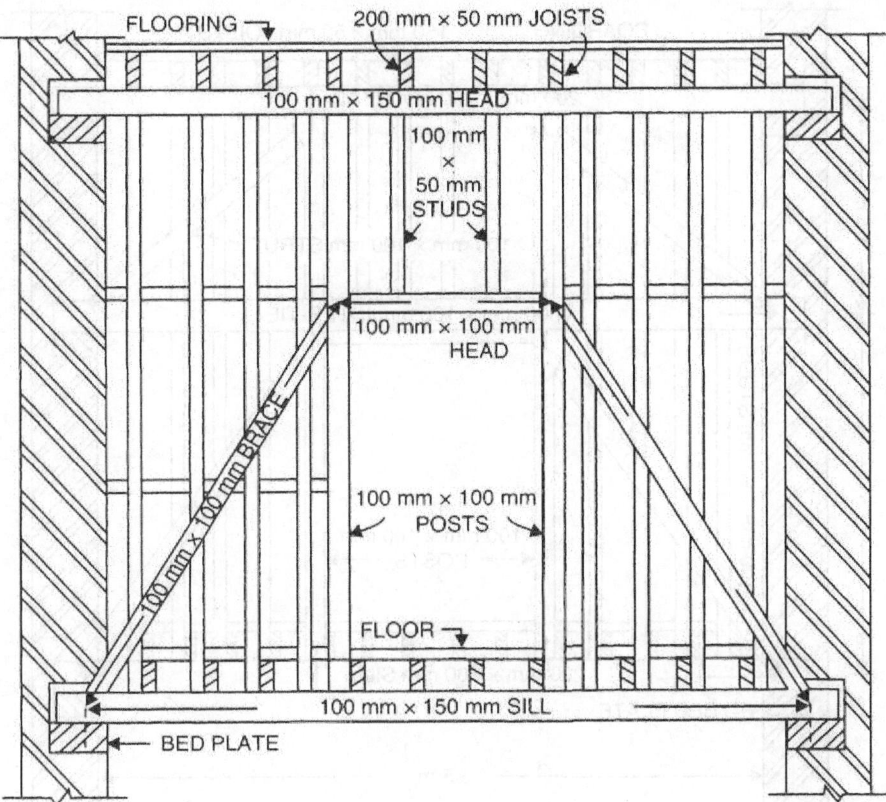

Fig. 8.26: Trussed partition.

For partitions which have to carry floor loads also, the trusses have to be designed suitably and the number of braces is increased (*see* Fig. 8.27). Further a horizontal member is added in between the sill and the head. This is called 'inter-tie'.

These partitions can also be covered with plywood or lath and plaster or wood boarding.

CAVITY WALLS

A cavity wall is a wall comprising two leaves or skins, each leaf being built of masonry unit and separated by a cavity and tied together with metal ties or bonding units to ensure that the two leaves act as one structural unit, the space between the leaves being either left a continuous cavity or filled with a non-load bearing insulating and waterproofing material. This type of construction is ideally suitable for places where prevention of dampness from exterior, high insulating value against heat and sound and economy are desired. In a country like India, where the climate is either hot and/or hot-humid, cavity wall construction is very desirable for conditions of comfort, health and preservation of the building against dampness. However, it is being used to a limited extent till today.

The two leaves of a cavity wall may be of equal thickness or the thickness of the inner leaf may be increased to meet the desired structural requirement. The inner and outer leaves of the wall should not be less than 100 mm in thickness throughout the height of the wall.

For a cavity wall to be effective, it is absolutely essential that the inner leaf is entirely disconnected from the outer leaf, except for ties. This is done by providing a cavity of uniform width, normally 50 mm but not exceeding 75 mm. It is, however, at the openings for doors and windows, where special precautions are necessary.

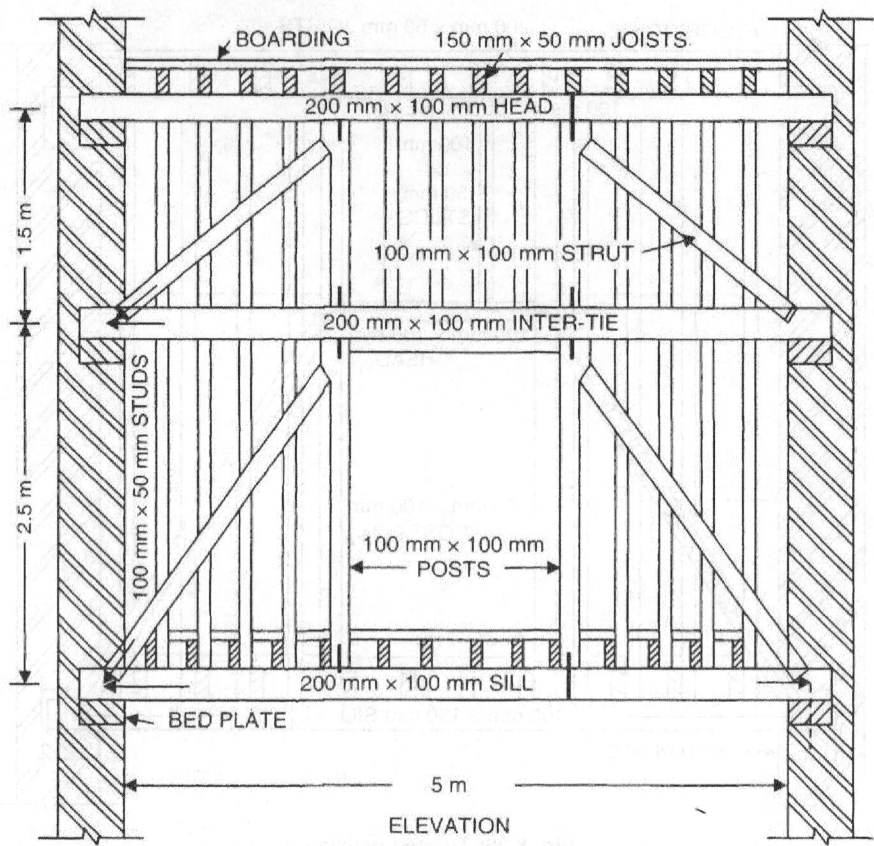

Fig. 8.27: Trussed partition to support floor loads.

The inner and outer leaves of the wall are securely tied together with suitable bonding ties of adequate strength. These ties should be of such a shape that water from the outer wall does not pass into the inner wall. Strong non-corrodable ties of wrought iron or mild steel thoroughly galvanised, dipped in hot tar and sand are used. Copper or bronze metal ties are used whenever excessive corrosion is anticipated. These ties can be of wire type, the ends of which are twisted together and turned downward to prevent the transmission of water from the outer face to the inner face. Flat sections can be used of twisted or loop shape (*see* Figs. 8.29 and 8.30). These ties should be placed at intervals not exceeding 1 metre horizontally and 40 cm vertically. Special bonding bricks of terracotta are also sometimes used as ties.

A cavity wall is built with an outer facing of specially selected face brick and the inner leaf is made of common bricks. The ties are placed wherever the joints coincide. The half brick walls are constructed in stretching bond. However at certain intervals half headers may be introduced to improve the appearance.

The cavity wall may extend down to the concrete foundation which means that water in the soil can pass into the inner face of the wall wherever the brickwork in exterior wall is not constructed properly. This would cause dampness in the floor. To prevent this, the cavity is taken only about 100 mm below the damp proof course of the external wall and below that level the wall is built of a solid section. If water is liable to collect, the course of bricks in the external face just below the damp proof course should be laid with hollow joints to enable this water to get out. These hollow joints may be left at about 1 metre interval along the length of the course. As an alternative small weep holes may be left.

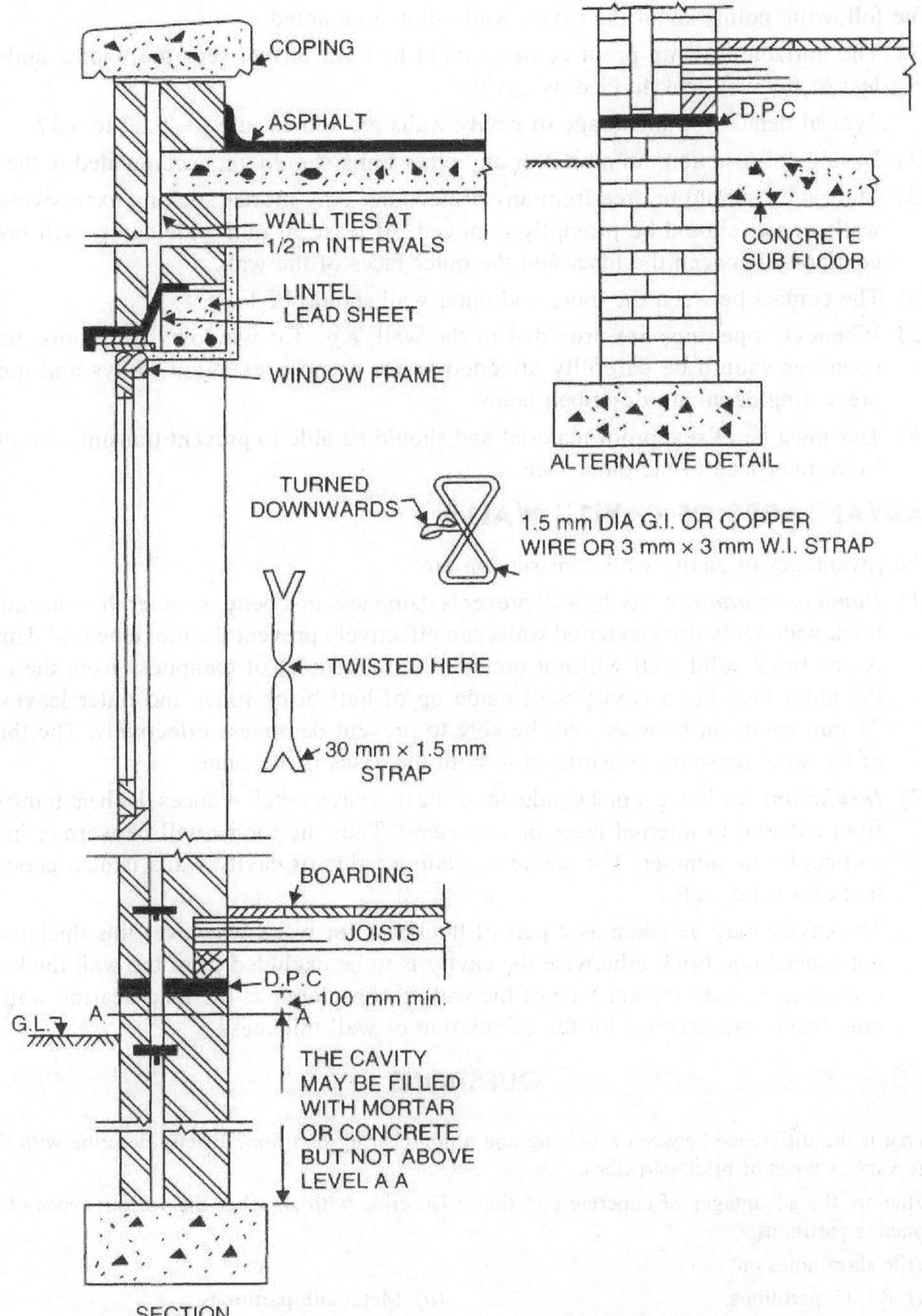

Figs. 8.28 to 8.31: Cavity wall details.

The damp proof course for these walls should be about 150 mm below the ground level.

Whenever wooden floors are used along with the cavity walls, it is essential to use only sound and seasoned timber otherwise dampness from the cavity may cause dry rot. The ends of joints bearing into the cavity walls should be creosoted and a similar treatment should be given to wall plates resting on these walls.

The following points about the cavity walls should be noted:

(1) The horizontal damp proof course should be built in two separate widths under each leaf of the wall and divided by cavity.

Typical details for anchorage of cavity walls are shown in Figs. 9.32 to 9.37.

(2) During construction, no mortar or any other thing should get accumulated in the cavity.

(3) The cavity should be free from any projections. Any mortar sticking excessively to the walls or ties should be promptly removed. At least 50 mm wide cavity will break the continuity between the inner and the outer faces of the wall.

(4) The contact between the inner and outer wall should be least.

(5) Whenever openings are provided in the wall, *e.g.*, for windows and doors, heads of openings should be carefully attended to for damp prevention. Ways and means of preventing damp are described below.

(6) Ties must be of rust proof material and should be able to prevent transmission of water from inner face to the outer face.

8.2 ADVANTAGES OF CAVITY WALLS

The advantages of cavity wall construction are:

(1) ***Damp prevention*:** A cavity wall prevents dampness in a better manner than an equivalent brick wall. Only thick external walls can effectively prevent the movement of dampness. A one brick solid wall will not prevent the movement of dampness from the outer to the inner face but a cavity wall made up of half brick inner and outer leaves and a 75 mm cavity in between will be able to prevent dampness effectively. The thickness of the solid masonry constructed in both the cases is the same.

(2) ***Insulation*:** Air being a bad conductor of heat, a cavity wall reduces the heat transmission from external to internal faces or *vice versa*. Thus the rooms will be warmer in winter and cooler in summer. The sound insulating value of cavity walls is also greater than that of a solid wall.

The cavity may be taken as a part of thickness for walls whenever this thickness does not exceed one brick otherwise the cavity is to be excluded from the wall thickness. In certain cases only thicker face of the wall is considered as the load bearing wall and is only taken into account for the calculation of wall thickness.

QUESTIONS

1. What is the difference between a bearing and a non-bearing partition? Briefly describe with sketches the various types of brick partitions.
2. What are the advantages of concrete partitions? Describe with sketches the various types of precast concrete partitions.
3. Write short notes on:
   (*i*) R.C.C. partitions. (*ii*) Metal lath partitions.
   (*iii*) Solid plaster partitions. (*iv*) Corrugated steel partitions.
4. What are the types of wooden partitions commonly used? Describe with sketches the construction of a trussed partition.
5. What are the advantages and disadvantages of cavity wall construction? Draw a section through an external cavity wall of a building showing clearly all details including the prevention of damp entry at the windows.

REFERENCES

1. **Capital Project, Chandigarh**: *Bye laws.*
2. **British Standard Specification**: *Metal wall ties No.* 1243 *of* 1945.
3. **Crockhill**: *Carpentry and joinery, Vol. III.*
4. **National Building Organisation**: *A symposium on building materials, Vols. I and II.*
5. **Dalzell and Townsend**: *Masonry, simplified.*
6. **Kay, N. W.**: *Modern building encyclopaedia.*
7. **Frederick, S. M.**: *Handbook of building construction.*
8. **Bureau of Indian Standards**: *National Building Code of India* 2005, *Part* 6, *Section* 4, *Masonry.*

CHAPTER

Arches and Lintels

An arch is a structure comprising of a mechanical arrangement of wedge shaped blocks, upholding each other by the mutual pressure of their own weight and maintained in equilibrium by the resistance of the supports or abutments and designed to support the superincumbent load of the wall. The materials used for the construction of arches are bricks, stones and concrete blocks. Steel and reinforced concrete arches are, however, built of a single unit and are of a rigid type. They are used to span bridges and are not covered in this chapter.

9.1 TERMS USED

(1) ***Arch Ring***: This is the curved ring of masonry forming the arch.

(2) ***Span***: This is the width of the opening which is covered by the arch.

(3) ***Voussoirs***: These are wedge shaped blocks of masonry of which the arch ring is made of.

(4) ***Key stone***: The central voussoir at the highest point of the arch.

(5) ***Extrados***: The outer curve of the arch.

(6) ***Intrados***: The inner curve of the arch.

(7) ***Soffit***: The inner or under surface of the arch.

(8) ***Crown***: The highest part of the extrados.

(9) ***Abutment***: The portion of the wall which supports the arch.

(10) ***Skewbacks***: These are the inclined or splayed surfaces of the abutments prepared to receive the arch and from which the arch springs.

(11) ***Springing points***: The points at the intersection between the skewbacks and the intrados.

(12) ***Springing line***: The horizontal line joining the two springing points.

(13) ***Springer***: The lowest voussoir immediately adjacent to the skewback.

(14) ***Haunch***: The lower half of the arch between the crown and a skewback.

(15) ***Rise***: The vertical distance between the springing line and the highest point of intrados.

z The triangular space between the back of the arch ring and a horizontal plane tangent to it at the crown.

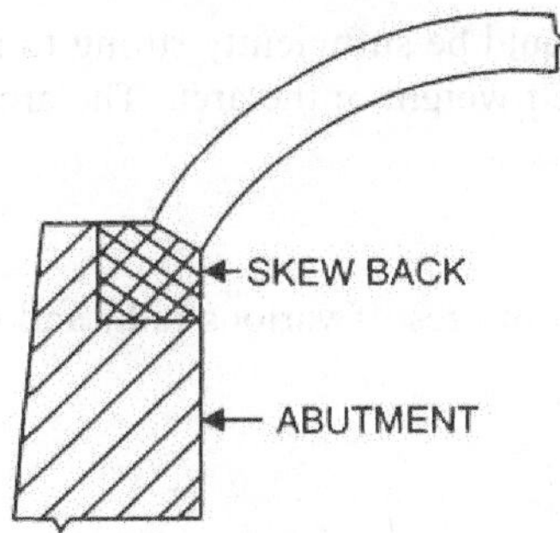

Fig. 9.1

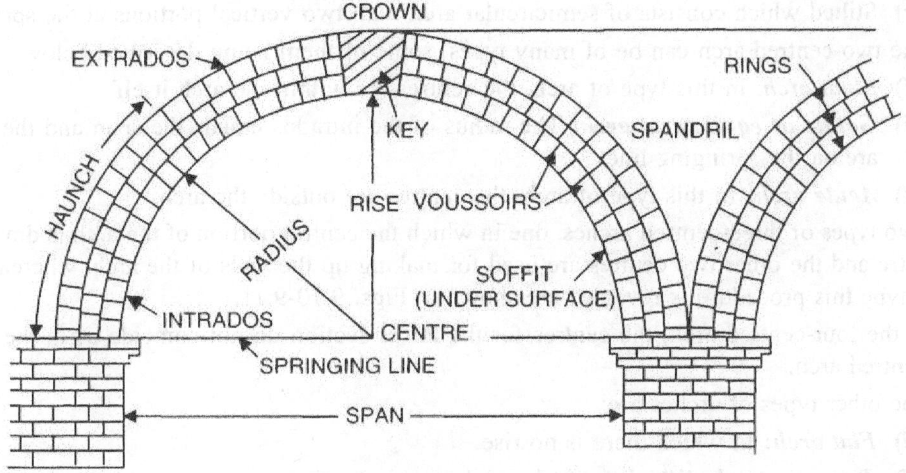

Fig. 9.2: Terms used in Archwork.

9.2 STABILITY OF ARCHES

The arch remains stable due to the friction between the surface of voussoirs and the cohesion of the mortar. The various ways in which an arch can fail are:

(1) The crushing of the material.

(2) The sliding of one voussoir on another.

(3) Rotation of some joint about an edge.

(4) Uneven settlement of abutments.

An arch will fail in crushing, if the thrust in some part of the arch exceeds the safe crushing strength of the material comprising the arch. For this the dimensions of the voussoirs should be adequate to withstand the pressure coming on them. Usually they are of uniform height, generally 1/12th of the span but not less than 230 mm. For large spans, the height of the voussoirs may be varied, less at the crown and increasing towards the skewback.

To safeguard against sliding of one voussoir on another, the angle between the line of resistance and the normal to any joint should be less than the angle of internal friction. Also the depth of the voussoirs should be adequate to resist the tendency of the joints to open and slide upon one another.

To prevent rotation or overturning, the line of the thrust or resistance should lie within the intrados and the extrados. Further the line of thrust may be made to cross each joint away from the edge to avoid crushing at an edge.

The abutments of the arch should be sufficiently strong to resist the thrust of the arch due to the superincumbent load and the self weight of the arch. The arch should be symmetrical to avoid unequal settlement of the abutments.

9.3 TYPES OF ARCHES

The intrados is a combination of arcs of various radii and centres or may have one arc with one centre.

One-centred arches are:

(*i*) Semicircular,

(*ii*) Segmental which is less than a semicircle,

(*iii*) Horse-shoe which includes more than a semicircle and

(*iv*) Stilted which consists of semicircular arch with two vertical portions at the springings.

The two-centred arch can be of many types, some of them being described below:

(*i*) *Blunt arch*: In this type of arch, the centres are within the arch itself.

(*ii*) ***Gothic or equilateral arch***: The radius of the intrados equals the span and the centres are on the springing lines.

(*iii*) ***Acute arch***: In this type of arch, the centres are outside the arch.

Two types of three-centred arches, one in which the central portion of the arch is drawn with one centre and the other two centres are used for making up the ends of the arch, whereas in the second type this procedure is reversed are shown in Figs. 9.10-9.11.

In the four-centred arch, the centres for the lower section do not coincide as in the case of three-centred arch.

The other types of arches are:

(*i*) ***Flat arch***: In which there is no rise.

(*ii*) ***Two cusp arch***: Which is used mainly for decorative purposes.

(*iii*) ***Elliptical arch***: Which has an elliptical shape for its intrados.

Some other types of arches, *e.g.*, a relieving arch which is used to take the major portion of the load coming over a lintel (*see* Fig. 9.15) are also used.

BRICK ARCHES

Brick arches may be divided into the following types:

(*i*) Gauged arches.

(*ii*) Axed or rough cut arches.

(*iii*) Rough brick arches.

(*iv*) Brick flat arches.

Primarily this classification is based on the type of bricks used in the arch construction. The varieties of bricks which can be used are:

(*a*) Ordinary standard bricks.

(*b*) Ordinary bricks cut to a wedge shape.

(*c*) Purpose made bricks. These are special bricks of different sizes suitable for various types of arch work.

(*d*) Soft bricks. These are made of diatomaceous earth and can be sawn and rubbed to the desired shapes.

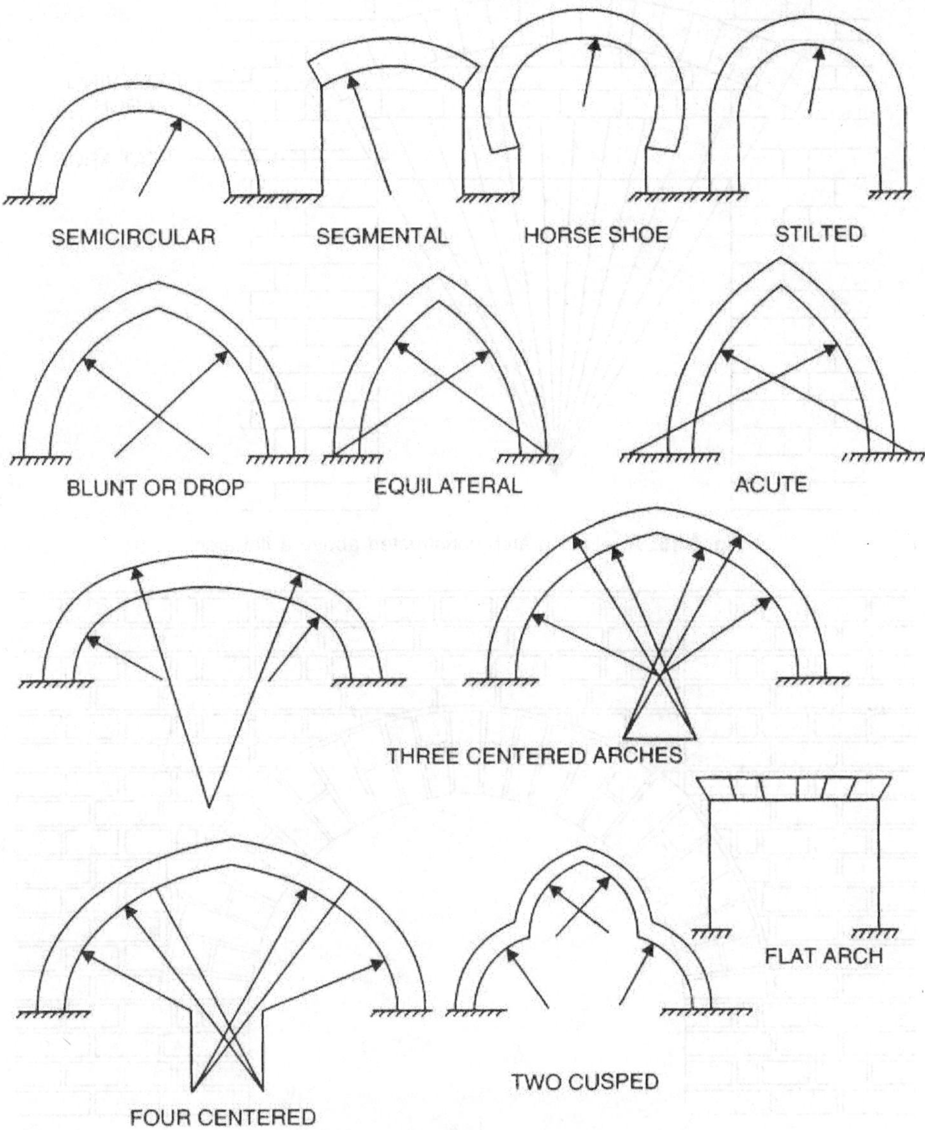

Figs. 9.3 to 9.14: Various types of arches.

9.4 GAUGED ARCHES

In this type of arches, bricks are cut and dressed accurately to a wedge shape so as to conform to the various voussoir sizes. Truly radial joints are present and they can be made as thin as possible. In order that the bricks are correctly shaped, a full scale drawing of the arch showing the various voussoirs and the joints is prepared and metallic templates are cut corresponding to each voussoir. The voussoirs are cut to this shape accurately.

The centering of the arch, as described later on, is erected and each voussoir fitted in its position with mortar. When all the voussoirs are placed in position, the key stone is fitted and hammered slightly. If very thin joints are desired, the voussoirs are dipped in mortar and then placed in position otherwise mortar can be applied with a trowel.

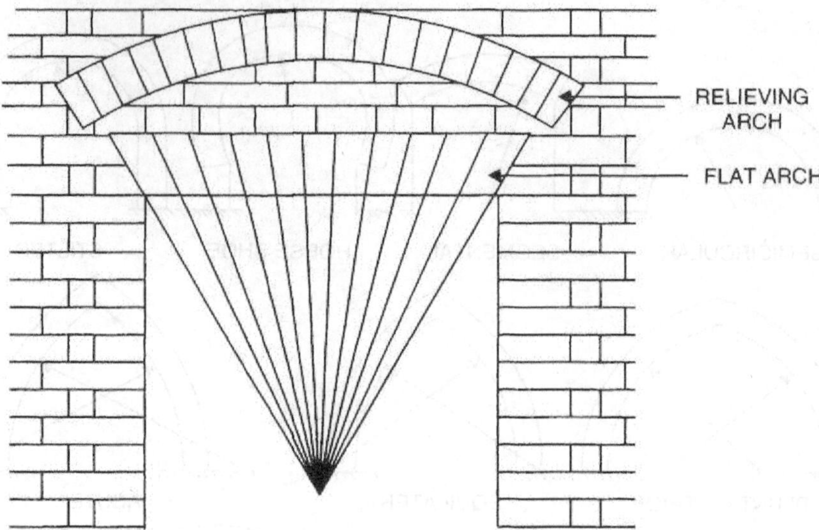

Fig. 9.15: A relieving arch constructed above a flat arch.

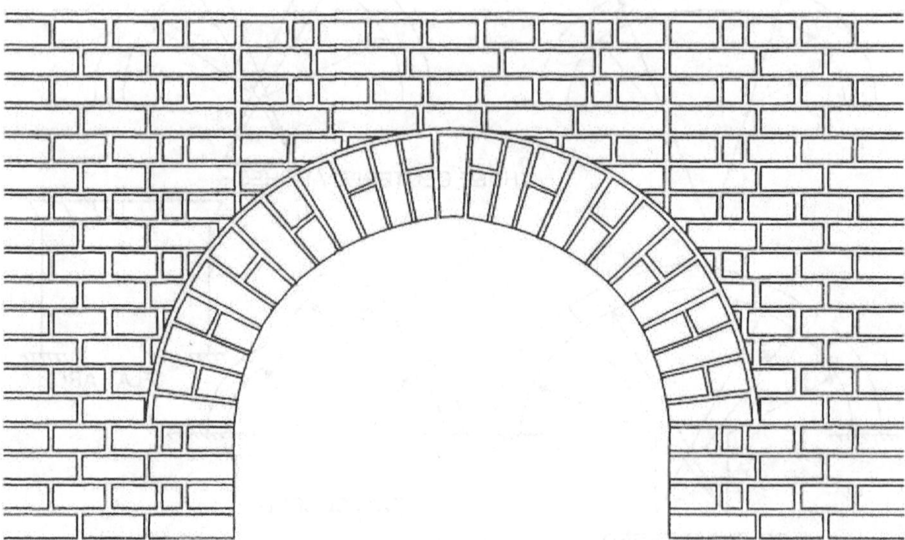

Fig. 9.16: Semi-circular gauged brick arch.

9.5 AXED ARCHES

The bricks are cut with a bricklaying axes to form wedge shaped voussoirs. They have a rough appearance at the joints. The construction procedure is the same as described above.

9.6 ROUGH BRICK ARCHES

These are made of ordinary bricks which are neither cut nor dressed to the wedge shape. As the length of the extrados is greater than that of the intrados, the joints near the extrados are wider. The appearance of these arches is not pleasing. They are mostly constructed in half brick rings so as to reduce the thickness of the joints. (*See* Fig. 9.17).

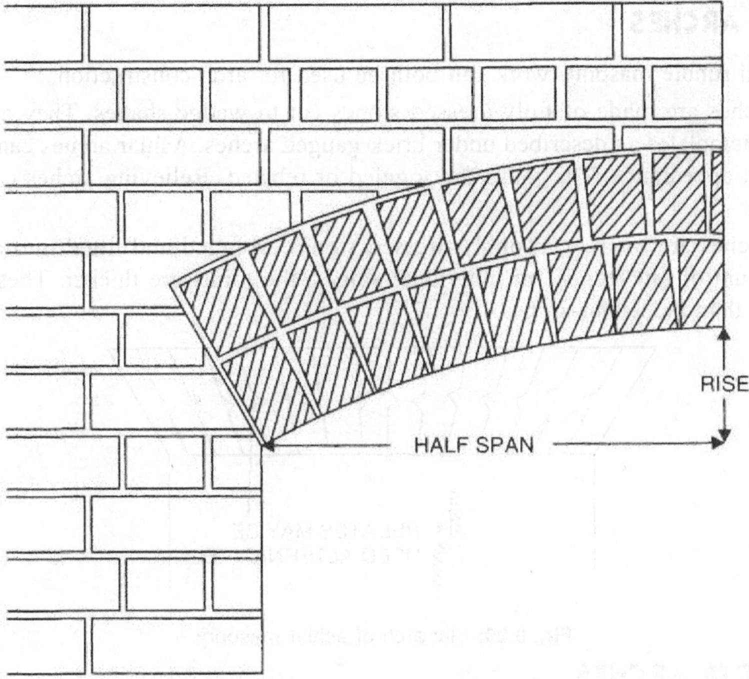

Fig. 9.17: Segmental rough brick arch.

9.7 BRICK FLAT ARCHES

These are constructed of specially cut bricks made to wedge shape and arranged in a manner so as to have flat under surface. All the joints are made to radiate from one common point. Either one brick thickness may be used or half brick may be used for thicker arches.

Whenever loads are excessive, segmental or semi-circular arches can be built above the flat arches so as to reduce the loads on the latter. (*See* Fig. 9.15).

Flat arches can also be used for facework and part of the wall thickness can be supported by a horizontal wooden section (lintel).

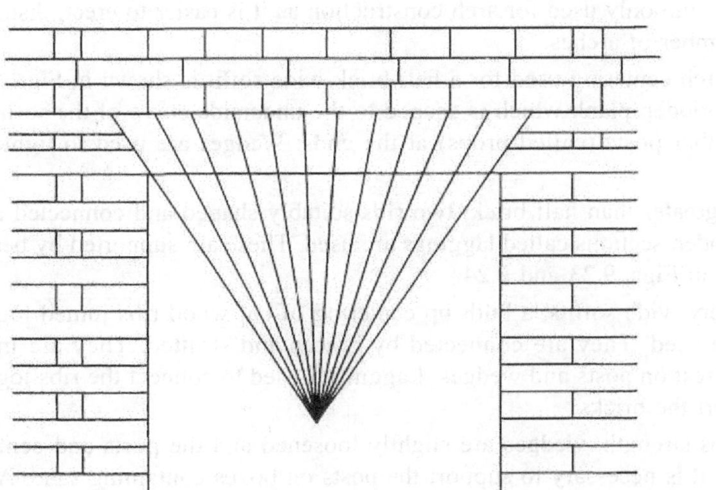

Fig. 9.18: Flat arch (gauged).

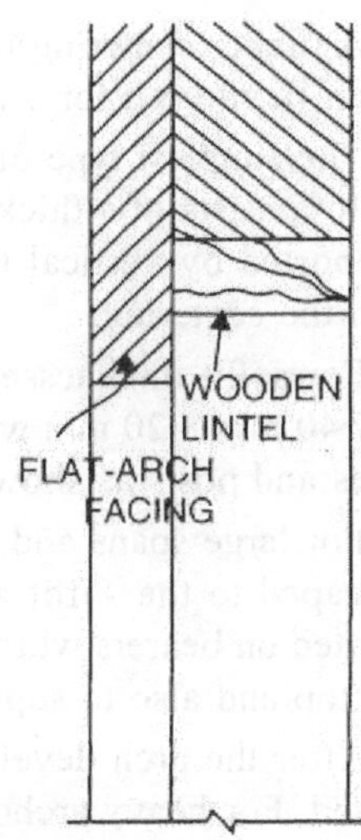

Fig. 9.19: Flat arch facing.

9.8 STONE ARCHES

Ashlar and rubble masonry work can both be used for arch construction.

Ashlar arches are made of fully dressed stones cut to wedge shapes. They are constructed with the aid of templates as described under brick gauged arches. Ashlar arches can also be made to give flat arch appearance. The joints are joggled or rebated. Relieving arches can be built for excessive loads.

Rubble arches are made of roughly dressed stones arranged and fitted into a definite arch shape. On account of unevenness of individual stones, the joints are thicker. These arches have lesser strengths than the ashlar arches.

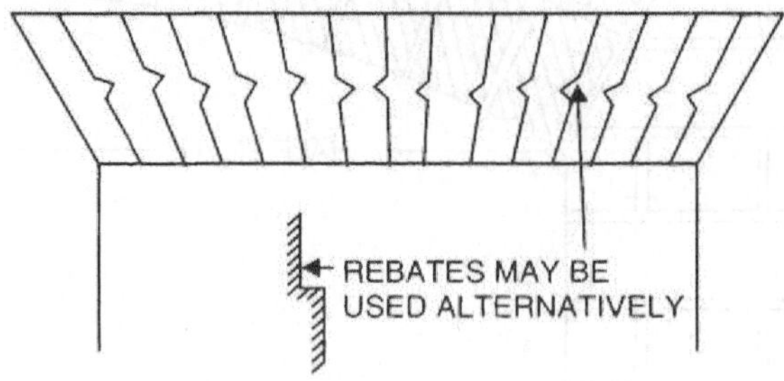

Fig. 9.20: Flat arch of ashlar masonry.

9.9 CONCRETE ARCHES

Cement concrete blocks are frequently used these days for arch construction. The blocks are similar to stone voussoirs and are of precast type. These arches are suitable where stones are not available and concrete work can be carried out easily. Arches of reinforced cement concrete are rarely used in buildings. They are used for carrying suspended floors or improving the appearance of very big buildings.

9.10 CENTERING FOR ARCHES

A temporary structure is necessary to support the arch till it develops strength. For very small works, mud masonry, made to the shape of the soffit of the arch is built and plastered, over which archwork is constructed. The temporary masonry work can be dismantled later on.

A timber centering is commonly used for arch construction as it is easier to erect, dismantle and can be re-used for a number of arches.

The simplest type of arch centering used for a half brick wide soffit is shown in Figs. 9.21-9.22. It consists of a thick wooden plank which is shaped to the underside curve of the soffit and is supported by vertical timber posts (called props) at the ends. Wedges are used to tighten or loosen the centering.

For soffit thicknesses greater than half brick, two ribs suitably shaped and connected at the top by 40 mm × 20 mm wooden sections called laggings are used. These are supported by bearers, wedges and posts as shown in Figs. 9.23 and 9.24.

For large spans and very wide soffits, a built up centering of cut wood ribs joined together and shaped to the soffit are used. They are connected by braces and strutted. They are in turn supported on bearers which rest on posts and wedges. Lagging is used to connect the ribs together at the top and also to support the bricks.

After the arch develops strength, wedges are slightly loosened and the posts and centering removed. For heavy arches, it is necessary to support the posts on boxes containing sand. A hole is kept plugged into the box. When the plug is removed, sand flows out and the post sinks slowly thereby relieving the arch centering.

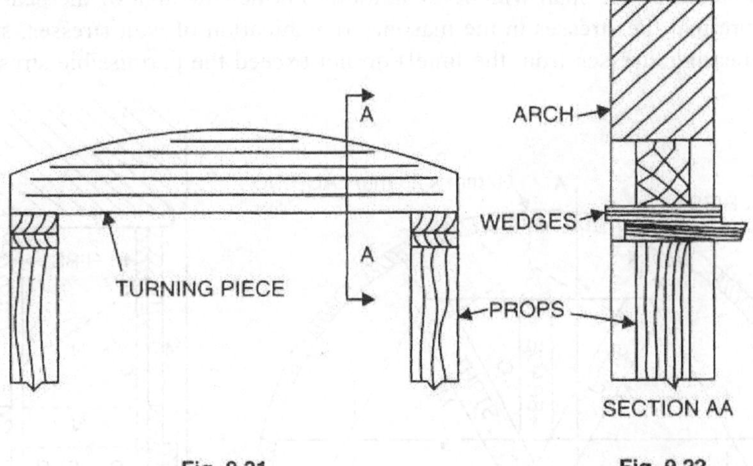

Fig. 9.21 **Fig. 9.22**

Fig. 9.21 and 9.22: Arch centering for half brick wide soffit arches.

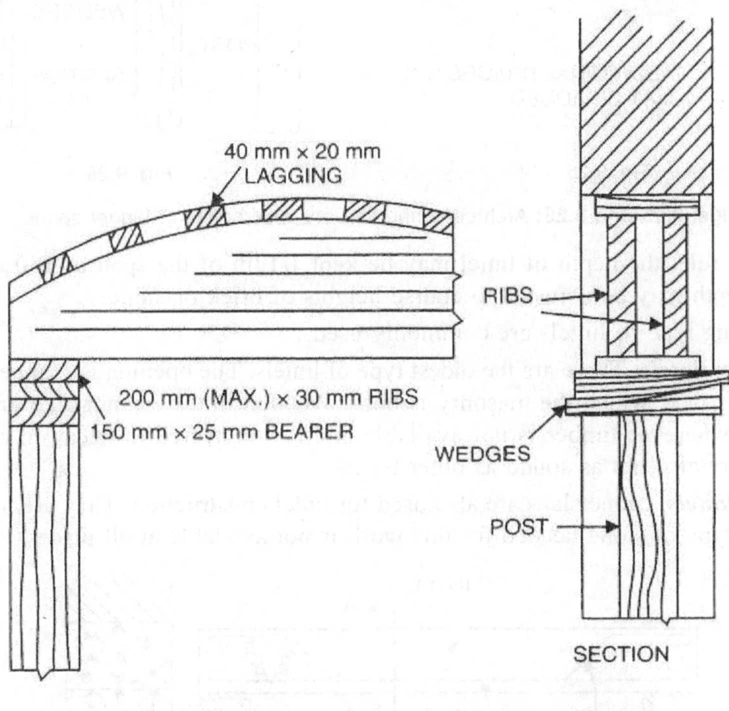

Fig. 9.23 **Fig. 9.24**

Figs. 9.23 and 9.24: Arch centering for wider soffit arches.

LINTELS

Arches are not preferred these days for bridging openings over doors and windows of a building, because they require strong abutments (walls) to withstand the arch thrust. Instead intels are commonly built to span the openings and they support the superimposed load by beam action. They are easy to build and the supporting walls need not be very strong. However, a proper bearing for lintel ends is very necessary. The length of bearing of lintel at each end should not be less than 9 cm or one-tenth of the span whichever is more. Further the area of the bearing should be sufficient to ensure that the stresses in the masonry (combination of wall stresses, stresses due to arch action and bearing stresses from the lintel) do not exceed the permissible stresses.

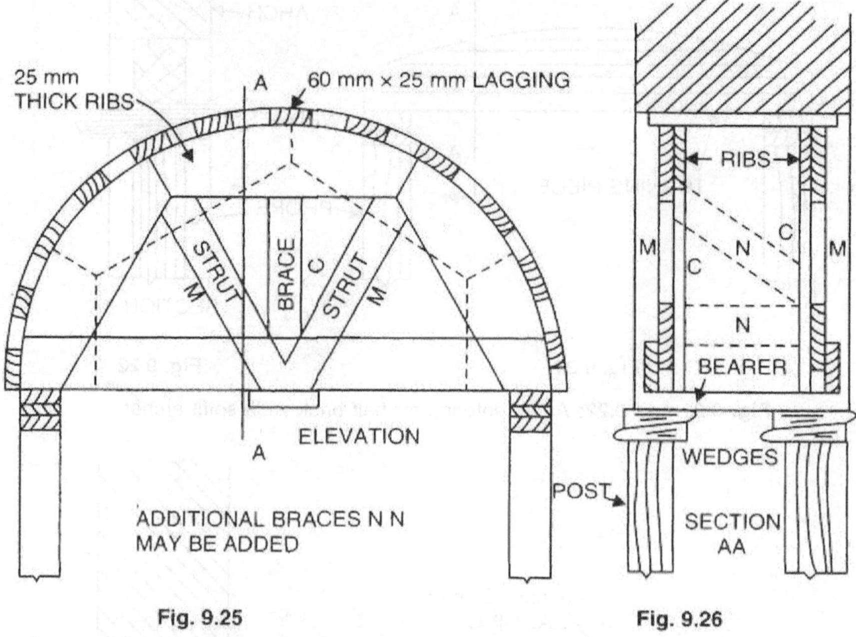

Fig. 9.25

Fig. 9.26

Figs. 9.25 and 9.26: Arch centering for very wide soffits or bigger spans.

As a thumb-rule, the depth of lintel may be kept 1/12th of the span or 150 mm whichever is greater. The depth may be adjusted to course heights of brick or stone.

The following type of lintels are commonly used:

(i) ***Wooden lintels*:** These are the oldest type of lintels. The opening is spanned by a wooden member over which the masonry is built. Wooden lintels cannot take greater load, are costly wherever timber is not available and are subjected to decay in course of time. They are also not as sound as other types.

(ii) ***Stone lintels*:** Stone slabs are also used for lintel construction. They are not widely used as the type of stone needed for this work is not available at all places.

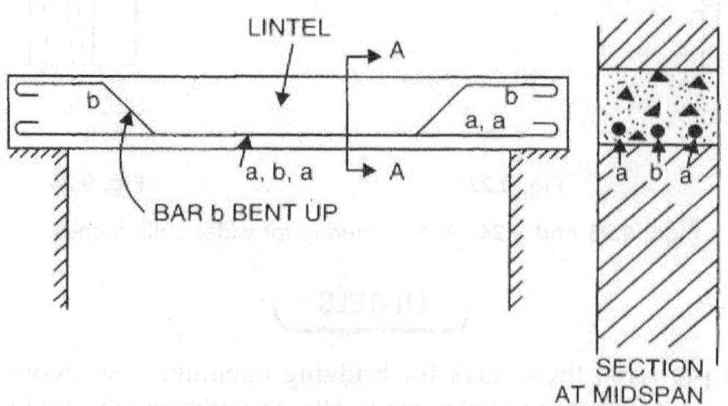

Fig. 9.27: R.C.C. Lintel.

(*iii*) ***Reinforced concrete lintels***: These are most commonly used these days. They consist of a rectangular concrete section, strengthened with steel rods. For small spans, one or two rods at the bottom are sufficient. For larger spans, some of the bars are taken to top at the ends and vertical bars (stirrups) are used in addition. (*See* Fig. 9.28). They may be cast in place or precast R.C. lintels may be used with advantage to accelerate the progress of work.

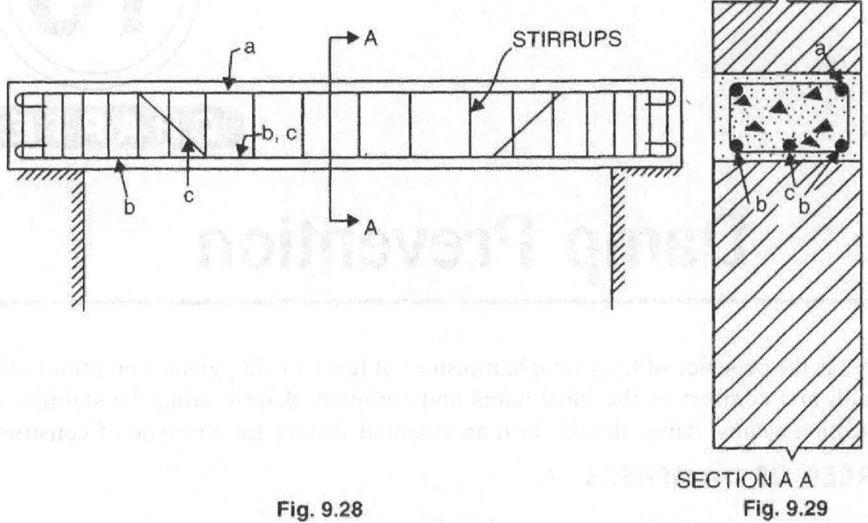

Fig. 9.28 **Fig. 9.29**

Fig. 9.28 and 9.29: R.C.C. Lintel.

(*iv*) ***Reinforced brick lintels***: These are used for comparatively small openings and consist of bricks laid in mortar and strengthened with steel bars. Ordinary brick masonry is not capable of taking loads over large openings.

QUESTIONS

1. What is meant by the term 'arch'? Briefly illustrate with sketches the various types of arches used in masonry construction.
2. What are the various types of brick arches? Briefly describe the construction of a 3 metre span segmental arch for $1\frac{1}{2}$ brick thick wall.
3. Define voussoirs, key stone, soffit, rise of the arch, extrados, skewback and haunch. Illustrate your answer with sketches.
4. What are the different types of lintels used? Sketch a reinforced concrete lintel to span an opening 2 metre clear in a masonry wall.

REFERENCES

1. **Braley E. L.**: *Brickwork.*
2. **Bombay P.W.D.**: *Specification Vols. I and II.*
3. **Huntington W. C.**: *Building Construction.*
4. **Mackey W. B.**: *Building Construction.*
5. **Frost W. and Boughton R. V.** *Modern Practical Brickwork.*

Damp Prevention

Dampness is the presence of hygroscopic moisture. It leads to unhygienic conditions affecting badly the health and comfort of the inhabitants and seriously deteriorating the stability of the building. Protection against damp should form an essential feature for any type of construction.

10.1 SOURCES OF DAMPNESS

The sources which create dampness in a building are:

(*a*) Geological and climatic causes:

(*i*) Rain penetration.

(*ii*) Ground moisture rise.

(*iii*) Condensation in buildings due to moisture in the atmosphere.

(*iv*) Drainage of the site.

(*v*) Orientation of the building.

(*b*) Structural causes:

(*i*) Water introduced in the structure during construction.

(*ii*) Defective construction.

(1) ***Rain penetration***: Climate of a country has a great bearing on the extent of dampness in a building. Some of the buildings can withstand a heavy rain shower of a short duration whereas most of the buildings fail to withstand continuous down-pour of rain for a number of days. Wherever such climatic conditions exist, dampness in the building is inevitable.

Direct penetration of rain through the entire area of the wall is rarely seen. Properly constructed walls offer considerable resistance to the penetration of moisture but a rapid penetration of moisture takes place through the capillaries between the mortar joints and the bricks or stones used to construct the wall. Porous external renderings prevent rain penetration to a large extent where as dense renderings usually crack and are, therefore, sources through which damp can pass. Projecting features of a building arrest water on them and transmit it inside.

Rain penetration is common on the roof components such as valley gutters and dormer windows. Similarly parapets and chimneys collect water. Leakage occurs through flat roofs also and water in this case may penetrate at a crack or at a joint. Hence inspection of the entire surface is essential in this case.

(2) ***Ground moisture rise***: The soil on which the building is constructed is an important factor for causing the damp or its prevention. Gravel and sandy soils allow water to pass through them very easily whereas soils of clayey nature create damp conditions due to heavy capillary rise. Further if a clayey stratum is covered by a sandy layer it is difficult to drain such soil.

Dampness in floors is primarily due to this reason. The rise of water also causes the soluble salts to be brought up and they get deposited on the walls at the floor level. A sort of darkened surface with irregular patches is created for some height above the floor level.

(3) ***Condensation in a building due to moisture in the atmosphere***: Except under very cold climates, condensation is not prevalent in this country. In the case of dampness due to condensation usually the whole area of walls, ceilings and floors is affected. Deliquescent salts cause condensation, *e.g.*, if sea water containing sodium chloride is used in the construction of a building during humid climate, the surfaces present a damp appearance. Similarly nitrates and chlorides of alkali metals cause dampness.

(4) ***Drainage of the site***: The elevation of a building site has a correlation with the sub-soil conditions and the strata beneath. Houses built on a higher ground can be drained easily and are hence less liable to dampness. Low lying sites cannot be easily drained and present somewhat water logged conditions if impervious soil is present underneath the building. While deciding upon the location of a building, it should be ensured that the site can be drained off easily.

(5) ***Orientation of the building***: Every building has some walls which are subjected to a constant splash of water and also some walls are getting less of direct sun's rays. This is due to geographical conditions which create rain falling at a place mostly in one particular direction. The walls which are in this direction and those which also get less sunshine during the hotter part of the day are liable to get damp.

(6) ***Water introduced in the building during construction.*** Walls while being built are in a very damp condition. This persists for a long period after the wall has been constructed. While the walls get dried, efflorescence of soluble salts results. If this dampness persists, it is to be rectified. Similarly if wet timber is painted, it cannot get dried and hence is liable to be attacked by vermin.

(7) ***Defective construction***: Improper construction of the various parts, *e.g.* fixtures in a building, joints in the roofs, throating of sills and coping, proper jointing of walls, etc., creates dampness by transmitting water. In every component of a building, there are specific points where damp is liable to find its way and it is essential that enough care is taken to see that proper construction is carried out. In the following chart various types of defective constructional causes for some important structural members are described. The extent of dampness along with the time of occurrence is shown in Tables 10.1 and 10.2.

10.2 EFFECTS OF DAMPNESS

(1) Corrosion of metals used in the construction.

(2) Plasters get crumbled and softened.

(3) The floor coverings loose adhesion with the floor bases.

(4) Timber, when in contact with damp gets rotten.

(5) The paints get blistered and bleached. Further they get disfigured.

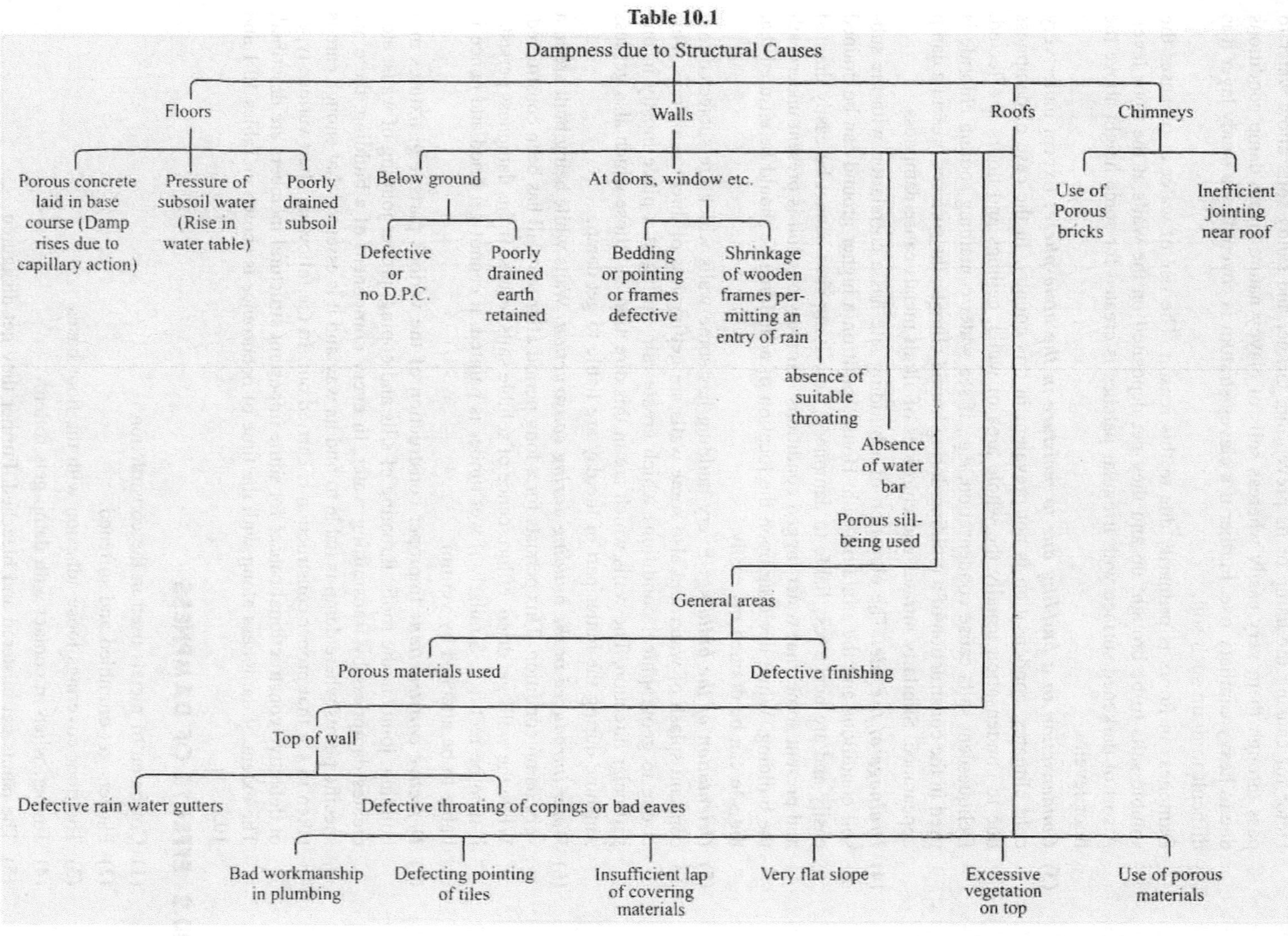
Table 10.1
Dampness due to Structural Causes
Floors
Walls
Roofs
Chimneys
Porous concrete laid in base course (Damp rises due to capillary action)
Pressure of subsoil water (Rise in water table)
Poorly drained subsoil
Below ground
At doors, window etc.
Use of Porous bricks
Inefficient jointing near roof
Defective or no D.P.C.
Poorly drained earth retained
Bedding or pointing or frames defective
Shrinkage of wooden frames permitting an entry of rain
absence of suitable throating
Absence of water bar
Porous sill-being used
General areas
Porous materials used
Defective finishing
Top of wall
Defective rain water gutters
Defective throating of copings or bad eaves
Bad workmanship in plumbing
Defecting pointing of tiles
Insufficient lap of covering materials
Very flat slope
Excessive vegetation on top
Use of porous materials

(6) The stones, bricks, etc. are disfigured on account of a coating of salts which collect on them.
(7) All electrical installations get deteriorated.
(8) Causes unhygienic conditions for the occupants of the building.

Table 10.2: Causes of dampness

Structural elements	*Extent of dampness*	*Timing of occurrence*	*Sources of damp*	*Other causes*
Wall, partitions, ceilings, floors	(*a*) Completely spread	After a certain rise in temp. and humidity of the outside air or of rooms only	Air	Condensation
	(*b*) In patches	During high humidity	Air	Condensation aided by deliquescent salts
	(*c*) One patch	—	From plumbing fitting	Leaking pipes or condensation or cold pipes
Wall	From floor level to some distance above	At all times of the year	Due to ground moisture movement	Absence or defect in D.P.C.
Chimneys	(*a*) From roof downwards	After rain	Moisture penetration.	Absence or defective D.P.C.
	(*b*) Patches with brown colour	Whenever chimney is used	Condensation inside flues.	—
Ceilings	(*a*) Patches	After rain	Leakage in roofs.	—
	(*b*) Wide spread	During cold weather	Condensation	—
Ground Floors	(*a*) Patches	Whenever ground is wet	Ground moisture movement	Defective base
	(*b*) Around edges	-Do-	-Do-	D.P.C. not effective

PREVENTION OF DAMPNESS

There are various methods of preventing dampness. Different types of materials are used for reducing the occurrence of dampness. The methods of damp prevention may be classified as under:

(1) Use of water proofing mixtures or integral treatment.
(2) Application of water proof surface treatment.
(3) Interposing a water proof membrane.
(4) Use of special constructional devices.

These are described below:

10.3 WATER PROOFING MIXTURES OR INTEGRAL TREATMENT

Synthetic compounds are available which can be added during the process of mixing of the constructional materials. Substances such as chalk, talc or fuller's earth fill the pores in the concrete or mortar mix and make it denser. The other type of chemicals are silicates, chlorides and sulphates which react chemically and fill the pores. Soaps and other fatty acid compounds such as calcium, sodium, ammonium stearates and oleates and petroleum oils are used as ad-mixtures in concrete to make it water repellent. The water soluble soaps are not water repellent in themselves but they react with hydrated lime to form insoluble compounds. Some compounds are available in commercial patented forms like Pudlo, Impero, Cico, etc. Some of them are used after diluting with water.

10.4 WATER-PROOF SURFACE TREATMENT

Surface treatments can be either external or internal, the former are considered to be more effective in preventing dampness. While the latter cannot eliminate dampness but can only reduce it to an acceptable degree.

External surface treatments consist of pointing joints in brick-work or stone masonry and plastering over brick walls. The inner walls are normally plastered. Paints, oils, waxes, soaps and silicate solutions, etc., can also be applied on the surfaces for treatment against damp. These, however, need renewals after a few years.

A thin film of a water proofing material can be applied to the surface of concrete after it is laid. These materials consist of silicates of aluminium, potassium, barium hydroxide and magnesium sulphate in alternate applications, soft soap and alum, lime and linseed oil, coal tar, bitumen, waxes, fats, etc. Some of these covering materials give a durable resisting surface to the concrete while others afford only a temporary protection. However, the tars and asphalts render the surface black which is not desirable.

10.5 INTERPOSING A WATER PROOFING MEMBRANE

This method consists of introducing a layer of water repellent material against the travel of damp. These are called Damp Proof Courses (D.P.C.). The materials used for the damp proof courses are:

(*i*) Flexible materials like bituminous sheet, plastic sheet, metal sheet etc.

(*ii*) ***Semi-rigid materials***: Like mastic asphalt.

(*iii*) ***Rigid materials***: Like slates, bricks, stones, dense cement concrete, etc.

For selection of the material for D.P.C., the following rules should be applied:

(*i*) The material should be durable and impervious.

(*ii*) It should be capable of withstanding the dead load of the wall and other superimposed loads without disintegrating itself.

(*iii*) It should not allow any movements in itself so that the walls resting on it do not crack.

(*iv*) For walls of 230 mm thickness any of the types described above may be used.

(*v*) For thicker walls or wider areas, a D.P.C. which has least number of joints should be used.

(*vi*) When water under pressure is to be retained, a jointless D.P.C. is desirable to prevent leakage.

(*vii*) For bridging cavities or joints, flexible materials should be used.

A brief description of some of the types of membrane damp proofing agents is given below:

(*a*) ***Concrete layers***: Cement concrete layer of 1 : 2 : 4 mix with water proofing agents is used as D.P.C. It does not permit the rise of water by capillary action but allows the water to pass through the cracks which occur in this layer due to shrinkage or minor settlement of the building. This can be used wherever damp is not excessive. For preventing the damp effectively, a layer of concrete at least 40 mm in thickness painted with two coats of hot bitumen are used. This type of D.P.C. is used as a horizontal plinth D.P.C.

(*b*) ***Mortar***: This can be used as a bedding layer for other types of D.P.C. and is made up of cement sand mix in the ratio of 1 : 3 with a slight addition of lime to increase the workability. For vertical D.P.C. a plaster of 1 : 3 cement sand mortar, at least 2 cm in thickness covered with two coats of hot bitumen is used.

(*c*) ***Bricks***: Dense bricks which absorb less than $4\frac{1}{2}\%$ of their weight of water are suitable for D.P.C. wherever the damp is not excessive. They should not be used for downward passage of damp prevention. The joints should be left open.

(*d*) ***Stones***: Dense stones, *e.g.*, granite, trap, slates, etc. are used as D.P.C. and should be laid in cement sand mortar. Two layers are at least used with joints breaking vertically.

(*e*) ***Hot laid bitumen***: This material is used on a bedding of concrete or mortar and should be applied so as to have a thickness of at least 3 mm.

(*f*) ***Mastic asphalt***: This provides a semi-rigid impervious D.P.C. Special care is needed in its laying.

(*g*) ***Metal sheets***: Lead, copper and aluminium sheets are used for D.P.C. Lead is to be protected from corrosion with an application of bitumen coating. This can with stand distortion and can be suitably lapped. At least 12 kg of lead per square metre should be used. Copper sheets are also quite durable and damp resistant but may stain the external wall surfaces. The copper sheets are not less than 3 mm in thickness. Aluminium sheets have to be protected with a layer of bitumen.

While laying a D.P.C. the following points should be kept in view:

(*i*) The D.P.C. should cover the full thickness of the wall and should not be recessed into the wall.

(*ii*) The mortar bed upon which the D.P.C. is laid should be even.

(*iii*) Each D.P.C. should be placed in correct position with another D.P.C. so as to provide a continuous protection.

(*iv*) Whenever D.P.C. are to be used to cover angle projection they should be laid continuously.

10.5.1 Use of Membranes in the Prevention of Damp at Various Positions in a Building

Basement: Vertical D.P.C. is applied to the enclosing walls of a basement. The horizontal D.P.C. is applied to the flooring of basement. There are various methods of applying vertical D.P.C. The horizontal D.P.C. of the flooring may be continued to the top (*See* Fig. 10.1) so that a sort of waterproof tank is formed. This prevents water from passing through the floors and walls adjacent to the earth and also intercepts the water which walls would absorb directly from the ground. Bituminous mastic or bituminous felt may be used along with concrete surfaces of the basement. The D.P.C.'s may be laid on the inner surface or on the outer surface of the walls or else they be incorporated into the wall thickness.

The bottom layer of concrete is covered with a 20 mm thick Bituminous mastic laid in two coats and a second layer of concrete is laid over it. The mastic is continued as a vertical D.P.C. A 115 mm outer wall is constructed from the horizontal D.P.C. to a little above ground level. The joints are raked out on the inner surface and bitumen layer is applied. After some time a second layer of bitumen is laid in such a manner so as to cover the joints of the first layer. Then the inner wall is built. In case excessive pressure of water is encountered, a drain can be laid outside the wall junction with bottom concrete. (*See* Fig.10.2).

If the bituminous treatment is to be given on the outside of the basement, the wall is built, joints are raked out on the exterior surface and two coats of bitumen applied as described above. The joint with the horizontal D.P.C. above the ground level is made perfect and the vertical D.P.C. is made inside the wall for a height between the horizontal D.P.C. for the ground floor and the ground level (*See* Fig. 10.3). This prevents an ugly appearance on the exterior.

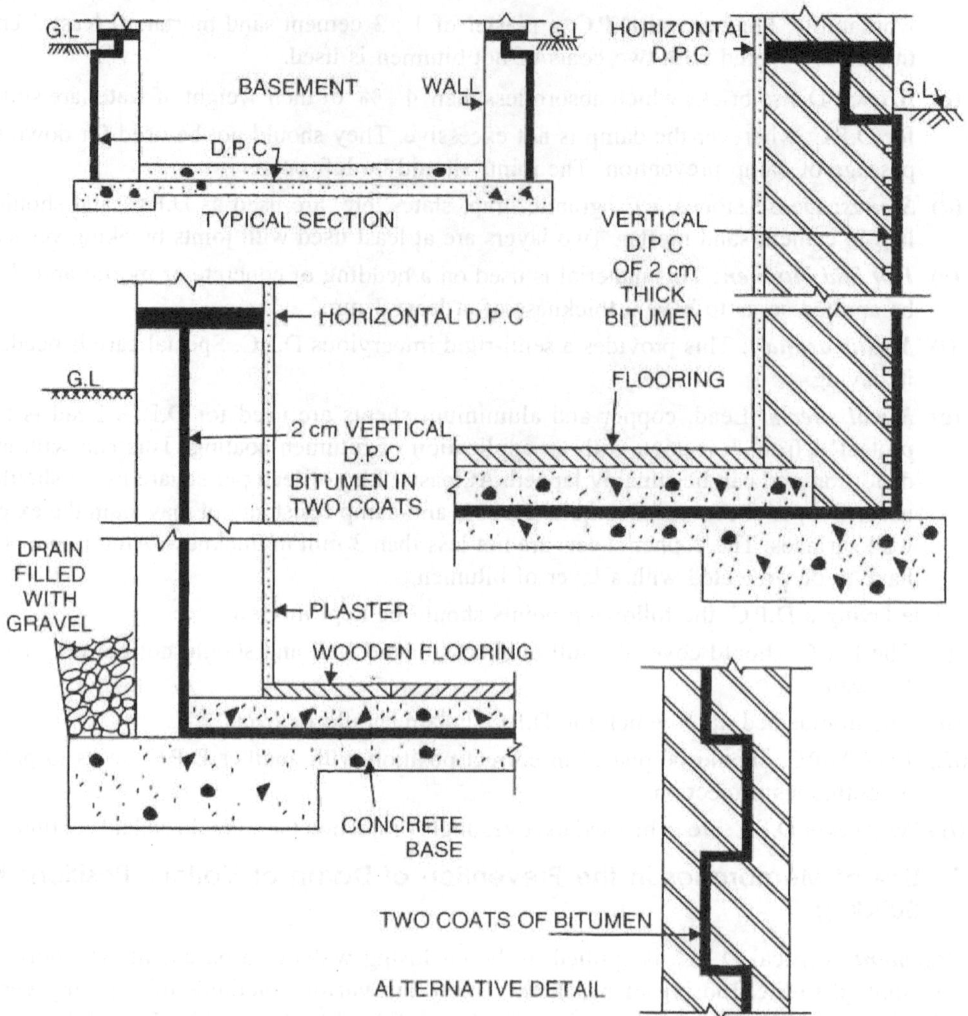

Figs. 10.1 to 10.4: Various methods for damp-proofing of basement.

Floor: The concrete floors can be treated with waterproofing layer of mastic bitumen or fibrous bitumen felt. This treatment is necessary when the subsoil water table is high and moisture is likely to rise through the flooring by seepage aided by capillary action of the soil. A bedding of concrete of 100 mm thickness is laid and allowed to set and dry for the required period. A priming coat of hot liquid bitumen is given and bituminous mastic is applied in two coats. If bituminous felt is used, the concrete surface should be levelled so as to prevent the tearing of the felt. Whenever excessive uplift pressures are expected, reinforcement should be provided in the concrete and the mastic or felt laid later on and covered with a concrete wearing coat. This treatment can be given to ground floors as well as to basement floors.

Horizontal D.P.C. for walls: In external walls the D.P.C. should be laid about 200 mm above the ground level. It is essential to have a vertical D.P.C. between the floor level and the D.P.C. level on the inside of external walls in this case. Some building codes prescribe that the horizontal D.P.C. should be at the plinth level below the flooring. In an internal wall, the D.P.C. is laid in level with the upper surface of concrete in the finished floor. The continuity of D.P.C. between the internal and external walls is obtained by inserting bitumenised bricks or cement concrete blocks.

Damp proofing in cavity walls: A horizontal D.P.C. is laid at about 200 mm above the ground level and is built of separate inner and outer slabs for the two walls. Dampness is also caused at the head of openings if proper protection is not afforded. Water which finds its way to the inner surface of the outer wall will drip on the opening, if not suitably treated. A layer of bituminous felt or lead or copper sheeting is stepped down from the inner wall and continued over the door or window frame. This covering should extend to at least 100 mm on either side of the opening. Lead sheet in the form of a trough should not be used as it collects water. Dampness can also occur at the window sills and lead or bituminous felt can be introduced in between the face of the sill and the inner wall covering.

Parapets: Water drains on the inner side of the parapet and finds way to the interior of the building. A D.P.C. just above the junction of the roof with the parapet wall gives an effective protection. Materials which form a continuous sheeting are preferred and flexibility is desirable. D.P.C. should extend through the full thickness of the wall including any plaster etc. A metal flashing may be provided on the exterior surface to improve appearance. A D.P.C. underneath the coping may be necessary where the bricks are not of good quality and are liable to get damaged on account of sulphate action. To reduce the number of joints in the coping, it should be cast monolithically. If any projection is present on a parapet, it should be protected by suitable flashings. Whenever it is anticipated that the plaster is going to crack, the plastering should be done on the interior surface only.

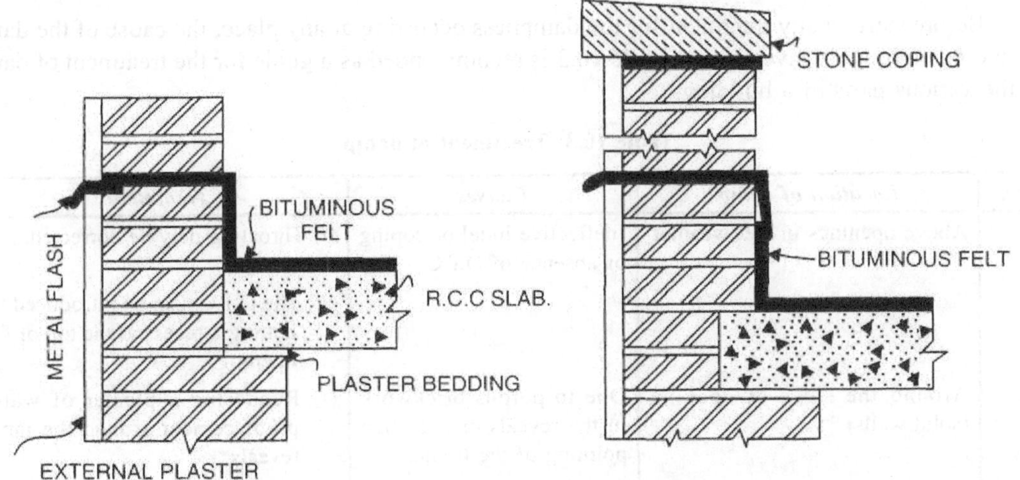

Figs. 10.5 to 10.6: Prevention of damp along parapets.

Grading of mud pushka with tiles: This method of damp-proofing flat roofs is best suited for hot and arid regions and is commonly adopted in Punjab, Haryana, Delhi, Rajasthan and U.P. In addition, this provides good insulation against heat as well. The various steps involved are:

1. Painting the top of roof slab uniformly with a layer of hot bitumen spread at the rate of 1.75 kg/sq m of the roof surface.
2. Spreading immediately coarse sand at the rate of 0.6 cub m of sand per 100 sq m of roof surface, when the bitumen is still hot.
3. Laying 100 mm (average) thick mud *pushka* prepared from puddled clay mixed with wheat bhusa at the rate of 8.0 kg of *bhusa* per cub m of clay. The required slope for the proper drainage of roof being given to this layer.
4. Plastering the consolidated layer of mud *pushka* with a 15 mm thick coat of mud-gobri mortar (3 parts of mud to 1 part of cow-dung).

5. Laying the brick tiles flat on the plastered surface and grouting the joints with 1 : 3 cement mortar.

10.6 SPECIAL CONSTRUCTIONAL DEVICES:

(i) ***Cavity wall construction***: For preventing the external walls from carrying damp inside, cavity wall construction can be adopted. The details are described in Chapter 8.

(ii) For solid walls at least $1\frac{1}{2}$ brick thickness should be used if lot of damp is anticipated.

(iii) The face bricks should be of best quality which can absorb very little moisture.

(iv) Openings and walls in general should be protected by the use of cornices, string courses, sills, plinth courses and canopies. These elements should be sloped away from the building and should have throating to prevent the water dripping on the surface.

(v) Roofs should be given an adequate slope to drain off water easily.

(vi) All plumbing fixtures, *e.g.*, down water pipes should be suitably fixed and it should be ensured that the water does not lead at their connections.

(vii) Special attention should be paid to the construction of joints and also to the junction of the roofs with parapet walls.

TREATMENT OF DAMP

Before starting any treatment for the dampness occurring at any place, the cause of the damp should be thoroughly investigated Table 10.3 is recommended as a guide for the treatment of damp in the various parts of a building.

Table 10.3: Treatment of damp

No.	*Location of damp*	*Causes*	*Treatment*
1.	Above openings in wet weather	Ineffective lintel or coping or absence of D.P.C.	(i) Throating may be corrected. (ii) The D.P.C. can be introduced by cutting a recess at the top of the openings.
2.	Around the sides of opening (solid walls).	Due to porous brickwork in the reveals or defective pointing of the frame.	(i) Rendering a plaster of water proof cement around the jamb reveals. (ii) Repointing the frame
3.	Along the underside of wooden sills in wet weather.	Defect in the design of setting of the sill.	A throating on the underside of the sill should be formed. Good slope should be given on the top surface of the sill outside. The sill should be rendered with plaster.
4.	Permanent dampness at or about the inside floor level on the walls of the rooms.	Absence or inadequateness of the horizontal D.P.C. or ineffective resistance to the horizontal movement of damp due to water getting collected on ground outside.	A D.P.C. is to be provided by suitably opening the wall. If the existing D.P.C. is inadequate, a double layer D.P.C. should be laid. To prevent the appearance of damp on the surface, the plaster is broken and the brick work is given a coating of neat cement slurry suitably water proofed and plaster applied.

(Contd.)

No.	Location of damp	Causes	Treatment
5.	On walls and floors below ground level.	(*i*) Site may be undrained. (*ii*) Floor is laid on sloping clay bed.	Disconnecting the floor from subsoil will help. A new thick layer of sandy soil is laid below the floor. The foundation of the floor may be built of rubble packing and then floor suitably finished. Suitable weep holes may be set in the walls. Special renderings can be given by stripping off the concrete topping and relaying it over the water proofing layer.
6.	On ceiling appearing suddenly in bad weather.	Bad eaves or choked rain water gutter.	Cleaning of gutters and repairing of eaves.
7.	In front of fireplaces at ceilings.	Defective lead work around chimney stacks.	Renewal of gutter lead work and constructing suitable fillets.

QUESTIONS

1. What are the various causes of dampness in a building? Give a detailed answer.
2. Describe briefly how dampness can occur in walls, chimneys, ceilings and floors of a building. What are the timings of their occurrence?
3. Write a short note on dampness due to structural causes. What are the bad effects of dampness?
4. Write a detailed note on the prevention of dampness by interposing a water proofing membrane in a structure.
   Name the various materials which can be used for this purposes.
5. A basement of a building is 3 metres deep and rectangular in plan of 6 m × 10 m in size. There is likelihood of dampness occurring in the inside of the basement. As the building is not yet constructed, explain briefly the various methods of damp prevention if the walls are to be built in brick masonry and a concrete floor is to be used.
6. Write short notes on the prevention of dampness in cavity walls and along parapets.
7. Write down the steps involved in damp proofing of flat roofs.

REFERENCES

1. **National Building Organisation**: *Dampness in building.*
2. **Dawes S.B.**: *Dampness in domestic buildings.*
3. **Capital Project Chandigarh**: *Building bye-Laws.*
4. **British Standard Specification 743 of 1951**: *Materials for damp proof courses.*
5. **British Standard 1097 of 1947**: *Mastic asphalt for damp proof courses.*
6. **British Standard 1418 of 1947**: *Mastic asphalts damp proof courses and tanking.*
7. **Blake E.G.**: *Damp walls.*

Wood Work Joints, Fastenings and Tools

Different pieces of timber have to be connected together in order to develop a framework which may be used in the construction of a building. Timber, being a natural product available in abundance in India, is used for the construction of doors, windows, roofs, partitions, beams, posts, etc. To increase the structural stability, to enhance the architectural beauty, to economise timber and to facilitate construction, joints form a very important part of carpentry.

It is well known that the joint is the weakest part of any structural member, hence it is desirable that utmost attention be paid to the construction and finishing of joints. The forces acting on wooden members can be of the following nature:

(*i*) ***Tensile***: This force will move the two parts of a joint apart.

(*ii*) ***Compressive***: A force of this nature will try to crush the fibres of the component parts.

(*iii*) ***Shear***: This will try to tear off the wooden fibres at right angles to their line of growth.

(*iv*) ***Torsion***: Any force trying to twist a member will tear it off crosswise.

Keeping in view these forces, the following points about the joints should be noted:

(*i*) The joints should be as simple as possible to save labour in construction. Further complicated joints present numerous surfaces or angles which can lodge verms and cause timber to decay.

(*ii*) The joint should be cut in such a way so as to weaken the connecting members as little as possible.

(*iii*) Each face abutting with the other should be cut in such a way that the face is perpendicular to the line of pressure acting on it.

(*iv*) Each part should be safe to withstand compressive forces acting on it. This means that the area of members being connected should be in proportion to the forces acting on them.

(*v*) The fastenings should be proportioned in such a way that they equal in strength to the members which they connect.

(*vi*) The fastenings should be placed in such a manner that there is sufficient resistance to the failure of a joint by the fastenings getting sheared or crushed through the timber.

Definitions

- (i) ***Chamfer***: Arris (edge) of timber plained off flat to form an angle usually 45°.
- (ii) ***Bevel***: If the angle of chamfer is not 45°, it is called bevel.
- (iii) ***Stopped chamfer***: When the chamfer is not continued to the end of the wooden piece but ends in another chamfer or slope, it is called stopped chamfer.
- (iv) ***Moulding***: Process of shaping various units of construction by hand or machine to produce a moulded section.
- (v) ***Bead***: Moulding consisting of a semi-circle and a quirk or bead returned; there quarters of a circle with quirk at each end.
- (vi) ***Bead jointed***: Term applied to two units of timber fitted at their sides, with a bead worked on one unit to cover the joint.
- (vii) ***Groove***: Recess formed in board or a piece of wood. The shape may be hollow, V-shape, semi-circular, etc.
- (viii) ***Rebating***: Cutting rectangular portion from a plank to receive another plank similarly cut.
- (ix) ***Mitring***: Joining two boards at an angle.

TYPES OF JOINTS

Joints may be classified as under:

1. ***Lengthening joints***: These connect members lengthwise in order to make use of small members.
2. ***Bearing joints***: These are the transverse joints.
3. ***Oblique shouldered joints.***
4. ***Side or widening joints.***
5. ***Angle joints.***

A brief discussion of each type is given below:

11.1 LENGTHENING JOINTS

Joints connecting two timbers may have to withstand tension or compression if the connected members are stressed likewise. Some joints may not have to stand these forces. This means that the design of the former type is to be done carefully and all members at the joint are selected in such a way that the forces are distributed evenly. It is not within the scope of this book to deal with the structural design of joints for which reference may be made to any standard book on design of timber structures. The various types of lengthening joints are:

- (i) ***Lap joint***: Method of joining two members by lapping one over the other and bolting the same is called lap jointing. Sufficient number of bolts are provided to withstand the forces. In this type of joint, the timbers are not cut but wrought iron straps are placed over them and the pieces are bolted. As the timber is not cut, a hundred per cent efficiency could be achieved. The joint is satisfactory for temporary construction. (*See* Figs 11.1 and 11.2).
- (ii) ***Fish joint***: This type of joint is more effective for transmitting only tension. The timbers are cut and iron plates are bolted to them. These plates are called 'fish' plates. (Figs 11.3 to 11.6). The strength of the joint depends upon the bolts and the fish plates. To increase this strength, keys are provided in the joint as shown in Figure 11.6. As an alternative the plates may be bent as shown in Figure 11.3.

(*iii*) ***Tabled joint***: The members of wood are cut with the shapes shown in Figures 11.7-11.9 and joined suitably. The joint is not very efficient as lot of material is cut. This joint is used in joinery work only.

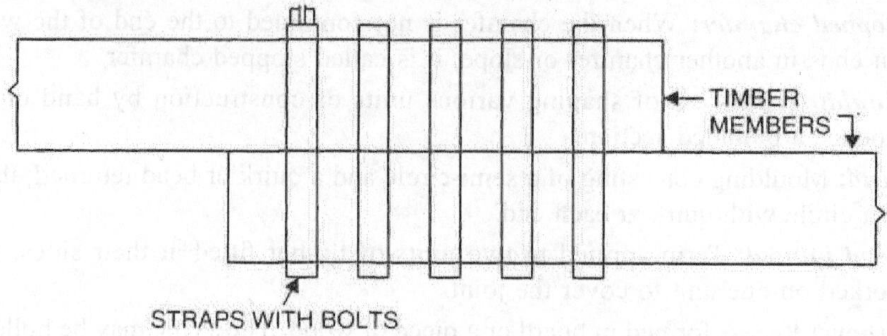

Fig. 11.1

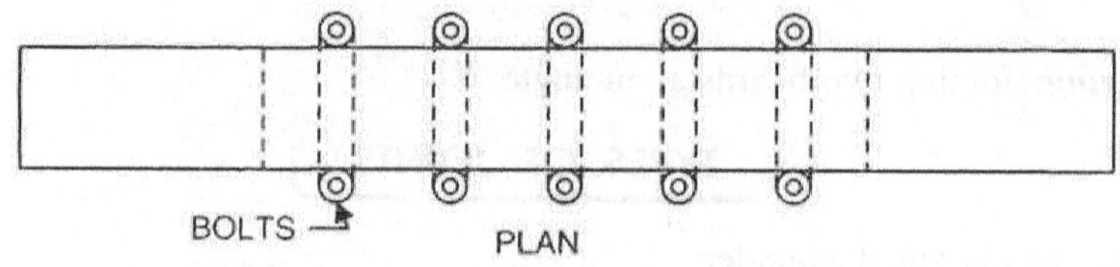

Fig. 11.2

Figs. 11.1 and 11.2: Plan and elevation of a lap joint with iron straps having bolt at ends.

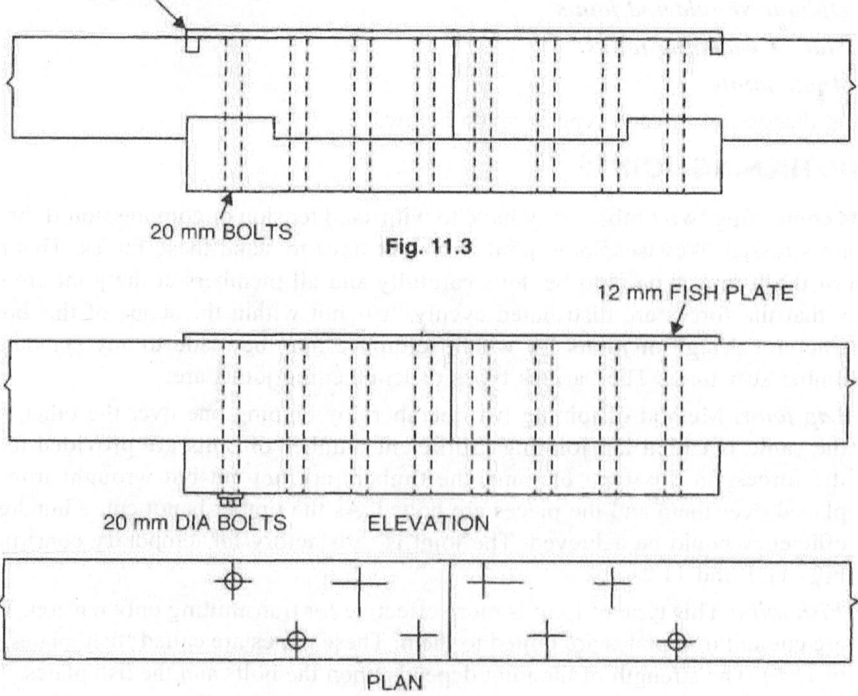

Fig. 11.3

Figs. 11.4 to 11.5

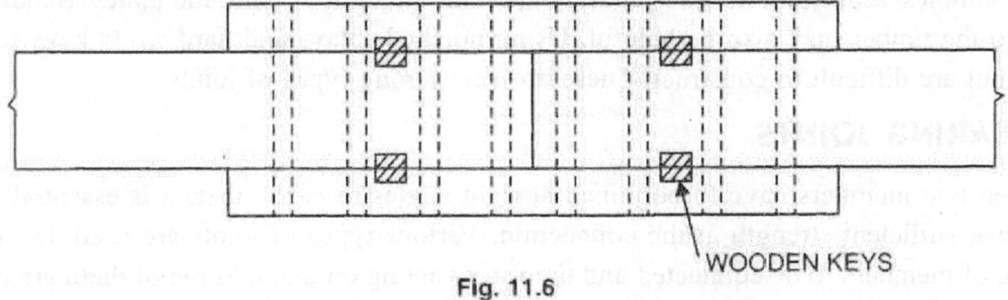

Fig. 11.6

Figs. 11.3 to 11.6: Various types of fish joints.

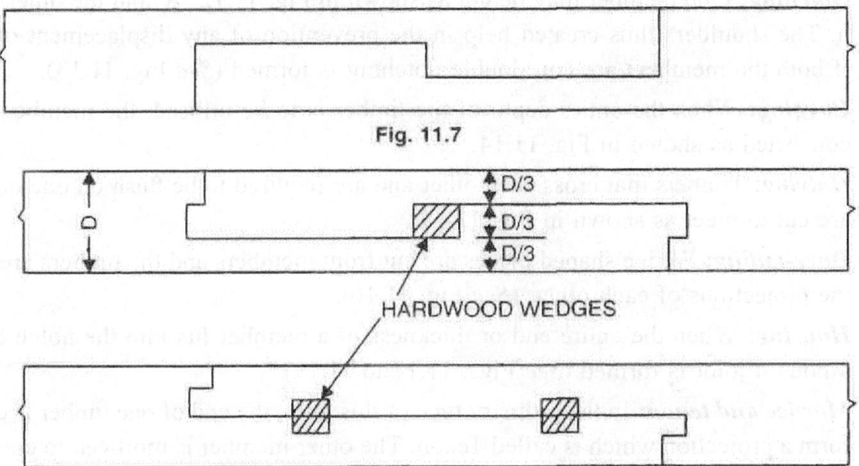

Fig. 11.7

Figs. 11.7 to 11.9: Various types of tabled joints.

(*iv*) ***Scarf joint***: These are used where strength as well as the shape of the joint are equally important. Different types of scarf joints are shown in Figs. 11.10 to 11.11.

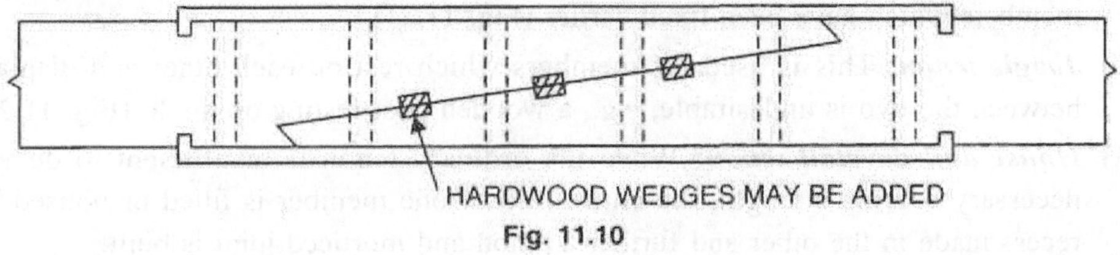

Fig. 11.10

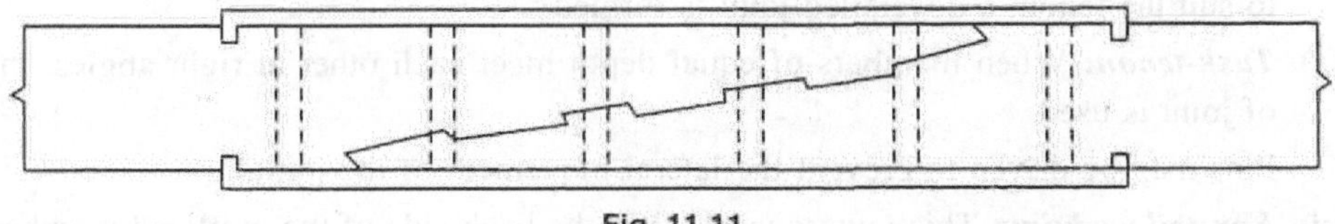

Fig. 11.11

Figs. 11.10 and 11.11: Typical scarf joints.

The simplest scarf joint depends entirely upon the strength of bolts and plates. Bending of the plates into the timber may also be helpful. Using notched splays and hard wood keys strengthen the joint but are difficult to construct. These are very strong types of joints.

11.2 BEARING JOINTS

When two members have to be joined at right angles to each other, it is essential that they should have sufficient strength at the connection. Various types of joints are used depending on the nature of members to be connected and the forces acting on them. Some of them are described below.

(*i*) ***Notching***: One member may be cut as shown in Fig. 11.12 'A' and the other fitted with it. The shoulders thus created help in the prevention of any displacement of the joint. If both the members are cut, double notching is formed (*See* Fig. 11.13).

(*ii*) ***Cogging***: When the entire depth of the timber is to be utilised, the members are to be connected as shown in Fig. 11.14.

(*iii*) ***Halving***: Timbers that cross each other and are required to be flush on one or two faces are cut to meet as shown in Fig. 11.15.

(*iv*) ***Dove-tailing***: Wedge shaped pieces are cut from members and the timbers are fitted into the projections of each other. (*See* Fig. 11.16).

(*v*) ***Housing***: When the entire end or thickness of a member fits into the notch of another, a housed joint is formed (*See* Figs. 11.17 to 11.19).

(*vi*) ***Mortice and tenon***: In the ordinary type of this joint, the end of one timber is cut so as to form a projection which is called Tenon. The other member is morticed to correspond to the dimensions of the tenon and the latter fitted into the former. This joint is commonly used. The two members may further be prevented from displacement by driving wedges or pins into the joint. (Fig. 11.20)

(*vii*) ***Chase morticing***: This is used when subsidiary members are connected to the main members which have been fixed earlier (Fig. 11.21).

(*viii*) ***Joggle tenon***: This is used for members which rest on each other and displacement between the two is undesirable, *e.g.*, a wooden post resting on a sill. (Fig. 11.22).

(*ix*) ***House and dovetail tenon***: When the ordinary tenon is insufficient to develop the necessary bearing strength, the entire End of one member is fitted or housed into the recess made in the other and further a tenon and morticed joint is built.

If the tenon is made of the shape of a dovetail and the mortice is cut in a similar fashion to suit the tenon, a dovetailed joint is formed.

(*x*) ***Tusk-tenon***: When members of equal depth meet each other at right angles, this type of joint is used.

Pins may be driven to prevent the lateral movement of the joint.

(*xi*) ***Fox-tail wedging***: This joint is used when the back side of the morticed member is not accessible. In this case the mortice is cut to a lesser depth than the member and is also slightly dovetailed. The tenon is cut and two sockets are made in the tenon in which wedges are inserted. The entire assembly is forced into the mortice. (*See* Fig 11.24).

(*xii*) ***Bridle***: In this, the members are slotted and tenoned as shown in Fig. 11.23. These joints are used in wooden trusses at the junction of struts and ties.

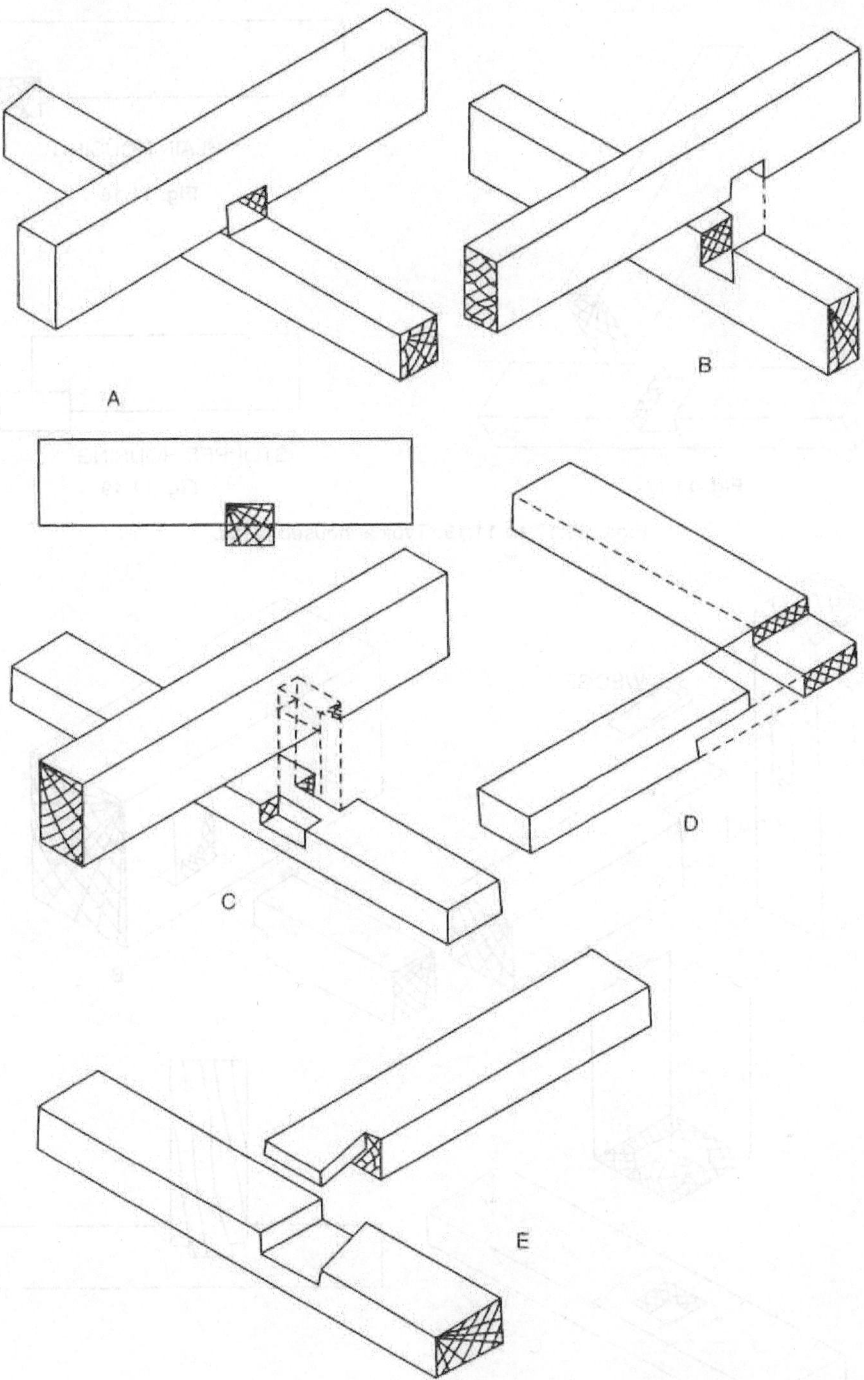

Figs. 11.12 to 11.16: Various types of bearing joints.
(A) Notching, (B) Double Notching, (C) Cogging, (D) Halving, (E) Dovetailing.

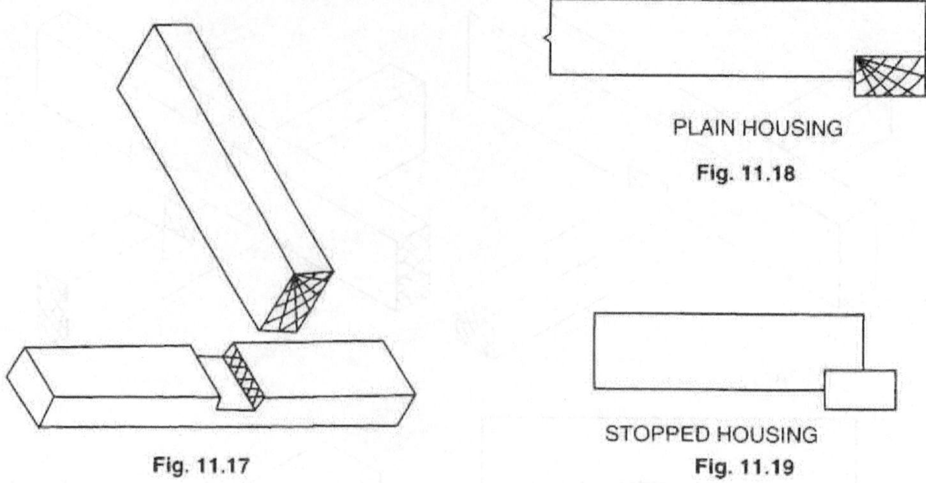

Figs. 11.17 to 11.19: Typical housed joints.

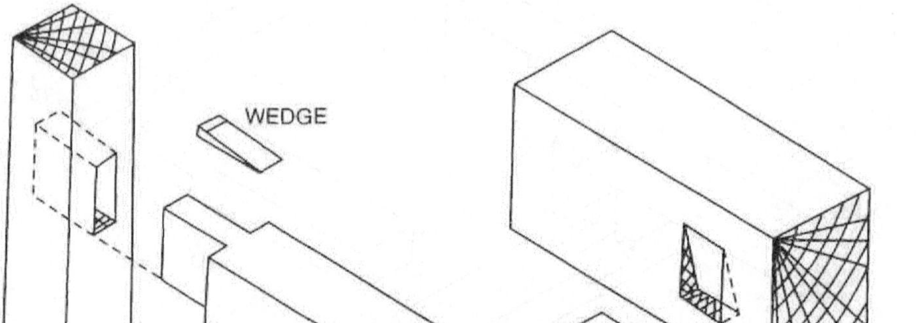

WEDGE

A

B

C

Figs. 11.20 to 11.22: Bearing joints. (A) Mortice and Tenon. (B) Chase Mortice, (C) Detail and Elevation of Joggle tenon joint.

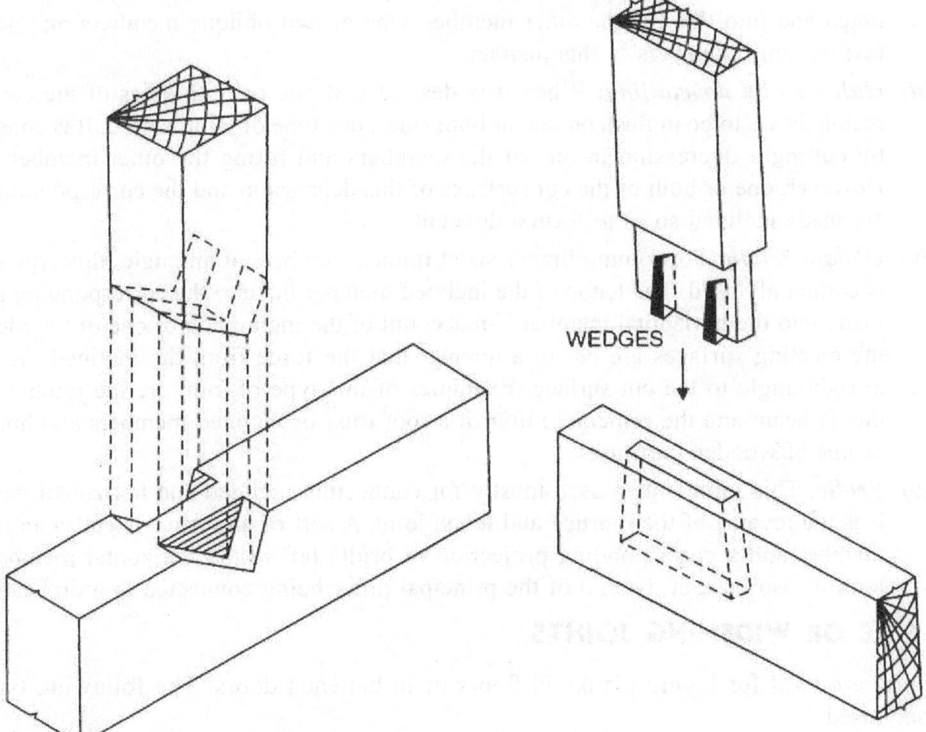

Fig. 11.23: A simple bridle joint

Fig. 11.24: Fox-tail wedging joint.

11.3 OBLIQUE SHOULDERED JOINTS

(i) *Mitre*: When two planks are to be joined at an angle, a mitre joint is formed. For right angled junctions, the two members are cut at 45° and joined. To prevent light from being seen through the widening of the joint, a wooden piece is fitted in the centre of the members and this is called mitred and feather joint.

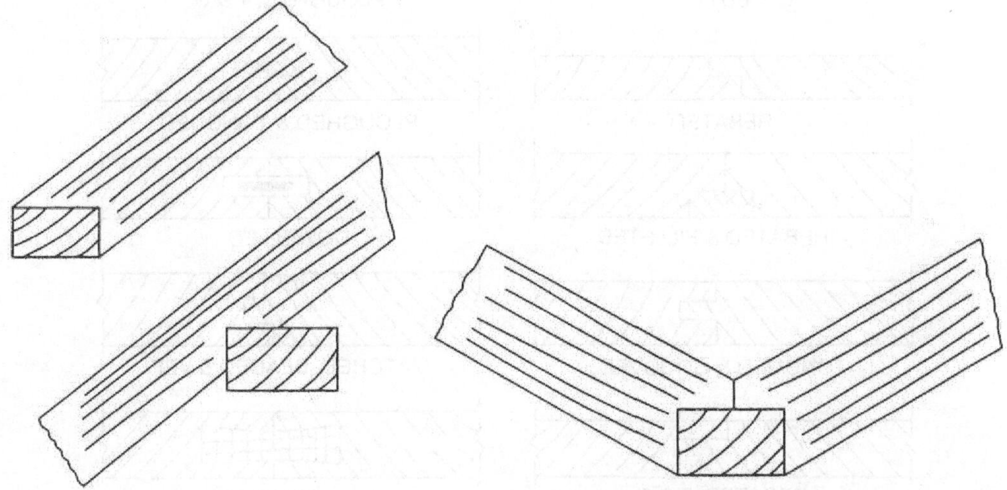

Figs. 11.25 to 11.27: Typical bird's mouth joints.

(*ii*) ***Bird's mouth***: The oblique member of this joint is cut in such a way so as to form a notch and into this fits the other member. One or two oblique members may join and rest on other members in this manner.

(*iii*) ***Halving and dovetailing***: When it is desired that one or both sides of the connected members are to be in flush on one or both sides, this type of joint is used. It is constructed by cutting a depression in one of the members and fitting the other member into it. However, one or both of the cut surfaces of this depression and the corresponding tenon are made inclined so as to form a dovetail.

(*iv*) ***Oblique tenon***: For joining bigger sized timber members at an angle, this type of joint is commonly used. The tenon of the inclined member fits into the corresponding mortice made into the horizontal member. On account of the inclination of one of the members, the meeting surfaces are cut in a manner that the force from the inclined member is at right angle to the cut surface. Examples of this type of joint are the joints between the tie beam and the principal rafter of a roof truss or inclined members and horizontal beams of wooden partitions.

(*v*) ***Bridle***: This joint is also used mostly for connecting inclined and horizontal members. It is the reverse of the mortice and tenon joint. A sort of a mortice is cut in an inclined member and a corresponding projection or bridle left in the horizontal member. This joint is also used at the end of the principal rafter being connected to a tie beam.

11.4 SIDE OR WIDENING JOINTS

These are used for laying planks in floors or in battened doors. The following types are commonly used:

(*i*) ***Butt***: The two members are simply joined side to side.

(*ii*) ***Rebated***: In this, depressions are cut in both the members and are joined lapping mutually.

(*iii*) ***Rebated and filleted***: A small depression is made in the connecting ends of both the members and other small piece fitted in.

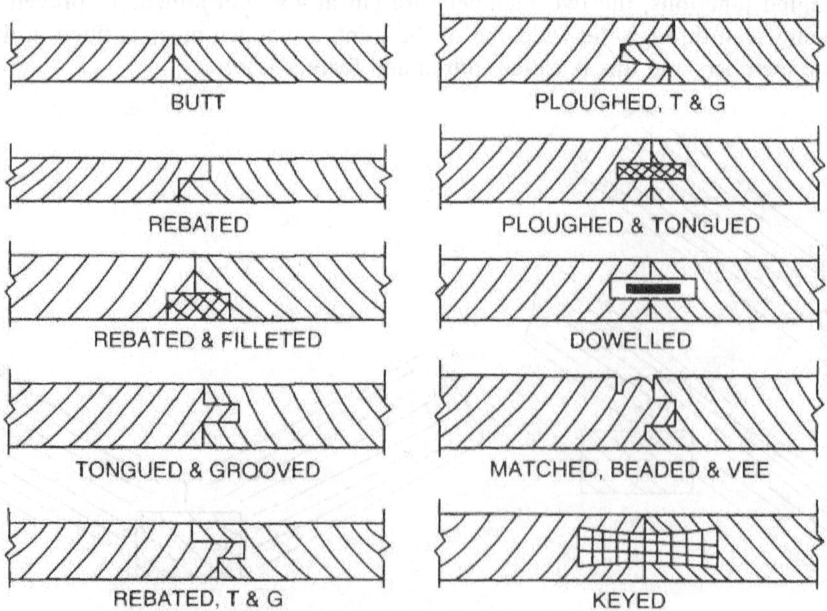

Figs. 11.28 to 11.37: Different types of side joints.

(*iv*) ***Tongued and grooved***: A depression (groove) is made in the side of one member and a corresponding projection (tongue) is cut in the other member. They are fitted tight.

(*v*) ***Rebated tongue and grooved***: In addition to forming a tongue and groove, a rebate is also given by cutting an additional width on one side of the tongue and making a similar shape for the other member.

(*vi*) ***Splayed, tongued and grooved***: This joint is similar to the joint (*iv*) except that the tongue has got one side cut at an angle.

(*vii*) ***Ploughed and tongued***: In this both the members have got similar grooves and a small wooden piece is fitted tight in between.

(*viii*) ***Dowel***: This is similar to the above type except that the additional small member may be of any other material (dowel) and may not fit tight.

(*ix*) ***Matched, beaded and vee***: This has got a tongue and groove arrangement and in addition has special moulding on one side and a V-shaped depression on the other.

(*x*) ***Keyed***: A dovetailed shaped key is used to fit in the depressions in the connecting members which are fitted tightly.

11.5 ANGLE JOINTS

These joints are generally used for connecting members parallel or at right angles to their grains. The following types are generally used:

(*i*) ***Butt***: The members are connected by just joining them together.

(*ii*) ***Rebate, butt and bead***: A rebate is formed in one of the members and the other member is fitted in it. To give it a better appearance, a moulding in the form of a bead is used.

(*iii*) ***Tongued and butt***: In this, the members are butted against each other and a small piece of wood is fitted into the corresponding depression of the two members.

(*iv*) ***Tongue and groove***: The members may be joined by fitting a projection of the two members.

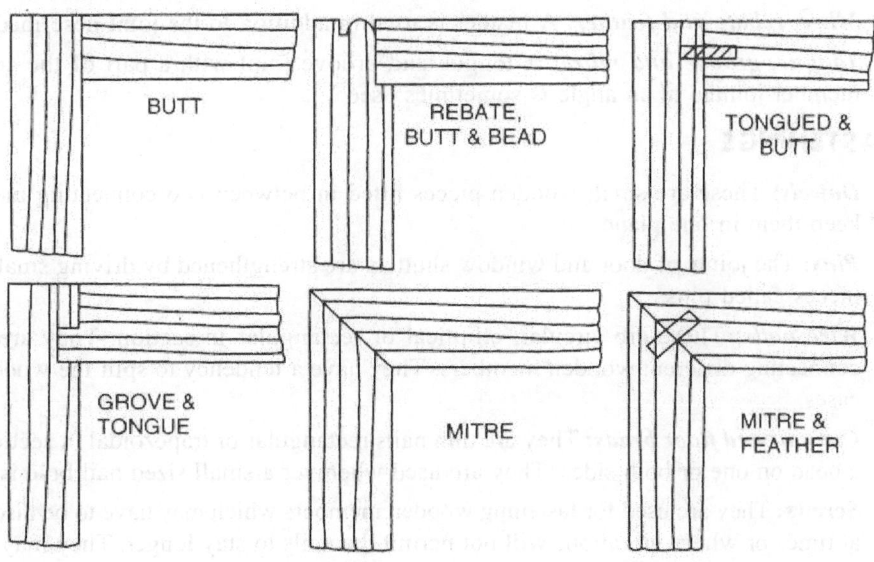

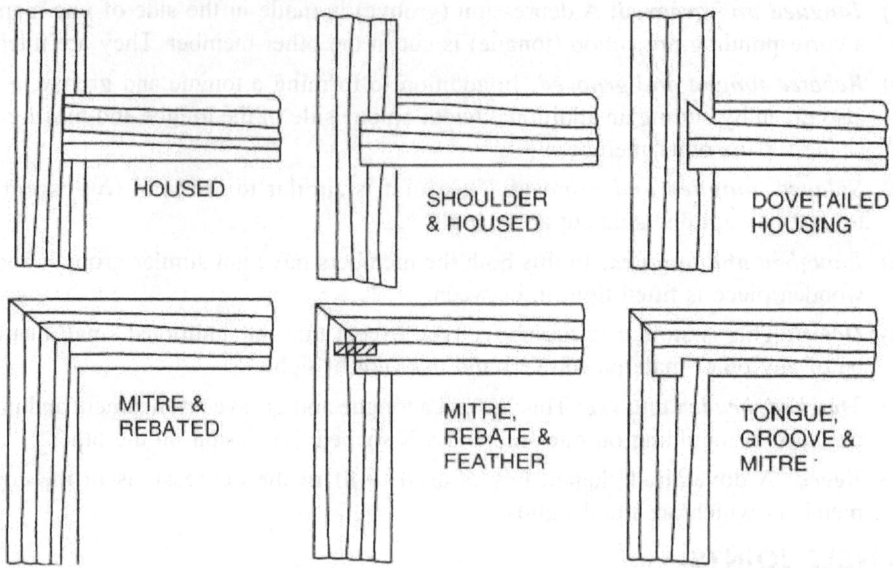

Figs. 11.38 to 11.49: Various types of angle joints.

(*v*) ***Mitre***: The members are cut at an angle at the joint and fitted.

(*vi*) ***Mitre and feather***: An additional small wooden member is inserted in the middle of the mitre joint.

(*vii*) ***Housed***: One member is completely fitted into the depression of the other.

(*viii*) ***Shouldered and housed***: Part of one member fits into a corresponding depression of the other member.

(*ix*) ***Dovetailed housing***: One member is fitted into the other by dovetailed joint.

(*x*) ***Mitre and rebate***: In addition to mitre, a rebate is used.

(*xi*) ***Mitre, rebate and feather***: A feather is used in addition to the joint described above.

(*xii*) ***Tongue, groove and mitre***: A tongue and groove joint with a part of the connecting member jointed at an angle is sometimes used.

11.6 FASTENINGS

(*i*) ***Dowels***: These are small wooden pieces fitted in between two connecting members to keep them in one plane.

(*ii*) ***Pins***: The joints of door and window shutters are strengthened by driving small wooden pieces called pins.

(*iii*) ***Wire nails***: These are circular, elliptical or rectangular in section. They are used for connecting different wooden members. They have a tendency to split the wood in some cases.

(*iv*) ***Cut nail and floor brads***: They are thin nails rectangular or trapezoidal in section having a head on one or both sides. They are used wherever a small sized nail hole is required.

(*v*) ***Screws***: They are used for fastening wooden members which may have to be disconnected at times or where vibrations will not permit the nails to stay longer. They may be round headed or counter sunk headed.

(*vi*) ***Bolts and coach screws***: Whenever spikes or ordinary screws cannot offer much resistance bolts have to be used. They are generally used for connecting large sized members.

(*vii*) ***Connectors***: Thin metal rings or corrugated sheet pieces can be driven into the members after abutting them. They have sufficient strength and do not damage the timber.

(*viii*) ***Dogs***: These are wrought iron U-shaped pieces with pointed ends. They are driven to connect the members.

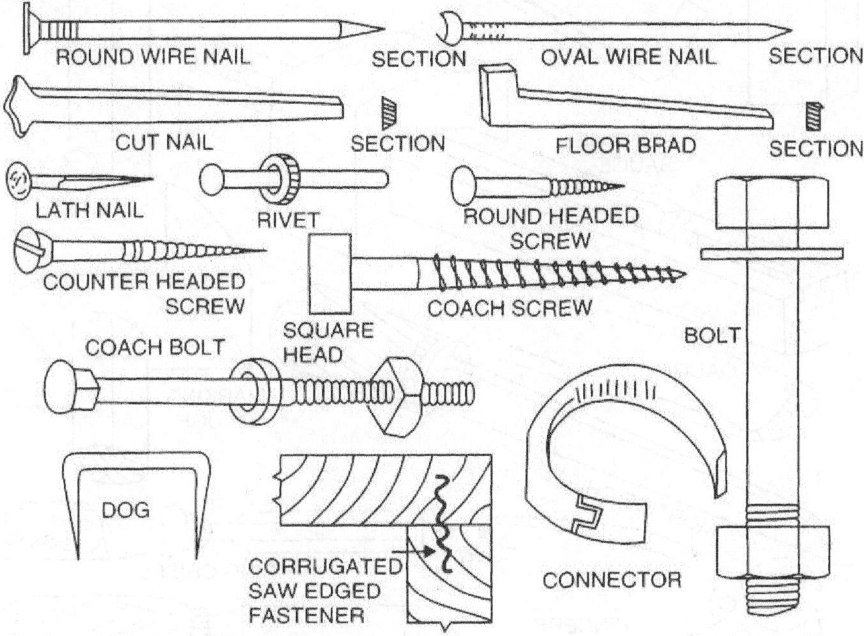

Figs. 11.50 to 11.63: Fastenings.

11.7 TOOLS FOR CARPENTRY WORK

Following are the types of tools used in carpentry work (Figs. 11.64 to 11.100).

(*i*) ***Marking tools***: These are used for setting out and adjusting carpentry work. They are try square, bevel, scribing knife, marking point, mortice gauge and marking gauge.

(*ii*) ***Cutting tools***: These are chisels and saws. The types of chisels commonly used are mortice chisel, paring chisel, firmer chisel etc. Crosscut saws, tenon saws, dovetailed saws, compass saws and coping saws are used for various types of cutting work.

(*iii*) ***Boring tools***: For driving holes ratchet and brace may be used to which various types of bits such as centre, shell, rose-countersunk, auger, screwdriver bits are used to get different shapes of holes. Augers and gimlets and bradwal are also used for driving holes.

(*iv*) ***Planning tools***: Various types of planes, *e.g.*, jack plane, rebate plane, trying plane, bead plane, etc. are used for smoothening the wood face or for creating small mouldings.

(*v*) ***Hammers and screw drivers***: Various types of these tools are shown in the figures.

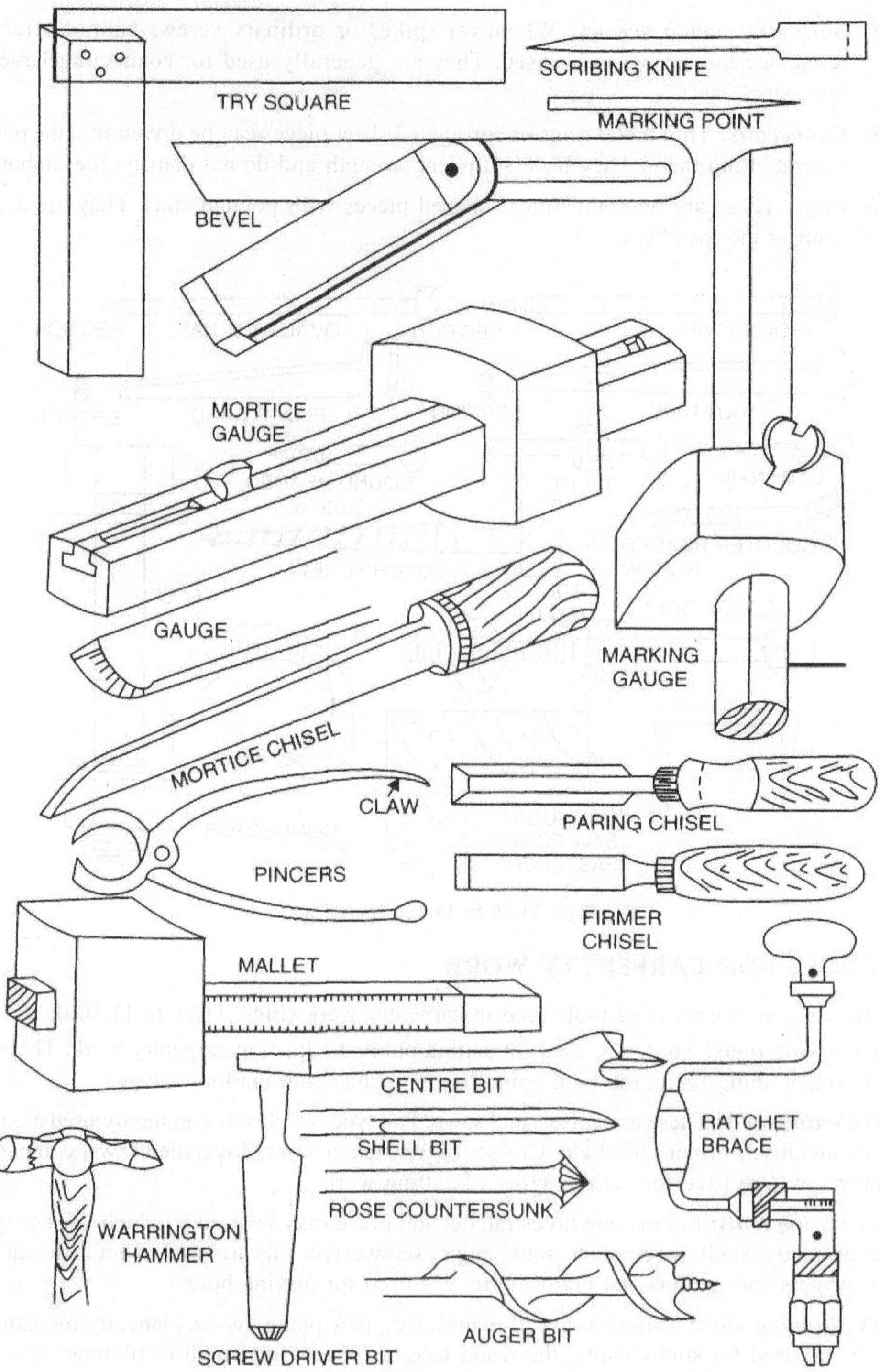

Figs. 11.64 to 11.82: Tools used in carpentry work.

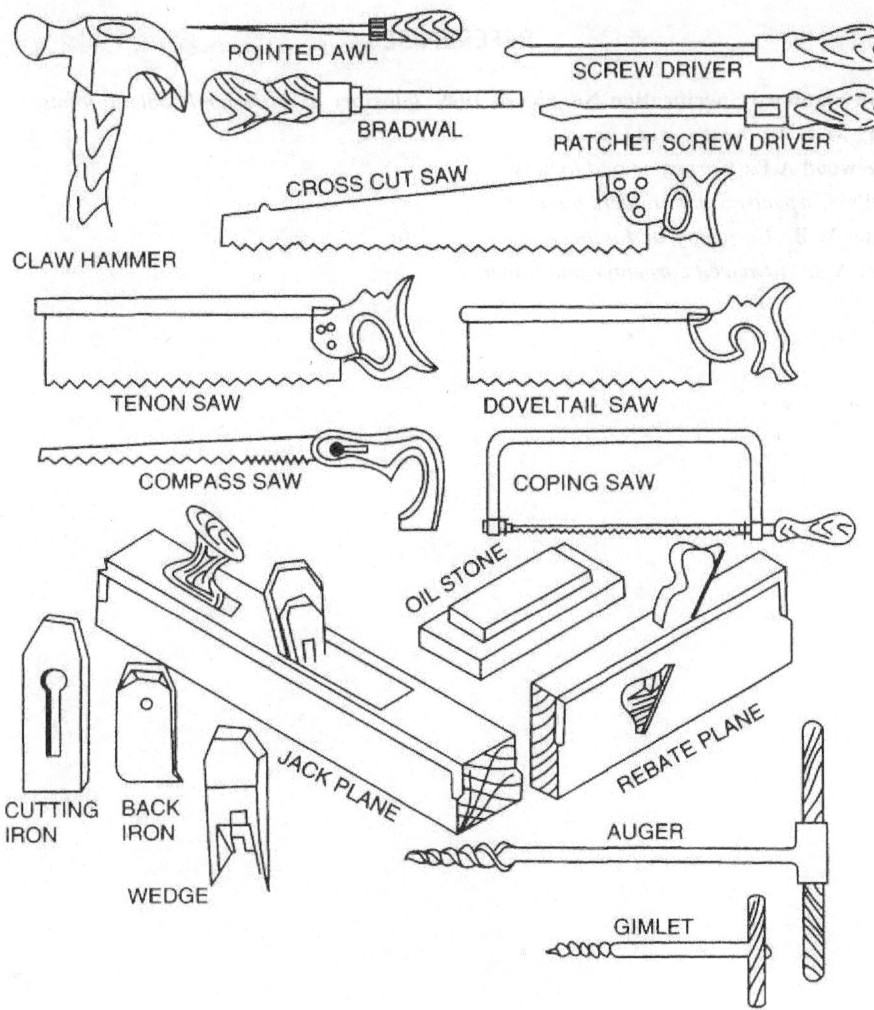

Figs. 11.83 to 11.100: Tools used in carpentry.

QUESTIONS

1. What are the various types of lengthening joints? Describe briefly how a scarf joint is made up so as to connect two members.
2. Write short notes on:
   (*i*) Notching. (*ii*) Cogging.
   (*iii*) Dovetailing. (*iv*) Halving.
3. Illustrate with sketches how a mortice and a tenon joint is made? What is meant by a chase mortice joint? When is fox-tail wedging used for a mortice and tenon joint?
4. Write short notes on:
   (*i*) Mitred joint. (*ii*) Bird's mouth joint.
   (*iii*) Oblique tenon joint. (*iv*) Bridle joint.
5. What are the various types of side joints used for joining wood planks? Illustrate with sketches.
6. What are the various types of fastenings and fixtures used in carpentry work? When are dowels and connectors used?

REFERENCES

1. **British Standard Specification No. 565 of 1949**: *Glossary of terms applicable to joints.*
2. **Telling M.T.**: *Carpentry and joinery.*
3. **Bridgewood A.E.**: *Carpentry and joinery.*
4. **Audel's**: *Carpenters and builders guide.*
5. **Dougles W.B.**: *Carpentry and joinery.*
6. **Emery A.S.**: *Advanced carpentry and joinery.*

Doors and Windows

12.1 LOCATION OF DOORS AND WINDOWS

12.1.1 Doors

Doors are provided to give an access to the inside of the room of a building. Hence they should be so located that freedom of movement in and out of the room is ensured keeping in view the maximum use of accommodation in the room and privacy of the occupants. Doors should preferably be placed in the corner of the room. In case a room has more than one door and they are to be located in the opposite walls of the room, the doors should face each other. This would provide a good ventilation and also create least interference with the inside arrangement of the room.

12.1.2 Windows

Distribution and control of day light, desired vision of outside, privacy, ventilation control and heat loss are some of the factors which have to be considered in the location of the windows. The pattern of air movement inside a room depends to a large extent on the angle at which air enters and leaves the room. It is desirable that the movement of air is downwards and across a room for proper ventilation. This means that windows should be located opposite to each other wherever possible. An attempt should be made to get the maximum amount of light into the room. Windows located on the northern side would be highly effective in this respect. Wherever windows have to be located in residential buildings situated just near public places, *e.g.*, a shopping centre or a cinema, privacy of the rooms on the ground floor is necessary. For this, the windows should be kept at a higher level say about two metres above the ground level outside. On similar considerations, the windows of bath rooms should be located at a higher level. Windows provide an effective means of getting a perspect of the outside surroundings. It is desirable to locate the windows above the floor level of a room in such a manner that the residents can conveniently look outside. A height of 0.75 to 1 metre for the window sill above the floor level is recommended.

12.2 SIZES OF DOORS AND WINDOWS

12.2.1 Doors

Internal doors of residential buildings should not, as a rule, be less than 0.9 m × 2 m. However the size of doors leading to bath rooms can be reduced to 0.75 m × 2 m. It should be aimed that two persons walking shoulder to shoulder should conveniently pass through the door. A size of 1.0 m × 2 m would be an average recommended standard. Doors in public places should be of larger size and their height increased accordingly. A common criterion for the sizes of the doors, used in India, is:

$$\text{width} = (0.4 \text{ to } 0.6) \text{ height}$$

Doors of garages should, as a rule, be not less than

$$2.5 \text{ m} \times 2.25 \text{ m}.$$

The following further points may be noted regarding the location, hanging and size of the doors:

(*i*) ***Width***: The aggregate clear width of doorways serving as exit for more than 40 persons shall not be less than 550 mm per hundred persons accommodated.

(*ii*). ***Hanging of doors***

(*a*) The doors shall be so hung and arranged that when opened, they shall not in any way diminish or obstruct the required width of passageway, stair or other means of exit.

(*b*) Building constructed facing streets should have door opening outside (*i.e.*, towards the street).

(*c*) Exit doors leading from rooms occupied by 15 persons or more shall be hung to swing in the direction of exit.

(*d*) No exit door shall open immediately on a flight of stairs, but a landing, the length and width of which are not less than the width of such door shall be provided between the door and the stairs. No riser should be located within 300 mm of an exit door.

(*iii*) ***Door fastenings***: Fastenings required on exit doors shall be such that the door can be readily opened from inside.

12.2.2 Windows

The size of windows would be governed by the total area of window space needed. The following criteria may be kept in view while designing a window.

(*i*) Breadth of window = $\frac{1}{8}$ (width of room + height of room).

(*ii*) There should be one square metre of window space for every 30 to 40 cubic metre of inside content of the room.

(*iii*) Glass area in the windows should be at least 15% of the floor area of the room. However, it is a good practice to provide 20% of the floor area as the total glass area of the windows. Continuous sash or one large opening in a room provides better distribution of light than a number of small windows.

12.3 MATERIALS

12.3.1 Doors

(*i*) ***Wood*** Wood is used in several forms for the construction of doors. Doors may be made of solid planks or built up of small individual pieces. Since wood is one of the most commonly available building materials in India, it is largely used for this purpose. Moreover, it can also take a good polish and can be given different type of mouldings to produce an attractive appearance. However, its life is less as it is easily attacked by vermin especially when in contact with walls built of mud masonry.

(ii) ***Glass*** This is used in panels of doors to admit light. Big openings in doors fitted with thick glass sheets give a very good appearance. Coloured glass, frosted glass, and beaded glass can be used for ornamental appearance.

(iii) ***Ply-wood*** This material is very useful for doors which have to be made flush.

(iv) ***Metals*** Steel framed doors are gaining popularity these days. Cast metal doors are relatively high priced. They are used mostly for monumental structures. Hollow metal doors made up of a metal frame and covered with sheet metal are selected for greater rigidity, life and fire resistance.

(v) ***Concrete*** Reinforced cement concrete frames can be used for doors in areas where there is a danger of disintegration of wooden frames by vermin.

12.3.2 Windows

The above-mentioned materials can also be used for the construction of windows. Use of metal in windows is progressively increasing. Aluminium and steel are mostly used. However bronze, stainless steel and galvanised steel are used for specific type of buildings.

12.4 DEFINITIONS

(i) ***Frame*** A group of members assembled and placed along the top, bottom and sides of an opening so as to form an enclosure and a support for a door or a window shutter.

(ii) ***Sill*** The horizontal member forming the bottom of the frame.

(iii) ***Shutter*** Planks panelled or otherwise which fit in the frame.

(iv) ***Style or stile*** Upright or vertical outside members of a shutter or framework.

(v) ***Top rail*** The topmost horizontal member of a shutter or framework.

(vi) ***Lock-rail*** Middle horizontal member of a shutter or framework wherein lock and bolt are fixed.

(vii) ***Bottom-rail*** The lower-most horizontal member of a shutter or framework.

(viii) ***Mullion*** The vertical member running through a shutter or a frame.

(ix) ***Jamb*** Vertical face of a window or door opening which supports the frame.

(x) ***Reveal*** External jamb of a door or window opening at right angles to the face of the wall.

(xi) ***Sash bar*** Light weight units of a frame which carry the glass within a window or a door-frame.

(xii) ***Sash-door*** Door whose upper part is glazed.

(xiii) ***Panel*** The area included between the rails.

(xiv) ***Hanging or hinged style*** The styles of a door, which is hung on one side, are called Hanging or Hinged styles.

(xv) ***Cross-rail*** Additional horizontal member connecting the two styles of a shutter or framework.

(xvi) ***Transom*** Horizontal dividing member in a window or a door frame.

(xvii) ***Threshold*** Wooden fixture fixed to the floor under a door-frame, thus enabling the door to be cut short enough to clear floor coverings on the inside.

(xviii) ***Rebate*** Depression cut in the frame to receive the door.

(xix) ***Louver*** An inclined piece of wood fixed in a framework.

12.5 GENERAL TYPES OF DOOR MOVEMENTS

(*i*) ***Swinging doors***: The shutters are hung to the door frame on hinges or butts fixed to one side of the shutter so that they swing on a vertical axis. These types are mostly used in residential buildings. The frame of these doors may be laid in flush with the outside or inside face of the wall so as to minimise the obstruction created by the opened shutter. They may be of either single swinging or double swinging type.

(*ii*) ***Sliding doors***: In these, the shutters can slide either upward, downward or sideway. These doors do not create obstruction in movements.

(*iii*) ***Rolling shutter doors***: This is an improvement on the sliding doors. They are generally made of steel and have the advantage of easy opening and closing. Moreover, they do not require much space. These doors are specially useful for garages or shops where large openings have to be provided.

(*iv*) ***Revolving doors***: These doors are used in places where frequent opening and closing of a door is to be avoided, *e.g.*, fish market. These doors hang on a central pivot and can rotate, whenever used, so as to close the opening automatically. Single shutter or double shutter revolving doors are also used in public places, *e.g.*, hotels and hospitals. These doors revolve about one side of the shutter and get automatically closed whenever pushed and left. They are suitable for air-conditioned buildings as they do not admit heavy draught of air when opened.

(*v*) ***Folding doors***: These doors may be of wood or metal. They can be constructed cheaply. They are used for large openings.

(*vi*) ***Collapsible doors***: They are made of light steel sections, mostly channels. They roll in small channels fixed at top and bottom wherein small rollers have been kept.

(*vii*) ***Telescopic doors***: They are used for very large openings. They are made up of five to fifteen leaves, generally centre parting. When opened, the shutters are stacked in pockets at the ends. These are operated by two motors located in the end pockets. The central leaves are driven by an endless chain operated by the motor. The other leaves are moved by a series of cables attached to the powered central leaf and arranged in such a way that all leaves arrive or close simultaneously.

12.6 DOOR FRAMES

(*a*) ***Wooden door frame***: External doors are fixed to frames. A door frame consists of two vertical pieces (posts) and one horizontal piece at top (head). Sometimes a horizontal member is added at the base also and this is called sill. The size of the members depends on the size of the door to be hung. Mostly they are 120 mm × 75 mm in section. They have a rebate of about 12 mm depth wherein the door shutter fits and one groove where the plaster abuts. The joints between the head and the post may be any one of the following types:

(*i*) Closed mortised and tenoned joint.

(*ii*) Haunched mortised and tenoned joint.

(*iii*) Pin and tenoned joint.

(*iv*) Double tenoned joint.

These are illustrated in the diagrams below:

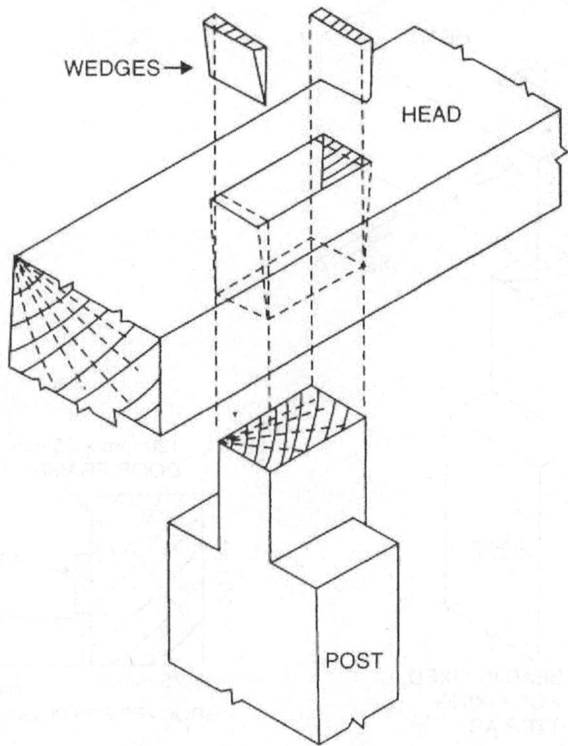

Fig. 12.1: Closed mortice and tenon joint between door frame head and vertical member.

Construction. The sections of suitable size are selected and plained. Rebates of required size are cut. The tenon and morticed joints are framed. The frame is temporarily fitted to check the correctness of the joints. The frame is opened and glue is applied at the contact faces of the joint. Pins of 10 mm diameter are inserted and cut to shape. In addition, wedges may also be inserted if desired. The frame is finally plained and smoothened with the aid of a sand paper.

Fixing:

(*a*) ***Before or during the construction of wall.*** The frame is erected vertical in plumb and secured tightly before any masonry around it is built. To ensure the proper fixing inside the wall, the head is projected beyond the frame dimensions (known as horns). In addition to this or as an alternative iron straps known as hold fasts of suitable dimensions are fixed to the frame.

To enable taking out of the frame for repairs or replacement after the building is constructed, wooden horns are not used. Hold fasts (Iron straps) which have got bolts running through the frame are used. The nut of the bolt is sunk into the frame and covered with putty. Any time removal is desired, the putty is scraped and the nut opened out.

(*b*) ***After the construction of wall.*** For expensive door frames which would get damaged during construction, the door frame is fixed into the wall after construction. Plugs of suitable size are driven into the wall as the masonry is constructed. They are sawn off true to the face of the jamb. The door frame is erected and nailed to these plugs.

(*c*) ***Steel door frame.*** These are made of steel channel sections and have depressions wherein the door shutter can fit in. These are welded at corners, and have suitable attachments, *e.g.*, fixing lugs, hinges, rubber shock absorbers, etc.

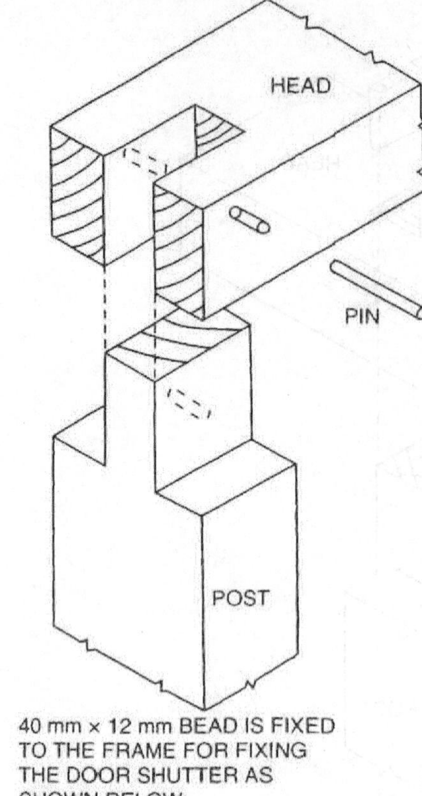

Fig. 12.2: Detail of pin and tenon joint.

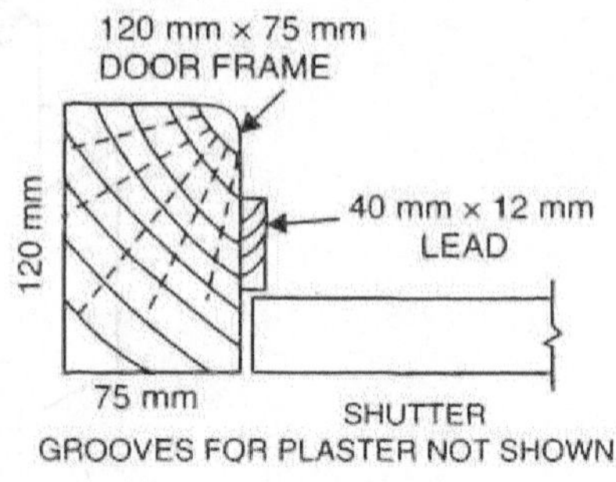

Fig. 12.3: Additional stop fixed to door frame for resting of the shutter.

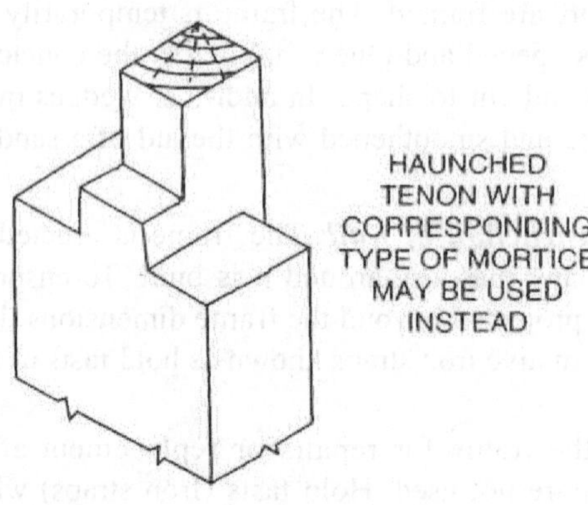

Fig. 12.4

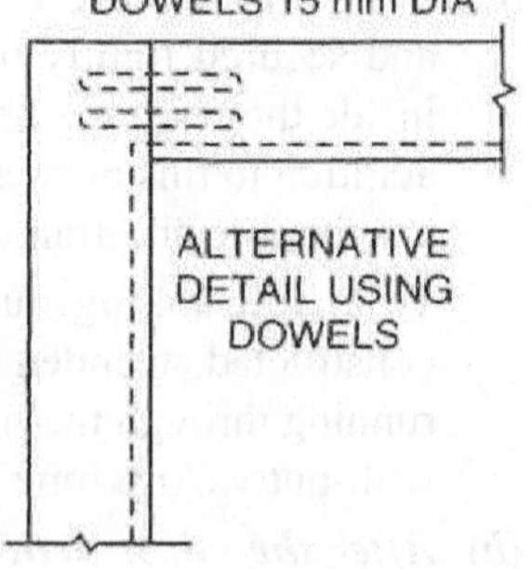

Fig. 12.5: Dowelled joint for door frame.

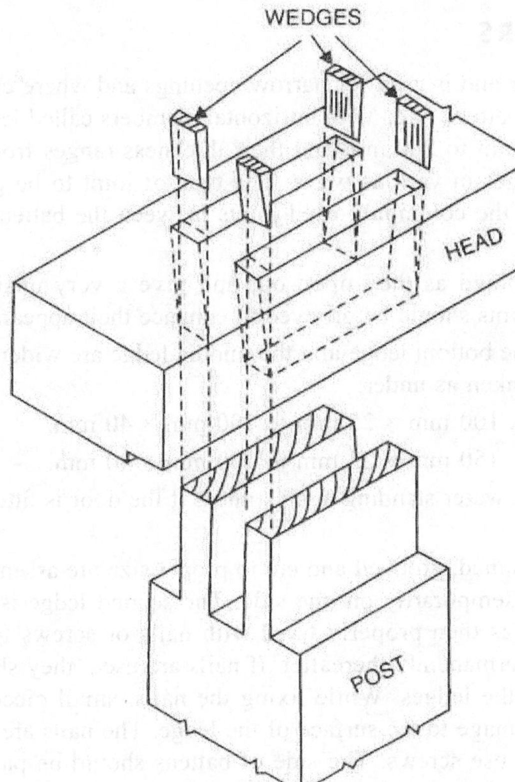

Fig. 12.6: Double tenon joint.

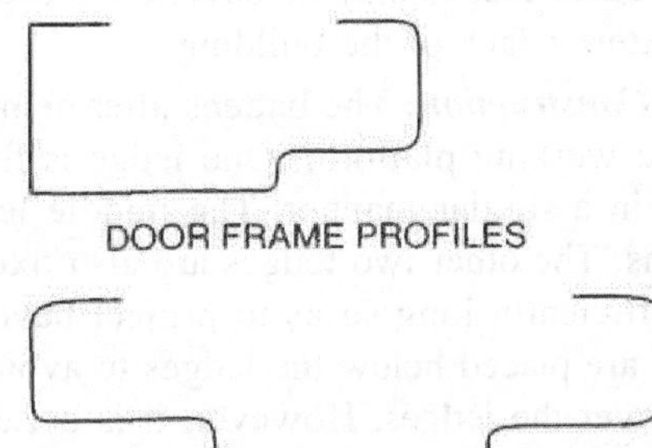

Fig. 12.7 and 12.8: Steel door frames.

TYPES OF DOORS

(1) ***Wooden doors***:

(*a*) Battened and ledged doors.
(*b*) Battened, ledged and braced doors.
(*c*) Battened, ledged and framed doors.
(*d*) Battened, ledged, braced and framed doors.
(*e*) Framed and panelled doors.
(*f*) Glazed or Sash doors.
(*g*) Flush doors.
(*h*) Louvered doors.
(*i*) Wire-gauged doors.

(2) ***Metal doors***:

(*a*) Hollow metal doors.
(*b*) Steel plate doors.
(*c*) Corrugated steel doors.
(*d*) Rolling shutters.
(*e*) Collapsible doors.

(3) ***Composite doors***:

(*a*) Metal covered doors.
(*b*) Revolving doors.
(*c*) Sliding doors.
(*d*) Telescopic doors.

12.8 BATTENED AND LEDGED DOORS

This is the simplest type of wooden door and is used for narrow openings and where cost is the main consideration. It consists of vertical battens fixed with horizontal members called ledges. The size of vertical battens varies from 100 mm to 200 mm and their thickness ranges from 20 mm to 40 mm. The battens are joined together in various ways. The type of joint to be given depends on the desired appearance. Some of the commonly used joints between the battens are shown in Figs. 12.15-12.19.

Square joints in battens should be avoided as they open out and give a very unsightly appearance. If they have to be used, square joints should be grooved to enhance their appearance.

Ledges are normally three in number. The bottom ledge and the middle ledge are wider than the top ledge. The section of ledges may be taken as under:

***Top ledge*:** 100 mm × 25 mm to 100 mm × 40 mm.

***Middle and bottom ledges*:** 150 mm × 25 mm to 200 mm × 40 mm.

These ledges may be bevelled to prevent water standing against them if the door is fitted on the exterior face of the building.

***Construction*:** The battens after being plained, grooved and cut to proper size are assembled on the working platform. One ledge is fixed temporarily on one side. The second ledge is also fixed in a similar manner. The middle ledge is then properly fixed with nails or screws to the battens. The other two ledges are also fixed permanently thereafter. If nails are used, they should be sufficiently long so as to project beyond the ledges. While fixing the nails, small pieces of wood are placed below the ledges to avoid damage to the surface of the ledge. The nails are then bent over the ledges. However, it is better to use screws. The side of battens should be painted after grooving and before fixing as otherwise when the joints open, a bad appearance is created.

***Hanging of door*:** The door is fitted to the frame with garnet hinges, which are generally 250 mm long and three in number.

***Fittings*:** The following fittings are generally used.

Bolts: 2 Nos. of 250 mm.

Handle: 1

Aldrop bolt and pad-lock: 1

Hooks: 2

Door stop: 1 No.

A very cheap type of a door would need a bolt on inside, a thumb latch and a hasp and staple with padlock on outside.

12.9 BATTENED, LEDGED AND BRACED DOORS

The battened and ledged doors, described above, has a tendency to droop at the 'nose'. This is particularly so in the case of wide doors. To prevent this tendency and also to strengthen the door, it is customary to insert sloping braces between the ledges. These braces must incline upwards from the hanging edge or else they will not be useful in preventing the droop. The position of the middle ledge should allow the braces to have the same inclination. The width of braces varies from 100 mm to 150 mm and are usually 25 mm to 40 mm thick.

***Construction*:** This door is made in a manner similar to the battened and ledged door. The joint between the ledge and the brace is shown in Fig. 12.12.

***Hanging of the door*:** This is hung by garnet hinges 200 to 250 mm long and three in number for a single leafed door and six for double leaved door.

***Fittings*:** Same as in the battened and ledged door.

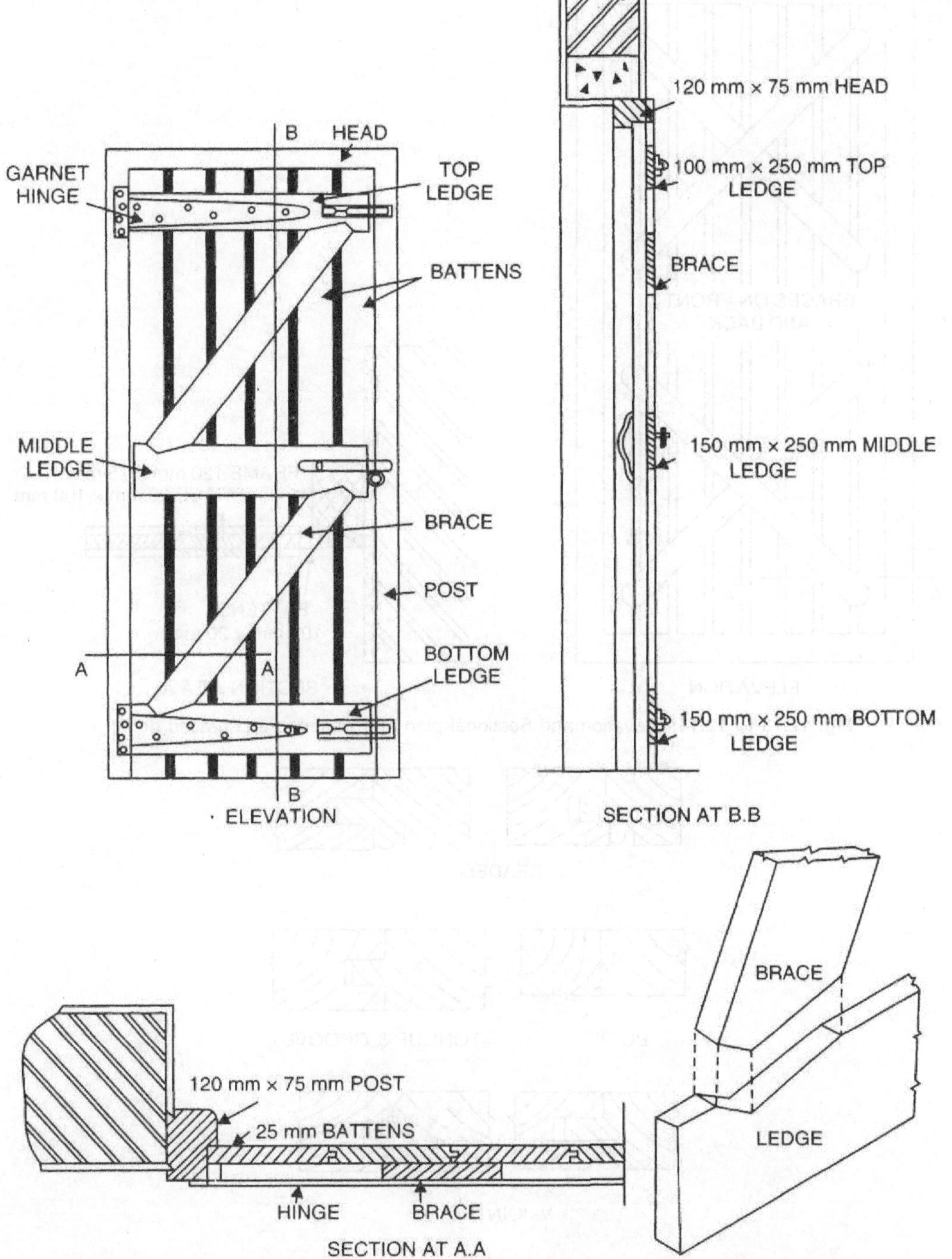

Fig. 12.9 to 12.11: Elevation, Vertical and Horizontal sections of a battened, ledged and braced door.

Fig. 12.12: Joint between brace and ledge.

Sometimes doors with battens and cross braces only are used to give a better appearance than a battened, ledged and a braced door (as shown in Figs. 12.13 and 12.14).

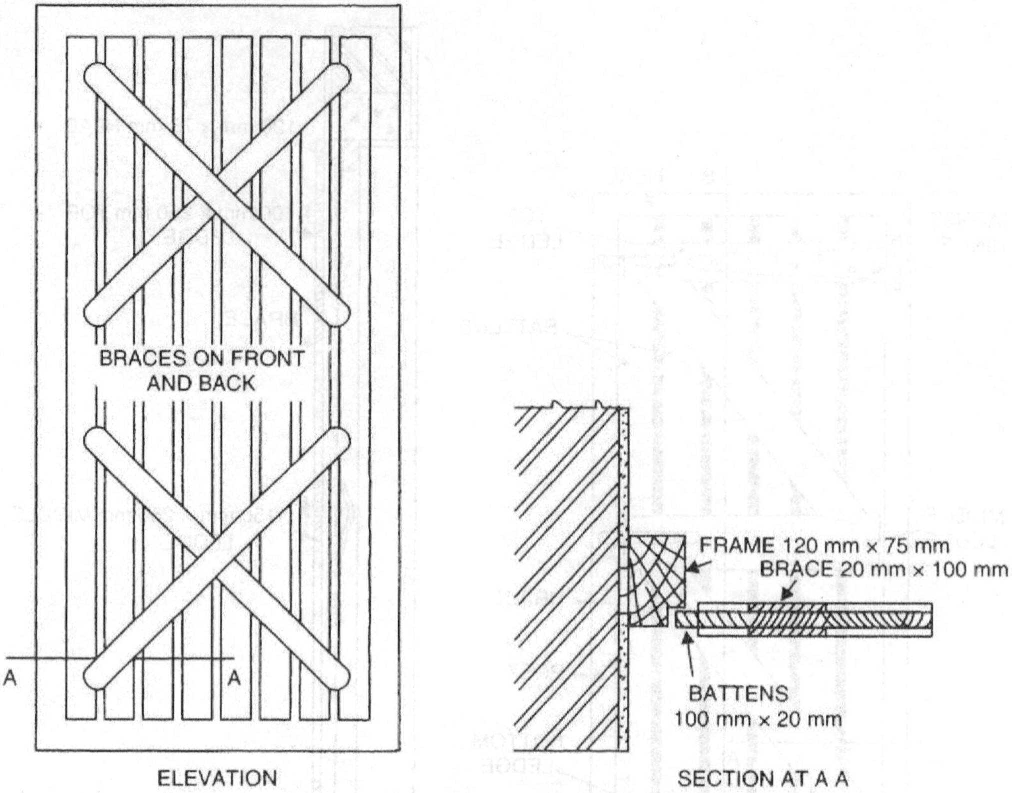

BRACES ON FRONT
AND BACK
A
A
ELEVATION
FRAME 120 mm × 75 mm
BRACE 20 mm × 100 mm
BATTENS
100 mm × 20 mm
SECTION AT A A

Fig. 12.13 to 12.14: Elevation and Sectional plan of a battened and braced doors.

BEADED
BUTT
TONGUE & GROOVE
V-JOINTED

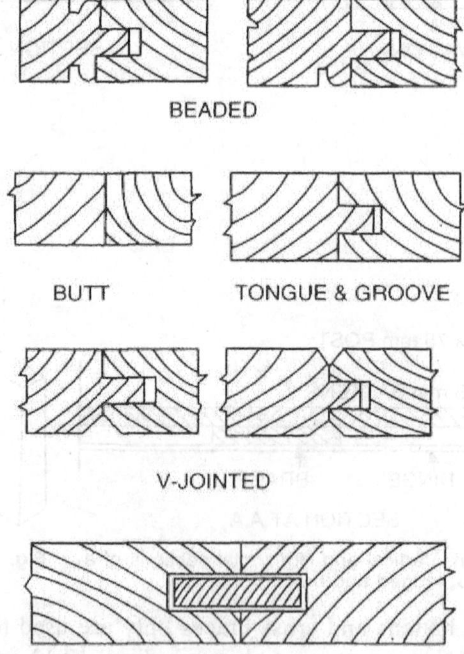

Figs. 12.15 to 12.19: Joints between battens of doors.

12.10 BATTENED, LEDGED AND FRAMED DOOR

This is superior in strength and appearance to an ordinary battened and ledged door. It consists of a framework of two vertical styles, bottom rail, top rail and one middle rail. The battens are fixed in this framework. The styles are, generally, 100 mm wide and 40 mm thick. The top and bottom rails are also of the same size. The middle rail is about 200 mm wide and 40 mm thick as the lock fittings are fixed to it. Battens are from 100 mm to 125 mm wide and about 15 to 20 mm thick.

The middle and bottom rails are mortised and tenoned into the styles. The upper and lower ends of the battens are let into the rails and the side battens are tongued into the styles.

Construction: The rails are fitted loosely into the styles. The battens are accurately fitted and slipped into the grooves of the styles and rails. The tenons are wedged and pinned firmly.

Hanging of the door: This is hung with heavy garnet hinges. Sometimes strap hinges or a Narmadi is used.

Fittings: The usual fittings are:
Strap hinges: 3 Nos.
Bolts: 2 Nos. of 250 mm.
Aldrop with padlock: 1 No.
Handle: 1 No.

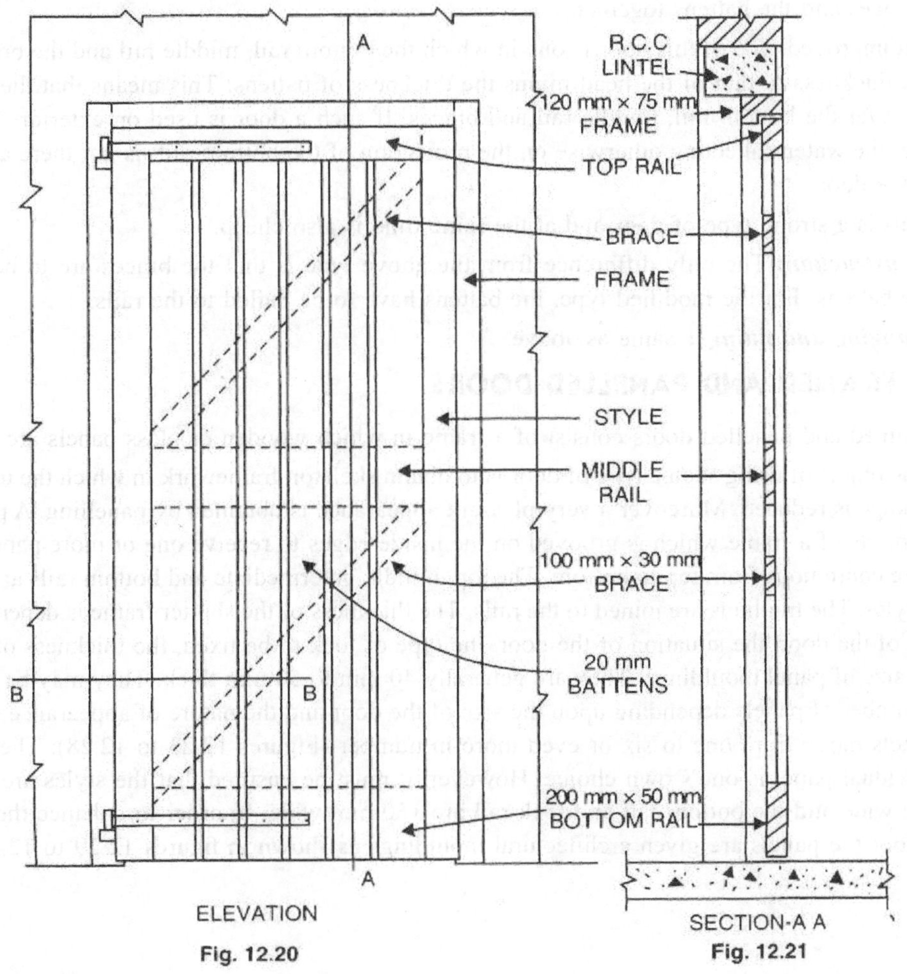

Fig. 12.20

Fig. 12.21

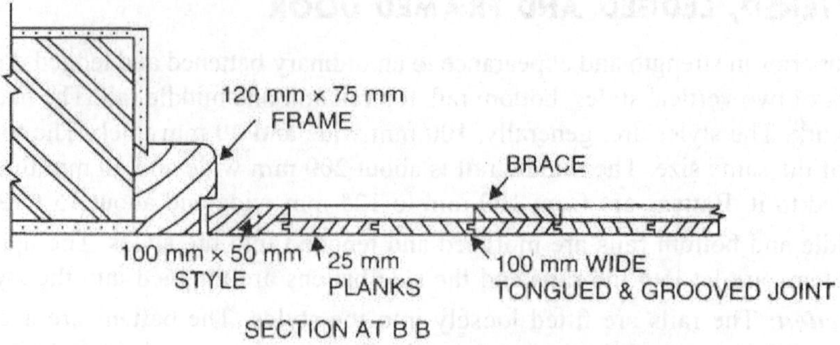

Figs. 12.20 to 12.22: Elevation, vertical and horizontal sections of a framed, ledged and braced door.

12.11 BATTENED, LEDGED, BRACED AND FRAMED DOOR

This door is stronger than the battened, ledged and framed type. In addition to the membes of that door, cross braces inclining upwards from the hanging side are used. The middle and bottom rails are mortised and tenoned into the styles and the braces are either housed into the rails at about 40 mm from the style or taken into the corners and tenoned into the styles. The size of braces is 15 mm × 120 mm. Generally the thickness of the top rail or the styles is equal to that of the braces and the battens together.

An improved type of this door is one in which the bottom rail, middle rail and the braces are equal to thickness to that of the head minus the thickness of battens. This means that the battens will run over the bottom rail, middle rail and braces. If such a door is used on exterior side of a building, the water collecting otherwise on the projection of the bottom rail, is not there and does not rot the door.

This is a strong type of door and at the same time is also cheap.

Construction: The only difference from the above type is that the braces are to be nailed with the battens. For the modified type, the battens have to be nailed to the rails.

Hanging and fittings: Same as above.

12.12 FRAMED AND PANELLED DOORS

Framed and panelled doors consist of a frame in which wooden or glass panels are fitted.

The object of using such a type of door is to obtain skeleton framework in which the tendency to shrinkage is reduced. Moreover a very pleasant appearance is obtained by panelling. A panelled door consists of a frame which is grooved on the inside edges to receive one or more panels. The styles are continuous from top to bottom. The top, middle, intermediate and bottom rails are joined to the styles. The mullions are joined to the rails. The thickness of the shutter frame is dependent on the size of the door, the situation of the door, the type of lock to be fixed, the thickness of panels and the size of panel mouldings. They are generally 40 mm to 50 mm thick. They may be divided into a number of panels depending upon the size of the door and the nature of appearance desired. The panels range from one to six or even more in number (Figures 12.23 to 12.28). The size of the individual panel is one's own choice. However, it must be ensured that the styles are at least 100 mm wide and the bottom rail and lock rail are 150 mm wide. In order to enhance the beauty of the door the panels are given architectural mouldings as shown in figures 12.29 to 12.33.

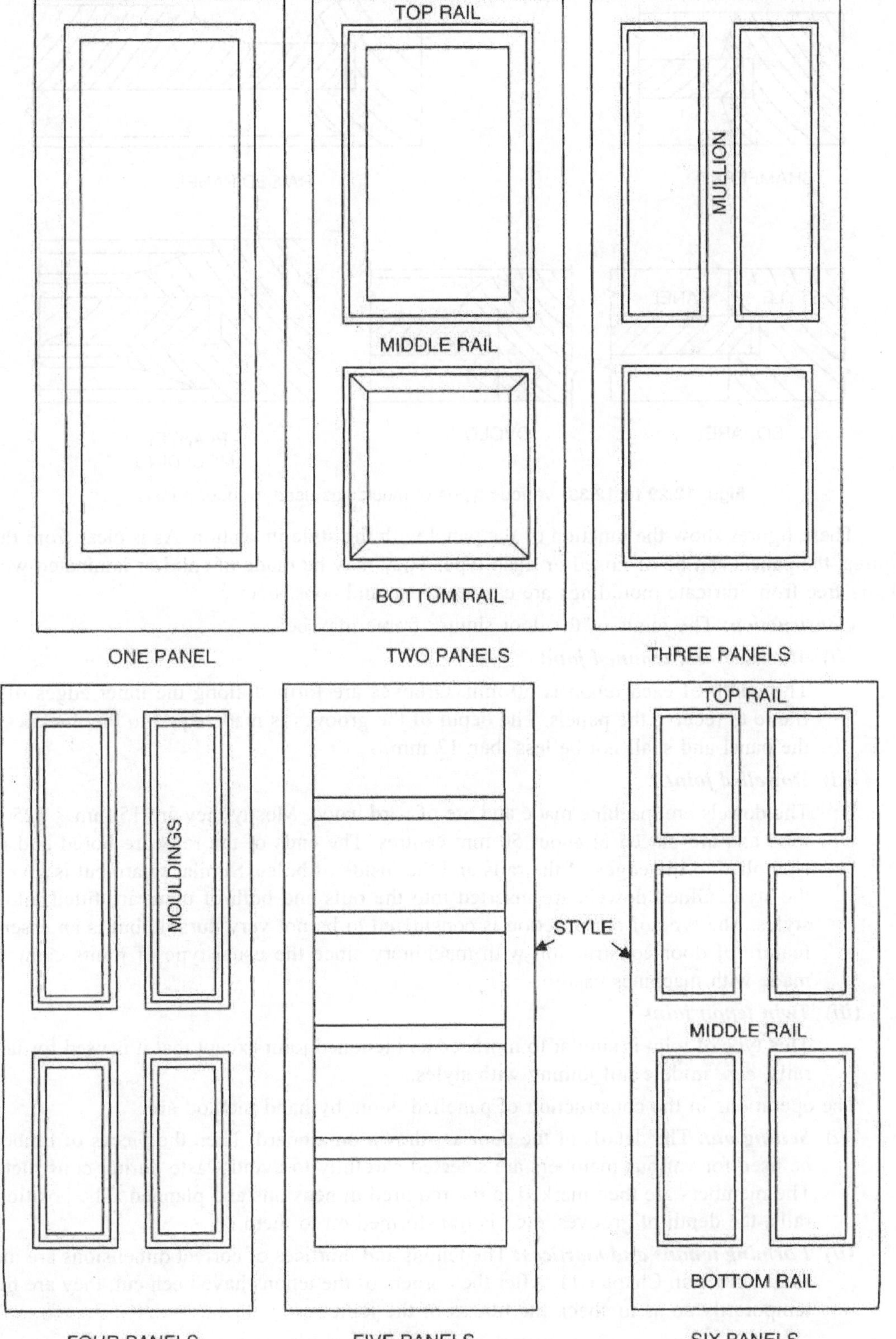

Figs. 12.23 to 12.28: Types of panelled doors.

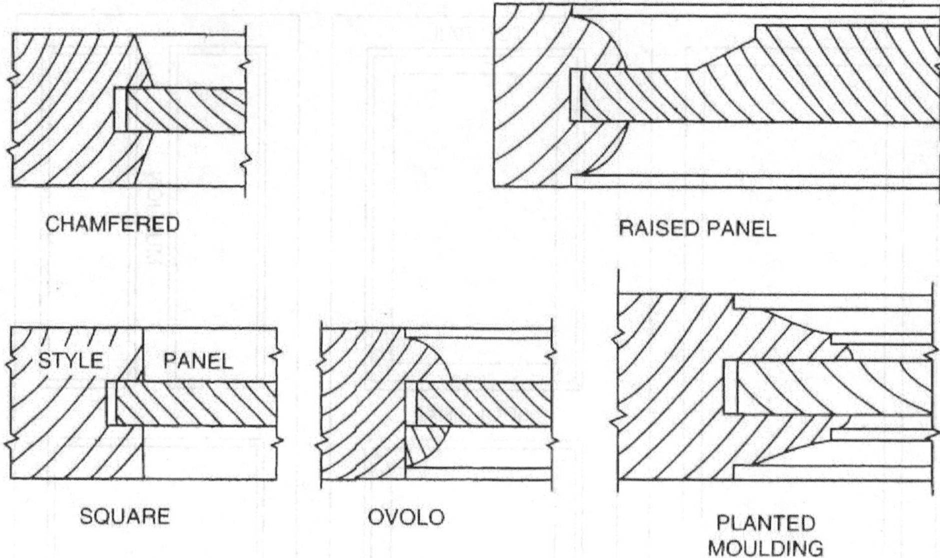

Figs. 12.29 to 12.33: Various types of mouldings used for door panels.

These figures show the junction of the panel with the style in section. As is clear from these figures, the panels can be of raised or flush type. They may be made of solid or laminated wood. Doors free from intricate mouldings are easy to clean and construct.

Construction: The joints of the door shutter frame may be:

(*i*) ***Mortised and tenoned joint***

The width of each tenon is 50 mm. Grooves are formed along the inner edges of the frame to receive the panels. The depth of the grooves is made equal to the thickness of the panel and shall not be less than 12 mm.

(*ii*) ***Dowelled joint***

The dowels are machine made and are of hard wood. Mostly they are 15 mm × 125 mm long and are placed at about 50 mm centres. The ends of the rails are bored and glue is applied to the edges of the rails and the inside of holes. Similar treatment is given to the style. Glued dowels are inserted into the rails and both of them are fitted into the styles. The type of construction is considered to be not very durable but is an essential feature of door construction with machinery since the usual type of joints cannot be made with machines easily.

(*iii*) ***Twin tenon joint***

This type of joint is similar to mortised and tenoned joint except that it is used for larger rails, *e.g.*, middle rail joining with styles.

The operations in the construction of panelled doors by hand method are:

(*i*) ***Setting out.*** The details of the door are drawn on a board. Then the pieces of timber to be used for various members are selected carefully to avoid waste during construction. The members are then marked to the required dimensions and planned. The position of rails, the depth of grooves, etc., is transformed on to them.

(*ii*) ***Forming tenons and mortices***: The tenons and mortices of correct dimensions are made as described in Chapter 11. After the corners of the tenons have been cut, they are fitted temporarily so as to check the fitness of the joint.

(*iii*) ***Making panels.*** The panels are next made to the required size and mouldings.

(*iv*) ***Gluing and fixing.*** The shutter frame is opened and glue is applied to all the places which come in contact with each other. The panels are placed in position and the frame is immediately re-assembled. Wedges or pins (*See* Fig. 12.43) are fixed into the joints.

(*v*) ***Cleaning.*** All the projections which are in excess to the required dimensions are sawn off. The shutter thus prepared is finally cleaned with a sharp plane followed by sand paper.

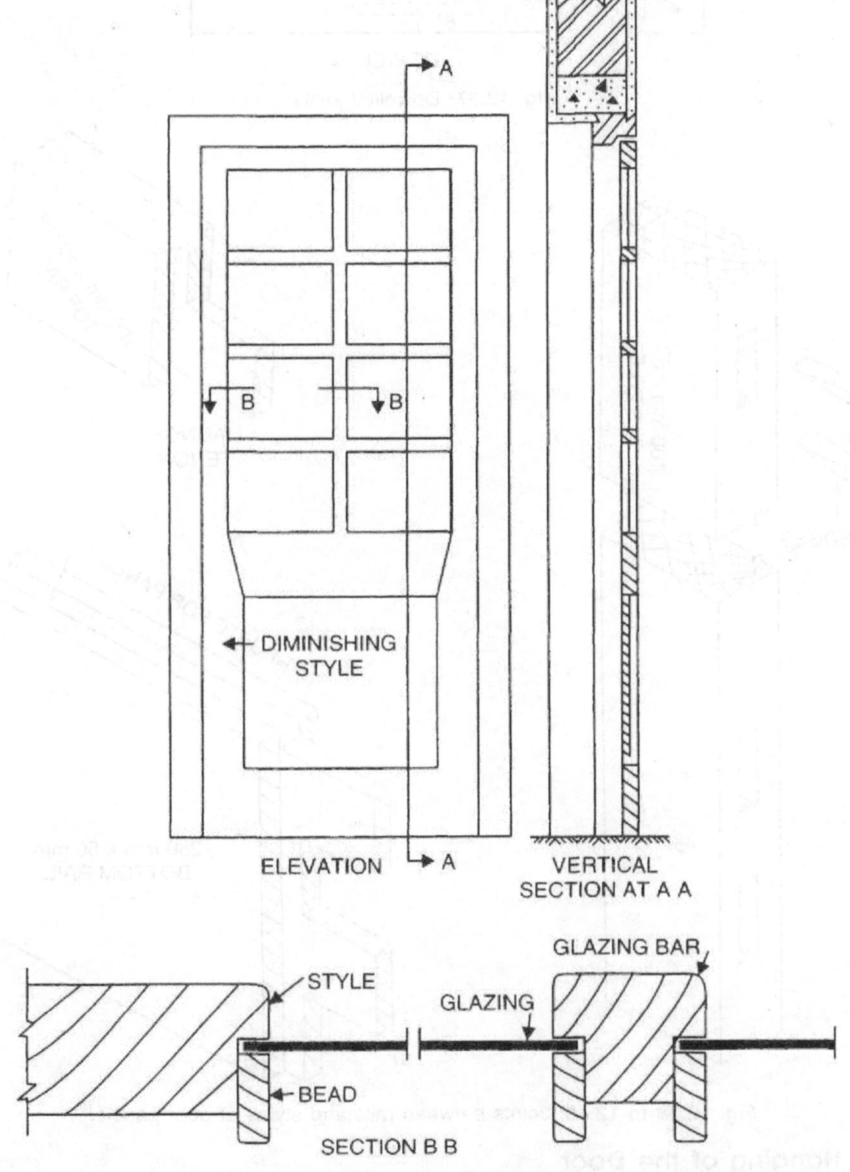

Figs. 12.34 to 12.36: Elevation, vertical and horizontal sections of a glazed and panelled door with diminishing style.

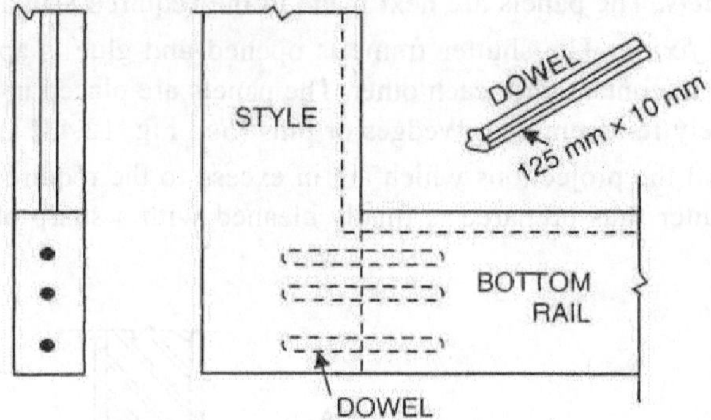

Fig. 12.37: Dowelled joints.

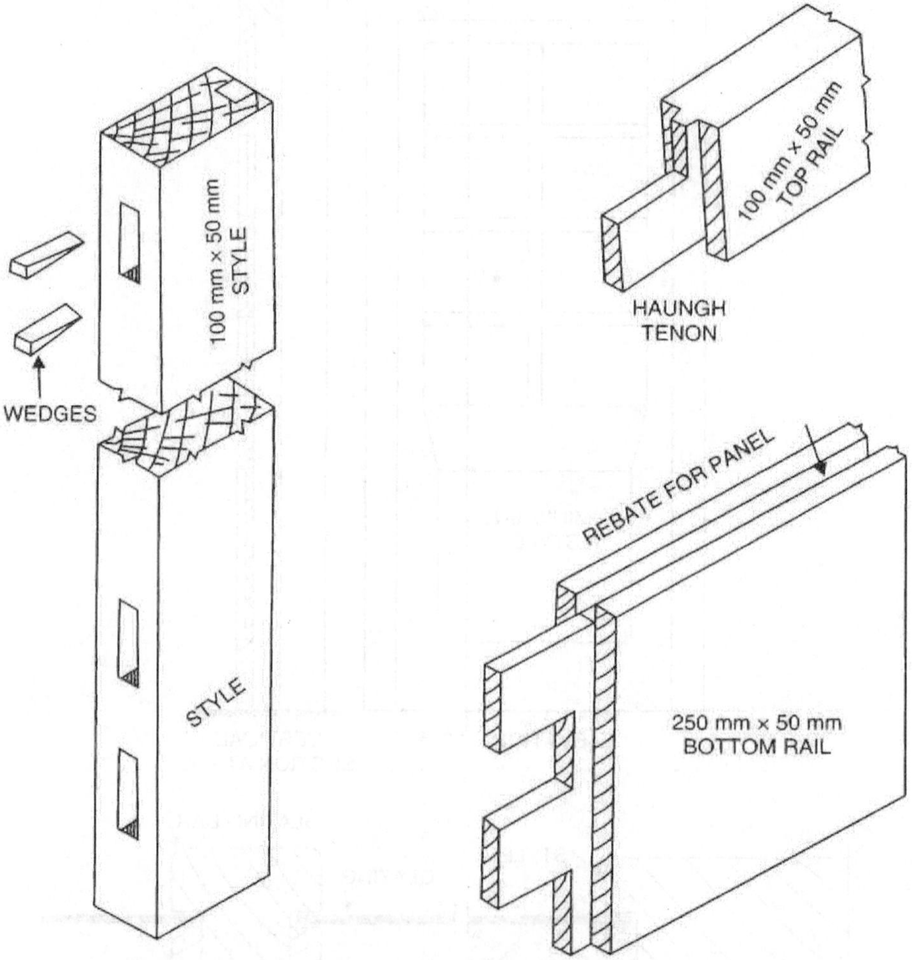

Fig. 12.38 to 12.40: Joints between rails and styles of door panels.

12.12.1 Hanging of the Door

These doors are hung with the aid of butt-hinges or rising-butt hinges.

Fittings: The following fittings are used:

Hinges	3 on each leaf
Bolts	3 Nos of 250 mm

Handles	3 Nos.
Door stops	2 Nos.
Aldrop and Padlock or mortise lock	1 No.

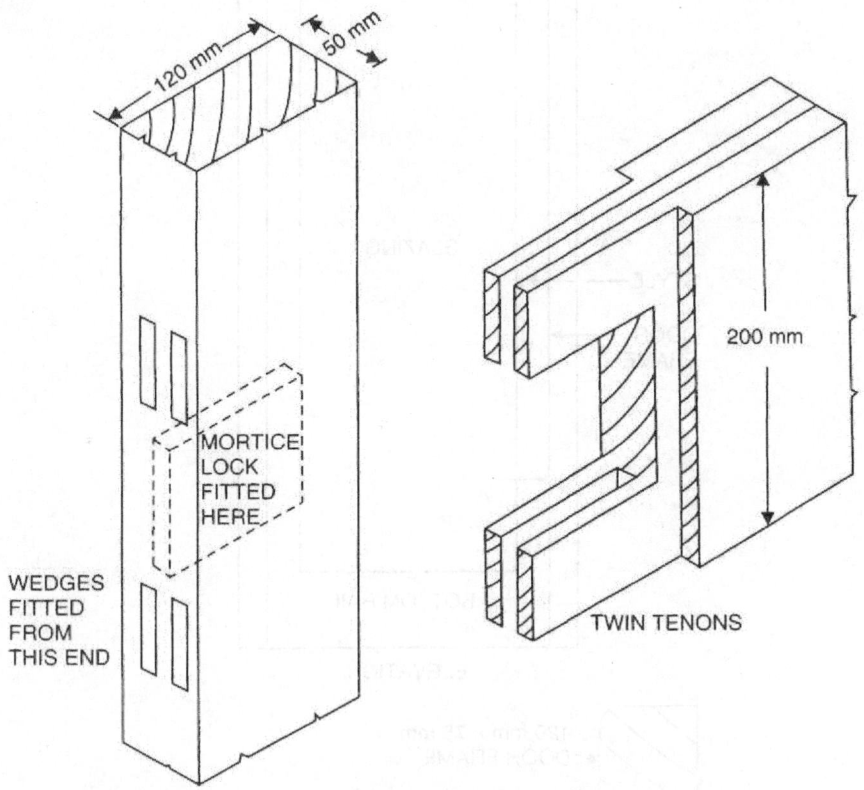

Figs. 12.41 and 12.42: A twin tenon joint for a Lock Rail and Style.

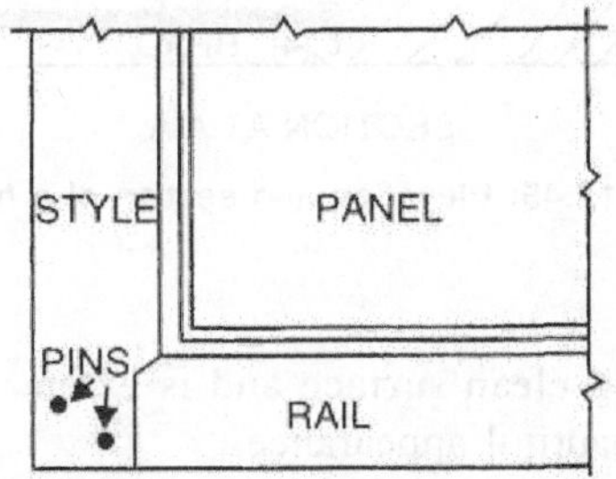

Fig. 12.43: A door joint between rail and style showing pins.

12.13 GLAZED OR SASH DOOR

When light is desired to be admitted inside a room through a door, its upper part may be glazed and the lower part panelled.

These are called partially glazed doors. The latest trend is, however, to provide fully glazed doors. Usually 5 to 8 mm thick polished plate glass is used. Special kind of glasses, *e.g.*, reinforced with wire mesh or toughened are being used to avoid shattering.

The glasses are fixed in the rebates of sash bar and they are secured with putty or by wooden beads fixed to the frame. The sash bars do not add to the strength of the door.

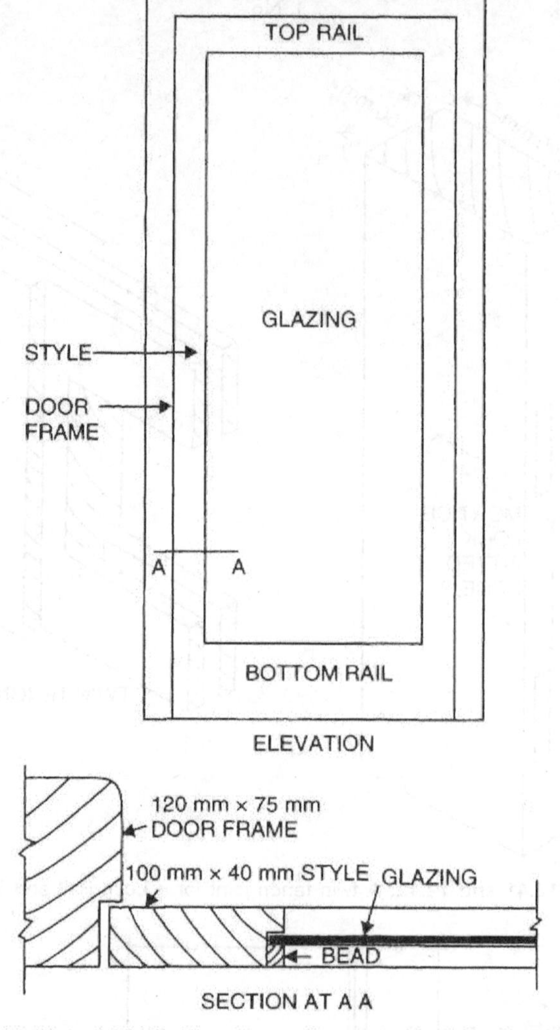

Figs: 12.44 and 12.45: Elevation and section of a fully glazed door.

12.14 FLUSH DOORS

This type of door provides a clean surface and is extensively used these days. The flush surface, when polished gives a beautiful appearance.

They are of two types:

(*i*) ***The solid or laminated core type*** (*See* Fig. 12.47 and for section see Fig. 12.49): The laminae are 15 mm thick and joined together in pieces. On the outside 6 mm plywood (3 plys) are fixed to the laminae. Edging strips are fitted, as shown, on the sides. This door is quite strong and sound insulating than the one described below. Sometimes instead of wooden laminae light weight material such as cork board, perforated wood resin boards and strips of compressed straw boards may also be used as a filling.

(*ii*) ***Skeleton framed type*** (See Figs. 12.46 and 12.48): This door consists of very light members. The styles, top and bottom rails are 40 mm × 25 mm in size, and 10 mm wide vertical and horizontal ribs are fitted into them. Each side is covered with 6 mm thick plywood. Special hinge blocks 150 mm × 40 mm × 25 mm in size are fixed. This type of construction is cheaper and saves wood.

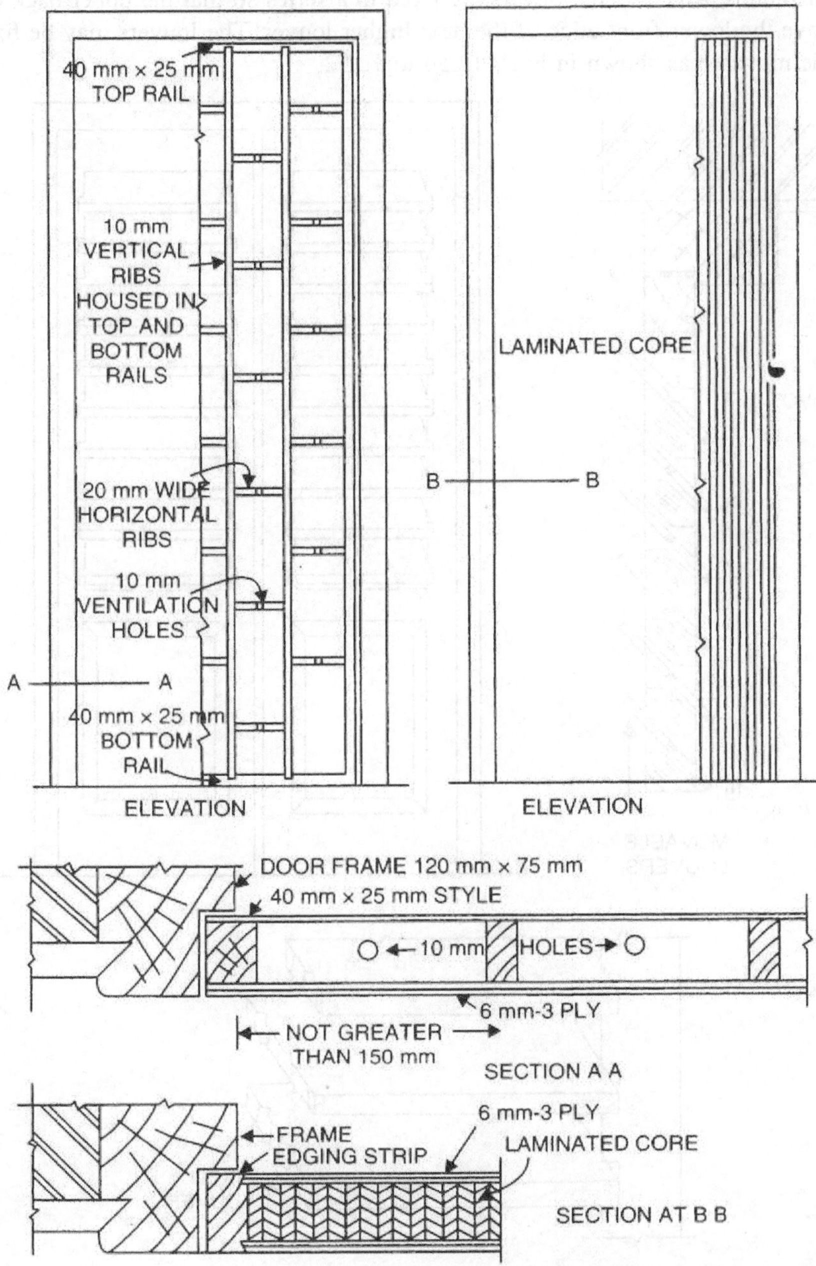

Figs. 12.46 to 12.49: Details of flush doors.

***Construction*:** In the laminated type, all the components are glued together. In the skeleton framed type the vertical ribs are housed 30 mm into the outer rails. The horizontal ribs are glued to the vertical ribs and styles. Ventilation holes are provided as shown for durability.

***Hanging and fittings*:** As for panelled type.

12.15 LOUVERED DOORS

This type of door allows sufficient amount of air circulation even when closed and at the same time maintains privacy. The louvers are fixed in a series so that the upper back edge of any louver is above the lower front edge of the next higher louver. The louvers may be fixed into the style or made movable as shown in Figs. 12.50 to 12.52.

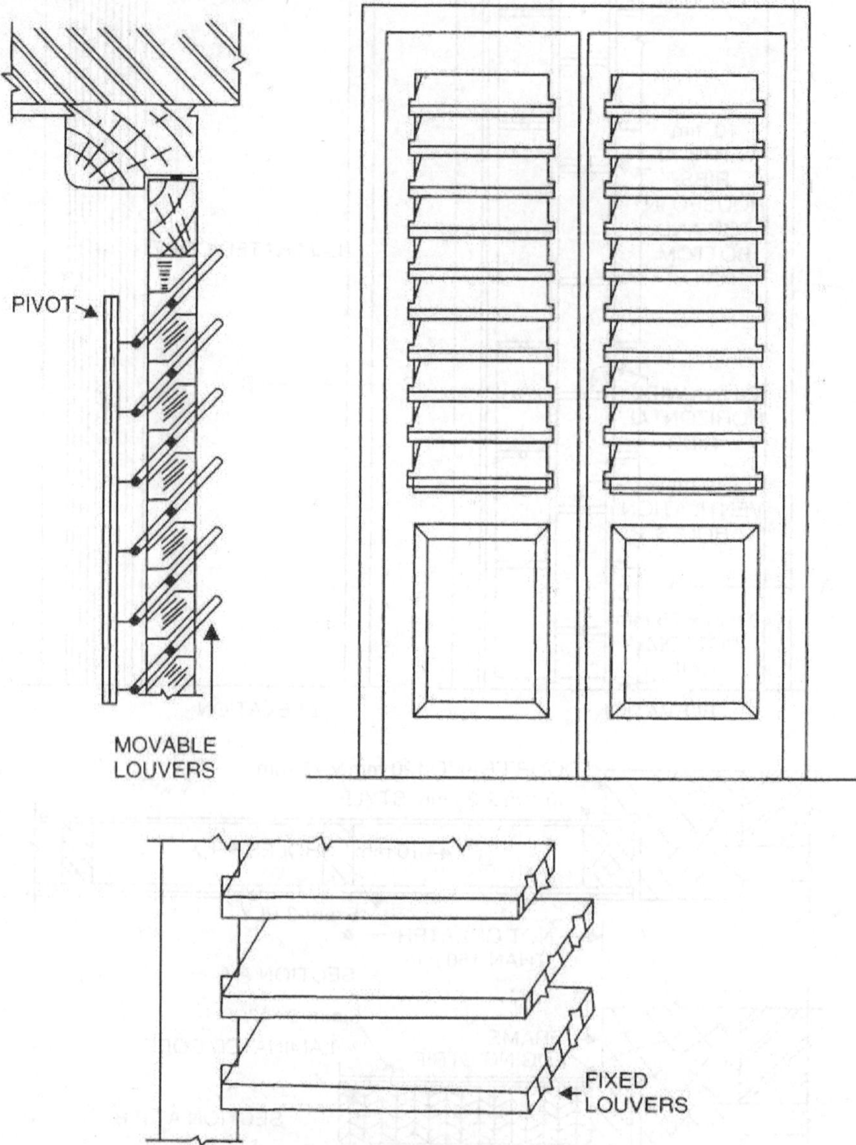

Figs. 12.50 to 12.52: Elevation and details of louvered doors.

12.16 WIRE-GAUGED DOORS

These doors enable air to pass into the room and do not allow flies or insects to come in. They are made of vertical and horizontal styles into which wire gauge is fixed. They may be hung on the same door frame or on a separate frame. The styles and rails shall be of the same size as that of the other door. Wire gauge shall be of best quality woven wire webbing. The entire area of the wire gauge within a panel shall not have any joint. Wire gauge is fixed to the leaf by rails and 20 mm × 20 mm heads are fixed. Double leaf wire gauge door shall close with the meeting styles abutting against each other. The leaves may close to such an extent that the junction projects outside from the face of the frame. The top of the frame and the still, if any, shall have a wedge shaped wooden projection.

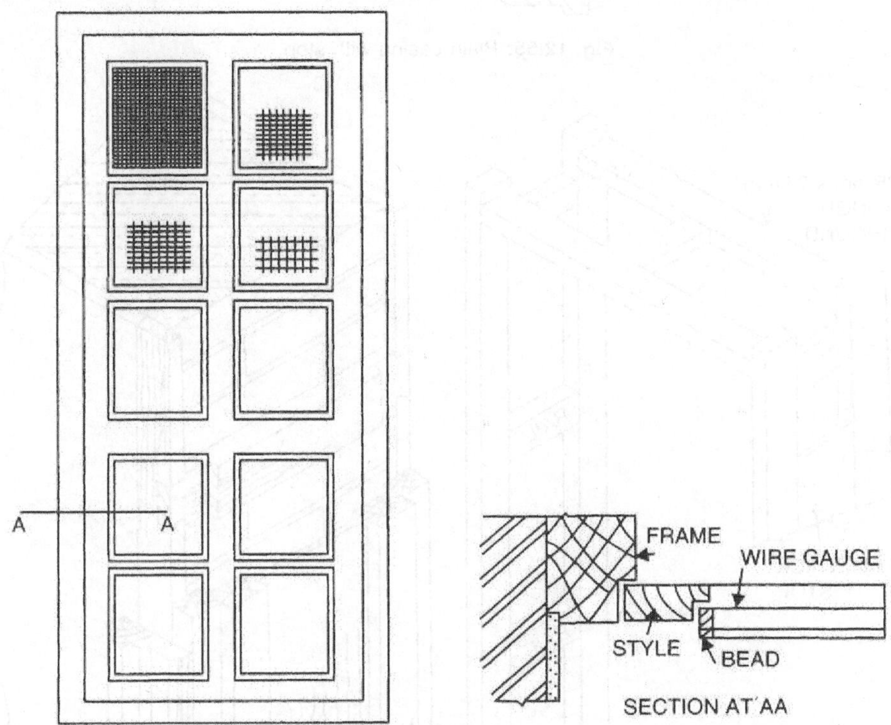

Figs. 12.53 and 12.54: Elevation and section of a wire gauge doors.

12.17 CLEARANCE UNDER THE DOORS

A clearance of 12 mm may be kept between the door and the flooring so that bottom of the door leaf clears off any carpet or other furnishing laid on the floor. This is done either by hanging the door on rising butt hinges or fixing a wooden threshold to the flooring. As an alternative a concrete flooring may be raised and sloped both ways under the door.

12.18 DOOR CASINGS

In important buildings internal doors are fixed to casings, or linings which provide a suitable finish to the openings. Following are the types of casings generally used:

(*a*) ***Plain casings*:** They are made of plain thick wooden boards (about 40 mm). The door leaf is fixed in rebate cut in this plank or a stop is fixed to the casing as shown in Fig. 12.55. The top casing is grooved to receive the vertical sides. This type of casing is suitable for walls less than one brick thick.

(*b*) ***Skeleton casings*.** This consists of vertical members to which thin boarding is fixed. Horizontal pieces are also fixed at intervals to give strength. These are used for thicker walls.

(c) ***Framed casings.*** This is made of panelled jamb and top member and is used where architectural beauty is the main consideration.

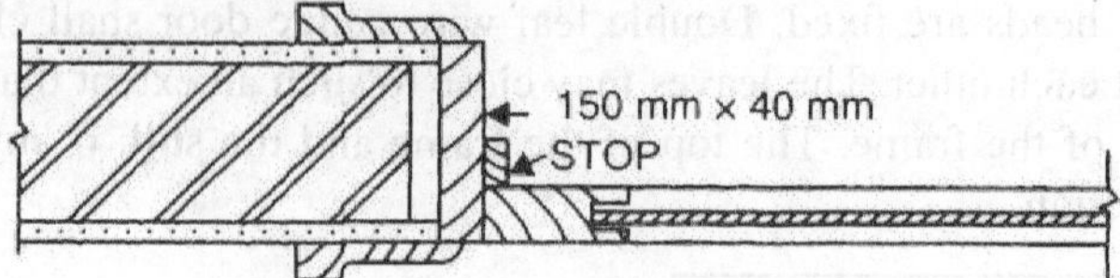

Fig. 12.55: Plain casing with stop.

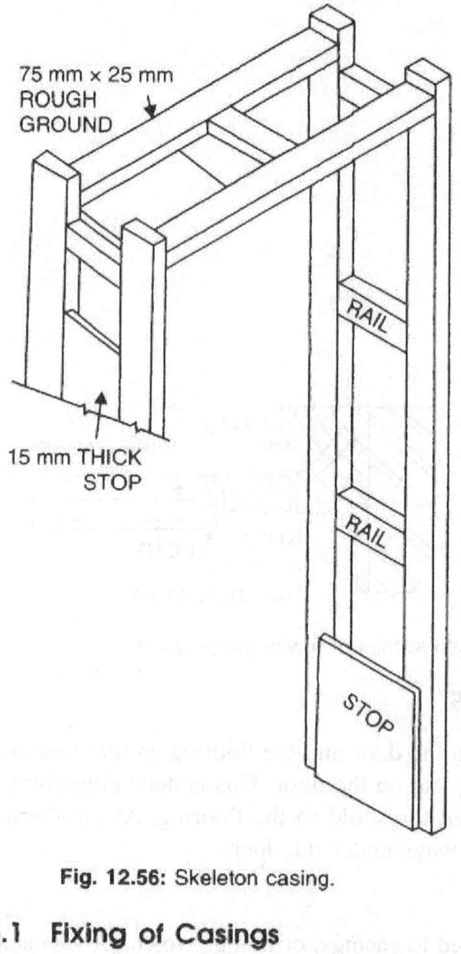

Fig. 12.56: Skeleton casing.

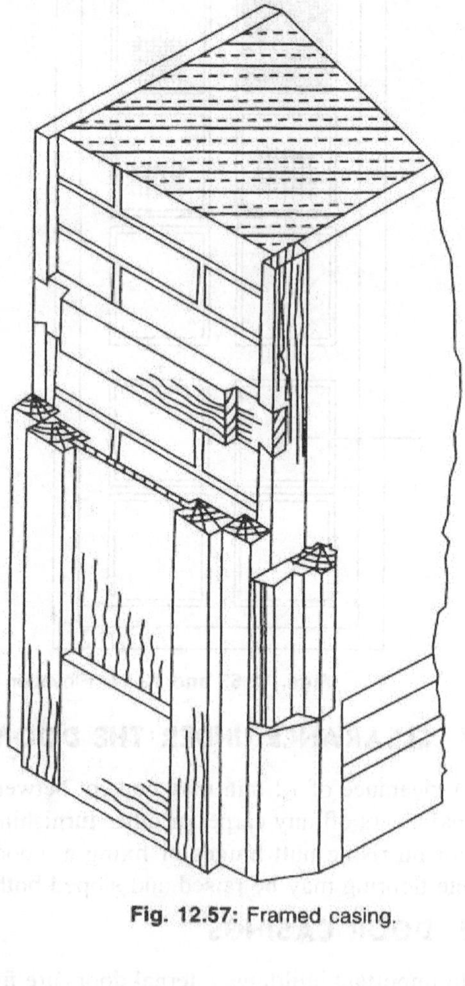

Fig. 12.57: Framed casing.

12.18.1 Fixing of Casings

The methods of fixing casings are as under:

(i) ***With pallets.*** 10 mm thick pallet pieces are fixed at regular intervals into the masonry and to this is attached the casing.

(ii) ***Plugs.*** Hard wooden plugs are fitted into the wall and casing is fixed to them.

(iii) ***Ground work.*** These provide continuous means of fixing the casings. The jamb grounds are fixed with plugs to the wall and cross pieces, which act as packing for wide casings, are fitted to these jamb grounds.

12.19 METAL DOORS

(*a*) ***Hollow metal doors.*** They are made of furniture steel and are shaped as wooden doors in appearance. The rails, styles and places where hinges, locks, etc., are fixed are strengthened by welding small T or I Sections inside. The styles are filled with some insulating material so that there is not much noise when they are opened or closed.

(*b*) ***Steel plate doors.*** Whenever a stronger door is needed, a framework of angle or channel steel sections is built. On this, steel plates of required thickness are welded.

(*c*) ***Corrugated steel doors.*** A frame-work of angles and channels is built up with suitable bracings. On this corrugated steel sheets are fixed vertically. If a stronger door is desired, an additional sheet can be placed and the gap between the two being filled with asbestos or any other fire-proof material.

(*d*) ***Rolling shutter doors.*** These are commonly used, these days for entrance to the garages, show-rooms and shops, where the doors width is large. They are available in varying widths, upto 6 m. The shutter rolls up and causes no obstruction to the floor space. Their operation is also easy. They are sufficiently strong and can be used in exposed situations.

(*e*) ***Collapsible door.*** The doors have the advantage that they do not require any frame or hinges to hang upon. These metallic doors are made from small rolled steel channel sections and strips. They are fitted with rollers at the bottom which slide in rails provided in the floor. When opened, they collapse to the sides. They are normally provided at the entrance of residential buildings, godowns or public buildings.

12.20 COMPOSITE DOORS

Metal covered doors. They consist of cores of seasoned non-resinous timber covered with a tight fitting sheet metal, *e.g.*, furniture steel, galvanised steel, cold rolled copper sheet, bronze, tin and lead combination (terne plate), etc. The cores are made of two or three layers of 25 mm boards and about 200 mm wide. One layer should be vertical and the other horizontal. The covering is made of 350 mm × 500 mm sheet preferably with double joint. The three-ply doors are very strong and two-ply doors give a moderate degree of protection.

In case cheaper doors of this type are needed, the same can be made by making a framework of wooden styles and rails and the inner spaces being filled with asbestos or other fire-proof material. This type has got considerable fire resistance qualities.

WINDOWS

Windows comprise of two parts, one frame which is fitted in the opening in the wall and the other sashes which are small shutter frames containing the glass and attached to the frame by hinges or retained by grooves. The type of window is determined by the type of window movement.

12.21 TYPES OF WINDOW MOVEMENTS

(*a*) ***Fixed type.*** This consists of a framework in which the panel or glass is fitted and the framework cannot move.

(*b*) ***Opening outside.*** In this the shutters open towards outside of a building.

(*c*) ***Opening inside.*** Here the shutters open inside the building.

(*d*) ***Top-hinged.*** The shutter is hinged at the top and can be moved outside.

(*e*) ***Bottom-hinged.*** The shutter is hinged at the bottom and can be moved inside.

(*f*) ***Pivoted.*** These window shutters rotate about a pivot which may be fixed to the window frame. The windows can rotate horizontally or vertically depending upon the position of the pivot.

(*g*) ***Sliding.*** In this type, the shutters can move inside the walls horizontally or vertically.

(*h*) ***Double hung.*** One shutter goes up and the other moves down.

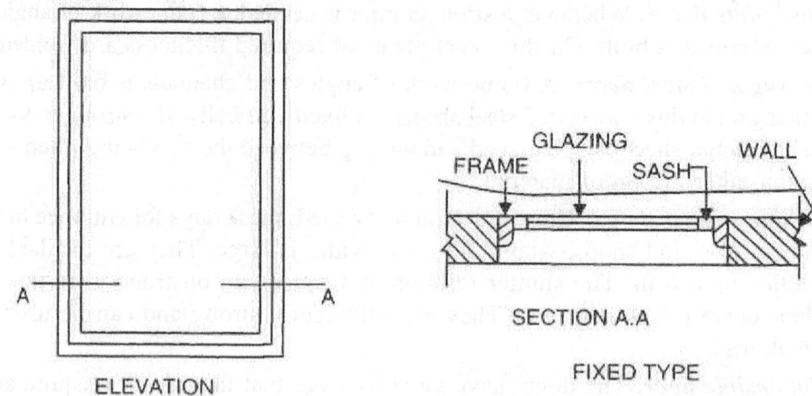

Figs. 12.58 and 12.59: Elevation and section of a fixed type window.

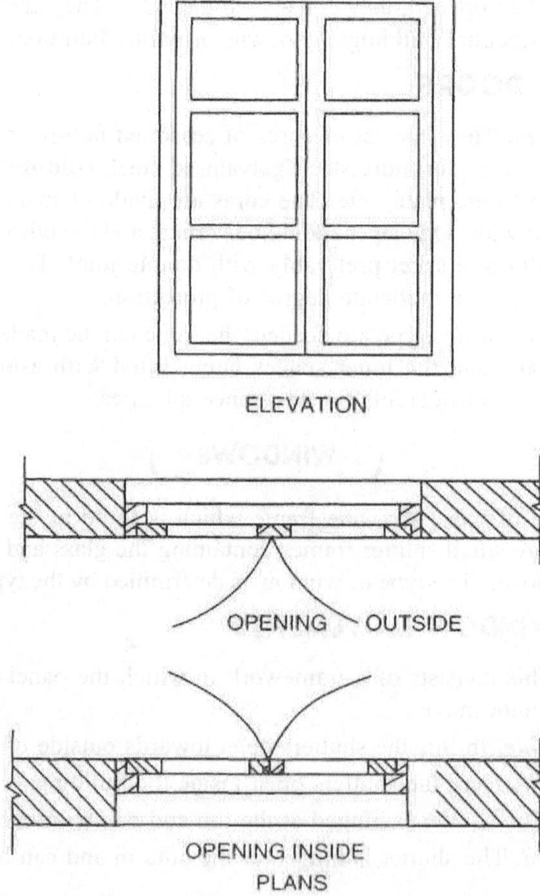

Figs. 12.60 to 12.62: Window elevation and plans with opening arrangements.

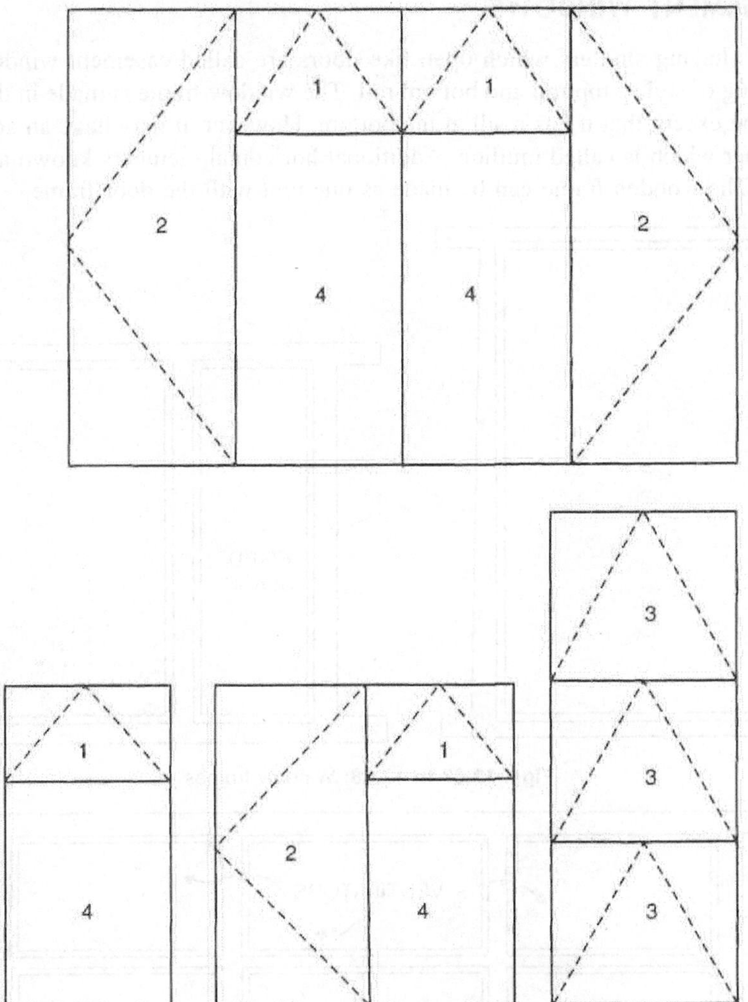

Figs. 12.63 to 12.66: Various types of window movements.
1. Top hung. 2. Side hung. 3. Bottom hung. 4. Fixed.

TYPES OF WINDOWS

(1) *Wooden windows:*
(*a*) Casement windows.
(*b*) Double hung windows.
(*c*) Sliding windows.
(*d*) Pivoted windows.
(*e*) Louvered or venetiated windows.

(2) *Metal windows:*
(*a*) Solid section metal windows.
(*b*) Metal covered and hollow metal windows.

(3) *Miscellaneous windows:*
(*a*) Composite windows.
(*b*) Bay windows.
(*c*) Clere storey windows.
(*d*) Corner windows.
(*e*) Dormer windows.

12.22 CASEMENT WINDOWS

Windows having shutters which open like doors are called casement windows. This has a frame consisting of styles, top rail and bottom rail. The window frame is made in the same manner as a door frame except that it has a sill at the bottom. However, it may have an additional central vertical member which is called mullion. Additional horizontal members known as transoms may also be used. The wooden frame can be made as one unit with the door frame.

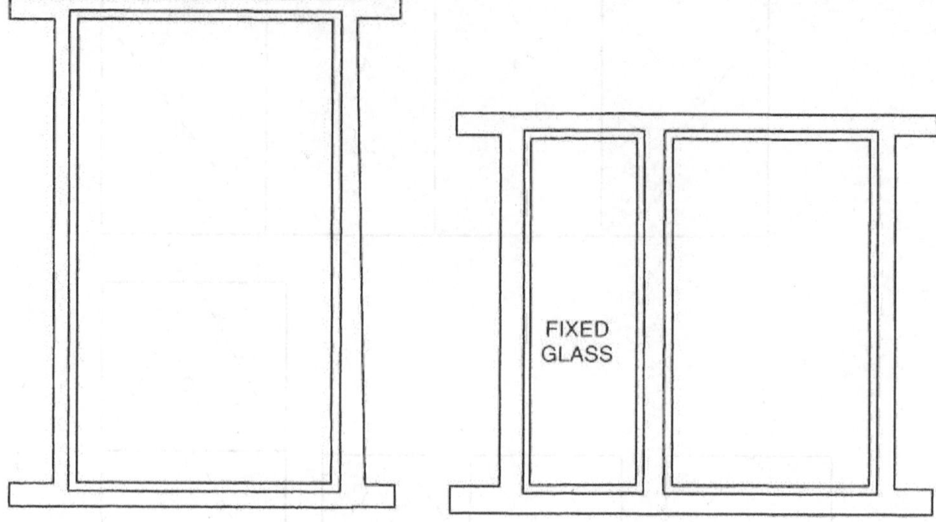

Figs. 12.67 to 12.68: Window frames.

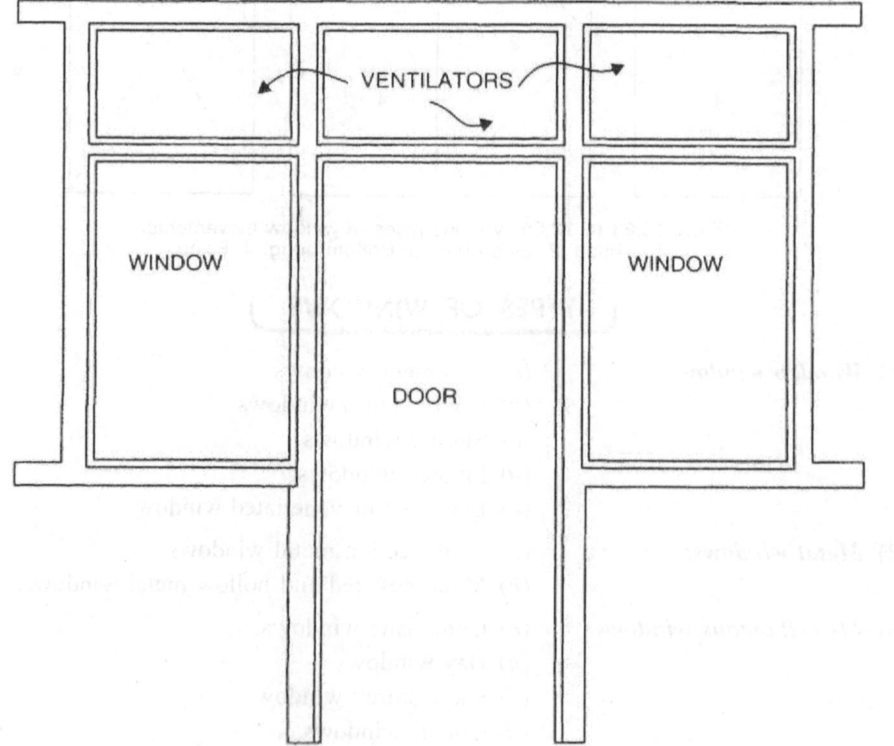

Fig. 12.69: A combined door and a window frame with ventilators

Heads, mullions and transoms are generally 100 mm × 60 mm to 120 mm × 75 mm in section. The size of sills may be more. The joints and the construction of the window frames is similar to the door frame.

The window frame is provided with iron bars or welded iron grills, particularly on the ground floor, as a safety measure against burglars. Wrought iron bars varying in diameter from 15 mm to 20 mm and spacing of 80 mm to 150 mm centre to centre are used in ordinary buildings. However, the present trend is to provide welded grills with beautiful patterns.

Sash or shutters which comprise of vertical and horizontal sash bars are also similarly constructed. Each shutter should not be greater than 0.6 m × 1.2 m in size. The sash bars have a size 40 mm × 30 mm or 40 mm × 35 mm. Small grooves are made in the frame and the rails. These are capable of stopping water which would otherwise proceed by capillary action into the room. It is essential that the windows should not admit rain water into the room. Hence the sills may be projected outside and throated as shown in the section of the window in Fig. 12.71.

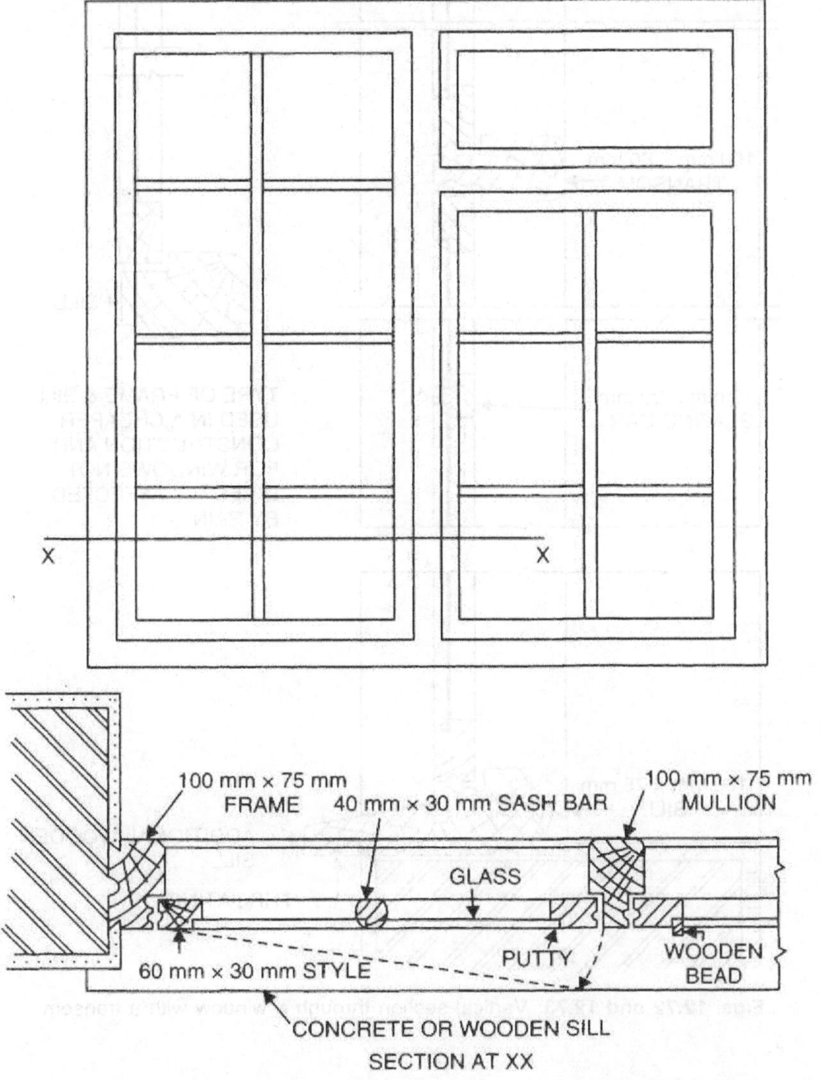

Figs. 12.70 and 12.71: Elevation and section of a casement window.

Fittings:	Hinges	— 4 Nos. of 100 mm.
	Bolts	— 3 Nos.
	Handles	— 2 Nos.
	Stops	— 2 Nos.

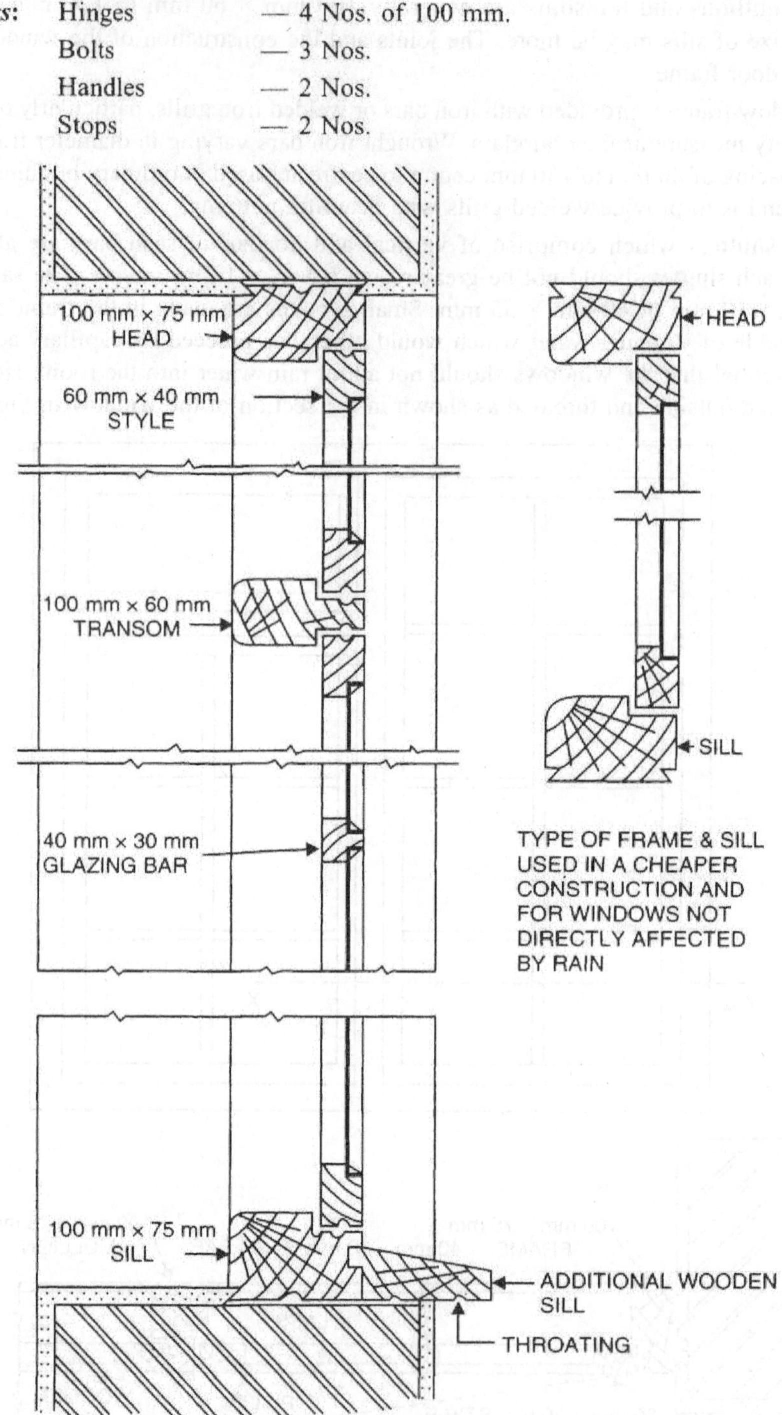

Figs. 12.72 and 12.73: Vertical section through a window with a transom.

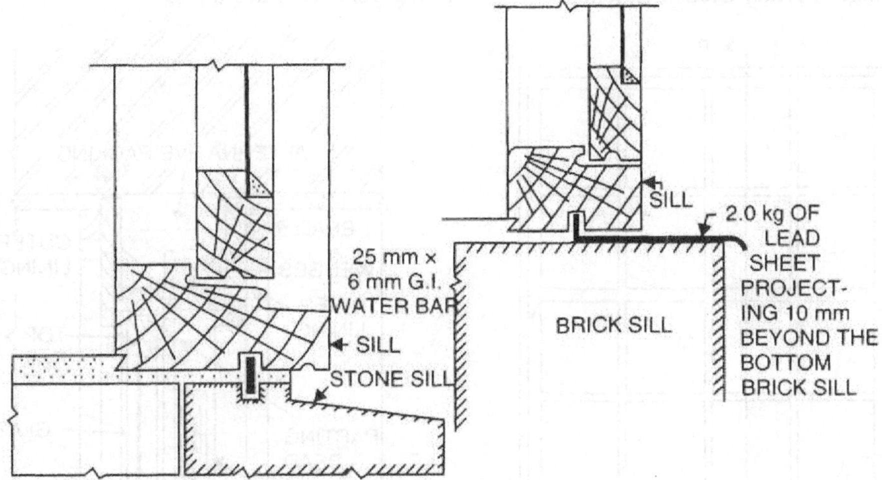

Figs. 12.74 and 12.75: Different arrangements of window sills showing methods of preventing entry of water.

12.23 DOUBLE HUNG

This type of window consists of a pair of shutters both of which can slide to facilitate cleaning. Ventilation can also be controlled effectively. The shutters move vertically within openings left in the frame and the walls. A pair of metal weights contained within the frame are connected to each shutter by means of cords or chains after being passed over pulleys which are fixed to the frame. The components of this type of a window are briefly described below:

(*i*) ***Frame***: This consists of two vertical members, a head and a sill. The vertical members are made of an inner lining, pulley style, back lining, and an outer lining. A small wooden section is used between the two weights to keep them apart. Small parting bead is provided to separate the two shutters when they are opposite to each other. The sill is made of solid wood.

(*ii*) ***Shutters***: These are of ordinary type with two vertical styles, a top rail, a bottom rail and sash bars. The usual thickness of the shutter shall be about 50 mm.

(*iii*) ***Fittings***:

Shutter fastener	— 2 Nos.
Shutter lift	— 1 Nos.
Shutter Handle	— 2 Nos.
Pulleys, Cord, Chains, etc.	
Tower bolts	— 4 Nos.

12.24 SLIDING WINDOWS

These windows are made of shutters which move horizontally or vertically on small roller bearings. The shutters are of usual size. Suitable openings in the frame are left to accommodate the shutters.

12.25 PIVOTED WINDOWS

This type consists of a frame and a shutter which is pivoted to allow it to open with the top rail swinging inside. The pivots are fixed slightly higher than the horizontal central line of the shutter so that the shutter can be self-closing. The frame is similar to that of a casement window except that it is not rebated.

Fittings: Pivots, Cleats, Latches and Ventilating gear, etc., are used.

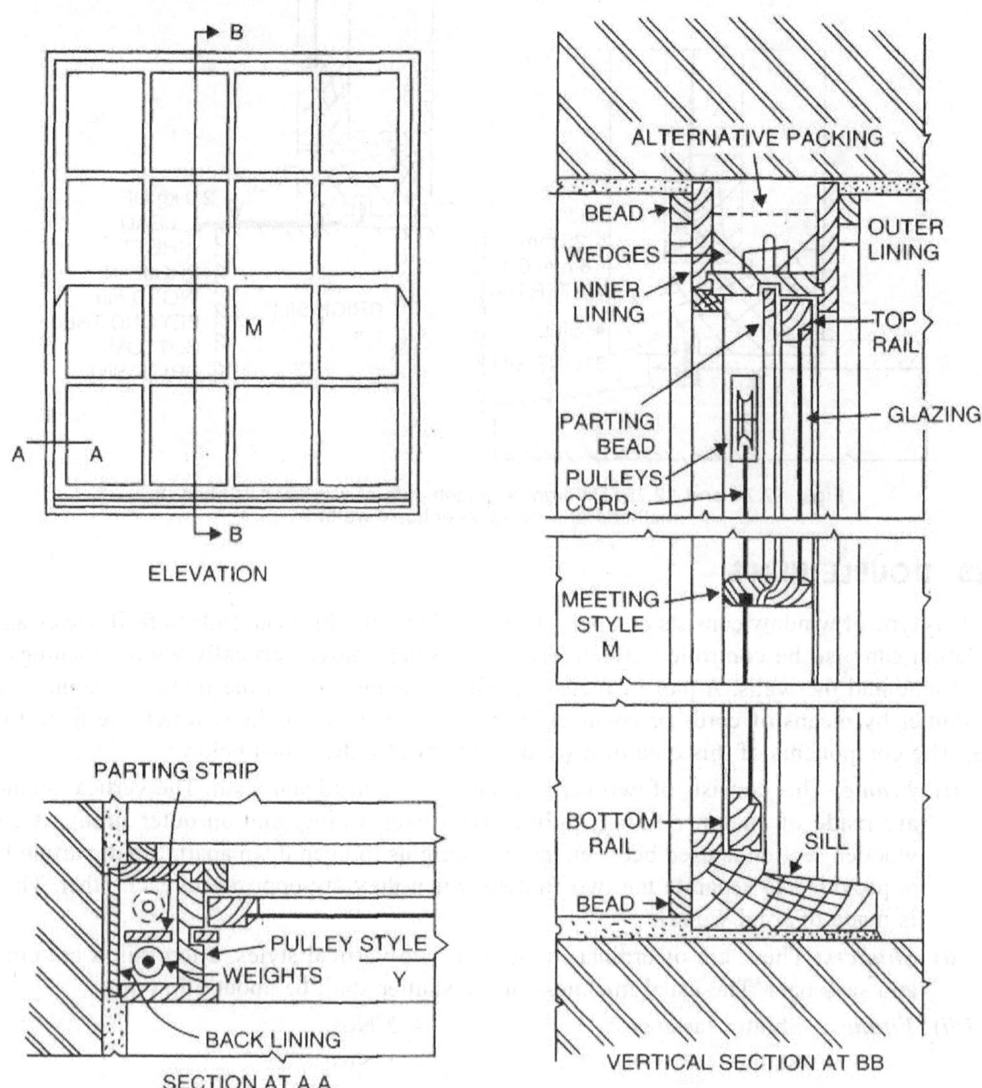

Figs. 12.76 to 12.78: Elevation and details of a double hung window.

12.26 LOUVERED OR VENETIATED WINDOWS

Whenever privacy as well as enough air is desired, louvered windows are used. These may be of fixed or moving type. The construction is similar to that of louvered doors.

12.27 SOLID SECTION METAL WINDOWS

They are made of light rolled steel sections which act as a frame. The glazing is fitted into this frame. The shutter containing the glass can have various types of movements, *i.e.*, it may be hinged at bottom, top, side or pivoted to move in a horizontal or vertical direction. Similarly horizontal or vertical moving shutters of double hung types are possible. The frame is fastened into the masonry in a groove with cement grout. Teak wood frame and sill can also be used to fit in the steel shutter to enhance the appearance.

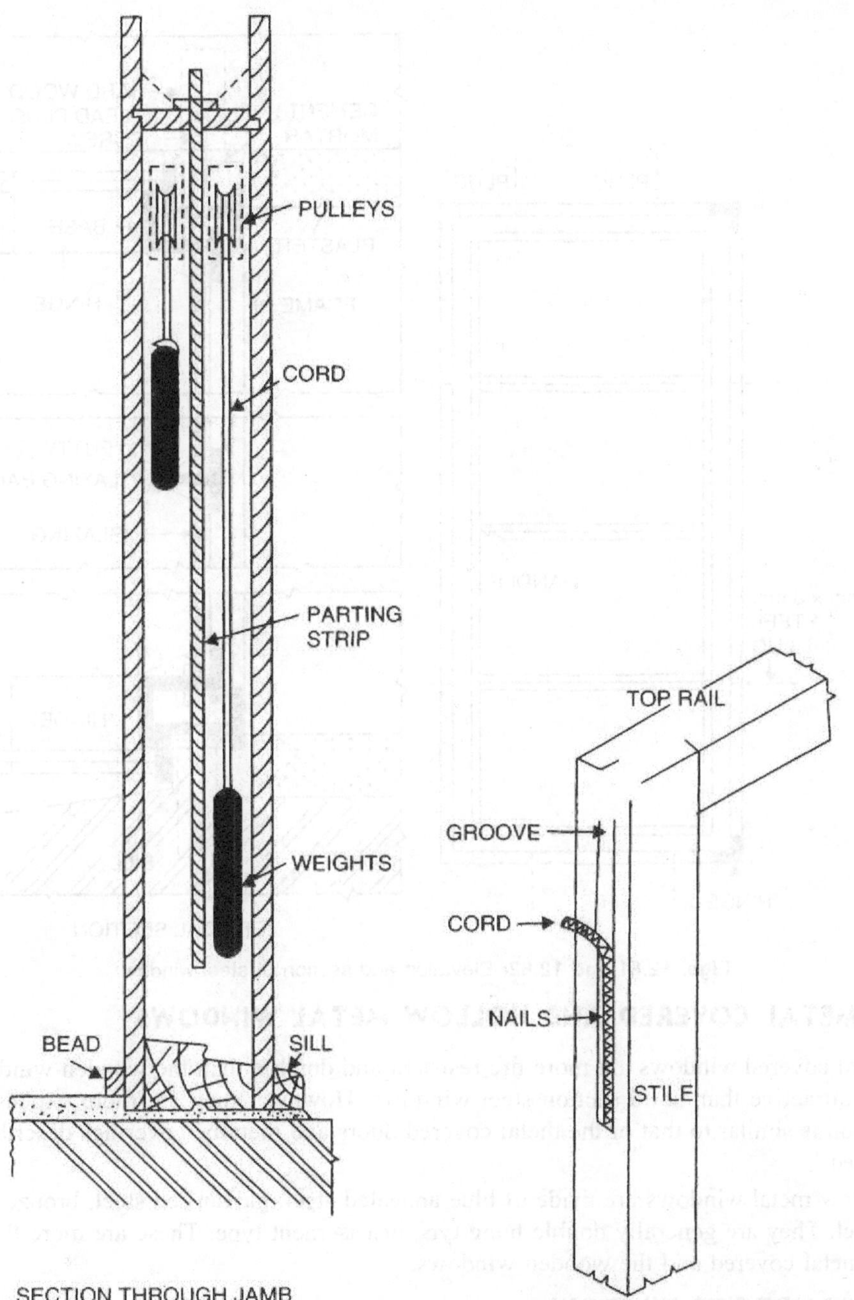

Figs. 12.79 and 12.80: Double hung window details.

Continuous windows which can be operated mechanically are used for big openings and are pivoted suitably at top.

Steel windows should receive one coat of primer before being installed. Then they should be painted with another coat of a durable paint and repainted after the glazing is fixed. The materials used for metal windows may be steel, bronze, aluminium, stainless steel, etc.

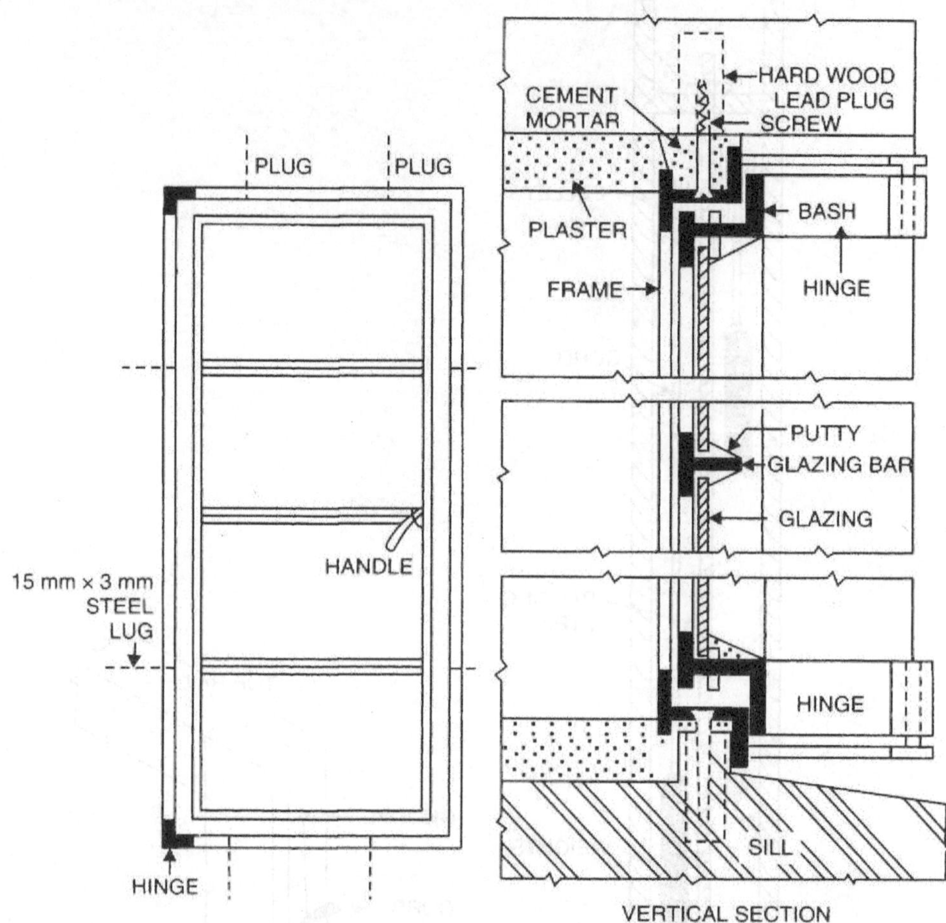

Figs. 12.81 and 12.82: Elevation and section of steel window.

12.28 METAL COVERED AND HOLLOW METAL WINDOWS

Metal covered windows are more fire resistant and durable than the wooden windows. They are more attractive than solid section steel windows. However, they are more expensive. Their construction is similar to that of the metal covered doors and metallic coverings described therein can be used.

Hollow metal windows are made of blue annealed steel, galvanised steel, bronze, copper or nickel steel. They are generally double hung type or casement type. These are more fire resistant than the metal covered and the wooden windows.

12.29 COMPOSITE WINDOW

Different type of movements can be obtained in the same window. This type of window can be adopted for a small room where only one large window is used. This can control movement of air and at the same time provide maximum light.

12.30 BAY WINDOW

This is a window which projects outward from the face line of the wall or a building. This gives extra floor area to the room and at the same time provides an additional space for admitting light and air into the room. A number of these windows might increase the architectural beauty of the entire house.

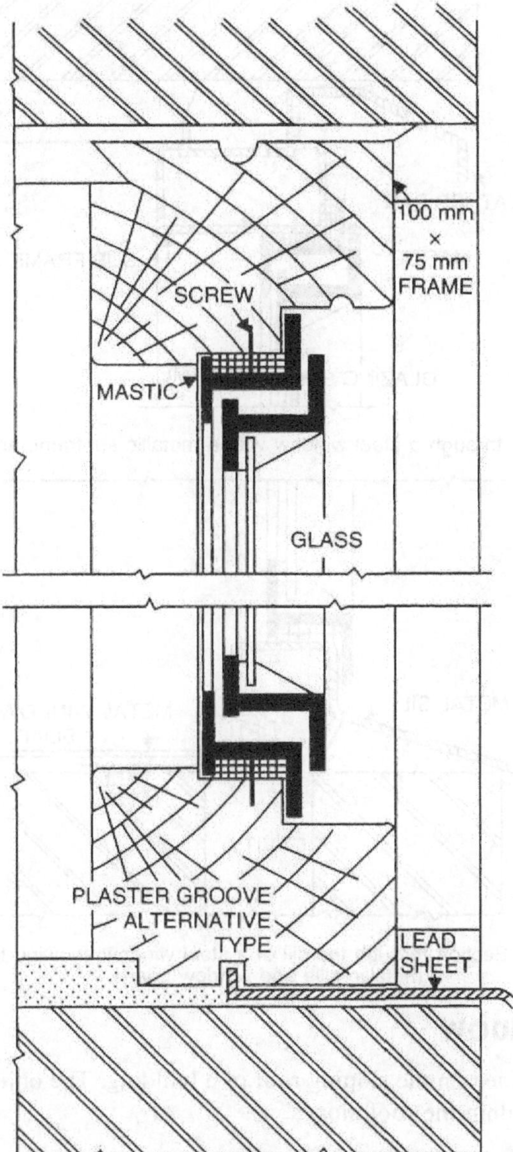

Fig. 12.83: Section through a steel window showing the use of an additional wooden frame surrounding it.

12.31 CLERE-STOREY WINDOW

It is fixed above the lean to roof but near the main roof and is of pivoted type. The shutter can be operated by a cord passing round a pulley. This window provides ventilation to the room whose front is blocked by a verandah roof.

12.32 CORNER WINDOW

This is placed in the corner of a building so that light and air can come from the two directions.

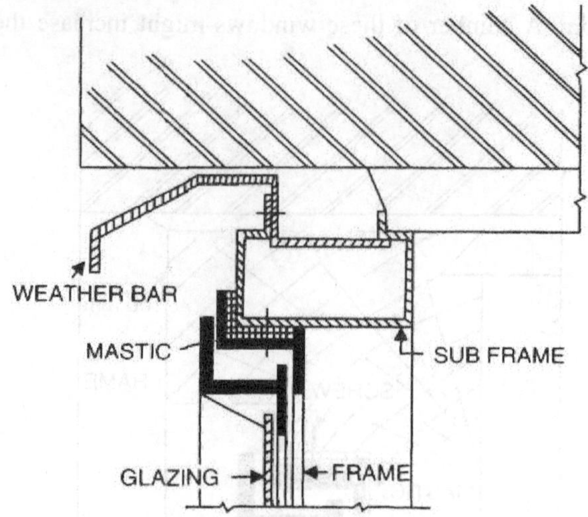

Fig. 12.84: Section through a steel window with a metallic subframe and a weather bar.

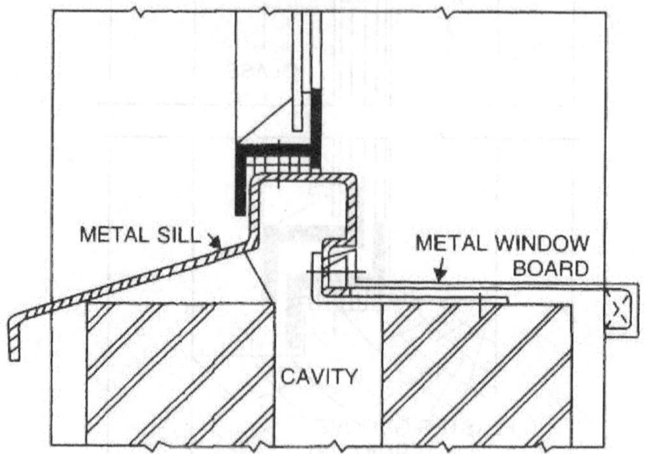

Fig. 12.85: Section through the sill of a steel window showing the use of metallic sills and window board.

12.33 DORMER WINDOW

This window is provided on the sloping roof of a building. The object is to provide light and air to the room built up within the roof slopes.

12.34 VENTILATORS

They are provided for ventilation purposes and can be combined with a window or a door frame. They may, as an alternative, be located near the ceiling of the room if the height of the room permits. They are usually pivoted type and can be operated with a pulley and a cord or have self-closing small bolts.

12.35 ARCHITRAVES

These are used to conceal joints between the door frames or casings and the plaster at the door. They consist of two vertical and one horizontal member which have mitred angle. These are nailed to the plugs and edges of casings. The size and design depends on the size of opening, the quality of timber and the general architectural effect desired. Generally they are about 100 mm × 40 mm to 60 mm × 20 mm.

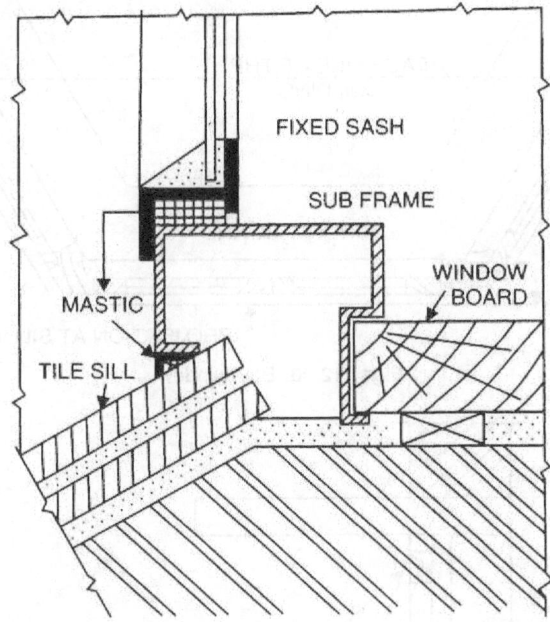

Fig. 12.86: Section through a steel window showing metallic subframe and tiled sill.

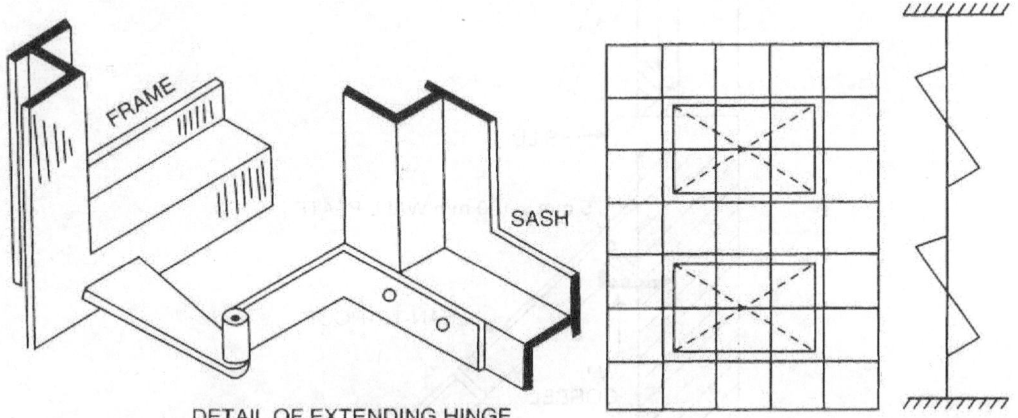

Fig. 12.87: Hinge for a steel window.

Figs. 12.88 and 12.89: A typical industrial steel window.

12.36 PICTURE RAILS

These are generally fixed at the top level of architrave and have various types of mouldings and finishes. They are fixed to the walls by plugs. They are used for hanging pictures or photographs. They are made of wood and are about 75 mm × 20 mm in section.

12.37 ANGLE BEADS

External angles of plaster are protected from damage by wooden beads 10 mm or 20 mm in size fixed to plugs in the walls.

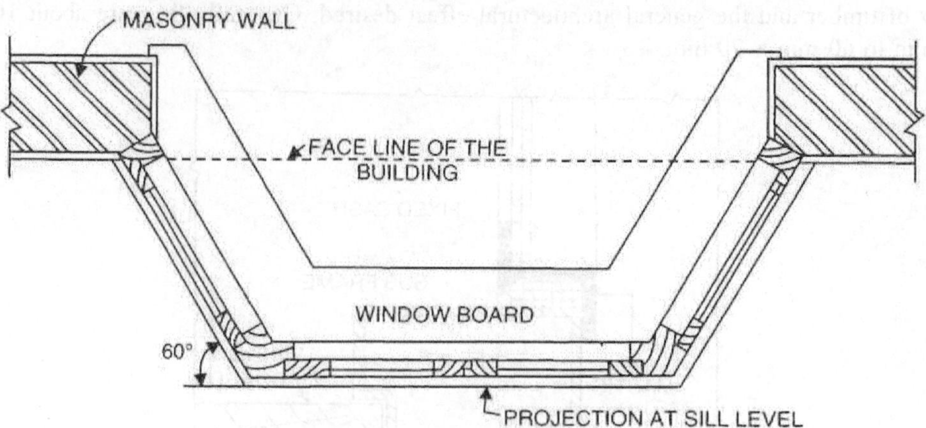

Figs. 12.90: Bay window.

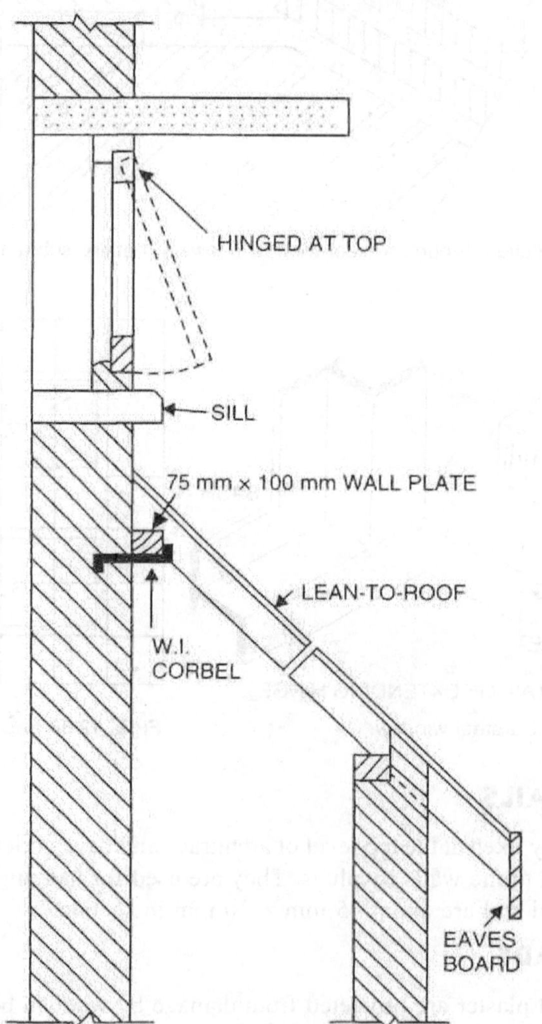

Fig. 12.91: Clere-storey window.

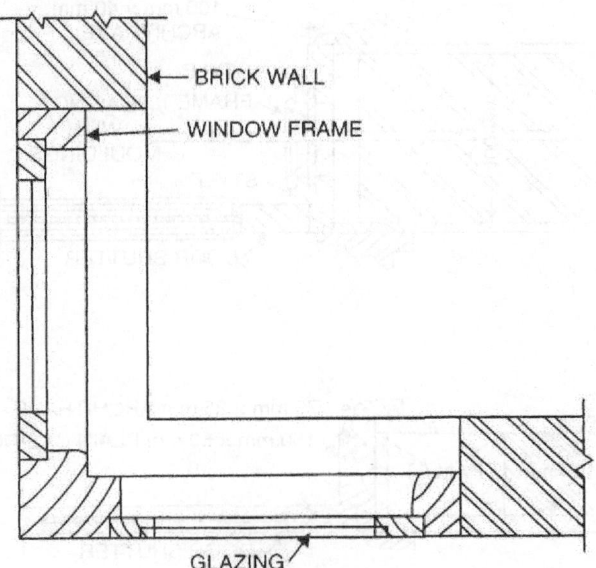

Fig. 12.92: Corner window.

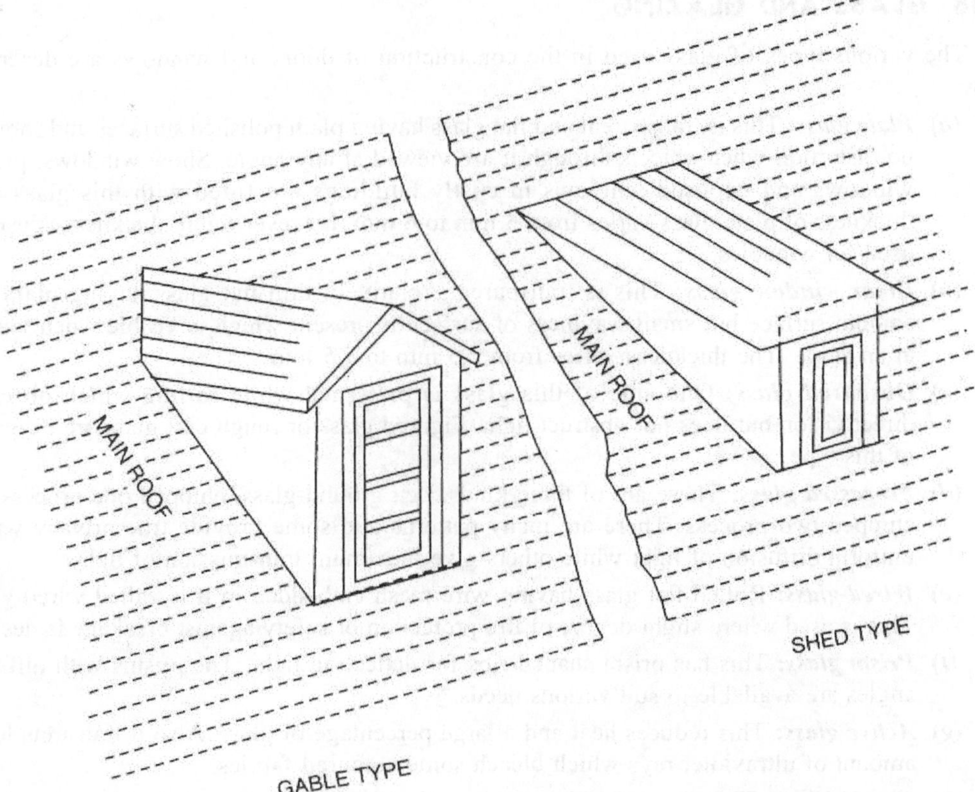

Figs. 12.93 to 12.94: Dormer windows.

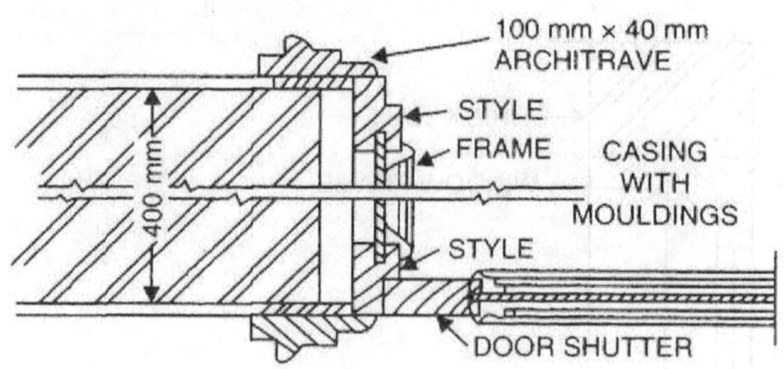

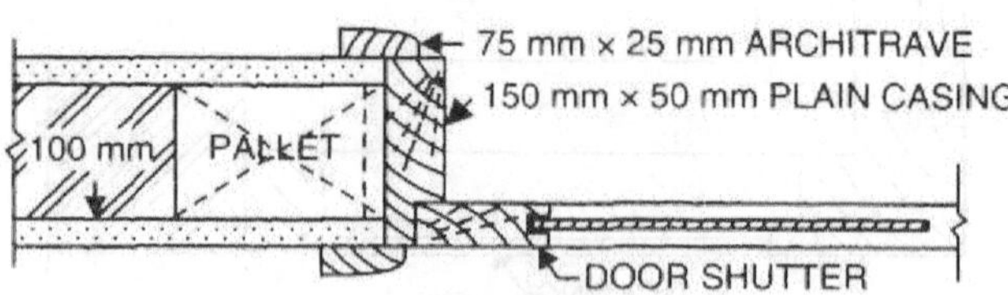

Fig. 12.95 to 12.96: Architraves.

12.38 GLASS AND GLAZING

The various types of glass used in the construction of doors and windows are described below:

(*a*) ***Plate glass***: This is transparent and flat glass having plain polished surfaces and showing no distortion when objects through it are viewed at any angle. Show windows, picture windows and exposed windows in costly buildings are fitted with this glass. The thickness of plate glass varies from 3 mm to 6 mm, however 6 mm thickness is usually used for windows.

(*b*) ***Clear window glass***: This is transparent, relatively thin flat glass having plain and smooth surface but small waviness of surface is present which is visible when viewed at an angle. The thickness varies from 2.5 mm to 7.5 mm.

(*c*) ***Obscured glass***: One side of this glass is patterned while rolling which obscures direct vision but does not obstruct light. Figured glass or rough cast glass are examples of this type.

(*d*) ***Processed glass***: These are of three kinds, *i.e.*, ground-glass, chipped one process and chipped two process. There are many patterns and some provide true privacy with a uniform diffusion of light while others give maximum transmission of light.

(*e*) ***Wired-glass***: Rolled flat glass having wire mesh embedded in it is called wired glass. This is used where slight degree of fire protection or safety against breakage is desired.

(*f*) ***Prism glass***: This has prism shaped ribs for deflecting light. The prisms with different angles are available to suit various needs.

(*g*) ***Active glass***: This reduces heat and a large percentage of glare. Also it transmits lesser amount of ultraviolet rays which bleach some coloured fabrics.

(*h*) ***Quartz glass***: This transmits ultraviolet rays and is used in hospitals where maximum benefit from sun's rays is desired for the health of the patients.

(*i*) ***Bullet-proof glass***: This is made of laminated plate glass which may break under impact but will not shatter.

Glazing: The process of placing glass within sash bars is called glazing. Rebates of about 6 mm depth are provided on one side of the sash bars to support the glass. After the glass is placed, it is kept in position by means of putty, glazing beads, strips or moulds of wood. Windows are sometimes provided with two thicknesses of glass with an air space between them. This is called ***"Double glazing"*** and is adopted in air-conditioned buildings.

12.39 FITTINGS FOR DOORS AND WINDOWS

(a) Hinges

(*i*) ***Butt hinges***: These are commonly used on doors and windows and are screwed to the edges of the shutters and rebates in the frame.

(*ii*) ***Back-flap hinge***: This is used where the shutters are thin and butt hinges cannot be fixed. These are fixed to the back side of the shutter and the frame.

(*iii*) ***Garnet hinge***: This is fixed to ledged and braced doors.

(*iv*) ***Strap hinge***: This is also used on ledged doors.

(*v*) ***Narmadi***: This is fixed on heavy doors. The pin on which the strap rotates is fixed into the wall or frame.

(*vi*) ***Parliamentary hinge***: This is fixed to a door which otherwise would obstruct the passage when open. This enables the door shutter to rest against the longitudinal profile of the wall.

(*vii*) ***Rising butt hinge***: This is used on doors which are desired to leave a clearance between the leaf and the floor when opened.

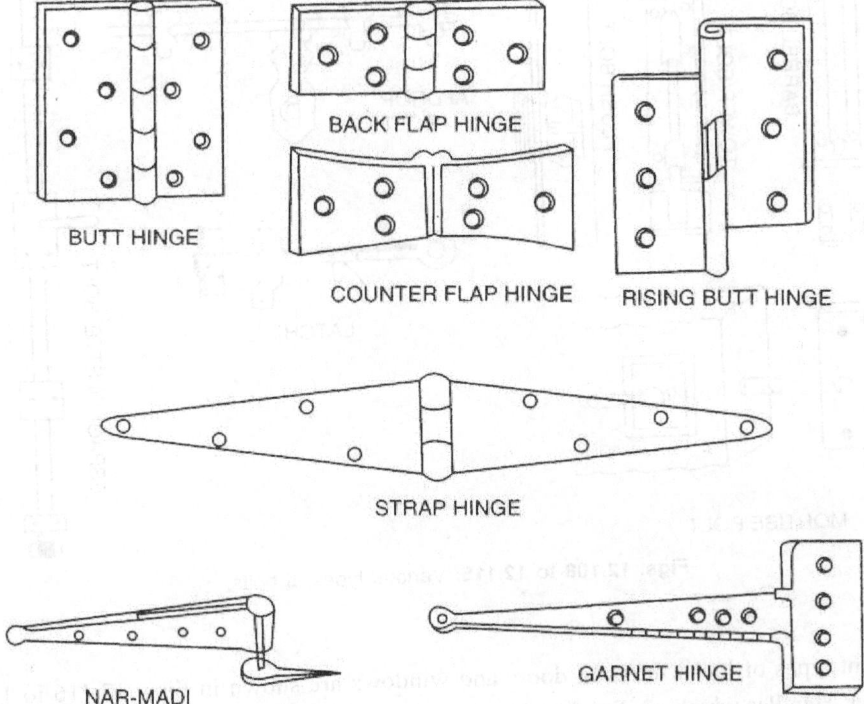

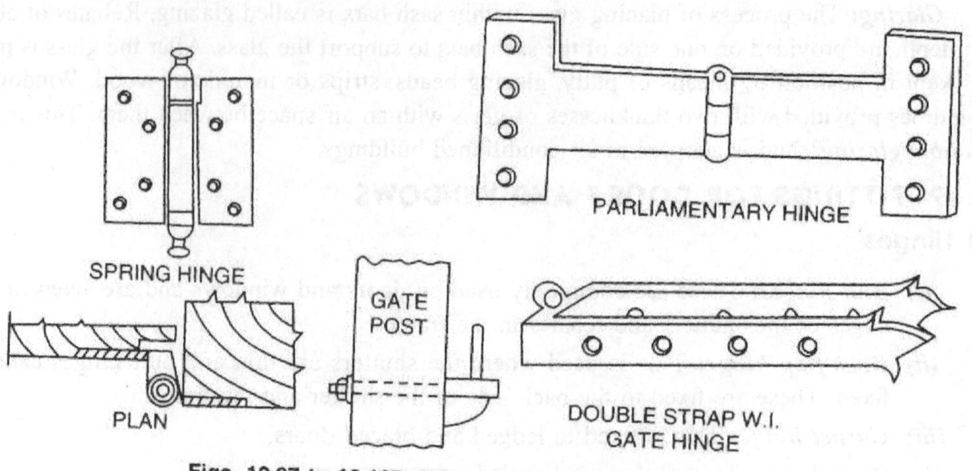

Figs. 12.97 to 12.107: Different types of hinges for doors.

(b) Bolts

Various types of bolts are shown in the Figs. 12.108 to 12.115.

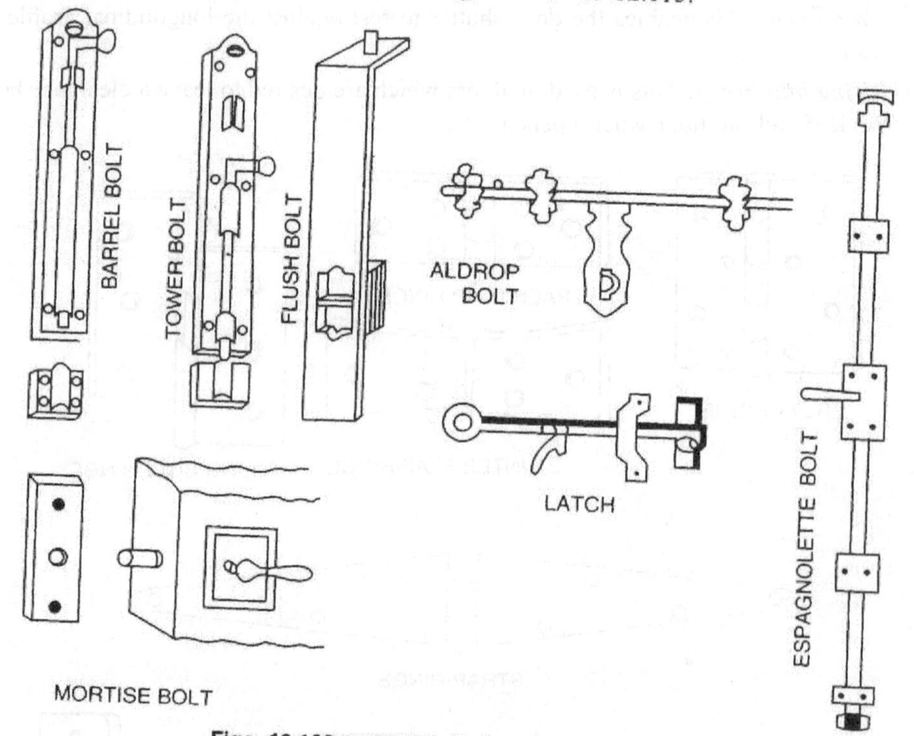

Figs. 12.108 to 12.115: Various types of bolts.

(c) Locks

Different types of locks fitted on doors and windows are shown in Figs. 12.116 to 12.120.

Handles. Handles which are fitted to doors and windows are shown in Figs. 12.121 to 12.124.

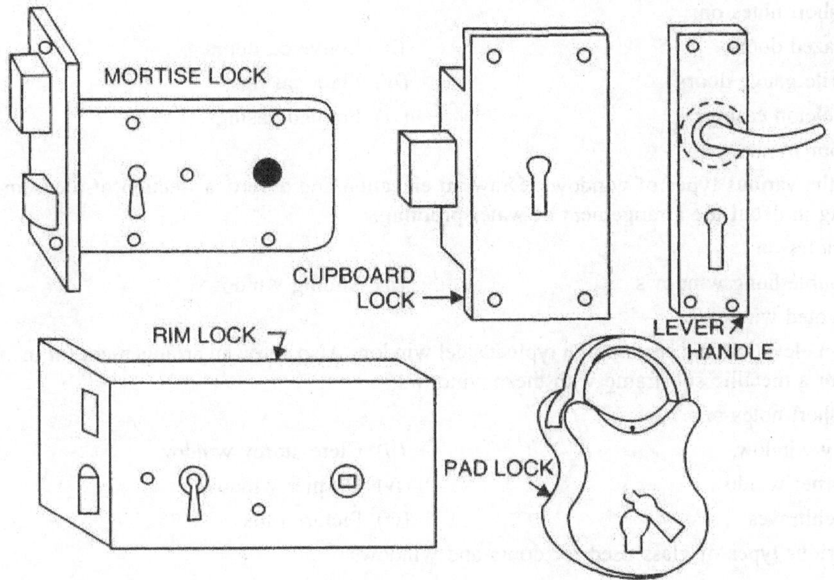

Figs. 12.116 to 12.120: Door locks.

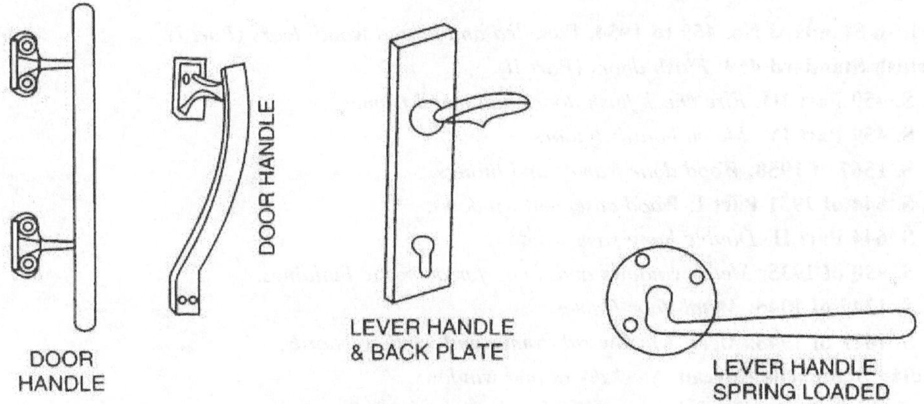

Fig. 12.121 to 12.124: Door handles.

QUESTIONS

1. What are the various points to be considered while locating the position of doors and windows? What are the commonly used sizes for them and how much window area is necessary for a building?
2. Name the various types of doors used in a building and also describe briefly the different types of door movements.
3. Sketch a ledged and braced door in elevation and section and describe its construction giving approximate sizes of members used. Also mention the fittings and fastenings used.
4. Write short notes on:

   (*i*) Ledged door. (*ii*) Various types of panelled doors.
5. Draw detailed sketches to show the constructional joints of a panelled door shutter. Roughly sketch the various types of door mouldings.
6. What are the advantages of flush doors. Draw a section of a flush door with skeleton frame as the interior support.

7. Write short notes on:
   (*i*) Glazed doors.
   (*ii*) Louvered doors.
   (*iii*) Write gauge doors.
   (*iv*) Plain casings.
   (*v*) Skeleton casings.
   (*vi*) Framed casings.
   (*vii*) Door frame.
8. Name the various types of windows. Draw an elevation and a vertical section of a casement window showing in detail the arrangement of water proofing.
9. Write notes on:
   (*i*) Double hung windows.
   (*ii*) Sliding windows.
   (*iii*) Pivoted windows.
10. Draw an elevation and section of a typical steel window. Also show an arrangement for using a wooden frame or a metallic sub-frame with these windows.
11. Write short notes on:
   (*i*) Bay window.
   (*ii*) Clere-storey window.
   (*iii*) Corner window.
   (*iv*) Dormer window.
   (*v*) Architraves.
   (*vi*) Picture rails.
   (*vii*) Various types of glass used for doors and windows.

REFERENCES

1. **British Standard No. 459 to 1954.** *Panelled and glazed wood doors (Part I).*
2. **British Standard 459**: *Flush doors (Part II).*
3. **B. S. 459 Part III**: *Fire check flush doors and metal frames.*
4. **B. S. 459 Part IV**: *Match boarded doors.*
5. **B. S. 1567 of 1958**: *Wood door frames and linings.*
6. **B. S. 644 of 1951 Part I**: *Wood casement windows.*
7. **B. S. 644 Part II**: *Double hung sash windows.*
8. **B. S. 990 of 1935**: *Metal windows and doors for domestic buildings.*
9. **B. S. 1245 of 1945**: *Metal door frames.*
10. **B. S. 1422 of 1948**: *Metal window subframes and window boards.*
11. **Indian Standards Bureau**: *Steel doors and windows.*
12. **National Building Organisation**: *Wooden doors and windows.*
13. **Urquhart, L. C.**: *Civil Engineering, handbook.*
14. **Frederick S. Merritt**: *Building construction handbook.*
15. **Mackey, W. B.**: *Building construction.*

Floors

GENERAL PRINCIPLES

Floors divide a building into different levels, one above the other for the purpose of creating accommodation within restricted space. Bottom-most floor of a building is called *ground floor* unless it is below the surrounding ground level when it is termed *basement floor*. The object of the ground floor is to have clean, smooth, impervious level and a durable surface. The top floors, in addition to the above objects have to be sufficiently strong so as to withstand the loads which come over the floor.

Two distinct components of the floor are:

(*i*) A subfloor or sometimes termed as base course.

(*ii*) Floor covering.

The object of providing a subfloor is to give proper support to the covering and hence loads are taken by this part. Moreover the subfloor is used to carry the covering and to make the flooring, in general, water-proof and sound-proof. A thin layer of another material may be used in between the two components.

The materials used for ground floor construction are:

(*i*) Bricks	(*ii*) Stone
(*iii*) Wood	(*iv*) Concrete

The materials used for the finishes of ground floors are:

(*i*) Bricks	(*ii*) Stones
(*iii*) Concrete	(*iv*) Tiles
(*v*) Wood	(*vi*) Terrazzo
(*vii*) Mosaic	(*viii*) Bitumen
(*ix*) Rubber	(*x*) Linoleum
(*xi*) Cork	(*xii*) Magnesite
(*xiii*) Glass	

Since ground floors rest directly on the soil, it is essential to have a proper drainage for the floor. This may consist of a system of drains suitably constructed below the floor and arranged to lead the water outside the building. However, this measure is not necessary in the normal type of construction. Generally a porous layer of an inert material is provided to prevent the rise of water into the floor. The material generally used is sand, crushed rock, gravel or cinder. In certain cases bituminous layers may assist in the general drainage of the surface.

13.1 SELECTION OF TYPE OF FLOOR

To select a suitable type of floor construction in a building certain points have to be kept in view. These are discussed on page 300 and apply mostly to upper floors. However, for ground floors, the selection of the type of the wearing surface is important and the various considerations governing the suitability of one over the other are given below:

(*i*) ***Initial Cost.*** The cost of the floor covering is a very important item. The expensive type of floor coverings are marble, rubber tiles and special clay tiles. Terrazzo, cork, tiles, linoleum, bitumen and slate form a less expensive group of floor coverings whereas concrete and brick are the cheapest.

However, the type of base course used with a particular covering has an important bearing on the cost of the floor, *e.g.*, if a wearing surface is to be provided on a wooden base, the cost would be different for different coverings as special treatment would be needed to fix the covering with the base in each case.

(*ii*) ***Appearance.*** Colour, texture and decorative value with regard to the architectural beauty of a room are the factors which have to be taken into consideration while selecting a type of floor for a room. Terrazzo, tiles and marble give a good appearance whereas a bituminous covering has an unattractive look.

(*iii*) ***Durability.*** Resistance to wear is an important factor for a floor covering. Further, resistance to temperature changes, humidity, disintegration and decay have to be taken into account. Tiles, terrazzo, marble and concrete floor coverings offer good resistance to these forces of disintegration whereas linoleum and rubber give less satisfactory resistance while cork and bituminous coverings have very low resistance. Wherever heavy floor traffic is not anticipated, bricks, wood blocks and bituminous mastic can be used.

(*iv*) ***Cleanliness.*** A floor should be non-absorbent and capable of being easily cleaned. All joints should be simple so that they can be made as water-tight as possible. Terrazzo, tiles, marble and slates provide surfaces which can be easily cleaned.

In addition to general cleanliness, a floor covering should have an effective resistance against the absorption of oil and grease which would otherwise give a very unsightly appearance. Greasy substances are not absorbed by tiles but affect terrazzo, concrete and bricks to some extent. Grease, oil and gasoline disintegrate bitumen and rubber floors to a considerable extent.

(*v*) ***Dampness.*** Resistance to damp is an important factor to be considered for ground floor construction. Wood, rubber, linoleum and cork coverings are not suitable in damp places whereas clay, tiles, bricks, concrete and terrazzo are suitable for use on floors which are subjected to dampness.

(*vi*) ***Indentation.*** It is desirable that indentation should not occur when loads are to be supported by a floor. In the case of lighter loads also *e.g.*, chairs, tables, etc., care should be taken that the floor covering does not get any imprint. Certain floor coverings do not get any mark on account of loads but some of the floor coverings get imprints which are not visible after the load is removed. Examples of the first type are concrete and other hard floor coverings whereas bitumen and rubber surfaces belong to the latter type.

(*vii*) ***Noiselessness.*** While this is not an important factor for ground floor construction, however care should be exercised to ensure that the floor is somewhat noiseless when travelled over. Wood, linoleum and bitumen are good in this respect whereas cork and rubber coverings are much superior.

(*viii*) ***Maintenance.*** To keep the surface in good condition, it is necessary to clean, repair and apply any other treatment from time to time. Tile, marble, terrazzo and concrete surfaces need less maintenance whereas wood blocks, mastics and corks may need frequent maintenance. However, concrete surfaces cannot be repaired easily while tiles of marbles or slates can be replaced quickly.

TYPES OF GROUND FLOORS

The various types of floorings used in ground floor construction are:

(*i*) Brick flooring (*ii*) Stone flooring
(*iii*) Concrete flooring (*iv*) Tiled flooring
(*v*) Wood-block flooring (*vi*) Terrazzo flooring
(*vii*) Mosaic flooring (*viii*) Bituminous flooring
(*ix*) Rubber flooring (*x*) Linoleum flooring
(*xi*) Cork flooring (*xii*) Magnesite flooring
(*xiii*) Glass flooring

A brief description of these is given below:

13.2 BRICK FLOORING

These are commonly used in cheap construction and where good bricks are available. Well burnt bricks of good colour and uniform shape are used. The filling, over which this floor is to be laid, should be well compacted. The level of the flooring top being known, the filling is excavated to the desired depth. This depth of excavation depends upon the type of bedding to be provided for the brick bearing surface. Generally two types of bedding are provided. In the first type, after excavation the surface is levelled and rammed carefully. On this bed, 75 mm layer of sand is spread over which a course of brick laid flat in mortar is built. This acts as a sublayer for the brick wearing course. As an alternative a lime concrete mix or lean cement concrete (one part of cement, six of sand and 18 of (aggregates 38 mm. size) of 100 mm to 150 mm in thickness is

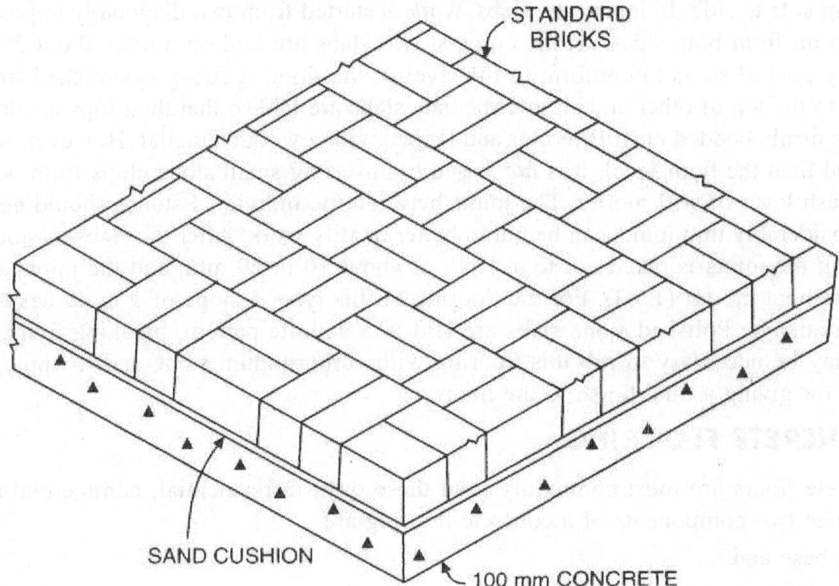

Fig. 13.1: Brick flooring.

laid on the compacted bed. The bricks for the wearing course are laid on edge breaking bond with each other. However, Herring bone pattern can also be used. The bricks are cemented by the use of cement or lime mortar and the joints are pointed to get good finish. For pointing, the mortar is raked out from the joints to a depth of 20 mm, and cement mortar (1 : 3) pointing carried out. The whole surface is cured with water for a suitable period. Whenever brick flooring is used just outside the building, the brick joints are simply grouted with dry sand, known as sand grouting.

13.3 STONE FLOORING

This type of work is also called paving. This consists of comparatively thin slabs of stone laid on concrete bedding. For damp and black cotton soil, a layer of sand or rubble is used as a cushion below the base. After excavating to the required depth, the earthen base is levelled, rammed

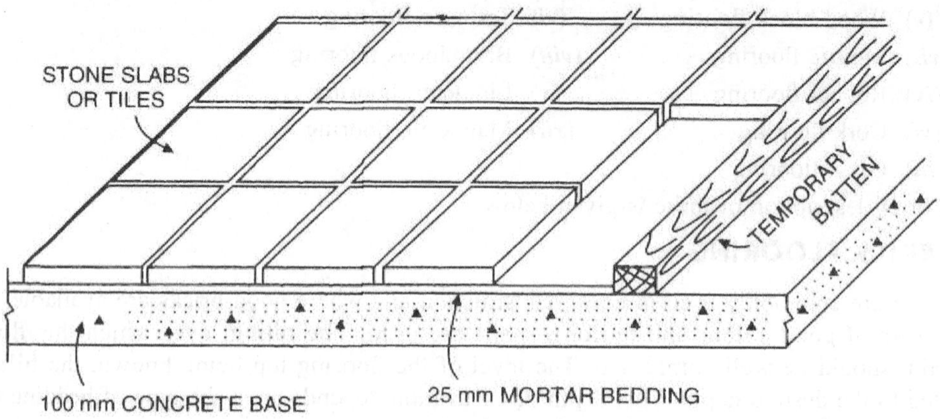

Fig. 13.2: Stone flooring.

and watered. Then a concrete layer 100 mm to 150 mm in thickness is laid and rammed thoroughly. The stone slabs used should be of a hard, durable and even quality and of as uniform thickness as possible. They should be finely chisel dressed on surface and should have their edges true and parallel from side to side. In laying the slabs, work is started from two diagonally opposite corners and brought up from both sides. At the corners, two slabs are laid on mortar about 25 mm thick and securely pushed so as to conform to the level of the floor. A string is stretched from the top of one slab to the top of other and all intermediate slabs are laid so that their tops touch the string. Each slab is firmly bedded on stiff mortar and tapped with a wooden mallet. However, when a slab gets lowered than the floor level, it is not raised by inserting small stone chips from below but is relaid on fresh layer of stiff mortar. The joints between the individual stones should not exceed 6 mm but considerably thin joints can be put in better quality work. After the slabs of stone are laid, the mortar in the joints is raked out to a depth of about 10 to 20 mm, and the joints are pointed flush with cement mortar (1 : 3). For the flooring of this type, a slope of 1 in 40 has to be given for proper drainage. Polished stone slabs are laid to a definite pattern, breaking joints with each other. It may be necessary to rub this flooring with carborundum stone and to apply a suitable wax polish for giving a final finish to the floor.

13.4 CONCRETE FLOORING

Concrete floors are most commonly used these days in residential, commercial and public buildings. The two components of a concrete flooring are:

(*a*) a base and

(*b*) a wearing surface.

The flooring can be constructed either monolithically, *i.e.*, base layer is laid and then immediately a concrete topping is provided or non-monolithically the topping is laid after the base has set. Monolithic construction has got certain disadvantages, which are:

(*i*) The topping is likely to get damaged due to subsequent building operations.

(*ii*) It is likely to develop hair cracks on account of small settlements which may occur in the base course immediately after it is laid.

(*iii*) When the surface of the topping gets damaged, it is very difficult to repair it.

(*iv*) The progress is slow since the topping can be laid only after the base course has sufficiently set to allow the workman to lay the top layer.

However the advantage of monolithic construction is that a smaller thickness is needed as the bond between the base course and the topping is good.

The type of floor finish generally used for the topping is either an ordinary concrete finish or a granolithic finish which is concrete made up of specially selected aggregates. The thickness of granolithic concrete should not be less than 12 mm. When the thickness is between 12 mm to 40 mm the finish should be laid monolithic with the base. When it is greater than 40 mm., this may be laid either monolithically or after the base concrete has hardened. Expected wearing of the surface and the nature of the base surface will determine the thickness of topping. If the thickness of topping is not greater than 40 mm for non-monolithic construction, the topping may be laid in one layer otherwise in two layers, the top layer being laid when the bottom layer is still in plastic condition. The grading of aggregates should be such that the floor should give a granolithic appearance.

Generally hard fine grained granite, basalt, limestone and quartizite stones are suitable for coarse aggregate. Preferably crushed aggregates should be used to get a granolithic finish. Granolated metal is sometimes used as coarse aggregates. The fine aggregates used are the natural or crushed sands and a suitable grading is given below:

I.S. Sieve	*Percentage by weight passing I.S. sieve*
10 mm	100
4.75 mm	90–100
2.36 mm	75–100
1.18 mm	55–90
600 micron	35–50
300 micron	10–30
150 micron	0–10

For ordinary concrete finish, the proportions of the concrete mix should be 1 : $1\frac{1}{2}$: 3. However, under controlled mixing conditions, 1 : 2 : 4 ratio with carefully selected aggregates may be used. For granolithic finish a mix of 1 : 1 : 2 is suitable but a careful selection of aggregates is essential.

Before laying the topping, the base concrete which may be about 100 mm in thickness is laid. The top surface of this base is finished roughly so as to get the desired bond between the base course and the topping. This concrete is generally of lean proportions and lime concrete is mostly used. For laying the topping, the surface of the base is cleaned and the work is planned in such a way that pouring can be done with ease. Whenever monolithic construction is desired, the wearing coat is placed within 45 minutes to 4 hours after placing the base.

For non-monolithic construction, the surface of the base concrete is brushed with a stiff broom and cleaned thoroughly. It is wetted the previous night and surplus water is brushed off. The floor is laid in rectangular panels made with the help of battens set on mortar beds. The panels used are not greater in size than 2 m × 2 m. Generally in exposed floorings 1 m × 1 m panels are used. These battens are laid to the desired level and slope. The concrete is poured into these bays after a grout of cement sand mix is applied to the base course. Concreting of alternate bays is done at one time and the intermediate bays are not laid earlier than 72 hours. This enables most of the initial shrinkage to take place. As soon as the layer of concrete is even, the surface is rapidly compacted by ramming or beating and screeded to a uniform level and line. This is followed by trowelling. The trowelling is done to give a level surface just after the laying of concrete. Further trowelling is done when the mix has stiffened to a point where a solid well compacted surface can be obtained without bringing up the slurry. Dusting of the surface with neat cement to facilitate trowelling or the application of a cement slurry to the surface should not be permitted.

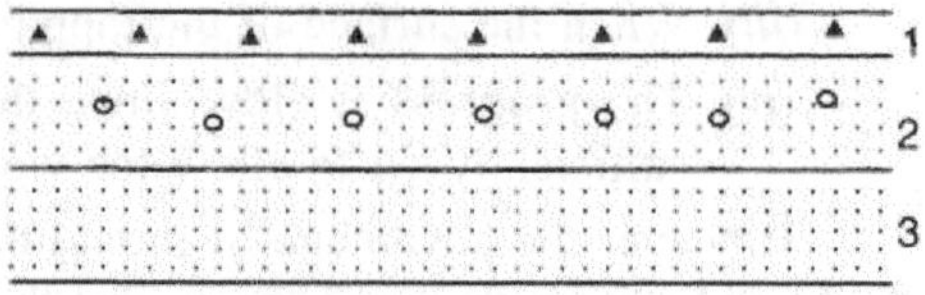

1. 40 mm CONCRETE TOPPING
2. 100 mm LIME OR LEAN CEMENT CONCRETE
3. SAND CUSHION

Fig. 13.3: Concrete flooring.

The floor thus laid is protected from sunlight, wind and rain for at least 12 hours but preferably 24 hours. It is then kept wet by spreading a layer of wet sand for a period of at least 7 days. As an alternative, special membranes can be used for curing. In the case of granolithic finish, the surface may be rubbed to get a better finish. The grinding is delayed sufficiently to prevent displacement of the aggregates. Power operated machines, fitted with abrasive stones, can be used for rubbing the floor surface. The surface is kept wet while it is rubbed. Any holes or depressions appearing on the surface are finished with an application of a thin cement grout which is allowed to set for a period of 3 days and then rubbed again.

Non-slippery surfaces can be obtained by incorporating suitable abrasives *e.g.*, silicon carbide, carborundum, aluminium oxide in granular form at the rate of 3 to 6 kg per square metre of the surface depending on the thickness of the topping.

For surface hardening, three coats of sodium silicate are given each being applied after the previous has dried.

In ordinary type of concrete flooring, the topping is replaced by only a thin coat of cement mortar 1 : 1 mix rendered on the base course which is generally of cement concrete. On lime concrete bases, thicker rendering is needed.

For getting a coloured finish, either coloured cements are used or to get the different colours, pigments as specified below may be used for every cubic metre of concrete in the topping.

Red colour	$\frac{1}{10}$ cu m red oxide of iron powder.
Terra-cotta colour	Burnt yellow cochre.
Black colour	$\frac{1}{20}$ cu m of manganese dioxide.
Buff colour	$\frac{1}{20}$ cu m of yellow cochre.

Floor paints in various colours are available and these give a hard and wear resistant surface which is also water-proof.

13.5 TILED FLOORING

Clay tiles of various shapes, sizes, thicknesses, colour and surface finishes are manufactured for use as surface covering for floors. Flooring tiles are set on a concrete base with mortar. Special bedding made up of emulsified bitumen and Portland cement is available for use over wood, steel and concrete bases. The concrete bedding is generally of 150 mm in thickness and is laid evenly with a slight rough surface at the top. After a period of 2 to 3 days, a mortar layer of 1 : 1 mix is spread on the concrete bed and the tiles are set truly and evenly with a thin paste of cement applied to their sides. They are slightly tapped till the cement oozes out through the joints to the top surface. This extra cement is wiped off and the joints are cleaned with saw dust. After 2 to 3 days, these joints are rubbed with carborundum stone to chip off all the projecting edges or surfaces. The whole surface is then polished with a very soft carborundum stone followed by a pumice stone. Finally the surface is washed with soap. This process of rubbing with carborundum stone is not suitable for tiles which have a glazed surface.

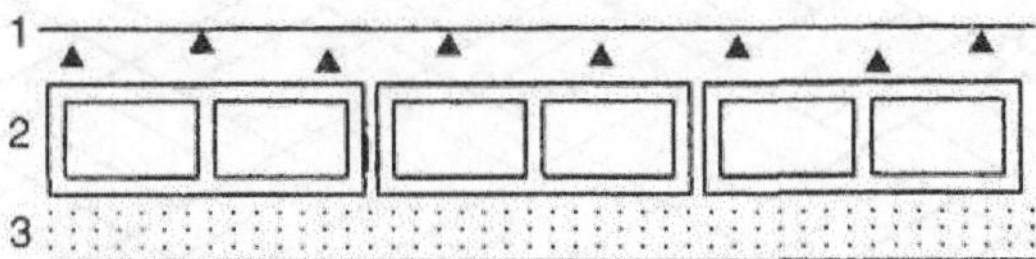

1. 50 mm CONCRETE TOPPING
2. HOLLOW TILES
3. SAND CUSHION

Fig. 13.4: Hollow tiled flooring with concrete topping.

Floors may be built of hollow tiles which act as a base and the same are covered with a layer of concrete. This type of floor ensures freedom from damp. A cushion of sand should be given below the tile layers, as shown in fig. 13.4.

13.6 WOODEN FLOORING

This type of floor construction is not extensively used but is popular for special purpose floor, *e.g.*, in auditoriums, hospitals, etc. The various types of wooden floorings used are:

(*i*) ***Strip flooring.*** This consists of narrow and thin strips of wood joined to each other by tongue and groove joint.

(*ii*) ***Planked flooring.*** In this type, wider planks are used and these are also tongued and grooved.

(*iii*) ***Heavy wood block flooring.*** These are made up of thicker pieces of wood cut in short lengths ranging from 5mm to 100 mm so as to form blocks which are set with the ends of the grains exposed.

(*iv*) ***Fabricated wood blocks.*** These consist of small square or rectangular blocks with tongue and groove joints on all sides.

Wooden flooring should have a concrete base or should rest on joists spanning across dwarf walls which are constructed at suitable intervals. For fixing the wooden floors on côncrete slabs, longitudinal nailing strips are provided. These have a bevelled section and are embedded in concrete at suitable intervals. To increase the hold of these strips with the concrete, nails or other fixtures can be driven into their sides. As an alternative to nailing strips, special concrete called nailing concrete may be used. The surface in contact with the concrete layer is treated with an application of bitumen or asphalt mastic which may be applied hot or cold. Special flooring nails are used for nailing down the floor coverings. Plank flooring should be laid with metal spacers 1 mm in thickness temporarily for providing expansion joints.

Strip flooring is used in thicknesses of 20 mm to 25 mm and width of 60 mm to 100 mm. Thicker floors of this type are used in warehouses, factories and gymnasiums. Planks for flooring are about 200 mm wide. Wood block flooring may have joists filled with bitumen for expansion.

For places where flooring is supported on dwarf walls, the intermediate space is left hollow. This provides an effective means of damp prevention and also can be used for dancing halls where a springing action is desired. In the construction of these floors, the joists rest on wall plate fixed on the dwarf walls and the boarding is nailed to the former. In order to prevent damp a 150 mm thick layer of lean concrete is spread on the earthen base. The hollow space between the ground and the floor is effectively ventilated.

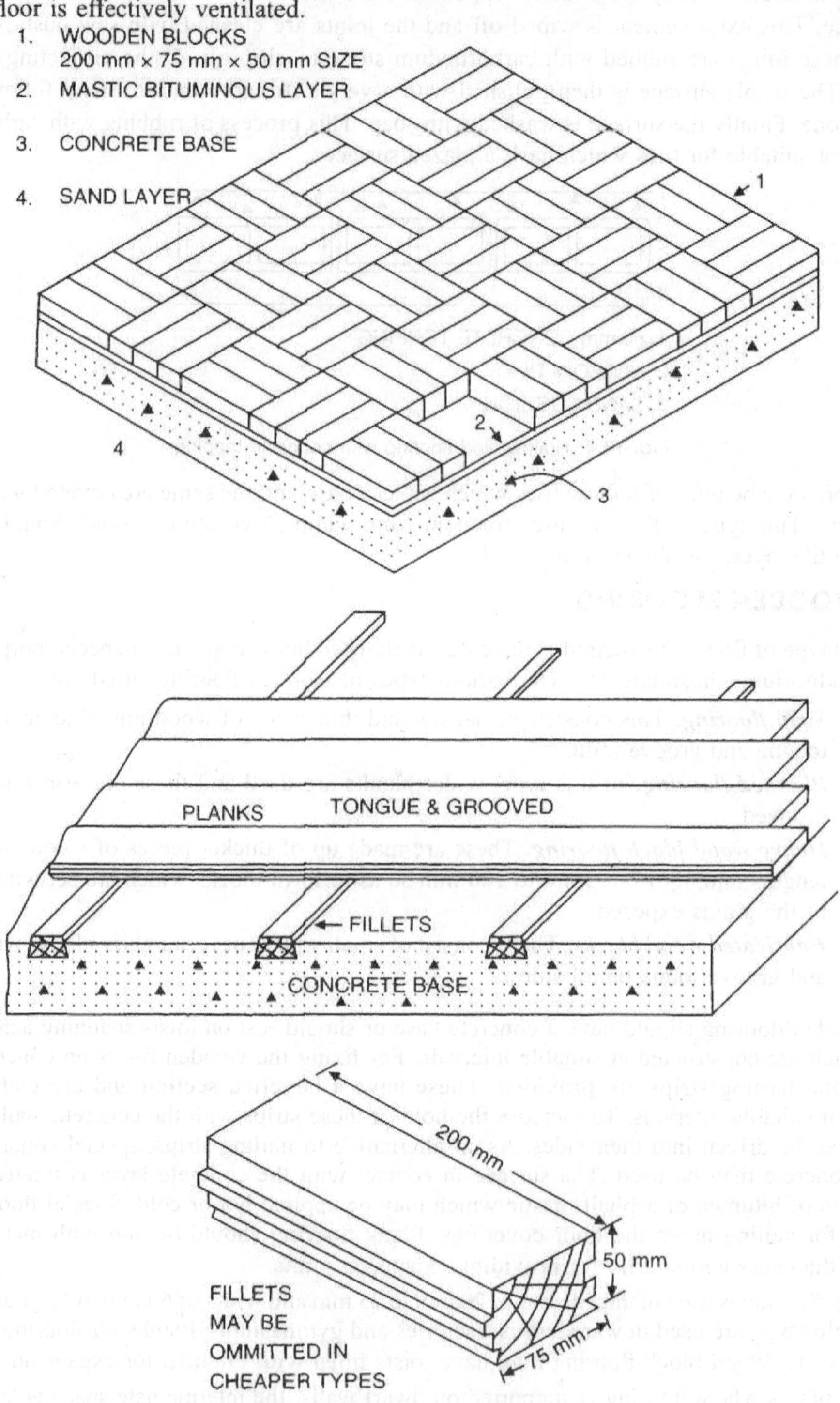

Fig. 13.5 to 13.7: Wooden flooring details.

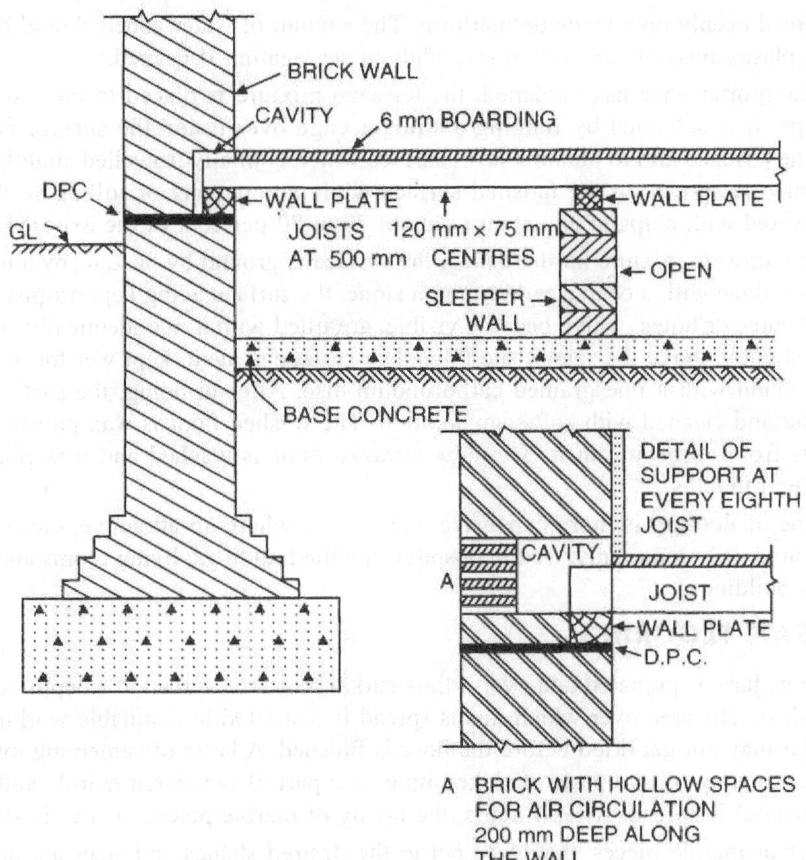

Figs. 13.8 and 13.9: Supported wooden flooring.

13.7 TERRAZZO FLOORING

Terrazzo is concrete containing marble chipping as an aggregate. The surface is rubbed to a fine finish. The base course is, as usual, of concrete. On this, a thin layer of sand is spread evenly and is covered by tarred paper. Over this a layer of rich mortar is spread. The mortar base should be at least 30 mm thick and should consist of 1 : 3 cement sand mixture. It should have only enough water to produce a mortar of as stiff consistency as possible so that it can be struck off with a straight edge. The mortar base can be placed directly on the concrete slab and can be bonded to it by cleaning the surface thoroughly. But in this case, slight irregularities, which develop in the base course on account of minor settlement would be transmitted to the top layer of terrazzo mixture. This mortar bedding is struck off about 10 mm to 5 mm below the finished floor level. Metal dividing strips of 20 gauge in thickness are inserted into the mortar base to form the desired pattern. These strips should be of a height slightly greater than the thickness of the mortar base and the terrazzo covering so that after rubbing of the surface, they come in level with the remaining surface.

The terrazzo mixture is made up of cement and marble chips in different proportions. Marble chips of sizes 3 mm to 6 mm can be used. Coloured cements can also be used to get the desired tint.

A uniform surface is absolutely essential and hence thorough mixing of the ingredients is desirable. The chippings should be thoroughly mixed dry followed by the addition of cement and remixing the ingredients. The materials should not be heaped so as to cause segregation but should be spread evenly on a made up platform. The amount of water added should be such so as to produce a plastic mixture which will stay without segregating if heaped.

After the mortar base has hardened, the terrazzo mixture is placed to the top level of the dividing strips. It is screeded by running a straight edge over it and the surface is rammed to consolidate the terrazzo and to obtain a level surface which is finally trowelled slightly. Additional aggregates may be spread on the finished surface while compacting or rolling so that the final surface is covered with chips to an extent of about 70 to 80 per cent of the exposed area.

After the terrazzo mixture has hardened, the surface is ground by hand or by a machine. The first grinding is done with a coarse carborundum stone, the surface being kept profusely wet while grinding. All pores or holes, which become visible, are filled with a fine identically coloured mix. This is kept slightly above the rest of the floor. The surface is again kept wet for at least 5 days and finally ground with a fine grained carborundum disc. After grinding, the surface is washed with hot water and cleaned with soft soap solution. The washed floor is wax polished to prevent the chippings from absorbing dust. The new terrazzo floor is washed and wax polished every week for several months.

This type of flooring is more expensive and is used where an attractive, clean and durable surface is desired. It is commonly used in hospitals, public buildings, living rooms and bath rooms of residential buildings.

13.8 MOSAIC FLOORING

A concrete base is prepared and over it lime surkhi mortar is spread to a depth of 5 mm to 80 mm and levelled. The area over which this is spread is restricted to a suitable working period so that the mortar may not get dried before the floor is finished. A layer of cementing material about 3 mm thick consisting of two parts of slaked lime, one part of powdered marble and one part of puzzolona material is spread. After 4 hours, the laying of marble pieces or tiles is started.

The tiles or marble pieces should be cut to the desired shapes and sizes and arranged into definite outline to get the desired pattern. They are broken into cubes or wedges and hammered into the mortar at the outer line of the pattern. The inner area is filled with coloured pieces of marble in the desired fashion. A stone roller about 300 mm in diameter, 450 to 600 mm long is passed over the surface gently, water being sprinkled every now and then to work up the cement between the marble pieces.

The surface, thus prepared is allowed to set for 24 hours and is rubbed with a pumice stone 200 mm × 250 mm × 70 mm fitted to a long wooden handle. The object is to polish the surface and to make it smooth and level. The floor is dried for about two weeks before use.

13.9 BITUMINOUS FLOORING

Bituminous floors are of two types:

(*a*) ***Bituminous tiles.*** These are made from bitumen, asbestos fibres and inert materials under pressure. They are square and vary in sizes from 200 to 450 mm and in thickness from 3 to 6 mm. They are used as a covering for wooden or concrete floors. On wooden floors, a bitumen saturated felt is first fixed to the wooden surface and the tiles are cemented to it. As an alternative, a mastic layer can be used for fixing. On concrete floors, the tiles are directly cemented to the floor. These tiles are resilient, non-absorbent, moisture proof and relatively cheaper. They are used in schools, offices and hospitals.

(*b*) ***Bituminous mastic.*** Asphalt mastic is a mixture of fine aggregates (sand) and natural or artificial bitumen. It can be mixed hot and laid in continuous sheets or pressed into blocks which can be used as a flooring. As an alternative it is mixed with a mineral oil and asbestos and applied cold.

For hot mix construction, the bitumen is broken into pieces of not more than 250 gm in weight and put into an iron pot which can be heated from underneath. While heating, the bitumen is stirred thoroughly so that the layer at the bottom may not get burnt. When the whole quantity is fused, sand or grit equal to twice the volume of bitumen is added gently and mixed thoroughly. This mixture is then ready for laying.

An ordinary concrete or wood base may be used for laying this mixture. If the wooden base is used, a tarred paper layer is introduced. Gauges are erected at the sides of the portion which is proposed to be laid with mixture. The mix is poured and spread by means of trowels into uniform thickness. Straight edges are used for giving a levelled surface. A thin layer of sand is spread over this and rubbed with a trowel. At the end of the day's work, the joints in the mastic are lapped as shown.

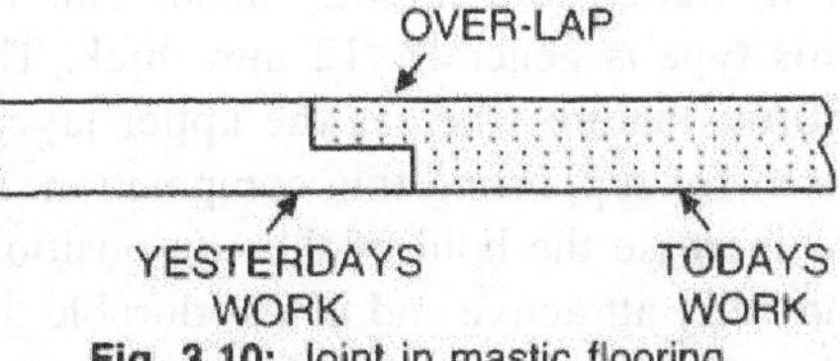

Fig. 3.10: Joint in mastic flooring.

Generally 25 mm thickness is sufficient for ordinary floor construction.

In the cold process, the ingredients are mixed without heating and a fluxing mineral oil is added to obtain a soft consistency suitable for laying. Pigments may be added for improving colour.

13.10 RUBBER FLOORING

Rubber flooring is made of pure rubber mixed with fillers such as cotton fibre, granulated cork or asbestos fibre and desired colour pigments. Suitable patterns can be incorporated. It is used in thickness from 3 to 9 mm. Many colours and patterns are available. It is cemented to concrete or wood in the same manner as linoleum. It is attractive in appearance and elastic. However, rubber floors are not resistant to oil, grease and gasoline. They are not commonly used for ground floors. Rubber flooring can be made in sheets or tiles of suitable size and thickness.

13.11 LINOLEUM FLOORING

This is used as a covering for wood and concrete floors. Plain and printed varieties of linoleum are available in thickness from 2 to 6 mm and in rolls which are about 2 m to 4 m wide. Linoleum tiles are also available for specific uses. The patterns made on the linoleum surface are durable but when they get tarnished sufficiently, the covering should be varnished. Linoleum coverings are laid on concrete and wood with a layer of saturated felt in-between to act as a cementing material. If cemented directly to the flooring, there is a likelihood of its getting torn on account of the movement in the floor boards due to expansion or contraction of the planks. Moreover, depressions which appear on the surface of the linoleum boards shrink and give an unattractive appearance. Plywood boards, if used, as a sub-base for the linoleum flooring, the depressions caused otherwise are not visible. Linoleum coverings are attractive, resilient, durable, cheap and can be cleaned very easily. Varnishing and waxing the surface would give a longer life and help in the cleaning of the surface. Linoleum, however, is not used for basements.

13.12 CORK FLOORING

These are suitable for use in churches, theatres, public libraries and other places where noiseless floor covering is desired. This should not be used in basement or floors where dampness prevails. They are available as cork carpet and cork tiles. Base may consist of a concrete floor or wooden flooring. The cork flooring is fixed in a manner similar to the linoleum floor covering. It may also be fixed directly to wooden floors. Cork tiled flooring is available in various shades and is made up of cork shavings compressed in moulds to a thickness of about 12 mm and baked subsequently for effective cementation of the individual components.

13.13 MAGNESITE COMPOSITION FLOORING

These are patent type of floor materials available for use on special occasions. Magnesites composition flooring consists of a dry mixture of magnesium oxide, asbestos or other inert material and a pigment. Liquid magnesium chloride is added on the job to form a plastic material which is trowelled to a smooth finish. This mixture sets to a hard surface in a few hours. Flooring of this type is generally 12 mm thick. Thicker floors are laid in two layers. The lower layer is of fibrous nature whereas the upper layer is harder. Base of wood, concrete or steel plates can be used for supporting this composition. For wooden bases, a foundation of metal lath is necessary to increase the hold of the composition on the smooth wooden surface. This type of flooring is not very attractive and is less durable than clay tiles, terrazzo and marble. However it is less noisy than these floors. It is used in schools and in office buildings.

13.14 GLASS FLOORING

Glass floors are used wherever it is desired to admit light into the basement through the top flooring. The glass blocks are fitted within frames of various types. They are of special shapes, if it is desired to transmit light at an angle to the farther areas in a room. Structural glass is available in the form of tiles or slabs and in thicknesses from 12 to 30 mm. The frame work is spaced close apart so that the glass can stand loads coming over it. Glass flooring are not commonly used.

UPPER FLOORS

13.15 SELECTION OF TYPE OF FLOOR

Slection of a suitable type of construction for upper floors of a building depends upon a number of factors, some of which are detailed below:

(1) ***Floor Loads.*** Light loads can be taken by any type of floor construction whereas heavy loads need special types of floors, e.g., steel joist with thin slabs of stone over them are suitable for light loads whereas for heavy loads thick reinforced concrete slab and beam construction is necessary. The use of flat slab construction will not be advantageous if only light loads are to be catered.

(2) ***General type of Construction.*** If beams, girders and columns are of wood, the floor construction will also be of wood whereas for steel framed building, the floors may be of steel and concrete type. Concrete framed structures should have R.C.C. floors. If brick walls are used to take the loads, flat slab construction is not possible.

(3) ***Plan of a Building.*** If a building is divided into rectangular panels which are nearly square in size, flat slab construction is possible. For short spans, simple slabs are economical whereas for long spans ribbed slabs will be satisfactory.

(4) ***Initial Cost.*** The initial cost of construction is directly dependent on the type of floor construction. As far as the direct cost is concerned, the various types of floors may be arranged in the following order:

(*i*) Wooden single floors or R.C.C. battened floors.

(*ii*) Light steel joist with timber planks

(*iii*) Double and framed wooden floors.

(*iv*) Steel joist with R.C.C. slab or jack arches.

(*v*) R.C.C. slab and beams.

(*vi*) R.C.C. ribbed slabs or flat slab construction.

(*vii*) R.C.C. ribbed slabs with clay tiles.

(*viii*) Hollow clay tile arches and steel beams.

However, wooden floors may be costlier at places due to the non-availability of timber and similarly concrete floors may cost more where wood or stone slabs are in abundance, *e.g.*, in hilly areas.

The indirect cost which accounts for the cost of the base course, special ceilings, increased cost of sub-structure to support a given flooring and the increased height of a building on account of the use of thicker flooring affects the type of floor to be selected. Examples are the use of special nailing strips for laying a wooden floor on a concrete surface. Wearing surfaces, *e.g.*, linoleum, cork, etc, need a smooth and strong base course of concrete or wood flooring. Tiles, marble and terrazzo will need concrete bedding. Plastered ceilings are needed for floors which have large projections underneath. Flat slabs on other hand do not need any additional ceiling. Whenever metal lath and thicker plaster layers are used for the construction of ceilings, the cost gets increased. Heavier type of floors will need stronger supporting structure. The use of flat slabs as flooring provides maximum room height. Concrete slabs supported on beams reduce the clear available floor height. Increased height means increased cost due to the cubic content of walls getting increased, the height of stairs being increased and heavier foundations being needed to support the increased loads.

(5) ***Function of the Building.*** The use to which a building is to be put determines the general type of construction needed. This factor, as already described under item (2), has an important effect on the selection of a suitable flooring. Moreover, the type of flooring loads, need for fire-proof construction, use of ceilings and other architectural features will limit the type of flooring which can be used under the given circumstances. Monumental or palatial structures will need a very decorative type of flooring and this may be the deciding factor for selection.

13.16 TYPES OF UPPER FLOORS

The various types of flooring used for the construction of upper floors are:

(1) **Wooden floors.**

- (*a*) Single joist.
- (*b*) Double joist.
- (*c*) Framed or triple joist.

(2) Steel joist and stone or precast concrete slabs.

(3) Jack arch floor of brick or concrete.

(4) R.C.C. slab floor and steel girders.

(5) R.C.C. beams and slabs.

(6) Flat slab construction.

(7) Beams and patent reinforcing slabs.

(8) Hollow tile ribbed floors.

(9) Hollow tile jack arch floors.

The details of each of them are given below:

13.17 WOODEN FLOORS; SINGLE JOIST

These are used for residential buildings where spans are short and the loads are comparatively lighter. They consist essentially of a set of wooden beams called joists and wooden boarding laid over them. These joists are spaced at about 300 mm centre to centre, spanning the room in the shorter dimension. They are generally of sufficient width for rigidity. A minimum width of 50 mm is taken as a guiding factor. The spans for this type of flooring are not generally greater than 4 metres. However in exceptional cases, spans up to 6 metres have been used.

The joists are supported on walls plates of 100 mm × 70 mm to 120 mm × 70 mm in size. The object of providing these wall plates is to distribute the loads from the joist to the wall below. Further they provide an easy means of bringing the upper edges of the joist to a horizontal plane for receiving the floor boarding. Wall plates are generally laid flush with the inner face of the wall and lengths as long as possible should be used. However if joists are a necessity, the same are provided by giving half lapped or scarfed joists. The joists rest on the wall plates with their ends nailed, cogged or notched. Single cogged or double cogged joist may be used. Corbels might be needed whenever the supporting width of the wall is not sufficient due to an adjacent flooring resting on it. Whenever joists of adjacent room run in the same direction, they may be overlapped and nailed to each other.

Joists for floors should not only be strong enough to support the loads for which they are designed but should also not deflect too much so as to cause the cracking of any plaster ceiling attached to them. A deflection greater than $\frac{1}{333}^{th}$ of the span is not permitted. The size of the joist is roughly calculated by the rule

$$\text{Depth (in mm)} = (\text{Span in m} \times 40) + 50 \text{ mm.}$$

When the length of the joist exceeds 3.5 m strutting of the joist with each other becomes necessary to avoid side bukling which is prominent in the case of deep joists. Herring bone strutting is the best type of bracing. This consists of a pair of inclined pieces of timber which are tightly fitted between the joists. These vary in size from 50 mm × 30 mm to 50 mm × 50 mm. The ends of these struts are nailed to the joist. At the end, joist wedges are incorporated between the joists and the wall. The second type of strutting is the solid strutting which is the simplest type and is used for cheap work. This consists of short lengths of wooden boards nailed to the joist in a continuous row. To strengthen this type of strutting, a circular rod about 12 mm in diameter is passed through the joist near the strutting and is held at the ends by nuts.

The planking consists of wooden boards about 40 mm thick and 100 to 150 mm wide. Narrow boards are used as the shrinkage will not appreciably open the joints and there is less tendency for the planks to distort. Care has to be taken in fitting the joints of the boards together. Various types of joints are used. The simplest type of joint is a plain butt joint in which two pieces of boards just about each other. Rebated joints are used for a better class of work and are specially suitable for end or heading joints. For good quality of work, tongued and grooved or feathered and grooved joints are used. Rebated tongued and grooved type is a superior type of joint. The boarding should be sufficiently long from one wall to the other in order to avoid heading joints. Heading joints when necessary are formed of square or butt type and should rest entirely on the joists below. Each plank is cut so as to have the joint centrally on the joist. Generally rebated and tongued joints are used. The individual planks are jammed together as they are laid by flooring cramps. Each board is secured in position by 60 mm screws screwed to each joist below. Nailing, if resorted to, should be done in a manner that the head of the nail is not visible. In the case of tongued and grooved joints nails can be driven from each tongue into the joist below. This is called secret nailing as the head of the nail is not visible. The length of the nails are driven below the surface by using a hammer and a punch. For secret nailing, oval wire nails should be used as they are less liable to split the tongues. Wherever gas pipes, cables, etc., are taken below the floors screws are used.

For surfaces which are subjected to a considerable wear, double planking is used. The first layer consists of a 20 mm thick roughly sawn square edged planks laid diagonally across the joists to avoid the joints of this layer coming right below the joints of the top layer. The upper layer consists of 20 to 25 mm thick hard wood planking and is laid at right angles to the direction of the joists.

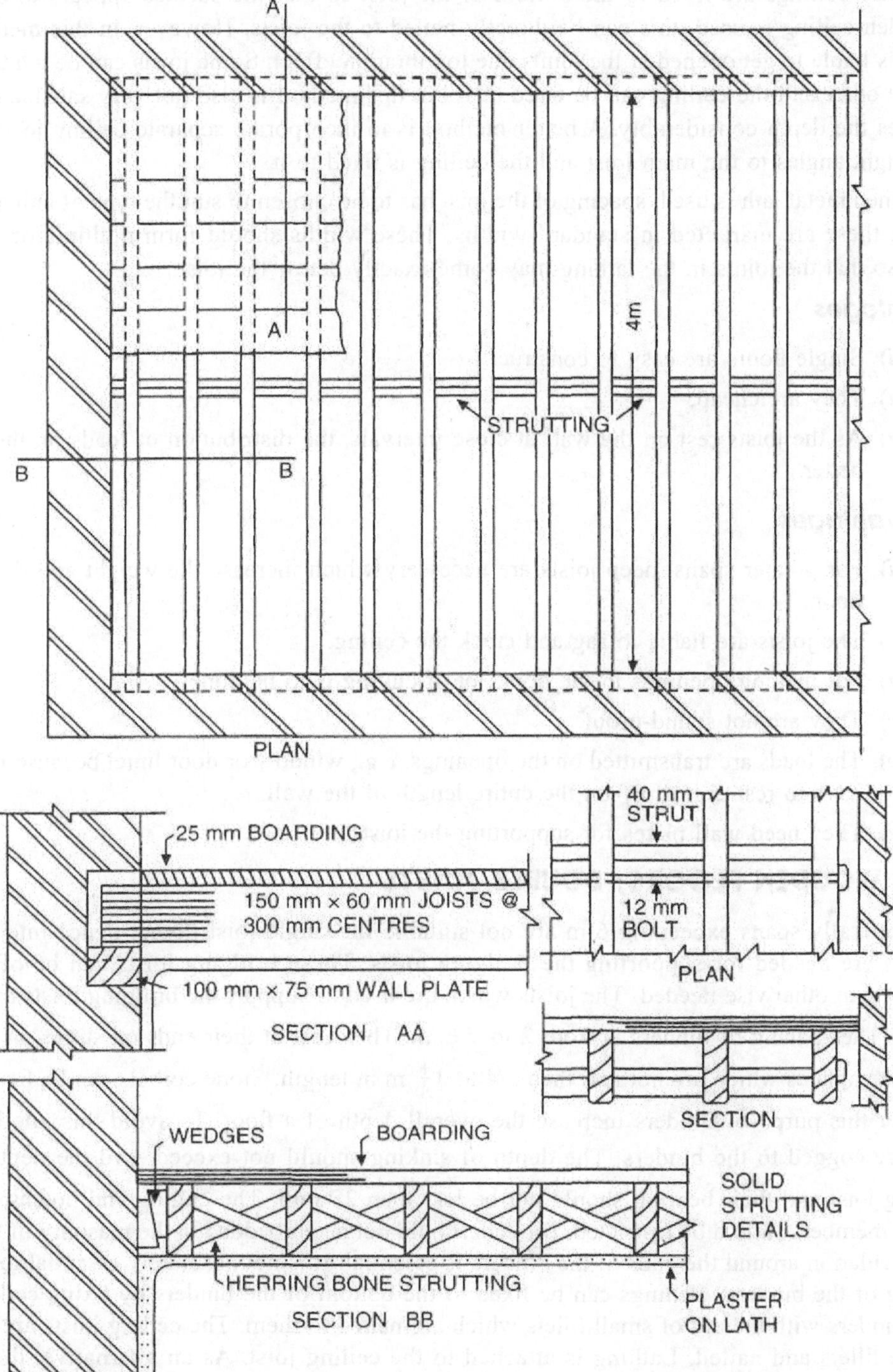

Figs. 13.11 to 13.15: Plan and sectional details of single wooden flooring.

On completion, the wooden floor planking is plained to a smooth and level surface. Hard wood floors are scrapped and rubbed smooth with sand paper. They are finally waxed and polished.

False ceilings are fixed to the bottom of the joist so that the surface appears to be plain. If wooden ceiling is used, this can be directly nailed to the joists. However, in this method, the ceiling is liable to get opened at the joints due to vibration effect. Some joists can be made deeper than the other and the ceiling can be fixed to it but this method is also not very satisfactory as it increases the depth considerably. A better method is to incorporate separate ceiling joists which run at right angles to the main joist and the ceiling is fixed to it.

When metal lath is used, spacing of the joist has to be chosen to suit the type of lath available because these are marketed in standard widths. These widths should form multiple of the joist spaces so that the joints in the lathing may come exactly below the joist.

Advantages

(*i*) Single floors are easy to construct.

(*ii*) They are cheap,

(*iii*) As the joists rest on the wall at close intervals, the distribution of loads on the wall is better.

Disadvantages

(*i*) For greater spans, deep joists are necessary which increase the weight and the cost of floor.

(*ii*) The joists are liable to sag and crack the ceiling.

(*iii*) For making openings in the floor, lot of cutting is to be done.

(*iv*) They are not sound-proof.

(*v*) The loads are transmitted on the openings, *e.g.*, window or door lintel because the joists have to rest uniformly on the entire length of the wall.

(*vi*) They need wall plates for supporting the joists.

13.18 WOODEN FLOORS; DOUBLE JOISTS

Generally spans exceeding 6 m are not suitable for single joist floors, hence intermediate supports are needed for supporting the bridging joists. These bridging joists can be of smaller sections than otherwise needed. The joists which are used to support the bridging joists are called binders. The spacing of binders is from 2 to $3\frac{1}{2}$ m. They rest at their ends on stones or wooden bearing templates which are not less than 3/4 to $1\frac{1}{4}$ m in length. Stone corbels can be fixed in the walls for this purpose. Binders increase the overall depth of a floor. To avoid this, the bridging joists are cogged to the binders. The depth of sinking should not exceed $\frac{2}{3}$rd the depth of the bridging joist and their bearing should not be less than 25 mm. The cutting and notching of the bearing members should be restricted. Binders should not be embedded in the masonry of the wall. Free circulation around the ends of the binders is essential. It is not absolutely essential to provide strutting of the binders. Ceilings can be fixed to the bottom of the binders by fixing ceiling joist to the binders with the aid of small fillets which are nailed to them. The ceiling joists are notched to these fillets and nailed. Lathing is attached to the ceiling joist. As an alternative, the ceiling may be directly fixed to the bridging joists. Whenever the section of the binders becomes very heavy it may be economical to replace the wood binders by rolled steel joists.

Advantages

(*i*) The loads are transmitted at specific points and hence window and door openings can be avoided.

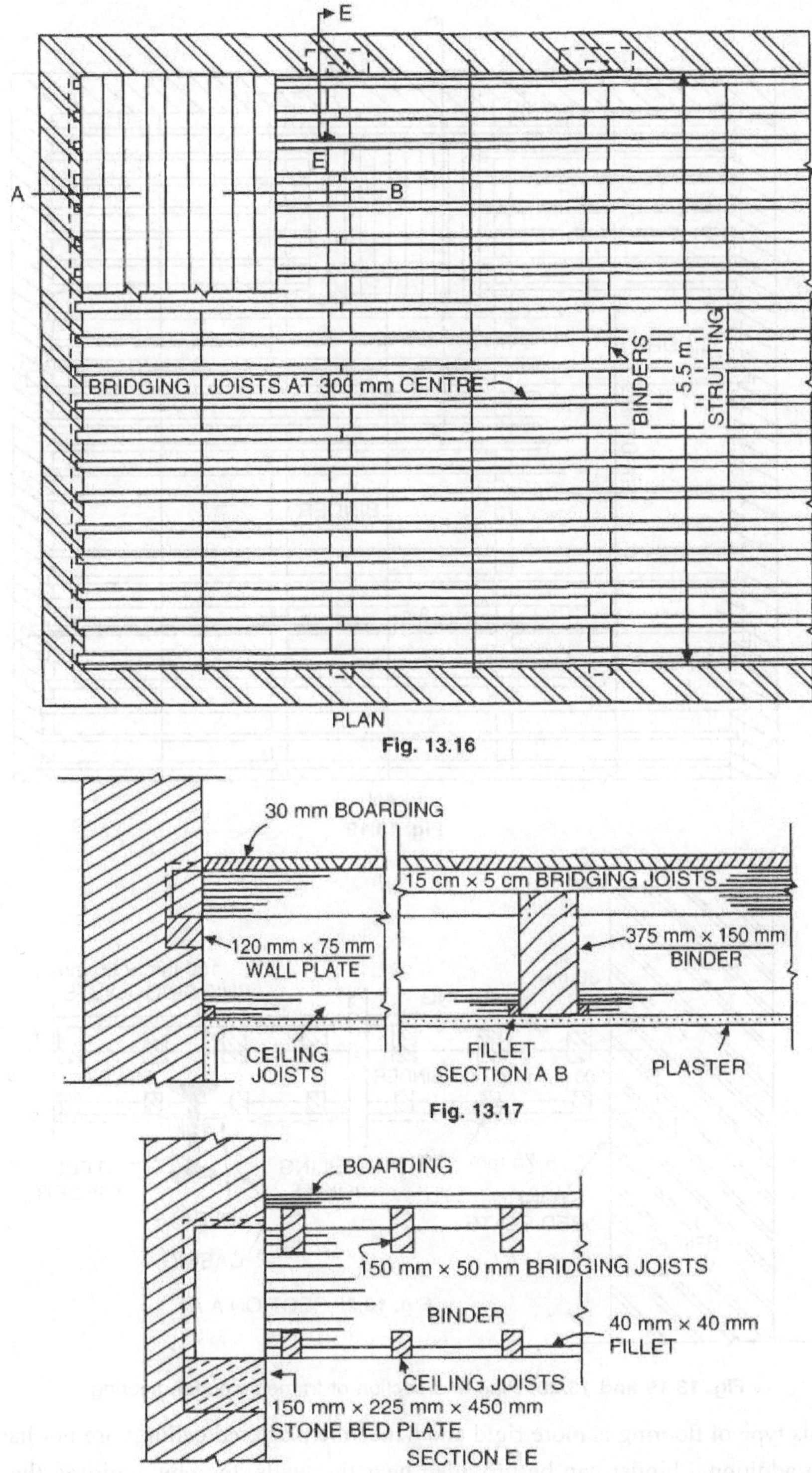

Fig. 13.16

Fig. 13.17

Fig. 13.18

Figs. 13.16 to 13.18: Plan and sections of double wooden flooring.

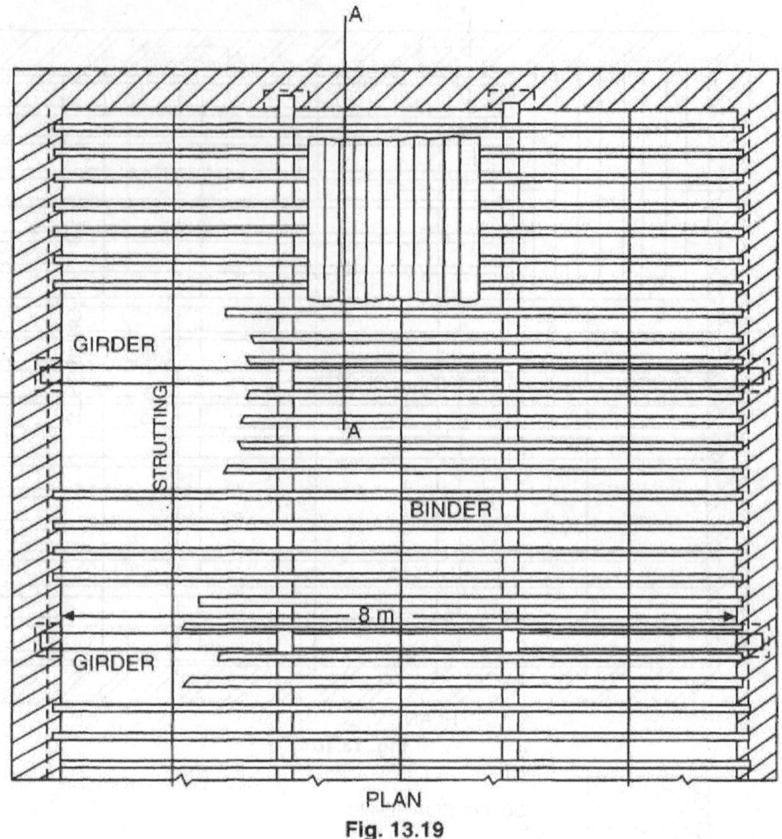

PLAN
Fig. 13.19

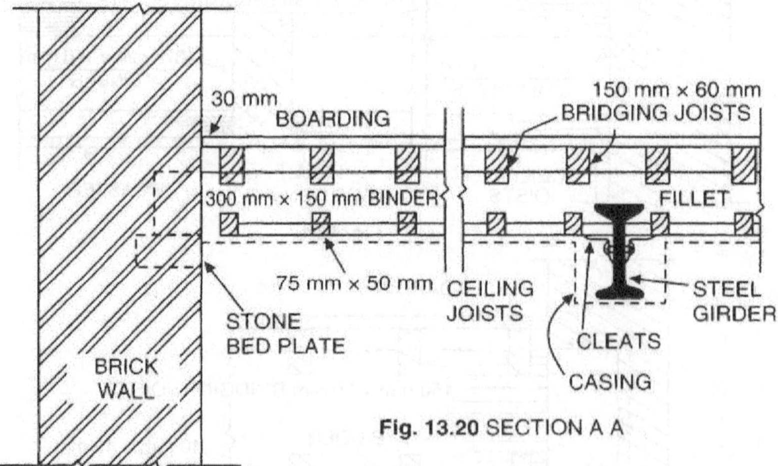

Fig. 13.20 SECTION A A

Fig. 13.19 and 13.20: Plan and section of framed wooden flooring.

(*ii*) This type of flooring is more rigid and hence the plastered ceilings are not liable to crack.

(*iii*) An additional binder can be provided near the walls, thereby avoiding the use of wall plates which are liable to decay on account of their contact with masonry under damp conditions.

(*iv*) It is more sound-proof.

Disadvantages

(*i*) The depth of floors is increased which reduces the available head room.

(*ii*) Extra labour is needed to join the various members together.

13.19 WOODEN FLOORS: FRAMED OR TRIPLE JOIST

For still greater spans (8 m or more) and heavier loads, the size of binders becomes exceptionally large and is, therefore, uneconomical. Hence if beams are placed to span the distance between the walls and the binders rest on them, the bridging joist in turn rest on the binders some economy is possible. These beams are termed as girders. The spacing of these girders depends on the type of girder used and the size of binders adopted. Steel girders or wooden girders can be used. For steel girders, the binders are notched over the top flange of the girder and are supported by two mild steel angle cleats secured to the web of the girder by rivet. In case wooden girders are used, the binders are connected to them by forming special type of joints, *e.g.*, tusk tenon joints. The binders should not come against each other as this will weaken the girder considerably. These should be staggered.

Ceilings can be attached directly to the binders or ceiling joists may be used. If the girders project below the binders also, the ceiling may be fixed to the binders and the girders suitably encased by covering with wood blocks (bearers, which may finally be finished with wooden lining or plastering).

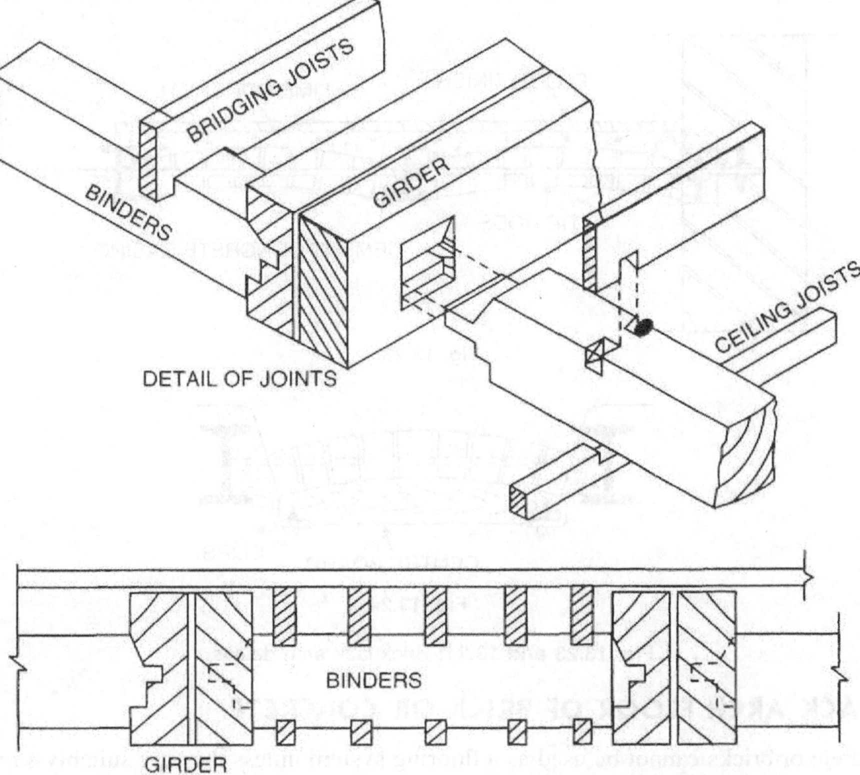

Fig. 13.21 and 13.22: Joint details of framed floors using wooden girders.

13.20 STEEL JOIST AND STONE OR PRECAST CONCRETE SLABS

Flagstone is available in abundance in certain parts of the country and is used for flooring. For small spans, steel joists are spanned on the shorter dimension of the room and are spaced at suitable intervals, depending on the availability of the maximum length of stones. Generally flagstone slabs are available in thicknesses of about 40 mm and widths of about 300 mm. These flagstones are inserted in the inter-space between the joists. They rest on the lower flanges of the joist. The portion above the flagstone is filled with light weight concrete made up of crushed brick or slag aggregate. If lime concrete is used, the steel joists should be encased in cement concrete so that it is not in direct contact with lime which would cause disintegration of the joists. Instead of light weight concrete, any other suitable filling can be filled in the inter-space between the joists. However the top of these joists should be kept about 10 mm lower than the concrete surface. A layer of suitable wearing coat may be laid as topping.

Instead of stone, precast coment concrete slabs of about 450 mm length and 50 mm thickness can be used in a similar manner.

For cheap type of construction, it may not be necessary to obtain flat surface at the bottom and hence the precast slabs or flagstone slabs can be laid on the top of these joists in such a way that all the heading joints come centrally over the joists. Where joists are needed, the underside of these slabs is pointed with cement. The top may be suitably finished by laying a wearing coat of concrete or other suitable materials.

For larger spans, it is necessary to provide a few girders on which these steel joists can rest.

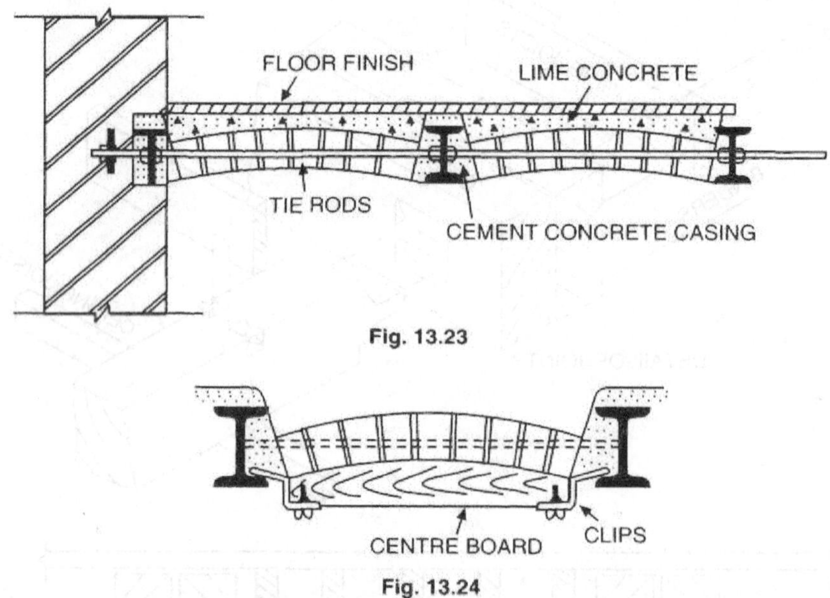

Fig. 13.23

Fig. 13.24

Fig. 13.23 and 13.24: Brick jack arch details.

13.21 JACK ARCH FLOOR OF BRICK OR CONCRETE

Concrete or bricks cannot be used as a flooring system unless they are suitably strengthened. Either reinforcement can be provided or arch action developed to bear the loads easily. Jack arch flooring consists of a series of arches built from small rolled steel girders. These arches have spans varying from 1.25 to 2 m and the rise is $\frac{1}{12}$th to $\frac{1}{16}$th of the span. Since these exert thrust at their ends, suitable arrangements for tying the individual spans together should be made. This is accomplished by running tie rods which are 20 to 25 mm in diameter and are spaced at 2 to 2.75 m apart. The ends of these rods are anchored into each wall with a stout washer. The end arch is supported on the wall by encasing rolled steel joists into the wall or by fixing an angle instead of rolled steel joist. The spandrils are filled with concrete. If a wooden flooring is to be used, two nailing strips of wood are fixed over each arch. The flooring is nailed in turn to them.

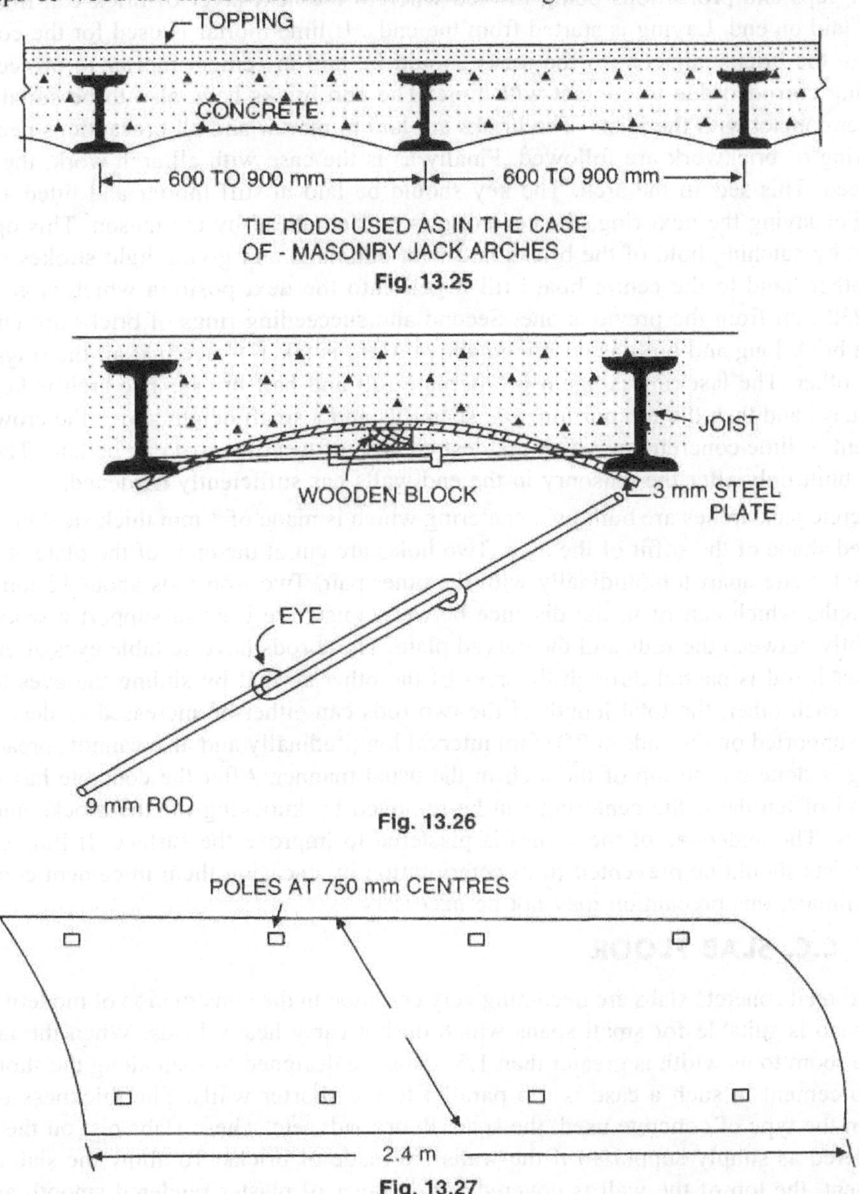

Fig. 13.25

Fig. 13.26

Fig. 13.27

Figs. 13.25 to 13.27: R.C.C. Jack arch details.

Brick jack arches are constructed by bringing up the arches against a concrete backing laid on the lower flanges of the joists. The centring for the arch consists of a segmental piece of wood 40 mm thick with a chord length equal to the span of an arch and conforming to the soffit. The ends of the segment are cut off slightly. The bottom ends of the centre board rest on the lower flanges of the joist and the curved surface forms the top. Sometimes a bent iron strap is attached to its ends to form a hook by which the centring board is suspended from the rolled steel joist as shown in figures 13.23 and 13.24. The brick arch is built over the centre board. In constructing the arch, the mason will sit on a plank laid across two joists and start laying one ring of bricks. In the first ring, the bricks have to be laid of one brick and half brick lengths alternately. This enables a series of gaps and projections being formed wherein the next layer of brick can fit in. All the bricks are laid on end. Laying is started from the ends. If lime mortar is used for the construction of the arch, the bricks adjacent to the joists should be laid in cement mortar to protect the joist from getting corroded due to contact with lime. The end bricks have also to be suitably cut for their proper contact with the joists. The bricks are laid in mortar and all precautions necessary for proper laying of brickwork are followed. Finally as is the case with all arch work, the key brick is introduced. This sets in the arch. The key should be laid in stiff mortar and fitted as much as possible. For laying the next ring, the centering is pushed ahead by the mason. This operation is carried out by catching hold of the bricks laid with one hand and giving light strokes of hammer with the other hand to the centre board till it gets into the next position which is at a distance of about 230 mm from the previous one. Second and succeeding rings of bricks are laid with all bricks one brick long and built as described above. Care is taken to see that all the rings interlock with each other. The last ring is again of alternate full and half bricks. The arch is kept wet for about ten days and then the top portion, *i.e.*, spandrils and a small height above the crown is filled with cement or lime concrete on which the desired type of wearing coat can be laid. These arches should be built only after the masonry in the end walls has sufficiently hardened.

Concrete jack arches are built on a centering which is made of 3 mm thick steel plates bent to the required shape of the soffit of the arch. Two holes are cut at the ends of the plate at a distance of about 3/4 metre apart longitudinally with the other pair. Two iron rods about 12 mm diameter and of lengths which can fit in the distance between joists are used to support a wooden block driven tightly between the rods and the curved plate. These rods have suitable eyes at one of their ends and each rod is passed through the eyes of the other so that by sliding the eyes towards or away from each other, the total length of the two rods can either be increased or decreased. The plate gets supported on the rods at 750 mm interval longitudinally and also cannot spread laterally. Concreting is done on the top of the arch in the usual manner. After the concrete has cured, say for a period of ten days, the centering can be removed by knocking off the blocks and opening out the rods. The underside of the arches is plastered to improve the surface. If lime concrete is used, the joists should be prevented from deterioration by encasing them in cement concrete. For very dry climate, this precaution may not be necessary.

13.22 R.C.C. SLAB FLOOR

Reinforced concrete slabs are becoming very common in the construction of modern buildings. A simple slab is suitable for small spans which do not carry heavy loads. When the ratio of the length of a room to its width is greater than 1.5, slabs are designed to span along the shorter width. The reinforcement in such a case is run parallel to the shorter walls. The thickness of the slab depends on the type of concrete used, the span, floor loads, etc. These slabs rest on the walls and are considered as simply supported if the walls are made of bricks. To allow the slab a freedom of movement, the top of the wall is covered with a layer of plaster rendered smooth and a thick coat of bitumen is applied to this plaster. The slab rests at its end on the wall. If the building is of reinforced concrete framed construction, it is necessary to build the slab monolithically with the supporting beams. Mixing, laying, finishing and curing of the concrete is described in chapter 18.

When rooms are nearly square *i.e.*, the ratio of the adjacent sides of the room is less than 1.5, the slabs are designed spanning in both directions and are called two-way slabs. The main reinforcement of the slabs runs parallel to both the sides of the room. At corners suitable mesh reinforcement is provided at top and bottom to allow for stresses which are created on account of the partial fixity of the corners.

Reinforced concrete slabs are easy to cast and provide a very smooth surface at the bottom which provides a good appearance and does not create difficulties in lighting arrangements. However, these are restricted to spans not greater than about 4 m for ordinary loads of residential buildings.

***Steel girder and R.C.C. slab floor*:** When the spans are larger, RCC slabs become uneconomical unless additional supports are introduced below the slabs. As a simple type of construction, steel girders can be introduced below the slabs to reduce the spans. These slabs may be built as individual units on the steel girder or can be laid as continuous slabs on the girders. The reinforcement is provided in the first type, on the underside of the slab whereas in the second case, the reinforcement has to be provided at the top of the slab over the beams, some reinforcement being taken continuously at the bottom. The amount of steel provided at the top of the slab is usually equal to that used at the bottom in the mid span of the slab. In some buildings it may be necessary to protect the steel beam from fire by encasing them in concrete or they may be encased in a covering of metal lath and cement plaster.

13.23 R.C.C. BEAM AND SLAB FLOOR

For larger spans and for heavier loads, R.C.C. beam and slab construction is becoming a very important type of floor construction which is commonly used for most of the important buildings. The slab is designed as a continuous slab monolithic over the beams, the reinforcement being provided as in the case of RCC continuous slabs laid on steel girders. The slab acts as a flange of the beam and as such is cast monolithically with the beams. Hence the size of the beam is reduced. The main reinforcement of the slab runs parallel to the short span but small amount of reinforcement is also provided parallel to the longer span for proper distribution of loads, etc. Two-way slabs can also be used for openings which are nearly square. The slab and beam type of construction is very economical. Sometimes the bottom projecting beams are covered by laying a false ceiling underneath it.

13.24 FLAT SLAB FLOOR

Whenever the use of beam is objectionable, flat slab construction is preferred. Flat slab floor is a slab directly supported on columns without any intermediate beams. The advantages of flat slab construction are:

(*a*) No projection of beam is visible and hence no additional ceiling is needed.
(*b*) More clear head room is available
(*c*) Better lighting facilities are available
(*d*) For heavier loads, thinner section of the slab is needed and
(*e*) The construction is easy.

The disadvantages are:

(*a*) The panels have to be nearly square
(*b*) At least three continuous panels are needed for economical construction; and
(*c*) It is not economical for lighter loads.

There are various types of flat slab construction. The two types generally used are:

(*i*) ***Two-way reinforced flat slab.*** In this type the reinforcement is parallel to the panel's length and breadth. The steel is at the bottom in the middle portion of the panels while it is at the top near the column edges and top of the column.

(ii) ***Four-way reinforced flat slab.*** In this system, reinforcement bands run from column to column diagonally in addition to the reinforcement provided parallel to the panel's sides.

Flat slabs are used in factories, ware-houses, commercial buildings etc. Sometimes the slab is thickened at the column head (which is always flared). This type of treatment is needed when very heavy loads are to be catered for. Such construction is called flat slabs with drops.

13.25 BEAMS AND PATENT REINFORCEMENT SLAB FLOOR

Generally round or square mild steel bars are used in the construction of RCC slabs. But for small slabs and light loads, it may be of great advantage to use patent reinforcements. The usual type of reinforcements are wire meshes, expanded metal lath or ribbed metal lath. Wire mesh is used for ordinary construction whereas the ribbed metal lath is used on light slabs poured without any formwork. The ribs span the distance between the supports and are rigid enough to take the loads. The mesh is so fine that the concrete will not run down but will form a substantial grip with the lath. The supporting girders may be of rolled steel joists of solid section or steel open web joist built up of two steel channels or T-section and inclined bars acting as a web. The rough lower side of the slab may be concealed and protected by plastering the ceiling. These joists are spaced at about 600 mm centre to centre. The slabs are generally 50 mm to 75 mm thick.

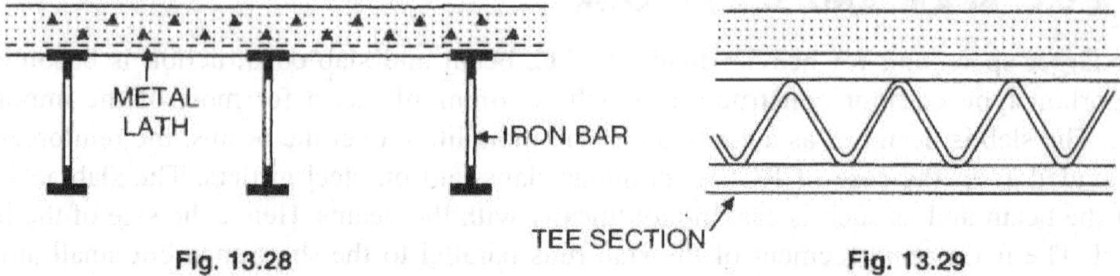

Fig. 13.28 Fig. 13.29

Figs. 13.28 and 13.29: Light metal girders to support patent reinforced slab.

13.26 HOLLOW TILED RIBBED FLOOR

Ribbed slab floors consist of a number of small beams spaced closely and cast monolithically with the slab. In the ordinary slab, the area of concrete below the neutral axis of the section does not contribute to the increase in strength of the section. Its object is only to hold the reinforcement in position. Hence if concrete below the neutral axis could be eliminated and only left in position at places where the steel reinforcing bars are placed, an economical construction would be feasible. This would make a slab with small ribs at the bottom. The steel is carried within the ribs and is about the same as needed for a solid slab except for small reduction due to the decreased weight of the slab which is to be carried now. Larger diameter bars are used as this space within the ribs is limited for accommodating the steel. Generally ribs are not less than 100 mm width. The slab between the ribs is 50 to 100 mm in thickness and is reinforced with wire mesh or with bars of small diameter running perpendicular to the ribs. The reinforcement is not bent over the ribs. For nearly square spans, the slab may be ribbed in both directions.

The cost of such ribbed slabs is very high due to intricate formwork needed for their construction. Hence the sides of the joists and the bottom of the slab are cast by placing hollow clay tiles, hollow gypsum tiles or sheet steel cores. Wooden formwork is built for the bottom of the ribs and is slightly wider than the width of the ribs so that the tiles can rest on it. The portion between the tiles is concreted and a slab is laid on it. The clay or gypsum tiles provide a smooth surface at the bottom which can be easily plastered.

The sheet steel cores can be made of thicker metal so that they can be removed later on. Non-removable types are made of very thin sheet steel and these are left in place. In this case, the metal lath for the plaster is fixed before the cores are set and is wired to the bars in the ribs. As an alternative, it may be fixed in the steel cores. Formwork for such type of construction is similar to the one needed for clay tiles.

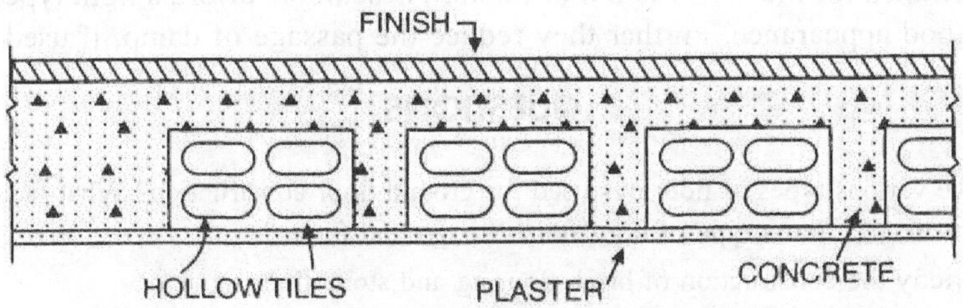

Fig. 13.30: Hollow tiled ribbed floor.

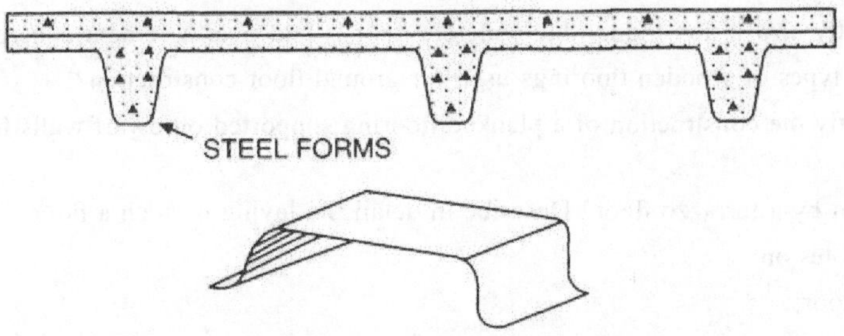

Figs. 13.31 and 13.32: Ribbed floor with steel forms.

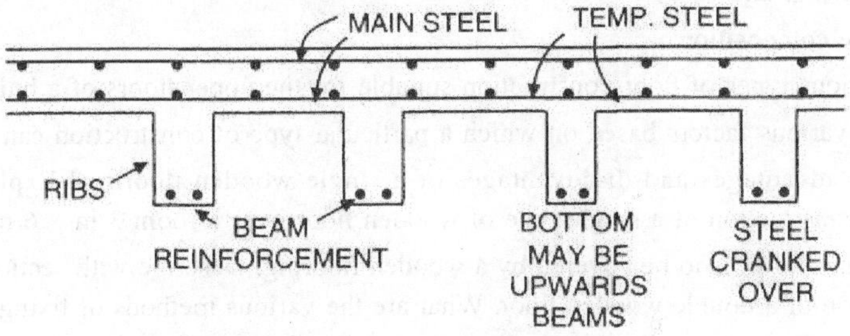

Fig. 13.33: Section through a R.C.C. ribbed floor.

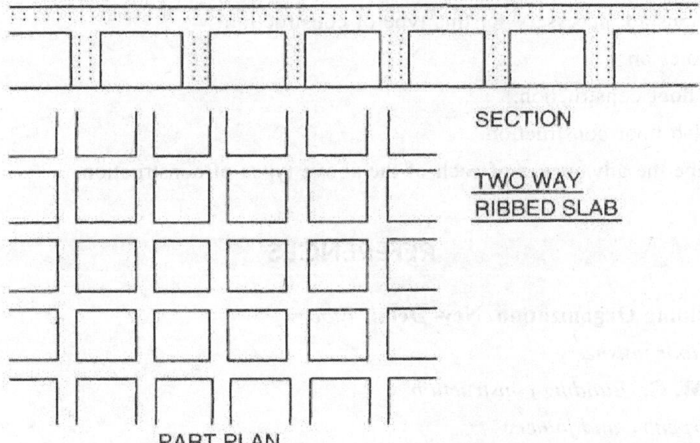

Fig. 13.34 and 13.35: Part plan and section through a two-way ribbed floor.

13.27 HOLLOW TILE JACK ARCH FLOOR

Instead of using bricks as individual units for building up a jack arch, hollow tiles of structural clay can be substituted for the construction of floors. These floors afford a light type of construction and have also good appearance. Further they reduce the passage of damp, if used in a roof.

QUESTIONS

1. What are the various types of floorings used for ground floor construction? What factors will you take into account in selecting a type of floor finish for ground floors?
2. Describe briefly the construction of brick flooring and stone flooring.
3. What is meant by conglomerate cement flooring? What are the advantages and disadvantages of monolithic type of floor construction?
4. Describe briefly how a non-monolithic concrete floor is laid over a cement concrete base.
5. What are the types of wooden floorings used for ground floor construction?

   Describe briefly the construction of a planked flooring supported on dwarf walls for ground floor of a building.
6. What is meant by a terrazzo floor? Describe in detail the laying of such a floor.
7. Write short notes on:

   (*i*) Mosaic floor;

   (*ii*) Bituminous flooring;

   (*iii*) Glass flooring; and

   (*iv*) Magnesite composition.
8. Name the various types of floor construction suitable for the upper floors of a building.

   What are the various factors based on which a particular type of construction can be adopted?
9. What are the advantages and disadvantages of a single wooden flooring? Explain with the aid of sketches the construction of a single type of wooden floor over a room 3 m × 6 m in size.
10. A room 5.5 m × 7.5 m is to be covered by a wooden flooring. Describe with sectional details and plan the construction of a double wooden floor. What are the various methods of fixing ceilings to wooden floors?
11. What is meant by a jack arch floor? Describe briefly the construction of a jack arch floor using bricks. What precautions are necessary for this type of construction?
12. Write short notes on:

    (*i*) Flat-slab floor construction;

    (*ii*) Ribbed slab floor construction.

    Briefly describe the advantages of each of the above types of construction.

REFERENCES

1. **National Building Organization, New Delhi**: *Floors.*
2. **Lea F. M.**: *Floor finishes.*
3. **Huntington W. C.**: *Building construction.*
4. **Crokhill**: *Carpentry and joinery.*
5. **Voss W. C.**: *Fire-proof construction.*
6. **British Standard 1187**: *Wood-blocks for floors.*

7. **British Standard 1286**: *Clay tiles for flooring.*
8. **British Standard 1344**: *Asphalt tiles for paving and flooring.*
9. **British Standard 1076, 1410 and 1451**: *Mastic asphalt flooring.*
10. **Mackey W. B.**: *Building construction Vols. II and III.*
11. **Turneaure F. E. and Maurer E. R.**: *Principles of reinforced concrete construction.*
12. **Sutherland, Hale etc.**, *Introduction to reinforced concrete design.*

Stairs

Stairs are provided in a building to afford a means of communication between the various floors. These are steps arranged in series and placed in an enclosure called staircase. Since they have to perform a very important function, they should be designed properly to provide maximum comfort, ease and safety. They are generally placed in the centre or to one side of a building. The location will depend upon the position of rooms and the type of approach needed to them. They are made into different units to facilitate construction and to economise space. Stairs should be properly ventilated and lighted. A building may have a number of staircases between the various floors depending on the number of people using the building. In public buildings, the staircases should be near the main entrance whereas in private buildings some privacy is essential.

14.1 MATERIALS

The selection of materials for the construction of stairs depends upon the availability of materials, funds available, desired life of the building, aesthetical importance and fire resisting quality expected. Stairs may be made either of timber, bricks, stones, mild steel, wrought iron or plain and reinforced concrete. Combination of two or more of these materials may be used in the same stair. Sometimes, a finishing of marble, mosaic or checkered, plaster on metal lath, etc. is also provided.

14.2 DEFINITIONS

(*i*) ***Tread*** : Flat or horizontal member which forms the upper surface of a step on which the foot is placed for ascending and descending.

(*ii*) ***Riser*** : The front vertical portion of a step to which the tread is connected.

(*iii*) ***Rise*** : Rise of a step is the vertical distance between the tops of two consecutive treads, and rise of a flight is the total height from floor to floor, floor to landing or landing to landing, as the case may be.

(*iv*) ***Going or run*** : Horizontal distance between the faces of any two consecutive risers.

(*v*) ***Flight*** : A continuous series of steps without an intermediate platform, extending from floor to floor, floor to landing, or landing to landing.

(*vi*) ***Landing*** : Landing is a platform between two flights provided to serve as a rest and, when required to make effective provision for turning a stair. This term is also applied to the portion of the floor adjacent to the top of a stair.

(*vii*) ***Nosing***	:	The projecting edge of a tread, usually moulded.
(*viii*) ***Line of noising***	:	An imaginary line connecting the nosing points parallel to the slope of the stair.
(*ix*) ***Hand rail***	:	A moulded member running parallel to the nosing line or landing. Rail of wood or metal on the side of stair fixed at about waist height at an incline or at level on a landing.
(*x*) ***Newel***	:	A post placed at the junction between a flight of stairs with a landing or floor.
(*xi*) ***Baluster***	:	Vertical members supporting the handrail and protecting the open side or sides of a stair.
(*xii*) ***Winders***	:	Radiating or angular steps which give a change of direction from one flight of steps to another.
(*xiii*) ***Strings or stringers***	:	Sloping members which support the steps in a stair.
(*xiv*) ***Walking line***	:	This is an approximate line on which people walk on a stair and is about 450 mm from the centre of the hand-rail in plan.
(*xv*) ***Pitch***	:	The angle between the pitch line and the horizontal.
(*xvi*) ***Pitch line***	:	The line joining the intersection of the face of each riser and the top of each tread.
(*xvii*) ***Stairs***	:	A set of steps and landing from one floor to another.
(*xviii*) ***Staircase***	:	Stairs together with the part of the building accommodating them.

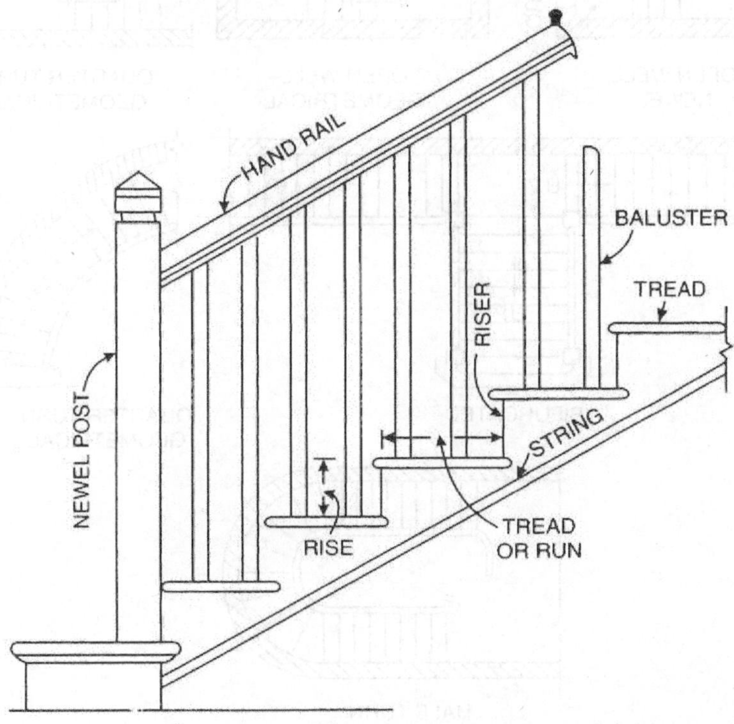

Fig. 14.1: Terms used for stairs.

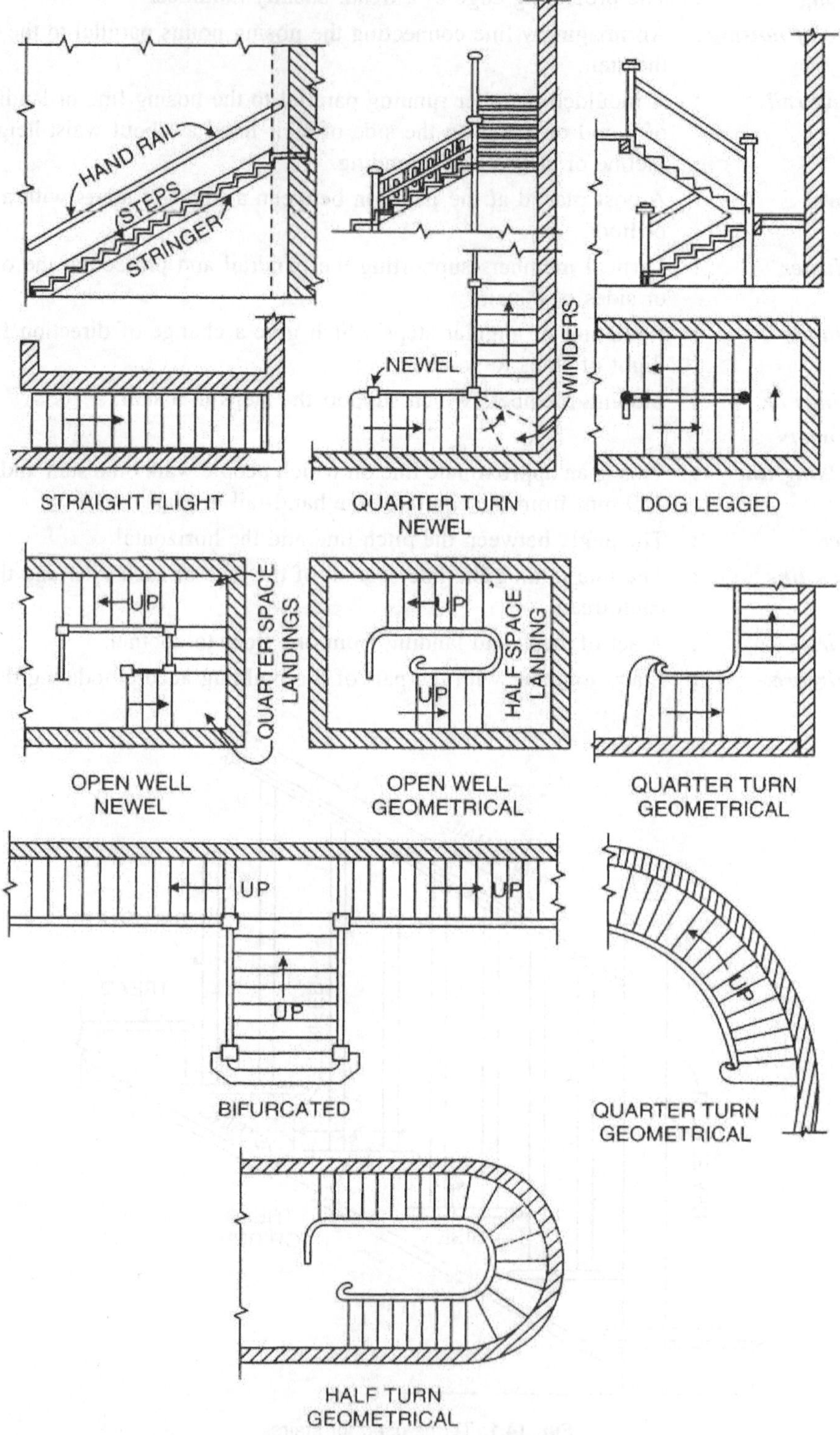

Figs. 14.2 to 14.13: Types of stairs.

14.3 TYPES OF STAIRS

The types of stairs are:

(*i*) ***Straight flight stairs.*** The stairs continue throughout their entire length in one direction only. This may consist of a single flight or a flight with one or two landings in-between.

(*ii*) ***Quarter turn stairs.*** When the direction of the flight is changed at right angles either to the left or to the right, quarter turn stairs are used. These are of two types. In the first type the change in direction is effected by introducing a quarter space landing. In the other type winders are used.

(*iii*) ***Half-turn stairs.*** In this, the direction of flight is reversed, *i.e.*, turned by 180° by the introduction of landings and or winders. A half space landing may be used.

A quarter space landing and winders may be used as shown in figure 14.5.

(*iv*) ***Three-quarter turn stairs.*** This has its direction changed three times with its upper flight crossing the bottom one.

(*v*) ***Bifurcating stairs.*** In this, the wide bottom flight is divided into narrow flights at the landings.

(*vi*) ***Geometrical or continuous stairs.*** The strings and handrails are continuous and are set out in accordance with geometrical principles.

They may be of the following types:

(*a*) ***Winding stairs.*** See figure 14.12.

(*b*) ***Half turn stairs*** with landings or newels.

(*c*) ***Spiral stairs.*** In this the steps project from a central post.

14.4 LAY-OUT OF STAIR-CASES

The lay-out of stair-cases is influenced by the space available for the construction of stairs and the height between the two floors.

A few examples are worked out below to illustrate the lay-out of staircases.

Problem 14.1. *The stairs of a residential building is to be located in a staircase measuring 1500 mm × 4500 mm. If the vertical distance between the floors is 3000 mm, draw a lay-out plan of the proposed stairs.*

Solution. Since the width of the stair-case is given as 1500 mm, it is not possible to take any turn in this limited space. Moreover the length of the stair is also 4500 mm, so the provision of a landing is also not feasible to climb a vertical height of 3000 mm. Thus straight flight stairs will be provided.

Assuming a rise of 200 mm,

number of risers required $(R) = \dfrac{300}{20} = 15$

No. of trends to be provided $(T) = R - 1$

$= 15 - 1 = 14$

It is proposed to keep the width of tread

$= 250$ mm

Space occupied by 14 treads $= 250 \times 14$

$= 3500$ mm

Space left for passage $= 4500 - 3500$

$= 1000$ mm

The layout is shown in Fig. 14.14.

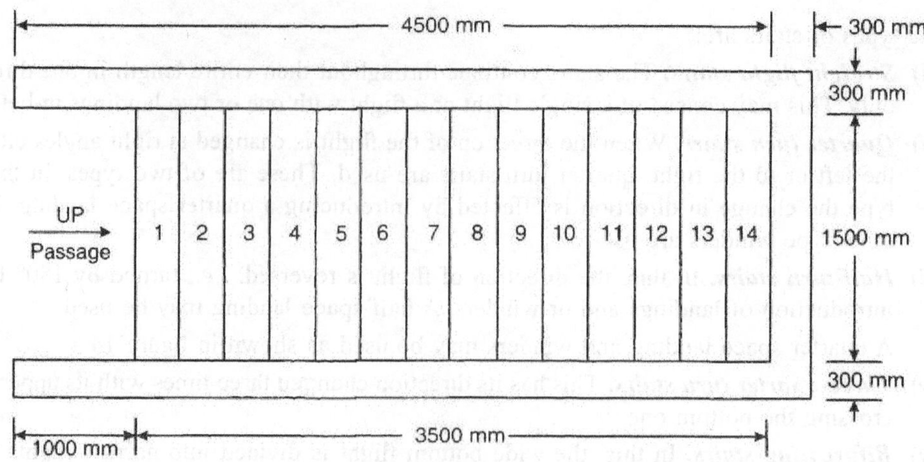

Fig. 14.14: Layout of straight flight staircase for Problem 14.1.

Problem 14.2. *A suitable stair is to be provided in an office building. The stair-case has the dimensions 2800 mm × 6000 mm. The vertical distance between the floors is 3600 mm. Two doors measuring 1200 mm each are provided in the longitudinal walls, facing each other and located 500 mm from the transverse wall.*

Solution. Assume, the width of the stair = 1300 mm.

Since the width of the stair case is 2800 mm, a half turn stair can be provided. The direction of the flight is reversed by the introduction of a half space landing and thus two flights shall be provided. The width of the landing will be kept equal to the width of the stair.

Space available in the centre of the stair-case

$$= 2800 - 2 \times 1300 = 200 \text{ mm}$$

Since this space is limited, a dog-legged stair will be provided.

$$\text{Height of each flight} = \frac{3.6}{2} = 1800 \text{ mm.}$$

$$\text{Assuming rise of 160 mm, number of risers required} = \frac{180}{16} = 11.25 \text{ say } 11$$

$$\text{Exact rise of each step} = \frac{180}{11} = 163.5 \text{ mm.}$$

Number of treads to be provided in each flight

$$T = R - 1 = 11 - 1 = 10.$$

Keeping the width of tread as 270 mm.

$$\text{Space taken by 10 treads} = 270 \times 10 = 2700 \text{ mm}$$

$$= 2700 \text{ mm}$$

$$\text{Space left for passage} = 6000 - 2700 - 1300 = 2000 \text{ mm.}$$

The arrangement of the steps and doors is shown in Fig. 14.15.

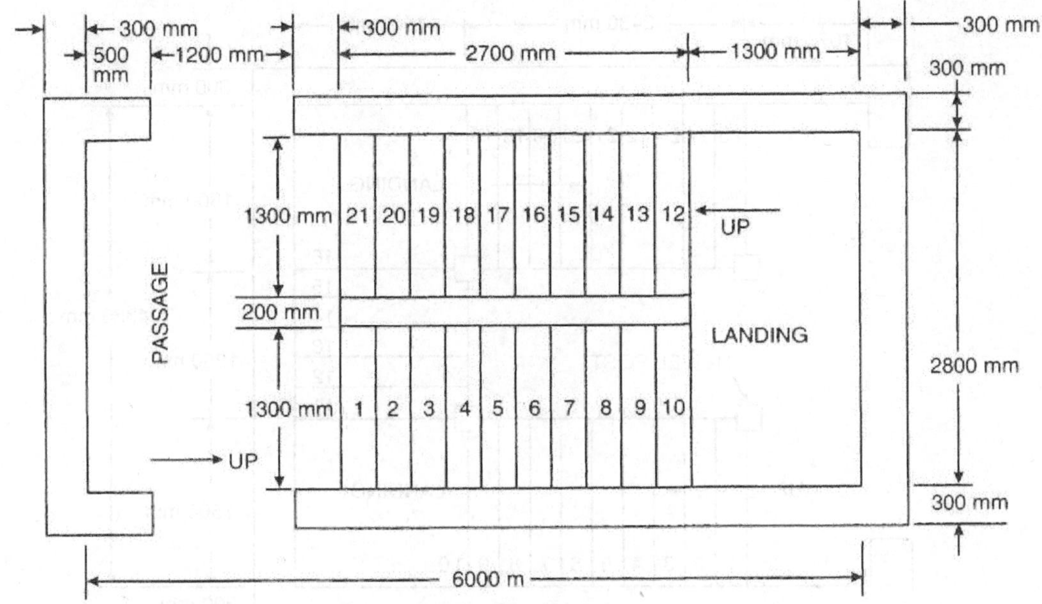

Fig. 14.15: Layout of a dog-legged half-turn staircase for Problem 14.2.

Problem 14.3. *A stair is to be located for an office in a stair-case hall measuring 4350 mm × 5000 mm. The vertical distance between the floor is 4000 mm. Draw a dimensional plan of the proposed stair.*

Solution. Since it is a public building, the width of the stair should be adequate for the number of people expected to use them.

Assume stair width = 1500 mm.

Space available in the centre of the stair-case

$$= 4350 - 2 \times 1500 = 1350 \text{ mm}$$

Hence an open well type of stair shall be provided.

Taking a rise of 160 mm, number of risers required

$$= \frac{4000}{160} = 25$$

If two flights are provided, number of risers shall be 12 and 13. However, being a public building, the number of steps in one flight should be restricted to 10. Hence provide 5 treads in the landing portion and make the stair of quarter space landing type. This will need to provide 9 treads each in the other two flights.

∴ Total number of rises needed to provide 23 treads in three flights = 23 + 3 = 26

$$\text{Hence, height of each riser} = \frac{4000}{26} = 154 \text{ mm.}$$

Adopting trend width of 270 mm, space occupied by 9 treads

$$= 9 \times 270 = 2430 \text{ mm}$$

$$= 2430 \text{ mm.}$$

The arrangement of steps is shown in Fig. 14.16.

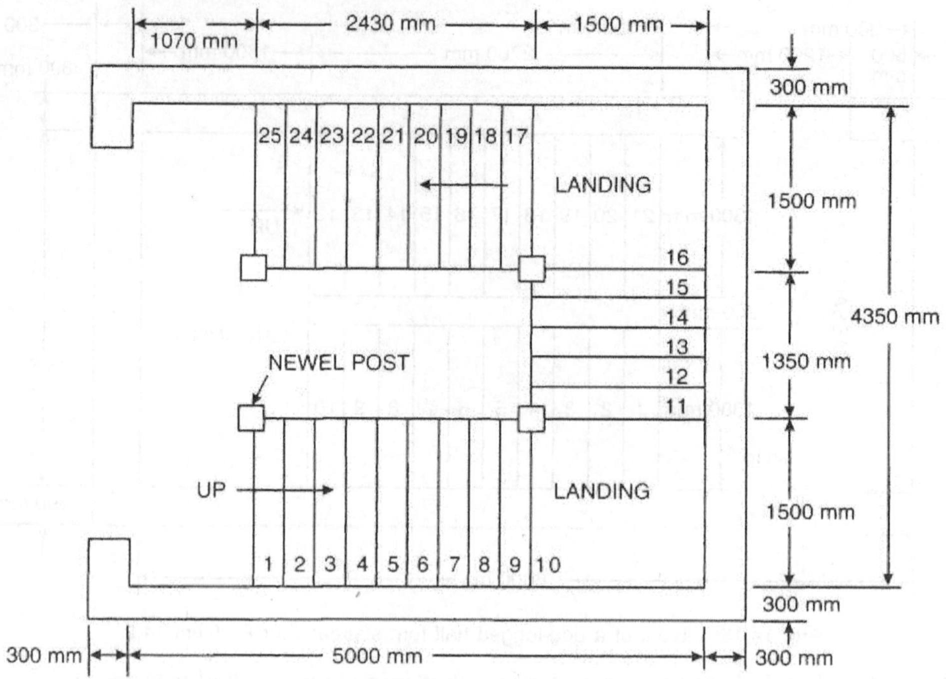

Fig. 14.16: Layout of an open well type staircase for Problem 14.3.

14.5 ESSENTIAL REQUIREMENTS FOR A GOOD STAIRCASE

For easy, quick and safe ascent or descent along the stairs, the following essential requirements should be fulfilled:

(1) The location should be such as to provide easy access to the occupants. The approaches should be convenient and spacious. The stair should be well lighted from the exterior, especially at turns. It should be well ventilated.

(2) Width of the stair should be adequate for the users without much crowd. The width of stair depends upon the type of building. About 1000 mm wide stair for residential building and 1350 mm for public buildings are considered the average values.

(3) The pitch or slope of the stair should be such as to prevent undue exertion and at the same time there should not be any wastage of space. The pitch of the stairs should be made flatter by introducing landings to make the ascent less tiresome. In general the slope of the stair should never exceed 40° and should not be flatter than 25°.

(4) ***Length of flight***: The number of steps in a flight should be restricted to a maximum of 12, from comfort point of view and a minimum of 3.

(5) Sufficient head room should be provided. The head room or the clear distance between the tread and the soffit of the flight immediately above it should not be less than 2200 mm. However a headroom of 2300 mm is desirable.

(6) ***Landing***: The width of landing should not be less than the width of stair.

(7) Construction of stair should be carried out with sound and fire-proof materials.

(8) All the risers and treads should be of uniform dimensions.

(9) Winders should be avoided as far as possible. If these are necessary, they should be of sufficient width.

(10) ***Step proportions:*** It is very essential that the design of steps should be very carefully worked out. The steps should neither be too widen or too short. Further the rise should not be excessive so as to cause inconvenience and fatigue to the user.

The following thumb rules may be followed:

(*i*) Rise + tread not < 400 and not > 450

(*ii*) 2 × Rise + tread = 600. It should be not < 580 and not > 630

(*iii*) Rise × tread = 4000 to 4100 approximately. It should not be > 5000

(All dimensions are in mm.)

(*iv*) Adopt the standard size of tread and riser as 300 mm and 140 mm and for each 25 mm deduction from tread, add 12 to 13 mm to riser.

The maximum riser and minimum tread for public and residential buildings are given below:

Type of building	*Maximum riser*	*Minimum tread*
Public building	150 mm	300 mm
Domestic building	190 mm	250 mm

WOODEN STAIRS

The wooden stairs are of the following types:

14.6 STRAIGHT FLIGHT STAIRS

This is used in places where restricted width is available for the location of the stair and the height between the floor is less.

(a) Steps. The thickness of the treads should not be less than 40 mm and that of risers 25 mm. The treads and the risers are connected by tongue and groove joint and screwed or nailed. The nosing should not project more than the thickness of the tread. It may be suitably finished at the ends by giving a scotia or cavetto moulding. A half round moulding may also be used.

(b) Stringers. The treads and risers are supported by two 350 mm × 50 mm stringers which are fixed to the walls. A 50 to 75 mm wide margin is provided and the upper edge of the stringer is suitably finished. For wider stairs, an intermediate support in the form of 100 mm × 75 mm wooden joist (bearer) is used. This is secured at the foot and the top of the stair. Small, shaped wooden pieces are fixed to the bearer. If more than one bearer are used, the stringers may be of lesser width. The ends of the steps are housed into the wall stringers to a depth of about 12.5 mm. The grooves are tapered and should permit hard wood wedges being put into them. These wedges enable the steps to be fitted tightly. Small triangular blocks may also be used. These are glued to the underside of the treads and risers.

(c) Hand rail. This is fixed to the wall with the aid of plugs. Special moulding may be used for giving a better appearance to the handrail. This is laid parallel to the slope of the stair.

14.7 HALF-TURN STAIRS

A brief description of one of the types called dog-legged stair is given below. This is so called because of its appearance in sectional elevation. It is conveniently used when the length of the stair-case is restricted and only two flights can be accommodated.

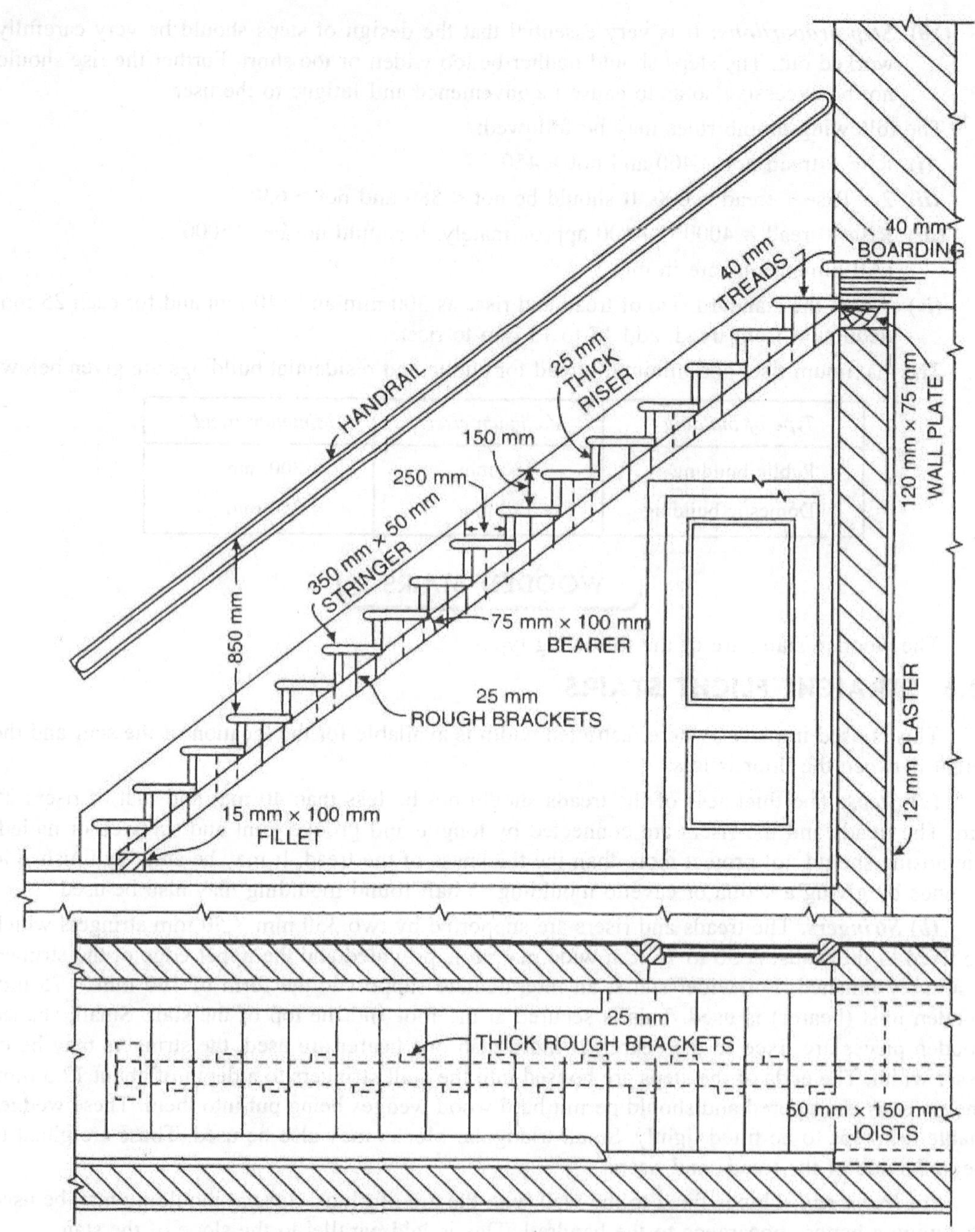

Fig. 14.17 and 14.18: Section and plan of a straight-flight wooden stair.

(a) *Steps.* They can be similar in construction to that of the straight flight-stair. However inclined risers may be used (*see* Figs. 14.19 and 14.20).

(b) *Stringers.* The outer ends of the steps are housed into the stringers which are slightly thicker than those fixed to the wall, *i.e.*, about 50 mm. The outer stringer may be of either one piece or of two pieces which are tongued and grooved together. The outer stringers are strengthened by newels at the foot and the head of each flight.

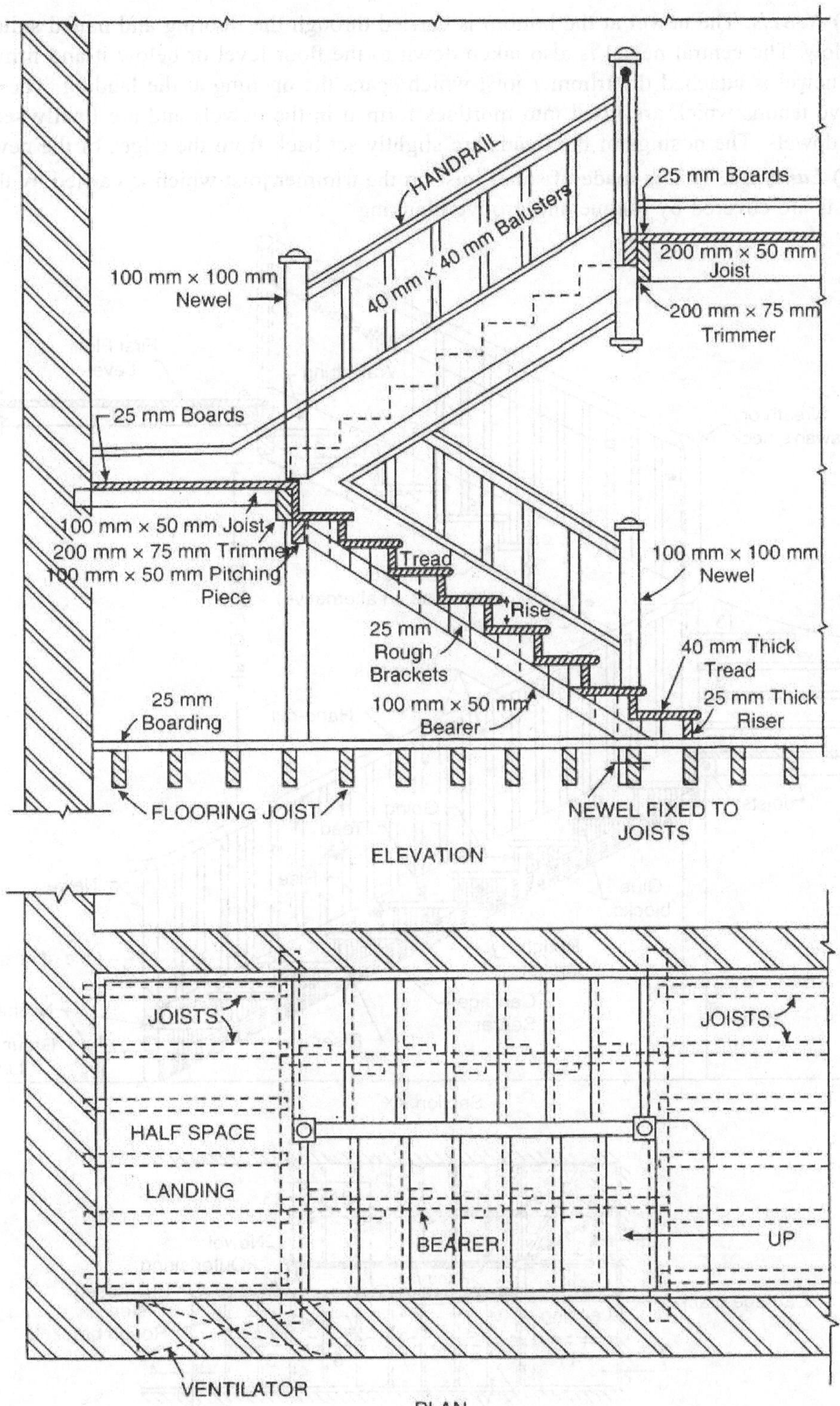

Figs. 14.19 and 14.20: Elevation and plan of a dog-legged wooden stairs.

(*c*) ***Newels.*** The newel at the bottom is carried through the flooring and nailed suitably to a joist below. The central newel is also taken down to the floor level or below it and firmly fixed. To this newel is attached the trimmer joist which spans the opening at the landing. The stringers may have tenons which are fitted into mortices formed in the newels and are finally secured by pins or dowels. The nosings of the treads are slightly set back from the edges of the newels.

(*d*) ***Landings.*** This is made of small joist on the trimmer joist which is carried by the walls. The joists are covered by tongue and groove planking.

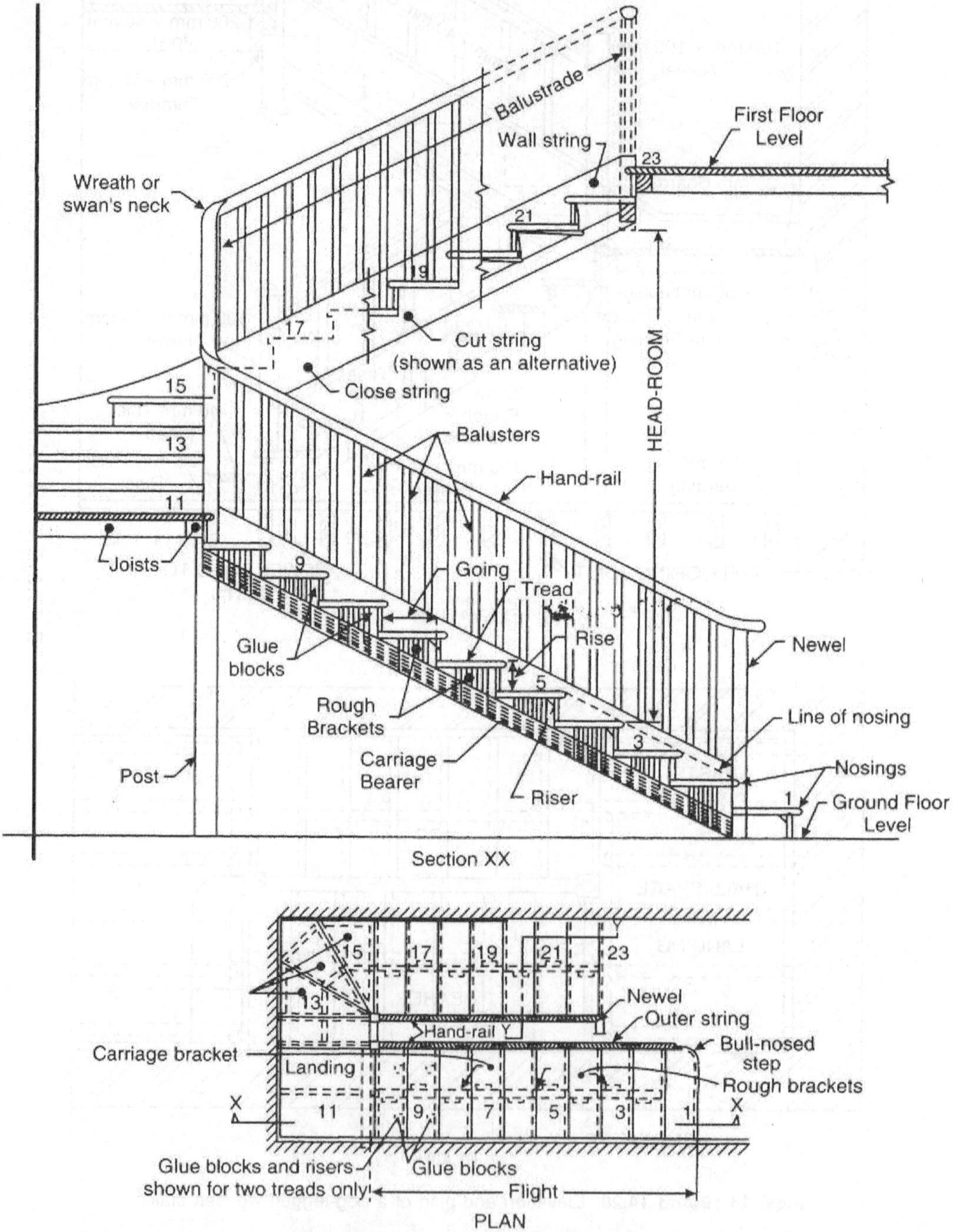

Balustrade
Wall string
First Floor Level
23
Wreath or swan's neck
21
19
17
Cut string (shown as an alternative)
Close string
15
13
11
Balusters
Hand-rail
HEAD-ROOM
Joists
9
Going
Tread
Rise
Glue blocks
Newel
5
Rough Brackets
Line of nosing
3
Carriage Bearer
Nosings
Post
Riser
1
Ground Floor Level
Section XX
15
17
19
21
23
13
Newel
Outer string
Hand-rail Y
Bull-nosed step
Carriage bracket
Landing
Rough brackets
X
X
11
9
7
5
3
1
Glue blocks and risers shown for two treads only
Glue blocks
Flight
PLAN

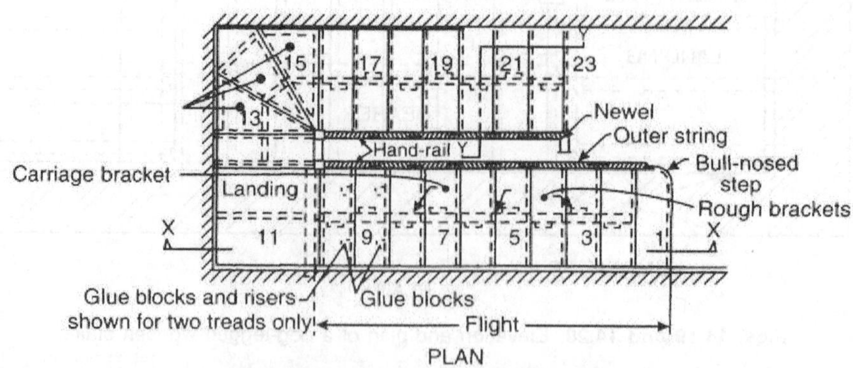

Figs. 14.21 and 14.22: Open well stair.

(*e*) ***Handrail.*** The handrail for the upper flight is housed, tenoned and dowelled to the two newels. The upper end of the lower handrail gets intersected by the upper outer stringer. As an additional measure, a handrail may be fixed to the wall for the portion of lower flight and is fixed as explained in straight flight stair. The handrail provided at the top landing is fixed to the newel at one end and to a half newel at the other end which in turn is plugged to the wall.

(*f*) ***Balusters.*** If square balusters are used, they should not be less than 50 mm wide and are spaced 75 to 100 mm apart. They are either housed or tenoned into the handrails and the stringers. A continuous groove is sometimes formed on the underside of the handrail and the balusters may be fitted into this, the grooves being finally filled in. For cheap work, the balusters may be nailed to the stringers and rails. Bronze or other type of metal balusters may be used for this type of stairs.

14.8 OPEN WELL STAIRS

This has an opening or well between the outer stringers. This gives a better appearance to the stair but needs greater width. The construction is similar to a dog legged stair except that the stringers are connected to the newels and rest on the joist at the end. The various components and the method of joining them is shown in Figs. 14.21 and 14.22.

14.9 SOME IMPORTANT COMPONENTS OF WOODEN STAIRS

Some important components of wooden stairs are described below in detail:

(1) ***Stringers.*** These are of the following types:

(*a*) ***Housed stringers.*** They have the slopes parallel to the slope of the stair. Grooves are cut in them to house the tread and the riser. The grooves are wide enough to enable the steps to be fitted in and also the wedges to be fixed. This type of stringer has been described on page 327 and shown in Figs. 14.23 and 14.24.

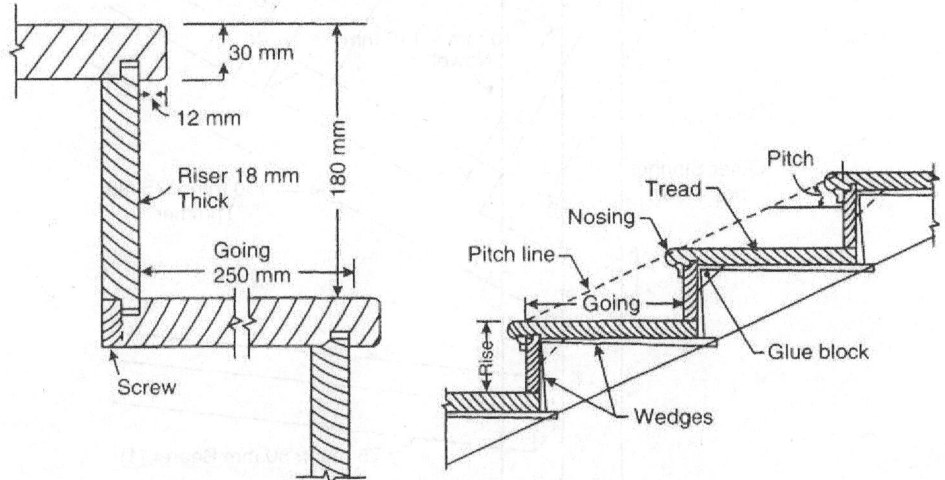

Figs. 14.23 and 14.24: Typical details for fixing of wooden steps.

(*b*) ***Cut stringers:*** They have one side-slope parallel to the stair slope whereas the other side has cut notches so that the steps can be fixed to them.

(*c*) ***Wreathed stringers.*** They are needed in geometrical stairs. They are in plan similar to the shape of the stair.

(*d*) ***Rough stringers.*** When the stringers need intermediate supports, rough stringers are used. They are also called bearers or carriers. (*see* Figs. 14.25 and 14.26). They also support the boarding or plaster which may have to be laid on the underside of the stair.

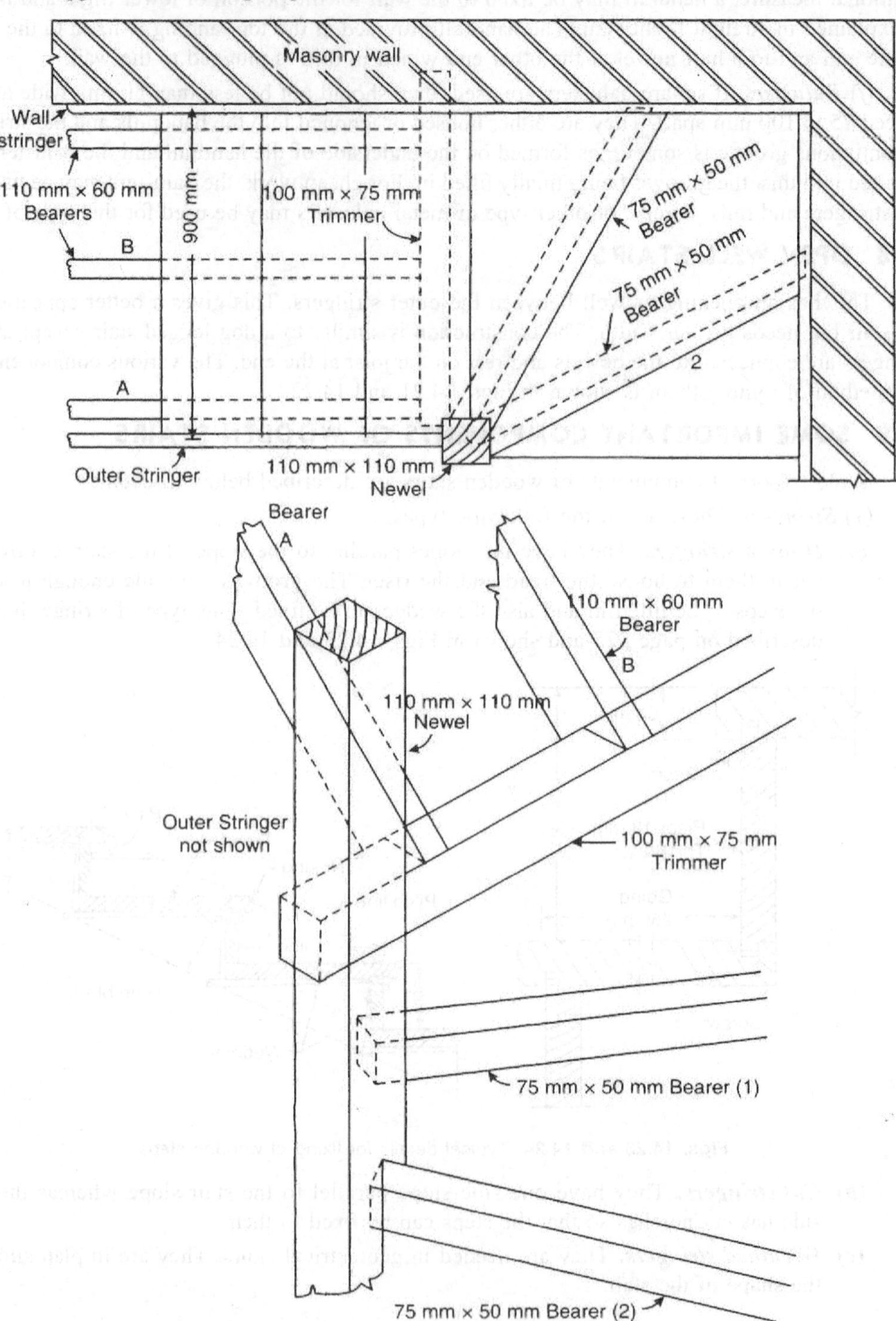

Figs. 14.25 and 14.26: Plan and detail of winding stair.

(2) ***Winders.*** It is desirable to have the treads of winders at the walking line equal and must be at least 230 mm. For narrow stairs, the treads of winders are supported on the risers which are cantilevered from the walls, the outer end being housed into the newel post. These risers have to be thicker than the remaining risers of the stairs. For wider stairs, since the bearers cannot be continued through the circular portion, the winders are supported on separate bearers which are placed below the risers and are built into the wall at one end and housed into the newels at the other end.

(3) ***Landings.*** A trimmer is placed across the walls of the staircase and bearers rest on it. Small joists are fixed at one end to the trimmer and supported on the wall at the other end. The planking is fixed to these joists. This forms a half spaced landing. In the case of a quarter spaced landing, these small joists rest at one end on the walls and at the other end on a beam (pitching piece) which is built into wall and the newel. This pitching piece may also be cantilevered from the wall. The planking is supported on the joists.

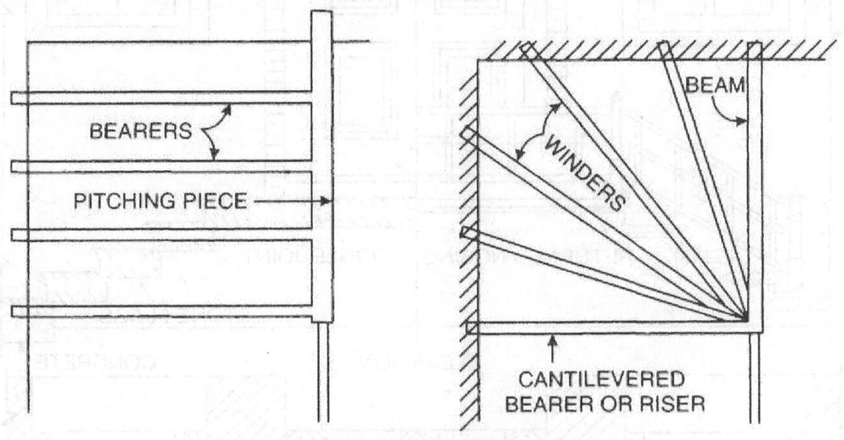

Figs. 14.27 and 14.28: Landings.

STONE STAIRS

Stone stairs are heavy and need substantial supports. They are not in much use these days. With use, they become slippery and dangerous.

14.10 STONE STEPS

These are used for threshold, approaches to basements, heating chambers, etc. Stone provides a very good material for use as steps because it is hard, durable and weather resistant. The following types of steps are used.

(1) ***Rectangular steps.*** These are cut of solid stone and the lower edge of one step is supported on the top back edge of the other. Rebated or checkered joints, if used, strengthen the stepping. The top front edge may have moulded noses. The steps may be built solidly into a wall and rest on the other side on a dwarf wall. This type of construction can be used for entrances to a building.

The landings may be built of stone slabs as shown in the figure which are joggled to each other.

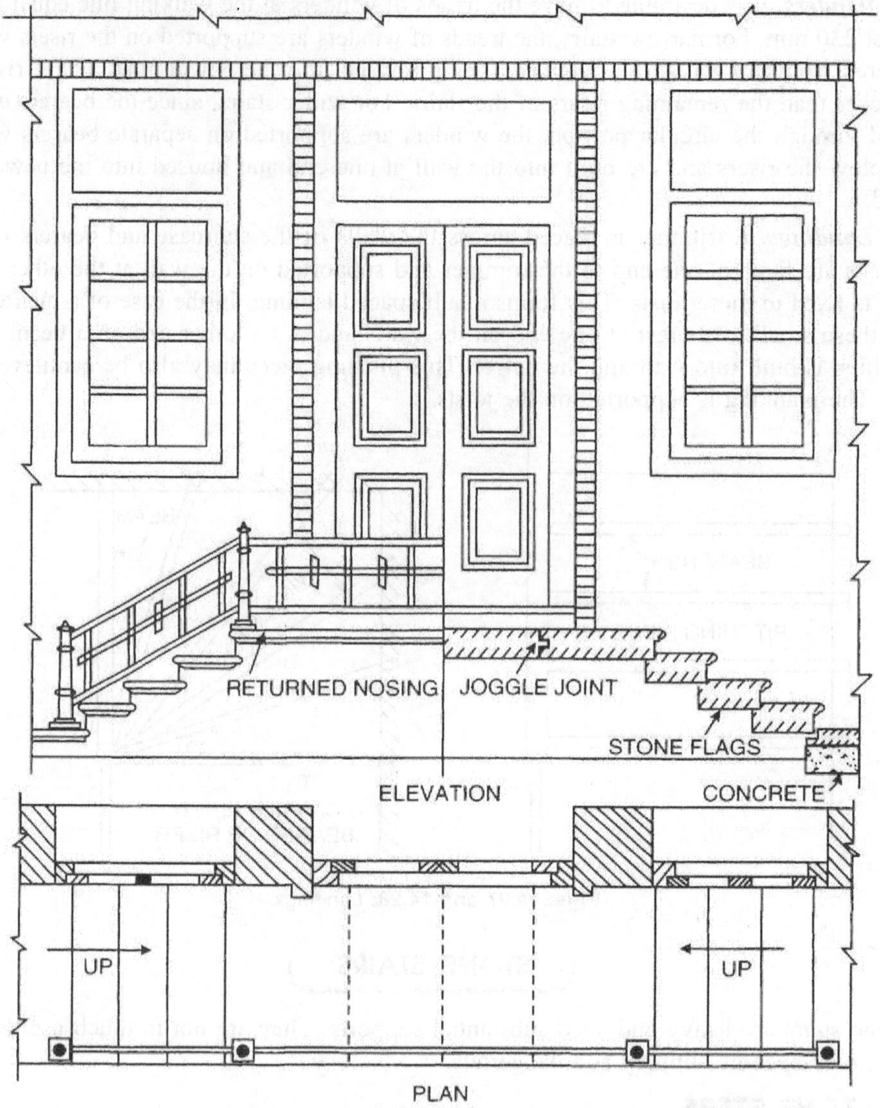

RETURNED NOSING
JOGGLE JOINT
STONE FLAGS
ELEVATION
CONCRETE
UP
UP
PLAN

Figs. 14.29 and 14.30: Built up stone steps for entrances to buildings.

LANDING SLAB
SLOPE
STEPS

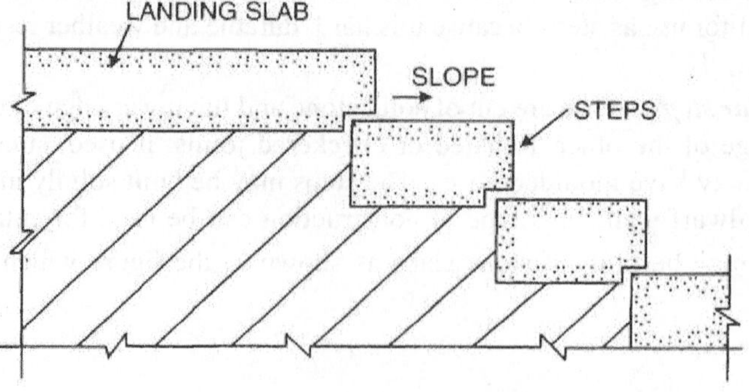

Fig. 14.31

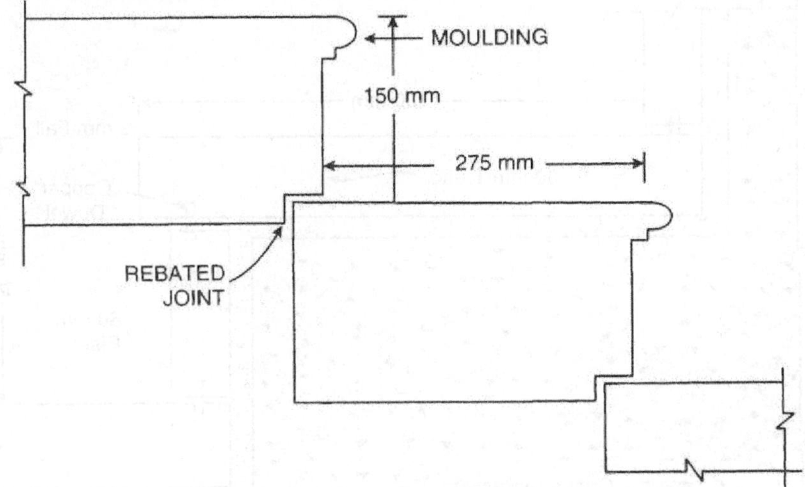

Figs. 14.31 and 14.32: Rebated square stone steps.

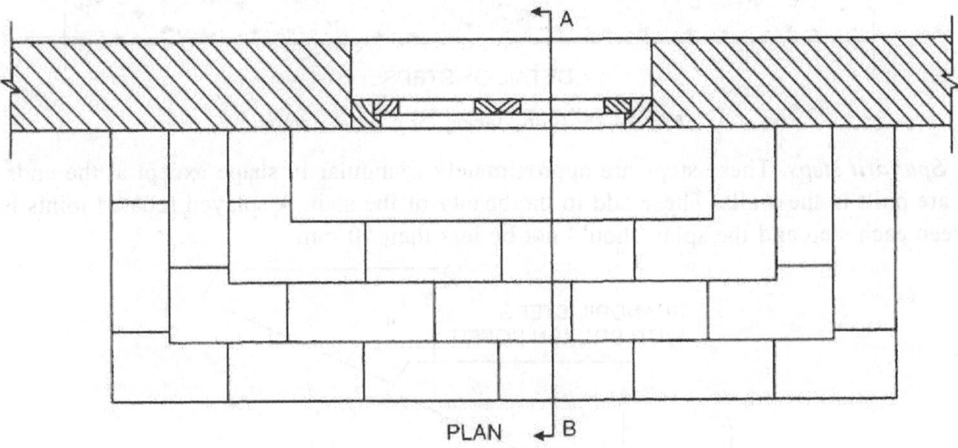

Fig. 14.33: Stone step entrance.

(2) ***Built-up steps.*** The tread and riser of each step are made of thin sawn slabs. The thickness of the treads, if supported on ends only should be at least 50 mm. They may be used as a facing for concrete or brick steps. In this case, the treads and the risers are bedded and jointed with cement mortar.

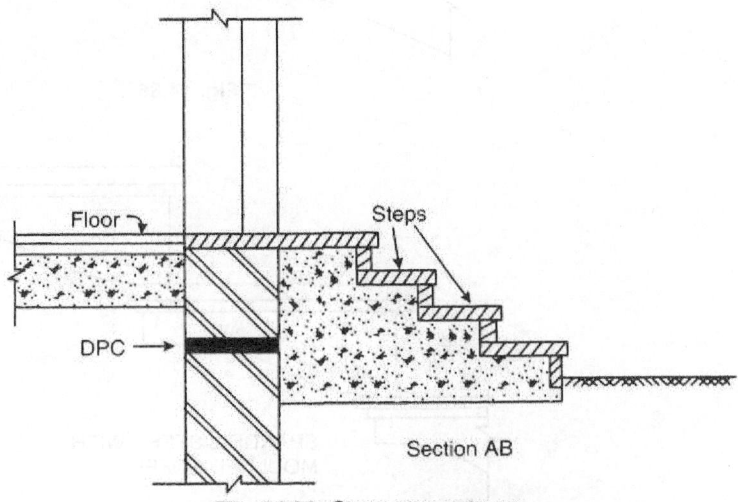

Fig. 14.34: Stone step entrance.

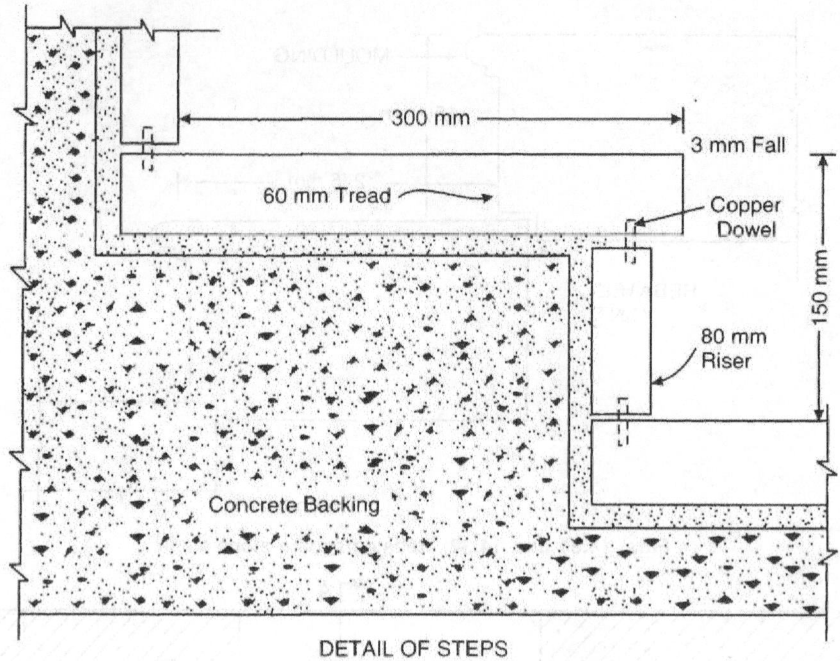

Fig. 14.35: Stone facing for concrete steps.

Spandril steps. These steps are approximately triangular in shape except at the ends where they are built in the walls. These add to the beauty of the stair. A splayed rebated joints is given between each step and the splay should not be less than 50 mm.

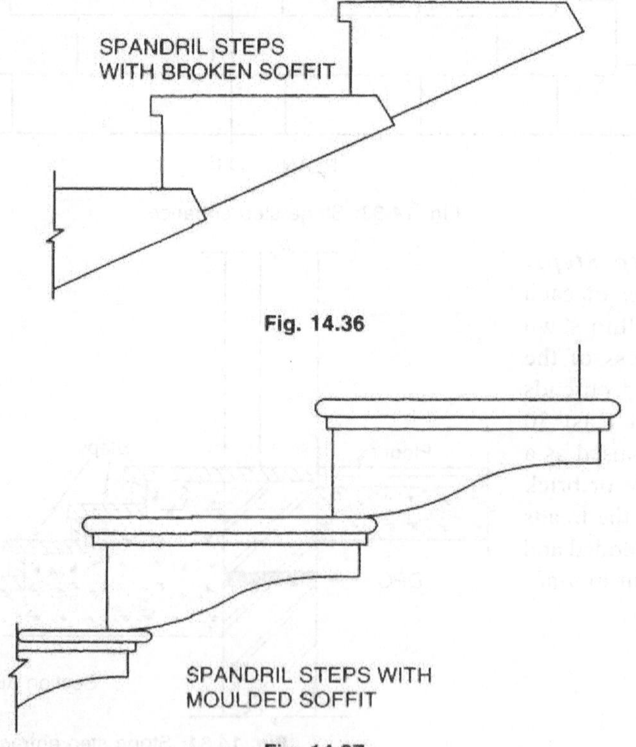

Fig. 14.36

Fig. 14.37

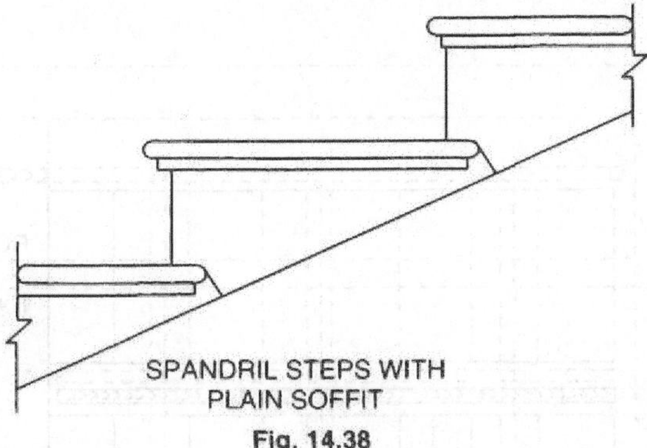

Fig. 14.38

Figs. 14.36 to 14.38: Spandril steps.

Stone stairs. These were formerly used in the case of commercial or public buildings. The steps generally used are of spandril type. They are either :

(*a*) ***Cantilevered from the wall.*** In this case, the steps are properly supported in the wall and are fixed with mortar. As already stated the section at the support is rectangular. The projection of the step should be limited to 15 mm. The minimum bearing of the step should be 200 mm.

(*b*) Built into the wall at both ends.

(*c*) Built into the wall at one end and the outer end beam supported on a steel I-Section. Special angle iron cleats are used at the connection of the stone steps with the I-Section.

(*d*) Built into the steel joist at both ends.

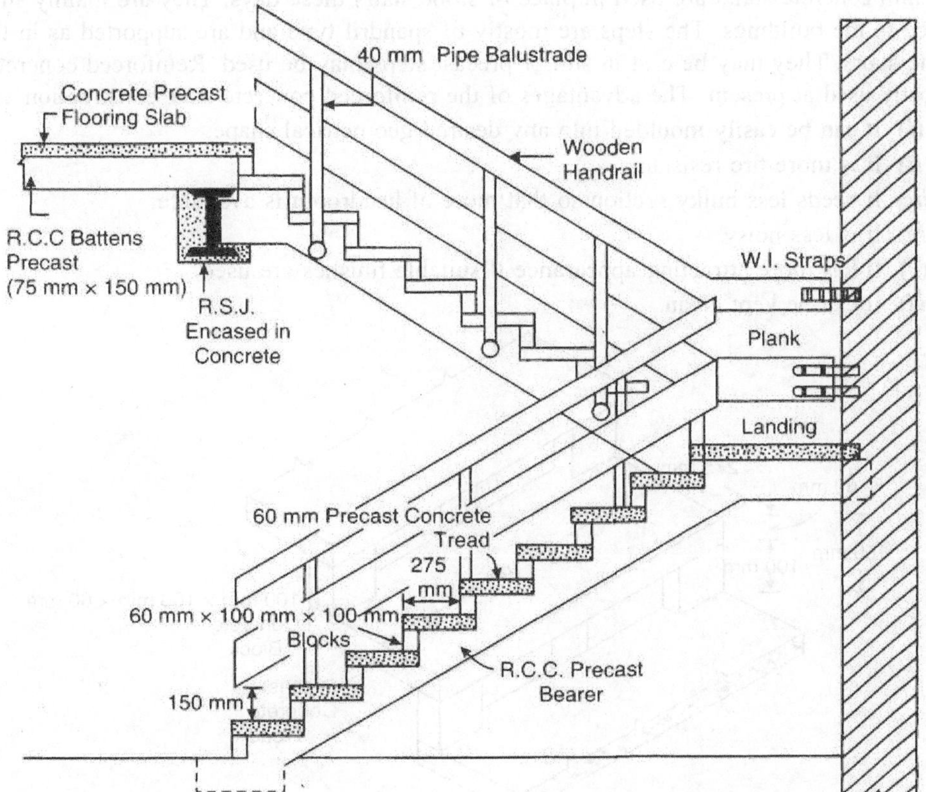

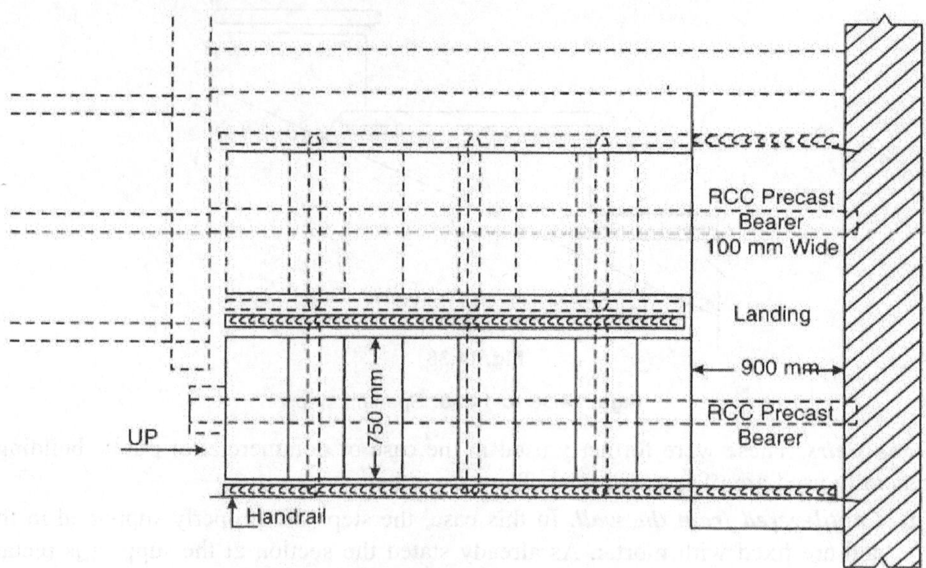

Figs. 14.39 to 14.40: Precast R.C.C. stairs.

CONCRETE STAIRS

Concrete stairs are becoming very popular these days because of their advantages such as strength, durability and fire resistance. Moreover, they can be constructed in many forms.

Plain concrete stairs are used in place of stone stairs these days. They are mainly suited for entrance to the buildings. The steps are mostly of spandril type and are supported as in the case of stone stairs. They may be cast in situ or precast steps may be used. Reinforced concrete stairs are mostly used at present. The advantages of the reinforced concrete stair construction are:

(*i*) It can be easily moulded into any desired geometrical shape.

(*ii*) It is more fire resistant.

(*iii*) It needs less bulky section so that more of headroom is available.

(*iv*) It is less noisy.

(*v*) It has more attracting appearance if suitable finishes are used.

(*vi*) It can be kept clean.

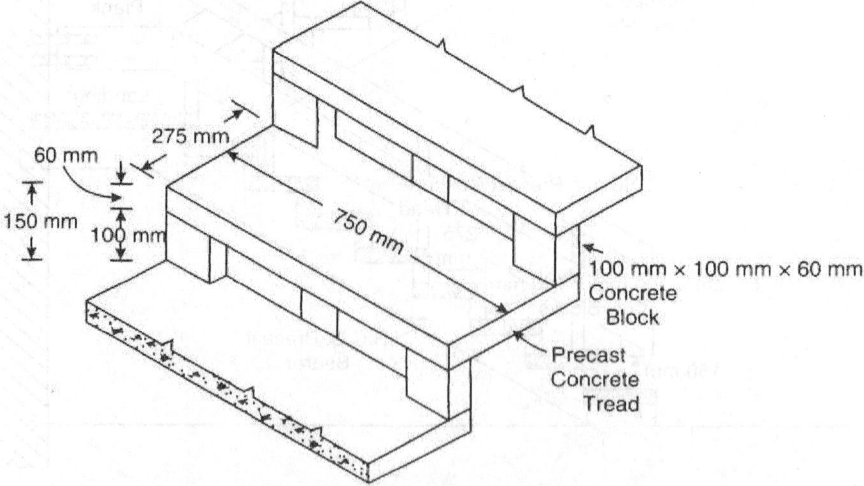

Fig. 14.41: Precast step details.

For small buildings, composite type of stairs envisaging the use of plain concrete and reinforced concrete are used. In such a type, the stringers can be of R.C.C. beams, the top slope being cut in a shape to suit the risers and the treads. The treads are of precast plain concrete slabs and the risers are of small, precast concrete blocks. One end of the stringer is resting on the base and the other rests on the wall. A precast concrete slab is used as landing. The handrails are made of wood and the balusters are of steel pipes.

R.C.C. stairs can be of three types:

(i) Cantilever type: In which R.C.C. steps are cantilevered from the wall. The reinforcement is placed at the top of the tread and a few bars are placed at the bottom of the riser, suitable cover being given.

(ii) Slab type: A stair in which the steps are designed to act as slabs between the two stringers or a wall and a stringer. The stringers act as inclined R.C.C. beams and are supported at the top and the bottom of the landing.

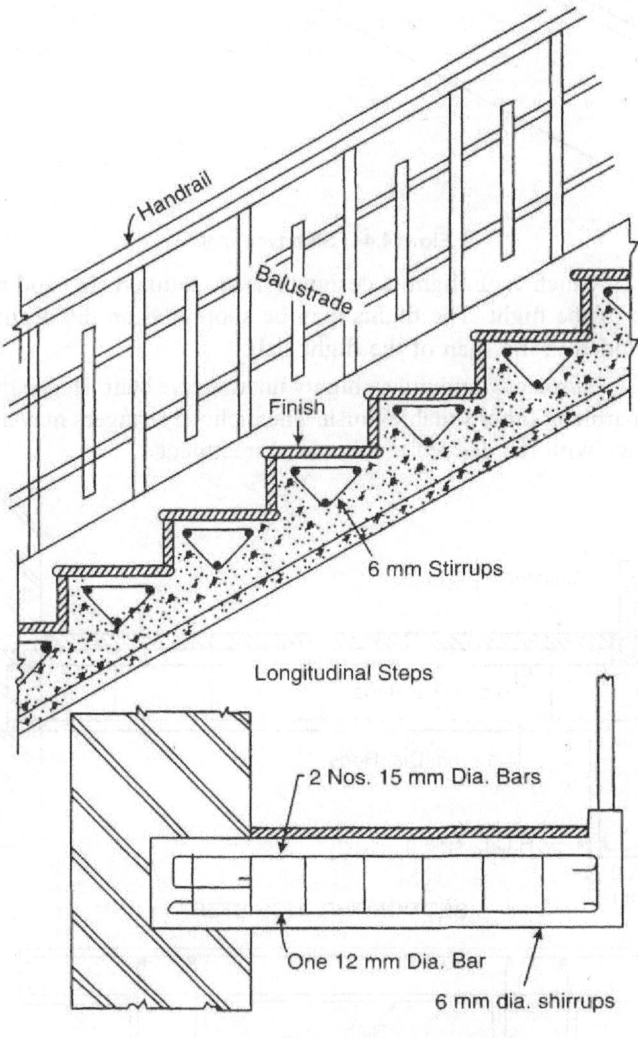

Fig. 14.42 and 14.43: R.C.C. cantilevered stairs.

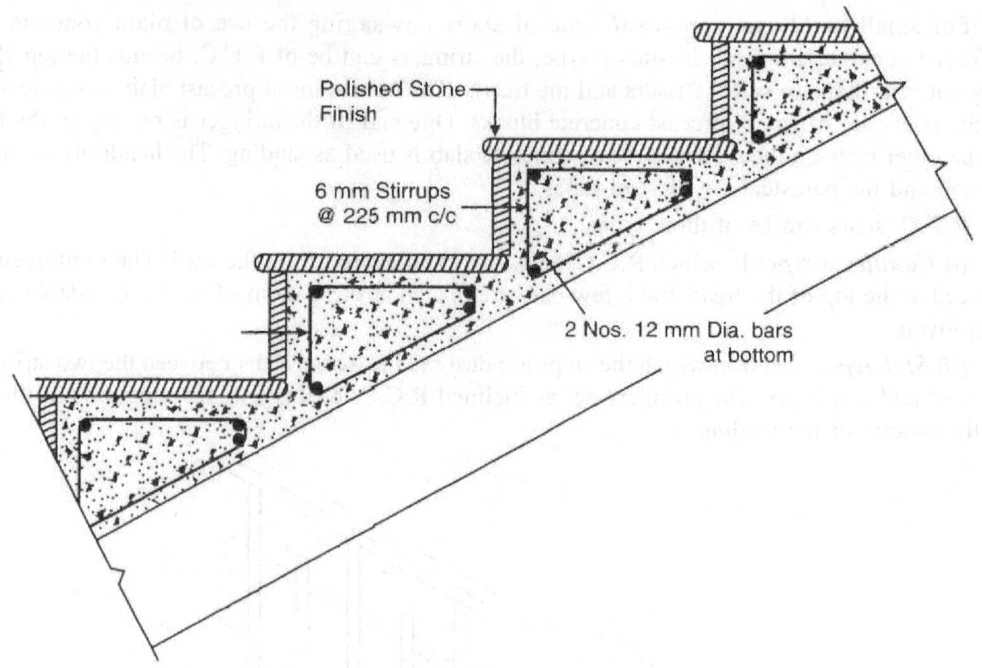

Fig. 14.44: Slab type stair.

(iii) Slab type: In which each flight is designed as an inclined slab and reinforcement is laid parallel to the stlope of the flight. The flights may be supported on the beams at the ends or the landings may be included in the span of the flight slab.

Unfinished R.C.C. surface would give a highly unattractive stair. Hence the steps are provided with tiles, terrazzo, marble or other suitable finish. The inclined stringers may also be given suitable covering in accordance with the desired architectural treatment.

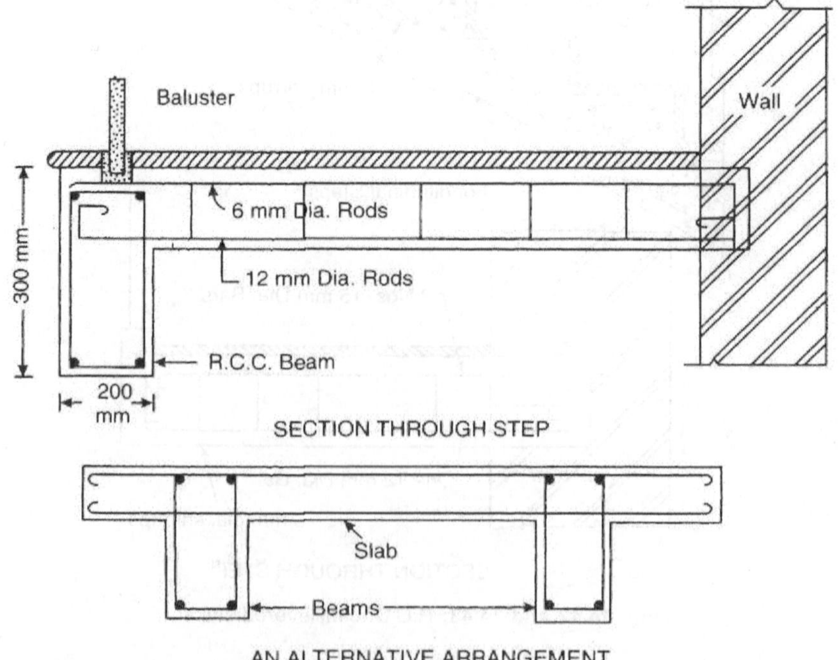

Figs. 14.45 and 14.46: Slab type stair.

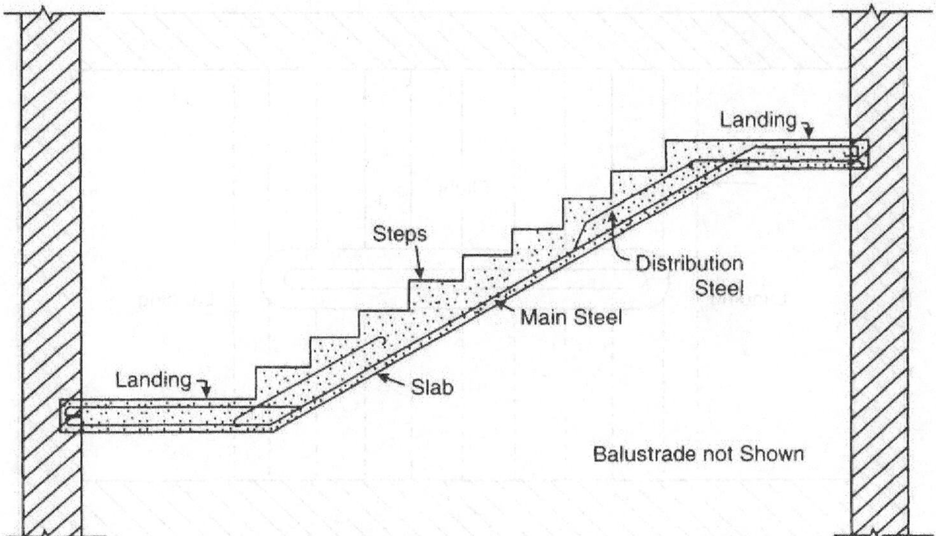

Fig. 14.47: R.C.C. slab stair.

It is essential that the steps are covered with a material which will not render the tread slippery, hence checkered grooves may be made on the surface or other suitable roughening devices may be used.

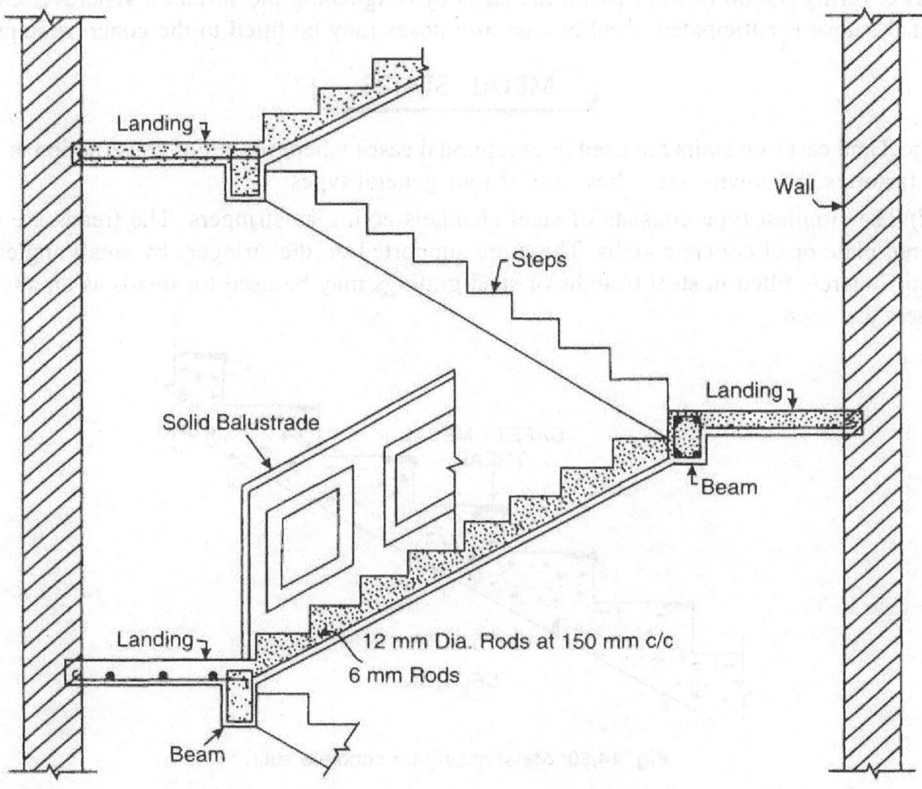

Fig. 14.48

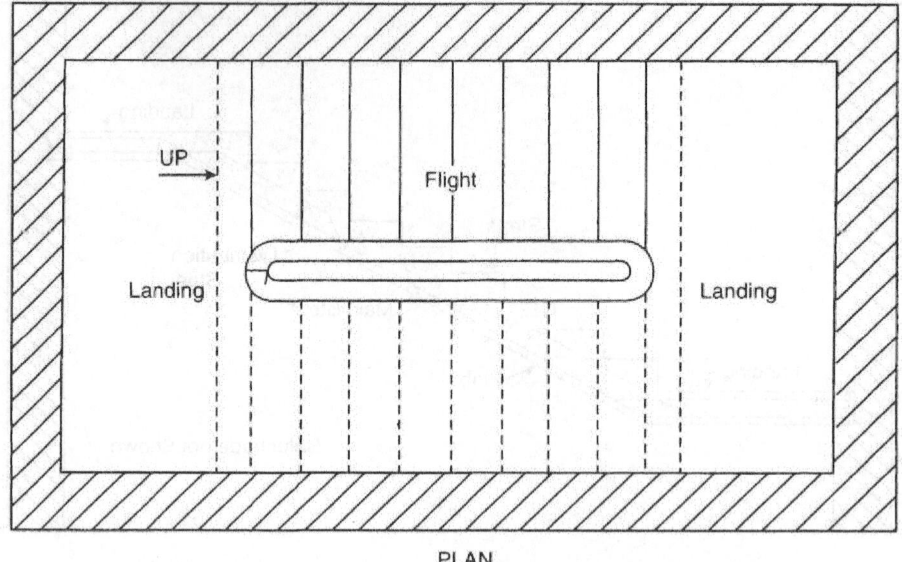

Figs. 14.48 and 14.49: R.C.C. continuous slab stair.

The nose of the steps should be so treated that it does not get cracked easily. Hence it is desirable that the concrete covering, if used, should be laid immediately after the steps are cast or else this covering should be keyed with the steps by roughening the surfaces. Wherever excessive wear of the nose is anticipated, steel or cast iron noses may be fitted to the concrete steps.

METAL STAIRS

Steel and cast iron stairs are used in exceptional cases where fire proof construction is desired, *e.g.* in factories, go-downs, etc. These are of four general types:

(i) The simplest type consists of steel channels acting as stringers. The treads are of steel checkered plate or of concrete slabs. These are supported on the stringers by small angles bolted to them. Concrete filled in steel troughs or steel gratings may be used for treads as an alternative. The risers are open.

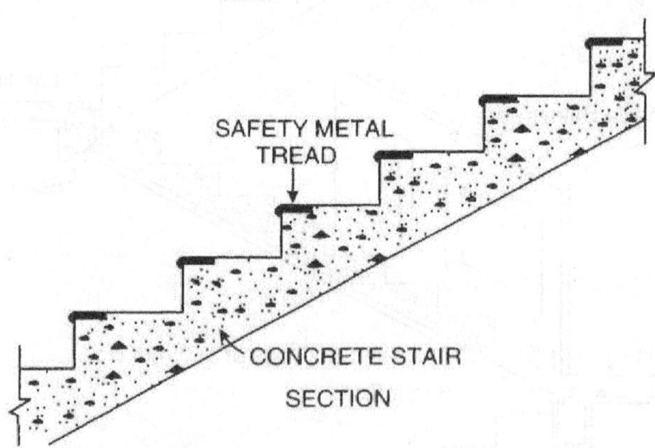

Fig. 14.50: Metal nosing for concrete stair.

(ii) This consists of steel stringers made of channels or plates and the treads are made of steel plates. They are finished in a suitable manner, *i.e.*, they may be covered with stone, marble, etc. They are used in important buildings.

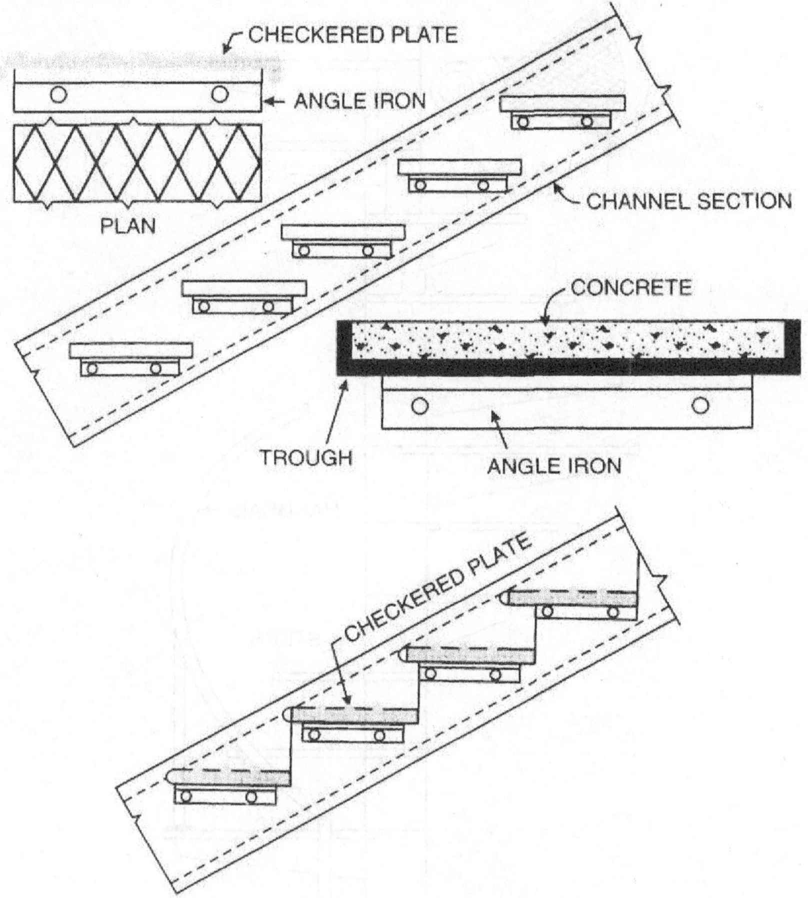

Fig. 14.51 to 14.54: Metal stair details.

(iii) In this type the stringers are of steel and risers of concrete. The concrete is suitably finished.

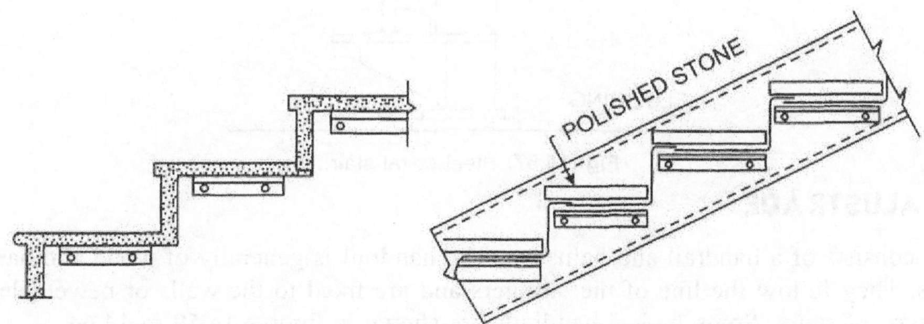

Figs. 14.55 and 14.56: Concrete or metal and stone stairs.

(iv) Spiral Type stairs. They are used where the space available is too small and the traffic is less. They are made of C.I. pipe newel fixed in the centre and around this C.I. steps are fixed. The steps have suitable checkered treads for safety.

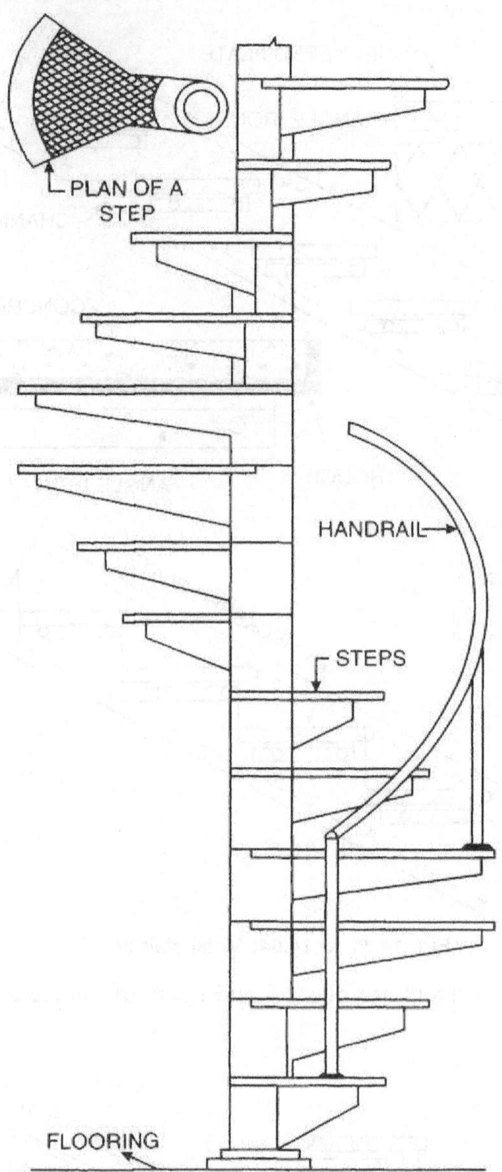

Fig. 14.57: Steel spiral stair.

14.11 BALUSTRADE

This consists of a handrail and balusters. The handrail is generally of wood and has special mouldings. They follow the line of the stringers and are fixed to the walls or newels depending upon the type of stairs. Some typical handrails are shown in figures 14.59 to 14.64.

These days handrails of chromium plated steel fixed on the top of polished wood are used. The handrail should be fixed at a height of 450 mm to 500 mm above the line of nosing. Handrail on landings should be at least 900 mm high.

Balusters. They are generally not spaced greater than 150 mm centre to centre. For wooden stairs, they are made of wood whereas for other types of stairs, balusters made of metal are used. They may have special shapes and mouldings carved on them or else they may be made of plain rectangular or pipe sections. Some of the types commonly used are shown in figures 14.59 to 14.64.

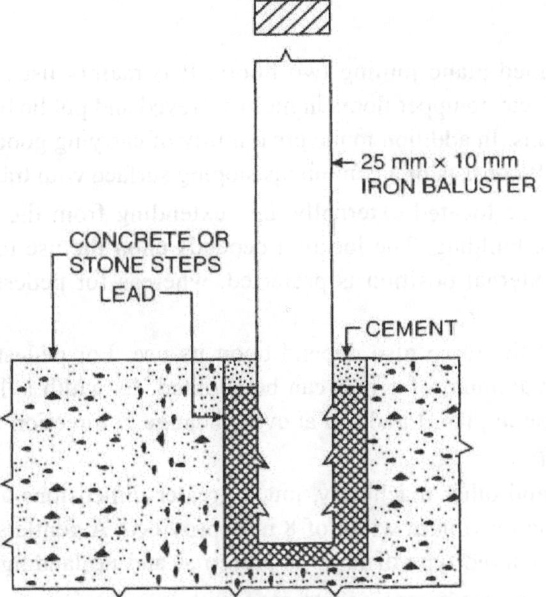

Fig. 14.58: Fixing of metal baluster in concrete.

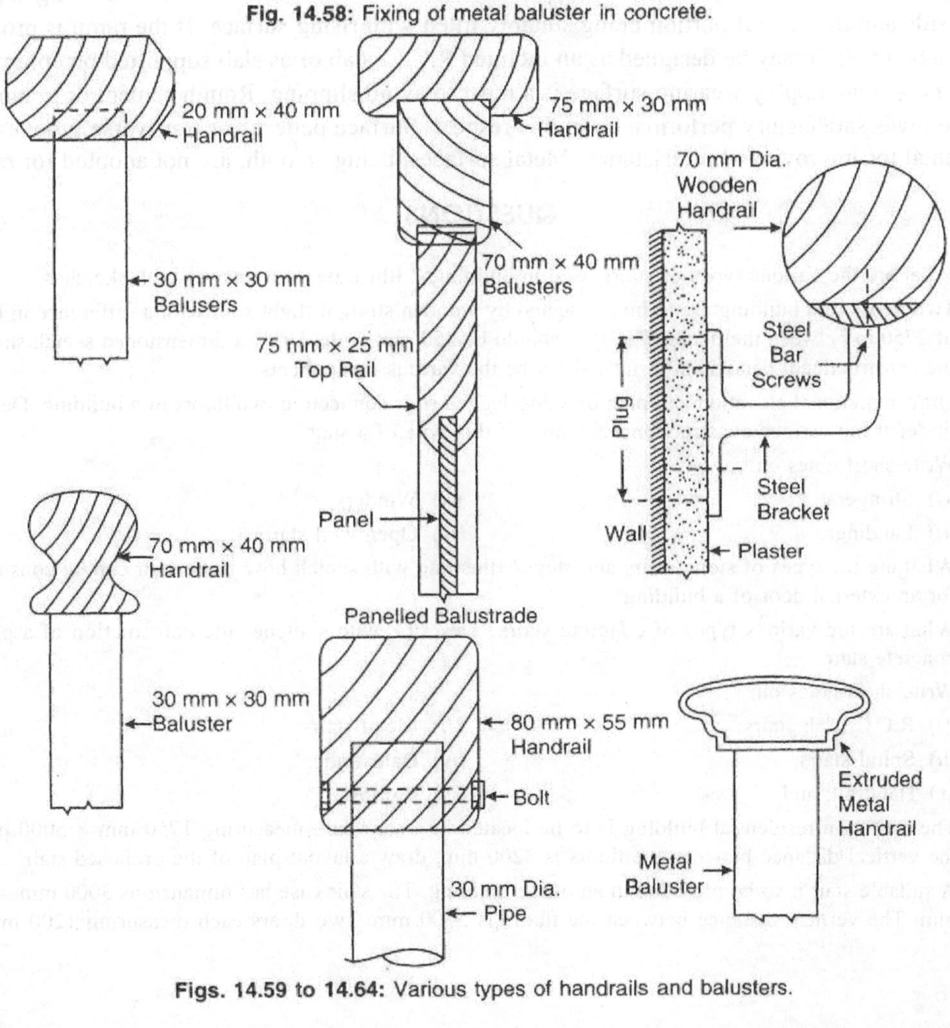

Figs. 14.59 to 14.64: Various types of handrails and balusters.

14.12 RAMPS

A ramp is an inclined plane joining two floors. It is mainly used for carrying machinery, equipment, trolleys, cars, etc. to upper floors in multi-storeyed and public buildings. It can, however, also be used by pedestrians. In addition to the great utility of carrying goods and pedestrians, ramps have an aesthetic value. It consists of a continuous sloping surface with transitions at the floor level.

Ramps may either be located externally, *i.e.*, extending from the general building line or internally, *i.e.*, inside the building. The location depends upon the use to which ramp is put. For carrying heavy loads, external position is preferred, whereas for pedestrians, it can be situated inside the building.

The dimensions of the ramp also depend upon its use. For pedestrian traffic, a minimum slope of 1 in 10 and a maximum of 1 in 6 can be adopted. Its width is limited to about 2 m and can be of dog-legged type in plan. Landings at every change in direction are provided with widths equal to that of the ramp.

For carrying cars and other machinery, much greater dimensions are required. A minimum width of 4 m is kept. The maximum width of 8 m is provided. A gentle slope of 1 in 10 is given. A straight flight type or curved type of ramp is preferred and no landing is necessary.

For external location, ramps are constructed with two retaining walls with rising tops on either side and the central portion being suitably filled with rising surface. If the ramp is provided inside a building, it may be designed as an inclined R.C.C. slab or as slab supported on columns or stringers. A non-slippery wearing surface is a must to avoid slipping. Rough cement or granolithic surface gives satisfactory performance in this respect. Surface patterns or transverse grooves may be formed for improving slip resistance. Metal surfaces, being smooth, are not adopted for ramps.

QUESTIONS

1. What are the various types of stairs used in buildings? Illustrate your answer with sketches.
2. Two floors of a building are to be connected by wooden straight flight stair with a difference in height of 2750 m between the floors. The stair should be 450 mm wide. Draw a dimensioned sketch showing the constructional details and briefly describe the various components.
3. Draw a sectional elevation and plan of a dog-legged stair connecting two floors of a building. Describe in detail the various constructional features of this type of a stair.
4. Write short notes on:
   (*i*) Stringers; (*ii*) Winders;
   (*iii*) Landings; (*iv*) Open well stairs.
5. What are the types of stone stairs and steps? Illustrate with sketch how stone stair can be constructed for an external door of a building.
6. What are the various types of concrete stairs? Describe with sketches the construction of a precast concrete stair.
7. Write short notes on:
   (*i*) R.C.C. slab stairs; (*ii*) Metal stairs;
   (*iii*) Spiral stairs; (*iv*) Balustrade;
   (*v*) Handrail; and (*vi*) Ramps
8. The stair of a residential building is to be located in a stair-case measuring 1750 mm × 5000 mm. If the vertical distance between the floors is 3200 mm, draw a layout plan of the proposed stair.
9. A suitable stair is to be provided in an office building. The stair case has dimensions 3000 mm × 7000 mm. The vertical distance between the floors is 3500 mm. Two doors each measuring 1200 mm are

provided in the longitudinal walls facing each other and located 500 mm from the transverse wall. Design and draw the stair details.

10. A stair is to be located for an office building 4.5 m × 5.25 m. The vertical distance between the floors is 3.8 m. Draw a dimensional sketch of the proposed stair.

REFERENCES

1. **Huntington W. C.**: *Building Construction.*
2. **British Standard 585**: *Wooden stairs.*
3. **Voss W. C.**: *Fire proof construction.*
4. **Telling M. T.**: *Carpentry and joinery.*
5. **Riley J. W.**: *Building construction for beginners.*
6. **Townsend**: *Stair Building.*
7. **IS: 1634–1973**: *Code of practice for design and construction of wood stair in houses.*

CHAPTER

Lifts and Escalators

LIFTS

For multi-storeyed buildings the installation of lifts is a must to avoid fatigue in climbing up the stairs and for quick vertical circulation between different floors. The provision of lifts in a building is a highly specialised job. However certain provisions are required to be made in the building layout and structures for accommodating lifts and other accessories such as operating devices. A vertical shaft with openings at the floor level is provided. The shaft is located at a suitable place *e.g.* by the side of the stair or within the open well of a stair. It may have sides built in masonry or concrete or metallic cage with suitable doors. The shaft extends below the ground floor or the basement floor, as the case may be, to accommodate the spring buffers for slow speed lifts and hydraulic buffers for high speed lift. Usually a machine room is located at the top of the lift shaft for housing equipment and accessories. However it can also be at the bottom, by the side or at the back of the shaft. The size of the machine room is normally 4 m × 3 m × 2.5 to 3 m. Its floor should be suitably designed to support the weight of the lift car, equipment, passengers, the balancing weight and the weight of motor with the winch arrangement. Floors of machine room should be designed to carry a load of not less than 500 kg/m^2 plus the load imposed by the equipment or any reaction from such equipment.

The shaft for the lift should be able to accommodate the lift car, balancing weight, and vertical guides for them. Previously collapsable doors were provided both for the lift car or cage and for the opening in the shaft in the floor level. Now a days flush doors of sliding type are provided. The doors at the floor level are fitted with electro-mechanical safety locking devices with special emergency lock release.

Fig 15.1 shows the different arrangements for 3,4,6 and 8 lifts.

The outline dimensions of machine room, pit depth, total headroom, overhead distance and sill are specified in Tables 15.1 to 15.4 as given below:

Passenger lifts	Table 15.1 and 15.1 A
Goods lifts	Table 15.2
Hospital lifts	Table 15.3
Service lifts	Table 15.1 and 15.1 A
Dumb Waiter	Table 15.4

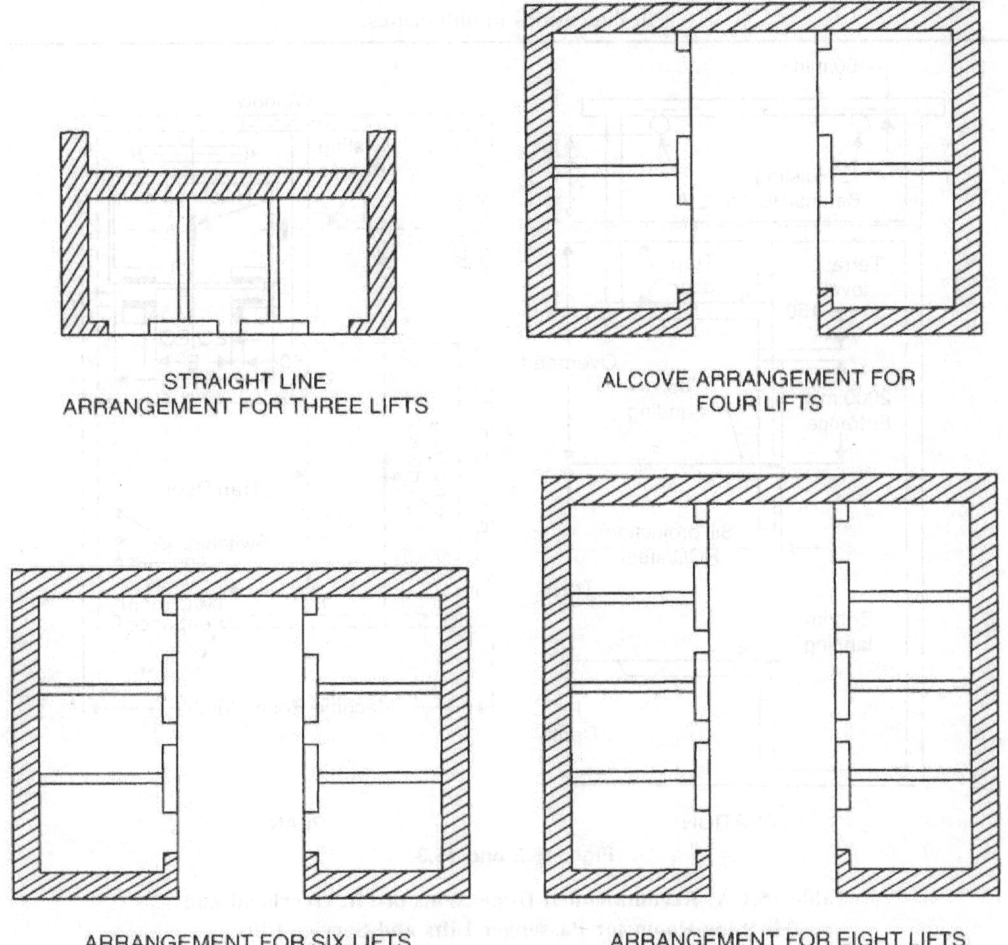

Fig. 15.1 Arrangement of Lifts

Table 1: Recommended Dimensions of Passenger Lifts and Service Lifts
(All dimensions in millimetres)

Load		*Car side*		*Lift well*		*Entrance*
Persons	*kg*	*A*	*B*	*C*	*D*	*E*
4	272	1 100	700	1 900	1 300	700, Min
6	408	1 100	1 000	1 900	1 600	700, Min
8	544	1 300	1 100	1 900	1 900	800
10	680	1 300	1 350	1 900	2 100	800
13	884	2 000	1 100	2 500	1 900	900
16	1 088	2 000	1 300	2 500	2 100	1 000
20	1 360	2 000	1 500	2 500	2 400	1 000

All dimensions in millimetres.

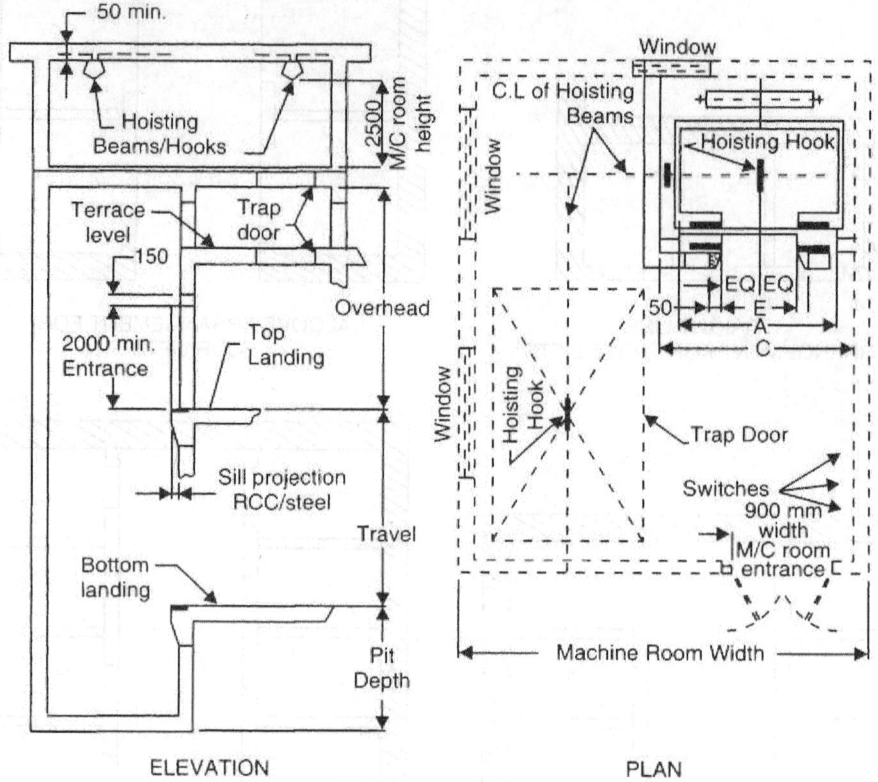

Figs. 15.2 and 15.3

Table 15.1 A: Recommended Dimensions of Pit, Overhead and Machine-Room for Passenger Lifts and Service Lifts

All dimensions in millimetres.

Speed in m/s	Upto 0.70	> 0.70 ≤ 1.00	> 1.00 ≤ 1.50	> 1.50 ≤ 1.75	> 1.75 ≤ 2.00	> 2.00 ≤ 2.50
Pit depth	1 350	1 500	1 600	2 150	2 200	2 500
Overhead	4 200	4 250	4 800	4 800	5 200	5 400
Machine-room Depth	D + 2 000			D + 2 500		
Machine-room Width	C + 1 000		C + 1 200		C + 1 500	

Notes 1. The total overhead dimension has been calculated on the basis of car height of 2.3 m.

2. In case of manually operated doors, clear entrance will be reduced by the amount of projection of handle on the landing door.

3. All dimensions given above for lifts having centre opening power operated doors with counterweight at rear, are recommended dimensions primarily for architects and building planners. Any variations mutually agreed between the manufacturer and the purchaser are permitted. However, variation in:

(*a*) Car inside dimensions shall be within the maximum area limits specified in accordance with accepted standards.

(*b*) Entrance width on higher side is permitted.

(*c*) Entrance width on lower side is permitted up to 100 mm subject to minimum of 700 mm.

4. Dimensions of pit depth and overhead may differ in practice as per individual manufacturer's design depending upon load, speed and drive. Recommended dimensions for pit depth, overhead and machine-room for different lift speeds are given in Table 15.1A. However, the pit depth and overhead shall be such as to conform to the requirements of bottom clearance and top clearance in accordance with the accepted standards.

Table 15.2: Recommended Dimensions of Goods Lifts (For Speeds Up to 0.5 m/s)

All dimensions in millimetres.

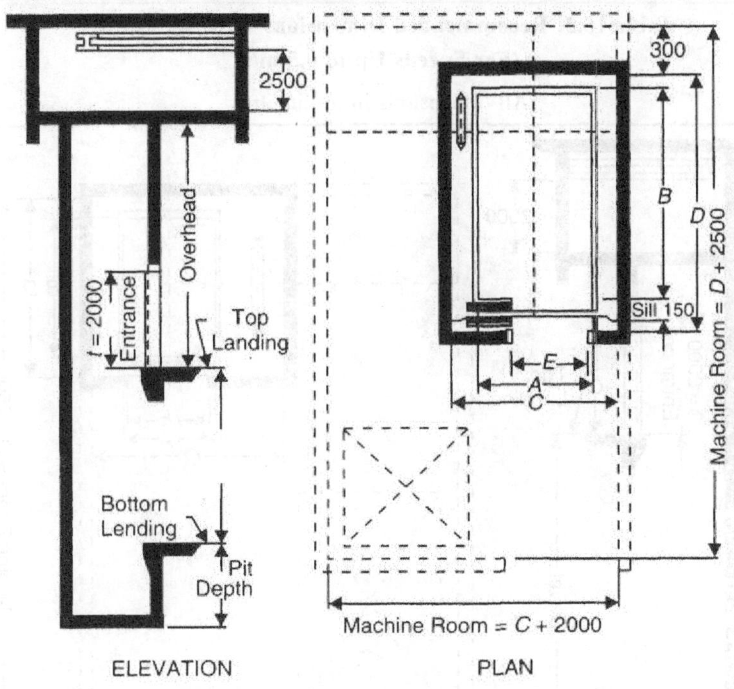

Figs. 15.4 and 15.5

Load	*Car Inside*		*Lift well*		*Entrance*
kg	*A*	*B*	*C*	*D*	*E*
500	1 100	1 200	1 900	1 500	1 100
1 000	1 400	1 800	2 300	2 100	1 400
1 500	1 700	2 000	2 600	2 300	1 700
2 000	1 700	2 500	2 600	2 800	1 700
2 500	2 000	2 500	2 900	2 800	2 000
3 000	2 000	3 000	2 900	3 300	2 000
4 000	2 500	3 000	3 400	3 300	2 500
5 000	2 500	3 600	3 400	3 900	2 500

Notes 1: The width of machine room shall be equal to be lift well width 'C' subject to minimum of 2500 mm.

2: The total headroom has been calculated on the basis of a car height of 2.2 m.

3: Clear entrance width 'E' is based on vertical lifting car-door and vertical biparting landing doors. For collapsible mid load clear entrance width will be reduced by 200 mm (maximum 1 800 mm).

4: All dimensions given above are recommended dimensions primarily for architects and building planners. Any variation mutually agreed between the manufacturer and the purchaser and permitted. However, variation in car inside dimensions shall be within maximum area limits in accordance with accepted standards.

5: Dimensions of pit depth and overhead may differ in practice as per individual manufacturer's design depending upon load speed and drive. However, the pit depth and overhead shall be such as to conform to the requirements of bottom clearance and top clearance in accordance with accepted standards.

Table 15.3: Recommended Dimensions of Hospital Lifts
(For Speeds Up to 0.5 m/s)

All dimensions in millimetres.

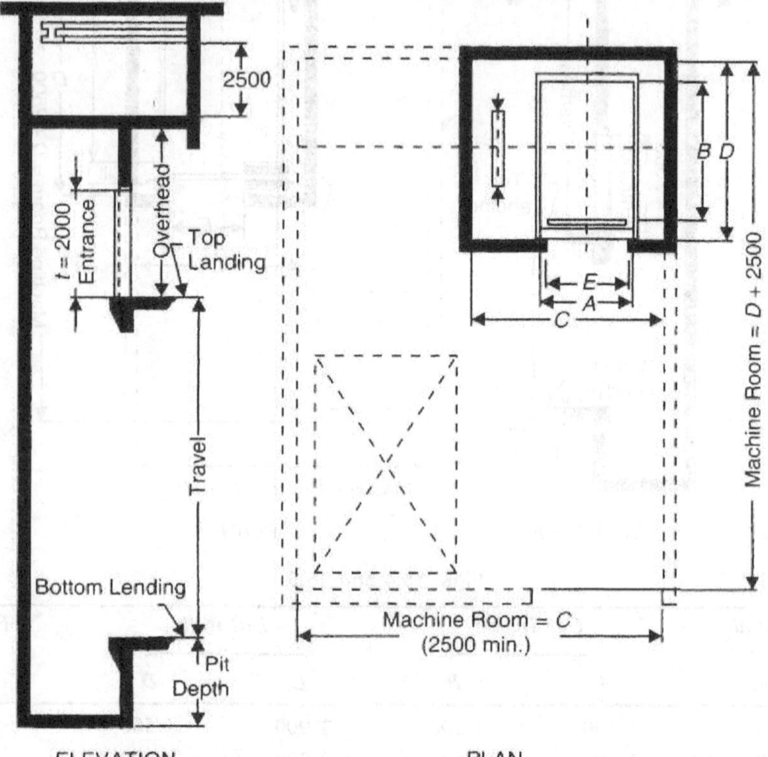

Figs. 15.6 and 15.7

Load		Car Inside		Lift well		Entrance
Persons	kg	A	B	C	D	E
15	1020	1000	2400	1800	3000	800
20	1360	1300	2400	2200	3000	1200
26	1768	1600	2400	2400	3000	1200

Notes 1: The total headroom has been calculated on the basis of a car height of 2.2 m.

2: In the case of manually-operated doors, clear entrance will be reduced by the amount of projection of handle on the landing door.

3: Although 15 persons capacity lift is not standard one, this is included to cover lifts of smaller capacity which can be used in small hospitals.

4: All dimensions given above are recommended dimensions primarily for architects and building planner. Any variations mutually agreed between the manufacturer and the purchaser are permitted. However, variation in car inside dimensions shall be within the maximum area limits in accordance with accepted standards [8-5(4)].

5: Dimensions of pit depth and overhead may differ in practice as per individual manufacturer's design depending upon load, speed and drive. However, the pit depth and overhead shall be such as to conform to the requirements of bottom clearance and top clearance in accordance with accepted standards [8-5(5)].

Table 15.4: Recommended Dimensions of Dumb Waiter

(For Speeds Up to 0.5 m/s)

All dimensions in millimetres.

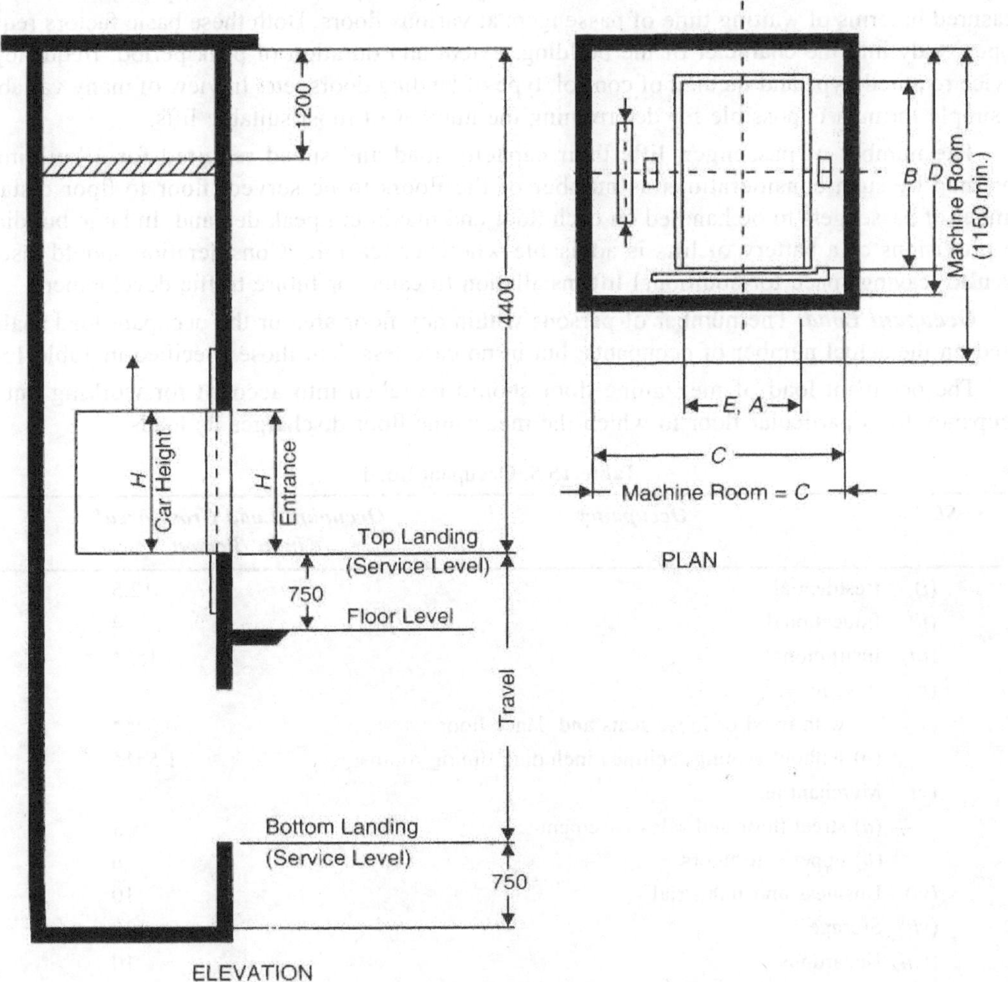

Figs. 15.8 and 15.9

Load	Car Inside			Lift well		Entrance
kg	A	B	H	C	D	E
100	700	700	800	1 200	900	700
150	800	800	900	1 300	1 000	800
200	900	900	1 000	1 400	1 100	900
250	1 000	1 000	1 200	1 500	1 200	1 000

Note: Entrance width '*E*' is based on assumption of provision of vertical biparting doors (no car door is normally provided).

15.1 DESIGN CONSIDERATIONS

15.1.1 Number of Lifts and Capacity

Two basic considerations, namely, the quantity of service required and the quality of service desired, determine the type of lifts to be provided in a particular building. Quantity of service gives the passenger handling capacity of the lifts during the peak periods and the quality of service is measured in terms of waiting time of passengers at various floors. Both these basic factors require proper study into the character of the building, extent and duration of peak period, frequency of service required, type and method of control, type of landing doors, etc. In view of many variables, no simple formula is possible for determining the number of most suitable lifts.

The number of passenger, lifts their capacity load and speed required for a building is governed by such considerations as number of the floors to be served, floor to floor distance number of passengers to be handled on each floor and maximum peak demand. In large buildings, the provisions of a battery of lifts is advisable wherever feasible. Consideration should also be given to leaving space for additional lift installation to cater for future traffic development.

Occupant Load. The number of persons within any floor area or the occupant load shall be based on the actual number of occupants, but in no case less than those specified in Table 15.5.

The occupant load of mezzanine floor should be taken into account for working out the occupancy for a particular floor to which the mezzanine floor discharges its loads.

Table 15.5. Occupant Load

Sl. No.	*Occupancy*	*Occupant Load Gross Area* in m^2/Person*
(*i*)	Residential	12.5
(*ii*)	Educational	4
(*iii*)	Institutional	15**
(*iv*)	Assembly:	
	(*a*) with fixed or loose seats and dance floor	0.6***
	(*b*) without seating facilities including dining rooms	1.5***
(*v*)	Merchantile:	
	(*a*) street floor and sales basement	3
	(*b*) upper sale floors	6
(*vi*)	Business and industrial	10
(*vii*)	Storage	30
(*viii*)	Hazardous	10

* The gross area shall mean plinth area or covered areas.

** Occupant load in dormitory portions of homes for the aged, orphanages, insane asylums, etc. where sleeping accommodation is provided, shall be calculated at not less than 7.5 m^2 gross area/person.

*** The gross area shall include in addition to the main assembly room or space, any occupeid connecting room or space in the same storey or in the storeys above and below, where entrance in common to such rooms and space and they are available for use by the occupants of the assembly place. No deductions shall be made in the gross area for corridors, closets or other sub-divisions; the area shall include all space serving the particular assembly occupaney.

Quantity of Service. The quantity of service is a measure of the passenger handling capacity of a vertical transportation system. It is measured in terms of the total number of passengers handled during each five minute peak period of the day. A five-minute base period is used as this is the most practical time over which the traffic may be averaged. The passenger handling capacity (*H*) for different occupancies, expressed in percent of the estimated population that has to be handled in the building in the five-minute peak period, should be approximately as follow:

Type of Building	*H (per cent)*
Office—Diversified (mixed) tenants	10-15
Office—Single tenant	15-25
Residential	7.5

Quality of Service. The quality of service is generally measured by the passenger waiting time at the various floors. The following is the guiding factor for determining this aspect:

Quality of Service or Acceptable Interval	*Rating*
20-25 seconds	Excellent
30-35 seconds	Good
34-40 seconds	Fair
45 seconds	Poor
Over 45 seconds	Unsatisfactory

Note: For residential buildings, longer intervals should be permissible.

Car Speed. The recommended car speed for the different types of lifts in different occupancies are given in Table 15.6.

Determination of Handling Capacity. The handling capacity is calculated by the following formula:

$$H = \frac{300 \times Q \times 100}{T \times P}$$

where

H = Handling capacity as the percentage of the peak population handled during 5 minute period.

Table 15.6: Car speed for lifts in different occupancies

Sl. No.	*Type of Lift*	*Occupancy*	*No. of Floors Served*	*Car Speed m/s*
(*i*)	Passenger lifts	Low and medium class flats (residential)	—	0.5
		Office building, hotels	4-5	0.5-0.75
		Large flats (high)	6-12	0.75-1.5
		Hospital	13-20	above 1.5

	Shops and departmental stores*	—	2-2.5
(*ii*) Goods lifts	Normal**	—	0.25-0.5
	Serving many floors	—	1
(*iii*) Hospital bed lifts	Short travel lifts in small hospitals	—	0.25
	Normal	—	0.5
	Long travel lifts in general hospitals	—	1

*The high speed is desirable especially where restaurant or other facilities are provided on the top floor as the traffic would at times demand a lift or lifts to be used entirely between the ground and the top floors.

**Slower speeds may be used for heavier loads.

Q = average number of passengers carried in a car,

T = waiting interval in seconds, and

P = total population to be handled during peak morning period (it is related to the area for which particular bank of lift serves).

The value of Q depends on the dimensions of the car. It may be noted that the car is not loaded always to its maximum capacity during each trip and, therefore, for calculating H the value of Q is taken as 80 percent of the maximum capacity of the car.

The waiting interval, T in seconds is calculated by the following formula:

$$T = \frac{RTT}{N}$$

where

RTT = round trip time, that is, the average time required by each lift in taking one full load of passengers from ground floor discharging them in various upper floors and coming back to ground floor for taking the fresh passengers for the next trip; and

N = number of lifts

Note: *RTT* is the sum of the time required in the following processes:

(*a*) Entry of passengers on the ground floor.

(*b*) Exit of the passengers on each floor of discharge.

(*c*) Door closing time before each time operation.

(*d*) Door opening time on each discharging operation.

(*e*) Acceleration periods.

(*f*) Stopping and levelling periods.

(*g*) Periods of full rated speeds between stops going up and

(*h*) Periods of full rated speeds between stops going down.

It is observed that the handling capacity is inversely proportional to waiting interval which in turn is proportional to *RTT*. Reducing the *RTT* of a lift from 120 to 100 seconds increases its handling capacity by 20 per cent.

The round trip time can be decreased not only by increasing the speed of the lift but also by improving the design of the equipment related to opening and closing of the landing and car doors, acceleration, deceleration, levelling and passenger movement.

These factors are given below:

(*a*) The most important factor in shortening the time consumed between the entry and exit of the passengers to the lift car is the correct design of the doors and the proper car width. For comfortable entry and exit for passengers, it has been found that most suitable door width is 1000 mm and that of car width is 2000 mm.

(*b*) The utilization of centre opening doors has been a definite factor in improving passenger transfer time, since when using this type of door the passengers, as a general rule, begin to move before the doors have completely opened. On the other hand, with a side opening door the passengers tend to wait until the door has completely opened before moving.

The utilization of centre opening door also favours the doors opening and closing time periods. Given the same door speed, the centre opening door is much faster than the side opening type. It is beyond doubt that the centre opening door represents an increase in transportational capacity in the operation of a lift.

An example illustrating the application of the above consideration is given below:

Gross area per floor = 1200 m^2

Net usable area per floor = 1000 m^2

No. of landings including ground = 15

Assuming population density = 10 m^2/person

$$\text{Population in upper floors} = \frac{14 \times 1000}{10} = 1400 \text{ persons}$$

Taking 20 passengers per lift with a speed of 2.5 m/s. The calculated value of RTT = 165 s

$$Q = 20 \times 0.8 = 16$$

(*a*) Taking No. of lifts, $N = 6$

$$T = \frac{RTT}{6} = \frac{165}{6} = 27.5 \text{ s}$$

$$H = \frac{300 \times Q \times 100}{T \times P}$$

$$= \frac{300 \times 16 \times 100}{27.5 \times 1400}$$

$$= 12.5\%$$

(*b*) Taking No. of lifts $N = 8$

$$T = \frac{165}{8} = 20.63 \text{ s}$$

$$H = \frac{300 \times 16 \times 100}{20.63 \times 1400}$$

$$= 16.66\%$$

15.1.2 Positioning of Lifts

A thorough investigation should be made for assessing the most suitable position for lift(s) while planning the building. It should take into account future expansions, if any. Though each building has to be considered individually for purposes of location of lifts, factors influencing the locations of passenger and goods lifts are given below.

The location of lifts may also conform to the travel distance requirements.

Arrangement of Lifts. The lifts should be easily accessible from all entrances to the building. For maximum efficiency, they should be grouped near the centre of the building. It is preferable not to have all the lifts out in straight line and, if possible, not more than three lifts should be arranged in this manner. It has to be kept in mind that the corridor should be wide enough to allow sufficient space for waiting passengers as well as for through passengers.

In some cases when there are more than three lifts, the alcove arrangement, is recommended. With this arrangement, the lift alcove lead off the main corridor so that there is no interference by traffic to other groups or to other parts of the ground floor. This arrangement permits the narrowest possible corridors and saves space on the upper floors. Walking distance to the individual lift is reduced and passengers standing in the centre of the group can readily see all the lift doors and landing indicators. The ideal arrangement of the lifts depends upon the particular layout of the respective building and should be determined in every individual case.

15.1.2.1 Passenger Lifts

Low and Medium Class Flats. Where a lift is arranged to serve two, three or four flats per floor, the lift may be placed adjoining a staircase, with the lift entrances serving direct on to the landings. Where the lift is to serve a considerable number of flats having access to balconies or corridors, it may be conveniently placed in a well ventilated tower adjoining the building.

Office buildings, Hotels and High Class Flats. It is desirable to have at least a battery of two lifts at two or more convenient points of a building. If this is not possible, it is advisable to have at least two lifts side by side at the main entrance and one lift each at different sections of the building for inter-communication. When two lifts are installed side by side, the machine room shall be suitably planned with sufficient space for housing the machine equipment. The positioning of lifts side by side gives the following advantages:

(*a*) all machines and switchgear may be housed in one machine room,

(*b*) the lifts can be inter-connected more conveniently from an installation point of view, and

(*c*) greater convenience in service owing to the landing openings on each floor being adjacent.

Shops and Departmental Stores. Lifts in shops and stores should be situated so as to secure convenient and easy access at each floor.

Hospitals. It is convenient to place the passenger lifts near the staircases. For buildings with more than 12 floors, it is recommended to have provision of 1 stretcher/service lift in addition to the passenger lifts.

15.1.2.2 Goods Lifts

The location of lifts in factories, warehouses and similar buildings should be planned to suit the progressive movement of goods throughout the buildings, having regard to the nature of processes carried out in the building, the position of the loading platforms, railway sidings, etc. The placing of a lift in a fume or dust laden atmosphere or where it may be exposed to extreme temperatures, should be avoided wherever possible. Where it is impossible to avoid installing a lift in an adverse atmosphere, the electrical equipment should be of suitable design and construction to meet the conditions involved.

15.1.2.3 Hospital Bed Lifts

Hospital bed lifts should be situated conveniently near the ward and operating theatre entrances. There shall be sufficient space near the landing door for easy movement of stretcher.

15.1.3 Shape and Size of Lift Car

The shape and size of the passenger lift car bears a distinct relation to its efficiency as a medium of traffic handling. The width of the lift well entrance is, in reality, the basic element in the determination of the best proportion. The width of the car is determined by the width of the entrance and the depth of the car is regulated by the loading. Centre opening doors are the most practicable and the most efficient entrance units for passenger lifts.

15.1.4 Positioning of Machine Room

The machine room should as far as possible, be placed immediately above the lift well as this has several advantages, such as reduced load on the building, lower capital cost of the lift, a smaller lift well for a given size lift car and reduced power consumption compared with a machine room in the basement, renewal of suspension ropes is less frequent and the cost of such renewals is less because shorter ropes are required and time taken for fitting them is less.

If a machine room on the lift well is impracticable for architectural or other reasons, the machine room may be placed below the lift well or in the basement, but guidance of a lift engineer should be followed on each instance, to minimize the disadvantage of its being so placed.

ESCALATORS

Escalators are deemed essential where the movement of people, in large numbers at a controlled rate in the minimum of space, is involved, for example, airports, railway stations etc. In exhibitions, big departmental stores and the like, escalators encourage people to circulate freely and conveniently.

As the escalators operate at a constant speed, serve only two levels and have a know maximum capacity, the traffic study is rather easy. Provided the population to be handled in a given time is known, it is easy to predict the rate at which the population can be handled.

For normal peak periods, the recommended handling capacities for design purposes should be taken as 3200 to 6400 persons per hour depending upon the width of the escalator.

The number of persons that may be theoretically carried by the escalator in one hour can be determined as follows:

(*a*) It is assumed that one step with an average depth of 0.4 m can carry 1 person for a step width of 0.6 m, 1.5 person for a step width of 0.8 m and 2 persons for a step width of 1 m.

(*b*) The theoretical capacity is given by

$$\frac{3600 \times (\text{rated speed in m/s} \times \text{k})}{0.4}$$

where $k = 1$, 1.5 or 2 for 0.6, 0.8 and 1.0 m step widths.

(*c*) Some values calculated as per the above are:

Step Width	*Theoretical Capacity in Persons/hour*		
	0.5 m/s *speed*	**0.65 m/s** *speed*	**0.75 m/s** *speed*
0.6 m	4 500	5 850	6 750
0.8 m	6 750	8 775	10 125
1.0 m	9 000	11 700	13 500

15.2 ESSENTIAL REQUIREMENTS

(1) Angle of inclination shall not be in excess of 30° from the horizontal excepting that with an escalator having a vertical rise not exceeding 6 m an angle up to 35° may be permitted.

The rated speed of the escalator shall not exceed 0.75 m/s for an angle of inclination up to 30° and 0.5 m/s for inclinations between 30° to 35°.

(2) The horizontal distance (measured at right angles to the direction of travel) between the balustrade interior panelling at lower points shall be equal to or lower than the horizontal distance measured at points higher up. The maximum distance between the balustrade interior panelling at any point shall be less than the distance between handrails.

(3) The parts of the balustrade facing the steps shall be smooth. Covers or strips not in the direction of travel shall not project more than 3 mm. They shall be sufficiently rigid and have rounded or bevelled edges. Covers or strips of such nature are not permitted at the skirting.

Cover joints in the direction of travel (in particular between the skirting and balustrade interior panelling) shall be arranged and formed in such a manner that the risk of trapping is reduced to a minimum.

Gaps between interior panels of the balustrade shall not be wider than 4 mm. The edges shall be rounded off or bevelled. The balustrade interior panelling shall have adequate mechanical strength and rigidity. When a force of 500 N is applied at any point of the panelling at right angles on a area of 2 500 mm^2. There shall be no gap greater than 4 mm and no permanent deformation (setting tolerances are permitted).

The use of glass for balustrade interior panelling is permitted with the approval of the Authority; further, provided it is splinter free one layer safety (tempered) glass and has sufficient mechanical strength and rigidity. The thickness of the glass shall not be less than 6 mm.

There shall be no abrupt changes in the width between the balustrades on the two sides of the escalator. Where a change in width is unavoidable, such change shall not exceed 8 per cent of the greatest width. In changing the direction of the balustrades resulting from a reduction in width the maximum allowable angle of change in balustrades shall not exceed it from the line of the escalator travel.

Where the skirting of the escalator is placed beside the steps the horizontal clearance shall not exceed 4 mm at either side and 7 mm for the sum of the clearances measured at both sides at two direction opposite points.

Where the building obstacles can cause injuries appropriate preventive measures shall be taken. In particular, at floor intersections and on criss-cross escalators, a vertical obstruction of not less than 0.5 m in height (not presenting any sharp cutting edge for example as an imperforate triangle) shall be placed above the balustrade decking. It is not necessary to comply with this requirement when the distance between the centre line of the handrail and any obstacle is equal to or greater than 0.5 m.

For escalators arranged adjacent to one another either parallel or criss-cross, the distance between the edges of the handrails shall not be less than 120 mm.

(4) *Handrails*

(*a*) Each balustrade shall be provided with a handrail moving in the same direction and at substantially the same speed as the steps.

(*b*) Each moving handrail shall extend at normal handrail height not less than 300 mm beyond the line of points of combplate teeth at the upper and lower landings.

(*c*) Hand or finger guards shall be provided at the point where the handrails enters the balustrade.

(*d*) The width of the handrail shall be between 70 mm and 120 mm. The distance between the handrail and the edge of the balustrade shall not exceed 50 mm. The distance between centreline of handrails shall not exceed the distance between the skirtings by more than 0.45 m.

(5) *Step Treads*

(*a*) The depth of any step tread in the direction of travel shall not be less than 40 cm and the rise between treads shall be not more than 22 cm. The width of a step tread shall be not less than 40 cm nor more than 102 cm.

(*b*) The maximum clearance between step treads on the horizontal run shall be 4 mm.

(c) The tread surface of each step shall be slotted in a direction parallel to the travel of the steps. Each slot shall be not more than 6.5 mm wide and less than 9.5 mm deep; and the distance from centre to centre of adjoining slots shall be not more than 9.5 mm.

(6) *Landing*

The landing area of escalators shall have a surface that provides a secure foot hold for a minimum distance of 0.85 m measured from the root of the comb teeth. Exempted from these are the combs.

(7) *Combplates*

There shall be a combplate at the entrance and at the exit of every escalator. The combplate teeth shall be meshed with and set into the slots in the tread surface so that the points of the teeth are always below the upper surface of the treads. Combplates shall be adjustable vertically.

(8) *Trusses or Girders*

The truss or girder shall be designed to safely sustain the steps and running gear in operation. In the event of failure of the track system it shall retain the running gear in its guides.

(9) *Step Wheel Tracks*

This shall be designed to prevent displacement of steps and running gear if a step chain breaks.

(10) *Driving Machine, Motor and Brake*

(*a*) The driving machine shall be connected to the main drive shaft by toothed gearing, a coupling, or a chain.

(*b*) An electric motor shall not drive more than one escalator.

(*c*) Each lift shall be provided with an electrically released, mechanically applied brake capable of stopping the up or down travelling escalator with any load up to rated load. This brake shall be located either on the driving machine or on the main drive shaft.

Where a chain is used to connect the driving machine to the main drive shaft, a brake shall be provided on this shaft. It is not required that this brake be of the electrically released type if an electrically released brake is provided on the driving machine.

(*d*) *Speed Governor*

Escalators shall be equipped in such a way that they stop automatically before the speed exceeds 1.2 times the rated speed.

QUESTIONS

1. When does it become necessary to provide lifts in a building? Give the components of a lift.
2. Draw sketches for the following types of lifts:
   (*a*) Passengers lift (*b*) Goods lift
   (*c*) Hospital lift, and (*d*) Dumb waiter.
3. Discuss the design considerations for the provision of lifts in a building.
4. What factors influences the positioning of passenger and goods lift in a building?
5. What is an escalator? Give the essential requirements for its provision.

REFERENCES

1. Bureau of Indian Standards, New Delhi, National Building Code of India 2005—Part 8, Section 5: Installation of lifts and Escalators.
2. **IS: 1860—1980:** Code of practice for installation, operation and maintenance of electric passenger and goods lifts.
3. **IS: 4666—1980:** Specification for electric passenger and good lifts.
4. **IS: 4591-1968;** Code of practice for installation and maintenance of escalators.
5. **IS: 14671-1999:** Code of practice for installation and maintenance of hydraulic lifts.

Roofs

The function of a roof is to give a protective covering to the building so that rain, snow or wind may not damage the building. It is erected at the highest part of the building and is made up of a framework over which a covering is laid. Roofs are constructed keeping in view the climate of the place and the materials available. Sometimes architectural considerations may also have an important bearing on the selection of roof type. The general types of roofs are listed below:

(*a*) Sloping roofs,

(*b*) Flat roofs,

(*c*) Shelled roofs, and

(*d*) Domes.

SLOPING ROOFS

The various shapes which can be given to roofs of this type depend on the area covered, materials available, type of lighting and ventilation needed inside, available appliances etc.

Some important types of sloping roofs are given as under:

(i) Shed roof: This is the simplest type and slopes only in one direction. It is used for smaller spans.

(ii) Gable roof: This roof slopes in two directions and is used commonly.

(iii) Hip roof: This slopes in four directions.

(iv) Gambrel roof: This roof also slopes in two directions but there is a break in slope.

(v) Mansard roof: This slopes in four directions but there is break in slopes.

(vi) Saw-tooth or north-light roof: This is used in factories, etc. where light is admitted from the glazing fixed on the steep sloping sides of the roof.

16.1 WOODEN SLOPING ROOFS

16.1.1 Terms Used

(i) Ridge: Apex line of a sloping roof.

(ii) Ridge piece: A member which runs horizontally at the highest level of the roof.

(iii) Hip: External angle of a sloping roof at which the roof slopes are turned down.

(iv) Common rafters: Members supporting the battens or boardings under the roof covering.

(v) ***Hips rafters:*** These are wooden members which form the hip of the roof and to which the common rafters are attached.

(vi) ***Eaves:*** These are the lower edges of the sloping surface of a roof.

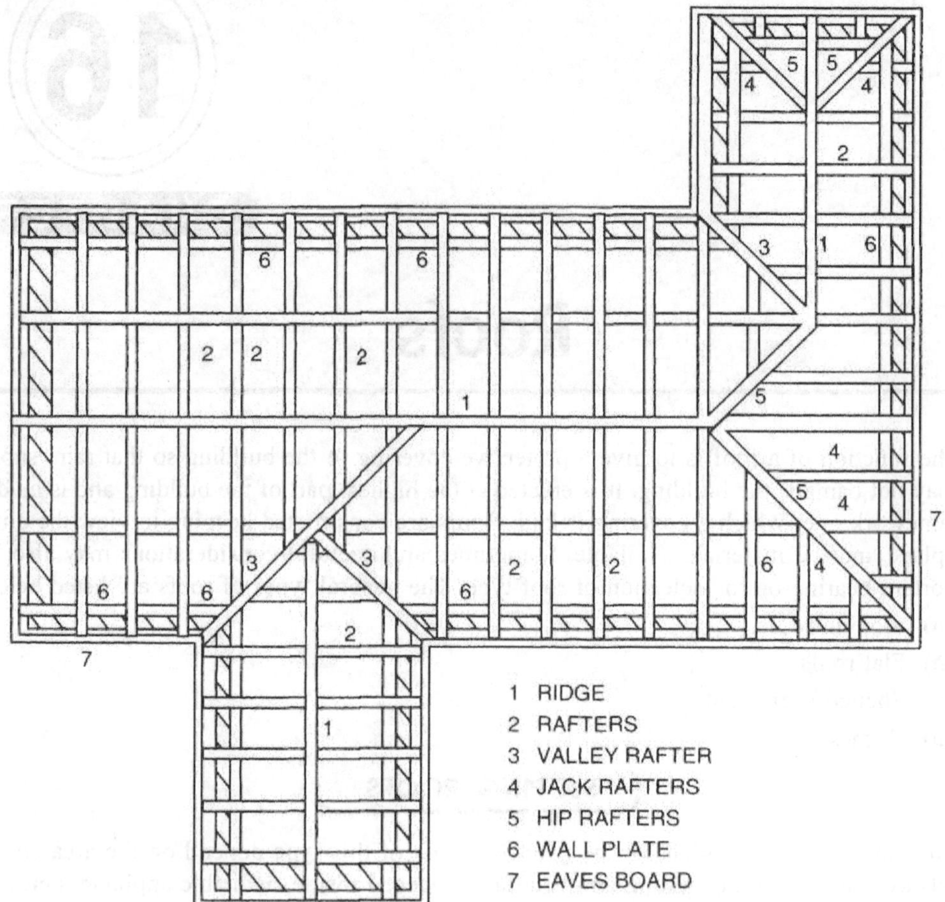

Fig. 16.1: Terms used in sloping roof construction.

(vii) ***Eave-board:*** Wooden board fixed along the eaves connecting the common rafters.

(viii) ***Gable:*** Wall which follows the slope of the roof from eave to ridge and covers the end of a roof.

(ix) ***Barge board:*** Wooden planks used to fix the ends of common rafters projecting beyond the sloping top of a gabled wall.

(x) ***Valley:*** A depression formed at the intersection of two sloping roofs at their junction.

(xi) ***Jack rafter:*** Short lengths of rafters fixed to the hip rafters and eave's board.

(xii) ***Valley rafter:*** Member at the intersection of two inverted slopes to which rafters are fixed.

(xiii) ***Dragon beam:*** Short member of wood which carries the foot of a hip rafter which is attached to the tie fixed across the angle formed by the intersection of two wall plates at the corner of a building.

(xiv) ***Purlins:*** These are members laid horizontally to support the common rafters. They transmit the loads to the trusses or walls.

(xv) ***Purlin cleat***: Short piece of timber bolted to the rafters of a roof truss for fixing the purlins.

(xvi) ***Truss***: A framework of triangles.

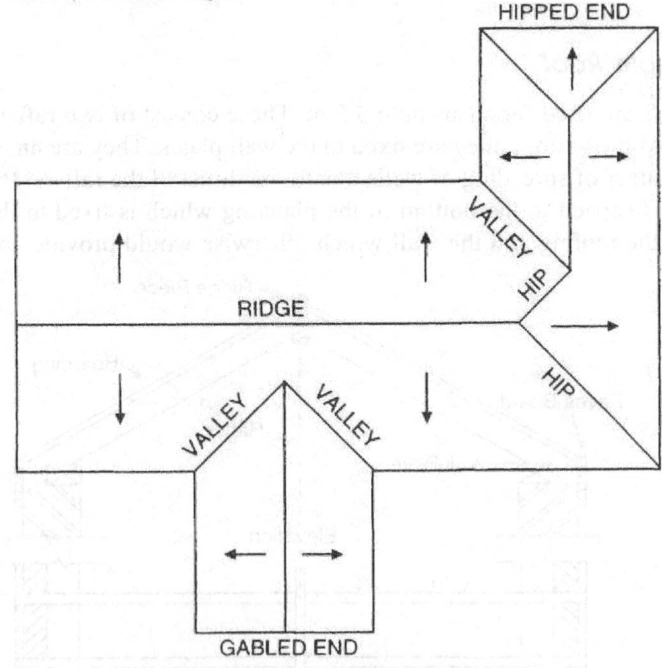

Fig. 16.2: Sloping roof junctions.

16.1.2 Types of Wooden Sloping Roofs

16.1.2.1 Lean to Roof

This is a roof which covers the verandah of a building and projects from the main wall of the building. This is suitable for spans upto 2.5 m. If, however, slightly greater spans are used, purlins may be added. On one side, the common rafters are supported on a wall plate which in turn rests on a projecting corbel from the wall. The lower side may also rest on the wall plate or may be fixed to a vertical post with bolts and nuts in case the verandah is exposed.

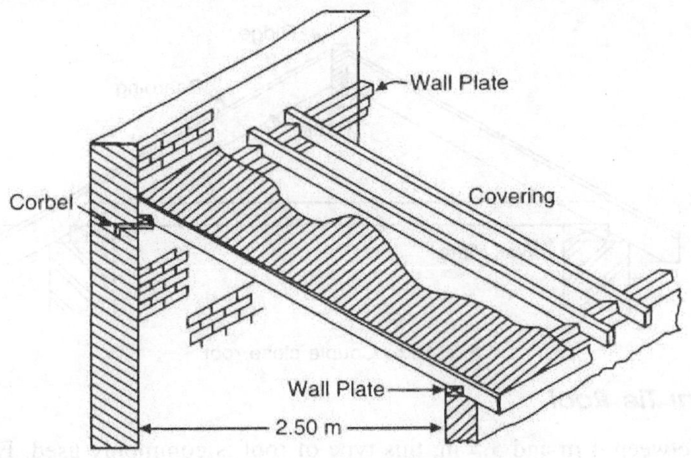

Fig. 16.3: Lean-to-roof.

The covering may be fixed to battons running across the rafters or a wooden boarding directly fixed to the rafters. It is essential that the leakage at the junction of the roof with the wall be prevented.

16.1.2.2 Couple Roof

Couple roofs are used for spans upto 3.5 m. These consist of two rafters with a central ridge piece at the top. At the bottom, they are fixed to the wall plates. They are unsuitable for large spans because of the danger of spreading of walls due to the thrust of the rafters. Brickwork between the common rafters is carried to the bottom of the planking which is fixed to the rafters. This leaves no gap between the roofing and the wall which otherwise would provide an inlet for birds, etc.

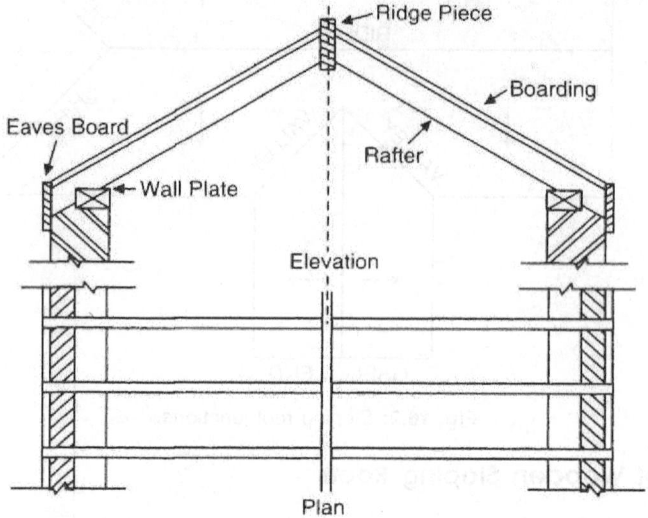

Fig. 16.4: Couple roof

16.1.2.3 Couple Close Roof

This consists of two rafters arranged as above and connected by a wooden member which acts as a tie. This horizontal tie prevents the outward spreading of the walls and can also act as a support for any ceiling which might be fixed beneath it. These roofs can be used for spans upto 4.5 m.

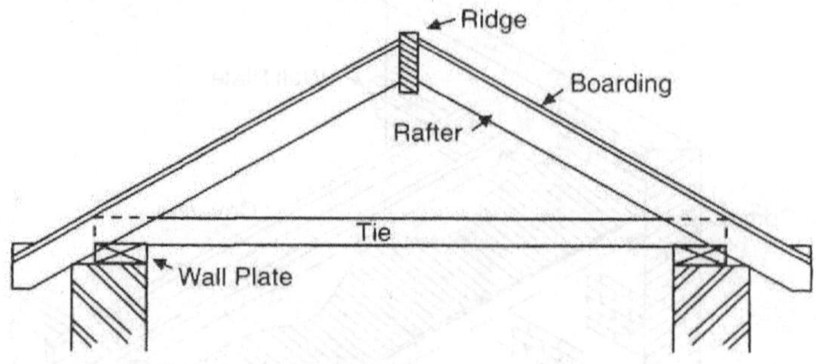

Fig. 16.5: Couple close roof.

16.1.2.4 Collar Tie Roof

For spans between 4 m and 5.5 m, this type of roof is commonly used. Each pair of rafters has a collar, *i.e.*, a wooden section of the same width as the rafter. This is fixed at a height of half to one-third of vertical height between the wall and the ridge. It is better to place it as low as possible for giving maximum strength to the roof. This collar subjects the rafters to excessive bending stresses and it is essential that the joint between the collar and the rafter be perfect. Any deflection of the rafter will try to spread the walls away. The collar is dovetailed with the rafter. Bolts may be used as an additional safety.

16.1.2.5 Collar and Tie Roof

When the roof span exceeds the previous given values, a combination of collar tie and couple close roof is used. In this, the rafters are supported by purlins which rest at the ends on walls. A collar and struts are used to support the purlins and the rafters. Advantage is taken of any partition wall, etc., to support the ceiling joist. This roof is used where purlins can be supported at the ends economically.

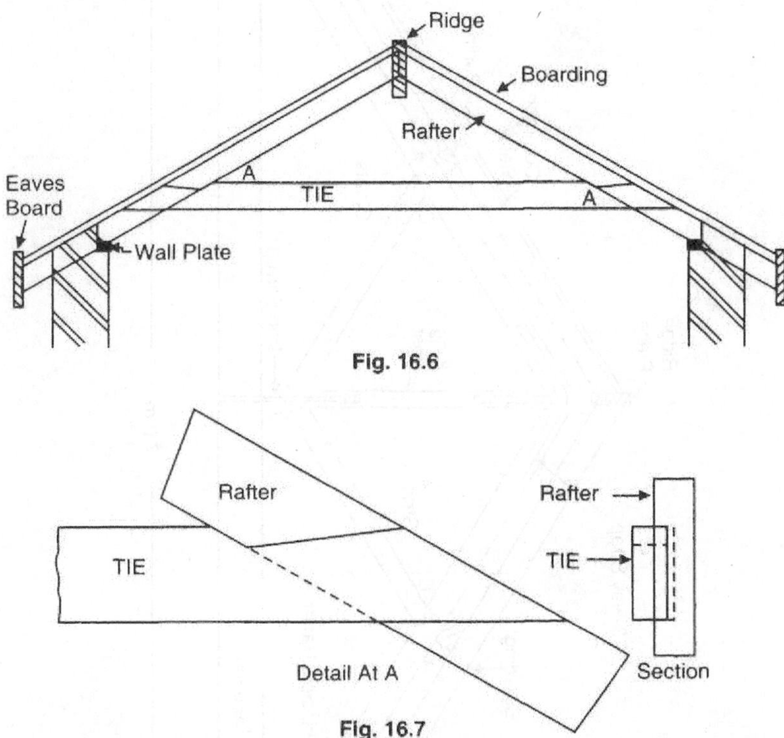

Figs. 16.6 and 16.7: Collar tie roof.

16.1.2.6 King-Post Truss

When the spans are large or intermediate supports for purlins and ties are not available, trusses are used. The total weight of the roof is carried on to the walls vertically. The framework of a truss is built in such a way that it does not alter its shape when loaded. Triangular shape of the frame offers greater rigidity. Members of trusses are subjected to direct stresses of compression or tension, the magnitude of bending stresses is negligible. To ensure this, the central lines of the connecting members should meet at one point and the loads should be applied at the apex points of the triangles.

A king-post truss is used for spans of 6 m to 9 m. The components of a king-post truss are:

(*a*) A central upright member called king-post.

(*b*) Inclined rafters called principal rafters.

(c) A horizontal connecting member called tie-beam.
(d) Two struts which give support to the rafters.
(e) Purlins running over the principal rafters from truss to truss.
(f) Ridge piece and common rafters for supporting the roof coverings.

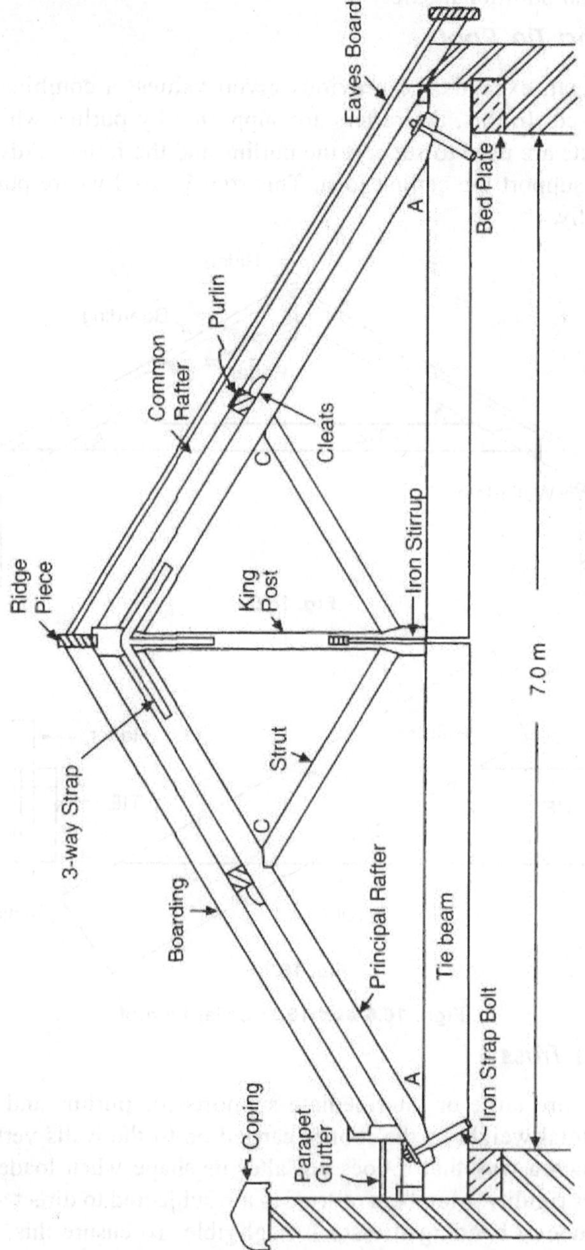

Fig. 16.8: Elevation of a king-post truss.

The weight of the ridge is carried down the principal rafters to the tie beam which prevents the feet of the rafters from spreading out. The weight carried by the purlin is placed in the centre of the principal rafter and is carried through the struts to the foot of the king-post. The load that is taken to the foot of the king-post is transmitted to the tie beam which in turn transmits it to the supports. In this manner, the principal rafters and the struts act as compression members whereas the tie beam and the king-post act as tension members. Hence the connecting joints have to be designed accordingly.

JOINTS AND FASTENINGS OF KING-POST TRUSS

The joint between the principal rafter and the tie is a bridle joint. A central tenon is formed in the tie and the foot of the rafter is shaped accordingly. The width of the tenon should not be greater than one-third of the beam width. The cuts of the notches on the tie beam are quite long and extend to the central line of the rafter. The bearing face of the notch is made at right angles to the central line of the rafter. The joint is further strengthened by passing a bolt between the tie and the rafter as shown. Alternatively a strap and a plate may be used to connect these (Fig. 16.10). It should be ensured that the principal rafter's central line and the central line of the tie beam meet at a point which lies on the vertical line passing through the centre of the supporting wall.

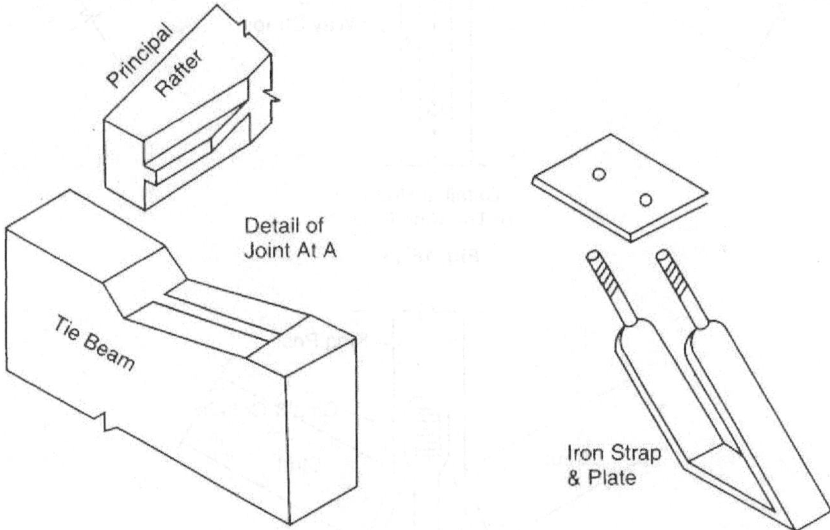

Fig. 16.9: Joint between principal rafter and tie beam. **Fig. 16.10**

The joint between the principal rafter and the king-post is made by cutting a tenon in the principal rafter and the corresponding mortice into the head of the king-post. The top edge of the tenon and the corresponding edge in the mortice are made horizontal whereas the lower edge is parallel to the slope of the rafter. The tenon is not greater than 30 mm in depth. The rafters and the king-post are connected by iron straps bolted together.

The joint between the king-post and the tie beam may be an ordinary mortice and tenon joint or a stub-tenon joint. The king-post and the tie-beam are connected by iron stirrup to strengthen the joint further as shown in Figs. 16.13 and 16.14.

Joints between the strut and the king-post or the rafter are also of mortice and tenon type. The joint at the head of the strut (*See* Fig. 16.15) has an oblique tenon which is housed into the rafter. The joint at the foot of the strut is of a single abutment tenon type. The king-post is shaped to form a good support for the struts and the lower edge of the tenon is cut horizontally. Double abutment tenon joint is also used.

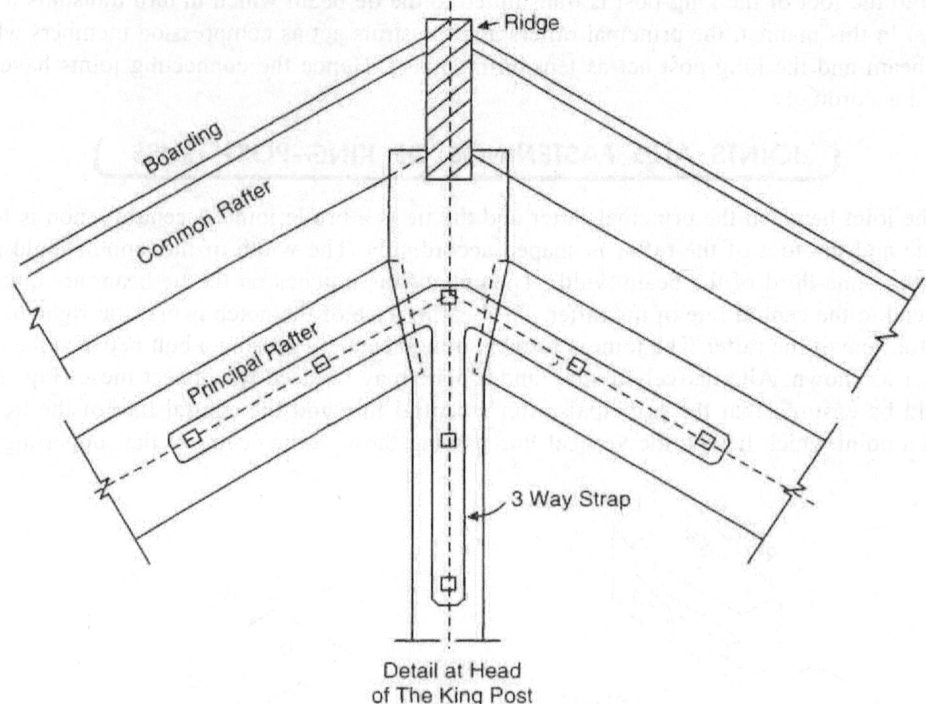

Fig. 16.11

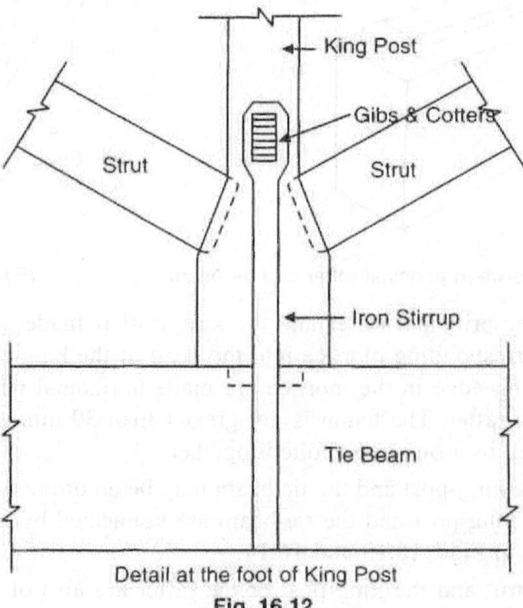

Fig. 16.12

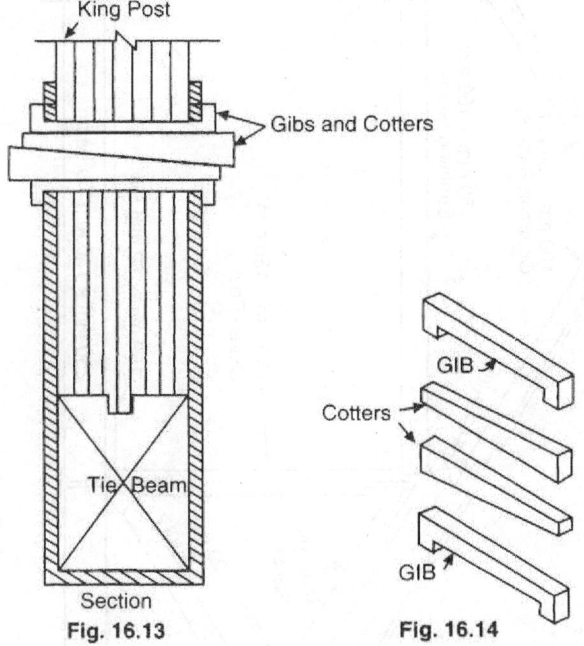

Fig. 16.13 **Fig. 16.14**

Figs. 16.13 and 16.14: Joint between king-post and tie-beam.

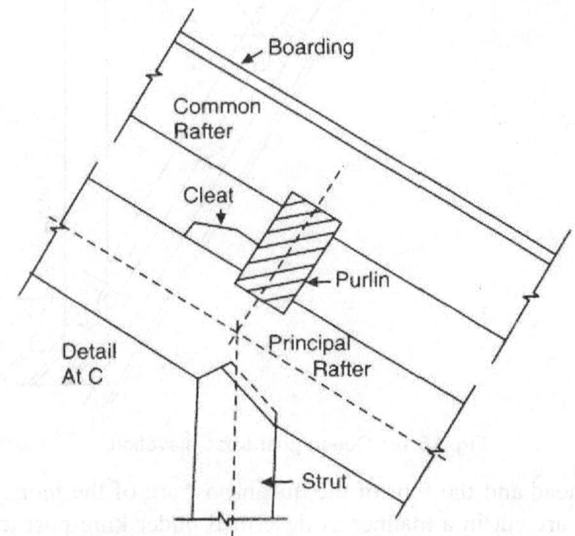

Fig. 16.15: Joint at the head of strut.

16.1.2.7 Queen-Post Truss

Queen-post trusses are used for spans upto 13.5 metres. It consists of two queen-posts, two principal rafters, struts, tie beam, straining beam, straining sill, purlins, etc., (*See* Fig. 16.16). The struts, as usual, will cause thrust on the foot of the queen-post and to resist this force a straining sill is fitted in-between the two posts. A straining beam is fixed in-between the heads of the queen posts to resist the thrust which the rafters give to the posts. The principal rafters, straining beams and struts are in compression whereas the queen-post and the tie beam are in tension.

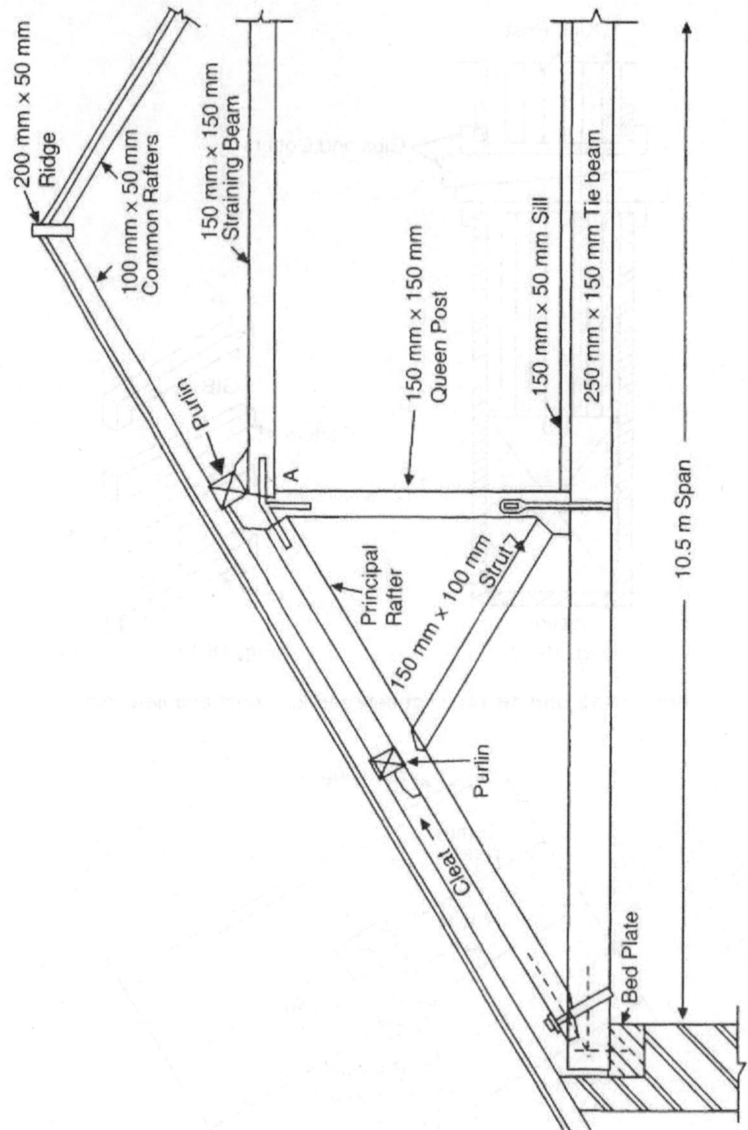

Fig. 16.16: Queen-post truss elevation.

The joints at the head and the foot of the queen-post are of the mortice and tenon type and the edges of the tenons are cut in a manner as described under king-post truss.

The usual sizes of the timber for various members of the king and queen post trusses are given in different handbooks. However the following rules may be used in determining the approximate sizes.

(*i*) Thickness of king-post truss in millimetres = 5 × Span in *m*

(*ii*) Thickness of queen-post truss in mm = 4 × Span in *m*

(*iii*) Area of cross-section of principal rafter in sq. mm = 1500 × Span in *m*

(*iv*) The section of king-post, queen post and struts should be square.

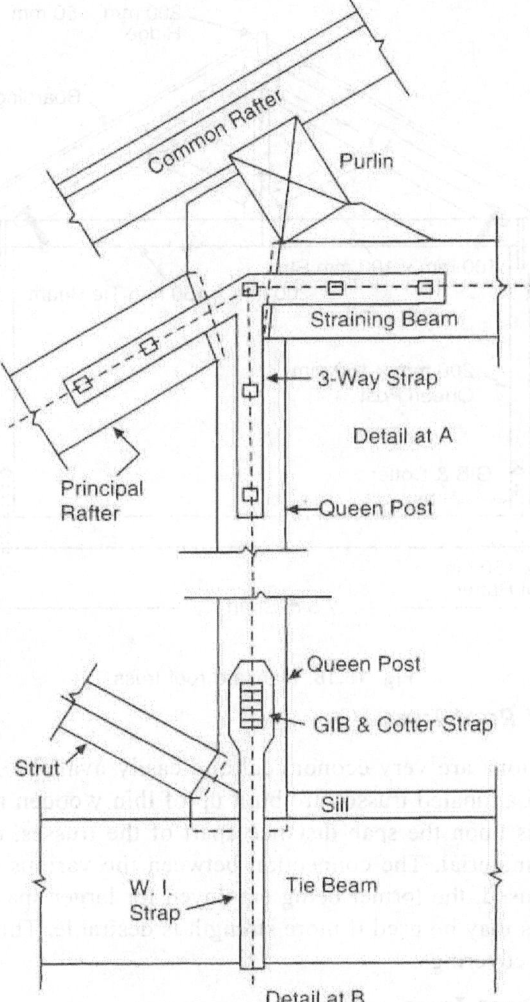

Fig. 16.17: Details of queen-post truss.

(*v*) The depth of the tie beam in centimetres should be from 5 to $6\frac{1}{2}$ times the thickness of the truss.

(*vi*) Depth of straining beam should be about $\frac{3}{4}$th the depth of tie beam.

16.1.2.8 Mansard Roof Truss

The advantage of this type of truss is that a room can be made inside the roof. This is a two-sloped truss and an approximate shape can be visualised by placing a king-post truss over a queen-post truss. The lower slope should not be steeper than 75° and the upper slope should not be greater than 30°. The construction of various joints is similar to the king and queen-post trusses.

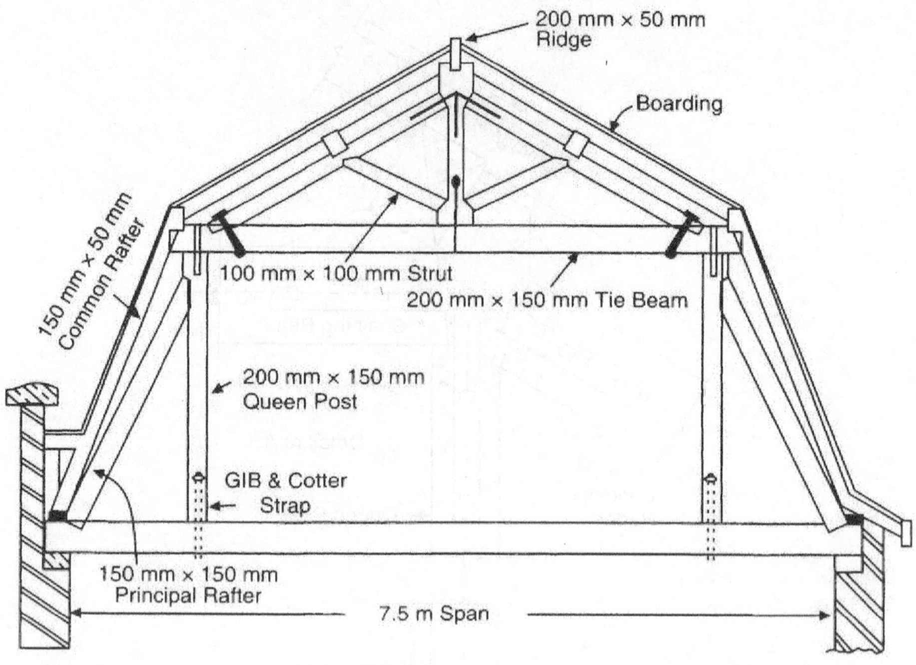

Fig. 16.18: Mansard roof truss.

16.1.2.9 Laminated Roof Truss

Thin wooden sections are very economical and easily available. Moreover they transmit lesser load to the truss. Laminated trusses are built up of thin wooden members. The design and size of members depends upon the span distance apart of the trusses, quality of timber and the weight of the covering material. The connection between the various members is very simple. Either bolts or nails are used; the former being employed for larger spans and the latter for small spans. Special connectors may be used if more strength is desirable. This type of roof is specially suitable for light weight coverings.

16.1.2.10 Bel-Fast Roof Truss

This is also called a Bowstring or latticed roof truss. It is made of thin sections of timber and can be used for big spans upto 30 m but lighter roof coverings are to be used. The central rise is about one-eighth of the span.

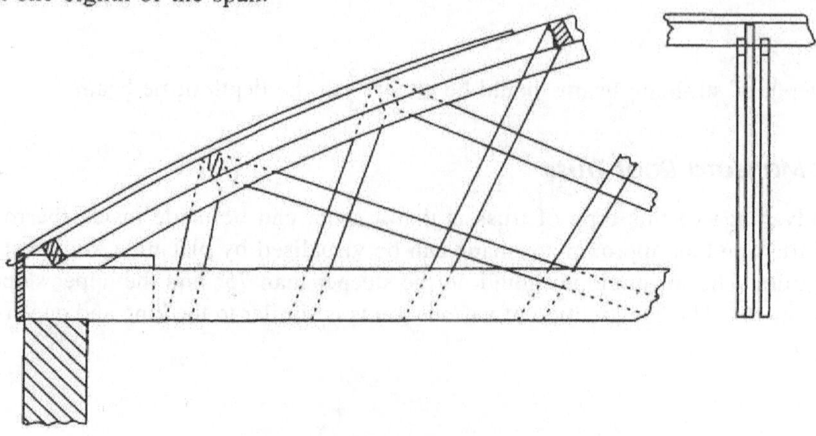

Fig. 16.19: Bel-fast roof truss.

16.2 COMPOSITE SLOPING ROOFS

Roof trusses built of timber and steel are termed composite roof trusses. The tension members if made of wood are very bulky because their section is reduced at the joints. Hence if steel bars are used for tension members, an economical construction is feasible. However special fittings have to be used where the wooden members join the steel members. The joint should be so designed as to enable the use of cast or forged fittings. The types of trusses of composite nature are shown in Figs. 16.20 to 16.23.

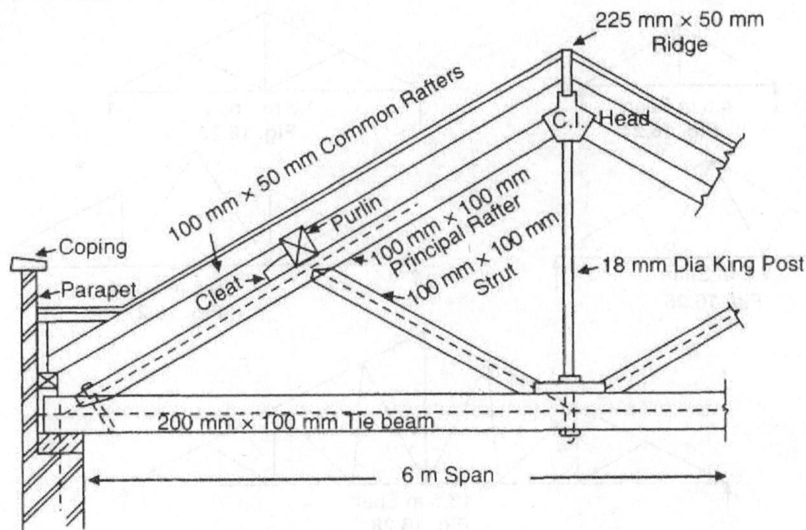

Fig. 16.20: Composite king-post truss.

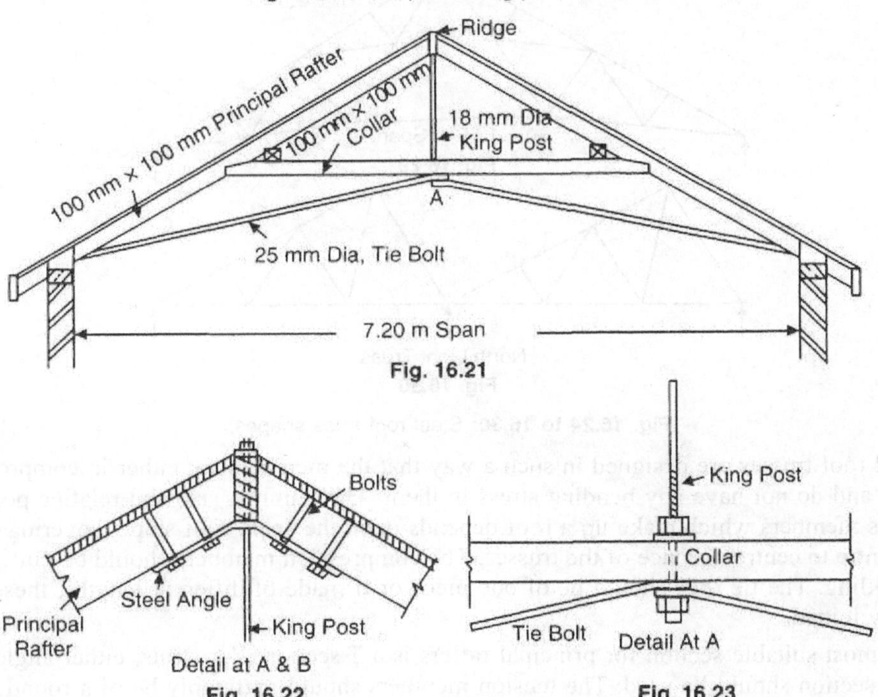

Figs. 16.21 to 16.23: Composite collar and tie truss with details.

16.3 STEEL SLOPING ROOFS

The various types of steel sloping roofs are as given below:

(*i*) Open trusses.

(*ii*) North light trusses.

(*iii*) Bow-string trusses.

(*iv*) Arched rib trusses and solid arched ribs.

The various shapes of these are shown in line diagrams in Figs. 16.24 to 16.30.

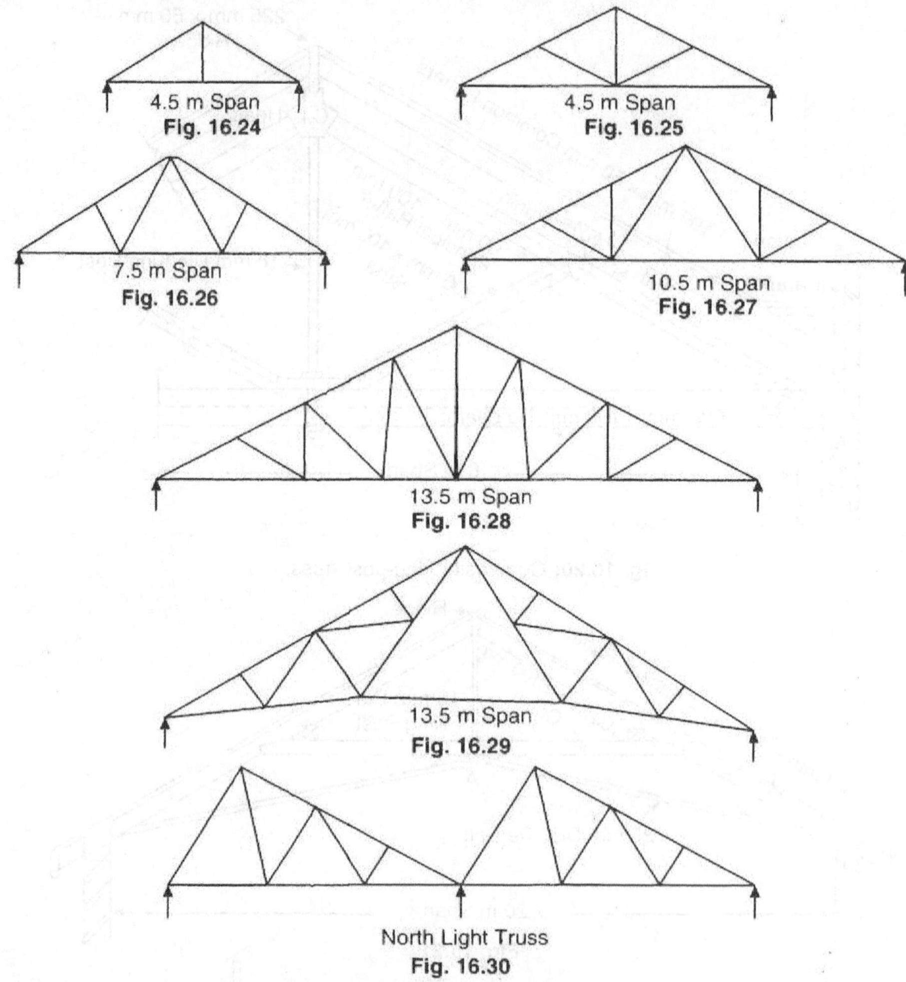

Fig. 16.24 to 16.30: Steel roof truss shapes.

Steel roof trusses are designed in such a way that the members are either in compression or in tension and do not have any bending stress in them. The number, size and relative position of the various members which make up a roof depends upon the span, roof slope, covering material and the centre to centre distance of the trusses. The compression members should be quite short to avoid buckling. The tie rods should be of one piece or if made of different lengths, these should be suitably joined.

The most suitable section for principal rafters is a T-section. For struts, either angle iron or a channel section should be used. The tension members should preferably be of a round or a flat section. The various members may also be built to two or more sections, *e.g.*, a principal rafter may be made of two angles placed side by side. In an ideal design all members of the structure should fail simultaneously. However due to practical considerations, small sections, *e.g.*, angles less than 50 mm × 50 mm × 6 mm are not used. Allowance should be made in the selection of sections for stresses imposed on account of the load not being applied along the central line of the member.

Joints: The various members are connected to each other by bolts, rivets and thin plates called gussets. The minimum pitch of the rivets should not be less than three times the diameter of the rivets. The maximum pitch is 150 mm for compression members and 200 mm for tension members. Further a minimum distance from the centre of the rivet to the edge of the member shall not be less than 25 mm for 15 mm diameter rivets. Generally 15 mm diameter rivets are used for smaller spans and 20 mm rivets for large spans. At least two rivets should be used for all connections. Gusset plates should be designed according to the forces coming on the members but should not be less than 6 mm in thickness.

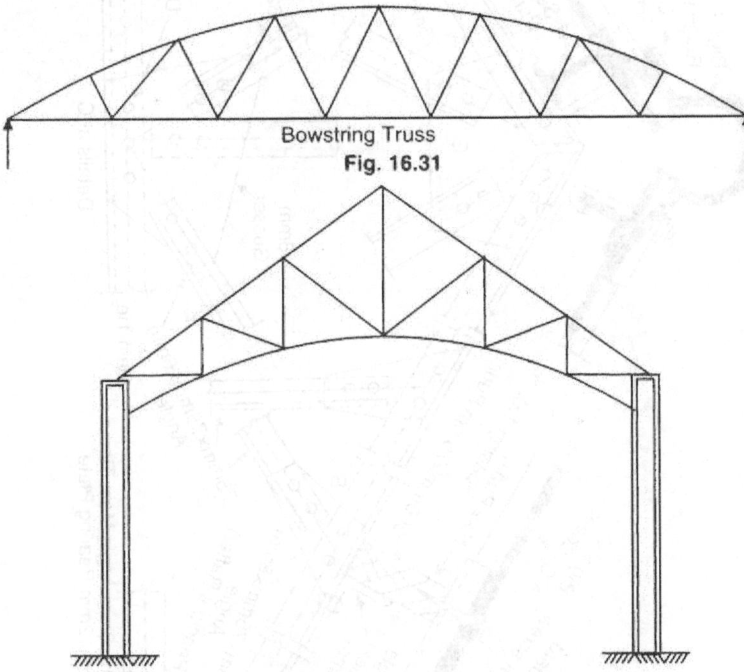

Fig. 16.31 and 16.32: Bow-string type steel roof trusses.

Some of the roof trusses are detailed below:

16.3.1 Trusses for Small Spans (Upto 7 m)

The ends of these small trusses rest on bed plates which may be of stone or concrete. The ends are further bolted down with rag bolts which hold the truss down.

For small spans, all members may consist of angles connected with gusset plates. The diagonal tension members are connected on one side of the gusset plate whereas the main tension member is connected on the other side. At the foot of the truss a short angle is fitted on both sides of the gusset which rests on a bearing plate. The holes in the bearing plate are made in such a fashion that they correspond to the holes in the angle cleats and the holes in the bed plates so that the rag bolts can pass through. The cleat holes are made slightly bigger than the diameter of the rag bolt to allow for the small variation in dimensions.

At the apex, the wooden ridge piece is secured by two bent plates and are bolted or rivetted to the main rafters. The purlins, if made of wood, are fixed to the rafters by angle iron cleats which are bolted or rivetted to both of them. Whenever joints occur at the purlins, longer cleats are used.

For this span, 15 mm diameter rivets are used.

16.3.2 Trusses for Large Spans

The arrangement of members for larger spans is shown in Figs. 16.33 and 16.34. The members may consist of more than one section. They may be two angles or channels or flats connected by a gusset with the other members at the joints. The section of the member can be reduced as the force in it gets decreased.

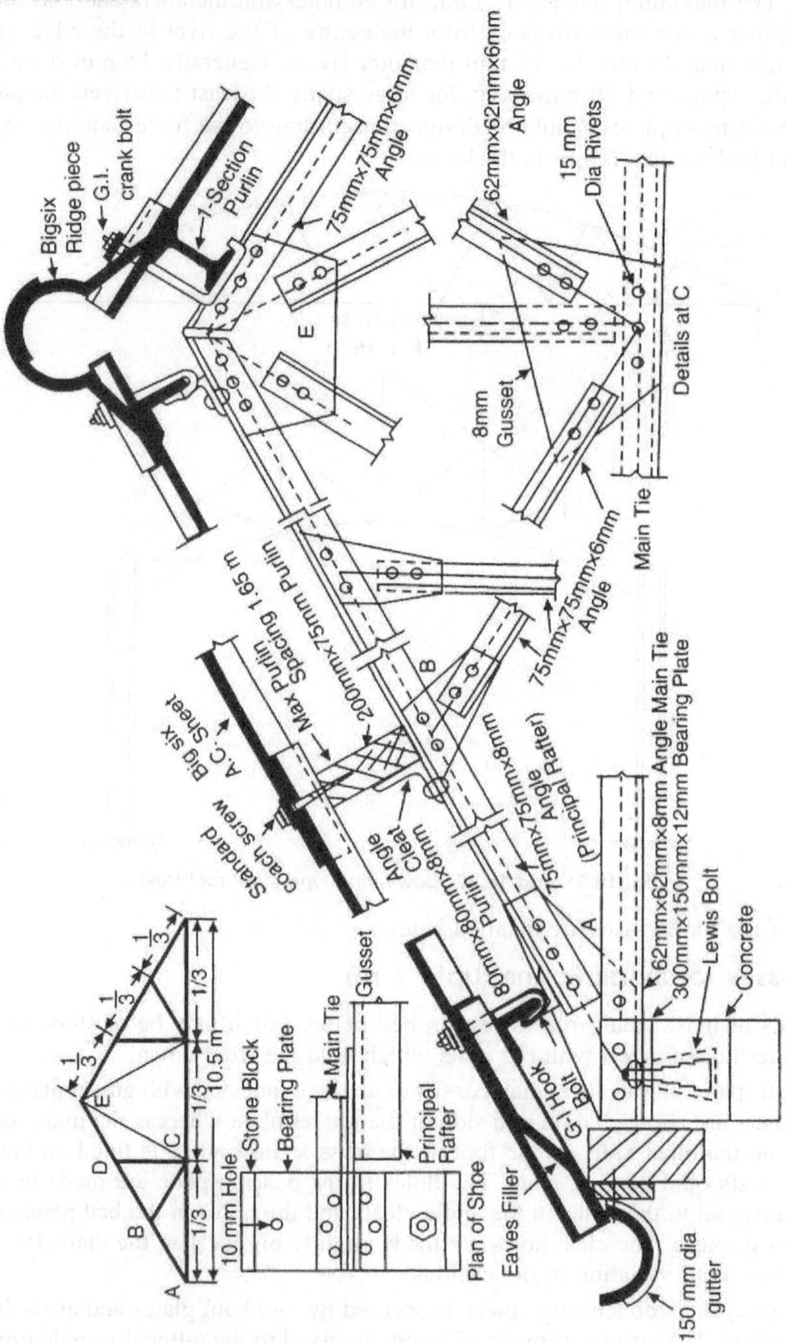

Figs. 16.33 and 16.34: Roof truss details.

The method of joining and construction is similar to the one described above except that bigger trusses cannot be rivetted completely in the fabricating shops. For this reason, the truss is rivetted in two portions at the shop and gusset plates at the connected ends are rivetted to them. The two halves are erected and finally rivetted at site.

Bigger bearing plates are used for these trusses and the cleats are connected to them by counter sunk rivets. Slotted holes are left in the bearing plates for fitting the rag bolts.

FLAT ROOFS OR TERRACED ROOFS

Generally roofs of this type are used in areas which have less rainfall or have no snow-fall during the year. Sloping roofs have a very good advantage that they can be drained easily but the drainage of flat roofs, is a big problem. Whenever materials which are not impervious or members frequently joined are used in the construction of roofs, drainage is difficult. Hence wooden roofs, as a rule, are made a little sloping and reinforced concrete roofs are flat. Moreover wooden roofs are constructed of small members which if arranged in a trussed form could be used for a larger span. Wooden flat roofs are used only for small spans.

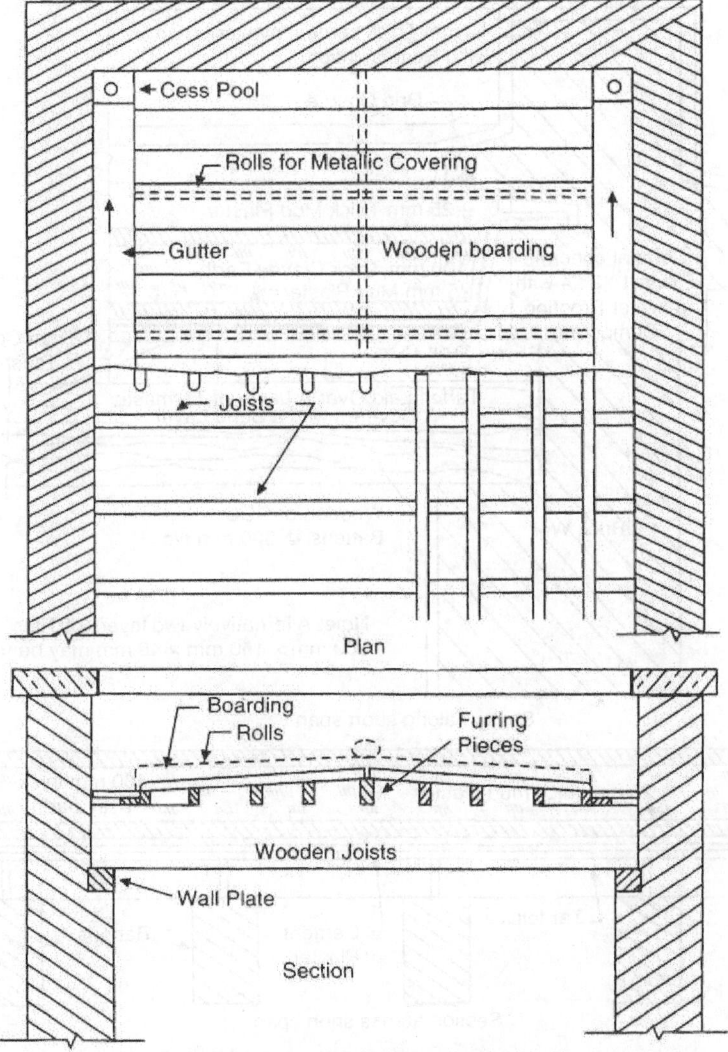

Figs. 16.35 and 16.36: Wooden flat roofs.

16.4 WOODEN FLAT ROOFS

The roofs are generally having a slope of less than 10°. They are in fact similar to wooden floors but have a steep slope on the top to drain off water. They consist of joists which are covered with cross girders or planking. The method of fixing the joists, the cross-girders and the boarding or planking is described under the chapter on wooden floors. However some important points concerning wooden flat roofs are noted below:

(*a*) The slope to be given to the roof should not be less than 1 in 50. However steeper slopes than this are desirable.

(*b*) The direction of slope is dependent on the position of outlet for water.

(*c*) The boarding should be laid parallel to the slope of the roof.

(*d*) The slope on the boarding is formed by fixing small tapering sections of wood (furrings) either parallel to or at right angles to the joists. The latter method is to be preferred.

(*e*) The roof should be given a suitable covering for water proofing and for prolonging its life.

A section of simple type of roof is shown in Figs. 16.35 and 16.36.

16.5 BATTENED FLAT ROOF WITH MUD TERRACE

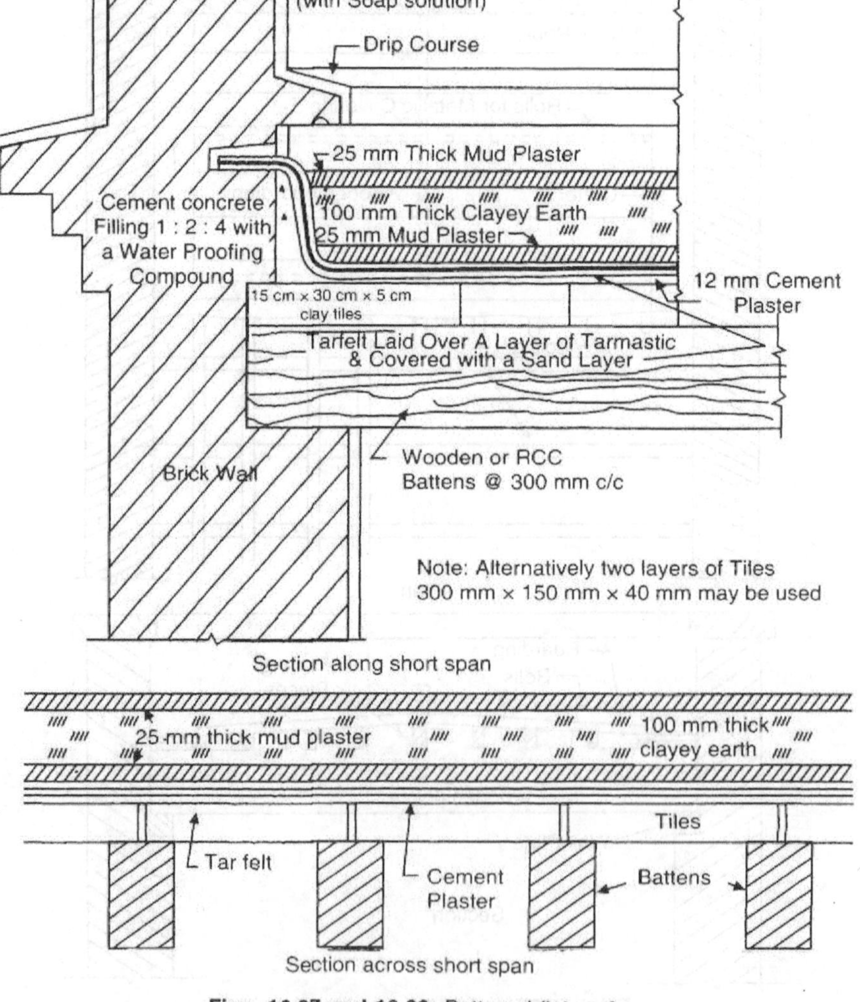

Figs. 16.37 and 16.38: Battened flat roofs.

This type is made up of wooden or R.C.C. joists (battens) (See Figs. 16.37 and 16.38) spaced close together and are supported on walls or in addition on an intermediate steel joist and are covered with brick tiles in one or two layers. These tiles are cement pointed and further are plastered at top. For water proofing this type of a roof, two coats of bitumen are applied on the plastered top. The tiles are 300 mm × 150 mm × 30 mm if laid in two layers or 300 mm × 150 mm × 50 mm if laid in one layer and are laid in such a way that the joints come over the centre of the battens. It is to be ensured that the joints are straight and fine. The second layer of tiles is laid on a 12 mm bed of mortar which is spread on the first layer of tiles. The joints in the top layer with respect to the bottom layer are broken. Tiles resting on walls should have a bearing of at least 110 mm and shall be tightly fitted into the brickwork of the parapet so as to leave no gap. After plastering the tiles, 100 mm of earth may be put and the same mud plastered. Wherever water is led away, special platform shall be built on the top of the earth to prevent scour.

16.6 MADRAS TERRACE ROOFING

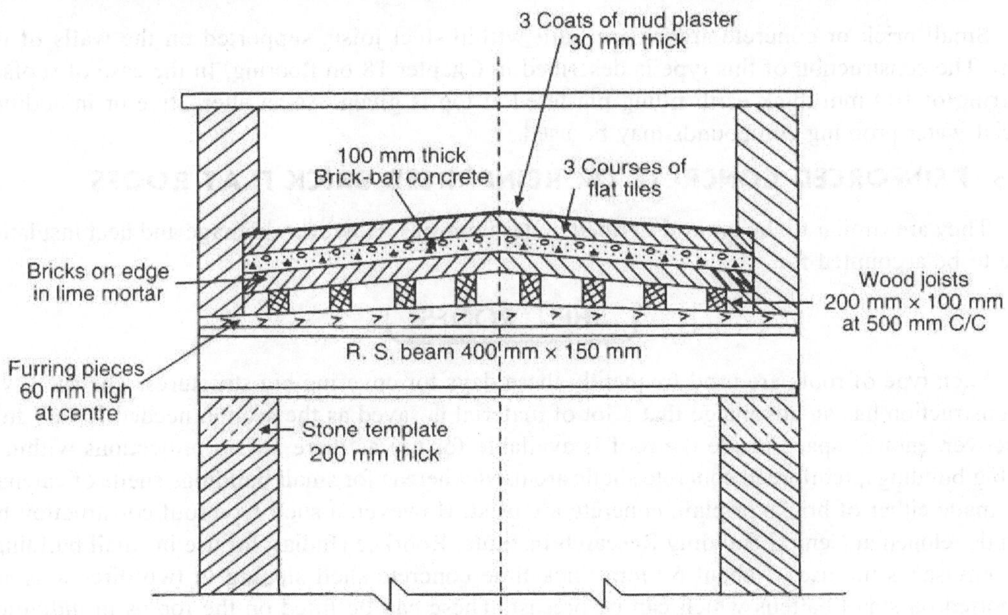

Fig. 16.39: Madras terrace roof.

In this type of roofing, rolled steel beams (400 mm × 150 mm) is placed over stone templates (200 mm) thick resting on the masonry walls. Furring pieces varying in height from 60 mm at the centre to 30 mm at the ends are placed over the rolled steel beams to give proper slope to the roof which is usually 1 in 30 and for support to the joists, wooden joists (200 mm × 80 mm) are placed on the furrings at a spacing of 400 to 500 mm. A course of well burnt terrace bricks (150 mm × 80 mm × 20 mm) is laid on edge diagonally across the joists of the roof frame in 1 : $1\frac{1}{2}$ lime-sand mortar. The bricks should be well immersed in water for 24 hours before use and a joint of 10 mm is provided. This is allowed to set completely. On the bricks, a course of 100 mm thick brick-bat concrete, comprising of three parts of brick-bats 20 mm and down gauge, one part of gravel and sand and 50 per cent by volume of lime mortar, is laid and beaten down to 75 mm thickness by wooden hand beaters. This is normally done for three days with frequent wetting the surface with water or till the wooden beater leaves no impression on the concrete layer. Curing is done for about 7 days. Then, three courses of Madras flat tiles are laid in mortar, breaking joints between the courses. The size of the tiles in the lower course is 150 mm × 100 mm × 18 mm and in the top layers 150 mm × 100 mm × 12 mm. Sometimes, the tiles of same size are also used for all the three courses. The joints of the top course are left open to provide a key for the mud plaster. Finally three coats of plaster of lime mortar are applied and rubbed to give a polished surface.

The thickness of the terracing is about 260 mm; as per details below:

Brick course on edge	100 mm
100 mm Brick-bat concrete beaten down to	75 mm
3 courses of flat tiles	50 mm
3 coats of plaster	30 mm
Total	270 mm

16.7 JACK ARCH FLAT ROOF

Small brick or concrete arches are built within steel joists supported on the walls of the room. The construction of this type is described in Chapter 18 on flooring. In the case of roofs, a covering of 100 mm thick earth filling plastered at top is given. As an alternative or in addition special water proofing compounds may be used.

16.8 REINFORCED CONCRETE OR REINFORCED BRICK FLAT ROOFS

They are similar to the types described under floorings. However drainage and heat insulation have to be accounted for.

SHELL ROOFS

Such type of roofs are used frequently these days for covering big structures. The shell type of construction has an advantage that a lot of material is saved as the section needed is very thin. Moreover, enough space inside the roof is available for use as there are no projections within it. For big buildings, reinforced concrete shells are used whereas for small buildings shells of catenary type made either of bricks or plain concrete are used. However, a shell type roof construction has been developed at Central Building Research Institute, Roorkee (India), for use in small buildings. This envisages the use of about 50 mm thick lime concrete shell sloping in two directions and supported on small battens which can be precast. These can be fitted on the rooms in individual units and supported on walls or joists as the case may be.

Shell type construction can be used for factories. Wherein north lighting is desired, saw-toothed shells are used. Details of this type of construction are given in chapter 19 on concrete construction.

DOMES

A dome is a roof of semi-spherical or semi-elliptical shape. They are constructed of stone or brick or concrete and are supported on circular or regular polygon shaped walls. The structure of the dome is such that within certain height and diameter ratios, very small thickness is needed. They are used where architectural treatment is needed, *e.g.*, in monumental structures or where roofs have to be built on buildings circular in plan. They need special type of construction procedure which is described in chapter 19.

QUESTIONS

1. What are the various types of sloping roofs? Describe briefly the construction of a lean-to-roof using wood as the main material.
2. Write short notes on:
   (*i*) Couple roof
   (*ii*) Couple-close roof
   (*iii*) Collar tie roof
   (*iv*) Collar and tie roof.

   Illustrate your answer with sketches showing construction details.
3. What is meant by the term "King-post-truss"? A room 7.5 m × 15 m in plan is to be covered by a wooden sloping roof. Show in detail the type of roof suitable with typical elevation, sections and joint connections.

   Approximately give the sizes of the various members.
4. When does it become necessary to use a "Queen-Post Truss" for the construction of a sloping roof? Give detailed elevation, sections, etc. of a Queen Post Truss for 12 m span.
5. Write short notes on:
   (*i*) Mansard roof truss
   (*ii*) Laminated roof truss
   (*iii*) Composite king post roof truss
   (*iv*) Bel-fast roof truss.
6. What are the advantages of steel roof construction? Draw neat line diagrams to illustrate the use of various roof trusses for different spans.
7. A room 10.5 m × 21 m in plan is to be covered by a steel roof. Draw detailed elevation and sections of a typical roof truss suitable for this span. Complete details of joints and supports should be given.
8. Write short notes on:
   (*i*) Wooden flat roofs
   (*ii*) Shell roofs.

REFERENCES

1. **Mitchel, G. A.**: *Building construction and drawing*, Vol. I.
2. **Riley, J. W.**: *Building construction for beginners*.
3. **Telling, M. T.**: *Carpentry and joinery*.
4. **Central Building Research Institute Roorkee**: *Corrugated shell roofs*.
5. **Bombay P.W.D. Specifications**.
6. **Punjab P.W.D. Specifications**.
7. **Greenhaulgh**: *Building construction*.

Roof Coverings

Roof covering is a material which gives a protective surface to the roofing structure. The function of the covering is only to prevent ingress or egress of heat and entry of moisture into the building. It does not withstand structural loads which are directly taken by the roofing elements. It is only to take loads to the extent over which it is supported on the roof members. There are various types of coverings depending on the character of the building, the type of roofing structure, local conditions, cost, etc. Mainly in India, thatch, wood, shingles, tiles, slates, bitumen, asbestos cement sheets and corrugated iron sheets are used as covering materials. In other countries bituminous felt, glass and copper or other sheet metal roof coverings are also used. A brief description of each covering is given in the following paragraphs:

17.1 THATCH

This is one of the most ancient types of roof coverings and is used mainly in rural areas. It is suitable for rural buildings, mainly because the cost is very low and thatch is abundantly available in those regions. Thatch either from straw or reed is used.

The supporting structure in the case of thatch may consist of ordinary truss work which is overlain with a network of battens so as to form a grid over which the thatch can be tied. Ordinarily the thatch is available in a sort of bundles which can be laid on this network and tied to it with the aid of wire or hemp rope. Each bundle is lapped sideways and also course after course so that the penetration of rain into the structure is resisted as much as possible. Thatch unit can be made on ground itself by tying it to network of 25 mm diameter bamboo spaced at 150 to 220 mm centres and making the unit in size of 1500 mm to 2000 mm by 1200 mm to 1500 mm. These units can thus be taken as a whole and laid over the wooden trusses or other types of supporting structures.

For adequate resistance against rain penetration the thatch covering should be at least 150 mm thick. It should have a good slope of 40° to 45° as a minimum. Thatch is liable to catch fire and as such it is desirable that it be made fire proof so as to resist fire hazards. Thatch can be dipped in a fire-resisting solution or the latter can be sprayed on to the roof after the thatch is laid. Various types of fire proofing materials are available and a lot of research work at the Forest Institute of India Dehradun and by other countries has been carried out. One of the formulae generally used for fire proofing is:

	Kg.
Sulphate of Ammonia	13
Carbonate of Ammonia (Lump)	6
Borax (Lump)	3
Boric Acid	3
Alum (Lump)	7
Water	225

It is also desirable to make thatch worm-proof so that it is not easily attacked by insects and birds do not gather inside it. For this reason it is treated with solutions of sodium bicarbonate and copper sulphate or other chemicals which are poisonous to the breeding of insects.

17.2 SLATING

Slate is a sedimentary rock which owing to intense pressure, volcanic upheavals and other internal structural changes has lost its original bedding and has acquired its essential characteristic

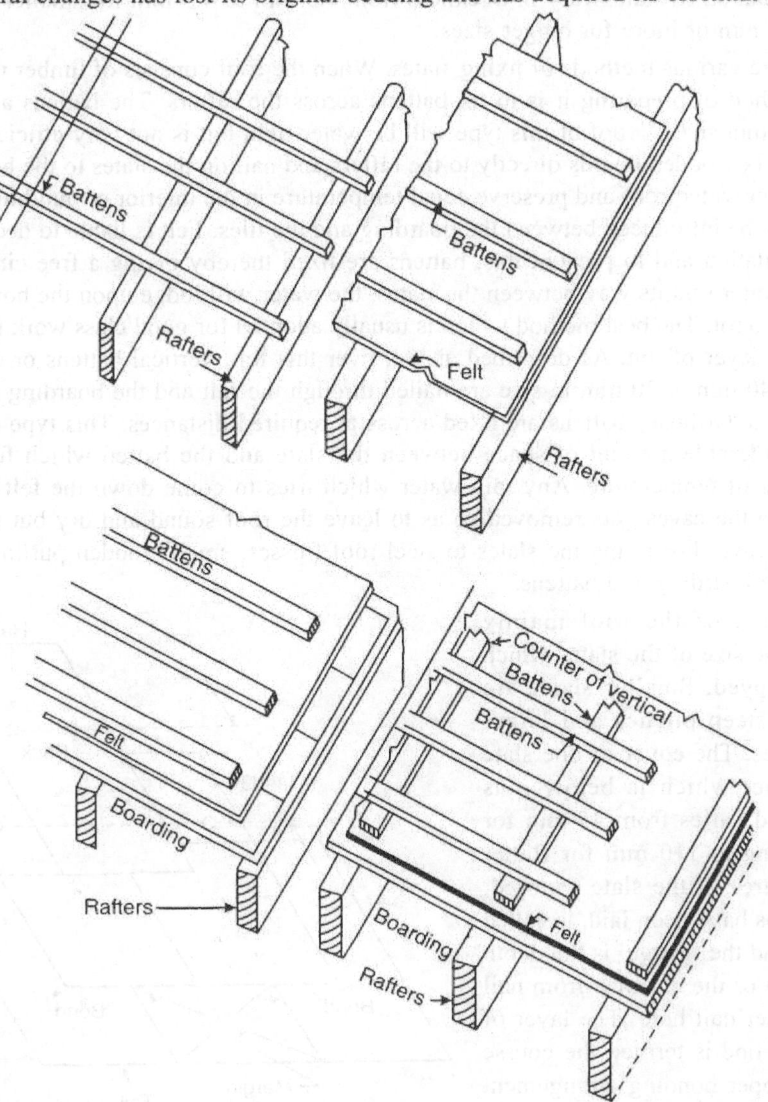

Figs. 17.1 to 17.4: Different types of supporting structures for slate roof covering

tendency to split into thin sheets. Chemically, slate is built of silica and alumina compounds which are inert and incorrodible and contribute largely to its permanent stability. The various colours of slates are due to the staining by absorbed products of metallic origin. The natural and most common colour is some shade of grey and the other colours are bluish-grey, purple, red, blue and black.

A good slate should be free from white patches which are generally of small dimensions and nearly round in shape. These patches are caused by the presence of iron pyrites and they decompose when exposed to weather. Slate should be even in colour and texture throughout. It must be hard and brittle but not tough, *i.e.*, should not break easily when thrown on ground. When struck it should give a ringing sound. It must be impervious to moisture.

The common sizes of slates vary from 600 mm × 350 mm to 250 mm × 125 mm. Sizes of random widths and of constant lengths or of random sizes are also available. Various names are given to slate sizes by tradesmen. The larger sizes are suitable for bigger roofs as they make fewer joints, require lesser nails, fewer battens and need lesser number of small pieces at hips and valleys. Regular sized slates vary in thickness from 1.5 mm to 4.5 mm for smaller sizes and from 10 mm to 12 mm or more for bigger sizes.

There are various methods of fixing slates. When the roof consists of timber rafters, the only cheapest method of preparing it is to fix battens across the rafters. The battens are generally of 50 mm × 20 mm size. A roof of this type will be water-tight but is not very efficient. The second method is to fix wooden boards directly to the rafters and nailing the slates to the boards. To make the roof further waterproof and preserve equal temperature in the interior of building, a bituminous felt layer may be introduced between the boarding and the tiles. Felt is liable to decay at times for want of ventilation and to prevent this, battens are fixed thereby giving a free circulation of air. Should any water find its way between the slates, the water will lodge upon the horizontal battens thereby tend to rot. The best method which is usually adopted for good class work is to have close boarding and layer of felt. As described above, over this felt, vertical battens or counter battens of 50 mm × 40 mm to 20 mm in size are nailed through the felt and the boarding over the centre of these rafters. Ordinary battens are fixed across at required distances. This type of construction gives a considerable amount of space between the slate and the batten which further prevents rapid changes of temperature. Any rain water which tries to come down the felt and below the slate batten to the eaves gets removed so as to leave the roof sound and dry but the cost of this work is excessive. For fixing the slates to steel roof trusses, small wooden purlins may be fixed to support the boarding and battens.

The pitch of the roof mainly determines the size of the slates which can be employed. Smaller slates are suitable for steep pitches and larger for flat pitches. The cover of one slate over the other which is below it is called lap and varies from 45 mm for vartical slating to 110 mm for flatter slopes. The area of the slate exposed, after the slates have been laid, is called the margin and the 'gauge' is the depth of the margin or the distance from nail hole to another nail hole. The layer of slates in one line is termed the course and has a proper bonding arrangement

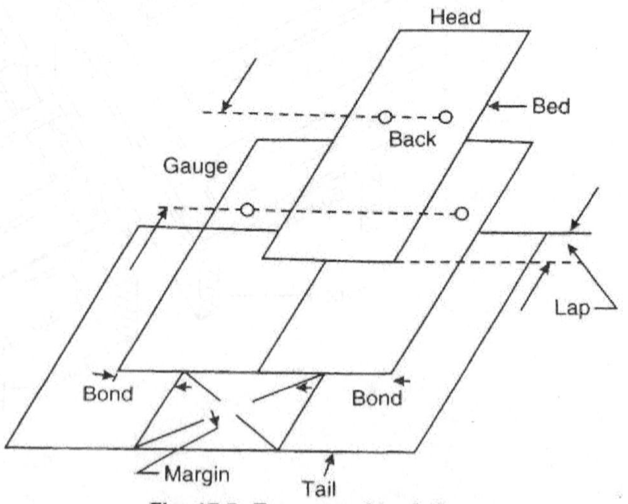

Fig. 17.5: Terms used in slating.

with the other course. Bond is an important factor in making the roof waterproof especially when the pitch is very flat.

Sometimes the gauge is irregular and is not related to the lap. Slates of random length are usually laid so that their length diminishes from the eaves to the ridge. The gauge may be regulated to produce the breaks on margin of the successive courses, but care must be taken to check that the lap is not reduced below a specified limit.

There are two methods of fixing slates depending upon the position of the nail holes. When the holes are placed near the centre it is termed centre nailing. The holes in this case are punched or drilled sufficiently above the centre of the slates to allow the nail to clear the top of the slates in the course below. This system gives a better hold to the slate and employs fewer slates in a given area. The nails are protected by only one thickness of slates so that if this breaks at any time the water will find its way in. Centre nailing is the method which is usually adopted.

When the holes are made below the top of the slate, the method is termed head nailing. The nail hole gets covered by two thicknesses of slates so that if one gets broken, it is still protected from weather by the other slate. The actual lap is 25 mm greater than the nominal one. More slates are needed in a given area of the roof and the lifting tendency on account of wind is more in this case.

Each slate should be secured with two nails drawn through the holes perforated for them into the batten or boarding below.

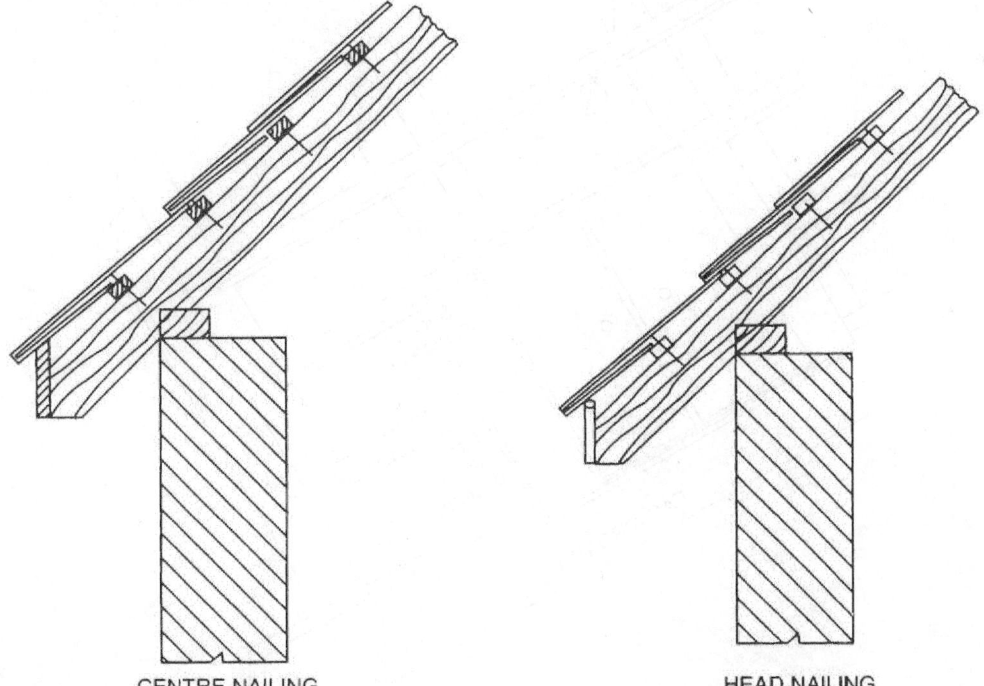

Fig. 17.6: Centre nailing method for fixing slates

Fig. 17.7: Head nailing method for fixing slates.

Copper, zinc or malleable iron nails are used. Composite nails made of an alloy of copper, zinc and tin are stiff and being cheaper are sometimes preferred. Cast yellow metal nails are very durable. They should be used where the atmospheric conditions are harmful to the other types of nails. The durability of copper nails is considered to be adequate for normal exposure and for slating; nails less durable than copper nails are also in use. Zinc nails should be used only in areas where the atmospheric pollution is very low and at a place away from the sea coast. The length of the nàils should vary according to the size and thickness of the slate. Where slates are centre-nailed longer nails are usually needed. If the nails are too long they will damage the battens. Slates may be holed either by hand or by machine and sometimes drilling by machine is preferred. The holes should be made from the front towards the back of the slates, a small counter sunk is thus left in one face of the slate which serves to take the head of the nail when the slate is thick. Care should be taken while making a hole by hand to ensure that there is no undue spalling. The distance from the slate edge to the centre of the hole should not be less than 30 mm. Usually for head nailing, the centres of the holes are 25 mm from the head of the slate. For centre nailing the holes are so placed that the nails will just clear the head of the slate underneath. In slates which are not of uniform thickness, the holes should be so located that the thicker end of the slate supports it. Thicker slates should be used in the lower course and the thinner ones on the ridges.

The first row of slates is called the doubling eaves course. This row is fixed all along the eaves before the main part of the roof is enclosed. But bottom edge of the slate projects about 75 mm over the facia board so that the rain water is thrown clear into the eaves gutter. This course is head-nailed and a 50 mm batten is fixed at such a distance from the outside of the facia board that the head of the slate rests on the centre of the battens. The second row is so laid that slates can be laid flush with eaves course at the bottom, each slate being nailed through the centre.

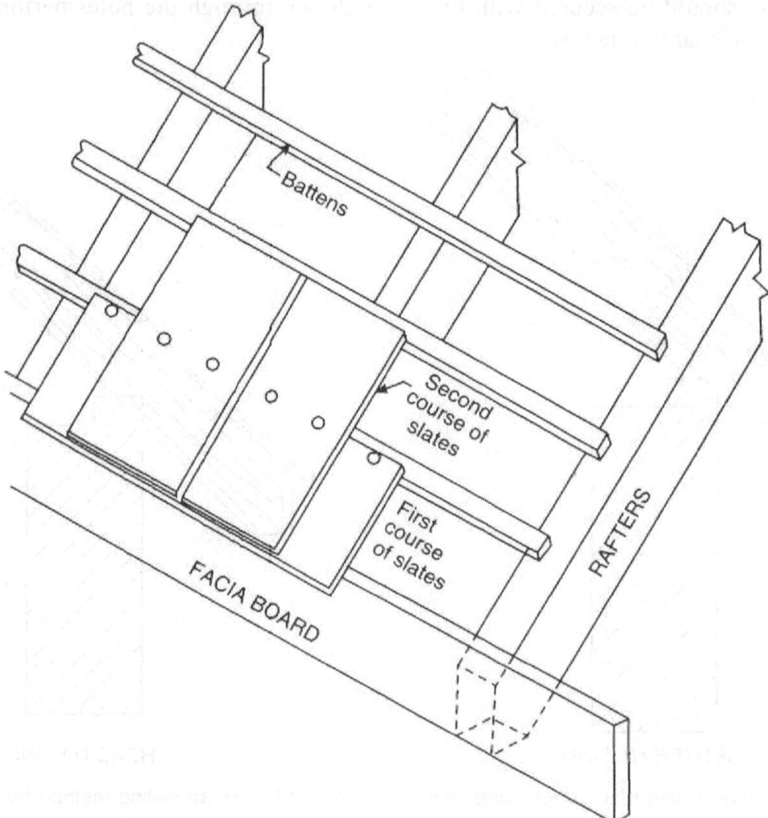

Fig. 17.8: A typical method of fixing slates at eaves.

Sometimes the double batten or the tilting fillet is placed behind the facia board. This method is better as otherwise the slates sometimes break, if the holes are too near the top of the slates.

At the gable end of the wall or at the verge, the side of the slate is allowed to overhang a distance of about 50 mm. The gap between the slate and the brick-work is pointed with mortar or suitable wooden moulding fixed.

The ridge course may also be of double slates so as to preserve a proper lap. The tilting fillet or thicker battens must be placed along the ridge piece. The last course is finished about 40 mm from the ridge. The gauge and the lap should be carefully worked out from the top of the ridge. The joint at the top should be perfect so that water may not drip in. Sometimes tiles are used for protecting the ridge. State ridges with wings on each side are also used. The wings are fixed with brass or copper screws.

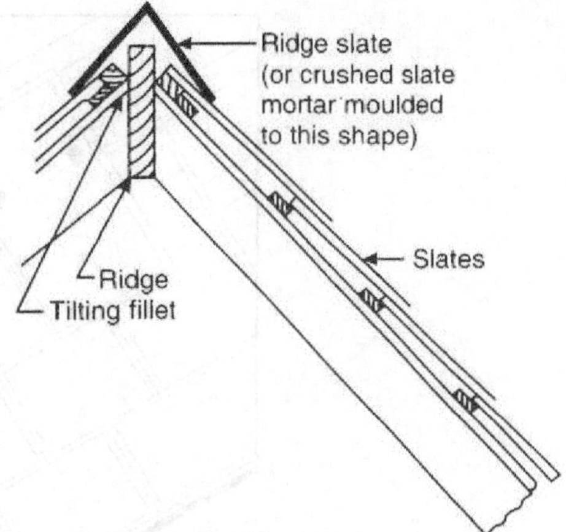

Fig. 17.9: Covering of ridge with slates.

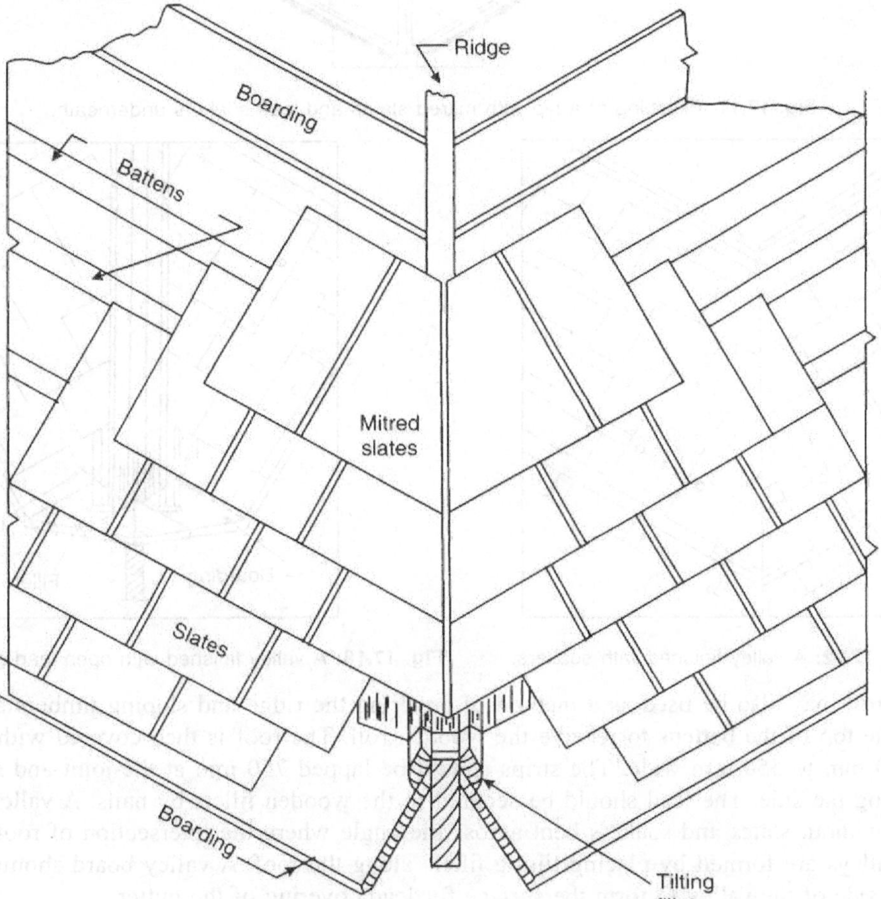

Fig. 17.10: Finishing of a hip with mitred slates and a secret gutter.

A similar treatment can be given for finishing the hips at the lower end. Usually a hip hook made of wrought iron is fixed so as to make a stop to prevent the lowest tile from slipping down. The hip may be formed by slates closely cut and mitred down the line of the hip to form a secret gutter along the hip. The hip may also be finished by placing soakers of lead across the angle.

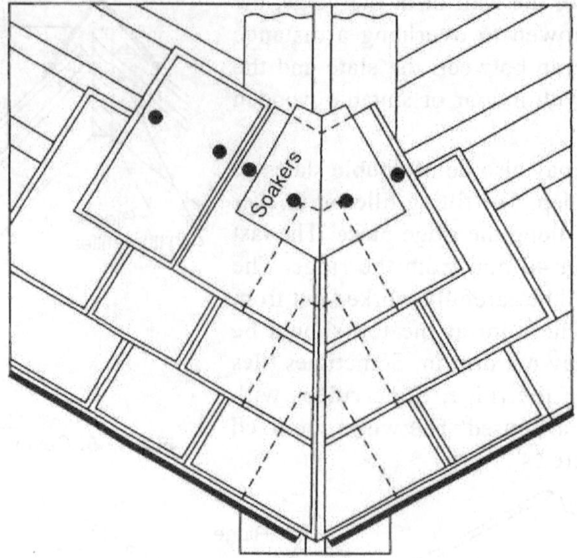

Fig. 17.11: Finishing of a hip with mitred slates and lead soakers underneath.

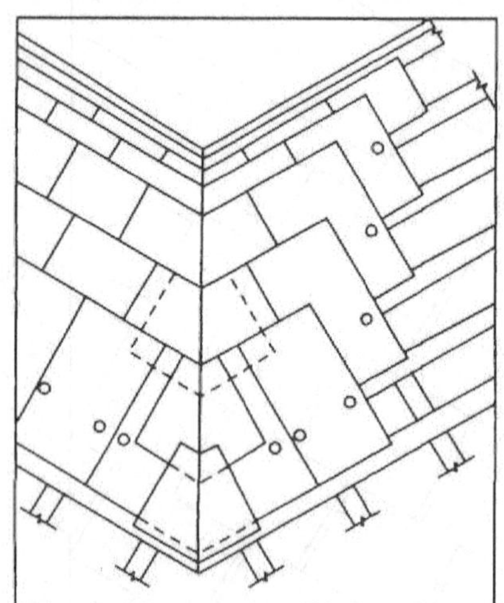

Fig. 17.12: A valley finished with soakers.

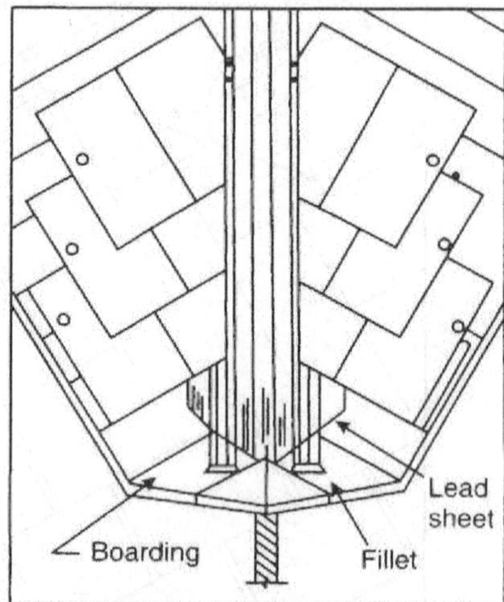

Fig. 17.13: A valley finished with open lead gutter.

A roll may also be used as a method of covering the ridge and sloping timbers are set up above the top of the battens to receive the wooden roll. The roof is then covered with strips of lead 500 mm to 550 mm wide. The strips should be lapped 750 mm at the joint and should be bent along the side. The lead should be secured to the wooden fillets by nails. A valley may be formed without slates and soakers bent across the angle where the intersection of roofs occurs. Open valleys are formed by placing tilting fillets along the roof. A valley board should be used on each side of the valley to form the surface for lead covering of the gutter.

Glass slates may be used in the roof to give proper light to the rooms below.

17.3 TILING

Tiles have been used from olden times as a means of covering the roof. Various shapes of different specifications have been improved from time to time for use. Clay tiles are mostly used after they have been burnt thoroughly in a fashion similar to bricks. Sometimes cement concrete tiles have also been used but their use is limited on account of excessive cost and the difficulties in their manufacture.

The preparation of roof for tiling is precisely the same as that for slating. The battens are usually of 40 mm × 20 mm to 25 mm in size and are set out to the required gauge depending upon the tile size. The tiles are hung to the battens with some projections which are made in the tile itself. Some courses of the tiles are also nailed to give added strength. Tiles are mainly kept in position by an interlocking action on account of their self-weight. Tiles are a better non-conductor of heat and cold and as such the use of battens only is not very objectionable. If boarding is used, the tiles have to be fixed on battens nailed to the boards or otherwise counterbattens may also be used and should be less than 40 mm × 15 mm in section.

In normal cases, the pitch of the common rafters or the surfaces on which the tiles are laid should not be less than 40° for plain tiles and 35° for single lap tiles. Where abnormal conditions may be expected, for example, near coasts or in areas where heavy snowfall occurs, these pitches may not ensure full protection from the weather. In such cases the pitch may be increased or the roof may be felted. Alternatively counter battens may be used with or without boarded background. The pitch of the outer surface of the tile gets further decreased due to the fact that each tile is tilted up at its lower edge by the tiles below and thus this slope decreases than the pitch of the common rafters.

17.3.1 Plain Tiles

Plain tiles measuring about 250 mm × 150 mm to 280 mm × 180 mm in size are used in thicknesses from 9 mm to 15 mm. They have continuous projections on one end or may have two small projecting nibs. These nibs or hand-made tiles should not be less than 9 mm wide and should have a depth of not less than 9 mm. Camber should not be less than 5 mm and not more than 10 mm. These values can be slightly decreased in machine made tiles. Two nails are provided, the centre of each nail hole being not less than 25 mm from the side of the tile and 15 mm from the under side of the nib.

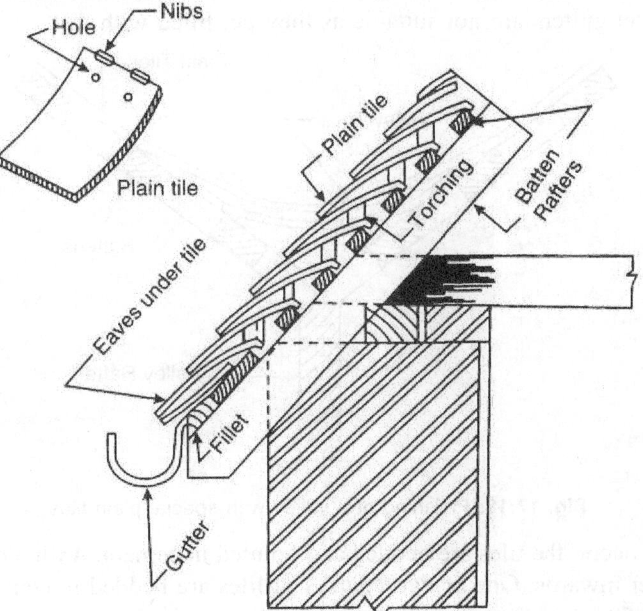

Figs. 17.14 and 17.15: Details of fixing of plain tiles.

The lap and gauge in tiling are quite important. The maximum gauge should be about 100 mm giving about 60 mm lap for a 260 mm tile. For a lesser gauge, *i.e.*, about 85 mm the number of tiles needed per unit surface of the roof area gets increased. Special tiles are made for the under course at eaves and top course at the ridges to avoid cutting. Special tiles are also made for the hips and valleys. Hip tiles may be in a bonnet shape or may have the shape of a half round or angular. The valley tiles may be in angular or rounded form. The bonnet hip tiles give a satisfactory finished appearance. There is no hard line at the intersecting plane and slight amount of latitude in use is allowed. The half round tile is suitable for ridges but is somewhat difficult to be used on the hips of small roofs. Half round tiles are bedded along their lower edges only and their unders are pointed. The angular hip tile is mainly used with hand made tiles to allow for any distortion or warping of the tile during burning.

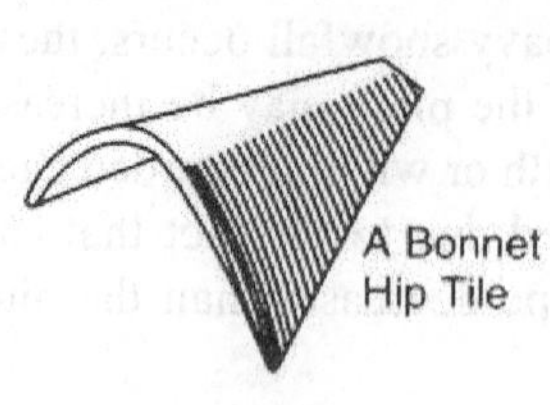

Fig. 17.16: A bonnet hip tile.

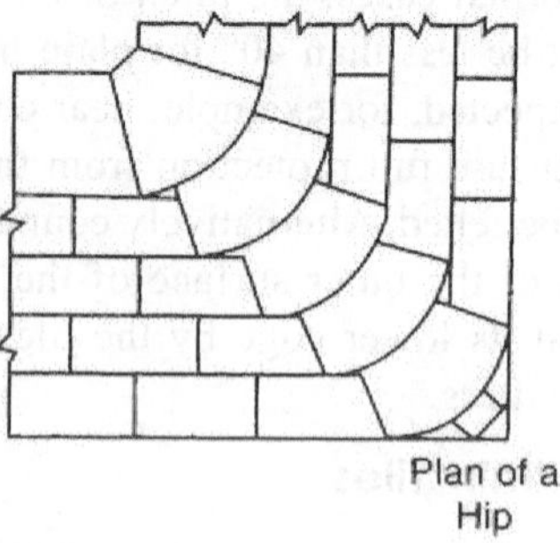

Fig. 17.17: Plan of hip corner using bonnet tiles.

Valley tiles made in angular or rounded form must be preferred to open or secret gutters as they preserve the homogeneous appearances of the roof, which is better from aesthetic considerations. Alternatively the valley may be laid with mixed tiles and the joints made waterproof by the use of soakers in a manner similar to slating. Whenever open or secret gutters are used, the lead sheet is suitably bent over the fillets and carried up over each side slope about 200 mm higher than the meeting place. Secret gutters are not suitable as they get filled with dirt.

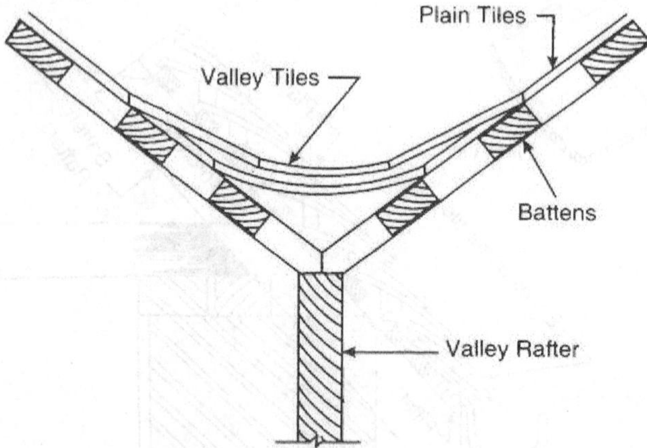

Fig. 17.18: Finishing of a valley with special plain tiles.

Where verges occur, the tiles are bedded and pointed in cement. A tilt along the edge is given so as to throw water inwards. One or two courses of tiles are bedded in cement along the wall. A projection of about 50 mm is given on the outside.

Whenever the gable has a parapet wall, the roof is kept satisfactorily finished with lead soakers and stepped flashing of a cement fillet.

17.3.2 Single Lap Tiling

This differs from the plain tiling in a manner that it overlaps only the tile immediately in the course below and not in the course which is next to one below it. They are not suitable to resist the weather owing to their being only one thickness of tiles in most parts of the roof except at laps. The breaking of one tile or some crack may cause leakage. To prevent leakage, some tiles are bedded with mortar and pointed. A better method would be to use boarding, felt and battens or counter battens.

17.3.3 Pan Tiles

This is one of the oldest types of clay tiles. The roof is covered by the use of bent tiles. There is saving in the quantity of materials and consequently in the weight of covering. This is on account of the fact that the lap of these tiles is only with the tile just below it. Precaution against leakage from the vertical joints is provided by the edge of the tile which turns over and completely covers the edge of the next tile in the same course. As the tile is practically segmental in section, water is diverted to the central channel immediately it falls and hence there is little chance of it leaking from the joints.

The tiles are about 300 mm to 350 mm long and are about 200 mm to 250 mm wide. The head or longitudinal lap varies from 75 mm to 100 mm according to the pitch of the roof and the degree of exposure. The two opposite corners at a joint may be cut to the depth of the lap so that reasonable cover is maintained between the tiles; otherwise four thicknesses of tiles would occur at the corners resulting in open joints due to tilting, etc. The joint of the bottom left hand corner of a tile with the top right hand corner of the tile below and to the left is called shouldering.

Pan tiles are nailed in a manner similar to that of plain tiling. They are fixed to tiling battens of about 40 mm × 10 mm size. At eaves, the pan tiles are bedded with mortar on plain tiles. Occasionally three or four courses of plain tiles are provided at 100 mm gauge at the eaves. The idea of this treatment is to distribute the flow of water from the channels of the pan tiles above and to prevent it from overshooting the gutter. Fillings at the ridge course are cut out of waste tiles. The ridge tiles are then bedded in a level line along the top of the pan tiles. Where an extra course of plain tile is bedded along the top of the rolls, an improved effect is achieved. Verges in pan tile roofs can be given easily. Two courses of plain tiles are bedded on brickwork to form an under clock and on this is placed a

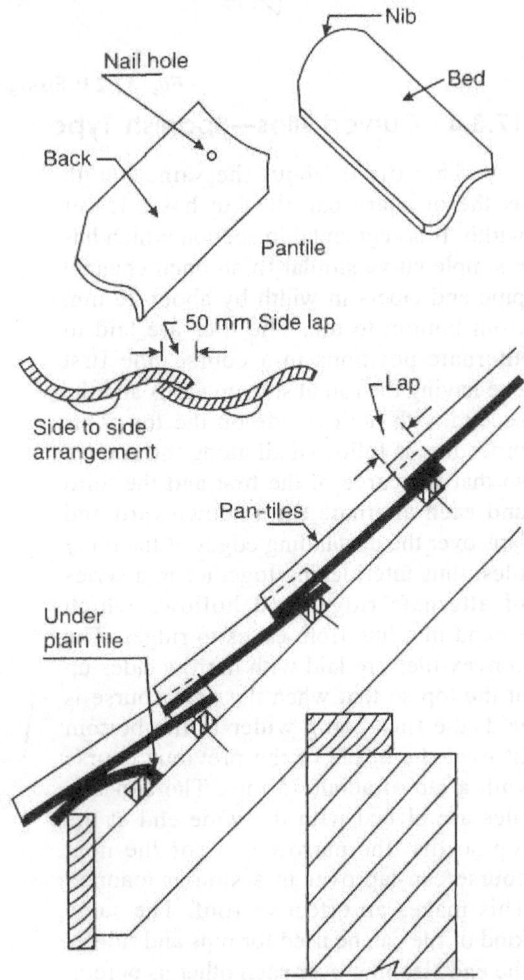

Figs. 17.19 to 17.22: Details of pan tiling.

roll to cover the up-turn of the pan tile. At exposed positions, the pan tiles can be made resistant to storms by fixing clips at each splayed joint so as to get additional strength. The valleys in this type of tiling work are finished with a lead gutter, the lead sheeting may be bent over the fillet.

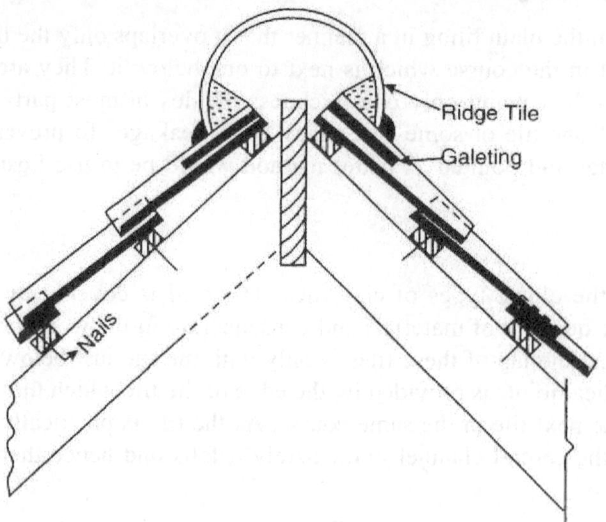

Fig. 17.23: Fixing of pan tiles at ridges.

17.3.4 Curved Tiles—Spanish Type

This tile is about the same length as the ordinary pan tile but has a lesser width. It is segmental in section which has a simple curve similar to an open channel pipe and tapers in width by about 20 mm from bottom to top. The tiles are laid in alternate positions in a course, the first one having the round side upwards and the second with hollow side on the top. This procedure is followed all along the course, so that the curve of the first and the third and each alternate tile is downward and laps over the upstanding edges of the other tiles, thus interlocking together in a series of alternate ridges and hollows which extend in a line from eaves to ridges. The convex tiles are laid with narrow sides up at the top so that when the next course is laid, the tiles being wider at the bottom fit over the heads of the previous course with a lap of about 75 mm. The concave tiles are placed with the wide end at the top so that the narrow ends of the next course can lap over in a similar manner. This makes an effective roof. The same kind of tile can be used for hips and ridges, the end slapping over each other as before.

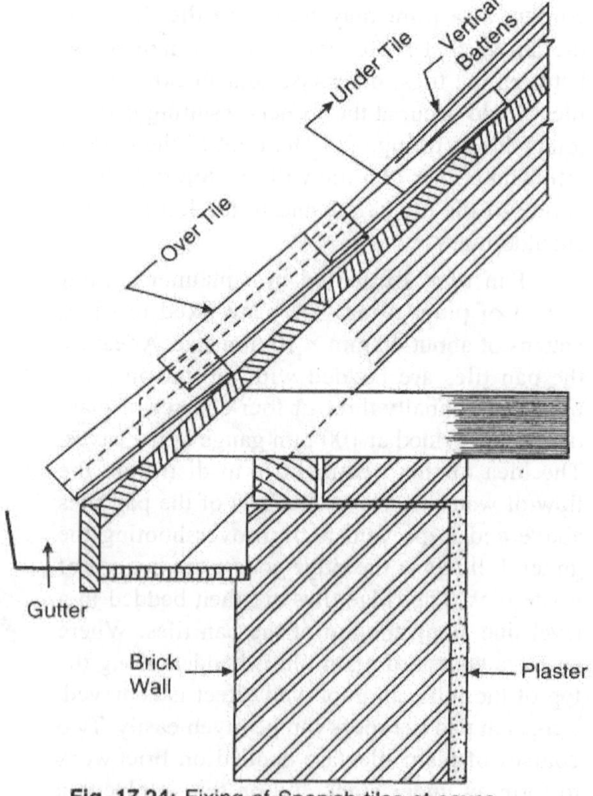

Fig. 17.24: Fixing of Spanish tiles at eaves.

A specially wide tile is used for the valleys, the course of each side resting on the edge of the valley tile. No pointing is needed with this type of covering as joints fit in completely together. Bedding with mortar is only needed on the hips and ridges at the sides of valleys. The minimum pitch with this type of roofing is about 35°.

17.3.5 Italian or Allahabad Tiles

These tiles with flat broad bottom have a beautiful shape. The under tile is a flat one and has a level depression with upturned flanges along the two edges. It measure about 230 mm at the interior end and about 260 mm at the wide end, the length being about 370 mm. The over tile is half round in section and has tapering from 120 to 160 mm. This allows the tile in the next course to fit in. Vertical battens are placed between the sides of the under tiles and to this the over tiles are secured with one nail. The horizontal battens of about 50 mm × 125 mm size are fixed with rafters, to support the vertical battens. Alternatively boarding nailed to rafters may be used for support. Half round ridges and hips are used in these tiles. They are bedded along their sides only and may be jointed in cement mortar. The pitch of the roof is about 35°. The over tiles may have interlocking devices for added strength.

Clay tiles make a good roof for residential houses. They are non-conductors of heat and help to prevent extreme changes of temperature within the building. When they are made of good materials and are burnt well, they form quite a durable covering. These have a good appearance and suit the surrounding architecture of urban and rural houses. These days they are available in different colours ranging from bright red to different shades of purple or even black.

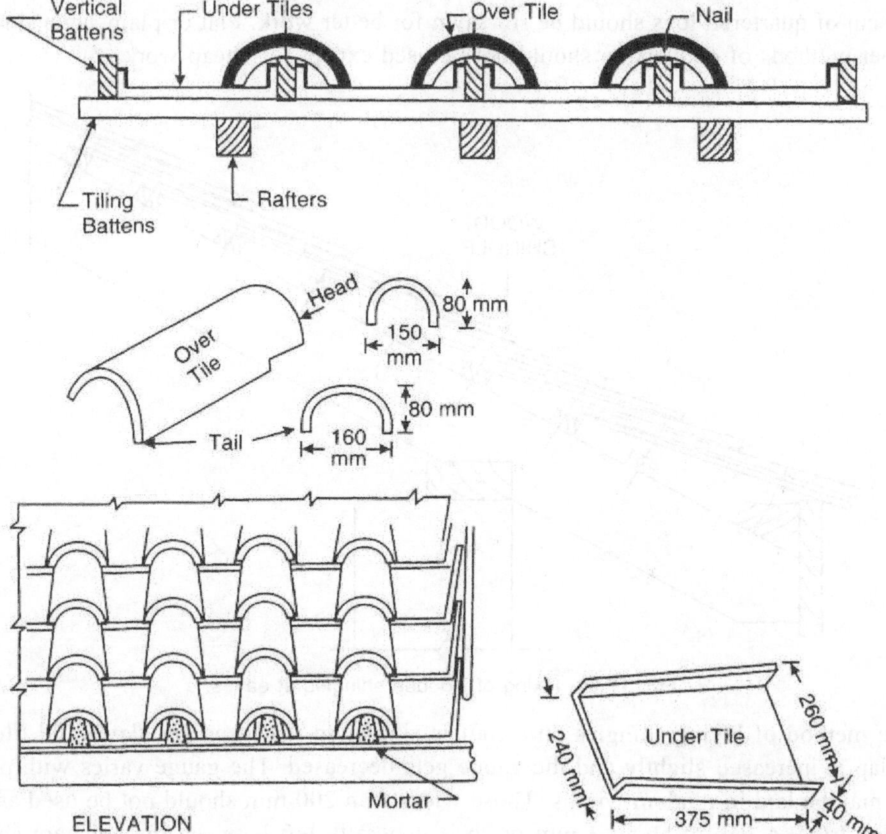

Figs. 17.25 to 17.28: Details of Italian tiles or Allahabad tiles.

The disadvantage of a tiled roof is that the tiles are heavy and being smaller in sizes need more laps which increase the total weight of the roof covering to an appreciable extent. The weight is further increased because of the necessity of placing the rafters closer in order to reduce the span of timbers and to throw off rain water effectively. The average weight of a tile roof is about 75 kg per square metre and thereby heavy timbers are needed to support them. Most tiles are porous and therefore absorb moisture to a great extent depending on their quality. The tiles are not so porous as to allow the water to drip freely to the other side but care should be taken while casting tiles that imperviousness is ensured as much as possible.

17.4 WOOD SHINGLES

Shingles are not used as a roofing material abundantly these days, but are only adopted in places where timber is cheaply available. Such type of construction is possible in the hilly areas. This type of covering is very light and consists of small units which give an artistic appearance when assembled on the roof slope. They do not get affected in their strength by rough use and can be replaced easily. However, they have got great disadvantage in being liable to catch fire. They are also liable to get split under the influence of heat.

Shingles should be cut of good seasoned timber which is not liable to get warped or shrunk. Shingles are cleft out of the log or may be sawn. Sawn shingles are generally used but they are slightly inferior as the timber fibres get cut and thereby the strength of the shingle gets reduced.

Shingles vary in length from about 330 to 400 mm and in width from about 60 mm to as much as 250 mm. They are about 9 mm thick at the tail and may taper at the head to lesser thickness. Shingles cut of quartered logs should be rift sawn for better work. Flat or plain sawn shingles got by cheaper methods of conversion should not be used except for cheap work.

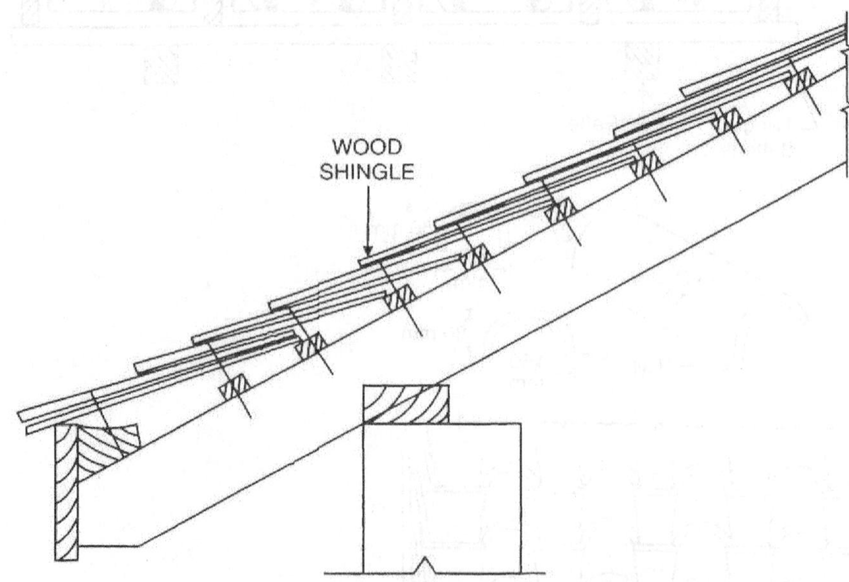

Fig. 17.29: Fixing of wooden shingles at eaves.

The method of laying shingles on a roof is similar to that used for slates and tiles except that the lap is increased slightly and the gauge gets decreased. The gauge varies with pitch. The shingles may be laid in random widths. Those wider than 200 mm should not be used as they get split when dried. A gap of about 3 mm or so is generally left between the adjacent shingles to allow for any swelling which may occur. These are fixed with copper nails or alternatively with galvanised iron nails. In olden days wooden pegs were used to fix the shingles but that method is not satisfactory and is not used now-a-days. Two nails are required to fix each shingle and this must be driven carefully so as not to split the wood. Generally centre nailing of the shingle is preferred to head nailing.

Whenever there are abrupt changes in weather, it is possible that shingles may get split near the centre and this crack in the centre will lead to a vertical joint in the course below and thereby would give a free passage for the rain water to pass. As a protection against this, it is advisable to lay the shingles with the joist across at a point about one-third of the distance across the shingle in the next course.

Shingles may be fixed to battens which are supported on rafters directly. Since shingles are unable to provide a substantial protection against heat or cold, they can be fixed to wooden boards and battens or directly to the boarding. Fixing along with the boarding will provide a sort of air gap in between the shingle and the board thereby reducing the ingress of heat or cold. When felt is not used along with the wooden boarding these boards can be fixed with a small gap in between the two so as to permit free access of air to the underside of the shingles so that drying of the roofing can take place easily after it is wet with rain.

Generally two thicknesses of shingles are used. At the eaves and at other places on account of lap, three thicknesses occur. Suitable eave course may be provided for better work. The ridges may be fixed with the aid of wider shingles each butt jointed. A split board covered with wood roll may be used alternatively. The ends of this board may be dovetailed. Lead covered ridges are also used. At the hips, the shingles are cut mitred and lead soaker can be used. For valleys, shingles with lead soakers are used. Alternatively a lead gutter with boards provided at each side to receive the lead sheets is used. Ornamental shingles are used for roofs so as to get a pleasing effect.

17.5 ASBESTOS CEMENT ROOFING

Asbestos cement is a material consisting of cement and asbestos. It is quite durable and is being adopted commonly as a roofing material. This type of roof covering is not heavy and hence a great reduction in the size of roof members is possible which gives a lot of economy. The roof covering is quite impervious and no moisture can pass through the sheets. Asbestos cement is perfectly fire and vermin proof. It does not rust and needs no protective coating of any other material. The cost of fixing is considerably lesser than the other types.

One of the disadvantages of this type of roof covering is that in colder climate, condensation is introduced on the underside of the roof where the buildings are occupied by human beings or by animals. Since most of the types of this covering material have got a single lap instead of having an additional lap over the unit which is one course below, there is always a possibility of some leakage to occur. But this is prevented by fitting tightly one sheet over the other. This material has also got a very low thermal conductivity and heats the room when it is used in summer or keeps the room very cool in winter. Another shortcoming of this material is that, it does not give a good attractive finish to the exterior of the roof. The roof covering of this material is not considered to be of a great aesthetic value.

Asbestos cement products for covering roofs are available in the form of slates, tiles and sheets.

17.5.1 A.C. Tiles

The tile may be of a diamond or of a rectangular size. These tiles can be laid on boards on the roof or directly on the battens which are nailed to the rafters. Usually the latter method is followed. The tiles which are laid in the diamond pattern have got sizes of about 380 mm square before the corners are cut off. They are supplied at the site with holes ready for nailing so that no cutting, etc. is to be done later on. These tiles have got a lap of 70 mm to 85 mm. At the eaves similar pieces of specially cut units are used. These are nailed to the tiling battens with two nails, the bottom edge projecting about 50 mm into the gutter or so. The larger pieces are then placed in flush at the bottom edge for the first course and nailed at each side. Two copper rivets with large flat ends are used with these tiles. The head of one rivet is slipped under the chamfered corner of the tile with the shank pointing upwards before the nails are driven. When the next tile is laid, the rivet gets gripped between the two and cannot move out. This method is followed throughout the roof until the opposite end is rigid. The subsequent courses are laid by slipping the hole in the tile over the projecting shank of the rivet and nailing the tiles at the sides as before. The ends of the rivets are then bent over to hold the tiles securely. This when completed, the whole surface of the roof is not only nailed tightly to the battens but each separate unit is also fastened to the course below so that no movement is possible. Special ridge tiles are used for finishing the ridge. These ridge tiles are lapped over the ordinary tiles which are carried to the top of the ridge line. A pitch of 30° is usually adopted for such type of roof.

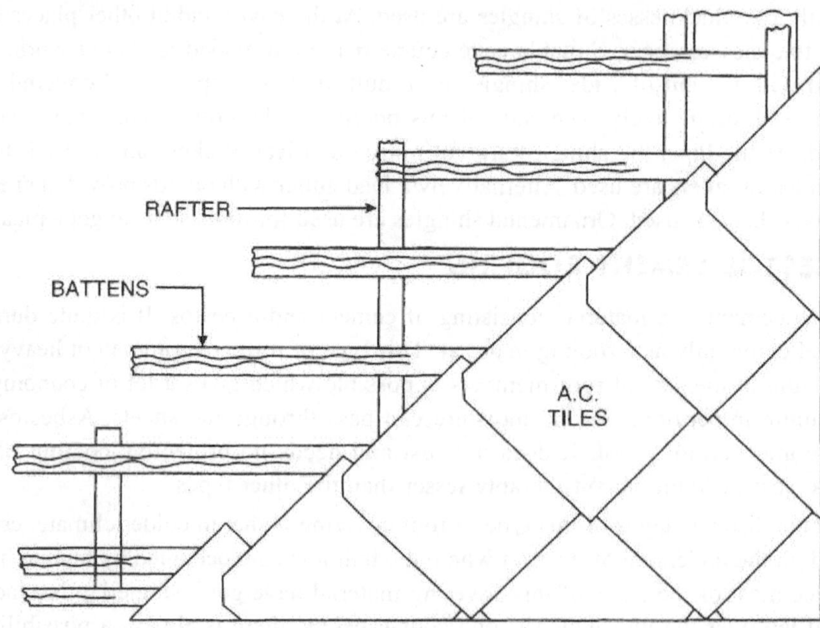

Fig. 17.30: Fixing details of A.C. tiles.

17.5.2 A.C. Slates

An alternative to the diamond pattern is the use of slates with square corners. Each slate is centrally nailed with two nails and in addition to this, one rivet is placed at the centre of the tail, the head of the rivet being pushed under the edge of the slates below when the last course is fixed. These rectangular slates vary in sizes from 600 mm × 330 mm to about 370 mm × 170 mm. They are fixed in the eave courses and other odd places.

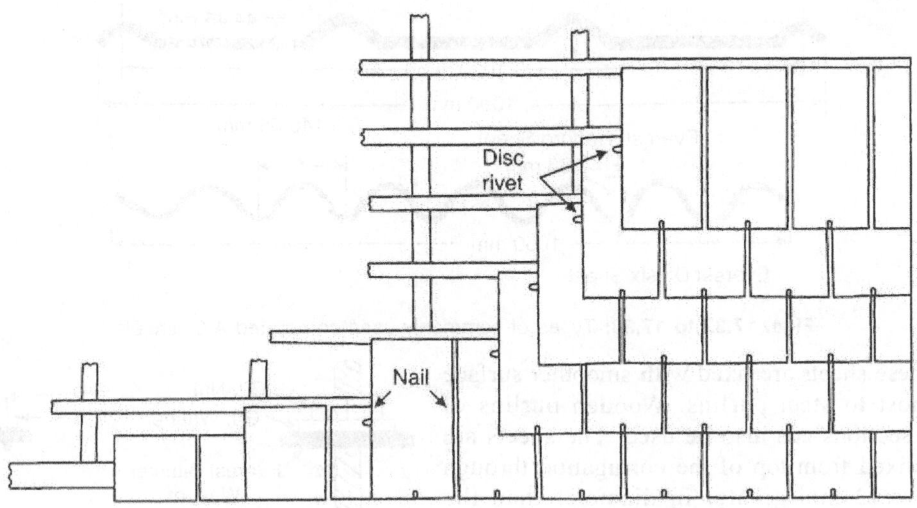

Fig. 17.31: Fixing of A.C. slate roof covering.

17.5.3 Corrugated Sheets

Table below gives details of commonly available corrugated sheets

Table 17.1 : Commonly available Corrugated Sheets

Dimensions	*Big Six Corrugated Sheets*	*Trafford Sheets*
Standard lengths in metres	1.52, 1.83, 2.13, 2.44, 2.74, 3.05	1.52, 1.83, 2.13, 2.44, 2.74, 3.05
Overall width in metres	1.05	1.09
Laid width in metres	1.01	1.02
Cover efficiency (allowing for side and end laps of 3.05 m sheet)	91% (approx.)	89% (approx.)
Purlin spacing in metres	1.68 (Max.)	1.68 (Max.)
Spacing of rails for side cladding in metres	1.98 (Max.)	1.98 (Max.)
Horizontal lap in metres	0.15	0.15
Actual cover of 3.05 m sheet as laid in sq. m.	2.92	2.94
9.29 sq. m. laid area requires area of sheeting (allowing for loss by side and end laps with 1.83 m sheet) in sq. m.	10.59 (approx.)	10.96 (approx.)
9.29 sq. m. laid area requires area of sheeting (allowing for loss by side and end lap with 3.05 m sheets) in sq. m.	10.22	10.59
Weight of sheets for 9.29 sq. m laid area in kg.	161.93	150.95

Straight and corrugated sheets of various sizes are available for being fixed on industrial roofing or on large spaces. At greater intervals, the corrugation may be small with a pitch of 70 mm and the thickness of the sheet being 3 mm. The width of such a sheet is about 750 mm and the length varies from 1250 mm to 3000 mm in increments of 150 mm. Wider corrugated sheets with deeper corrugations as a pitch of about 125 mm are also used. The corrugated sheets may also have flat portions in between the corrugations (Trafford Sheets). The thickness of this type of sheet may be as much as 6 mm. These sheets have got an end lap of 150 mm and a side lap varying from 50 mm to about 120 mm. The purlin spacing varies from 1000 mm to 1500 mm (normally the maximum spacing is 1250 mm).

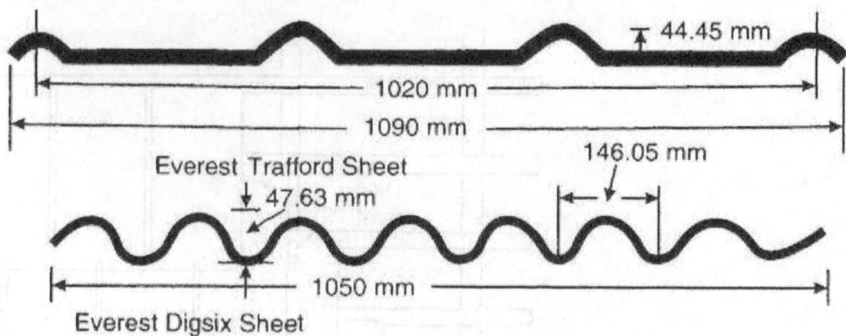

Figs. 17.32 to 17.33: Types of commonly used corrugated A.C. sheets.

These sheets are fixed with smoother surface uppermost to steel purlins. Wooden purlins of normal sections can also be used. The sheets are always fixed from top of the corrugation through the holes (2 mm greater in diameter than the screw) drilled to receive 7 mm diameter screws which are about 110 mm long. They are driven into the wooden purlins and water tightness is assured by the use of asbestos cup washer and a lead coated washer above it. The sheet is secured at six positions, two at the head, two at the bottom and two at the intermediate purlin. With steel purlins, hook bolts are used along with the washers to keep the sheets in position. At the eaves suitable eave fillers which fit in the respective corrugation of the asbestos cement sheets are used. These are used to prevent undue draft into the roof and entry of birds, etc. The asbestos cement ridge tile is provided for covering the joint at the apex of the roof. These ridge pieces are made of two parts each overlapping the other.

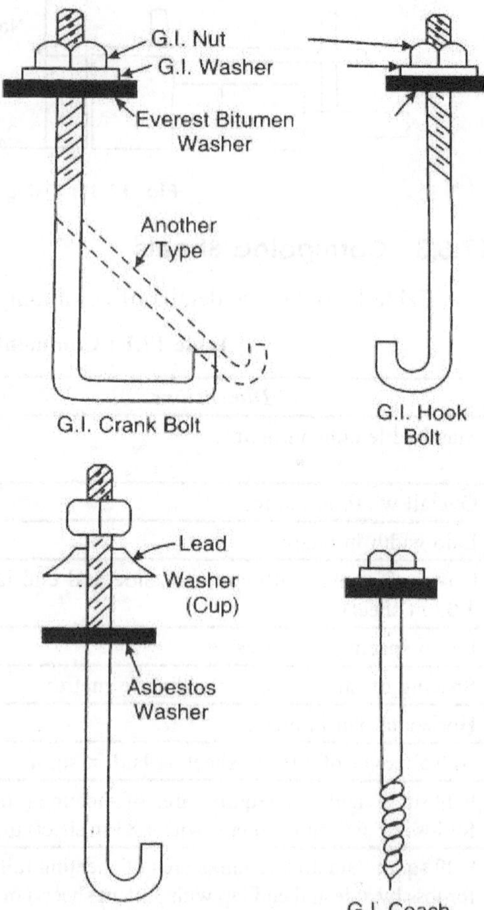

Figs. 17.34 to 17.37: Typical fixing of bolts and screws used with corrugated A.C. sheet.

While fixing the sheets, the purlin spacing and the length of the sheets are first checked to see that the arrangement will provide the specified overhang at the eaves and proper laps are obtained. An eaves course is laid first and the work started at the left hand end of the building so that the side laps have a better protection from the rain driven by wind. The first sheet is laid uncut but the remaining sheets in the bottom row should have the top left hand corners cut or 'mitred'. The sheets in the second or other intermediate rows should have both the left hand corner and the bottom right hand corner cut, with the exception of the first sheet in each row which should only have the left hand corner cut. The last or top row should have the bottom right hand corner cut, with the exception of last sheet which is laid uncut. Whenever sheets are laid from right to left, the whole process of cutting described above is reversed. 150 mm end laps are suitable for roof slopes greater than $21\frac{1}{2}$ degrees. For lesser slopes the laps should be increased accordingly. Sheeting should be laid at the end opposite to the direction of prevailing wind and rain. The overhang at the eaves must not be greater than 370 mm. Mitring of sheets is essential to get a proper fit wherever four corners meet. The top edges to eaves must extend 75 mm beyond the centre line of timber purlins or 40 mm beyond the back of steel purlin.

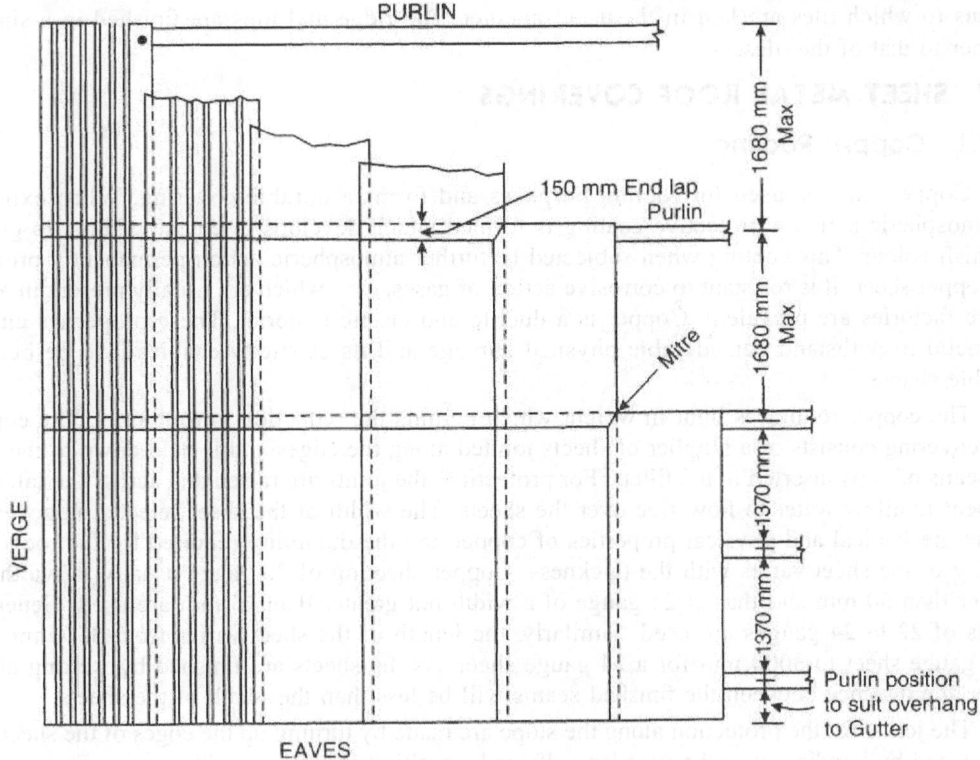

Fig. 17.38: Details of fixing A.C. corrugated sheets.

17.6 GALVANIZED CORRUGATED IRON ROOFING

Corrugated iron roofing is widely used, although it does not have a very good appearance. The main reason for its use is that it is very durable, light and fire proof. It is also easy to alter such a type of roof covering. Galvanised iron is usually manufactured in sheets which are corrugated or bent in a series of parallel depressions from end to end. The purpose of corrugation is mainly to give to thin iron sheets additional strength as they are only supported on purlins at fixed intervals. Moreover this corrugation helps in proper discharging of water away from the sheet. The idea of galvanising the iron sheet with zinc is to protect it from rusting in wet weather. This coating of zinc may get damaged only in atmospheres which are contaminated with fumes of factories containing acidic materials. One of the difficulties with this type of material is that it transmits heat and cold easily. It is liable to create condensation problems on the inner side in colder climates.

The laying of corrugated iron sheets is very simple. The sheets are fixed to purlins which are spaced to 2000 to 2500 mm centres. The sheets are nailed to these timbers with galvanised nails or screws. The washers have to be used to prevent water from passing through this hole. Each sheet is fixed with one screw in the central corrugation and one at each end. The screws are always placed on the top of the ridge to prevent infiltration of water. These screws are not driven until the adjacent sheet has been lapped over the first. About 75 to 150 mm lap should be allowed at the ends of the sheet. This would give a sort of air gap between the sheet and the plank and prevent condensation.

For preventing lot of heat getting transmitted from the galvanised iron sheet, tiles can be laid over this type of covering. Battens of 40 mm × 40 mm size are laid over the ridges of the corrugations at about 450 to 500 mm centres and are securely screwed from the sheets into the purlins below. Between these battens, the corrugated sheet is also fastened to the purlins by 75 mm long screws. The horizontal battens of about 50 mm × 15 mm size are fixed to the sloping battens to which tiles are laid in the usual manner. The ridge and hips are finished in a similar manner to that of the tiles.

17.7 SHEET METAL ROOF COVERINGS

17.7.1 Copper Roofing

Copper sheet is used for roofing purposes and forms a durable covering. When exposed to atmospheric action a protective coating is formed which develops in course of time to give a greenish colour. This coating when subjected to further atmospheric action permanently protects the copper sheet. It is resistant to corrosive action of gases, etc., which are usually present in areas where factories are prevalent. Copper is a ductile and elastic material. These properties enable the metal to withstand considerable physical damage and its elasticity enables it to be bent to suitable shapes.

The copper roofing is light in weight which permits the reduction in roof sizes. The copper roof covering consists of a number of sheets jointed along the edges and is held down to the roof by means of clips inserted in the fillets. For protection, the joints are raised and across the fall they are bent to allow water to flow free over the sheets. The width of the sheet selected is governed by the mechanical and physical properties of copper and the durability required for the roof. The rigidity of the sheet varies with the thickness. Copper sheeting of 26 gauge usually in width not greater than 50 mm and that of 21 gauge of a width not greater than 65 mm are used. Generally sheets of 22 to 24 gauges are used. Similarly, the length of the sheet varies from 3500 mm for a 22 gauge sheet to 3000 mm for a 24 gauge sheet. As the sheets are finished by welting at the edges, the distance between the finished seams will be less than the width of the sheets.

The joints for the protection along the slope are made by turning up the edges of the sheet and are formed by bending over the wooden rolls or by welting into the standing seam. The wooden roll is used where the traffic is expected on a roof and the standing seam is adopted where it is not liable to get damaged. The joints across the fall are made by welted seams. Where the pitch is greater than 60 degrees a single welted seam is used and whenever the pitch is less than 60 degrees a double welted seam is essential. For slopes between 60 degrees and 10 degrees, no drips are essential. Where the slope of the copper covering is less than 10 degrees, drip should be used to accelerate the flow of water.

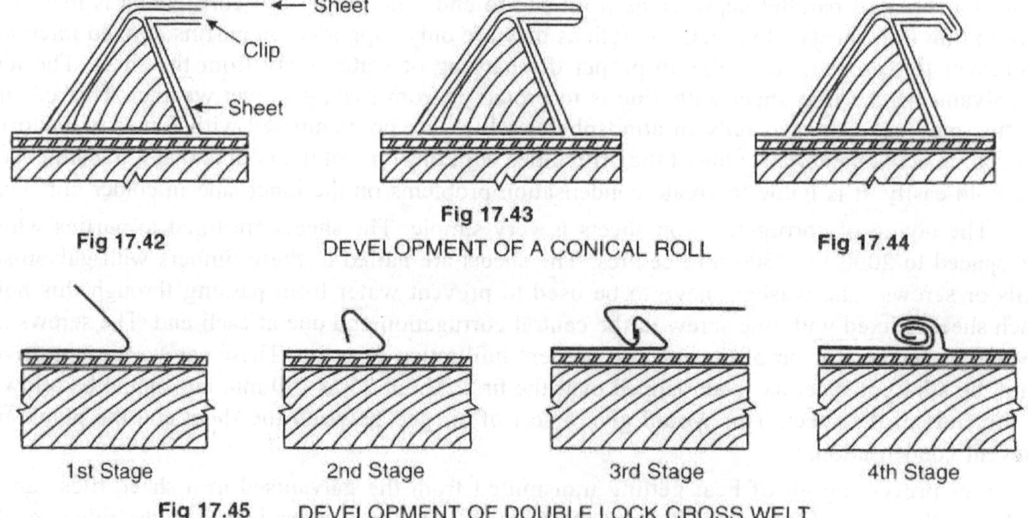

Fig 17.42

Fig 17.43
DEVELOPMENT OF A CONICAL ROLL

Fig 17.44

Fig 17.45 DEVELOPMENT OF DOUBLE LOCK CROSS WELT

Figs. 17.42 to 17.45

In any case the slope of the copper sheet when laid should not be less than 50 mm in a length of 3000 mm. The copper covering surface should be laid to even slope and provided with the smooth finish. Care should be taken to prevent the copper sheet from getting into direct contact with concrete by interposing a felt layer. Dovetailed wooden battens of 50 mm × 25 mm size are set flush with the surface and run across for standing seam joints. Whenever wood rolls are adopted the spacing can be increased. Underlayers of felt should be provided to concrete roofs to reduce condensation and on timber roofs to reduce noise. This felt is laid with butt joints. This felt may be fixed by copper nailing or other means to guard against wind, etc., during work, concrete roofs should be coated with two layers of bitumen before the felt is laid. The heads of the nails securing the boarding should be flushed below the surface and boarding plained to a smooth finish. Copper nails are used for fixing as iron nails would result in the decomposing of sheets due to electrolytic action.

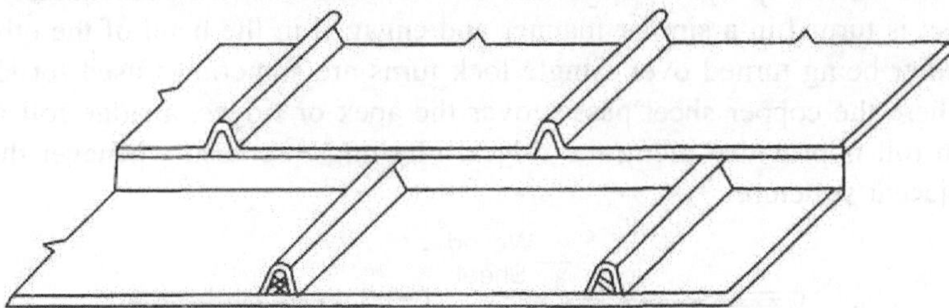

Fig. 17.46: Copper roofing plan using staggered rolls.

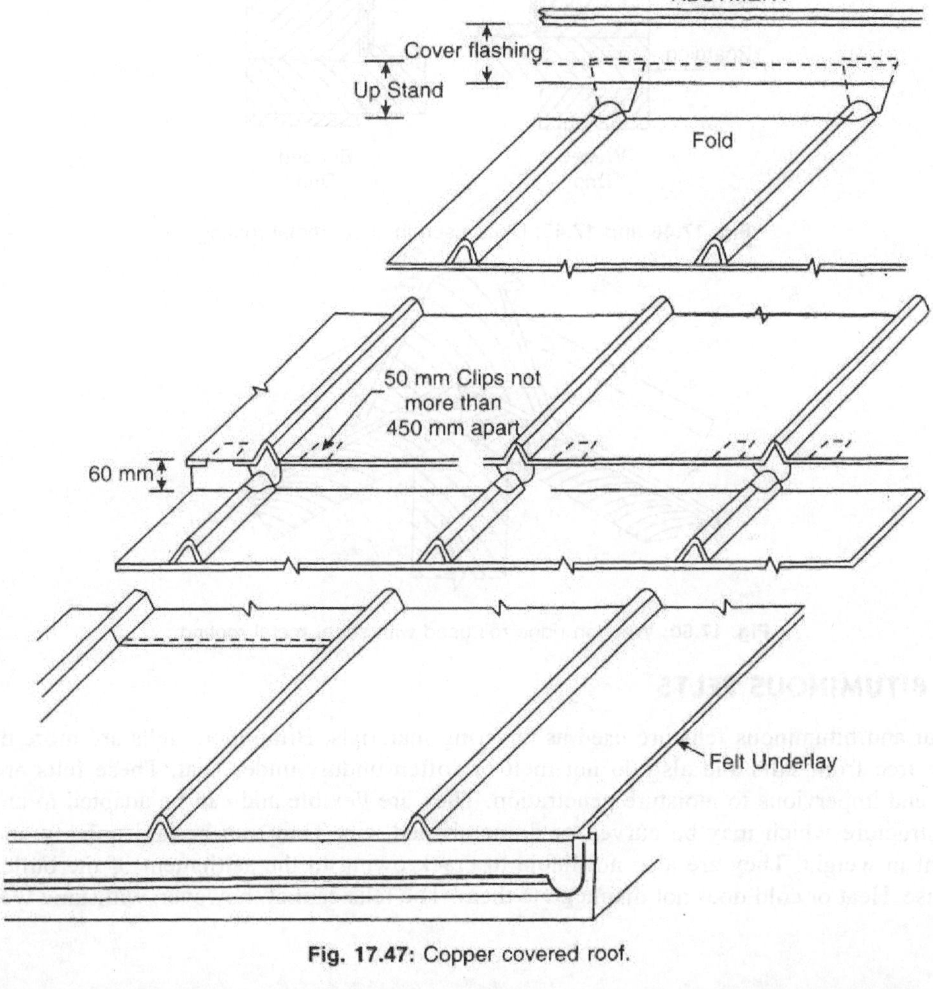

ABUTMENT

Cover flashing

Up Stand

Fold

50 mm Clips not more than 450 mm apart

60 mm

Felt Underlay

Fig. 17.47: Copper covered roof.

Although the expansion and contraction of copper due to changes in temperature are small, yet provision is to be made of these movements. Joints should be formed as stated earlier with the aid of the wooden rolls of standing seams. Wood rolls may consist of rectangular or square wooden sections. Copper clips of about 50 mm width are placed under the rolls at about 900 mm centres. The rolls are secured by copper or brass screws and are spaced at a distance equal to about 75 mm less than the width of copper sheet. The development of the roll is to be done carefully and different methods are adopted, some of which are shown in Figs. 17.39 to 17.45. Welted standing seams are used whenever the roofs are not subjected to traffic. The sheets are first welted end to end and then the stirrups are placed on the roof and the standing seams formed by means of pliers. The height of the finished joint is about 20 mm. Welts are used for end to end joints. In the first stage the edges are turned up by about 25 mm and are turned down subsequently. The edge of the adjacent sheet is turned in a similar manner and engaged in the bend of the other sheet so as to form a welt after being turned over. Single lock turns are sometimes used for sheets fixed on steep slopes. Where the copper sheet passes over the apex or ridges, a ridge roll 40 mm higher than the wooden roll is provided along the ridges. Flashings are used whenever the roll is to be jointed to an adjacent structure.

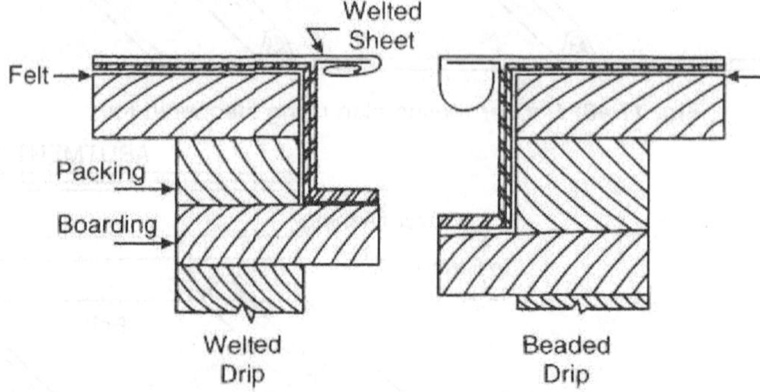

Fig. 17.48 and 17.49: Drips used in sheet metal roofing.

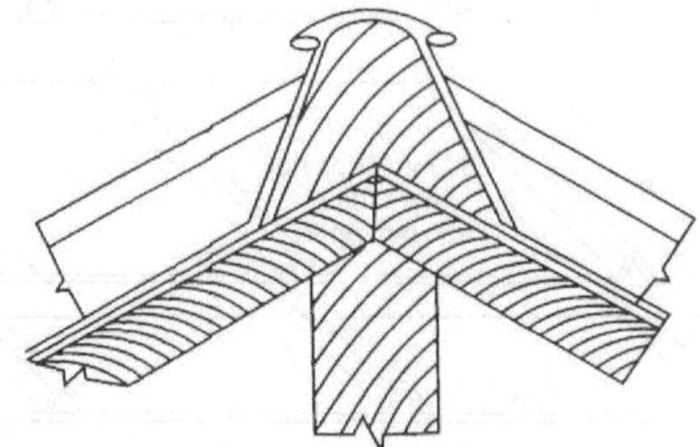

Fig. 17.50: Wooden ridge roll used with sheet metal roofing.

17.8 BITUMINOUS FELTS

Tar and bituminous felts are used as covering materials. Bituminous felts are more durable and are free from salts and also do not melt or soften unduly under heat. These felts are non-porous and impervious to moisture penetration. They are flexible and can be adapted to any type of substructure which may be curved or dome shaped, etc. They can be laid quickly and they are light in weight. They are also not liable to crack owing to the settlement of the building or otherwise. Heat or cold does not disintegrate them. The felts cannot, however, withstand wear and tear of the traffic. Its appearance is unattractive and it does not provide a self-supporting structure with the result that it needs a substructure which should be very durable.

The surface to which the roof is to be laid should be arranged to ensure a rapid flow of storm water in the required direction. For flat roofs, the finished surfaces should have a fall of not less than 1 in 60 and the gutters should have considerable slope and provision for that should be met while designing the substructure. The substructure may consist of wooden boardings which are duly supported on joists. The concrete substructure should be treated and prepared so as to avoid loss of adhesion and cracking. A primer coat may be given to reduce porosity and to ensure proper adhesion.

On timber substructure, the first layer of the felt should be secured by nails spaced at about 50 mm centres along laps and about 20 mm from the exposed edges. Whenever additional security is needed, extra nails are added midway between the laps. The subsequent layers of felts are bonded as described below.

All concrete or similar substructure should get a bituminous coat primer after the surfaces have been brushed thoroughly and dried up. The first layer of roofing felt is bonded hot and each substructure layer is bonded similarly. The bonding compound is heated to a temperature not in excess of that required to ensure an effective bond. The first length of the first layer of the felt is laid is position and firmly rolled back half way. The bonding compound is spread on to the roof as the felt is rolled out, pressure being applied to the felt so as to engage the compound while it is still hot. Any surplus compound is squeezed out and the edge of the felt is smoothened as work proceeds. The other half of the length is then rolled up and the process repeated. Subsequent lengths are laid in a similar manner. The bonding compound may be spread in the form of roll or applied by means of brushes.

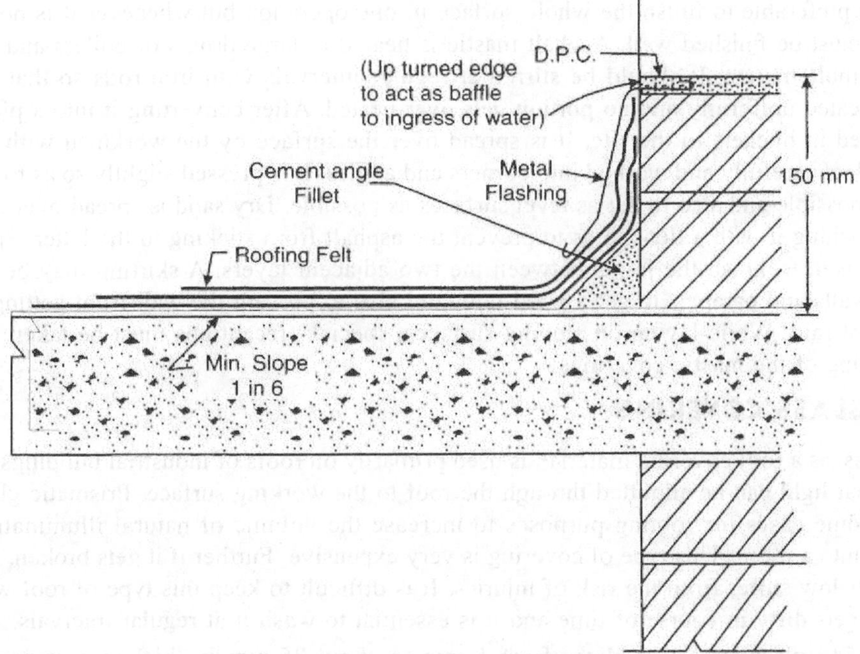

Fig. 17.51: A typical detail of bitumen roofing felt covering.

Layers should be made to follow each other in such a way that minimum of incomplete work is exposed to weather. Whenever a finish of cement concrete or tiles, etc. is applied to the surface, the top layer should be given a dressing of hot compound applied with the help of brush.

The felts are laid with lap joints at least 50 mm wide at the sides and 75 mm at the ends. When a single layer of bituminous felt is laid on the sloping timber substructure the joints should be raked with bituminous lap cement (a bitumen cut back applied cold) and a band of lap cement 50 mm wide applied along the joints to cover the exposed nails. At junction with walls of parapets, the felt should be taken up the angle fillet upto a height of at least 150 mm above the finished surface of the roof.

17.9 ASPHALT MASTICS

Two kinds of asphalts are used in building operations, *i.e.*, natural and artificial. The artificial product consists of bituminous material which is mixed with a filling of sand, chalk and other ingredients. It is not a true asphalt mastic which is made by the process of nature.

Asphalt mastics are quite impervious. They can be easily repaired and are not highly inflammable. They can withstand stress that arises from unequal settlement or other defects. Since this material when laid in a heated state is very plastic, it can be laid over an irregular shaped surface.

As a general rule, mastic asphalt is primarily used on flat roofs but can also be employed on sloping or vertical surfaces if desired. The materials that have to be reduced to plastic state by the application of heat are liable to soften or sag when they are subjected to the hot rays of the sun.

Before the asphalt mastic is laid, all dust and rubbish must be carefully removed from the surface of the roof so as to increase the adhesion. Asphalt is generally laid on concrete but this material can be applied to a boarded covering if necessary. The boards must be fixed close together so that no open joint is left through which the asphalt can sink. The roof may be covered with sheets of felt to minimize the danger of asphalt falling through the joints into the space below.

It is preferable to finish the whole surface in one operation but whenever it is not possible, the joint must be finished well. Asphalt mastic is heated in large drums or boilers and made into a plastic molten state. It should be stirred at regular intervals with iron rods so that the whole mass is heated uniformly and no portion gets over-heated. After converting it into a plastic state, it is carried in buckets to the site. It is spread over the surface by the workmen with the aid of wooden float carefully and worked into corners and angles. It is pressed slightly so as to make it as dense as possible and also to get as level surfaces as possible. Dry sand is spread over the surface before finishing it with a float so as to prevent the asphalt from sticking to the latter. Special care must be taken to finish the joints between the two adjacent layers. A skirting may be run along the side walls and parapets to get a good joint and also to prevent the wall from getting wet with splashes of rain. While laying on sloping surfaces, special precautions must be taken to prevent the dragging of the mastic coverings.

17.10 GLASS COVERING

Glass, as a roof covering material, is used primarily on roofs of industrial buildings, factories, etc., so that light can be admitted through the roof to the working surface. Prismatic glass is also used in some cases for roofing purposes to increase the volume of natural illumination that is admitted into a room. This type of covering is very expensive. Further if it gets broken, the people working below suffer from the risk of injuries. It is difficult to keep this type of roof water-tight. The roof gets dirty in course of time and it is essential to wash it at regular intervals.

Roofing glass is cast in plates from 5 mm to about 25 mm in thickness. It generally has got ribs which serve the purpose of providing better illumination than plain glass. The fixing of glass was previously done with the aid of plastic putty and wooden or other types of beads. This arrangement is not satisfactory. Putty gets cracked after sometime and results in a leaky roof. Moreover, use of putty needs constant maintenance. Various type of fixtures with glass plates have been devised. One of them using glazing bars is shown in Fig. 17.52. In this type, glass sheets are bedded on soft asbestos cords throughout the length of the bar and held in position by extruded lead capping which is secured to the stem with small brass bolts. A special stop at the foot of the bar is used to prevent the glass from slipping down the roof. Each pattern has a condensation channel underneath so that the moisture on account of the condensation on the inside of the glass can run off to the bar and escape by the channel to the gutter. A part of the bar projects above the surface of the roof thereby permitting planks to be laid across the sheet when maintenance is needed.

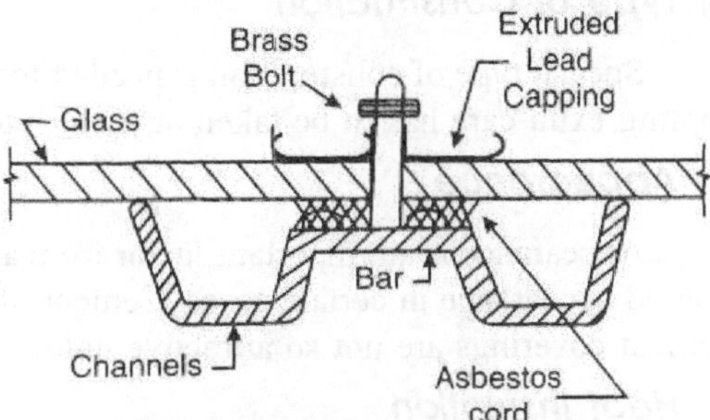

Fig. 17.52: A typical glazing bar of improved type used for glass roofs.

17.11 SELECTION OF A SUITABLE ROOF COVERING

The various factors which have to be kept in mind while selecting a particular type of roof covering for a given building are briefly discussed below:

1. *Initial Cost*

The initial cost of roofing material varies from time to time and also at a given place. Slate roofing is cheaper in hilly areas and similar is the case with wood shingles. Clay tiles are cheaper and can be manufactured only at places where the bricks are made. While considering the initial cost of a covering material, the additional cost on account of the weight of the supporting members must be considered. Materials which are heavier need stronger supporting structure which adds to the cost.

2. *Maintenance*

The cost of maintenance forms one of the main items while deciding the type of particular roofing material. Wood shingles and tiles need less frequent repairs than that of thatch roofing. Asbestos, roofing sheets, slates, etc. need occasional repairs and very little maintenance, while still lesser care is needed for built-up copper sheet metal roofing.

3. *Slope of the Roof*

Thatch, slates and tiles need steeper slopes to prevent water penetration, whereas corrugated sheet and sheet metal roofing need flatter slopes.

4. *Durability*

The life of the roofing material is important to determine its economical value. It depends on many factors and varies from place to place, but under normal conditions clay tiles, slate, copper, etc. have got a longer life. Asbestos cement roofs and wood shingles are classified as having medium life whereas thatch has got the least life.

5. *Resistance to Fire*

Fire resistance is equally important in any type of structure. Thatch roof covering has got the least resistance and clay tile, slate or sheet metal roof offers greater resistance to fire.

6. *Weight of the Roof Covering*

The heavier the weight of a particular roof covering the stronger supporting structure is needed which adds to the cost. Clay tiles and slates are the heavier types of roof coverings. Asbestos cement roof covering and wood shingles have got a lighter weight whereas sheet metal coverings are much less in weight.

7. Type of Construction

Special type of construction is needed for certain types of roof coverings, *e.g.*, for sheet metal roofing extra care has to be taken at joints, etc. Clay tiles do not need much attention.

8. Appearance

Appearance is an important factor from architectural point of view. Clay tiles or shingles give a good appearance in certain cases. Cement tiles are satisfactory for industrial buildings. Asbestos cement coverings are not so attractive unless specially treated.

9. Heat Insulation

In tropical countries insulation against heat is very important, as transmission of heat through the roof makes the rooms very hot and unsuitable for living.

Asbestos cement sheets, unless used in double layers, have got a very low resistance to transmission of heat and similar is the case with corrugated metal roofing. Clay tile or thatch gives adequate protection against thermal effects.

QUESTIONS

1. Describe in detail the various points to be considered in the selection of a suitable type of roof covering.
2. What are the essential qualities of a good slate to be used as roof covering? Give the sizes of slates used. Describe the various methods employed for fixing of slates.
3. What are the various types of tiles which can be used for roof covering? Discuss the merits and demerits of each one of them.
4. What is asbestos cement? What are the disadvantages of asbestos cement covering? Write short notes on:
   (*i*) A. C. Tiles.
   (*ii*) A. C. Slates.
5. Write short notes on:
   (*i*) Thatched roof.
   (*ii*) Corrugated sheets.
   (*iii*) Sheet metal roof coverings.
   (*iv*) Bituminous felts.
   (*v*) Asphalt mastic.
   (*vi*) Glass covering.
   (*vii*) Madras terrace roofing.

REFERENCES

1. **Miller, J.**: *Slating and tiling.*
2. **Banmgarten, R.H.**: *Manufacture of concrete roof tiles.*
3. **Blauie, E.G.**: *Roof coverings.*
4. **British Standards Institution:** *Standard code of practice for bitumen-felt roof slating and tiling.*
5. **Bovne, A.E.**: *Architects' working details.*
6. **British Standards Institution**: *Asbestos cement sheet roof covering.*
7. **Mackey W.B.**: *Building Construction.*
8. **British Standards Institution**: *Asbestos cement sheet roof covering, copper covering for roofs.*
9. **British Standards Institution**: *Standard cpde of practice for bitumen-felt roof covering.*

Steel Work

Steel is used to a large extent in modern multi-storeyed buildings. Construction in steel work is more economical for buildings which are more than ten storeys in height. Generally a framework of heavy steel sections is erected to take all loads and partitions or walls are built of other constructional materials. Structural steel and concrete combination may be used in building work. A typical example would be to have columns of structural steel and flooring systems of reinforced concrete. Due to ease in erection and heavy loads which structural steel can take, it is commonly adopted for factory buildings.

18.1 SECTIONS USED

Various types of sections and shapes are used for building work. Most of them are made by rolling. The white hot ingot of steel is reduced under the action of a steam hammer or in a cogging mill from where these are taken to the finishing rolls. Plate sections are formed by passing and repassing the metal through the rolls till the desired thickness is obtained. Other sections are made with the aid of cylindrical rolls having circumferential grooves which give the required shape to the metal.

The common types of sections used in structural steel work are shown in Figs. 18.1 to 18.22 and are described below.

Plates: Plates may be of any size or thickness but generally they are not rolled to thicknesses less than 5 mm and greater than 30 mm. The maximum area of a rolled plate is limited to 30 sq m. Larger plates have a tendency to be thicker at the centre than around the edges. Plates lesser than 4 mm in thickness are denoted as sheets. Sometimes plates are marked with different patterns which may be pressed into them, *e.g.*, checkered plates. Common uses of plates in building construction are as webs and flanges of deep beams, column flanges, column bases, etc.

Flats: These are rolled as in the case of plates, but are much longer in lengths and have shorter widths. The widths vary from 18 mm to 50 mm the minimum and maximum thicknesses vary from 3 mm to 80 mm. Flats are costlier than plates but are put to a considerable use in building construction. Flat section may have one rounded side with greater thickness than the remaining section and this is called bulb ball. These are used only in specific instances.

Angles: Angle section are the most commonly used in different components of a steel framed building. Indian Standard Specifications lay down the sizes of angle sections. Most common types are angles with equal legs and with unequal legs. The equal angles vary from 20 mm × 20 mm to 200 mm × 200 mm in size and 3 mm to 5 mm in thickness. The term 20 mm × 20 mm denotes that the widths of legs are 20 mm overall. Unequal angles vary in size from 20 mm × 30 mm to 220 mm × 100 mm and thickness from 4 mm to 20 mm. Special angles with a square toe, round backed, acute, square root and bulb types are also available. These are used to a limited extent in special constructional features.

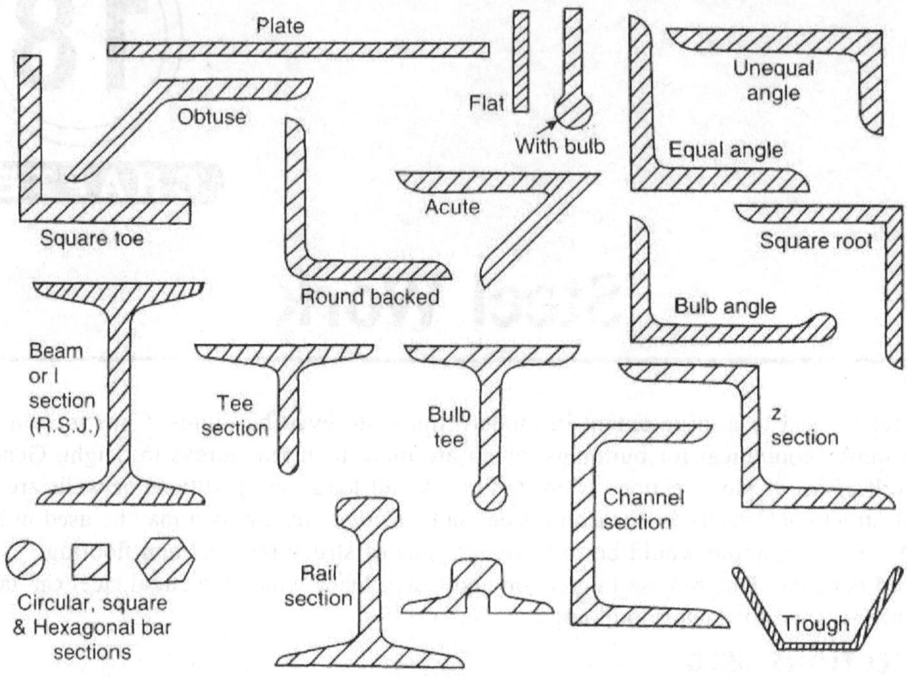

Figs. 18.1 to 18.22: Sections in steel work.

T-Sections: These are used for roof trusses and for certain built up columns. They are designated by the width of the stem, width of the table and by the thickness. The standard sizes vary from 40 mm × 40 mm to 150 mm × 150 mm with thicknesses from 6 mm to 8 mm. Special T-Sections with bulbs etc. are also used to some extent.

Channels: Channels are used for beams, columns and other small jobs. They are designated by the depth, flange width and weight per unit length. The size varies from 50 mm × 75 mm × 3 kg to 420 mm × 100 mm × 30 kg. Whenever stronger channels of lesser depth are required, these are specially cast with greater thicknesses than given by the Standards.

Joists: Rolled Steel Joists or I-Sections are most commonly used for beams and columns. They are denoted by the flange width, overall depth and weight per metre run.

Miscellaneous sections: Z-Sections, rail-sections, troughs, bars etc. are used to a limited extent in steel work for a building.

The British Standard lays down the basic properties of structural steel. Two qualities of steel are specified.

No. 1 quality: Steel made by the Open-Hearth Process or Acid-Bessemer process. Sulphur and Phosphorus contents not exceeding 0.06 per cent.

No. 2 quality: This is similar to No. 1 with the addition of copper from 0.2 to 0.5 per cent.

The tolerances in the dimensions are as under:

Lengths:
(*a*) When cut lengths are specified: ± 25 mm,
(*b*) When minimum lengths are specified: + 50 mm – 0.
(*c*) When exact lengths are specified: ± 3 mm.

Rolling margins, over and under the specified weight of $2\frac{1}{2}$ to 5 per cent are permissible.

Specifications for high tensile steel having a tensile strength of 5 kN/cm^2 are also given.

18.2 CONNECTIONS

Steel members are fastened together by means of rivets, bolts or by welding. Modern tendency is to go in for welding. Rivets are fitted at site or in the fabrication shop the latter being called as shop rivets and the former as field rivets.

Rivetting: Common practice is to connect the members by rivets which are permanent fastenings and cannot be removed unless the rivet is destroyed. Rivets used in building construction are made of soft steel with a tensile strength (breaking) of 3.5 to 4.0 kN/cm^2 with an elongation of not less than 26 per cent on a gauge of 8 diameters. Wrought iron rivets are sometimes favoured for site work as wrought iron has the property of retaining malleability even longer than that of steel, after being forged. Rivets generally have a hemispherical button-shaped head and a cylindrical shank. In driving a rivet, it is heated red hot and placed in a hole through the members which have to assist it in connecting. The projecting end is then upset to form a head with the aid of a pneumatic or hydraulic riveter. A corresponding projection left in the riveter gives the desired shape to the rivet head. While driving, the other side of the rivet is held in place by a dolly. As the rivet cools, it contracts thereby tightening the grip. Hand rivetting is used only in small structural jobs or in the field when only a few rivets are to be driven. Red-hot rivets are hammered down to the final shape.

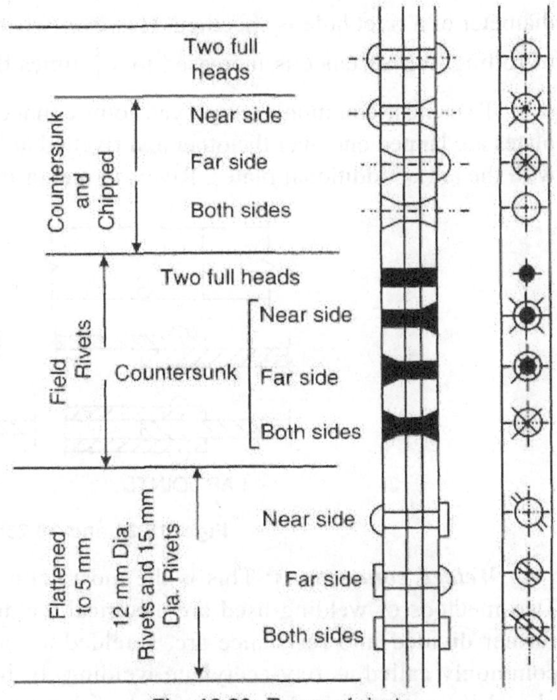

Fig. 18.23: Types of rivets.

Holes for rivetting may be punched, drilled or bored. For small works, rivet holes are generally punched. Drilled holes are always preferred as the holes are truly circular and cause less damage to the surrounding metal. The diameter of the hole is kept 1.5 mm larger than the rivet shank so that it could be inserted easily.

Rivet heads of various shapes, semi-circular button heads are commonly used. Whenever more clearances are desired, a flattened head is made. Counter-sunk heads are used when least projections are desired. They may have a flush face which is obtained by chipping. The conventional method of denoting various types of field and shop rivets is shown in Fig. 18.23.

Cutting of rivets may become necessary at times. This is accomplished by chipping off the head with a pneumatic chipper and then the rivet is driven out with a pin-maul. Sometimes the heads are burnt with an acetylene torch but care has to be taken to prevent damage to the surrounding metal.

The centre to centre distance between the adjacent rivets on the same rivet line is called pitch. Generally minimum pitch in structural steel work is specified to be three times the diameter of the finished rivet. However, the absolute minimum pitch is twice the diameter of the rivet plus 10 mm. Specifications lay down the minimum pitch as under:

10 mm	dia	rivets	– 40	mm	pitch
12.5 mm	dia	rivets	– 50	mm	pitch
17.5 mm	dia	rivets	– 60	mm	pitch
20.0 mm	dia	rivets	– 70	mm	pitch
25.0 mm	dia	rivets	– 80	mm	pitch
27.5 mm	dia	rivets	– 95	mm	pitch
30.0 mm	dia	rivets	– 110	mm	pitch

The maximum pitch should not be greater than 16 times the thickness of the thinnest plate with a maximum of 300 mm. Special rules for maximum pitch are also given for angle connections, tension members, etc.

A suitable edge distance, *i.e.*, the margin between the edge of the hole and the edge of the plate has to be provided for the prevention of failure. A minimum edge distance of $1\frac{1}{2}$ times the diameter of a rivet hole is specified. However where the edge has been formed by shearing instead of rolling, this distance is increased to $1\frac{3}{4}$ times the diameter of a rivet hole.

Two most commonly employed joint connections are (1) Lap joint in which the connected plates are lapped one over the other and rivetted and (2) Butt joint in which the plates are connected with the aid of additional plates. Rivets in one or more rows may be used to connect the members.

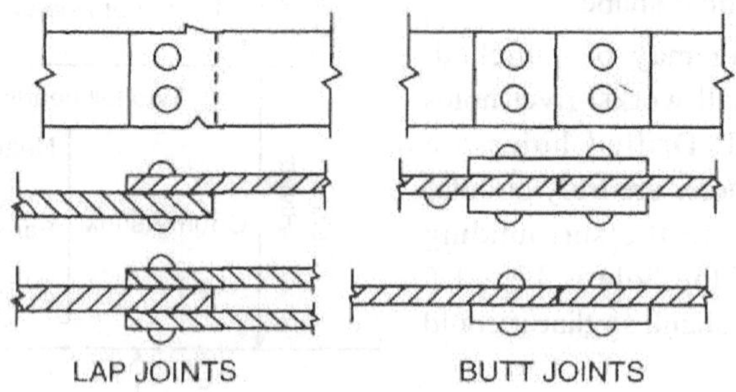

Figs. 18.24 and 18.25: Joint connections.

Welded connections: This is the most common method of connecting structural members. Two methods of welding used are electrical welding and gas welding. The electrical welding is further divided into resistance arc, shielded arc and atomic hydrogen welding. Gas welding is commonly called as oxy-acetylene welding. In both the methods, the pieces to be welded are placed in contact and edges are melted so that the metal flows from both the pieces together and when cooled, they get jointed with the aid of a weld. For satisfactory jointing, additional metal is supplied with the aid of a metallic rod which can be used as an electrode.

Bolt connections: For temporary connections, bolt joints are used. The diameter of the holes is kept 1 mm larger than the external diameter of the bolt threads. Bolt connections are also used wherever rivetting is difficult may be due to lesser number of rivets needed or difficulty in connecting.

18.3 STRUCTURAL MEMBERS

18.3.1 Steel Beams

Beams include girders, lintels, etc. A simplest beam would consist of a single rolled steel joist section or an angle section for carrying little loads. Whenever a beam has to take greater loads, compound sections are used. Compound beams consist of two or more single steel joists connected together through bolts and separators which hold them in position. Separators are placed at 1500 mm to 2000 mm apart.

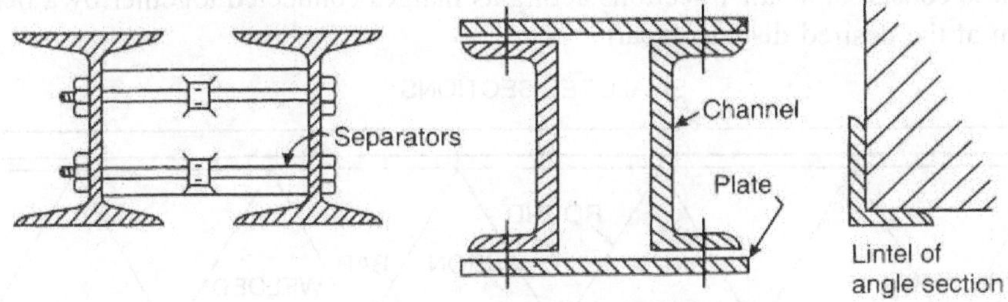

Figs. 18.26 to 18.28: Steel beams.

For still heavier beams or girders, two channels may be used back to back or spaced apart and their flange area increased by the addition of one or more plates at top and bottom.

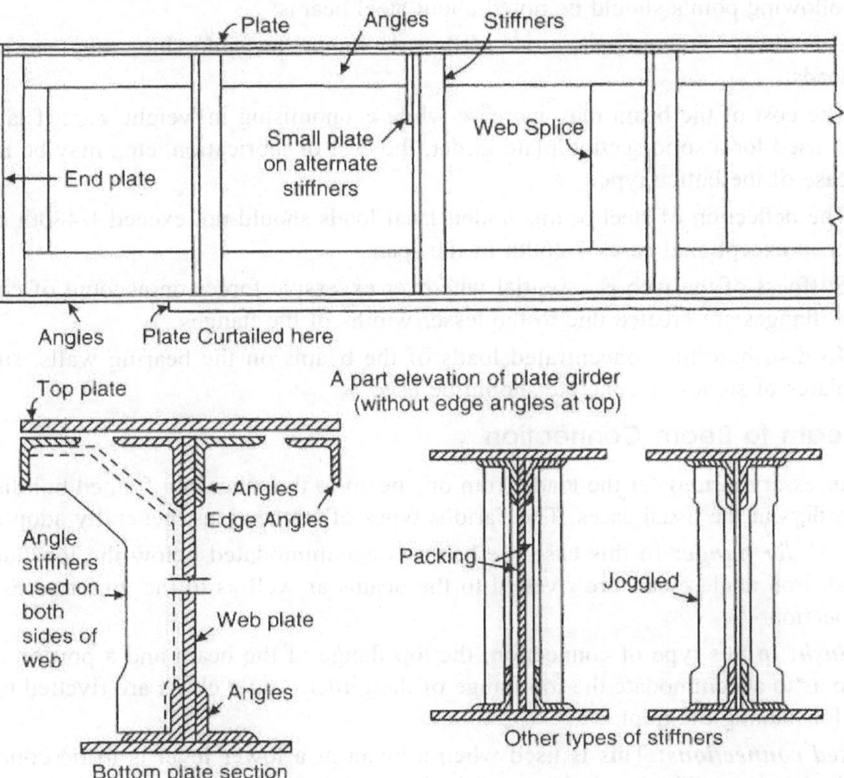

Figs. 18.29 to 18.32: Plate girder and its details.

Plate girders are used when very large loads have to be carried. A plate girder is a built up beam consisting of top and bottom flanges made up of angles and plates. The web consists of one or more steel plates. All the individual pieces are rivetted or welded together. Since plate girders are deep beams, it is necessary to stiffen them laterally to prevent buckling under high compressive forces in the top flange. Angles or T-sections are used as stiffners. They are rivetted to the plate girder at suitable intervals and may be bent so as to form joggle type or packings may be used to keep them straight.

18.3.2 Open Web Beams

Beams carrying light loads and where the shearing forces are not excessive, open web beams are used. These consist of small T-sections acting as flanges connected together by a bent iron bar to keep them at the desired distance apart.

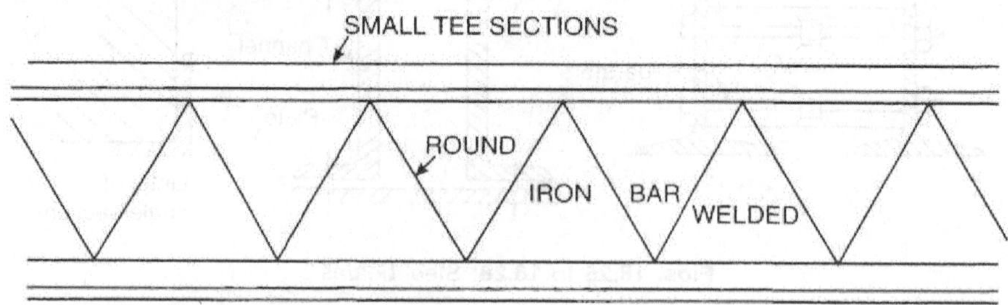

Fig. 18.33: Open web beam.

The following points should be noted about steel beams:

(*i*) It is always economical to use as deep beams as possible since they can take higher loads.

(*ii*) The cost of the beam may increase while economising in weight. *e.g.*, if lattice girder is used for a solid section plate girder, the cost of fabrication, etc., may be more in the case of the lattice type.

(*iii*) The deflection of steel beams under usual loads should not exceed 1/480th of the span or in exceptional cases 1/260th of the span.

(*iv*) Stiffness of the web is essential wherever excessive forces on account of compression in flanges are created due to the lesser widths of the flanges.

(*v*) To distribute the concentrated loads of the beams on the bearing walls, suitable bed plates of stones or concrete should be used.

18.3.3 Beam to Beam Connection

It is necessary to transfer the loads from one beam to the other in a framed building or from beams to girders in the usual cases. The various types of connections, generally adopted, are:

***Undergirder flange*:** In this case the beam is accommodated below the top flange of the girder. Small iron angle cleats are rivetted to the beams as well as to the girder so as to form a strong connection.

***Top-flush*:** In this type of connection, the top flange of the beam and a portion of the web is cut off so as to accommodate the top flange of the girder. Angle cleats are rivetted to the beam and girder for making the joint.

***Blocked connections*:** This is used when a beam at a lower level is to be connected to a girder at a higher level. The lower flange of the beam is cut off to accommodate the lower flange of the girder, the webs of the two being connected with angle cleats.

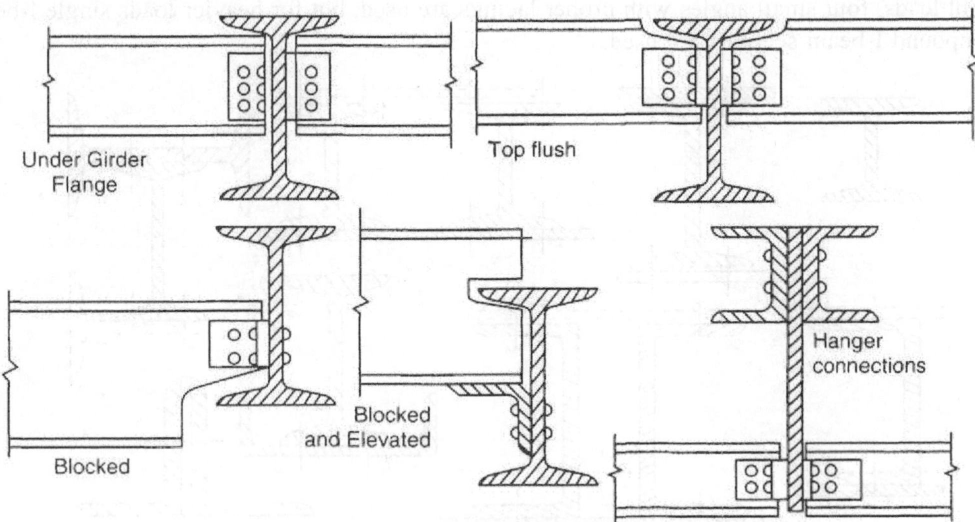

Figs. 18.34 to 18.38: Beam to beam connections.

***Blocked and elevated*:** This type of connection is suitable for beams meeting at higher levels with the girders. A suitable recess to fit in the top flange of the girder is cut from the beam and the beam rests on an angle cleat which in turn is bolted to the girder.

***Hanger connections*:** These are used to connect beams and girders at different levels. The connection is made with the aid of a plate and angle cleats or bolt and rods may be used.

18.4 COLUMNS

Individual small columns may consist of rolled steel joist or a rolled steel joist with two flange plates. Bigger columns can be made by rivetting together two or three rolled steel joists so as to form a compound column. Other forms of compound columns commonly used are a combination of rolled steel joists and channels, two angles, channels with flange plates or lattice bracing, four angles and a web plate and four angles joined together with lattice braces. Whenever angles or channels or joists more than one in number are used without flange plates, it is necessary to inter-connect them with the help of lacings and batten plates which make the column to act as a combined unit.

The types of lacings used are:

(*a*) ***Single lacing*:** These consist of small flat iron rivetted to the column components in a zigzag manner.

(*b*) ***Batten plates*:** These are thin plates or flat sections rivetted at suitable intervals at right angles to the axis of the columns.

(*c*) ***Double lacing*:** This consists of small flat irons rivetted to the column components in a cross manner.

(*d*) ***Z-type lacing*:** This comprises of small angle sections rivetted to the column, one in the horizontal position and the other inclined.

The choice of the type of column depends upon the load to be taken by the column, dimensions of the neighbouring parts, fixing conditions, fire proof treatment essential, architectural treatment desired, etc. It should be ensured that all connections are of simple type and it is easy to fabricate the individual parts. Columns can fail by bending along the axis having least moment of inertia. Hence it is desirable to have columns having same moment of inertia along both the axis. Generally for light loads, four small angles with proper lacings are used, but for heavier loads single I-beam or compound I-beam sections are used.

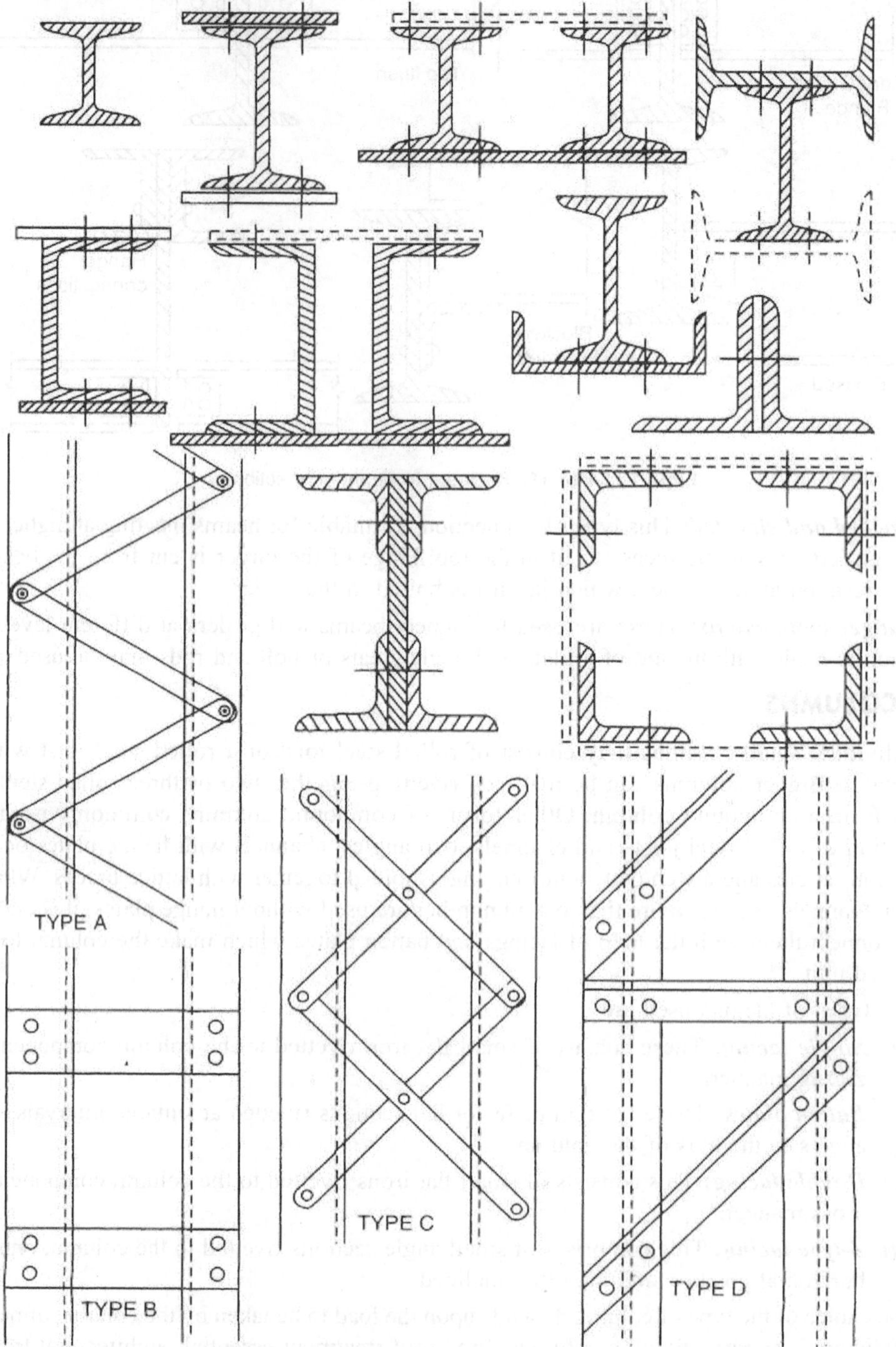

Figs. 18.39 to 18.52: Different types of columns.

Splicing of columns: Steel columns are made of constant sections for every two storey heights of a tall building. It is possible to change the sections of the column for different storeys on account of the reduced load which a column has to take in the upper portions of a building. This means that columns of two sections have to be joined together. Columns of different sections are connected with the aid of spliced plates and additional plates are used for proper connections. Sometimes bearing plates of steel are interposed to ensure a proper bedding for the upper column. These splices are commonly made about 0.75 m above the floor level so that they do not interfere with the beam and other connections at the floor levels.

18.5 COLUMN AND BEAM CONNECTIONS

Generally two types of connections are used, the framed connection type and the seated type. In the framed connection type, two angles are rivetted to the beam and the column. These angles are connected to the web of the girder or beam and the flange of the column. A small angle may be added at the bottom of the beam to seat the same on the column while erecting it. In the seated type of connection, two more angle sections are used to connect the beams with the columns. They are rivetted to the top and bottom flange of the girders. (*See* Figs. 18.53 to 18.54). For bigger column sections, stiffners in the form of two small angles rivetted to the column flange are fitted beneath the beams (*See* Figs. 18.55 and 18.56).

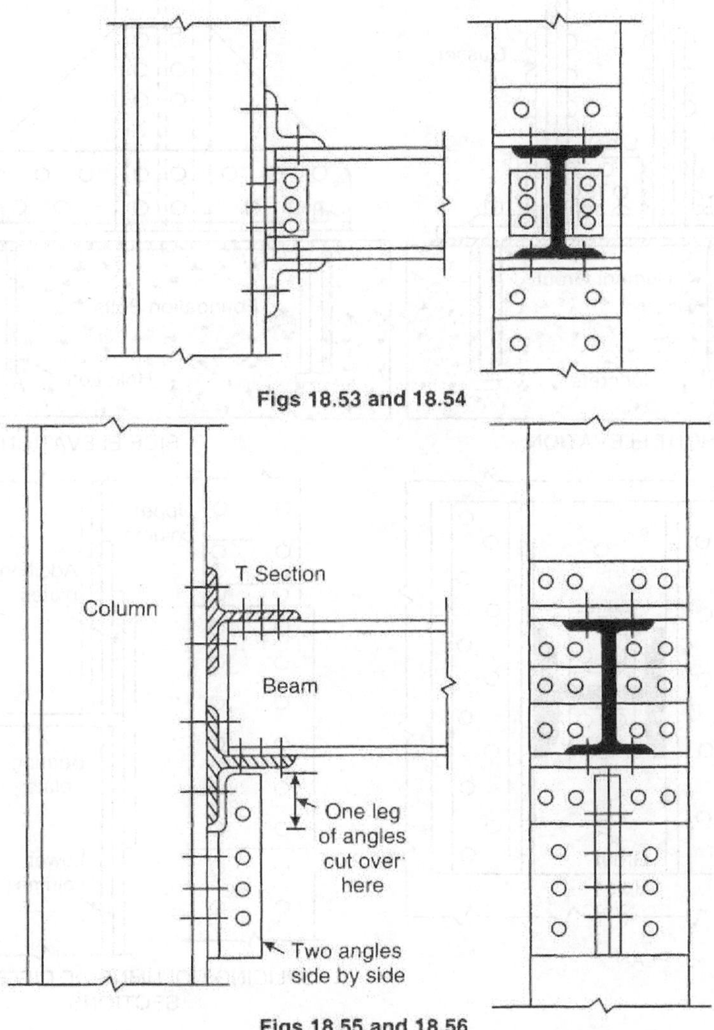

Figs. 18.55 and 18.56: Column and beam connections.

18.6 COLUMN BASES

The load at the lower end of a column has to be transferred to a concrete footing which in turn transmits it to the ground below. The bottom end of the steel column has to be supported on a special base to distribute the load from the column to the concrete footing which otherwise would get crushed. Column bases are of the following types.

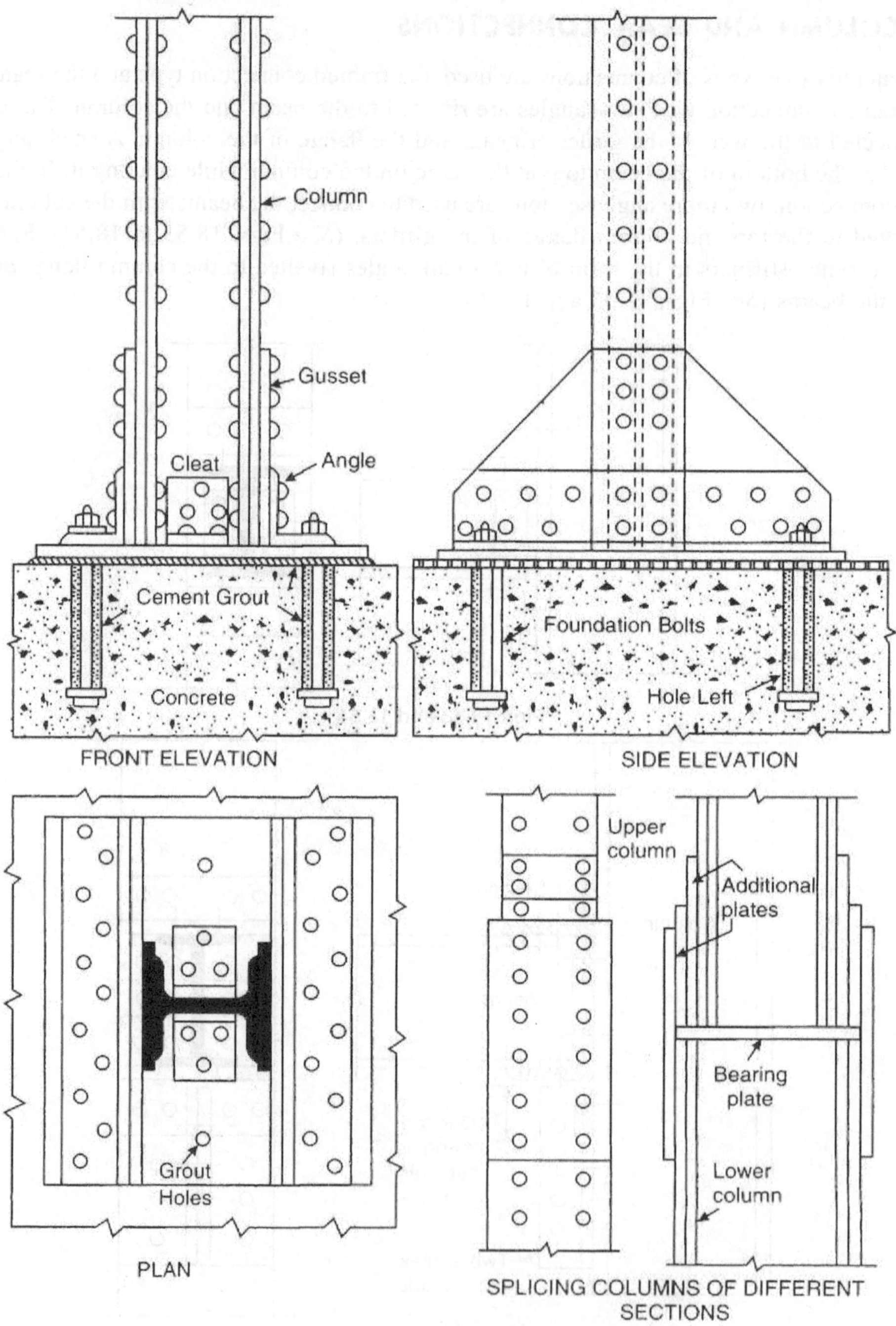

Figs. 18.57 to 18.60: Column bases.

(1) ***Built up bases***: Made from structural steel sections.

(2) ***Cast iron or cast steel bases***: These are specially cast for different types of columns.

(3) ***Steel grillage***: This consists of rolled steel joists connected together and laid in tiers.

Built up column bases are mostly used and are suitable for light and moderate loads. The base is made up of steel plates which are connected to the column with the aid of angle cleats through gusset plates. A thin cement grout is spread between the top of the concrete surface and the bottom of the steel base plate. The angles and the base plate are fixed to the concrete base with the aid of foundation bolts which are placed in suitable holes left for them.

18.7 FRAMED STEEL BUILDINGS

Buildings which are of industrial nature or consist of many storeys are made up of a steel framework which support all the loads. The steel framed building is similar to reinforced concrete framed building. In this case the columns, beams and girders consist of steel sections. The partitions and exterior walls are constructed of other light and fire proof materials. Tall buildings have to be adequately braced to resist the forces of wind and earthquakes.

18.8 FABRICATION OF STEEL WORK

This means preparing steel work for erection and includes all work necessary to layout, cut, drill, rivet or weld the steel sections. Most of the work is carried out in the fabrication shop and the work in the field is to be reduced as much as possible.

The first stage is to prepare the template according to the shape of the final job. Templates may be made of wooden strips showing location of all holes and cuts. For gusset plates, card board templates may be used.

As steel is procured from the mill, it is stored in a stock-yard. All materials should be straight and if necessary they are straightened by pressure unless they have to be of curvilinear forms.

Cutting is affected by shearing, cropping or sawing. Gas cutting by mechanically controlled arcs is also used for mild steel. For high tensile steel, gas cutting is permitted under special care so as to remove all hardened material by machining later on. Plates and angles are cut cold in one stroke by shearing. Beams and channels are usually cut to the desired lengths in factories but if shearing is to be resorted to, at least 25 mm of the material is wasted. A better method is to cut the beams with saws or in exceptional cases, gas cutting may be resorted to.

The next step is to lay out all the material which includes marking the steel either with the aid of templates or directly. The centres of the holes to be drilled are marked with a punch. Holes are drilled, punched or bored. All holes should generally be drilled for better work as they are true circles, are accurately centred and cause less damage to the surrounding metal. Holes are always drilled whenever thick sections are encountered. Drilled holes are generally not more than 2 mm larger than the nominal diameter of the rivet or the bolt. When punching is resorted to, the holes should be punched 6 mm less in diameter than the final desired value and pneumatic reamers are used for making the holes to proper size.

When the components of a member are ready, they are assembled and held in position temporarily by shop bolts. These are longer than necessary and have packing washers to save time in tightening. At least two bolts should be put in one part of a member. All parts assembled should be in close contact and all bearing stiffners should bear tightly at top and bottom. No drifting should be permitted except to draw the parts together and no drift should be larger than the normal diameter of the rivet. Drifting should not distort the metal or enlarge the holes.

The assembled parts are then rivetted. All rivetting should be done by hydraulic or pneumatic pressure. Bigger rivets are heated. Heaters commonly used in the fabrication shops burn oil and give a steady flame. Electric heaters are also sometimes used. Rivets of diameter less than 10 mm may be driven cold. Each rivet should completely fill the hole and form a head of standard size. All loose, burnt or badly formed rivets should be cut out and replaced.

Column splices, butt-joints of struts and compression members depending upon contact and stress transmission should be accurately machined and close butted over the whole section. Whenever sufficient gussets and rivets are provided to transmit the entire load, the column ends need not be machined.

The whole steel work with the exception of rivets, bolts, nuts and machine faces after being thoroughly cleaned should be given one coat of red lead paint. All machined faces should be coated with a mixture of white lead and tallow. Surfaces which are to be held in contact by rivetting or bolting should be painted before assembly and the parts brought together while they are still wet. All portions of steel work which are inaccessible after rivetting should be given two coats of red lead paint.

The steel work should be temporarily shop erected so that accuracy of fitness may be checked before despatch. All parts are inspected for defects, if any.

18.9 ERECTION OF STEEL WORK

Steel work is erected with the aid of derricks, slings, guys, cranes, etc. All parts of steel work should be erected according to drawings keeping in view the plumbness of columns and placing of other parts. During erection the steel work should be securely bolted and when necessary temporarily braced to provide for all loads to be carried by the structure during erection including that of the erecting equipment, etc. All parts should be finely rivetted only after they are properly placed in position and final alignment is checked. The steel work is finally painted.

18.10 DESIGN USING LIMIT STATE METHOD

18.10.1 General

The design in steel may be done using the limit state method, which is generally based on the following basic aspects/principles:

(*a*) It makes use of the plastic range of material for the design of structural members and incorporates load factors to take into account the variability of loading configurations.

(*b*) It considers the good performance of steel in tension compared to compression and specifies variable factors. It takes into account this variance by defining limit states which address strength and serviceability.

(*c*) According to this method, a structure or part of it, is considered unfit for use when it exceeds the limit state beyond which it infringes any one of the criteria governing its performance or use.

(*d*) The two limit states are classified as the ultimate limit State and Serviceability limit State. The limit states take care of the safe operation and adequacy of the structure from strength point of view. The criteria which are used to define the ultimate limit state are yielding, plastic strength, fatigue, buckling, etc. Serviceability limit state takes care of the performance and behaviour of the structure during its service period. Deflection, vibration, drift, etc are considered as serviceability criteria.

(*e*) Limit state method considers the critical local buckling stress of the constituent plate element of a member. This method has provisions to enhance resistance of plate elements against local buckling by suitably reducing the slenderness ratio. Hence, it is possible to develop the full flexural moment capacity of a member subjected to flexure or the limit State in flexure for a beam. in limit State method, based on slenderness ratio of the constituent plate elements, a member can be classified as Slender, Semi-compact,

Compact and plastic. This section classification becomes essential as the moment or load capacities of each of these sections take different values depending upon these classifications.

In this method, the factored loads, in different combinations, are applied to the structure to determine the load effects. The latter are then compared with the design strength of the elements.

This is expressed mathematically as:

The effects of

$$\gamma_L \cdot Q_k \leq \left[\frac{1}{\gamma_m}\right]$$

where

$\gamma_m = \gamma_f \cdot \gamma_{m_1} \cdot \gamma_{m_2}$

= Partial safety factor for material.

γ_L = partial factor for loads.

γ_f Factor that takes into account the inaccuracies in assessment of loads, stress distribution and construction tolerances.

$\gamma_{m_1}, \gamma_{m_2}$ = Factors that take into account, uncertainties in material strength and quality, and manufacturing tolerances respectively.

Q_k = Specified nominal load induced based on design stipulations.

σ_y = Yield strength of the material.

QUESTIONS

1. What are the uses to which steel is put in Building Construction? What are the various sections used?
2. What are the various methods of connections? Give details of rivetted and welded connections.
3. Write short notes on:
   (*i*) Steel beam.
   (*ii*) Beam to beam connection.
   (*iii*) Columns.
   (*iv*) Column and beam connection.
   (*v*) Column bases.
4. Write a short note on the fabrication and erection of steel work.
5. Give the general principles for the design of steel structures using limit state method.

REFERENCES

1. **I.S. 800–1984:** *Code of practice for general construction in steel* (*Second revision*).
2. **Voss, W.C.**: *Fire-proof construction.*
3. **Huntington, W.C.**: *Building construction.*
4. **Feber, Oscar**: *Structural steelwork.*
5. **Stewart, D.S.**: *Practical design of simple steel structure.*
6. **Reynold T.J. and Kent, L.E.**: *Structural steelwork.*
7. **British Standard Institution**: *Building materials and components for housing.*
8. **IS: 2062:1999:** *Specification for steel for general structural purposes* (*Fifth revision*).

Concrete Construction

Plain, reinforced, prestressed and precast concrete members are being abundantly used in the construction of buildings these days. This construction, which has withstood the test of time, is very durable and economical for most of the buildings. The construction involving either precast units or cast-in-situ building units made of concrete may be adopted.

Plain concrete is a mixture of cement, water, fine and coarse aggregates and any approved admixture. It possesses the property of hydraulicity or ability to harden under water.

As plain concrete by itself has little tensile strength, it has to be reinforced in structures subjected to tensile forces. The reinforcing material is usually steel.

Prestressing is the technique of introducing forces in a structural member, so as to produce stresses opposite to that which will occur during its service.

19.1 ADVANTAGES OF REINFORCED CEMENT CONCRETE (R.C.C) CONSTRUCTION

Reinforced cement concrete construction has a number of advantages over other types of construction. Some of these are listed below:

1. ***Higher strength***: Reinforced cement concrete has a higher compressive strength than masonry construction. The reinforcement imparts tensile strength. Plain concrete itself has a low tensile strength, which is about 8 to 12 per cent of its compressive strength. Its shear strength, measured as inclined tension, is about 8 to 10 per cent of its compressive strength. However the strength of concrete varies considerably due to variations in the following factors:

(*i*) Quality of materials and grading of aggregates

(*ii*) Water-cement ratio

(*iii*) Cement content

(*iv*) Age of concrete; and

(*v*) Methods of mixing, placing, compacting and curing.

2. ***Good workability***: Concrete can be placed easily in the forms and compacted. The workability of concrete is influenced by the water content, the size, the shape and grading of the aggregate, the ratio of coarse and fine aggregates and by the use of any admixtures.

3. ***High durability***: R.C.C. has a very high resistance to the forces of disintegration due to natural causes, such as weathering, action of water containing chemicals, temperature changes

including freezing and thawing, alterations in moisture content and chemical changes. Sound cement, durable aggregates, an optimum water-cement ratio, a dense grading of aggregates, careful mixing and placing, thorough compaction and adequate curing increase durability.

4. *Fire resistant*: Concrete is a good insulator and has a fairly good fire resistance owing to the presence of pores and also water which is present in it in a chemically combined state. This water is removed only when the temperature exceeds about 300°C. However when the exposed concrete is dehydrated at this temperature, it attains greater insulation capacity and higher resistance to heat penetration. The fire resistance of concrete depends mostly on the quality of the aggregates used. Broken brick aggregates make the concrete more fire resistant then broekn stone aggregates. Broken stone aggregates are better than gravel.

However, the steel in reinforced concrete, being a very good conductor of heat, looses it strength rapidly with rise in temperature. The steel reaches the danger point at about 250°C, and its strength is reduced to half at about 400°C. Therefore to protect the steel, an adequate concrete cover should be given to it. This cover generally varies from 13 mm in slabs to 40 mm in columns.

5. *Shrinkage of concrete provides good bond between the concrete and reinforcement*: When concrete sets and hardens in air, it shrinks, partly due to chemical action of the colloids produced by the water reacting on the cement, and partly due to the physical drying out of the mass. The advantage of shrinkage of concrete is that in shrinking, the concrete grips the reinforcement tightly, thus preventing slipping. Thus shrinkage is the main cause of the bond between the concrete and the reinforcement. Shrinkage of concrete is almost directly proportional to the amount of water used in the mix.

6. *Ease of construction*: Concrete can be placed in any shape with ease. The quickness and ease of construction in the case of concrete work is a great advantage.

7. *Improved appearance*: The appearance of concrete surfaces can be improved by applying various types of finishes to concrete.

8. *Light weight construction possible*: The weight of concrete can be reduced by adding gas forming or foaming agents during mixing which produces a cellular structure on hardening. This type of concrete can be used where lightness of the structure is desired e.g. partition walls. Light weight aggregates can be used in place of usual aggregates.

9. *Higher resistance to weathering by making air-entrained concrete*: One of the outstanding developments in concrete technology is the purposeful entrainment of air in concrete either by using air-entraining Portland cement or by adding some air-entraining agent directly to concrete at the time of mixing.

Air-entrained concrete has higher resistance to frost action, sulphate attack, internal stresses caused by the heat of hydration, and the disruptive forces caused by frequent wetting and drying and heating and cooling. It has greater workability, plasticity and cohesiveness. It increases water tightness, reduces bleeding and segregation and permits reduction in the water, cement and fine aggregate content of the concrete.

Good concrete, for economical results, can be made with materials, a bulk of which are generally obtainable near the site. Clean sand, strong coarse aggregates and good Portland cement are necessary for making good concrete. These materials are to be mixed in the proportions which will produce concrete having the necessary strength and other properties combined with the greatest economy. Concrete is graded as per Table 19.1. M refers to the mix and the number is the compressive strength of 150 mm size cube at 28 days, expressed in N/mm^2 or $\frac{\text{kg}}{10\,\text{cm}^2}$.

Table 19.1: Grades of Concrete

Group	*Grade Designation*	*Specified Characteristics Compressive Strength of 150 mm Cube at 28 days in N/mm²*
Ordinary concrete	M 10	10
	M 15	15
	M 20	20
Standard concrete	M 25	25
	M 30	30
	M 35	35
	M 40	40
	M 45	45
	M 50	50
	M 55	55
High strength concrete	M 60	60
	M 65	65
	M 70	70
	M 75	75
	M 80	80

19.2 MATERIALS USED FOR MAKING CONCRETE

Ordinary concrete is used for lean concrete bases and simple foundations for masonry walls. Grades M 25 and higher are adopted for reinforced concrete and prestressed concrete.

19.2.1 Aggregates

Aggregate is a general term applied to those inert materials which, when bound together by cement, from concrete. Aggregates for concrete are divided into three categories:

(*a*) Fine aggregate

(*b*) Coarse aggregate; and

(*c*) All in aggregate.

The most important functions of the fine aggregates in concrete are to assist in producing workability and uniformity in the mixture. Fine aggregate also assists the cement paste to hold the coarse aggregate particles in suspension, thus promoting plasticity in the mixture and preventing the possibility of segregation of the paste and coarse aggregate. Coarse aggregate is used mainly for providing bulk to the concrete. It is frequently used in two or more sizes to increase the density of the mix. All in aggregates contain both fine and coarse aggregates. Aggregates may be either from natural sources such as sand, gravel or stone, crushed or uncrushed or artificial aggregates e.g. broken brick, air-cooled blast furnace slag, light weight aggregates, etc.

In the selection of aggregates for good concrete work, it should be ensured that the aggregates are absolutely clean. They should be free from organic matter and other impurities. The aggregates used must be sound and must resist the action of weather, particularly of frost. Some aggregates contain silicious materials which may combine with the alkalies present in cement so as to produce disruption which causes disintegration of aggregates. Certain aggregates contain readily oxidizable materials which expand on oxidation and cause disruption. These types of aggregates are unsuitable for concrete work.

The grading of aggregates is very important for getting good quality concrete. If the grading of aggregates is such as will create a large proportion of voids when the concrete is compacted, a concrete of poor quality is obtained. The material as obtained from the quarry is to be suitably adjusted for concrete work, *i.e.*, grading has to be done according to definite specifications for different types of work. The aggregates should not be very fine as large amount of water is needed for their mixing and the concrete gets weakened. For best results the proportion of very fine material should not exceed 10%. The ratio of the quantity of fine aggregates to the coarse aggregates is very important and has got a great bearing on the strength of concrete in the final stage.

The maximum size of the particles which is going to be used in a particular portion of a building will depend upon the nature of work and the reinforcement used. The maximum size generally used is 5 mm less than the minimum distance between reinforcement or is taken as the minimum cover between the face of work and the reinforcement. For thin walls, slabs, etc., the maximum size of the particles should not exceed 1/5th of the thickness of walls. For mass concrete work, the maximum size of particles may be increased in such a limit that they can be used easily. The use of large aggregates will reduce the quantity of cement needed for a given strength as the surface area of the aggregate particles gets decreased, but larger sizes are unsuitable for being worked easily. Large sizes known as plums or displacers are used in mass aggregate work to save materials but they have to be suitably packed in concrete. These plums should be sound and clean and should not be placed very near to one another or to the exposed face.

The shape of the aggregate particles will affect the workability and hence the strength. Aggregates composed of roughly spherical particles produce the most workable concrete. The longer particles produce a harsh workable mix and hence it is a common practice to add some water to the mix. This has the effect of increasing the ratio of water to a given quantity of cement and hence decreases the strength of concrete.

19.2.2 Cement

The cement used may be any of the following types conforming to accepted standards:

(*i*) 33, 43, 53 grade of ordinary Portland Cement

(*ii*) Rapid hardening Portland Cement

(*iii*) Portland Slag Cement

(*iv*) Portland Pozzolana Cement (fly ash based)

(*v*) Portland Pozzolana Cement (Calcined clay based)

(*vi*) Hydrophobic Cement

(*vii*) Low heat Portland Cement

(*viii*) Sulphate resisting Portland Cement

A part of the Ordinary Poartland Cement may be replaced by any of the following mineral admixtures, provided uniform blending with cement is ensured

(*i*) Pozzolanas

(*ii*) Fly ash (Pulverised fuel ash)

(*iii*) Rice husk ash

(*iv*) Metakaoline having fineness between 700 to 900 m^2/kg

(*v*) Ground granulated blast furnace slag.

19.2.3 Reinforcement

The reinforcement may be any of the following:

(*a*) Mild steel and medium tensile steel bars

(*b*) High strength deformed steel bars

(*c*) Hard-drawn steel wire fabric

(*d*) Grade A of structural steel

19.2.4 Admixtures

Admixtures are substances, other than cement, aggregate and water, which are added to concrete in order to improve its qualities either during its placing or in its subsequent performance.

The important modifications obtained in the properties of concrete by the use of admixtures are:

(*i*) Improvement in workability

(*ii*) Reduction in bleeding

(*iii*) Retardation or reduction of heat evolution

(*iv*) Retardation of initial stiffening or increased Setting time

(*v*) In grout mixtures, improvement of penetration and pumpability and the reduction of segregation.

(*vi*) Acceleration of the rate of development of strength at early ages.

(*vii*) Increase in durability or in resistance to deterioration under special exposure conditions.

(*viii*) Decrease in capillary flow of water

(*ix*) Decrease in permeability to liquids

(*x*) Increase in the bond between reinforcement and concrete

(*xi*) Prevention of shrinkage or creation of slight expansion in concrete for certain special uses; and

(*xii*) Economy in the use of cement.

Admixtures can be broadly divided into the following categories according to the kind of materials constituting them or to the characteristic effects produced by them:

(*i*) Workability agents

(*ii*) Pozzolanic materials

(*iii*) Retarders

(*iv*) Accelerators

(*v*) Water repelling agents

(*vi*) Air-entraining agents

(*vii*) Gas forming agents

(*viii*) Natural cementing materials; and

(*ix*) Miscellaneous

Commercial admixtures may contain materials belonging to two or more of the above groups.

19.2.5 Water

Water to be used in mixing concrete must be clean and free from injurious amounts of oils, acids, alkalies, salts, sugar, organic materials or other substances that may be deleterious to concrete or steel. Potable water is generally considered suitable for mixing concrete.

As a guide, the following concentrations represent the maximum permissible values:

(*a*) To neutralise 100 ml of water, using phenolphthalein as an indicator it should not require more than 5 ml of 0.02 normal NaOH.

(*b*) To neutralise 100 ml of water, using mixed indicator, it should not require more than 25 ml of 0.02 normal H_2SO_4.

Percentage of solids permissible as specified in the National Building Code 2005 should not exceed the following values:

Sl. No.	*Solids*	*Permissible limit (Max.)*
1	Organic Organic solids = (total solids–ignited residue)	200 mg/*l*
2	Inorganic (ignited residue)	3000 mg/*l*
3	Sulphates (as SO_3)	400 mg/*l*
	Chlorides (as *Cl*)	2000 mg/*l* for concrete not containing embedded steel and 500 mg/*l* for reinforced cement concrete work
4	Suspended matter	2000 mg/*l* The pH value of water should generally be not less than 6.

Sea water is not generally suitable considered for building work because of the presence of harmful salts.

19.3 WORKABILITY OF CONCRETE

The concrete mix proportions chosen should be such that the concrete is of adequate workability for the placing conditions of the concrete and can be properly compacted with the means available.

Suggested ranges of workability of concrete for some placing conditions, are given in Table 19.2.

Table 19.2: Workability of concrete

Placing conditions	*Degree of Workability*	*Slump mm*
Blinding concrete; Shallow sections; Pavements using pavers	Very low	compacting factor 0.75–0.80
Mass concrete; Lightly reinforced sections in slabs, beams, walls, columns; Floors; Hand placed pavements; Canal lining; Strip footings	Low	25–75
Heavily reinforced sections in slabs, beams, walls, columns	Medium	50–100
Slipform work; Pumped concrete	Medium	75–100
Trench fill; *in-situ* piling	High	100–150
Tremie concrete	Very High	By flow

19.4 DURABILITY

The durability of concrete depends on its resistance to deterioration and the environment in which it is placed. The resistance of concrete to weathering, chemical attack, abrasion, frost and fire depends largely upon its quality and constituent materials. Susceptibility to corrosion of steel is governed by the cover provided and the permeability of concrete. The quality and durability of concrete depends upon adequate cement content and a low water-cement ratio.

One of the main characteristics influencing the durability of any concrete is its permeability. With strong, dense aggregates, a suitably low permeability is achieved by having a sufficiently low water-cement ratio, by ensuring as thorough compaction of the concrete as possible and by ensuring sufficient hydration of cement through proper curing methods. Therefore, for given aggregates, the cement content should be sufficient to provide adequate workability with a low water-cement ratio so that concrete can be completely compacted with the means available.

Rapid hardening cement attains its strength at an earlier age than ordinary Portland cement and hence is economical in some cases but is more costly. High alumina cement is more resistant to deteriorating effects of weathering agencies. Coloured cements or white Portland cement may be used depending on the type of finish desired. For good work, the specifications of Portland cement must be checked regularly.

CONCRETE MIX DESIGN

The mix proportions should be selected to ensure that the workability of fresh concrete is suitable for the conditions of handling and placing, so that after compaction it surrounds all reinforcements and completely fills the formwork. When concrete is hardened, it should have the required strength, durability and surface finish.

In specifying a particular grade of concrete, the following information should be included:

(*i*) Type of mix *i.e.* design mix concrete or nominal mix concrete
(*ii*) Grade designation
(*iii*) Type of cement
(*iv*) Maximum nominal size of aggregate
(*v*) Minimum cement content (for design mix concrete)
(*vi*) Maximum water-cement ratio
(*vii*) Workability; and
(*viii*) Mix proportion (for nominal mix concrete)
(*ix*) Exposure conditions
(*x*) Maximum temperature of concrete at the time of placing
(*xi*) Method of placing
(*xii*) Degree of supervision

In appropriate circumstances, the following additional information may be specified

(*i*) Type of aggregate
(*ii*) Maximum cement content and
(*iii*) Whether an admixture shall or shall not be used and the type of admixture and the condition of use.

19.5 MIX DESIGN METHODS

The determination of the proportions of cement, aggregates and water to attain the required strength can be made as follows:

(*a*) By adopting nominal concrete mix, such concrete mix is called 'Nominal mix concrete'.

(*b*) By designing the concrete mix, such concrete mix is called 'Design mix-concrete'.

(*c*) By other methods such as Minimum Voids method and Fineness Modulus method.

These methods are explained below:

19.5.1 Nominal Mix Concrete

Nominal mix concrete is used for concrete of M 20 or lower. The proportions of materials for nominal mix concrete are given in Table 19.3.

Table 19.3: Proportions for nominal mix concrete

Grade of Concrete	*Total Quantity of Dry Aggregates by Mass per 50 kg of Cement. To be Taken as the Sum of the Individual Masses of Fine and Coarse Aggregates. Max*	*Proportion of Fine Aggregate to Coarse Aggregate (by Mass)*	*Quantity of Water per 50 kg of Cement. Max.*
(1)	(2) kg	(3)	(4) litres
M 5	800	Generally 1 : 2	60
M 7.5	625	but subject to	45
M 10	480	an upper limit	34
M 15	350	of 1 : $1\frac{1}{2}$ and a	32
M 20	250	lower limit of	30
		1 : $2\frac{1}{2}$	

Note: The proportions of the fine to coarse aggregates should be adjusted from upper limit to lower limit progressively as the grading of the fine aggregates becomes finer and the maximum size of coarse aggregate becomes larger. Graded coarse aggregate shall be used.

Example. For an average grading of one aggregate the proportions shall be 1 : $1\frac{1}{2}$, 1 : 2 and 1 : $2\frac{1}{2}$ for maximum size of aggregates 10 mm, 20 mm and 40 mm, respectively.

The cement content of the mix specified in Table 19.2 for any nominal mix should be proportionately increased if the quantity of water in a mix has to be increased to overcome the difficulties of placement and compaction, so that the water-cement ratio as specified is not exceeded.

19.5.2 Design Mix Concrete

The guidelines for concrete mix design is covered in IS : 10262–1982. The mix should be designed to produce the grade of concrete having the required workability and a characteristic compressive strength not less than the appropriate values.

Concrete should be of satisfactory quality in both the fresh and hardened states. The choice of proportions will be governed by both these conditions. If the plastic concrete is not satisfactory, it cannot be properly compacted and its structural value is greatly reduced. Therefore the mix design is generally carried out for a particular compressive strength of concrete, coupled with adequate workability so that fresh concrete can be properly placed and compacted.

The basic assumption made in the mix design is that the compressive strength of workable concrete is, by and large, governed by the water-cement ratio. Thus another convenient relationship applicable is that for a given type, shape, size and grading of aggregates, the amount of water determines its workability.

However, there are various other factors which affect the properties of concrete, for example, the quality and quantity of cement, water and aggregates, batching, transportation, placing, compaction, curing, etc.

19.5.2.1 Data for mix design

The following basic data are required for design of a concrete mix:

(*i*) Characteristic compressive strength of concrete at 28 days (F_{ck})

(*ii*) Degree of workability desired

(*iii*) Limitations on the water/cement ratio and the minimum cement content to ensure adequate durability.

(*iv*) Type and maximum size of aggregate to be used; and

(*v*) Standard deviation (s) of compressive strength of concrete.

19.5.2.2 Target strength for mix design

The concrete mix should be designed for a somewhat higher target average compressive strength ($\bar{f}_{ck}$) so that not more than the specified proportion of test results fall below the characteristic strength. The margin over the characteristic strength depends upon the quality control (expressed as standard deviation) and the accepted proportion of results of strength tests below the characteristic strength ($\bar{f}_{ck}$), expressed by

$$\bar{f}_{ck} = f_{ck} + t \cdot s$$

where $\bar{f}_{ck}$ = target average compressive strength at 28 days

s = standard deviation, as per Table 19.4 for the desired degree of control.

t = a factor, given by Table 19.5.

Table 19.4: Suggested values of standard deviation

Grade of Concrete	*Standard deviation for different degree of control in N/mm²*		
	Very good	*Good*	*Fair*
M 10	2.0	2.3	3.3
M 15	2.5	3.5	4.5
M 20	3.6	4.6	5.6
M 25	4.3	5.3	6.3
M 30	5.0	6.0	7.0
M 35	5.3	6.3	7.3
M 40	5.6	6.6	7.6
M 45	6.0	7.0	8.0
M 50	6.4	7.4	8.4
M 55	6.7	7.7	8.7
M 60	6.8	7.8	8.8

Table 19.5: Values of t.

Accepted proportions of low results	*t*
1 in 5	0.84
1 in 10	1.28
1 in 15	1.50
1 in 20	1.65
1 in 40	1.86
1 in 100	2.33

19.5.2.3 *Selection of mix proportions*

(i) ***Water-cement ratio.*** Since different cements and aggregates of different maximum size, grading, surface texture, shape and other characteristics may produce concretes of different compressive strength for the same free water-cement ratio, the relationship between strength and free water-cement ratio should be established for the materials actually to be used. In case such data

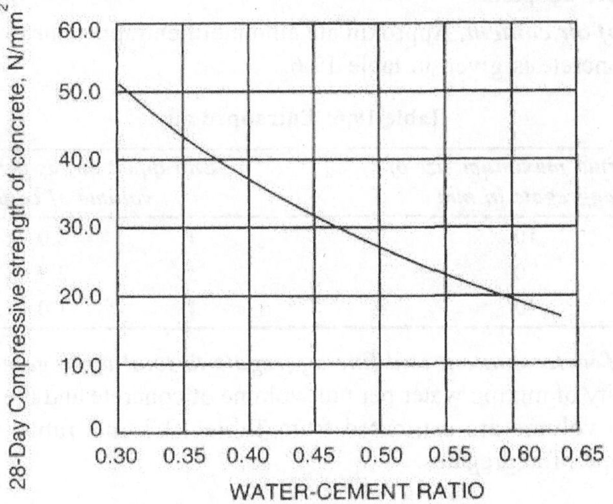

Fig. 19.1

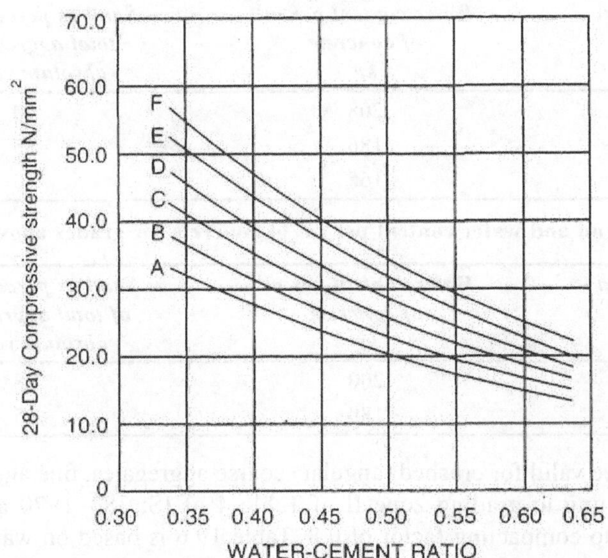

28 Day strength of Cement, Tested According to IS: 4031-1968

A = 31.9-36.8 N/mm² (325-375 kg/cm²)
B = 36.8-41.7 N/mm² (375-425 kg/cm²)
C = 41.7-46.6 N/mm² (425-475 kg/cm²)
D = 46.6-51.5 N/mm² (475.525- kg/cm²)
E = 51.5-56.4 N/mm² (525-575 kg/cm²)
F = 56.4-61.3 N/mm² (575-625 kg/cm²)

Fig. 19.2

is not available, the preliminary free water-cement ratio, by mass, corresponding to the target strength may be selected from Fig. 19.1. As an alternative, the water-cement ratio may be determined from the curves shown in Fig. 19.2 corresponding to the 28 days cement strength to be used for the purpose. The free water-cement ratio selected according to the above two methods is then checked against the limiting water-cement ratio for the requirements of durability and the lower of the two values adopted.

(ii) Estimation of air content. Approximate amount of entrapped air to be expected in normal (non air-entrained) concrete is given in table 19.6.

Table 19.6: Entrapped air

Nominal maximum size of aggregate in mm	*Entrapped air, as percentage of volume of concrete*
10	3.0
20	2.0
40	1.0

(iii) Selection of water content and fine aggregate to total aggregate ratio. For the desired workability, the quantity of mixing water per unit volume of concrete and the ratio of fine aggregate to total aggregate by volume are estimated from Table 19.7 and Table 19.8, for the nominal maximum size and type of aggregates.

Table 19.7: Sand and water content/m^3 of concrete for grades upto M 35

Nominal maximum size of aggregate mm	*Water content per m^3 of concrete kg*	*Sand as percentage of total aggregate by absolute volume*
10	208	40
20	186	35
40	165	30

Table 19.8: Sand and water content per m^3 of concrete for grades above M 35.

Nominal maximum size of aggregate mm	*Water content per m^3 of concrete kg*	*Sand as percentage of of total aggregate by absolute volume*
10	200	28
20	180	25

The above tables are valid for crushed (angular) coarse aggregates, fine aggregates consisting of natural sand conforming to grading zone II of Table 4 of IS: 383–1970 and workability of concrete corresponding to compacting factor of 0.8. Table 19.6 is based on water-cement ratio of 0.6 and table 19.7 for water-cement ratio of 0.35 (by mass). For other conditions of workability, water-cement ratio, grading of fine aggregate, and for rounded aggregates, certain adjustments in the quantity of mixing water and fine to total aggregate ratio should be made, as given in Table 19.9.

Table 19.9: Adjustment of values in water content and sand percentage for other conditions

Change in condition stipulated	*Adjustment required in*	
	Water content	*Percentsand in total aggregate*
For sand conforming to grading Zone I, Zone III or Zone IV of table 4 of IS : 383–1970	0	+ 1.5 per cent for zone I – 1.5 per cent for zone III – 3.0 per cent for zone IV
Increase or decrease in the value of campacting factor by 0.1	± 3 per cent	0
Each 0.05 increase or decrease in free water-cement ratio	0	± 1 per cent
For rounded aggregate	– 15 kg/m^3	– 7 per cent

(iv) Calculation of cement content. The cement content per unit volume of concrete may be calculated from the free water-cement ratio and the quantity of water per unit volume of concrete. The cement content so calculated should be checked against the minimum cement content for the requirements of durability and the greater of the two values adopted.

(v) Calculation of aggregate content. The total aggregate content per unit volume concrete can be calculated by the following relations:

$$V = \left[W + \frac{C}{S_c} + \frac{1}{p}\frac{f_a}{S_{fa}} \right] \times \frac{1}{1000} \quad \text{and} \tag{19.1}$$

$$V = \left[W + \frac{C}{S_c} + \frac{1}{1-p}\frac{C_a}{S_{ca}} \right] \times \frac{1}{1000} \tag{19.2}$$

where V = absolute volume of fresh concrete which equals gross volume (m^3) minus the volume of entrapped air

W = mass of water (kg) per m^3 concrete

C = mass of cement (kg) per m^3 of concrete

S_c = Specific gravity of cement

p = ratio of fine aggregate to total aggregate by absolute volume

f_a, C_a = total masses of fine aggregate and coarse aggregate (kg) per m^3 of concrete respectively; and

S_{fa}, S_{ca} = Specific gravities of saturated surface dry fine aggregate and coarse aggregate respectively.

(vi) Combination of different coarse aggregate fractions. Coarse aggregates of different sizes should be combined in suitable proportions so as to result in an overall grading conforming to IS : 383–1973, for the particular nominal maximum size of aggregate.

(vii) Calculation of batch masses. The masses of the various ingredients for concrete of a particular batch size is then calculated.

(viii) Trial mixes. The calculated mix proportions are then checked by means of trial batches. Workability of the trial mix is measured and carefully observed for freedom from segregation and bleeding and its finishing properties. Other trial mixes are prepared by adjusting water content or varying the free water-cement ratio by ± 10 per cent. The mix corresponding to the desired characteristics is then selected.

Problem 19.1.

Design a concrete mix of M 30 grade from the following data

(1) *Design conditions*

(a) Characteristic compressive strength required in the field at 28 days	*30 N/mm²*
(b) Maximum size of aggregate	*20 mm (angular)*
(c) Degree of workability	*0.70 compacting factor*
(d) Degree of quality control	*Good*
(e) Type of exposure	*Mild*

(2) *Test data for material*

(a) Cement used:	*Ordinary Portland cement*
(b) Specific gravity of cement	*3.15*

(c) *Specific gravity*

(i) Coarse aggregate	*2.60*
(ii) Fine aggregate	*2.60*

(d) *Water absorption*

(i) Coarse aggregate	*0.5 per cent*
(ii) Fine aggregate	*1.0 per cent*

IS Sieve Size mm	*Coarse aggregate fraction (Per cent passing)*		*Percentage of different fractions*			*Remarks*
	I	*II*	*I 50 per cent*	*II 50 per cent*	*Combined 100 per cent*	
20	100	100	50	50	100	Conforming
10	0	57	0	28.5	28.5	to table 2
4.75		7.5		3.75	3.75	of IS:
2.36		0		0	0	383–1970

(*ii*) Fine aggregate

IS Sieve size	*Per cent passing*	*Remarks*
4.75 mm	100	Conforming to grading
2.36 mm	100	zone III of Table 4 of
1.18 mm	93	IS : 383–1970
600 micron	60	
300 micron	12	
150 micron	2	

Solution.

(*i*) *Target mean strength of concrete*

For a tolerance factor of 1.65 and using Table 19.3, the target mean strength for the specified characteristic cube strength

$$= 30 + 6 \times 1.65 = 39.9 \text{ N/mm}^2$$

(ii) Selection of water-cement ratio

From Fig. 19.1, the free water-cement ratio required for the target mean strength of 39.9 N/mm^2 is 0.375. This is lower than the maximum value of 0.65 prescribed for mild exposure (See Appendix A of IS : 456—1978).

(iii) Selection of water and sand content

From table 19.6, for 20 mm nominal maximum size aggregate and sand conforming to grading zone II, water content per m^3 of concrete = 186 kg and sand content as percentage of total aggregate by absolute volume = 35 per cent.

For change in values in water-cement ratio, compacting factor and sand conforming to zone III, the following adjustment is required in accordance with table 19.9.

Change in condition	*Adjustment required in*	
	Water content per cent	*Percentage sand in total aggregate*
For decrease in water-cement ratio by (0.06—0.375) *i.e.* 0.225	0	–4.5
For decrease in compacting factor (0.8 – 0.7) *i.e.* 0.10	–3	0
For sand conforming to zone III of table 4 of IS : 383 – 1970	0	–1.5
Total	– 3%	– 6%

Therefore, required sand content as percentage of total aggregate by absolute volume

$$= 35 - 6 = 29 \text{ per cent}$$

$$\text{Required water content} = 186 - \frac{186 \times 3}{100} = 186 - 5.58$$

$$= 180.42 \ l/m^3$$

(iv) Determination of cement content

$$\text{Water-cement ratio} = 0.375$$

$$\text{Water} = 180.42 \text{ litres}$$

$$\text{Cement} = \frac{180.42}{0.375}$$

$$= 481.12 \text{ kg/m}^3$$

This cement content is adequate for mild exposure condition, according to Appendix A of IS: 456–1978.

(v) Determination of coarse and fine aggregate content

From table 19.6, for the specified maximum size of aggregate of 20 mm, the amount of entrapped air in the wet concrete is 2 per cent. Then applying equations

$$0.98 \text{ m}^3 = \left(180.42 + \frac{481.12}{3.15} + \frac{1}{0.29} \cdot \frac{f_a}{2.60}\right) \times \frac{1}{1000}$$

and

$$0.98 \text{ m}^3 = \left(180.42 + \frac{481.12}{3.15} + \frac{1}{0.71} \cdot \frac{C_a}{2.60}\right) \times \frac{1}{1000}$$

or

$$f_a = 488 \text{ kg/m}^3$$

$$C_a = 1194 \text{ kg/m}^2$$

The mix proportions then become

	Water	Cement	Fine aggregate	Coarse aggregate
	180.42	481.12	488 kg	1194 kg
or	0.375	1	1.01	2.48

(vi) Actual quantities required for the mix per bag of cement

The mix is 0.375 : 1 : 1.01 : 2.48 (by mass).

For 50 kg of cement, the quantity of materials are worked out as below:

(*a*) Cement = 50 kg

(*b*) Sand = 50.5 kg

(*c*) Coarse aggregate = 124 kg $\begin{bmatrix} \text{Fraction I} = 62 \text{ kg} \\ \text{Fraction II} = 62 \text{ kg} \end{bmatrix}$

(*d*) Water

(1) For water-cement ratio of 0.375, quantity of water = 18.75 litres

(2) Extra quantity of water to be added for absorption in case of coarse aggregate, at 0.5 per cent by mass = + 0.62 litres

(3) Quantity of water to be deducted for free moisture present in sand, at 1 per cent by mass = – 0.51 litres

(4) Actual quantity of water to be added = 18.75 + 0.62 – 0.51 = 18.86 litres

(*e*) Actual quantity of sand required after allowing for mass of free moisture = 50.5 + 0.51 = 51.01 kg

(*f*) Actual quantity of coarse aggregate required:

(1) Fraction I = 62 – 0.31 = 61.69 kg

(2) Fraction II = 62 – 0.31 = 61.69 kg

Therefore the actual quantities of different constituents required for the mix are:

Water : 18.86 litres

Cement : 50.00 kg

Sand : 51.01 kg

Coarse aggregate

Fraction I = 61.69 kg

Fraction II = 61.69 kg

19.5.3 Other Methods of Mix Design

19.5.3.1 Minimum voids method

This method aims at evolving a mix which has the minimum percentage of voids so as to obtain a dense concrete. This means that the fine aggregates should be in sufficient quantity to fill the voids in the coarse aggregates. The voids of each aggregate having been determined earlier, the cement is added in sufficient quantity to fill the voids in the fine aggregates. Sufficient water is added to make the mix workable.

19.5.3.2 Fineness modulus method

It has been found that the aggregates can be suitably defined in terms of grading by a term called "Fineness Modulus" which is a numerical index of fineness. It gives an idea of the mean size of particles in the entire body of aggregates. It is obtained by adding the percentages by weight of materials retained on each of the 10 Indian Standard sieves ranging from 76.2 mm to I.S. No. 15 sieve and dividing the sum by 100. As a result of considerable experience, certain values of fineness modulus for fine and coarse aggregates and mixed aggregates have been found to give good workability with a minimum quantity of cement. This value varies with the size of the particle in the body of aggregates. If the fineness modulus of a given sample does not correspond to these values, the mixture will need more cement for a given water/cement ratio and workability. Any increase in the values of fineness modulus of mixed aggregates from this value will result in a harsh mix while a decrease in the value will give an uneconomical mix for fixed water/cement ratios.

After having mixed the fine and coarse aggregates so as to obtain the desired fineness modulus of combined aggregates, the proportioning of the concrete can be done by actual field trials.

The actual quantity of cement required per unit volume of coarse aggregates and for a desired slump must be checked with the same trial mixes.

19.6 MIXING OF CONCRETE

The simplest method used for concrete mixing is by manual means. It is done only on small works where the quantity of concrete is insufficient to warrant the installation of machinery, the working area is less or when the distance from the mixing position to the placing area is small. This method has also to be adopted in places where machinery cannot be used due to excessive noise produced by it, for example, near hospitals, etc. When hand mixing is adopted, 10 per cent extra cement is added. The usual procedure for mixing concrete manually is to carry out this operation on a level, clean and water tight platform. The aggregates and cement are stored adjacent to this platform in a proper manner. Coarse aggregate is laid evenly in a layer over the platform and the fine aggregate is spread over it. The cement layer is then spread over uniformly at the top. In mixing the materials, care should be taken that each shovel contains some quantity of each of these materials. A suitable method of mixing is that the whole mass is mixed thoroughly 2 to 3 times while in dry state and again 2 to 3 times when water is added. Care should be taken to ensure that mixing is continued until the concrete is uniform in colour and consistency.

Mechanical mixing is adopted these days to a great extent. The following types of mixers are usually adopted:

1. ***Non-tilting types***: These are made in sizes from about 1/5 cu m to 5 cu m. In most cases the drum rotates in one direction and the concrete is discharged by a hopper. The direction of rotation of the drum may be reversed to discharge the concrete.

2. ***Tilting drum mixers***: These are generally made in smaller sizes but a few larger machines up to 3 cu m capacity have been made which are capable of producing about 60 cu m of concrete per hour.

3. ***Pan mixers***: These mixers are efficient but have small capacity.

4. ***Continuous mixers***: These mixers measure the ingredients by volume and once they have been set, their measuring is done more or less automatically. The power consumption of these mixers is small and these are capable of producing 10 to 20 cu m of concrete per hour.

5. ***Truck mixers***: These are used when the mixed concrete is to be carried to a long distance. They are used in conjunction with a central batching plant. Water is carried in a separate tank so that it can be mixed in the truck when it is near the depositing area.

The ingredients in dry state are filled into hoppers or skips attached to the mixer. The coarse aggregates are placed first at the bottom followed by fine aggregates and cement. These materials get admitted into the drums in the reverse order. No material should be put in the skip unless they have been mixed and used immediately. The materials are loaded into the skip after being correctly measured for large works and for a good quality of concrete. It is essential to have weighing machines with measuring devices for weighing the concrete ingredients correctly. Many types of weigh batching plants are available. These are swinging hopper types, horizontal hopper types and simple weigh batchers. They can be completely mobile or partially mobile and the hopper may be elevated or may have bottom discharging arrangements. The capacity of the plants is sufficient to give about 7 cu m of concrete per hour. For large works, fixed batching plants are used. They can be fed with bulk cement instead of cement available in sacks. They are economical only in large-scale construction as they need a lot of time for erection.

After all the materials have been put in the drum, mixing should continue until the mixture is uniform. The strength of concrete is effected by the duration of the mixing process. About $1\frac{1}{2}$ minute is necessary to give good strength and mixing up to 2 minutes improves the quality of concrete. There is no advantage in mixing beyond two minutes. For hydrophobic cement, the mixing time may be taken as $2\frac{1}{2}$ to 3 minutes. Some machines have a device for discharging the mixed materials automatically after they have been mixed for a sufficient time. Sometimes it is specified that the materials should be mixed in a dry state before adding water. This may not be practicable with large batching and mixing plants. When the mixing plant is to be closed for more than one hour it should be cleaned.

19.7 TRANSPORTING CONCRETE

After mixing the concrete, it has to be delivered to the final position which it has to take in a building unit. The need of transporting concrete at the earliest possible time is essential because concrete should be placed and compacted before the initial set of cement takes place. While transporting, it should be ensured that the containers are water-tight and that the loss of water and cement is avoided to a great extent. Segregation of the particles must be avoided at all costs as this would result in bad concrete. If it is unavoidable during transit, concrete should be remixed before being placed.

Good concrete may get spoiled by mishandling during transportation from the mixer to the point of incorporation. This is more true for areas where greater lead of concrete is essential. The methods of transportation usually adopted are:

1. Use of iron pans which are carried by men

These pans have got a small capacity and are usually of such a size that one man can lift and walk easily over the required distance. For raising concrete over some height so as to place it on the top floors of a building, ladders are erected and men stand at various levels. The same pan is being transferred from one person to the other quickly. This method of transporting concrete is only adopted for small and medium size buildings in India as it works out to be the cheapest method with such type of construction.

2. Wheel barrows

Ordinary wheel barrows hold about 2 cu m of concrete. The wheels may be fitted with pneumatic tyres to prevent the segregation of concrete on account of series of jerks which the wheel barrow is likely to experience while travelling over uneven surface.

3. Concrete carts

These vary in capacity from 7 to 14 cu m and they are designed for easy pushing, tipping and discharging. Suitable wooden platforms are laid for their easy travel.

4. Hoists used with barrows

For tall structures, hoists may be used either to lift the barrow to the required height or to discharge the concrete into the hopper from which it is carried further. Some sites do not permit the use of external hoists, *e.g.*, in the case of steel buildings. In confined areas the hoist must be erected within the building and built up in sections. The topmost will rise a few metre above the working level. The guide rails are held up against each floor level by studs. A vertical hoist can frequently be arranged in multi-storeyed buildings within the space allowed for stairs or lift wells. Otherwise temporary openings have to be left in the floor panels to allow for the hoists. Cranes may be used with hoist mainly to lift skips of concrete from the site of the mixer to the building where the concrete is to be placed.

5. Lorries

Lorries are used whenever the distance is large. Such a case occurs when concrete is to be deposited at a place where it cannot be mixed due to the non-availability of space. General mixing plant placed at some distance from the site of work is erected. Special lorries suitably designed are used for transporting concrete. The concrete is covered to protect it from the sun, wind or rain during transport. The time to transport should not exceed more than 20 minutes when mixed concrete is to be carried.

6. Chutes

When concrete is to be placed below general ground level, the mixer can be placed on an upper level and concrete discharged to the lower level through a small chute of corrugated iron or timber, leading to barrows, carts or other transport facility or directly to the final position where it has to be deposited.

7. Pumping of concrete

Pumping of concrete is economical and successful where the size of aggregate does not exceed 50 mm and concrete is of medium workability. The control of aggregate gradings and the batching of mix have to be done effectively. The advantages of this type of work are that no traffic zone for transporting concrete within the working radius of about 600 metres from the pump is created. About 8 to 25 cu m of concrete per hour can be delivered. The system is very flexible as the pipe lines can be easily changed. Handling of concrete between the mixer and the point of incorporation is avoided which makes a better quality of concrete. The plant consists of a pipe line and a pump. The type of pump will vary slightly with the consistency of concrete and length of the pipe line. For a continuous discharge, it is essential for the mixing plant to be of an adequate capacity. These pumps are operated by electric power or other means. The pipes are of about 150 mm internal diameter for large capacity pumps and about 100 mm diameter for smaller machines. Concrete can be delivered up to a height of about 40 m vertically or 2000 m horizontally. The power required to operate a pump at full range capacity is about 35 h.p. This type of transportation is rarely resorted to for building work.

19.8 PLACING OF CONCRETE

The quality of concrete depends amongst other factors, on the care with which it is placed in the final position. Concrete must be deposited quickly in its final position to avoid rehandling and its lateral movement after deposition should be reduced as much as possible. It should be placed in even layers each of which is compacted before the next layer is placed. Each layer must be placed before the previous one has set. Layers must be of such thickness that a good quality concrete is obtained depending on the type of work. After a break in the work, fresh concrete should be joined to the one which has already set. All formwork must be cleaned before any concrete is poured. Surface of the forms may be treated to prevent adhesion of concrete. The reinforcement must be fixed before concreting is started and it should be ensured that it does not displace during concreting. Except for slabs and large works, the concrete may be placed into the shutters by shovels. While depositing concrete with the aid of iron pans, it must be ensured that concrete is not dropped from greater heights as otherwise segregation would result.

19.9 COMPACTING CONCRETE

The purpose of compaction is to remove air bubbles from concrete mass to a great extent. Under-compaction results in a lot of air voids in concrete while over-compaction may lead to segregation. Compaction is carried out by hand or mechanical means. Manual compaction is carried out by rodding, tamping or ramming. Rodding is used for the vertical section, tamping for slabs and ramming for heavy reinforced concrete. Rodding consists of a rod of wood or metal being pushed into the concrete and moving it up and down until the concrete is thoroughly worked into empty places and completely filling the corners and other awkward positions. Slabs should be tamped with tamping rod which serves the double purpose of compacting and finishing concrete to the required level. Heavy masses of reinforced concrete work are rammed with wooden or steel rammer, the face of which is usually not less than 150 mm square. Ramming is continued till the slurry of cement and water starts to appear on the upper face of the concrete. Honey-combing must be avoided in good concrete work. This is due to inefficient tamping or due to leakage of cement grout from the shuttering. The ramming should not be heavy as otherwise there is a tendency for concrete to rise against the side of the shutters and also to exert excessive pressure on the shuttering face.

Compaction, by the use of mechanical vibration is becoming more common and useful these days. The advantage with mechanical vibration is that lesser water/cement ratio can be used which can give good concrete and also economise cement. The principle of compaction by vibration is that when a concrete mass is vibrated, the surface structure between the particles gets considerably reduced and thereby gives a sort of fluid nature to the concrete. As soon as the vibration is stopped such fluid properties disappear. The heavier particles sink rapidly to the bottom. Hence concrete which is very workable should not be compacted by vibration. The higher the frequency of vibration the lesser the time needed for compaction. For higher frequency, the power consumption increases rapidly and also the wear and tear of the equipment increases disproportionately. The minimum frequency for compaction appears to be 3600 r.p.m. Drier mixes or thick slabs can be compacted with larger amplitude or with larger frequency. A larger amplitude leads to more difficulties in finishing the surface. The usual types of vibrators used are internal, external, table and surface types. The internal vibrators consist of a long rod like vibrating head which has a motor and vibrations are given to this rod. Basically they are of two types, the rigid type in which the motor is rigidly attached to the head of casing or vibrating unit and the flexible type in which the vibrating head is mounted on the flexible shaft driven by an external motor or the vibrating head is mounted on a flexible hose feeding compressed air to the motor located in the vibratory heads. Vibrating heads are available in various sizes, small sizes being used in mass concrete work and heavier sizes for reinforced work. External vibrators are designed for being attached to the formwork which contains concrete. They are operated by an electric or an air motor. Table vibrators employed for vibrating precast units are available in lengths up to 4 m. These are not used in building construction work otherwise.

Where the sections of concrete members are thin and they are reinforced, vibration is to be done carefully. A higher water/cement ratio would be necessary in such work. Vibration saves time and labour in placing concrete. Good surface finishes are also produced and side shutters can be easily removed. The slump should never exceed 50 mm but lesser slump can be used when compaction can be thoroughly done. In the construction of walls, it must be ensured that the formwork is strong so that it does not get loosened on account of vibrations.

Vibrations should be stopped when the water comes to the top of concrete. Longer periods of vibration are required to compact concrete made with angular aggregates than that of the rounded aggregates. Vibrators have to be firmly attached with the shuttering so that energy may not be lost in vibrating the formwork only. For general work the internal vibrator is well known and is usually adopted. The advantages of this type are that the vibrator actually gets into the concrete and vibrations are therefore directly transmitted. It is easily moved from point to point. The operator can also observe the behaviour of concrete under vibration. These vibrators should be used vertically and should penetrate to the full depth. They should be inserted and withdrawn slowly and removed from concrete as soon as cement paste starts to appear on the surface.

19.10 CURING CONCRETE

The setting of cement in concrete takes place on account of hydration. Hydration continues for a great length of time but its rate decreases. Hence concrete must be kept moist till it gains sufficient strength. The mixing water is usually sufficient for initial hydration but most of the quantity of water is lost by evaporation at the surface. This is to be prevented by curing.

The period of curing depends on the type of work, the type of cement, and the climatic conditions. Normally for ordinary Portland cement, the concrete must be cured at least for a period of 7 to 10 days. This period may be reduced in the case of rapid hardening cement.

Curing may be done either by moist curing or membrane curing. The generally adopted method is to apply periodically water from a hose to exposed parts of concrete or by covering the face of walls with sacks, hessian or coconut matting, etc. Horizontal surfaces can be kept wet by a covering of wet sand, earth, saw-dust or by covering with water to a depth of 30 mm or so. This water in small ponds is kept within small clay bunds. These days large flat surfaces are sprayed with patented compounds or bituminous emulsions, and sometime the surface is covered with water proof papers. These papers consist of two sheets of white, tough, durable kraft paper joined together by a bituminous material.

A layer of calcium chloride spread on the concrete absorbs the moisture from the atmosphere and is helpful in curing concrete but the process is expensive.

The purpose of these coatings is to prevent water evaporation. They are applied as soon as concrete is finished. Bituminous layers are only advisable where the exterior finish is not of much importance.

A process of steam curing is widely used these days to cure precast pipes and other precast units. It accelerates the rate of gain of strength, reduces the time required for curing and enables early removal of formwork, thereby increases the rate of production of the units with a resultant saving in cost.

REINFORCEMENT

The reinforcement should be any of the following conforming to accepted standards:

(*a*) Mild steel and medium tensile steel bars

(*b*) Hot-rolled deformed bars

(*c*) Cold-twisted bars

(*d*) Hard-drawn steel wire fabric, and

(*e*) Rolled steel made from structural steel.

The modulus of elasticity of steel should be taken as 200 kN/mm^2.

All reinforcement should be free from loose mill scales, loose rust and coats of paints, oil, mud or other coatings which may destroy or reduce bond.

19.11 REINFORCED CONCRETE COLUMNS

Reinforced concrete columns can be of various shapes. These may be of square, rectangular, hexagonal, octagonal or circular sections. The main constituents of R.C.C. columns are the longitudinal bars which take the main load. These are tied by lateral steel bars which act as ties. These ties prevent the buckling of these longitudinal bars and keep them in position with respect to each other. Sometimes these ties are made in the form of spiral and this is called spirally reinforced columns. This spiral has got the additional advantage of providing extra strength to the column. Other types of columns usually are composite columns which consist of simple sections of steel pipes or joist sections which are encased in reinforced concrete work. Columns at the corners of

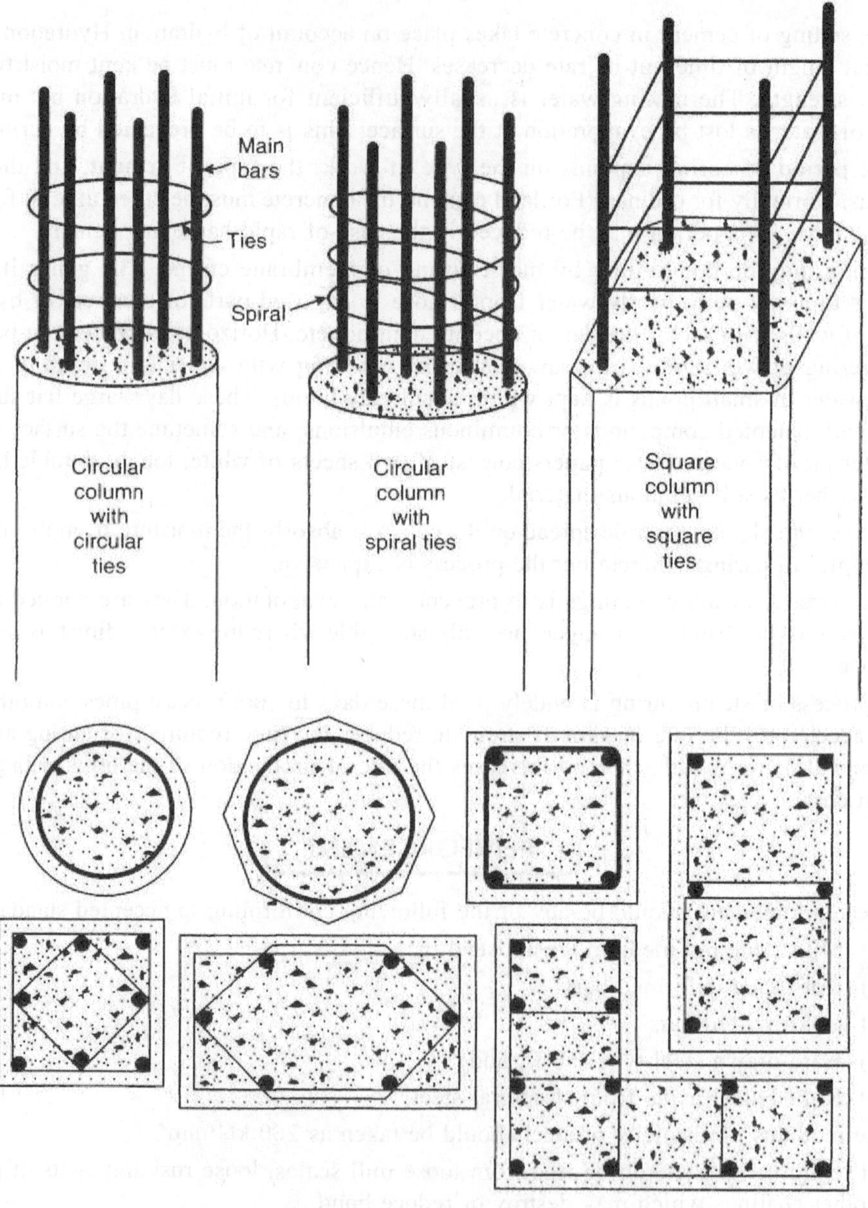

Fig. 19.3 to 19.12: Typical details of R.C.C. columns.

buildings may be 'L' shaped. For bigger columns the arrangement of ties is to be suitably varied to account for the size of the column. The load of the column is distributed between the concrete and longitudinal bar. The total area of the longitudinal bar is not less than 0.8% and not greater than 8%. The minimum diameter of such reinforcement is not less than 12 mm and not greater than 50 mm. A 40 mm cover of concrete is at least provided on all sides of the main steel. The minimum diameter of lateral ties is 5 mm and the maximum size usually adopted is 12 mm as bending would become difficult for larger sizes. The pitch of the lateral steel is not greater than 48 times the diameter of transverse bars or the least dimension of the column or 16 times the diameter of the longitudinal steel. The volume of lateral reinforcement is not to be less than 4% of the total volume of column. The longitudinal steel is suitably lapped whenever necessary. The spiral reinforcement of the column is held in place by at least 6 vertical bars.

After the column shuttering has been fixed, the reinforcement is checked and separated from the shuttering with the aid of small fillets so as to provide suitable cover to the steel. The first lift of concrete is then deposited and the shuttering is raised by another stage. The operation of placing concrete in each lift is continued but the rate of vertical progress is not to exceed about one metre in half an hour. The depth of concrete placed in a single layer is frequently specified and is not to exceed one metre. The risk of aggregate dropping towards the bottom of the shuttering or the presence of cavities being left, is minimised, whenever the bottom layer is allowed to set before the next layer is laid. Settlement is accelerated by ramming and tamping or by the use of vibrators. In the case of deep single deposits of concrete, hammering outside would assist to consolidate concrete if it is combined with the internal tamping. This ramming should be done rarely in the corners. It is preferable to have men who do this ramming process at the top of the column. The concrete is deposited centrally in the column. In the case of tall columns it is not permissible to drop the concrete vertically from the top of the box. It should be fed in at a convenient height through a chute extending to the centre of the column. The concrete should not be discharged directly from long chutes into the column box or else honey-combing may result. The concrete in the column is usually stopped a few centimetres below the level of the beams running into the columns. The portion of the column thus left is concreted with the beams. The consistency of concrete in column is sometimes varied. Drier concrete is used in the upper portions, as it is advantageous while placing continues.

Concreting of columns of different storeys is to be carried out very accurately. The central lines of column must be accurately one above the other for columns which are placed in different floors. This must be checked at every stage of construction. Since the loads in the top storey are less, the section of column also gets reduced correspondingly. The steel from the column in the lower storey is bent so as to get accommodated in the column of the top floor.

The steel or cast iron core of a composite column must be thoroughly encased in a reinforced concrete cover so that the load is carried by the concrete, the longitudinal bar and the core. The cross sectional area of the core should not be greater than 20% of the gross area of the column.

19.12 R.C.C. BEAMS AND GIRDERS

Reinforced concrete beams are being commonly adopted in building construction these days. The strength of a R.C.C. beam depends on the composite action of concrete and steel, the former taking all compressive forces whereas the latter takes all the tensile stresses.

This beam in its simplest form is rectangular. Steel is provided at the bottom in the form of bars to take tensile forces. Concrete in the lower portion does not take any force but merely keeps the bars in position. Rectangular beams are adopted wherever the slabs resting on the beams are of a different material or are not cast monolithically with the former. Since concrete in the lower portion is not useful it can be omitted and only that much quantity can be left which is sufficient to keep the steel in position. The slab is cast monolithically with the beams and if it extends on both sides of the beams, a T-beam is developed. This beam is more economical. Whenever the slab extends on the side of the beam only and is monolithic, it is termed as L-beam and is used at the end walls of rooms. Due to the limitations of the size of the beam it may become necessary to place steel bars at the top also to take additional compressive forces. This type of beam is termed as doubly reinforced beam and is not economical due to excessive cost of steel.

The R.C.C. beams described above may therefore consist of steel rods running along the length of the beam and placed at bottom or both at top and bottom. Additional steel has to be provided in the vertical direction or in the form of bent bars to take care of the shearing stresses created in a beam. The bars in the vertical direction (stirrups) take a U-shape and are connected to the longitudinal main bars at one side and to additional longitudinal holding bar at the other end. For larger beams, these U-shaped bars are suitably altered in shape so as to keep them in contact with as many longitudinal main bars as possible and thereby keeping them in correct position and also giving the desired area per unit length of the beam to resist the shearing forces. The spacing of vertical stirrups or inclined bars is suitably varied according to the shearing forces set up at a given section. Usually the spacing is least at the supports of the beam as the shearing forces are greatest there. The bars which carry the forces must have an adequate grip with concrete. For this reason, the ends are hooked or extra lengths are provided to account for the gripping action. Alternately, deformed bars shown in Figs. 19.13 to 19.16 may be used as they have a greater holding power with concrete.

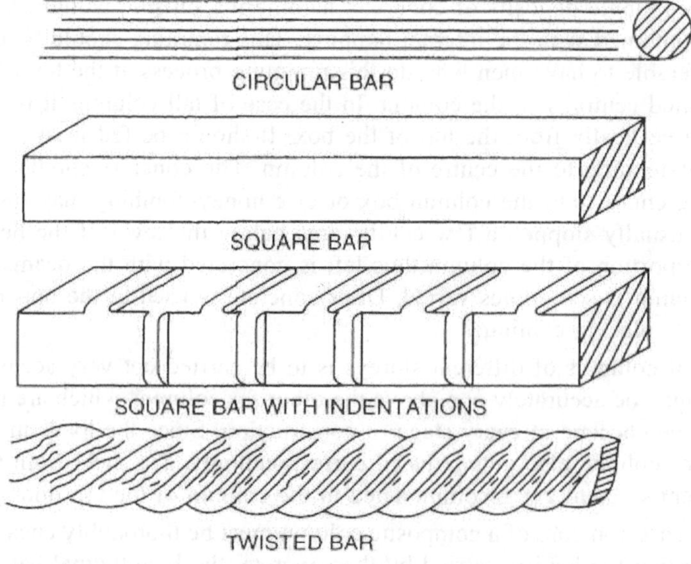

Figs. 19.13 to 19.16: Different types of bars used in R.C.C. work.

For economical reasons the beams may be continuous over several supports. They may also be made monolithic with columns over which they rest and with the connecting girders. (See Figs. 19.17 to 19.21). The main reinforcement of such beams has to be provided carefully as tensile stresses occur on the top face at the supports of these beams and in the bottom face at the mid-spans. The steel has to be provided on these faces at these points. The bars have to be suitably extended into the adjacent pannel to give proper laps. They are also anchored into the columns. (See Figs. 19.22 to 19.26).

19.13 REINFORCED CONCRETE SLABS

To span small openings and distances between beams, R.C.C. slabs are used. These slabs have steel bars placed at the bottom surface to take care of the tensile forces developed in them. For small spans, no steel is provided at the top but for greater spans and whenever this slab rests on masonry walls a part of the bottom steel which is greatest at the centre of the span, is taken to the top at the supports. Additional bars are provided across the length of the span to take care of the temperature stresses.

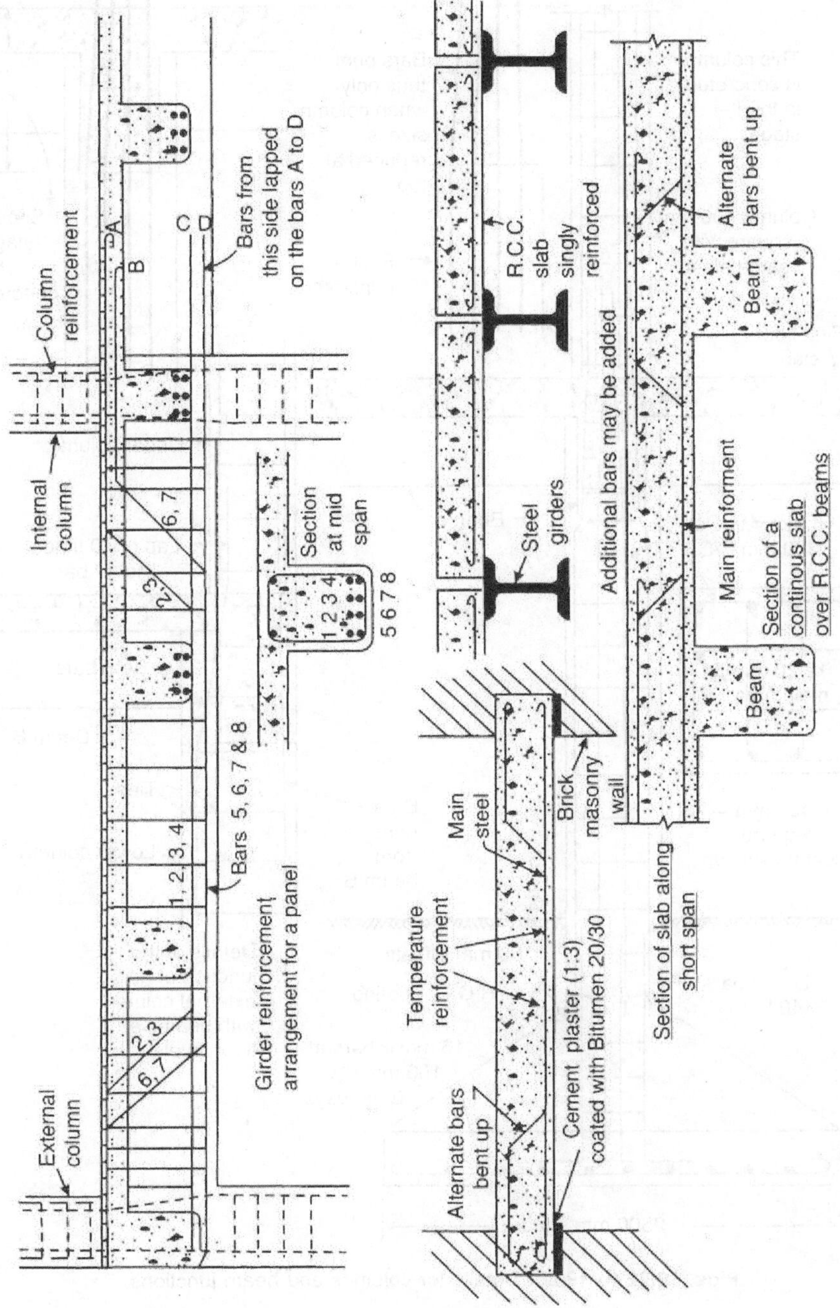

Figs. 19.17 to 19.21: Details of R.C.C. Beams and Slabs.

Slabs may be built monolithically with the beams and may also be continuous over a number of spans. In the latter case, tensile stresses occur over the support and as such additional steel is to be provided in the top surface of the slab over this area. The bottom steel is taken continuous over all the spans and its quantity over the supports may be reduced since very small tensile stresses are created in the bottom of the slab at this point.

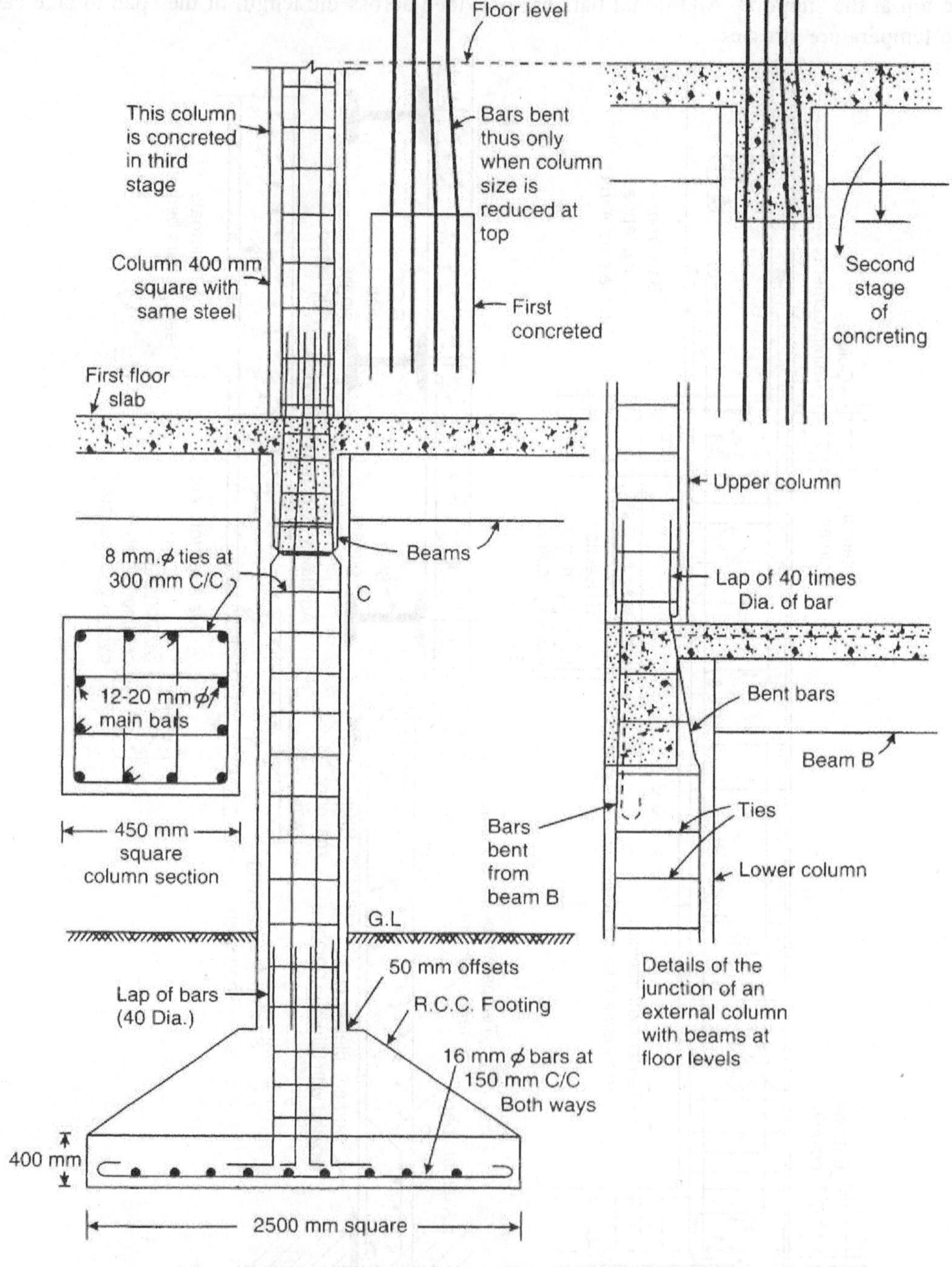

Figs. 19.22 to 19.26: Details for columns and beam junctions.

For small spans the slabs are simply supported over small rolled steel joints which carry the loads in turn. For bigger spans, monolithic construction is desirable as this is going to economise the quantity of R.C.C. work. Where the R.C.C. beams intersect the slab it should be ensured that no congestion of reinforcement is created as otherwise concreting would be difficult.

The usual type of reinforcement adopted for slab work consists of round, square or cold twisted or deformed bars but for small spans wire mesh or XPM *Jali* may be used. Where the length of the room is not greater than or equal to twice the width of the room, it is economical to put the steel bars in both the directions. This type of construction is called a two-way slab which is shown in Figs. 19.27 to 19.31. If such a slab is built within masonry walls additional bars have

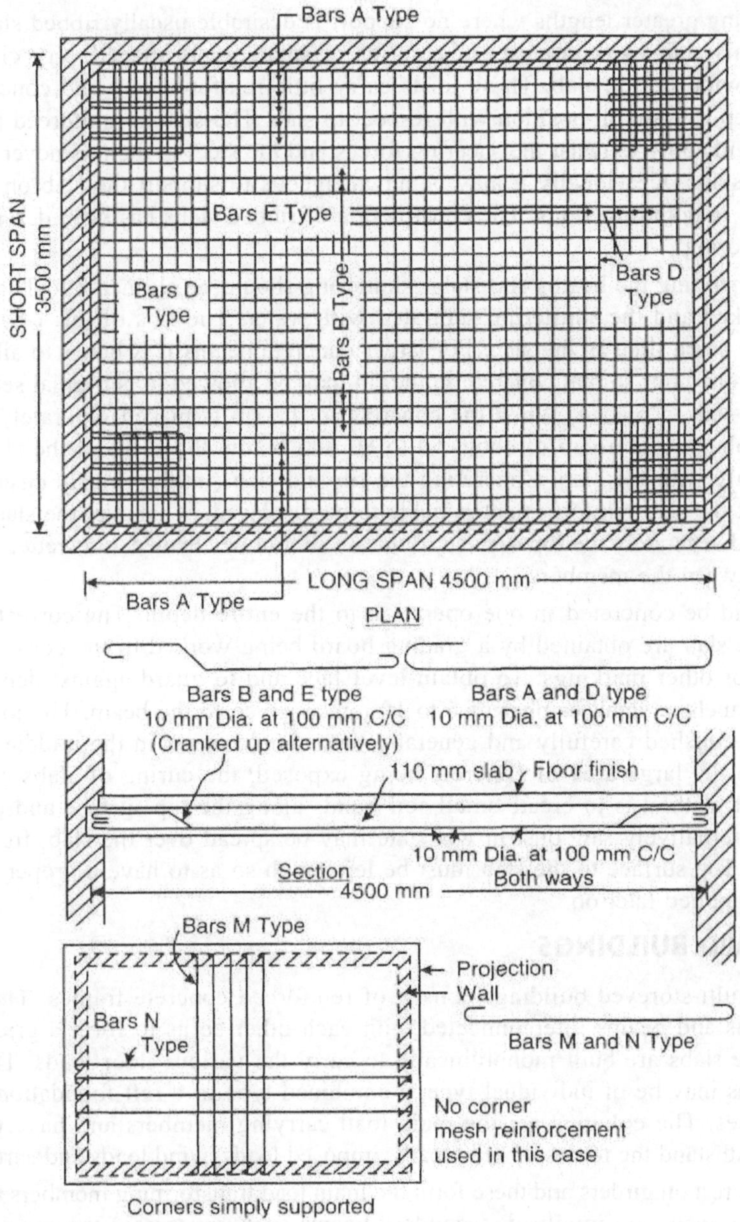

Figs. 19.27 to 19.31: Two-way slab details.

to be provided over the corners to account for the excessive stress created therein due to the restraining effect of the masonry at the top. The reinforcement of a two-way slab runs from corner to corner and additional steel should be provided to take care of temperature stresses. Whenever beams are not desirable for supporting the slabs, the slab in turn may be carried over columns which have top ends flared out. The slab in such cases gives a smooth surface, is economical for heavier loads, gives better lighting and is clean. This type of slab is called a flat slab. Such construction can also be done by thickening the area of the slab over the columns which are called drops. Flat slab construction is usually adopted in factories, big parking areas, restaurants, etc. In this type of work the distance between the columns has to be suitably arranged so as to get an economical construction.

For spanning greater lengths where no support is desirable usually ribbed slab construction is followed. This slab work consists of a series of small joists spaced very closely and cast monolithically with the top slab. These joists carry little reinforcement and concrete in them is sufficient to keep the bars in position with respect to slab. The slab is reinforced as usual. These ribs are made wide only to resist the shearing forces and the steel in them is never more than that of an ordinary slab. Occasionally it may be advantageous to support the slab on four sides and provide the ribs in both directions. The construction of ribbed slabs has been described under the chapter on "Flooring".

Before concreting the beams and slabs, shuttering should be cleaned out. The reinforcement should be checked and the shuttering saturated with water. The ribs of the T-beams should be concreted at the same time as the slab. For larger concrete beams it is better to allow an interval of not less than one hour so that concrete in the rib may be allowed to get initial settlement before concrete in the slab in placed. When the concrete in the rib is placed separately, the rib is not concreted completely but a gap of about 50 to 80 mm below the soffit of the slab is left to be concreted directly with the slab. Concreting the rib and slab simultaneously ensures monolithic construction but the slab reinforcement is liable to be displaced. Whenever the slab and the beam rib are concreted separately the top surface of the rib should be cleaned and wetted so as to ensure a good joint between the members.

Slab should be concreted in one operation to the entire depth. The correct top level and thickness of the slab are obtained by a grading board being worked to the correct height of the adjacent slabs or other markings. To obtain level face and to guard against depressions in the middle of the panels, a small camber of 5 to 10 mm is given to the beam. The joint in the day's work should be finished carefully and generally they should occur in the middle of the span of a beam. Due to the large area of concrete being exposed, the curing of slabs should be done effectively. Best method is to create small soil bunds along the top surface and pond the water within them. Alternatively saw dust in wet state may be spread over the slab, frequent watering being essential. Top surface to the slab must be left rough so as to have a proper bond with any finish which is applied later on.

19.14 FRAMED BUILDINGS

Modern multi-storeyed buildings consist of reinforced concrete frames. These frames are units of columns and beams interconnected with each other so as to form a grid of the beams and girders. The slabs are built monolithically to carry the various floor loads. The foundations of these columns may be of individual type, a combined type or a *raft foundation depending on* the circumstances. The columns are the main load carrying members and have to be designed adequately to withstand the forces of dead loads, imposed loads, wind loads and earthquake forces.

The beams rest on girders and these form the main load transforming members to the columns. Whenever larger spans are involved, secondary beams spanning across the main beams can be introduced to take care of the load distribution, the slab being spanned across these secondary beams.

It is difficult to concrete such a building in one operation and hence construction joints are to be provided at intervals. These joints are so located and constructed so as not to reduce the strength of the building as a whole. The joints in slabs, beams and girders should be vertical and at proper places where the shearing forces are least. The columns are concreted in suitable heights so as to give a proper lap with the sides of beams and columns which are going to come in the upper storeys. The section through the multi-storeyed buildings showing the slabs, girders and beams is

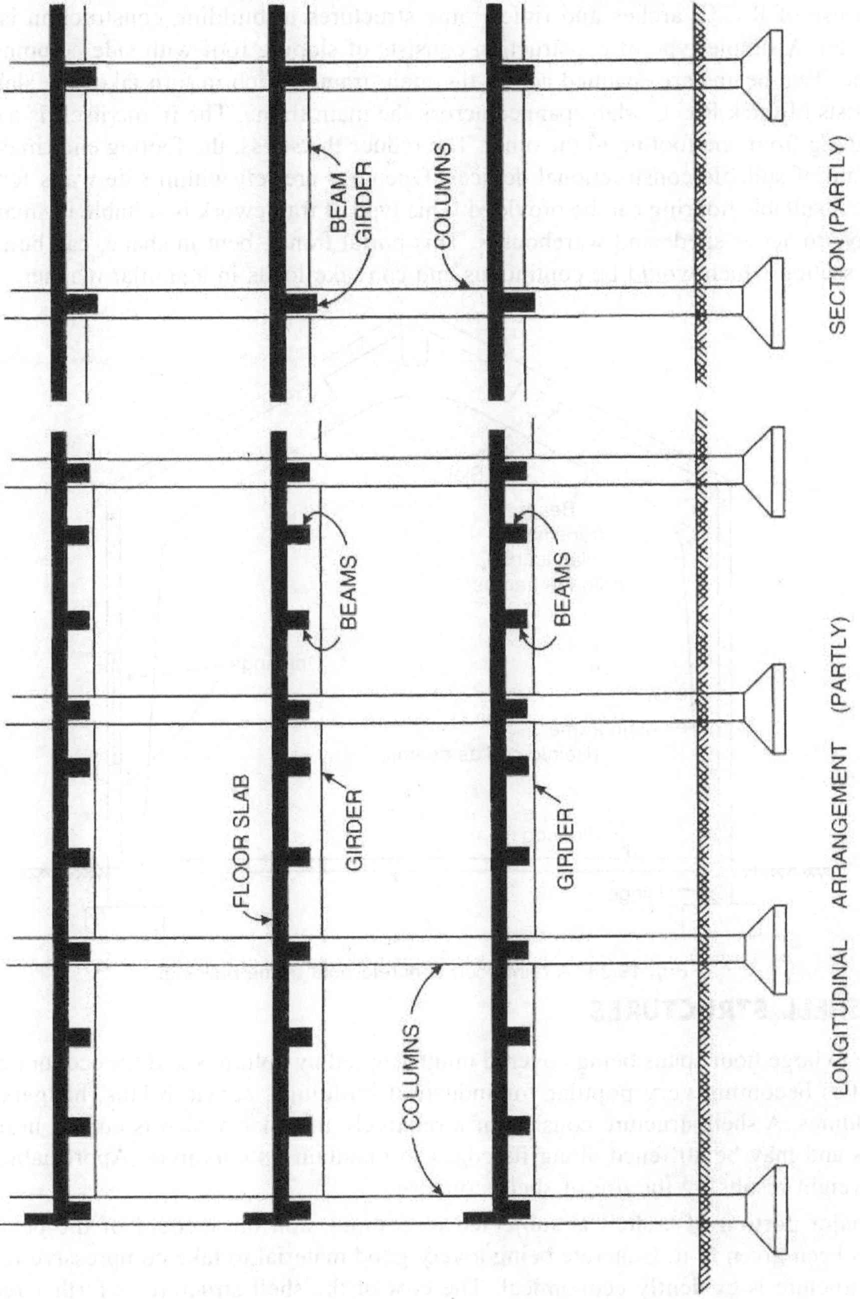

Fig. 19.32 and 19.33: R.C.C. framed building.

shown in Figs. 19.32 and 19.33. In multi-storeyed buildings, to have rooms of smaller sizes within the floor area, it is necessary to create partitions which are of light weight materials and are of small thicknesses. Provisions are to be kept for lifts so as to transport people from one floor to the other.

The concreting of these buildings is done in a similar manner as described for individual units.

19.15 REINFORCED CONCRETE ARCHES AND RIGID FRAMES

The use of R.C.C. arches and rigid frame structures in building construction is becoming very popular. A simple type of construction consists of sloping roof with side columns built in a bent shape. Two beams are spanned across the main frame which in turn takes the slab. The side wall consists of thick R.C.C. slab spanned across the main frame. The frame itself is a continuous beam running from one footing to the other. The reduce the stress, the footing ends may be hinged with the aid of suitable constructional devices. Openings are left within side walls for doors and windows. A suitable flooring can be provided. This type of framework is suitable in small buildings which have to act as sheds and warehouses. This portal frame, bent in shape, can be replaced by any arch section which would be continuous and can take loads in a similar manner.

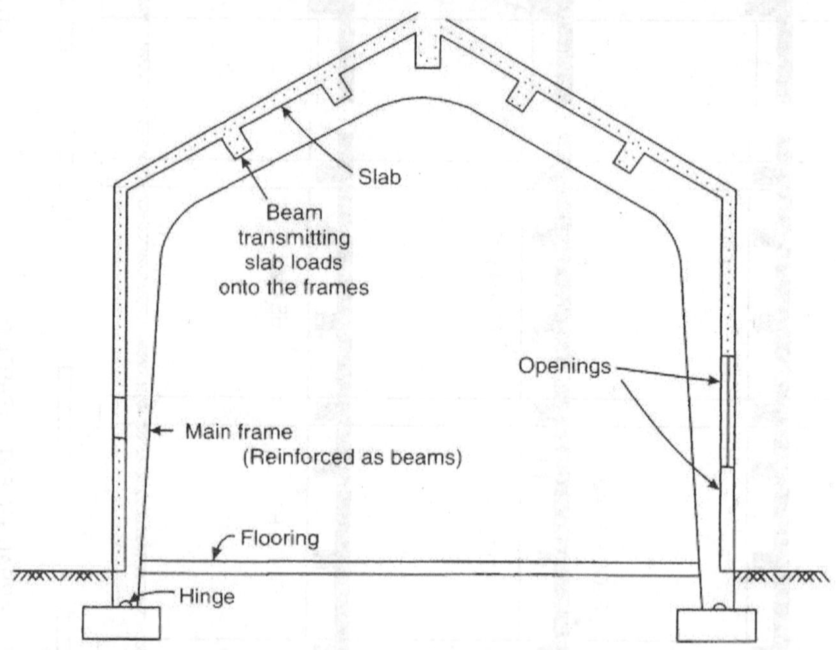

Fig. 19.34: A reinforced concrete rigid frame building.

19.16 SHELL STRUCTURES

Due to large floor spans being covered uninterrupted by columns and for economical reasons, shell roof is becoming very popular for industrial buildings, research labs, hangers and other large buildings. A shell structure consists of a relatively thin slab which is curved in one or both directions and may be stiffened along its edges to maintain its curvature. Appreciable reduction of dead weight results by the use of shell structure.

A major portion of a shell is subjected to compression on account of the peculiar shape which has been given to it. Concrete being a very good material to take compressive force, its use in shell structure is evidently economical. The cost of the shell structure is further reduced as a number of them have to be constructed and the centering can be repeated every now and then. The maintenance cost of the shell structures is least. The economy of steel is an advantage. The floor space can be advantageously used because larger spans can be created with small sized columns.

A good ventilation is available in the greater space volume which is created in a shell structure. A better appearance and good reflecting surfaces are possible. The construction of a shell can be carried out rapidly if proper care is taken to execute the job.

Many varieties of shell structure have been built and various shapes are always being developed. Shells of double curvature have been built in many forms such as domes of circular, elliptical or polygonal shapes, concoidal forms providing north lights, hyperbolic paraboloids and others. The shapes of the majority of shell roofs are of single curvature consisting of either long or short barrels. In the latter case, the shell has a long radius which is variable because the shell is supported on long span with arch ribs which are closely placed. The span of the shell between arch ribs is small compared with the span of the supporting ribs. In long barrels the cylinder itself becomes the supporting member between columns. The length of the cylinder is usually greater than its radius of curvature. The cylinder may be used single or in a series.

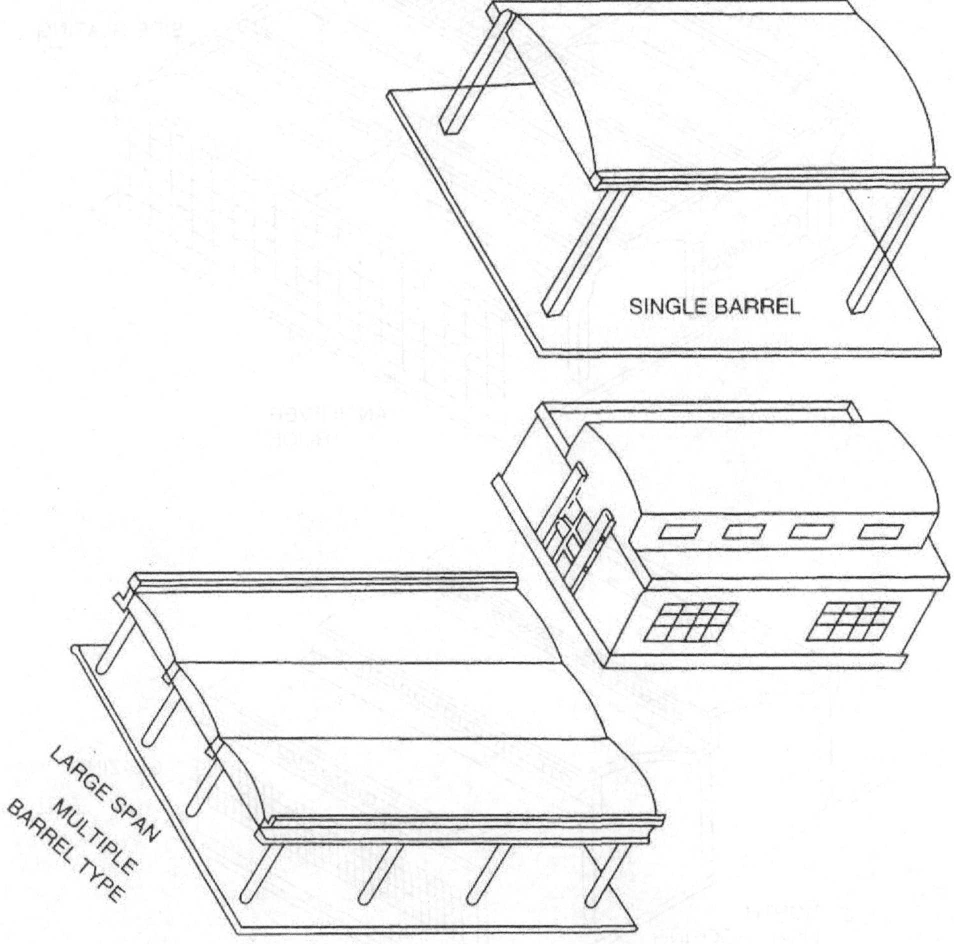

Figs. 19.35 to 19.37: Types of shell structures

Shells are used for covering the floor spaces in factories and as such they have north light arrangements. These shells have got edge beams at the ends and provision is kept for lighting in the sloping walls of the roof. North light shell construction is very economical. Similarly cantilevered roof type shells are used for factory areas. In this type of construction, the shell roof is cantilevered out from a series of columns. For rigidity, ribs are introduced across the length of the shells for spanning this type of structure.

Special shapes are being developed which include the doubly curved shells, shells in the form of the spherical triangles, etc.

Similarly corrugated shells are used for economical reasons. Figs. 19.40 and 19.42 show the twin cteisphon shell which is of corrugated type and has been built at the Central Building Research Institute, Roorkee. The depth of corrugation determines largely the rigidity of the arch ring. This depth must be increased as the span increases. Generally 200 mm depth of corrugation is adopted for spans up to 10 m and may be increased to as much as 600 mm for a span of 25 m. Multi-span corrugated shells are used to cover greater areas. These corrugated shells can be used for small low cost houses. They are provided with suitable ventilators and windows.

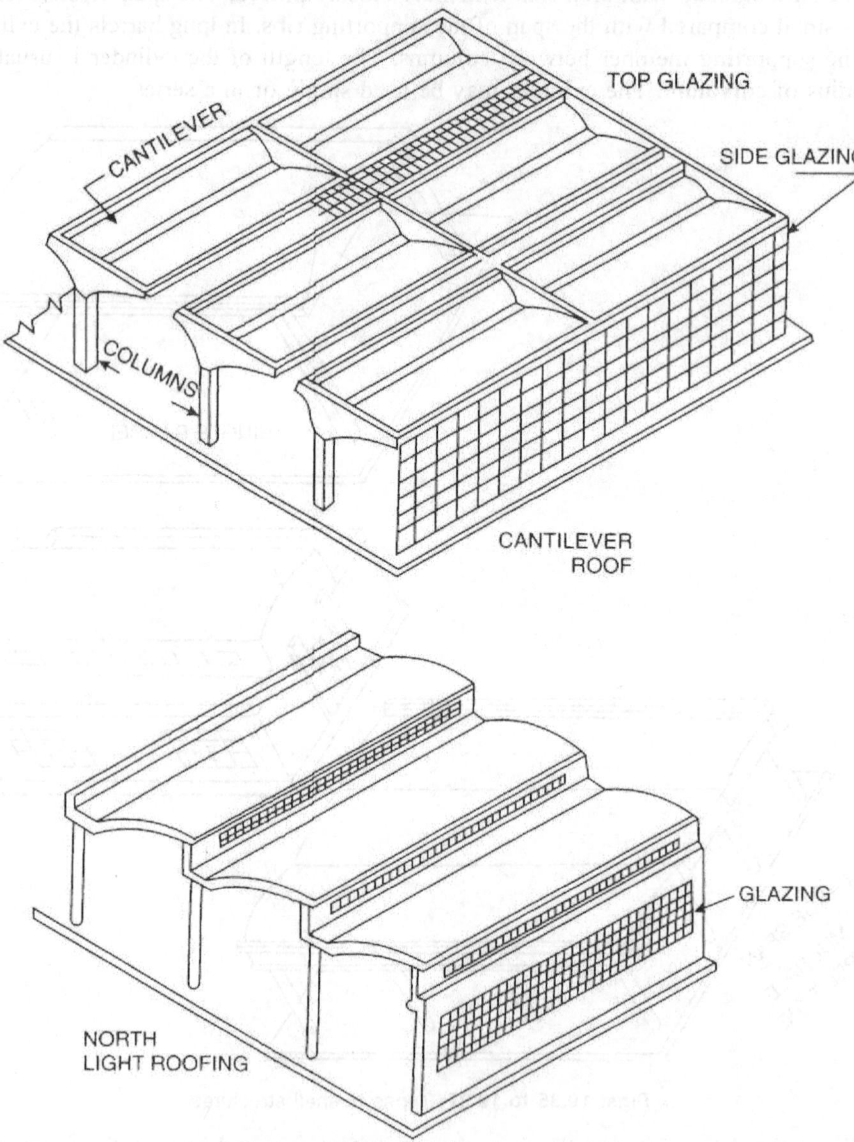

Figs. 19.38 and 19.39: Types of shell structures.

The construction of shell roofs is not difficult but they demand careful designing and handling of formwork. After a shell structure is designed the most important factor that governs its cost is the type of formwork which is used. The formwork structure must be so designed that repeated use

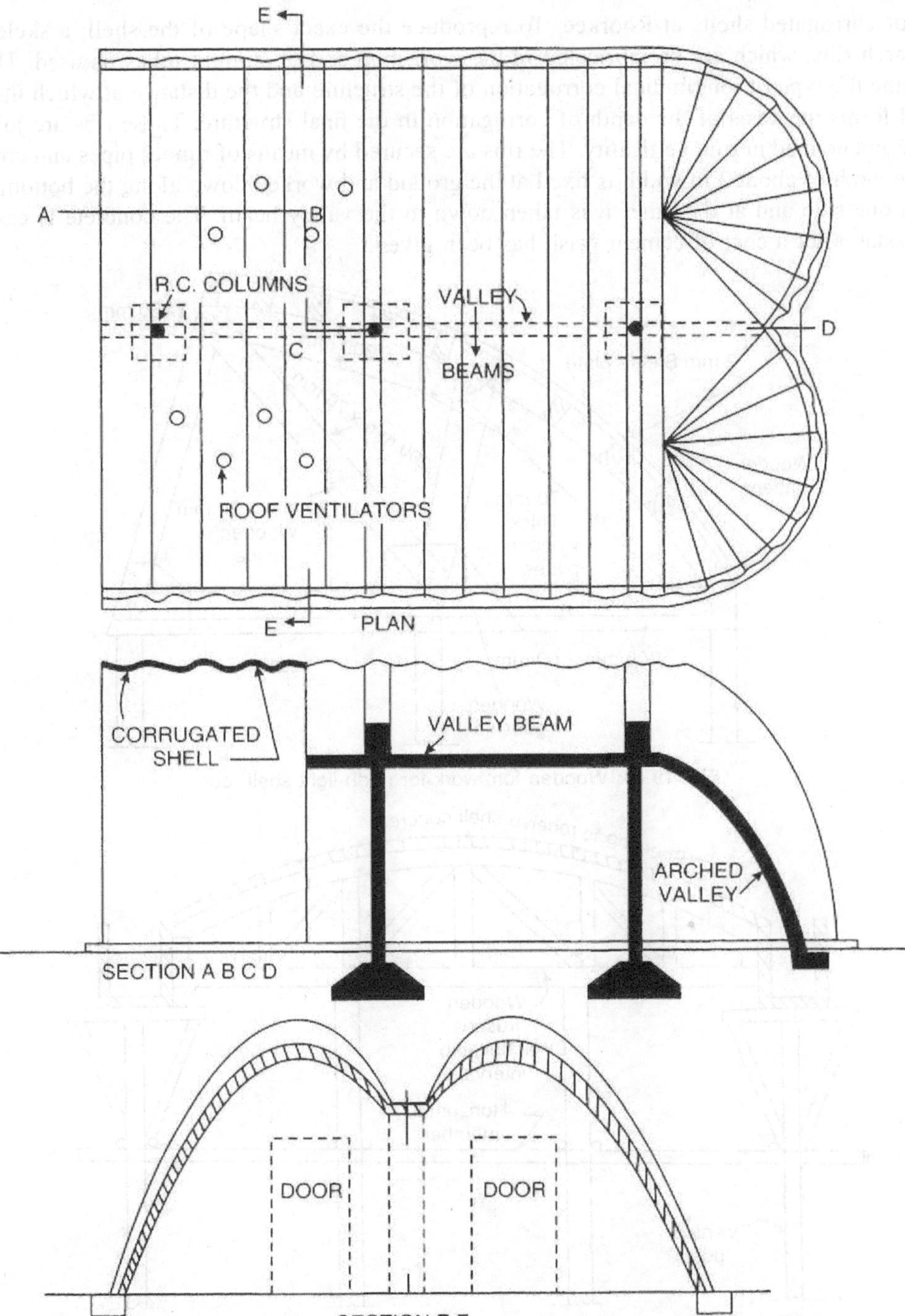

Figs. 19.40 to 19.42: Details of cteisphon shell roof (corrugated type).

of the same form becomes possible. In case larger units of formwork are adopted they lead to practical difficulties in handling them carefully. Mobile forms are generally used and these become specially convenient when the shell structure consists of one or more unbroken barrels with the smooth unobstructed units. Patent formwork which may consist of tubular steel fittings is increasingly used in the shell construction in other countries. Similar type of formwork has been used for corrugated shells at Roorkee. To reproduce the exact shape of the shell, a skeleton of a few arch ribs, which are prefabricated in the workshop and of straight tubes, is used. The ribs determine the type of longitudinal corrugation of the structure and the distance at which these are erected forms the basis of the depth of corrugation in the final structure. These ribs are joined at the site and erected in true verticality. The ribs are secured by means of runner pipes and couplers. Hessian cloth of about 2 m width is fixed at the ground and worked down along the bottom of the ribs on one side and at the other it is taken down to the valley beam. The concrete is cast over this hessian after a coat of cement wash has been given.

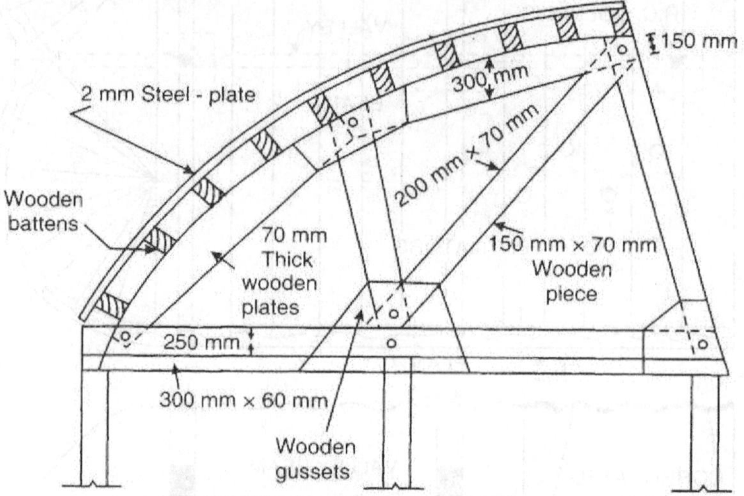

Fig. 19.43: Wooden formwork for north light shell roof.

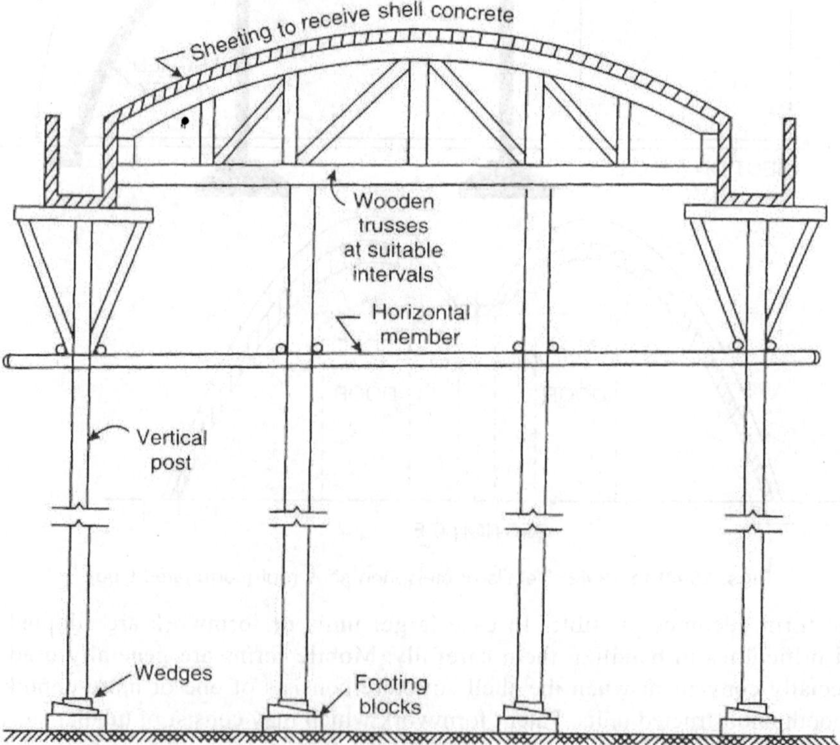

Fig. 19.44: Wooden formwork for barrel shell roof.

Wooden formwork can also be used for shell structures. The type of formwork for a north light shell is shown in Fig. 19.43 and for a barrel shell is shown in Fig. 19.44. The essential features of this type of formwork consist of wooden trusses which have been built to the required shape and are spaced at suitable intervals.

Suitable batten or sheeting is laid on these trusses at the top. The curved surface may be formed by fitting 2 mm steel plates to these battens. The top surface on which the concrete is placed should be covered with a thin coat of crude oil or a special solution so that the concrete when laid will not stick to the forms. The removal of the forms is thereby facilitated. The vertical posts or *ballies* supporting the formwork usually rest on wooden wedges which are supported on wooden planks or otherwise so as to prevent the settlement to a great extent.

Concreting of shells may be done in one thickness or in two or more layers in case the cover to the steel cannot be given properly. Richer mixes of concrete are usually adopted. Concreting is started from the edge beams or the end frames. Whenever concreting is restarted, thick slurry of cement is applied at the joints after cleaning them thoroughly. Suitable markings should be erected at intervals so that the exact thickness of the shell is obtained. The reinforcement should be placed carefully and its exact position may be marked with chalk on the formwork.

PRESTRESSED CONCRETE CONSTRUCTION

Prestressed concrete is one in which permanent internal stresses are deliberately introduced, usually by tensioned steel, to counteract, to the desired degree, the stresses caused in the member in service. For the preparation of concrete, any one of the following cements may be used:

(*i*) Ordinary Portland cement
(*ii*) Portland slag cement with not more than 50 per cent slag content
(*iii*) Rapid-hardening Portland cement; and
(*iv*) High strength ordinary Portland cement.

All aggregate, coarse and fine, should be of good quality and conform to accepted standards. The maximum size of the aggregate should in no case the greater than one-fourth the minimum thickness of the member or 5 mm less than the spacing between the cables, strands or sheathings where provided, subject to a maximum of 40 mm. Admixtures may be added. However, use of any admixture containing chlorides in any form is prohibited.

The prestressing steel may be any one of the following:

(*i*) Plain hard-drawn steel wire
(*ii*) Cold-drawn indented wire
(*iii*) High tensile steel bar; and
(*iv*) Uncoated stress relieved strand.

The concrete should be in grades designated as per Table 19.9.

Table 19.9: Grades of concrete

Grade designation	*Specified characteristic compressive strength at 28 days, N/mm²*
M 30	30
M 35	35
M 40	40
M 45	45
M 50	50
M 55	55
M 60	60

For pretensioned prestressed concrete, the grade of concrete should not be less than M 40.

For prestressed concrete construction, only 'design mix concrete' should be used. The cement content in the mix should preferably not exceed 5200 kN/m^3.

Prestressed concrete is more economical for building construction wherever large construction is involved or whenever the handling costs are less and also where it is difficult to erect shuttering. Prestressed concrete has been used in the construction of building components like columns, beams, slabs, etc. Large areas can be covered without the use of interior columns by prestressing. Similarly small units are also being used for economical construction.

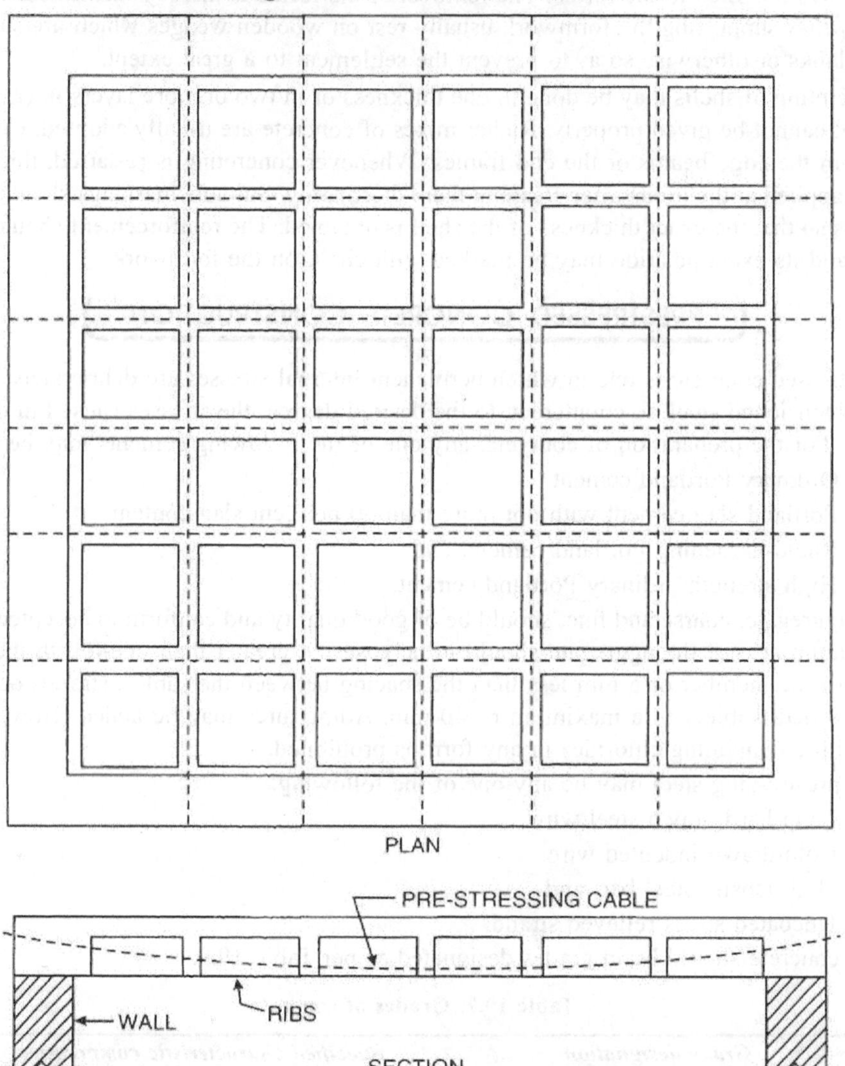

Figs. 19.45 and 19.46: Prestressed slab with ribs containing prestressed cables in both directions.

Basically two methods for prestressing concrete are used, one is the pre-tension method and the other is the post-tension method. In the pre-tension method the high tensile steel is placed in the formwork so that it gets stretched. When the concrete is hardened it is released and transfers all the forces to concrete. This type of a member when used to bear the forces of bending can stand much greater loads as the stresses which are introduced in the concrete under load are opposite to that of the stresses created by prestressing, thereby large saving in sections are possible. Since the dead weight of the structural members gets decreased considerably it is possible to use lighter sections and thereby more economy is possible. In the ordinary reinforced concrete work we cannot economically make use of higher strength of the concrete as the steel needed for such type of work would be excessive whereas in the case of prestressed concrete we can use a higher strength of concrete and thereby economise in the section.

The other method of prestressing is by applying the pull to the cable when concrete is set and transferring the forces of the pull on to the concrete member through suitable anchorages which are embedded at the ends of the concrete member. This method is gaining wider application for types of works which involve large sections to be used, but it is not so economical in the case of buildings when smaller sections have to be built but of a greater number. For building units the pre-tension method described earlier is used.

In all prestressed work the units are made earlier and are placed in the building without the aid of any formwork, etc.

Prestressed concrete is used for slab work in various forms. These slabs are not very economical but have been adopted in building construction. These slabs have been prestressed as continuous units. Alternatively they have been prestressed in both the directions so that they can be used for large spans. Similarly projecting ribs can be added to these slabs to further economise their construction (Figs. 19.45 and 19.46). The only possible construction in the case of slabs is to have precast pre-tension slabs thereby bringing saving in the formwork. The economic spans of slabs which are precast are generally more than 6 m to 7 m. For smaller spans, slabs can be economically built of reinforced concrete. When spans are greater than 10 m, slabs even though precast would become too heavy for economic lifting and as such ribs as described above or beams have to be provided. The lift slab type of construction which has resulted from an effort to reduce the cost of concrete construction by eliminating the expenses of bottom shuttering can be of prestressed type. These slabs are cast and prestressed at the ground level and then lifted into position by hydraulic jacks mounted on the top of the columns. Prestressing of these slabs reduces the thickness, controls deflection and makes lifting more economical.

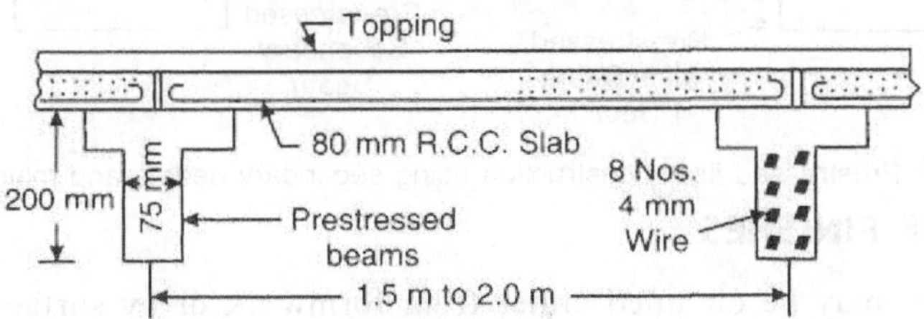

Fig. 19.47: Prestressed beams with R.C.C. slabs.

Prestressed floor or roof construction consists of small prestressed beams which are spaced at 1 m to $1\frac{1}{2}$ m centres and are covered with precast reinforced slabs. For spans up to 7 m or so the beams are precast and pre-tensioned in a factory. They are generally of a T-section with a flange of about 120 mm in width. Depth varies from 120 to 300 mm depending upon the span. The slabs may be slightly reinforced if the span is more than 1 m and the loads are excessive. The prestressed beams of smaller sections than specified above having a sort of I-shape are used for floor slab construction. These beams are spaced at about 300 to 450 mm centre to centre and hollow clay tiles are supported on the bottom flange of the beams. The top portion is then covered with a concrete topping (see Fig. 19.48). For spans greater than 8 m post-tension primary beams are placed 7 to 8 m apart and precast secondary beams are laid on the top of the bottom flange of these primary beams. These secondary beams are kept close with respect to each other and are covered with reinforced or unreinforced precast slabs as described above. For large spans, the beams may be of greater dimensions. These beams are usually of post-tension type and cast in situ. These beams have a slope given to their top so that they have more depth at the centre of their span. I-beams are economical but for greater economy the upper flange of the I-beams may be of greater width. Beams can also be of continuous type and prestressing can extend throughout the length.

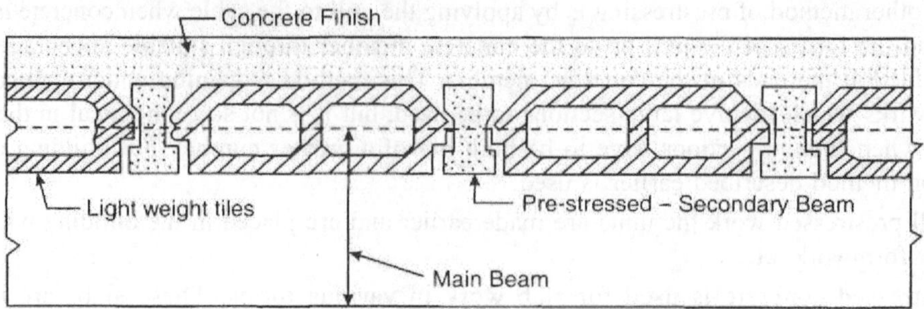

Fig. 19.48: Prestressed beams with a hollow clay tile floor.

Although it is usually not necessary to prestress building columns, sometimes it may be economical to do so due to the excessive bending forces which the column has to resist. Prestressed concrete piles have also been built of large length as they are more economical. The prestressing technique has been extended to the use of shells, stairs, etc.

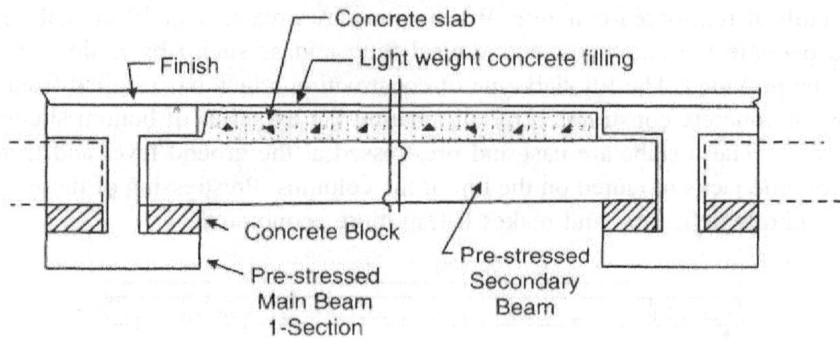

Fig. 19.49: Prestressed floor construction using secondary beams and main beams.

19.17 CONCRETE FINISHES

Surface finishes may be obtained either from formwork or by surface treatments or by applying finishes.

The formwork finishes are the patterns which are created on the surface of concrete merely by the type of formwork used and no additional treatment is given. If proper care is not taken in the selection of a good formwork, the grains and many defects in the wooden boarding are reproduced. Twisted or warped wooden boards create uneven surface. Similarly honey-combing can be caused by the leakage of cement paste through the joints of the boardings. The effect of rough timber can be created deliberately by having boardings of unequal thicknesses.

The boards may be chamfered at the edges so as to create small elevated lines on the surface of concrete. Fillets can be nailed over the concrete formwork to give an appearance of grooves along the surface of concrete. A smooth surface can be obtained by lining the formwork with thin steel sheets or plywood. To prevent the reproduction of grains, plywood should be painted. Pleasing effects can be obtained by using linings of hessian inside the forms. For all types of linings, care should be taken that the joints of linings are tight.

Surface treatments are sometimes adopted for giving a better appearance to concrete. The simplest form of surface treatment is to remove the outer film of cement from the concrete and thereby expose the coarse aggregate. With all exposed aggregate finishes, it should be ensured that the concrete is well mixed and the external film is removed within 48 hours of concreting. The surface can be scrubbed with stiff brushes and water. Care should be taken to check that concrete is sufficiently hard so as to prevent the dislodging of the bigger particles. Retarding agents which delay the setting of cement can be applied to the internal face of the formwork and the outer surface can be demoulded easily later on. Similarly by washing the outer surface with a dilute solution of hydrochloric acid, the thin film can be easily removed. A bush hammer can be worked against the face of hardened concrete so as to create rough surface. Similarly the surface can be punched with chisels.

Applied finishes may consist of paints or application of thin coats of mortar. Over these coats of mortar a number of surface treatments can be given. These include placing of pebbles or creating ornamental finishes.

19.18 JOINTS

Construction joints may be as strong as concrete in other portions of structural members or may be a source of weakness due to faulty construction. A badly formed joint may allow water to enter and subsequently lead to defects. The first step in preparing the construction joints is to remove any laitance, scum or other loose concrete. The exposed surface of the old concrete should be thoroughly wet and given a grout of neat cement and an application of 1 : 2 cement sand mortar immediately before placing the new concrete. The grout and mortar application should be kept as thin as practicable as otherwise the line of the joint will appear on the finished surface. A solution of hydrochloric acid and water is sometimes used in place of hammering to remove the cement on the exposed face at the construction joint but care has to be taken to remove all these acids completely before the grout is applied. As far as the profile to the constructional joint is concerned, the keyed joint is better indicated in Figs. 19.50 to 19.53. Suitable boards should be kept for the slabs and bars should be allowed to pass through.

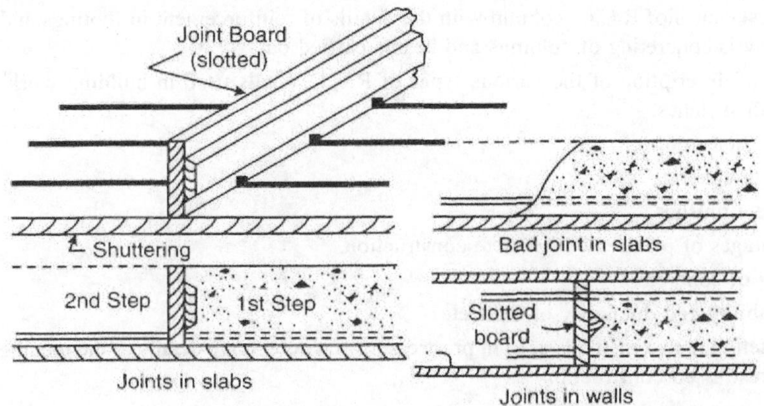

Figs. 19.50 to 19.53: Joints in slabs and walls.

Construction joints should be reduced to a minimum and their position should be predetermined after careful consideration. The amount of work to be done each day should be decided and this should determine the position for the joints. Primarily the joints should be at the section where the shearing is minimum and the face of the joint should be perpendicular to the compressive forces in the member. Thus for the beams, the joint should be at the centre of the span or at the most within the middle third. The face of the joint should be vertical. If the joint in the slab is made normal to the main reinforcement, it should be placed at the centre of the span but if it is placed parallel to the main steel it may be placed anywhere so long as it is not nearer than 8 times the thickness of the slab from the edge of the parallel beam designed as T-beam. When concreting columns, the constructional joint should be horizontal and placed a few mm below the beam intersection. In framed buildings the constructional joint should be at the joint of counter-flexture or alternatively at a place of maximum shear.

The necessity of introducing permanent joints in reinforced structure is to allow for shrinkage, temperature changes, settlements, etc. Joints in retaining walls and floors laid on solid ground must be defined correctly. Joint should be placed where concrete in one operation can be completed and thus the joining becomes obligatory. Joints are also placed where the marked reduction in the width of the slab occurs. The simplest type of joint between panels of slabs laid directly on the ground may be a dry-joint. The stopping board at the end of one period of concreting is removed, the strip of water proof-paper may be placed at the end before placing the concrete in the adjacent panels. In place of water-proof paper, felt or bitumen sheeting may be used. Sometimes a gap of 5 to 10 mm is left between adjacent panels and these gaps are left open or filled with bitumen or a mixture of sand and bitumen. The joint in the roofs should be water-tight and copper strips or water bars are sometimes placed. For simpler work, rubbed or coated joints may be given. The joints in the external walls of buildings must be protected from the effects of weather and as such a simple joint behind an external column is used. A more effective joint is by incorporating a copper strip, ends of which are greased and are free to move whenever movement starts to take place.

QUESTIONS

1. What are the advantages of reinforced cement concrete construction? Describe briefly one of the methods used for designing a concrete mix.
2. Describe briefly the steps in the preparation of concrete with the aid of modern mechanical appliances.
3. What are the various methods for transporting concrete?
   What is the advantage of compaction of concrete? Briefly describe the various methods of compacting concrete.
4. Draw neat sketches of R.C.C. column with the details of reinforcement in footings and junctions with beams. How is concreting of columns and beams carried out?
5. Write a brief description of the various types of R.C.C. shells used in building work. Illustrate your answer with sketches.
6. Write short notes on:
   (*i*) Joints in concrete work.
   (*ii*) Concrete finishes.
   (*iii*) Advantages of prestressed concrete construction.
   (*iv*) Curing of concrete.
   (*v*) Workability and durability of concrete.
7. Explain briefly the two methods used in prestressing concrete. Give details of the members of a building utilising prestressed construction.

REFERENCES

1. **Reynolds, C.E.**: *Concrete Construction.*
2. **H.M.S.O., London**: *Military Engineering, Vol. XIV Concrete.*
3. **Lakeman, A and Davison R.T.**: *Concrete Houses and Small Garages.*
4. **Bangarten, R.H. and Childe, H.L.**: *Manufacture of Concrete Roofing Tiles.*

5. **Concrete Association of India**: *Corrugated Concrete Shell Roofs.*
6. **Indian Construction News**: *Prestressed Concrete as applied to Buildings—Symposium held at C.B.R.I.*
7. **Magnel, G.**: *Prestressed Concrete.*
8. **H.M.S.O., London**: *Concrete Roads.*
9. **C.B.R.I., Roorkee**: *Manufacture of Precast Doubly-curved Shell Elements for Roofs, Floors, etc.*
10. **Bureau of Indian Standards**: *National Building Code of India Part 6, Section 5-Concrete.*
11. **IS: 1343 : 1980**: *Code of Practice for Prestressed Concrete (first revision)*
12. **IS: 1026–1982**: *Recommended guidelines for Concrete mix design.*

Prefabricated (Precast) Construction

Prefabricated building is the completely assembled and erected building, of which the structural parts consist of prefabricated individual units or assemblies using ordinary or controlled materials.

Prefabricated construction is a new technique and is desirable for large scale housing programmes, yet this has to take a firm hold in the country. This type of construction serves the following aims:

(*i*) Prefabrication is used to effect economy in cost.

(*ii*) This results in improvement in quality because components can be manufactured under controlled conditions.

(*iii*) The speed of construction is increased since no curing period is necessary.

(*iv*) Prefabrication helps in the use of locally available materials with required characteristics; and

(*v*) Use of materials which, due to their innate characteristics like light-weight, easy workability, thermal insulation, non-combustibility, etc. effect economy and improve quality.

20.1 MATERIALS OF CONSTRUCTION

While choosing the materials for prefabrication, the following special characteristics are to be considered:

(*a*) Easy availability;

(*b*) Light weight for easy handling and transport, and to economise on sections and sizes of foundations;

(*c*) Thermal insulation property;

(*d*) Easy workability;

(*e*) Durability in all weather conditions;

(*f*) Non-combustibility;

(*g*) Economy in cost, and

(*h*) Sound insulation.

The materials used in prefabricated components can be various and the modern trend is to use concrete, steel, treated wood, aluminium, cellular concrete, light weight concrete, ceramic products, etc. However, this chapter pertains to prefabricated concrete elements.

20.2 MODULAR CO-ORDINATION, ARCHITECTURAL TREATMENT AND FINISHES

20.2.1 Modular Co-ordination

The basic module is to be adopted. Basic module is the fundamental module used in modular co-ordination, the size of which is selected for general application to building and its components. The value of the basic module chosen is 100 mm for maximum flexibility and convenience. The symbol used for basic module is M. After adopting this, further work is necessary to outline suitable range of multimodules with greater increments, often referred to as preferred increments. A set of rules as detailed below are adequate for meeting the requirements of conventional and prefabricated construction. These rules relate to the following basic elements:

(*a*) The planning grid in both directions of the horizontal plan shall be:

(*i*) 15 M for industrial buildings

(*ii*) 3 M for other buildings

The centre lines of load bearing walls should preferably coincide with the gridlines;

(*b*) The planning module in the vertical direction shall be 2 M for industrial buildings and 1 M for other buildings

(*c*) Preferred increments for sill heights, doors, windows etc. shall be 1 M; and

(*d*) In the case of internal columns, the grid lines shall coincide with the centre lines of columns. In case of external columns and columns near the lift and stair wells, the grid lines shall coincide with centre lines of the column in the top-most storey.

20.2.2 Architectural Treatment and Finishes

While deciding the type of architectural treatment and finishes for prefabricated buildings, the following points should be kept in view:

(*a*) Suitability for mass production techniques;

(*b*) Recognition of the constraints imposed by the level of workmanship available;

(*c*) Possibility of using different types of finishes;

(*d*) Use of finishes and architectural treatment for the creation of a particular architectural character in individual buildings and in groups of buildings by the use of colour, texture, projections and recesses on surfaces, etc;

(*e*) The incorporation of structural elements like joists, columns, beams, etc. as architectural features and the treatment of these for better overall performance and appearance;

(*f*) Satisfactory finishing of surfaces; and

(*g*) Use of light weight materials to effect economy in the structural system.

Some of the normally acceptable methods of finishes are:

(*a*) Concrete surface moulded to design, shape;

(*b*) Laid-on finishing tiles fixed during casting,

(*c*) Finishes obtained by washing, tooling, grinding, grooving of hardened concrete,

(*d*) Exposed aggregates; and

(*e*) Other integral finishes.

20.3 COMPONENTS OF PREFABRICATED CONSTRUCTION

The preferred dimensions of precast elements are as follows:

(a) Flooring and Roofing Scheme

Precast slabs or other precast structural flooring units:

(1) ***Length.*** Nominal length shall be in multiples of 1 M;

(2) ***Width.*** Nominal width shall be in multiples of 0.5 M; and

(3) ***Overall Thickness.*** Overall thickness shall be in multiples of 0.1 M.

(b) Beams

(1) ***Length.*** Nominal length shall be in multiples of 1 M;

(2) ***Width.*** Nominal width shall be in multiples of 0.1 M; and

(3) ***Overall depth.*** Overall depth of the floor zone shall be multiples of 0.1 M.

(c) Columns

(1) ***Height.*** Height of columns for industrial and other building in multiples of 1 M.

(2) ***Lateral dimensions.*** Overall lateral dimension or diameter of columns shall be in multiples of 0.1 M.

(d) Walls

Thickness. The nominal thickness of walls shall be multiples of 0.1 M.

(e) Staircase

Width. Nominal width shall be in multiples of 1 M.

(f) Lintels

(1) ***Length.*** Nominal length shall be in multiples of 1 M;

(2) ***Width.*** Nominal width shall be in multiples of 0.1 M; and

(3) ***Depth.*** Nominal depth shall be in multiples of 0.1 M.

(g) Sunshades/Chajja Projections

(1) ***Length.*** Nominal length shall be in multiples of 1 M.

(2) ***Projection.*** Nominal length shall be in multiples of 0.5 m.

PREFABRICATION SYSTEMS

The word 'system' is referred to a particular method of construction of buildings using prefabricated components which are inter-related in functions and are produced to a set of instructions. With certain constraints, several plans are possible, using the same set of components. The degree of flexibility varies from system to system. However, in all the systems there is a certain order and discipline.

The following characteristics, among others, are to be considered in devising a system:

(*a*) Effective utilisation of spaces;

(*b*) Straight and simple walling scheme;

(*c*) Limited sizes and numbers of components;

(*d*) Limited opening in bearing walls;

(*e*) Regulated locations of partitions;

(*f*) Standardized service and stair units;

(*g*) Limited sizes of doors and windows with regulated positions;

(*h*) Structural clarity and efficiency;
(*i*) Suitability for adoption in low rise and high rise buildings.
(*j*) Ease of manufacturing, storing and transporting;
(*k*) Speed and ease of erection; and
(*l*) Simple jointing system.

The system of prefabricated construction depends on the extent of the use of prefabricated components, their materials, sizes and the technique adopted for their manufacture and use in building.

20.4 TYPES OF PREFABRICATED COMPONENTS

The prefabricated concrete components which can be used are:

(*a*) Reinforced prestressed concrete channel units,
(*b*) Reinforced/Prestressed concrete slab units,
(*c*) Reinforced/prestressed concrete beams,
(*d*) Reinforced/prestressed concrete columns,
(*e*) Reinforced/Prestressed concrete hollow core slab.
(*f*) Reinforced concrete waffle slabs/shells,
(*g*) Reinforced/prestressed concrete wall elements, and
(*h*) Hollow/solid blocks and battens
(*i*) Precast planks and joists for flooring and roofing
(*j*) Precast joists and trussed girders
(*k*) Light weight/cellular concrete slabs
(*l*) Precast lintels and *chajjas*,
(*m*) Large panel prefabricates
(*n*) Reinforced/prestressed concrete trusses
(*o*) Reinforced/prestressed roof purlins
(*p*) Precast concrete L-panel unit
(*q*) Prefabricated brick panel unit
(*r*) Prefabricates sandwich concrete panel
(*s*) Precast foundations

The elements may be cast at the site or off the site.

20.5 TYPES OF PREFABRICATED SYSTEMS

20.5.1 Open Prefab System

There are two categories of open prefab systems depending on the extent of prefabrication used in the construction as given below:

(*a*) ***Partial prefabrication open system.*** This system basically uses precast roofing and flooring components and other minor elements like lintels, *chajjas*, kitchen sills in conventional building construction. The structural system could be in the form of *in-situ* framework or load bearing walls.

(*b*) ***Full prefabrication open system.*** In this system almost all the structural components are prefabricated. The filler walls may be of brick block masonry or of any other local by available material.

20.5.2 Large Panel Prefabrication System

This system is based on the use of large prefab components. The components used are precast concrete large panels for walls, floors, roofs, balconies, staircases, etc. The casting of the components could be at the site or off the site.

Depending upon the extent of prefabrication, this system can also lend itself to partial prefab system and full prefab system.

20.6 PRECAST WALLS

Based on the structural functions of the walls, the walls could be classified as:

(*a*) load bearing walls,

(*b*) non-loading bearing walls, and

(*c*) shear walls.

Precast concrete walls can be classified as follows, based on construction:

(*a*) ***Homogeneous Walls*** which could be solid, hollow or ribbed; and

(*b*) ***Non-Homogeneous Walls.*** These could be composite or sandwich panels.

Based on their locations and functional requirements the walls are also classified as:

(*a*) external walls, which can be load bearing or non-load bearing depending upon the lay-out and are usually non-homogeneous walls of sandwiched type to impart better thermal comforts; and

(*b*) internal walls providing resistance against vertical loads, horizontal loads, fire, etc. and are normally homogeneous walls.

20.7 PRECAST FLOORS

Depending upon the composition of units, precast flooring units could be homogeneous or non-homogeneous.

(*a*) Homogeneous floors could be solid slabs, cored slabs, ribbed or waffle slabs.

(*b*) Non-homogeneous floors could be multi-layered ones with combinations of light weight concrete or reinforced/pre-stressed concrete, with filler blocks.

Depending upon the way the loads are transferred, the precast floors could be classified as one way or two ways systems.

One way system transfers loads to the supporting members in one direction only. The precast elements which come under this category are: channel slabs, hollow core slabs, channels and tiles system, light weight cellular concrete slabs, etc.

Two way systems transfer loads in both the directions imparting loads on the four edges. The precast elements under this category are room sized panels, two way ribbed or waffle slab systems, etc.

20.7 STAIRCASE SYSTEMS

Staircase system could consist of single flights with in-built risers and treads in the element only. The flights are normally unidirectional transferring the loads to supporting landing slabs or load bearing walls.

20.8 BOX TYPE CONSTRUCTION

In this system, room size units ae prefabricated and erected at site. Toilet and kitchen blocks could also be similarly prefabricated and erected at site.

This system derives its stability and stiffness from the box units which are formed by four adjacent walls. Walls are jointed to make rigid connections among themselves. The box unit rests on plinth foundation which may be conventional type or precast type.

JOINTS

The design of joints should be made in the light of their assessment with respect to the following considerations:

(*a*) ***Feasibility.*** The feasibility of joint shall be determined by its load-carrying capacity in the particular situation in which the joint is to function.

(*b*) ***Practicability.*** Practicability of joint shall be determined by the amount and type of material required in construction; cost of material, fabrication and erection and the time for fabrication and erection.

(*c*) ***Serviceability.*** Serviceability shall be determined by the joints/expected behaviour to repeated or possible overloading and exposure to climatic or chemical conditions.

(*d*) ***Fire-Rating.*** The fire rating for joints of precast components shall be higher or atleast equal to connecting members.

(*e*) ***Appearance.*** The appearance of joints shall merge with architectural aesthetic appearance and should not be prominent compared to other parts.

The following are the requirements of an ideal structural joint.

(*a*) It shall be capable of being designed to transfer the imposed load and moments with a known margin of safety;

(*b*) It shall occur at logical locations in the structure and at points which may be most readily analysed and easily reinforced;

(*c*) It shall accept the loads without marked displacement or rotation and avoid high local stresses;

(*d*) It shall accommodate tolerances in elements;

(*e*) It shall require little temporary support, permit adjustment and demand only a few distinct operation to make;

(*f*) It shall permit effective inspection and rectification;

(*g*) It shall be reliable in service with other parts of the building; and

(*h*) It shall enable the structure to absorb sufficient energy during earthquakes so as to avoid sudden failure of the structure.

Precast structures may have continuous or hinged connections subject to providing sufficient rigidity to withstand horizontal loading. When only compressive forces are to be taken, hinged joints may be adopted. In case of prefabricated concrete elements, load is transmitted *via* the concrete. When both compressive force and bending moment are to be taken, rigid or welded joints may be adopted; the shearing force is usually small in the column and can be taken up by the frictional resistance of the joint. Here load transmission is accomplished by steel inserted parts together with concrete.

When considering thermal shrinkage and heat effects, provision of freedom of movement or introduction of restraint may be considered.

20.9 JOINING TECHNIQUES

Joining techniques/materials normally employed are:

(*a*) Welding of cleats or projecting steel.

(*b*) Overlapping reinforcement, loops and linking steel grouted by concrete.

(*c*) Reinforced concrete ties all round a slab.

(*d*) Prestressing.

(*e*) Epoxy grouting.

(*f*) Bolts and nuts connection, and

(*g*) A combination of the above.

MANUFACTURE OF PRECAST CONCRETE ELEMENTS

A judicious location of precasting yard with concreting, initial curing (required for demoulding),storage facilities, suitable transporting and erection equipments and availability of raw materials are the crucial factors which should be carefully planned and provided for effective and economic use of precast concrete components in construction.

The manufacture of the components can be done in a centrally located factory or in a site precasting yard set up at or near the site of work.

20.10 FACTORY PREFABRICATION

Factory prefabrication is resorted to in a factory for the commercial production for the manufacture of standardized components on a long term basis. It is a capital intensive production where work is done throughout the year preferably under a closed shed to avoid effects of seasonal variations. High level of mechanization can always be introduced in this system where the work can be organized in a factory-like manner with the help of a constant team of workmen.

The basic disadvantage in factory prefabrication is the extra cost incidence of transportation of elements from plant to site of work where sometimes even the shape and size of prefabricates get limited due to lack of suitable transportation equipment, road contours, etc. The organized labour of permanent nature with regular benefits lead to huge establishment cost which add to ultimate cost of production.

20.11 SITE PREFABRICATION

In this scheme, the components are manufactured at site or as near the site of work as possible.

This system is normally adopted for a specific job order for a short period. The work is normally carried out in open space with locally available labour force. The equipment machinery and moulds are of mobile nature.

The site prefabrication may be either semi-mechanised or fully mechanised. In a semi-mechanised system, the work is carried out in open space with locally available labour force; whereas in a fully mechanised system, the work is carried out under shed with skilled labour. The equipment used will be similar to the one of factory production.

Though there is a definite economy with respect to cost of transportation, this system suffers from basic drawback of its non-suitability to any high degree of mechanization and no elaborate arrangements for quality control. Normal benefits of continuity of work is not available in this system of construction.

20.12 PROCESSES OF MANUFACTURE

The various processes involved in the manufacture of precast elements may be classified as follows.

20.12.1 Main Process

(*a*) Providing and assembling the moulds, placing reinforcement cage in position for reinforced concrete work, and stressing the wires in the case of prestressed elements;

(*b*) Fixing of inserts and tubes, where necessary (for handling)
(*c*) Pouring the concrete into the moulds;
(*d*) Vibrating the concrete and finishing;
(*e*) Curing (steam curing, if necessary); and
(*f*) Demoulding the forms and stacking the precast products.

20.12.2 Auxiliary Process

Process necessary for the successful completion of the process covered by the main process:

(*a*) Mixing and manufacture of fresh concrete (done in a mixing station or by a batching plant);
(*b*) Prefabrication of reinforcement cage (done in a steel yard or workshop);
(*c*) Manufacture of inserts and other finishing items to be incorporated in the main precast products;
(*d*) Finishing the precast products; and
(*e*) Testing of products.

20.12.3 Subsidiary Process

All other work involved in keeping the main production work to a cyclic working:

(*a*) Storage of materials;
(*b*) Transport of cement and aggregates;
(*c*) Transport of green concrete and reinforcement cages;
(*d*) Transport and stacking the precast elements;
(*e*) Repairs and maintenance of tools, tackles and machines; and
(*f*) Generation of steam, etc.

For the manufacture of precast elements all the above processes shall be planned in a systematic way to achieve the following:

(*a*) A cyclic technological method of working to bring in speed and economy in manufacture.
(*b*) Mechanization of the process to increase productivity and to improve quality.
(*c*) The optimum production satisfying the quality control requirements and to keep up the expected speed of construction aimed.
(*d*) Better working conditions for the people on the job; and
(*e*) To minimize the effect of weather on the manufacturing schedule.

The various stages of precasting can be classified as in Table 20.1 on the basis of the equipments required for the various stages. This permits mechanization and rationalization of work in the various stages. In the precasting, stages 6 and 7 given in Table 20.1 form the main process in the manufacture of precast concrete elements. For these precasting stages there are many technological processes to suit the concrete product under consideration which have been proved rational, economical and time saving. The technological line or process is the theoretical solution for the method of planning the work involved by using machine complexes. Figure 20.1 illustrates diagramatically the various stages involved in a plant process.

Table 20.1: Stages of precasting concrete products

Sl. No.	*Precasting Stage No.*	*Name of Process*	*Operations Involved*
(*i*)	1	Procurement and storage of construction materials	Unloading and transport of cement, coarse and fine aggregates, and steel, and storing them in bins, silos or storage sheds
(*ii*)	2	Testing of raw materials	Testing of all materials including steel
(*iii*)	3	Design of concrete mix	Testing of raw materials, plotting of grading curves and trial of mixes in laboratory
(*iv*)	4	Making of reinforcement cages	Unloading of reinforcement bars from wagons or lorries and stacking them in the steel yard, cutting, bending, tying or welding the reinforcements and making in the form of a cage, which can be directly introduced into the mould
(*v*)	5	Applying form release agent and laying of moulds in position	Moulds are cleaned, applied with form release agent and assembled and placed at the right place
(*vi*)	6	Placing of reinforcement cages, inserts and fixtures	The reinforcement cages are placed in the moulds with spacers, etc. as per data sheet prepared for the particular prefabricate.
(*vii*)	7	Preparation of green concrete	Taking out aggregates and cement from bins, silos, etc, batching and mixing
(*viii*)	8	Transport of green concrete	Transport of green concrete from the mixer to the moulds. In the case of precast method involving direct transfer of concrete from mixer to the mould or a concrete hopper attached to the mould this prefabrication stage is not necessary.
(*ix*)	9	Pouring and consolidation of concrete	Concrete is poured and vibrated to a good finish
(*x*)	10	Curing of concrete and demoulding	Either a natural curing with water or an accelerated curing using steam curing and other techniques. In the case of steam curing using trenches or autoclaves, this stage involves transport of moulds with the green concrete into the trench or autoclave and taking them out after the curing and demoulding elements from the mould. In the case of pre-tensioned elements cutting of protruding wires also falls in this stage. In certain cases the moulds have to be partly removed and inserts have to be removed after initial set. The total demoulding is done after a certain period and the components are then allowed to be cured. All these fall in this operation
(*xi*)	11	Stacking of precast elements	Lifting of precast elements from the mould and transporting to the stacking yard for further transport by trailer or rail
(*xii*)	12	Testing of finished components	Tests are carried out on the components individually and in combination to ensure the adequecy of their strength
(*xiii*)	13	Miscellaneous	(*a*) Generation of steam involving storing of coal or oil necessary for generation of steam and providing steam pipe connection up to the various technological lines (*b*) Repair of machines used in the production

Table 20.2: Precasting methods

Sl. No.	*Precasting Method*	*Where used*	*Dimensions and weights*	*Advantages and remarks*
(*i*)	Individual mould method. Precasting method which may be easily assembled out of bottom and sides, transportable, if necessary. This may be either in timber or in steel using needle or mould vibrators and capable of taking prestressing forces	(*a*) Rib slabs, beams, girders, window panels, box type units and special elements	No limit in size and weight.	(*a*) Strengthening of the cross section possible (*b*) Openings are possible in two planes
		(*b*) Prestressed railway sleepers, parts of prestressed girders, etc	Depending on the equipment used for demoulding, transporting and placing.	
(*ii*)	Battery form method (The shuttering panels may be adjusted into the form of a battery at the required distances equal to the thickness of the concrete member)	Interior wall panels, shell elements, reinforced concrete battens, rafters, purlins and, roof and floor slabs	Length: 18 m Breadth: 3 m Mass: 5 tonnes	Specially suitable for mass production of wall panels where shuttering cost is reduced to a large extent and autoclave or trench steam curing may be adopted by taking the steam pipes through the shuttering panels
(*iii*)	Stack method	Floor and roof slab panels	Length: Any desired length Breadth: 1 to 4 m Mass: 5 tonnes	For casting identical reinforced or prestressed panels one over the other with separating media interposed in between
(*iv*)	Tilting, mould method (This method is capable of being skipped vertically using hydraulic jacks)	Exterior wall panels where special finishes are required on one face or for sandwich panel	Length: 6 m Breadth: 4 m Mass: 5 tonnes	Suitable for manufacturing the external wall panels
(*v*)	Long line prestressing bed method	Double tees, ribbed slabs, purlins, piles and beams.	Length: Any desired length Breadth: 2 m Height: 2m Mass: Up to 10 tonnes	Ideally suited for pretensioned members
(*vi*)	Extrusion method (Long concrete mould with constant cross section concreting and vibration will be done automatically just as in hollow cored slab casting)	Roof slabs, foam concrete wall panels and beams cross-section where connecting and vibrations are done automatically just as in hollow cored slab casting.	Length: Any desired length Breadth: Less than 2 m Height: Less than 3 m	May be used with advantage in the case of unreinforced blocks, foam concrete panels

The various accepted methods of manufacture of precast units can be broadly classified into two methods:

(*a*) The *'Stand Method'* where the moulds remain stationary at places, when the various processes involved are carried out in a cyclic order at the same place, and

(*b*) The *'Flow Method'* where the precast unit under consideration is in movement according to the various processes involved in the work which are carried out in an assembly-line method.

The various accepted precasting methods are listed in Table 20.2 with details regarding the elements that can be manufactured by these methods.

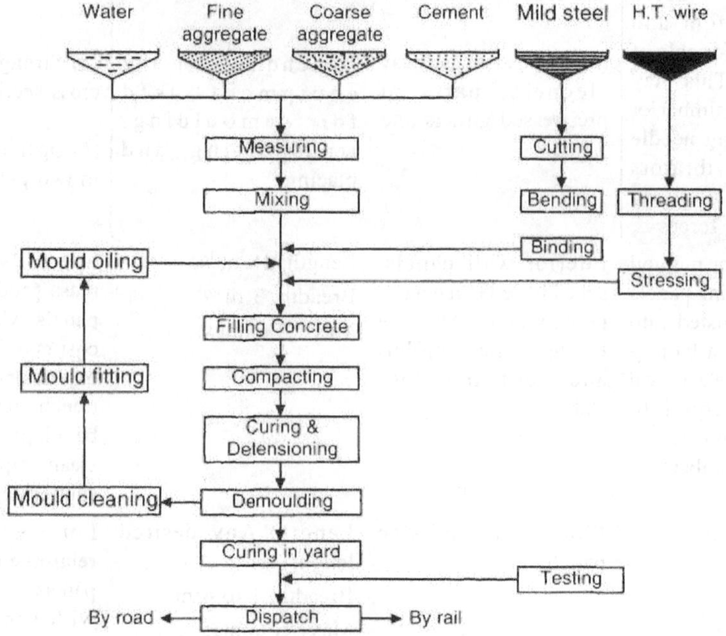

Fig. 20.1: Plant process

20.13 PRESERVATION AND STORAGE OF MATERIALS

Storage of materials is of considerable importance in the precasting industry, any mistake in planning can greatly influence the economics of production. From experience in construction, it is clear that there will be very high percentages of loss of materials as well as poor quality due to improper storage and transport. So, in a precast factory where everything is produced with special emphasis on quality, proper storage and preservation of building materials especially cement, coarse and fine aggregates, is of prime importance.

20.13.1 Storage of Cement

Storage of cement can be effected either in specially erected storage sheds where cement can be stored in the form of bags or in silos where it is stored loose.

20.13.2 Storage of Coarse and Fine Aggregates

The coarse and fine aggregates can be stacked either in open or in bunkers. In the case of open storage, the 'Parallel-Boxes' method with dividing walls up to about 3 metres in height, is considered to be most convenient and economical. The dividing walls can be made up of precast R.C.C. retaining walls or steel or timber panels inserted between the columns. In planning this method of storage, the following points shall be kept in mind:

(*a*) The stored aggregate shall be protected from mixing up with the local earth, clay or coal; and

(*b*) The various bins or boxes shall be properly designated about the size and type of material to be stored. Mistakes occurring due to dumping of one class/size of aggregates in the wrong bin should be avoided.

Yet another method of open storage is by heaps under which a tunnel is provided with conveyor belt system to extract from the heap whatever material is required for preparation and mixing of concrete.

In planning the storage of coarse and fine aggregates, bins, silos, etc. shall have a minimum storage capacity and shall be designed to suit the supply requirements of the factory. As far as batching silos are concerned, 2 to 4 hours storage capacity shall be provided.

20.14 MOULDS

Moulds for the manufacture of precast elements may be of steel, timber, concrete and plastic or a combination thereof. For the design of moulds for the various elements, special importance should be given to easy demoulding and assembly of the various parts. At the same time rigidity, strength and watertightness of the mould, taking into consideration forces due to pouring of green concrete and vibrating, are also important.

Slopes of the mould walls. For easy demoulding of the elements from the mould with fixed sides, the required slopes have to be maintained. Otherwise there is a possibility of the elements getting stuck up with the mould at the time of demoulding.

20.15 CURING

20.15.1 Accelerated Hardening

In most of the precasting factories, it is economical to use faster curing methods or artificial curing methods, which in turn will allow the elements to be demoulded much earlier permitting early re-use of the forms. Any of the following methods may be adopted:

(a) By heating the aggregates and water before mixing the concrete. By heating of the aggregates as well as water to about 70°C to 80°C before making the concrete mix and placing the same in the moulds, sufficiently high earlier strengths are developed to allow the elements to be stripped and transported.

(b) Steam curing. Steam curing may be done under high pressure and high temperature in an autoclave. This technique is more suited to smaller elements. Alternatively this could be done using low pressure steam having temperature around 80°C. For light weight concrete products when steam cured under high pressure, the drying shrinkage is reduced considerably. Due to this reason, high pressure steam curing in autoclave is specified for light weight low densities ranging from 300 to 1,000 kg/m^3. For normal heavy concretes as well as light weight concretes of higher densities, low pressure steam curing may be desirable as it does not involve using high pressures and temperatures requiring high investment in an autoclave.

(c) Steam injection during mixing of concrete. In this method low pressure saturated steam is injected into the mixer while the aggregates are being mixed. This enables the heating up of concrete to approximately 60°C. Such a concrete after being placed in the moulds attains high early strength.

(d) Heated air method. In this method, the concrete elements are kept in contact with hot air with a relative humidity not less than 80 per cent. This method is specially useful for light concrete products using porous coarse aggregates.

(*e*) ***Hot water method.*** In this method, the concrete elements are kept in a bath of hot water around 50°C to 80°C. The general principles of this type of curing are not much different from steam curing.

(*f*) ***Electrical method.*** The passage of current through the concrete panels generates heat through its electro-resistivity and accelerates curing. In this method, concrete is heated by an alternating current ranging from 50 volts for a plastic concrete and gradually increasing to 230 V for the set concrete. This method is normally used for massive concrete products.

(*g*) ***Consolidation by spinning.*** Such a method is generally used in the centrifugal moulding of pipes and such units. The spinning motion removes excess water, effects consolidation and permits earlier demoulding.

(*h*) ***Pressed concrete.*** This method is suitable for fabrication of small or large products at high speed of production. A 100-200 tonnes press compresses the wet concrete in rigid moulds and expels water. Early handling and a dense wear resistant concrete is obtained.

(*i*) ***Vacuum treatment.*** This method removes the surplus air and water from the newly placed concrete as in slabs and similar elements. A suction up to about 70 per cent of an atmosphere is applied for 20 to 30 minutes per centimetre thickness of the units.

(*j*) ***Consolidation by shock.*** This method is suitable for small concrete units dropped repeatedly from a height in strong moulds. The number of shocks required to remove excess water and air may vary from 6 to 20 and the height of lift may be up to as much as half the depth of the mould.

After the accelerated curing of the above products by any of the above accepted methods, the elements shall be cured further by normal curing methods to attain full final strength.

The curing of the prefabricated elements can be affected by the normal methods of curing by sprinkling water and keeping the elements moist. This can also be done in the case of smaller elements by immersing them in a specially made water tanks.

20.16 STACKING DURING TRANSPORT AND STORAGE

Every precaution shall be taken against overstress or damage, by the provision of suitable packings at agreed points of support. Particular attention is directed to the inherent dangers of breakage and damage caused by supporting other than at two positions, and also by the careless placing of packing (for example, not vertically one above the other). Ribs, corners and intricate projections from solid section should be adequately protected. Packing pieces shall not discolour, disfigure or otherwise permanently cause mark on units or members. Stacking shall be arranged or the precast units should be protected, so as to prevent the accumulation of trapped water or rubbish, and if necessary to reduce the risk of efflorescence.

The following points shall be kept in view during stacking:

(*a*) Care should be taken to ensure that the flat elements are stacked with right side up. For identification, top surfaces should be clearly marked.

(*b*) Stacking should be done on a hard and suitable ground to avoid any sinking of support when elements are stacked.

(*c*) In case of horizontal stacking, packing materials must be at specified locations and must be exactly one over the other to avoid cantilever stress in panels.

(*d*) Components should be packed in a uniform way to avoid any undue projection of elements in the stack which normally is a source of accident.

20.17 HANDLING ARRANGEMENTS

Lifting and handling positions shall be clearly defined particularly where these sections are critical. Where necessary special facilities, such as bolt holes or projecting loops, shall be provided in the units and full instructions supplied for handling.

For precast prestressed concrete members, the residual prestress at the age of particular operation of handling and erection shall be considered in conjunction with any stresses caused by the handling or erection of member. The compressive stress thus computed shall not exceed 50 per cent of the cube strength of the concrete at the time of handling and erection. Tensile stresses up to a limit of 50 per cent above those specified shall be permissible.

20.18 TRANSPORT

Transport of precast elements inside the factory and to the site of erection is of considerable importance not only from the point of view of economy but also from the point of view of design and efficient management. Transport of precast elements must be carried out with extreme care to avoid any jerk and distress in elements and handled as far as possible in the same orientation as it is to be placed in final position.

Transport inside the factory. Transport of precast elements moulded inside the factory depends on the method of production selected for the manufacture as given in Table 20.2.

Transport from stacking yard inside the factory to the site of erection. Transport of precast concrete elements from the factory to the site of erection should be planned in such a way so as to be in conformity with the traffic rules and regulations as stipulated by the authorities. The size of the elements is often restricted by the availability of suitable transport equipment, such as tractor-*cum*-trailers, to suit the load and dimensions of the member in addition to the load-carrying capacity of the bridges on the way.

While transporting elements in various systems, that is, wagons, trucks, bullock carts, care should be taken to avoid excessive cantilever actions and desired supports are maintained. Special care should be taken at location of sharp bends and on uneven or slushy roads to avoid undesirable stresses in elements.

Before loading the elements in the transporting media, care should be taken to ensure that the base packing for supporting the elements are located at specified positions only. Subsequent packings must be kept strictly one over the other.

20.19 ERECTION

In the *'erection of precast elements'*, all the following items of work are meant to be included:

(*a*) Slinging of the precast element;
(*b*) Tying up of erection ropes connecting to the erection hooks;
(*c*) Cleaning of the elements and the site of erection;
(*d*) Cleaning of the steel inserts before incorporation in the joints, lifting up of the elements, setting them down into the correct envisaged position;
(*e*) Adjustment to get the stipulated level, line and plumb;
(*f*) Welding of cleats,
(*g*) Changing of the erection tackles;
(*h*) Putting up and removing of the necessary scaffolding or supports;
(*i*) Welding of the inserts, laying of reinforcements in joints and grouting the joints; and
(*j*) Finishing the joints to bring the whole work to a workman like finished product.

20.20 AUTOCLAVED CELLULAR CONCRETE

The manufacture of cellular concrete products differs from that of dense concrete in certain respects as given below:

(*a*) The manufacture of cellular concrete being a highly controlled process has to be done in a factory;

(*b*) The principal raw material are cement or lime and fine materials (silicious sand, fly ash, granulated blast furnace slag);

(*c*) The silicious material is ground finely in a ball-mill and the slurry is prepared with predetermined quantity of cement or lime and water. Gas generating materials and harmless additives are also added in the required amount before the concrete is poured into the moulds;

(*d*) The cellular concrete is cast in standard moulds and the various components are cut to the required size before it is autoclaved.

(*e*) Curing is done in autoclaves at high temperatures (180°C to 200°C) and at high pressure (7 to 15 kgf/cm^2). The components are taken out after they are fully autoclaved.

(*f*) Each slab is provided with tongue at one side and groove at the other or any other provision is made to transfer load from one unit to another; and

(*g*) In view of the above, there will be some changes in the stages of manufacture given in Table 20.1.

20.21 EQUIPMENT

General. The equipment used in the precast concrete industry can be classified into the following categories:

(*a*) Machinery required for quarrying of coarse and fine aggregates.

(*b*) Conveying equipment, such as belt conveyors, chain conveyors, screw conveyors, bucket elevators, hoists, etc;

(*c*) Concrete mixing machines.

(*d*) Concrete vibrating machines.

(*e*) Erection equipment, such as cranes, derricks, hoists, chain pulling blocks, etc.

(*f*) Transport machinery, such as tractor-*cum*-trailers, dumpers, lorries, locomotives, motor boats and rarely even helicopters.

(*g*) Workshop machinery for making and repairing steel and timber moulds.

(*h*) Bar straightening, bending and welding machines to make reinforcement cages.

(*j*) Minor tools and tackles, such as wheel barrows, concrete buckets, etc; and

(*k*) Steam generation plant for accelerated curing.

In addition to the above, pumps and soil compacting machinery are required at the building site for the execution of civil engineering projects involving prefabricated components.

Each of the above groups can further be classified into various categories of machines and further to various other types depending on the source of power and capacity.

20.22 MECHANIZATION OF THE CONSTRUCTION AND ERECTION PROCESS

The various processes can be mechanized as in any other industry for attaining the advantages of mass production of identical elements which in turn will increase productivity and reduce the cost of production in the long run, at the same time guaranteeing quality for the end-product. On the basis of the degree of mechanization used, the various precasting factories can be divided into three categories:

(*a*) With simple mechanization,

(*b*) With partial mechanization, and

(*c*) With complex mechanization leading to automation.

In simple mechanization, simple mechanically operated implements are used to reduce the manual labour and increase the speed.

In partial mechanization, the manual work is more or less eliminated in the part of a process. For example, the batching plant for mixing concrete, hoists to lift materials to a great height and bulldozer to do earthwork come under this category.

In the case of complex mechanization leading to automation, a number of processes leading to the end-product are all mechanized to a large extent (without or with a little manual or human element involved). This type of mechanization reduces manual work to the absolute minimum and guarantee the mass production at a very fast rate and cheap price.

PREFABRICATED STRUCTURAL UNITS

Some of the structural members of a building which can be constructed with the help of prefabricated units are described below briefly.

20.23 WALLS AND COLUMNS

As described in chapter 7, walls and columns made of hollow block masonry are advantageous as they are easy to construct, are cheaper and have got a great thermal insulation effect. They considerably save mortar compared to brickwork since the number of joints are less. Internal plastering is reduced since a good finish can be obtained with one coat only; instead of two.

These blocks are built in concrete, the forms of which can be reused a number of times. Special types of forms are used keeping in view the shape of the hollow space to be left in the blocks.

20.24 LINTELS

Prefabricated RCC lintels can be used conveniently over window and door openings, thereby accelerating the speed of construction by eliminating curing period.

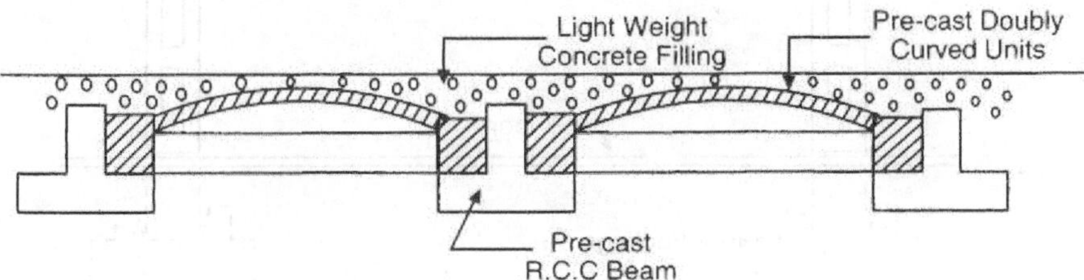

Fig. 20.2: Doubly curved precast concrete unit.

20.25 DOOR AND WINDOW FRAMES

Precast concrete door and window frames can also be built. Steel bars of about 4 mm to 40 mm in diameter run through them. Suitable hard wood blocks are used for fixing hinges, etc. to them. Concrete is vibrated on a table vibrator and is then poured into the mould, thereby forming a durable mix.

20.26 ROOFING AND FLOORING ELEMENTS

Prefabricated reinforced concrete battens and plain concrete tiles can be used for roofing and flooring for flat roofs, instead of wooden sections and brick tiles. For sloping roof, precast reinforced and prestressed concrete triangulated trusses can be used.

Plain concrete or lightly reinforced concrete can be used in the form of precast shells for roofing. Plain concrete, doubly curved shells have been developed at C.B.R.I.; Roorkee. While making them, a suitable frame is built over which hessian cloth is spread. This cloth has to be given a sort of sag and internal tension is controlled by the depth indicating frame. Over this, a thin frame of about 25 mm height is placed to retain the wet concrete and regulate its thickness. A piece of chicken mesh is next placed on the mould to act as reinforcement. Cement concrete is poured over the hessian and manually compacted. The frame is made to rest to permit the hessian to sag with the wet concrete. Moulds are set up for the edge beams and suitable steel rods are placed within these moulds. These beams are then cast. The beam sides are demoulded after an hour of casting. The unit is cured for about two days and is inverted for use. The final shape of the unit, as placed on small T-beams, is shown in Fig. 20.3. The top surface of the roof thus formed is wavy and as such the depressions are filled with concrete or earth so that level surface may be obtained. These units can be of 1000 mm × 1000 mm to about 3000 mm × 3000 mm in size.

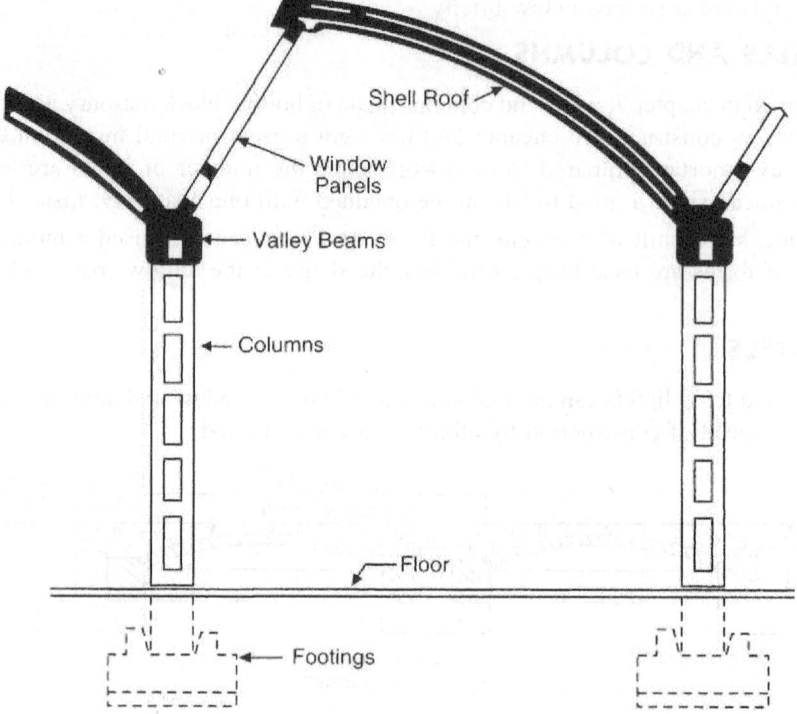

Fig. 20.3: Precast RCC Northlight shell roof construction.

20.27 STAIRS

Treads of pre-cast plain concrete slabs and the risers of small, precast concrete blocks can be used for construction of stairs. The details are shown in Fig. 14.41 in chapter 14.

QUESTIONS

1. What aims are served by prefabricated construction? What special characteristics are to be considered while selecting materials for prefabricated construction.
2. Discuss the concept of modular co-ordination in precast construction. What points should be kept in view while selecting the architectural treatment and finishes for prefabricated construction.
3. (*a*) Give the components of prefabricated construction.
   (*b*) What characteristics should be considered while devising a system?
4. Name the different types of prefabrication systems. What components can be used in each case?

5. What consideration govern the provision of joints in prefabricated construction? List the joining techniques normally employed.
6. Describe the procedure used for the manufacture of precast concrete elements.
7. List the stages of precasting concrete products. Draw a flow chart for the plant process.
8. Describe briefly the various methods used for curing concrete elements.
9. List the equipment used in prefabricated construction. Discuss the mechanisation of construction and erection processes.
10. Describe the various structural members of a building which can be prefabricated.

REFERENCES

1. **Bureau of Indian Standard IS**: *National Building code of India,* 2005 *Part* 6 *Section* 7*-Prefabrication and systems building.*
2. **IS: 3201 : 1988**: *Criteria for the design and construction of precast trusses and purlins* (*First revision*)
3. **IS: 6332 : 1984**: *Code of practice for construction of floors and roofs using precast doubly-curved shell units (first revision)*
4. **IS: 11447 : 1985**: *Code of practice for construction with large panel prefabricates.*

Formwork

Concrete construction in the modern building work has assumed a lot of importance and is being used to a great extent. Concrete, being a plastic material in the green state, has to be kept within an enclosure till it gains strength. This temporary structure which has to be built for any concrete member is called formwork. Timber is the most common material used for formwork construction. It is easy to work with and is also cheap.

Timber to be used should be free from knots and should have coarse grains. Hard woods are not used as they are difficult to be worked and nailed. Partially seasoned timber is best for formwork. Too dry timber will tend to swell with moisture while green timber will tend to dry and shrink in hot weather causing ridges on the concrete surface. Timber of a particular size should be finished to a uniform thickness so that all the pieces match together.

The use of steel forms has been on the increase in recent years. They can be obtained for all types of structures and in finished forms. The initial cost of such forms is very high but they can be repeatedly used for a number of times. Hence they are suitable for large works only and not for small structures involving the use of a particular form only once or twice. Steel forms can be easily erected and do not need many adjustments as they are available in standard sizes. Their erection is very mechanical. Another advantage of steel forms is that they give smooth surfaces which require very little finishing work later on.

21.1 REQUIREMENTS OF FORMWORK

(1) *Formwork must be strong*: This is very essential as it has to bear the weight of concrete which weighs about 2.4 kN/m^3 and exerts a lot of pressure on the members surrounding it while it is in the wet state. Resistance to this outward push is achieved with the aid of adequate bracings and by the use of wires which get embedded in the concrete and are fastened to the outer side of the formwork.

(2) *Formwork must be smooth*: It is always desirable to have smooth surfaces on all concrete structures. This is especially desirable on exposed concrete. All the impressions which are visible on the surrounding formwork will get transformed to the concrete.

(3) *Formwork should be true*: This means that the formwork should be erected in such a manner that all the faces of concrete are true with respect to the drawings. For example, the side formwork of footings and walls must be absolutely vertical. To get a true surface, it is also essential that no cracks are present within the formwork as concrete is likely to flow through them thereby causing a rough surface on the outside.

(4) *Formwork must be easily removable*: Formwork should be constructed in such a manner that it can be removed easily without damaging it. It should be constructed in such a way that the nails or screws driven through it do not interfere with its removal.

21.2 TOLERANCES FOR FORMWORK

The formwork should be designed and constructed to the shapes, lines and dimensions shown on the drawing within tolerances given below:

(*a*) Deviation from the specified dimensions of cross-section of columns and beams	– 6 mm + 12 mm
(*b*) Deviation from dimensions of footings	
(*i*) Dimensions in plan	– 12 mm + 50 mm
(*ii*) Eccentricity	0.02 times the width of the footing in the direction of deviation but not more than 50 mm
(*iii*) Thickness	± 0.05 times the specified thickness

The tolerances apply to concrete dimensions only, not to positioning of vertical reinforcing steel or dowels.

21.3 FOOTING FORMS

There are various types of forms used for footings depending upon the type of construction. Some of these are described below:

(1) *Wall footing forms*: These are the simplest to be built and consist merely of a heavy plank put in position to the required line. One side is set to the line initially and held with stakes spaced at about 2000 mm apart and the other side is set from this with the aid of spreaders at a correct distance equal to the width of the footing. These spreaders are knocked out as concreting reachès them. For footings up to one metre in depth the sides are made of 25 mm thick timber pieces built into panels about 2500 mm long with battens in the centre to hold the boards together. These panels can be held to line by 50 mm × 100 mm wailings which can be braced to stakes or sides of the panels (See Fig. 21.1).

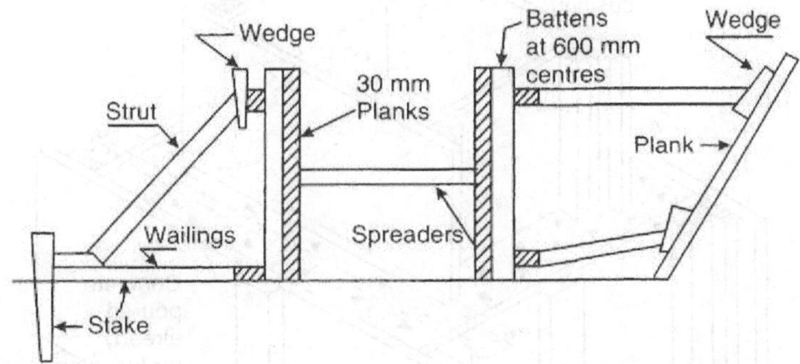

Fig. 21.1: A typical form for a wall footing.

(2) *Column footing forms*: They are of the following types:

***(a) Square or rectangular*:** The ends are built to the exact dimensions of the footing with a cleat at each end of the footing from the outside. The other two sides are built about 300 mm longer than the dimensions of the footing. The side and end panels are made on the carpenter's bench with holes bored for the wires, if they are to be used. Four panels are assembled by butting the end panels against the end cleats of the sides and are nailed. The corner should be braced back to stakes or to the sides of the excavation. The intermediate bracing should be done in similar manner or the cleat wired as shown in Fig. 21.2.

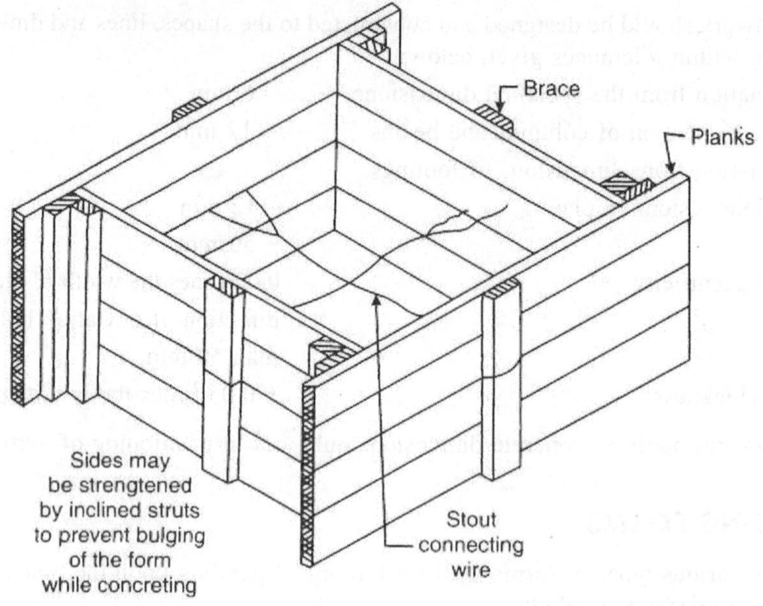

Fig. 21.2: A typical column footing form.

***(b) Stepped footing forms*:** These are of similar construction as described above, the largest and the lowest forms are placed first and the concrete is allowed to set a little before placing the next form. The upper form can then be held by placing a 50 mm × 100 mm wooden piece nailed to the top ends of the lower form.

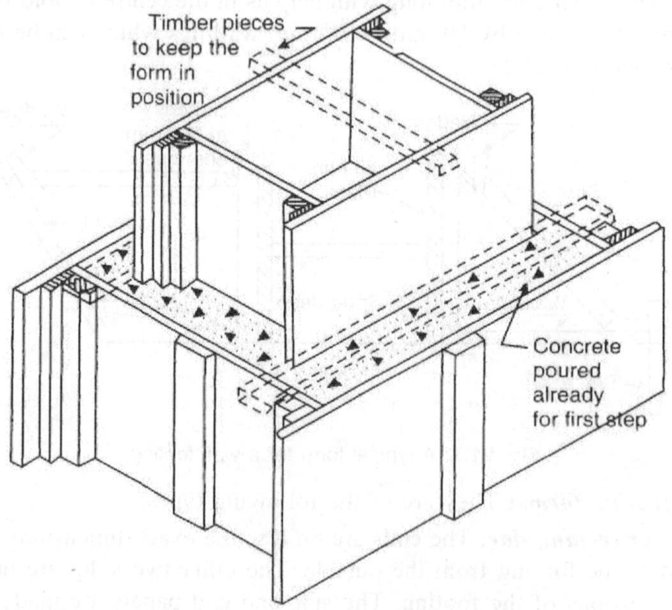

Fig. 21.3: A stepped column footing form.

*(c) **Sloping side footing form***: The two ends are cut to the exact dimensions of the footing, first laying out boards of approximately equal lengths, putting on the cleats, marking the sloping lines and then sawing off through the sloping line. The two sides should not be cut to the slopes but the board can be left of different lengths. The sloping box is always set on a very shallow vertical box of the same bottom dimensions. The vertical box is set on the ground and braced. The reinforcing steel is then placed and the sloping box set on the top of the first one. It is essential to prevent the lifting of the sloping sides due to concrete pressure. To prevent this uplift, wire can be embedded in the centre of the lower portion and fastened to the reinforcing steel or to any other heavily anchored member below. The other end of the wire is twisted around a piece of timber fitted to the top of the form. A few bags of sand, cement or heavy stones can be laid on the outside of the form over its sloping sides to prevent uplift.

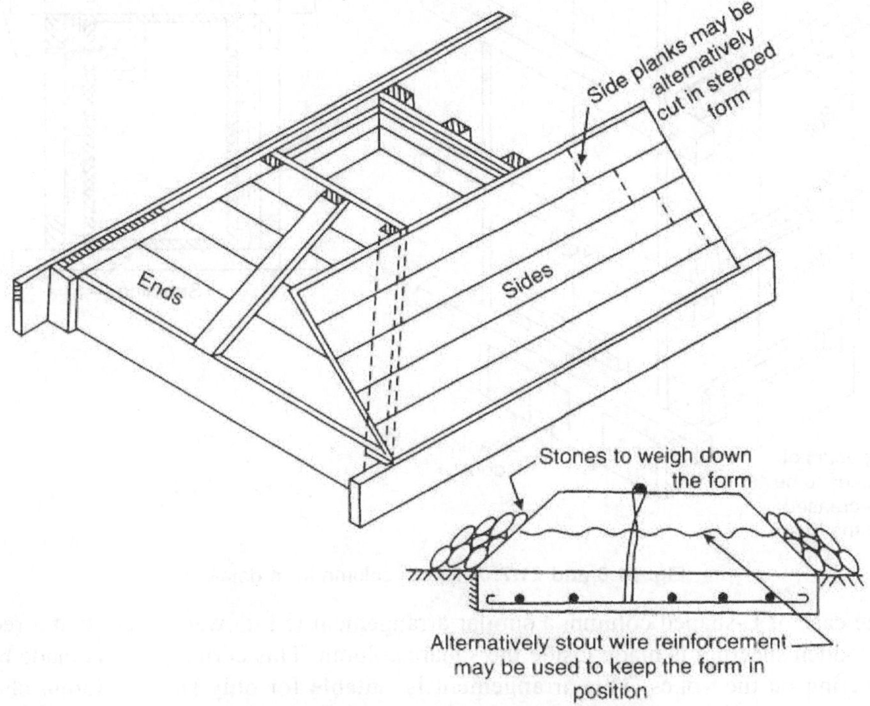

Fig. 21.4 and 21.5: Sloping footing form details.

21.4 COLUMN FORMS

A column form consists of two ends and two sides each built as a unit or panel, the height of the panel being the storey height less the slab thickness and the floor sheeting. The two ends are built in width equal to the dimension of the column plus twice the thickness of the sheeting, the ends of the yokes and sheeting being in flush. The two side panels are similarly made but the width of the sheeting will be the same as the dimensions of the column and the yokes will project 200 mm to 250 mm beyond the sheeting at each end. The spacing of yokes will be similar on both ends and sides. Bolt holes are bored in the side yokes about 3 mm larger than the size of the bolt. Their spacing is calculated from the size of the column and timber used allowing about 40 mm for wedging.

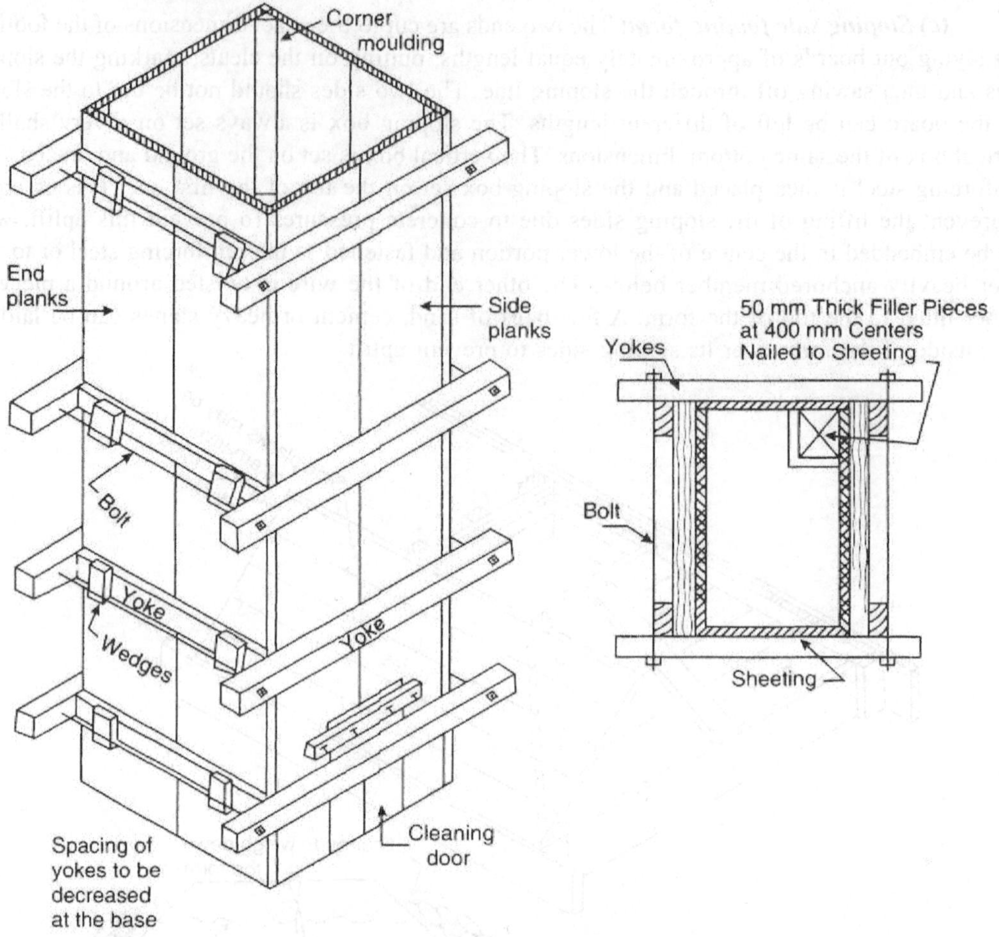

Fig. 21.6 and 21.7: A typical column form details.

In the case of L-shaped column, a similar arrangement is followed except that a rectangular block of wooden sheeting is made inside the square column. This corner piece is made by nailing timber sheeting on the yokes. This arrangement is suitable for only small columns. For bigger columns, the formwork itself is made into L-shape and bolts are fixed at the corner as shown in Fig. 21.8.

Wooden forms of round columns are used to a very less extent. The sheeting is cut into small widths so that it can be joined easily into a circular shape. The yokes have two different arrangements one of which is shown in Fig. 21.9.

21.5 WALL FORMS

The wall form consists of timber sheeting supported by vertical studs and horizontal wailers. The studs are first set to line and temporarily braced. The sheeting is wailed lightly breaking joints preferably at the studs. If the wall is narrow and heavy, reinforcement is to be placed; one side of the wall form is completed first and the other side placed in position after the reinforcement has been laid. Spacers are used at about one metre distance vertically and knocked out when the concreting reaches to that level. If the wall is wide enough for a man to work inside, both sides may be brought up together.

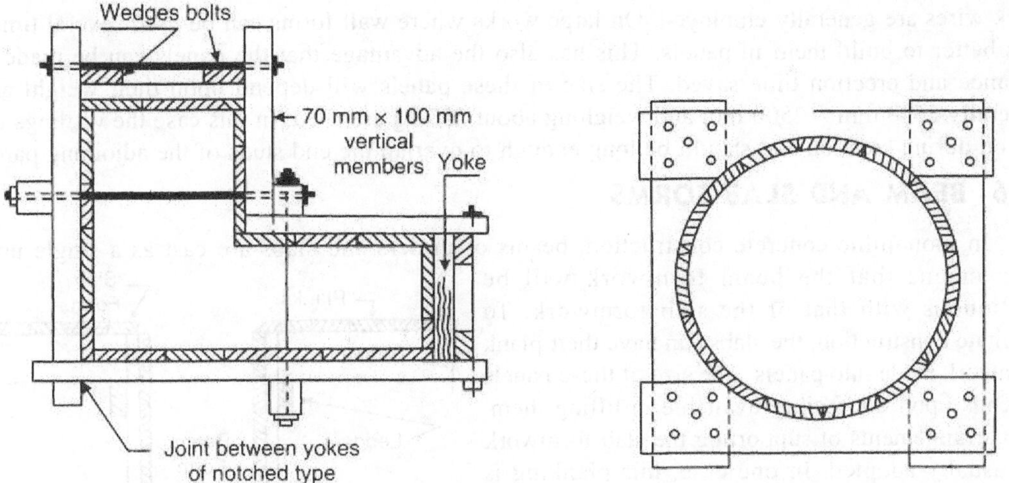

Fig. 21.8: A typical L-shaped column form.

Fig. 21.9: Form for a circular column.

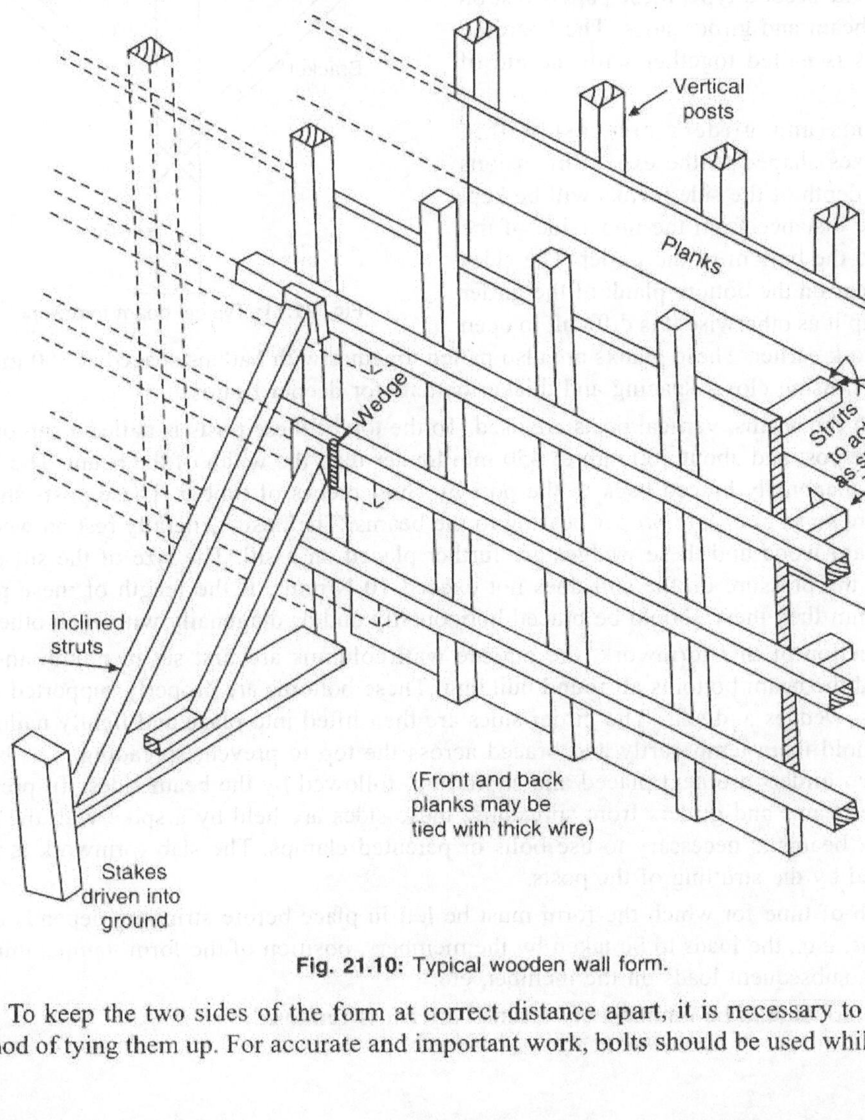

Fig. 21.10: Typical wooden wall form.

To keep the two sides of the form at correct distance apart, it is necessary to employ some method of tying them up. For accurate and important work, bolts should be used while for ordinary work wires are generally employed. On large works where wall forms can be used several times, it is better to build them in panels. This has also the advantage that the panels can be made in advance and erection time saved. The size of these panels will depend upon their weight and generally 2500 mm × 2500 mm and weighing about 250 kg are used. In this case the wailings are put on during erection and should be long enough to overlap the end studs of the adjoining panel.

21.6 BEAM AND SLAB FORMS

In monolithic concrete construction, beams or girders and slabs are cast as a single unit. This means that the beam formwork will be continuous with that of the slab formwork. To facilitate construction, the slabs can have their plank formwork made into panels. The size of these panels depends upon the facility available in lifting them. Two arrangements of supporting the slab formwork are usually adopted. In one case, this planking is supported by means of a wooden centering on the floor below. In the second type, these panels rest on the top of the beam and girder sides. The boarding of these panels is nailed together with the aid of battens.

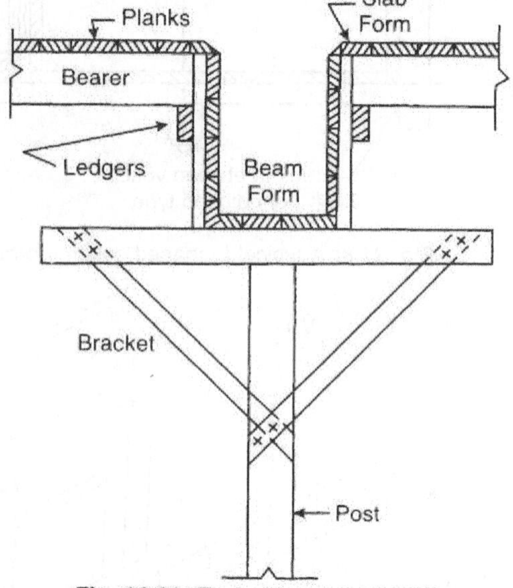

Fig. 21.11: Typical beam formwork

The beams and girders are cast within rectangular boxes shaped to the exact dimensions of the rib. The depth of the side planks will be kept the same as the distance from the underside of the slab sheeting to the bottom of the girder. The sides should never bear on the bottom plank of the girder but must overlap it as otherwise it is difficult to open the side formwork earlier. These planks are also nailed together with battens spaced at 300 mm to 450 mm centres, using closer spacing and thicker battens for deeper beams.

To support the beams, vertical posts are used. To the top of these posts is nailed a cap of the same size as the post and about 400 mm to 450 mm longer than the width of the beam. The ends of this cap are diagonally braced back to the post by short pieces of timber. These posts should be cut off square so as to ensure proper bearing to the beams. The posts generally rest on wedges which are of hard wood and these wedges are further placed on a sill. The size of the sill is so calculated that the pressure on the soil does not exceed 10 N/mm^2. If the length of these posts exceeds 2500 mm then these should be braced horizontally and/or diagonally with each other.

In the erection of this formwork, the outside wall columns are first set in plumb and on these are placed the beam bottoms all round building. These bottoms are properly supported with the aid of posts, wedges and sills. The girder sides are then lifted into place and lightly nailed to the bottom to hold them temporarily and braced across the top to prevent spreading. The beam bottoms between girders are next placed and shored up, followed by the beam sides. To prevent the sides of the beams and girders from spreading, these sides are held by a spike with the post cap. For deeper beams is necessary to use bolts or patented clamps. The slab formwork is next erected followed by the strutting of the posts.

The length of time for which the form must be left in place before stripping depends upon many conditions, *e.g.*, the loads to be taken by the members, position of the form, temperature of the atmosphere, subsequent loads on the member, etc.

All formwork should be immediately cleaned after it is removed.

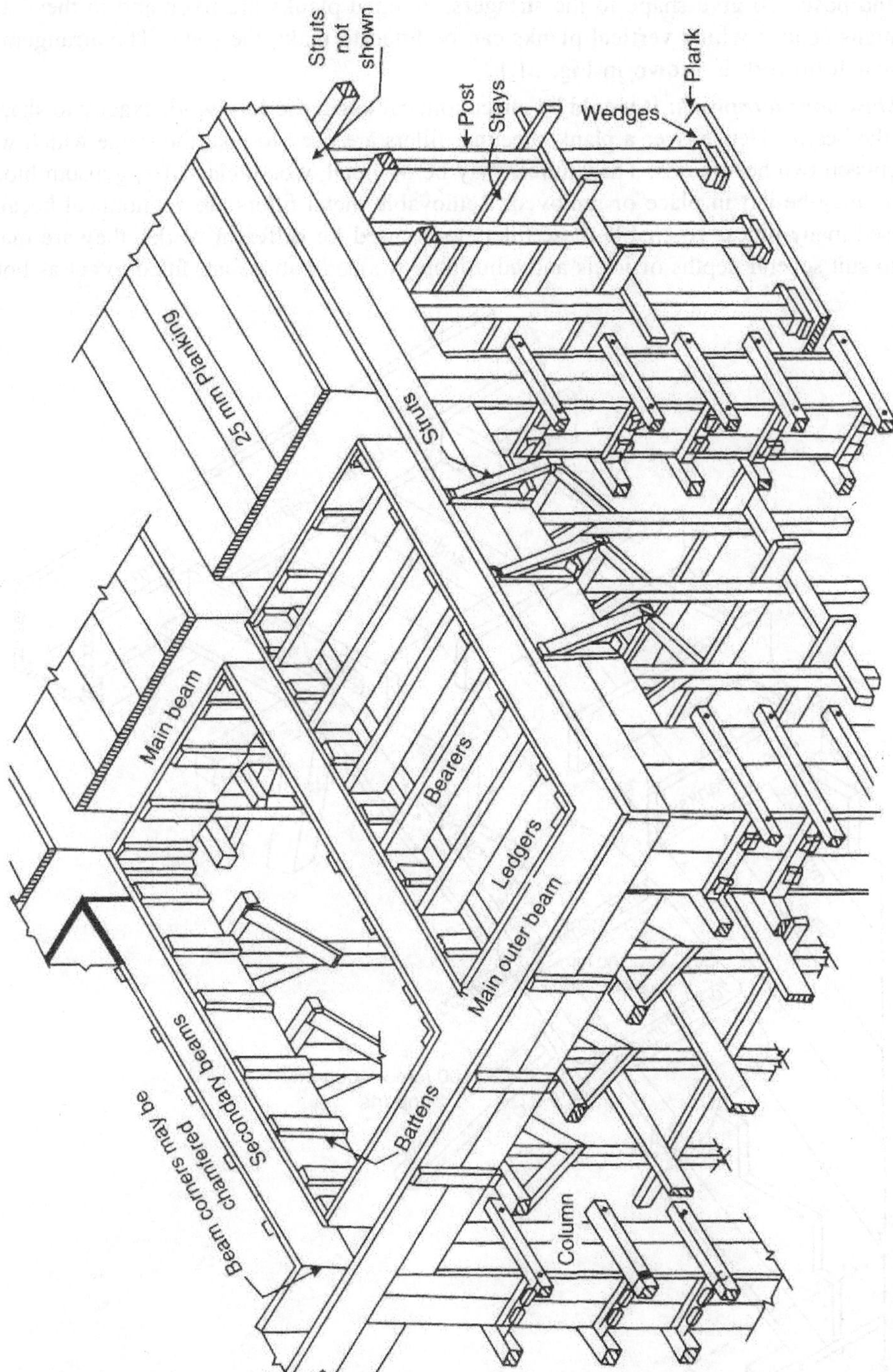

Fig. 21.12: A detailed layout of the beam and slab formwork.

21.7 OTHER TYPES OF FORMWORK

Stair forms: Stairs are designed to be self-supporting longitudinally from floor to landing or across wall to wall. The sloping slabs are supported by planking which is further supported on battens and posts. To give shape to the stringers, inclined planks are fixed and to these are fitted small battens against which vertical planks can be fitted to make the steps. The arrangement of a typical stair formwork is shown in Fig. 21.13.

***Ribbed floor formwork*:** It would be uneconomical to cut the formwork exactly to shape of the webs of the beams. Hence over a plank sheeting, fillers are used to take the space which would be left in between two beams ribs. These fillers may be of metal, wood, clay tiles, gypsum blocks, etc. The fillers may be left in place or removed. Removable metal fillers are economical because they can be used many times. To enable these fillers to be used for different works, they are made deep enough to suit several depths of joists and adjustable wooden battens are fitted to act as bottom.

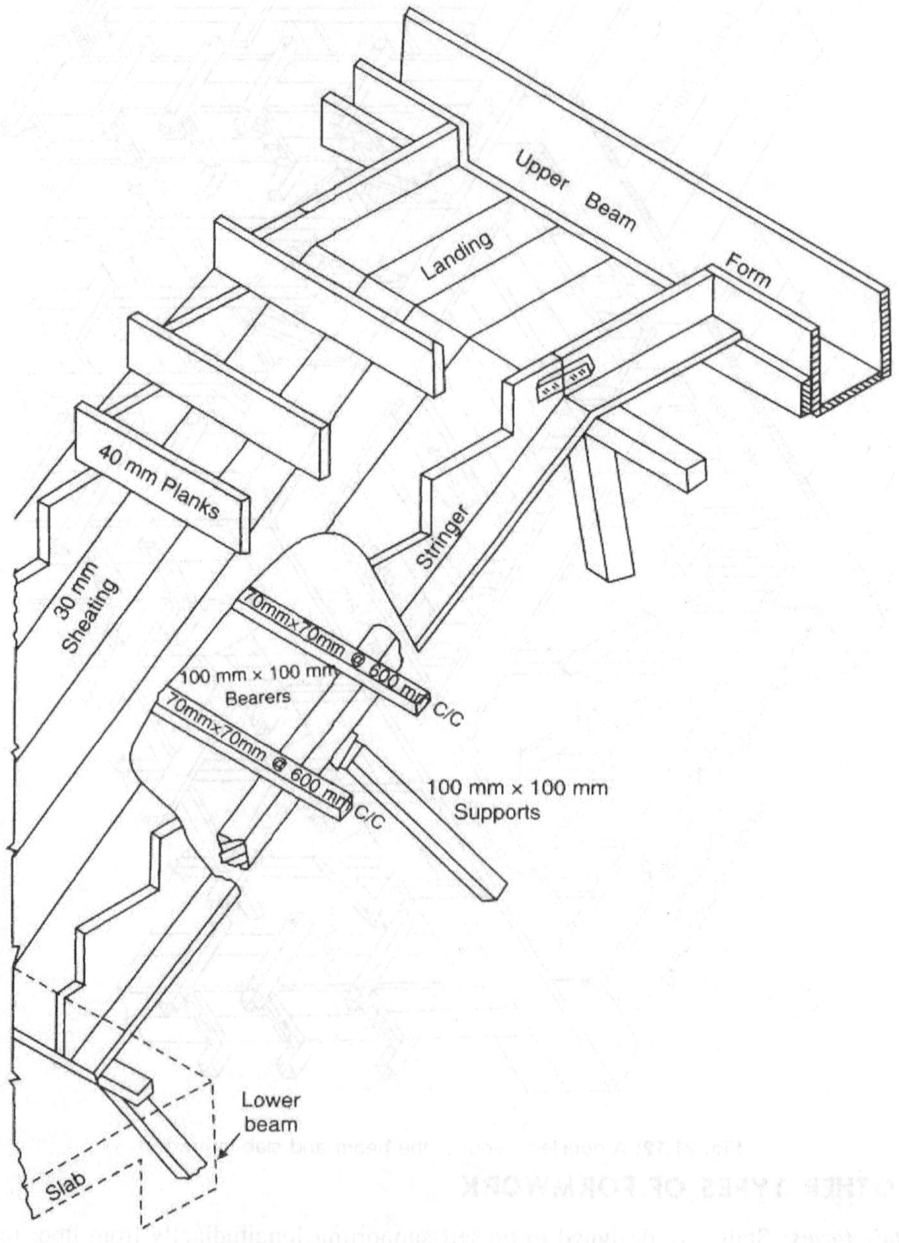

Fig. 21.13: A typical wooden form for a stair.

21.8 STEEL FORMWORK

The use of steel formwork is increasing in modern building construction. Wherever the work is large, these forms are economical and also possess the advantages specified earlier. Steel formworks are very useful for round columns and pile caps since very good shape can be obtained by their use. They are also used for flat slab construction. Slab panels are made up of 500 mm square plates with ribs. These need shoring from bottom as a support. Various systems of joists have been developed which will span the walls without any vertical support. The advantage is that no load is taken by the floor below and workmen can carry out their work without hindrance. Telescopic forms of these joists are becoming increasingly popular. Lattice girder made of very thin sections can also be used as supports. Light wall panels about 500 mm in size are made up of sheet metal with steel angle stiffners fastened together by clamps, wedges, etc. They are made in standard sizes so that any length of the wall can be made and adjustable pieces can be used for odd lengths. Two or three courses are used one above the other, concreting 1500 to 2000 mm on the first day and 500 to 1500 mm on succeeding days, the forms being stripped after 24 hours.

21.9 FORM LINING

These may be needed for the purpose of providing smooth surfaces and special markings on the concrete surface. Ordinary planks always leave ridges and fins and reproduce their grains even though the work is carried out with exceptional care. In order to minimize these defects and reduce the cost of finishing, smooth surfaces are obtained by the use of paper, sheet metal, plywood, hard-board as the linings. Water-proof paper of great toughness and durability is used and is available in rolls. Sheet metal of 22 or 23 gauge either galvanised or black with greased surface is used. Plywood is a very satisfactory type of lining and is available in several thicknesses. As plywood has got greater strength, it may be possible to space form boards at 50 mm to 200 mm apart depending upon the load so as to economise the cost of formwork. Grainless boards have been used recently and they have better adaptability than even plywood.

21.10 REMOVAL OF FORMS (STRIPPING TIME)

In normal circumstances (generally where temperatures are above 20°C), and where ordinary portland cement is used, forms may be struck after the expiry of following periods:

(*a*) Walls, columns and vertical faces of all structural members	24 to 48 hours, as may be decided by the engineer-in charge
(*b*) Slabs (props left under)	3 days
(*c*) Beam soffits (Props left under)	7 days
(*d*) Removal of props to slabs:	
(1) Spanning upto 4.5 m	7 days
(2) Spanning over 4.5 m	14 days
(*e*) Removal of props to beams and arches:	
(1) Spanning upto 6 m	14 days
(2) Spanning over 6 m	21

For rapid hardening cement 3/7 of the above period will be sufficient in all cases except vertical sides of slabs, beams and columns which should be retained for 24 hours.

After removal of formwork, all concrete is carefully inspected and any defective work or small defects are either removed or repaired before concrete has thoroughly hardened.

QUESTIONS

1. What are the essential requirements of a good formwork?
2. Give the details of the various types of formwork for footings. Explain your answer with sketches.
3. Write what you know about:
   (*a*) Column formwork.
   (*b*) Wall formwork.
   (*c*) Stair formwork.
   (*d*) Ribbed floor formwork.
4. Write short notes on:
   (*a*) Steel formwork.
   (*b*) Form linings.
   (*c*) Stripping time.

REFERENCES

1. **Wynn, A.E.**: *Formwork for Concrete Structures.*
2. **Mitchel, G.A. and Mitchell, A.M.**: *Building Construction*, Vol. II.
3. **Indian Concrete Journal**: *Various Issues.*
4. **Journal of the American Concrete Institute**: *Various Issues.*
5. **IS : 6461 (Part V)–1972**: *Formwork for concrete.*

Plastering and Pointing

Plastering is the method of covering various components of a building with a plastic material to form a durable surface. This material may have such a surface as to form a decorative finish or may be treated further. The object of external plastering is to cover the surface to enable it to resist the atmospheric influences, particularly the penetration of rain. The object of internal plastering is to provide a smooth surface in which dust and dirt cannot lodge. This surface is also not liable to be affected by vermin and forms a good surface for taking colour wash, distemper or a paint.

The prime consideration in plastering is that it should adhere to the background and should continue to do so while changes of weather frequently occur. It should also be possible to apply the plaster during all weather conditions. For economic reasons, the materials adopted and the methods of application should be cheap. The type of plaster to be used at a particular location will depend upon the rainfall, the weather conditions, presence of frost, the amount of finish that it will have to take and the appearance.

Plasters usually consist of a binding material, fine aggregates and water. In special cases certain additives are added to effect an improvement in the plasters as far as its adhesiveness, life etc., are concerned.

22.1 USED TERMS

Some of the terms used in plastering are given below:

Background

The surface to which the first coat of plaster is applied.

Blistering

This is the development of one or more local swellings on the finished plastered surfaces.

Cracking

This is the development of one or more fissures in the plaster usually but not always localised in relation to and arising from the movements in the background or the surrounding structure.

Crazing

This is the development of a series of hair cracks on the finished surface plastered usually in a haphazard manner.

Dado

The lower part of the plastered wall where the plaster treatment is varied from the upper portion more often to give a resistant finish.

Dots

Small patches of plaster laid on the background used in fixing screeds, etc.

Dubbing Out

This is the process of filling in hollow spaces in a solid background before the main body of the plaster is applied.

Finishing Coat

The layer in which plaster is laid over the whole area is called a coat. In three coats work, it is the final layer of plaster which is laid. It is also called the setting coat, face or skimming coat.

Flaking

The scaling away of patches of plaster due to the absence or failure of the adhesion with the previous coat.

Float

This is a tool used in plaster work normally consisting of a flat wooden board with a wooden handle grip. Cross-grained float is the float having grains of wood running across the face of its blade. A two handled float used for levelling surfaces is called a *Darby*. The float used for scratching the surfaces to form a key with the next coat has got a few nails protruding from its surface.

Gauging

The mixing of various constituents of plaster is called gauging.

Grinning

This is appearance on the surface of the plaster of the pattern of joints or similar patterns in the background.

Grounds

This term is applied to the wooden strips fixed to the background and to which primary finishing may be secured.

Hacking

The roughening of solid backgrounds to provide suitable key for plastering is called hacking.

Keys

These are the openings or indentations on the surface of under coat or background on which plastering material can be pressed in a manner that it will form a mechanical bond with the surface so prepared.

Laitance

This is the thin layer consisting of fine cement particles which form a screen on the surface of the fresh laid concrete when subjected to excessive trowelling.

Pattern Staining

This term is applied to the type of staining which occurs when the two sides of a composite structure are constantly exposed to different temperatures. The cause of such staining is the preferential deposition of suspended particles on the surface of those portions of the structure which on account of higher thermal conductivity are cooler.

Peeling

Dislodgment of plaster work form the background.

Under Coats

These are plaster coats, the main function of which is to provide for the application of succeeding coats. The first coat is applied to the building surface and is also called the rendering coat or the pricking up coat when it is applied to lath work. The second coat is applied to bring the first coat to a true and finish surface.

22.2 TOOLS USED IN PLASTERING

Laying trowel. This is used for spreading or laying material and for trowelling so as to get the desired finish. The blade is made of thin tempered steel and is of about 300 mm × 100 mm in size. For good work, two types of trowels are used. The first one with a stiff plate is used for applying the rendering and the second has got a thin blade which has a slightly springing action and is used exclusively for finishing coat.

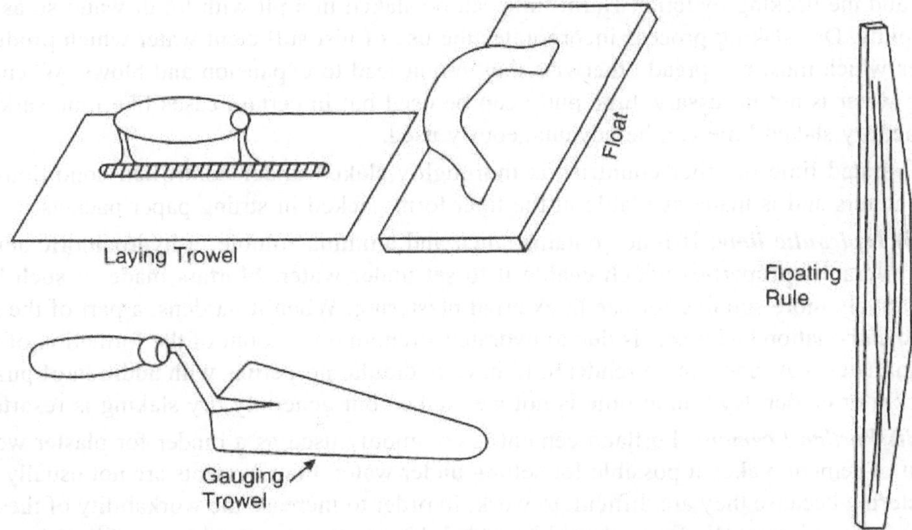

Figs. 22.1 and 22.4: Some of the tools used in plastering.

Skimming float. This is similar to the laying trowel but is made of wood. It is used for the finishing or the setting coat. The grains of the blade run length-wise. This varies in size from 300 mm × 100 mm to about 330 mm × 110 mm and is about 10 to 12 mm in thickness.

Devil float. This is similar to the above type except that it has got nails projecting by about 3 mm from the surface. It is used for scratching the plastered surfaces so as to form a key with the subsequent coat.

Felt float. A skimming float is covered with a layer of soft felt and this is used for securing plain surfaces.

Gauging trowel. This is used for gauging small quantities of materials on other tools for applying plasters to mouldings or to awkward places like corners, etc. This is available in sizes varying in length from 150 mm to 450 mm, the ends of the blades are pointed or bull-nosed.

The other types of tools which are commonly used are joint rules, straight edges, scratchers, brushes, spirit levels, plumb bob, set squares, etc.

22.3 MATERIALS USED FOR PLASTERING

(a) Binders

(i) Hydrated lime. When dry hydrated fat lime is mixed with water, it forms a solution of calcium hydroxide. When a mix in which lime is suspended is brought up to a high suction background, one part of the water is drawn away while remaining part, *i.e.*, the free water, can evaporate. The mix then falls short of water and part of the dissolved lime crystallises. This process of drying and crystallisation gives the lime mix its initial strength shortly after plastering. Crystals of calcium hydroxide are formed. Some part of the lime forms silicates in contact with silica contents of the plaster while other parts of lime form carbonates by absorbing carbon dioxide from the atmosphere. The carbonation is usually very slow and depends on the per cent of calcium hydroxide present. The porosity of the mix is an important factor. Thus in the case of lime mortar at early stages, a porous type can be carbonated greater than that of a denser mix. The effect of the relatively slow carbonation is more effective near the layers farthest away from the surface, *i.e.*, in contact with backing. At this layer the carbonate remains in a very weak condition for a long period after the plastering has been applied and this leads to weakening in adhesion between the plaster and the backing material. Burnt lime can be slaked in a pit with lot of water so as to form a lime putty. Dry slaking process incorporates the use of just sufficient water which produces dry particles which must be spread otherwise they might lead to expansion and blows. Whenever the binding water is not necessary, lime putty can be used but in certain cases like lime surkhi, sand and lime, dry slaked lime can be advantageously used.

Hydrated lime in other countries is thoroughly slaked under controlled conditions by the manufacturers and is made available in the final form packed in strong paper packets.

(ii) Hydraulic lime. If lime çontains silica and alumina soluble in hydrochloric acid it has certain hydraulic properties which enable it to set under water. Mortars made of such lime are comparatively more suitable for use in external plastering. When it hardens, a part of the strength is due to carbonation and a part is due to hydrated strength on account of the formation of silicates and aluminates. Fat lime can be rendered to have hydraulic properties with addition of puzzolanas like surkhi or cinder. Hydraulic lime is not wet slaked but generally dry slaking is resorted to.

(iii) Portland cement. Portland cement is commonly used as a binder for plaster work. The addition of cement makes it possible for setting under water. Pure cements are not usually suitable for plastering because they are difficult to work. In order to increase the workability of these mixes small amounts of hydraulic limes should be added. Up to 10 per cent addition of lime by weight is considered harmless to the strength of the mix. Improvement in the workability of cement plasters can also be made by adding air entraining agents. These might reduce the strength of the mix but they help to a great extent in the easy application of plastering. Water repellent agents can also be used to prevent the passage of moisture into the plaster.

(b) Sand

The only aggregate used in plasters is sand. This forms the greatest proportion of the plaster mix. The sand particles should be so graded that the binder can act as an adhesive and not merely as a filler. The object of sand is also to reduce the shrinkage of the binders. The porosity and strength of the plaster depend to a large extent on the quality of sand. The grading of sand and the workability of the plaster mix have a great effect on the adhesion of the plaster with the backing. Sand should be free from any organic matter, or harmful impurities. The maximum size of the particles should be limited to $\frac{1}{3}$ to $\frac{1}{5}$ of thickness of the plaster coat. The grading of sand normally recommended for different plaster coats is given in Table 22.1.

Table 22.1: Grading of sand for plaster work

SAND		% passing I.S. Sieve No.							
		IS Sieve	*4.75 mm*	*2.36 mm*	*1.18 mm*	*600 micron*	*300 micron*	*150 micron*	*Silt and Clay*
Plastering (Under coats)	Min.		100	98	80	30	5	0	Not more than 5%
	Max.		100	100	95	85	50	10	
Plastering (Finishing coats)	Min.		100	100	95	30	5	0	
	Max.		100	100	100	85	50	10	
External rendering	Min.		100	90	70	40	5	0	
	Max.		100	100	100	85	50	10	

(c) Surkhi

This is a material obtained by powdering burnt clay or bricks. The use of surkhi is mainly dependent upon the fine fractions consisting of secondary clay minerals and the presence of silica in such a state that their reaction with lime proceeds with adequate speed. The rate of strength if governed by the fineness of grading. Surkhi used in plaster should be all passing IS : 1.18 mm sieve and with at least 50 per cent of it passing through 2.36 mm sieve. The soluble contents of salts should not be greater than 2 per cent by weight.

(d) Cinder

Cinders or ashes containing lime give moderate hydraulic properties. They should be clean and must pass completely through an IS : 2.36 mm sieve. Excessive unburnt carbon particles make it impossible for the cinders to be used in mortar in a wet condition. In places where exposure to moisture is expected the combustible content of the cinders should be below 10 per cent by weight. In internal application the combustible content can be greater and may be as large as 20 per cent.

Other admixtures can be added to different mixes of plasters so as to increase the workability, to increase water penetrating quality or to increase the strength. Such mixtures or compounds are available in packed containers and can be readily mixed at site.

22.4 DESIGN CONSIDERATIONS FOR PLASTERING

The following features have to be kept in view while selecting a suitable plaster and applying the same.

(i) Combinations of Under-Coats and Finishing Coats, etc.

Under-coats and finishing coats should have a good relation with each other and also with the background as regards density, strength, liability to expansion or contraction during setting or afterwards. Bond failure between the successive coats may be caused by the defects in the background or under-coats, presence of unset films of plaster produced by premature drying, salt formation at interface, excessive relative moisture movement or combination of these.

For obtaining a good mix the plastering mix must have definite properties which make it suitable for particular conditions. The properties of different types of mixes are briefly reported below:

(a) Mix based on cement and lime or hydraulic lime for strength. Mixes containing cement, lime and sand have got high workability and therefore are easy to apply. They can be used reasonably after long working time. They have adequate early strength to withstand modern building conditions. For setting, they need moisture and rapid drying in the initial stages must be avoided. The drying and shrinkage of the first coat must be completed before the second coat is applied to avoid shrinkage cracks. When the cement quantity is increased the workability gets decreased correspondingly. Weaker mixes containing little of cement or hydraulic lime should not be used as an under-coat for a stronger finishing coat. They are useful for application to non-rigid backgrounds. Mixes of lime and cement are not suitable for trowelled finished coats as the shrinkage on drying creates a tendency to surface crazing.

(b) Mixes based on lime gauged with a gypsum plaster. These are workable and the working time depends on the type of gypsum plaster used. Gypsum plaster expands on setting and tends to limit the drying and shrinkage of lime. They are used for finishing coats.

(c) Mixes based on gypsum and anhydrite plasters. Basically they are divided into two main groups, *i.e.*, those which contain Plaster of Paris and retarded gypsum plaster and those containing anhydrated calcium sulphate plasters. The semi-hydrated plasters (Plaster of Paris) set quickly whereas the other type is slow setting. The setting of these plasters is accompanied by expansion which is variable in amount but when the setting is complete, no further movement takes place during drying. Hence it is not necessary to wait for long period so as to allow the undercoat to dry before the subsequent coats are applied, but they should be sufficiently strong.

No gypsum or anhydrous plaster should be permitted to remain under persistently damp condition after it has set as this causes weakness and disintegration. Most of these plasters can be used without admixture for the finishing coats, and when used in such form, expansion during setting becomes an important factor. They set to a hard strong finish which may exert considerable force on the undercoat or background which should on that account be strong. These types of plaster are relatively free from surface crazing.

(ii) Number of Coats

For best work, plastering should be applied in three coats. The first coat is responsible for creating the bond between the backing and plastering. Thickness suggested for this coat is about 3 mm. The mortar for this coat should be thrown on the surface and should not be worked but left to rest. A cement mix of 1 : 2 proportions is suitable. The second coat is intended to fill up the irregularities on the back of the wall. A thickness of about 10 mm is sufficient for this coat. This surface should be levelled with a trowel. The third coat is usually the skimming coat and mainly serves to give a good appearance.

Wood lath and metal lathing require three coats for a satisfactory finish. Brick-work and monolithic concrete work if executed carefully will normally be sufficiently good to allow the use of a single undercoat.

(iii) Type of Surface Finish

The usual practice is to produce the smooth finish for internal plaster work but with special techniques a variety of textures can be created. Texture finishing can be obtained in the plaster itself or by application of a textured paint. Sand finishing or other type of finishing can be given to the external plaster.

(iv) Resistance to Abrasion

As far as the hardness is concerned, cement finishes have the greatest hardness. These are followed by anhydrite plasters, anhydrous gypsum plasters, retarded semihydrite gypsum plasters and lime plasters in order of decreasing hardness. Addition of lime decreases the hardness, while gypsum or cement whenever added to the lime plaster increases their hardness. Resistance to abrasion is directly related to the relative hardness as described above.

(v) Suitability for Further Decoration

The characteristics of the plaster affect the final finish to be done. The degree of alkalinity of the plaster, the dryness and its background at the time of painting are the factors which have to be considered from this point of view.

(vi) Corrosive Effect on Metals

The corrosive action should only occur during the initial drying periods and subsequently during the periods of heavy condensation. Plaster containing uncarbonated lime, for example, lime and cement mixes and gypsum plasters containing a substantial admixture of lime have a protective effect on iron and steel but are likely to corrode lead when they are damp for longer periods. Gypsum plasters may have an acid reaction, if they contain salts which are added to accelerate their setting. Suitable metallic or paint coatings or other types of precautions may be given to steel to prevent corrosion.

(vii) Effect of Atmospheric Conditions

Plastering should not be continued during frosty weather. Suitable precautions should be taken if the weather is very cold.

In certain localities condensation may be excessive and that may spoil the finish of the plaster work. This is specially true for gypsum and anhydrate plasters. Under hot-dry conditions the applied plaster becomes dry before the setting has taken place completely. This will be true in all anhydrous calcium sulphate plasters or cement containing rapid hardening mixture.

(viii) Effect of Vermin

This liability of plaster to vermin attack is dependent on the amount of cracking present. Freedom from cracking or serious crazing apart from that created by the structural movement can be avoided by the use of good workmanship.

(ix) Fire Resistance

Application of internal or external plasters can increase the fire resistance of all structural components. The effect of plastering a one brick wall is not of much value because without plastering this wall has a good resistance. Plaster finishes on thinner walls or partitions increase their fire-resistance qualities considerably.

(x) Accoustical Properties

The effect of the plaster finish on the sound absorption is not much. Internal plaster finish of special types may be used to increase the sound absorption.

22.5 BACKGROUND FOR PLASTERING

The durability of plastering depends not only on the properties of the mixture itself but also its adhesion with the background. A good background must be plain enough for suitable application of plaster and should have enough strength. The usual types of backgrounds that are used for plaster work are:

(a) Solid backgrounds: These include brick-work, heavy clay blocks, in-situ concrete, concrete blocks, etc.

(b) Lathing: These include wood laths, expanded metal lathing, wire meshes, etc.

***(c) Boards and slabs in non-mortar construction*:** Under this, slabs of gypsum plaster, fibre boards, wood-wool slabs, etc., are included.

The main characteristic of the solid background is that the nature of the surface presented for plastering is taken care of primarily by conditions other than those of plastering. Care is needed in such a case for proper preparation. The normal method of construction of solid background introduces varying amounts of water and as a result needs adequate drying before the application of plaster is given. The amount of water contained in concrete placed *in situ* is similar to that of brick-work and similar precautions must be taken in that case also.

Local projections are more serious than depressions and these should be broken before plastering. Depression can be dubbed out if necessary. For three coat plastering the projections should not exceed 10 mm above the general surface and the depressions may not be greater than 20 mm. For two coat plastering they should be 5 mm less than these values. The efflorescence of the bricks should be removed before plastering by dry brushing. In the case of concrete either excessively smooth surface is left by shuttering or imperfections on account of shuttering occur or greasing defects are created. Whenever concrete surfaces are to be plastered, rough sawn shuttering boards should be used. Mechanical keys may be applied to the inside face of the shuttering so as to help in the bonding of the plaster later on. Alternatively metal mesh be left adhering to the concrete. The addition of hair to the first undercoat of the plaster may help in obtaining a good bond.

Lathing materials provide a strong mechanical key for the application with the undercoat. Wood laths are not suitable for cement sand finishes because of the risk of shrinkage. Boards and slabs have the advantage that a uniform thinner coat of plaster can be given. They have got relatively plain surfaces which give good results with two coats or a single coat plastering. Special treatments may be needed for soft type of boards so that plaster adheres to them properly.

22.6 APPLICATION OF THE PLASTER

It is preferable to apply plaster mix by throwing it with great force against the walls. As this job involves great effort, the mortar is sometimes applied in practice by pressing it thereon. In the latter case adhesion of the plasters is lesser and the plaster may also be more porous. The applied plaster must be levelled and smoothened. This is done by drawing a screed across the surface after the application of mix. When it is a backing or an undercoat the surface is scratched with a suitable trowel after the mix has hardened to a certain extent so as to get a good bonding with the coat to be applied later on. The plaster should be finished with wooden floats before it had started setting. If it is done too early the binder gets drawn, thereby resulting in the cracking of the plaster. The plastered surface is to be kept moist after application when the mix is of hydraulic type.

Slow setting under higher humidity improves the strength of the plaster.

The processes of laying the plasters of different mixes are briefly reported below:

22.6.1 Lime Plastering

The process for laying the lime plasters includes the selection of the suitable mix and application of various coats. The slaking of the lime should be carried out in a large vessel in which stirring can be done fairly well. This vessel is partly filled with water to a depth of 300 mm and the lime added to cover the bottom of the vessel to a uniform layer of about 150 mm. Stirring should be commenced immediately and care should be taken that no lime gets exposed above the water surface. As the mix thickens more water should be added. The resulting milk of lime is then run through I.S. sieve 600 micron into a suitable container. It should be allowed to mature for a period of about 2 weeks and will flatten up to lime putty. It should not be allowed to dry out during this time. This operation is needed for quick lime only. In the case of hydrated lime, it should be mixed with water and stirred to a thick creamy consistency. This should remain undisturbed for a period not less than 16 hours and should be prevented from drying out.

The materials should be in the proportions of one part of cement, one part of lime, six parts of sand or one part of cement, two lime and 9 parts of sand depending on the degree of workability, hydraulic strength and rapidness of hardening desired. For greater workabilities greater quantities of lime are used. For the first undercoat, the coarse stuff should be prepared by mixing one volume of lime putty with six volumes of sand. Water should be added to give stiff consistency and the mix should be allowed to stand for some time and prevented from drying. Immediately before it is used it should be thoroughly mixed with cement in the given proportions. All the mixed material should be used within two hours of the addition of cement. For application on lath work hair may be added in the proportion of 4 kg to one cubic metre of the coarse mix. In the case of hydrated lime-cement mix, initial mixing dry with sand in the chosen proportions, either manually or in a mechanical mixer may be done. Sufficient water should be added to make the mix workable.

The corase stuff should be stiff enough to cling and hold when laid and should be more stiff for ceilings than for walls. On walls this material should be laid in long spreads upward and across overlapping. Sufficient pressure should be applied to force it into intimate contact with the background. The material should be laid as uniformly as possible. It should be scratched after it is firm. Before applying the second undercoat the first undercoat should be cleaned and may be washed by passing a damp brush over the surface. The necessary screeds should then be formed and the intervening spaces floated in by coarse stuff till it is flushed with the screeds. It is then ruled thoroughly with a floating rule. When the surface is true it is scratched so as to form a bond with the third coat. Adequate time is left before the third coat is given.

The finishing coat may be made from non-hydraulic or some hydraulic lime prepared as described earlier. The mix should be of one part of lime putty, one part of sand, with $\frac{1}{4}$ part of suitable anhydrous gypsum plaster. It should be used within thirty minutes of the addition of the gypsum plaster. For smaller areas Plaster of Paris may be substituted, but in that case the plaster should be applied quickly. Alternatively two parts of lime putty may be mixed with three parts of sand and thus used along with or without gypsum plaster addition. Normally two applications of finishing coats are applied, the first application is laid leaving a coarse surface to receive the second and final application.

22.6.2 Cement Plastering

Plastering mix containing cement and sand may be used without the addition of lime. Whenever damp conditions are prevalent maximum resistance to abrasion is required, a cement sand mix should be employed. Water repellent agents may be added if excessive dampness is occurring. Mixes containing sand and cement only have a dry early strength and the drying and shrinkage sets up a considerable stress both in the applied coat and in the background. For this reason the background is to be specially rigid and must be of enough roughness. These types of plasters are suitable for application on solid background such as brickwork and to suitably finished concrete surfaces. They are not suitable on semi-rigid background.

It may be preferable to use a cement lime mix of suitable proportion to improve the working qualities of the plaster.

Coloured finish is obtained by the addition of white or coloured cement combined with coloured and graded sand. Before the application of cement plaster coats on the surface of the background it should be cleaned with water in such a manner that a slight amount of moisture remains on the surface. The first undercoat usually of the proportion 1 : 3 cement and sand should be stiff enough to hold to the wall or the ceiling. On soffits the mixture should be laid in long even spread outward from the mason overlapping each trowelful and using sufficient pressure to force it into key with the background. On walls the mix should be laid in long even spreads upward and across. The material should be laid in uniform thickness as far as possible, the average thickness not being greater than 20 mm. This coat should be allowed to stand till scratching for keying with the next coat is done.

For getting a proper thickness of the plaster and a level surface it is desirable to take suitable precautions. At places about 150 mm below the ceiling and about 150 mm from the edge or angle of the wall a small pat of the mix is laid to form a sort of pyramid about 100 mm square. A piece of wood or lath is pressed firmly to this so as to form the first point from which the other surface of the plaster can be checked.

The thickness of this pat, which is levelled, is normally about 10 to 20 mm so that it clears all the protruding surfaces of the walls. With the aid of plumb bob a dot (a small level plastered area) can be made correctly at the lower end. When these dots have been correctly finished with respect to each other, another pair of similar dots is made on the other end of the wall. In between these two series of dots plaster can be laid with the aid of strings or otherwise. The distance between the intermediate dots should be less than the length of the checking straight edge so that the surface of the plaster gets checked completely later on.

The next process is the forming of screeds. The material is laid vertically between the dots in strips which are about 70 to 100 mm in width and carefully levelled off with the floats to the thickness of the dots. The wall is now divided by screeds into a series of vertical bays. When the screeds are hard enough to withstand pressure at the surface the spaces on the walls are refilled with plaster and finished to a uniform level and true surface. Similarly careful arrangements have to be made for getting a proper thickness of the surface whenever there are excessive projections.

Before applying the second undercoat, the first coat is swept clean of any dust or loose particles. Screeds may again be formed and material laid in a similar manner. The average thickness of the second undercoat should not exceed 10 mm. As the surface is formed it is keyed by the use of comb to get a regular pattern of wavy lines.

The proportion of the finishing coat may be similar to that of undercoats. Coloured or white cement may however be used only in the finishing coats. The grading of the aggregate should be appropriate for the type of finish desired. No lime is added to mix containing coloured cement. The proportion of water must be constant for each batch and no additional water is added after initial mixing. The mix is applied with a wood skimming float and afterwards finished to a true and even surface. The surface is then finished with an under float and depressions are filled in. Overworking will tend to bring too much fine material to the surface and should be avoided to reduce the tendency of surface crazing.

22.6.3 Plaster Applied to Metal Lathing

Metal lathing is a term used to cover a wide variety of materials available in sheet form. They are classified as (*a*) Plain expanded, (*b*) Ribbed expanded with ribs forming an integral part of the sheets or attached subsequently, (*c*) Perforated, (*d*) Dove-tailed.

Lathing which is commonly used in the domestic building is "plain expanded type". This type is used mainly for ceiling construction or partition work, by fixing it to timber supports or sometimes to steelwork. The following points are to be noted for fixing the metal lathing while being used as a base for plasters:

(*a*) The lathing should be fixed and stretched tightly as much as possible.

(*b*) The correct gauge should be used in relation to the distances at which supports are spaced. For plain expanded metal lathing the weight should not be less than 15 kg per sq m over normal supports spaced at 350 mm. The mesh of the lathing should be 5 mm to 10 mm (measured in the shorter direction).

(*c*) The long dimensions of the sheets should span across the bearers.

(*d*) In fixing directly to timber supports suitable strips of hard wood should be fixed between the lathing and the supports to ensure continuity of the plaster key behind the lathing.

(*e*) All sheets must be overlapped not less than 25 mm at the sides and the ends. Over-laps should always occur on solid supports.

(*f*) The sides of the sheets must be wired together to come between supports, with galvanised wire.

(*g*) When metal lathing is applied to timber supports it should be fixed by galvanised nails or screws driven across the grain of the timber at about 100 mm centres.

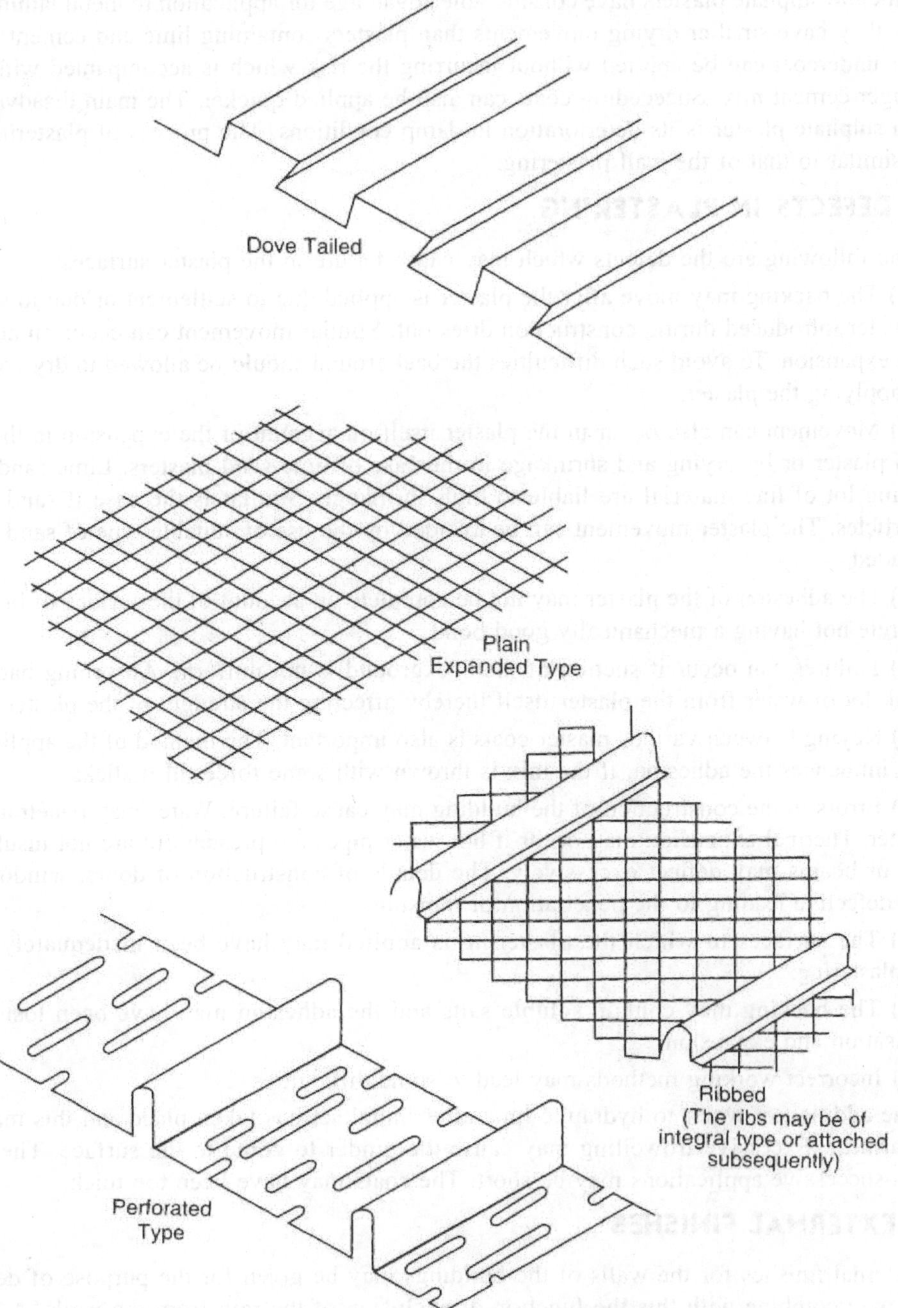

Fig. 22.5 to 22.8: Various types of metal laths.

Two main types of plastering may be used on metal lathing that is those having undercoats containing cement lime and those containing calcium sulphate plaster. Generally two coats are essential for getting a good finish. The undercoat mix adopted for metal lathing is 1 : 2 : 9 of cement, lime and sand mix. Hair may be added to help in the application to the extent of 4 kg per cubic metre. For plastering on metal lathing it is not necessary to use a stronger mix than 1 : 1 : 6 of cement, lime and sand.

Calcium sulphate plasters have considerable advantage for application to metal lathing chiefly because they have smaller drying movements than plasters containing lime and cement. Hence a stronger undercoat can be applied without incurring the risk which is accompanied with the use of stronger cement mix. Succeeding coats can also be applied quicker. The main disadvantage of calcium sulphate plaster is its deterioration in damp conditions. The process of plastering on the lath is similar to that of the wall plastering.

22.7 DEFECTS IN PLASTERING

The following are the defects which may cause failure in the plaster surfaces:

(1) The backing may move after the plaster is applied due to settlement or due to shrinkage as the water introduced during construction dries out. Similar movement can occur on account of thermal expansion. To avoid such difficulties the background should be allowed to dry completely before applying the plaster.

(2) Movement can also occur in the plaster itself on account of the expansion in the case of gypsum plaster or by drying and shrinkage in the case of lime sand plasters. Lime sand plasters containing lot of fine material are liable to high shrinkage. Similar is the case if sand contains clay particles. The plaster movement can be avoided by the use of suitable type of sand which is well graded.

(3) The adhesion of the plaster may not be complete on account of the surface of brick-work or concrete not having a mechanically good bond.

(4) Failures can occur if suction of the background is not uniform. Absorbing background will suck lot of water from the plaster itself thereby affecting the strength of the plaster.

(5) Keying between various plaster coats is also important. The method of the application of the mix influences the adhesion, if the mix is thrown with some force till it sticks.

(6) Errors in the construction of the building may cause failure. Water may penetrate behind the plaster. Thermal expansion may occur if hot water pipes are present but are not insulted. The column or beams may deflect excessively. The details of construction of doors, windows, etc., may be defective leading to the penetration of moisture.

(7) The surfaces to which the plastering is applied may have been inadequately cleaned during plastering.

(8) The backing may contain soluble salts and the adhesion may have been lost by their crystallisation and expansion.

(9) Incorrect working methods may lead to some difficulties.

The addition of water to hydraulic lime after initial set has taken place and this may prove to be harmful. Excessive trowelling may cause the binder to come to the surface. The interval between successive applications may be short. The coats may have been too thick.

22.8 EXTERNAL FINISHES

External finishes for the walls of the buildings may be given for the purpose of decoration only or may combine with this the function of exclusion of the rain from the walls. A rendered finish to an external wall is cheaper and can be used over variety of structural materials which may not themselves be suitable for exposure. In making the decision for the selection of a suitable finish, its specifications and colour, the following factors have to be kept in view:

(1) Appearance required.

(2) Maintenance necessary for giving the satisfactory appearances.

(3) Degree of protection against rain penetration to be provided by the finish.

(4) Severity of exposure to atmospheric and climatic agencies affecting durability.

(5) Time of year during which the finish is applied.

(6) Environment and use of the building.

(7) Background materials on which the finish is to be applied; and

(8) Cost.

22.8.1 Types of Finishes

The types of finishing treatments usually adopted for external surfaces are the following:

***(a) Pebble-dash or dry-dash*:** This is the finish in which small pebbles or crushed stones of suitable sizes are thrown on to a freshly applied final coat of mortar and left exposed. The pebbles are sometimes lightly pressed or tapped into the mortar after throwing. This is not suitable for weaker types of bricks or light weight concrete. It can be suitable for rural areas as far as aesthetic appearance is concerned. It can resist good amount of aesthetic deterioration.

***(b) Rough cast*:** This is a finish in which the final coat containing a proportion of fairly coarse aggregate is thrown as a wet mix and is left in a wet condition. The coarseness of the texture depends mainly upon the type and size of the coarse aggregate.

***(c) Scraped finishes*:** In this type of finish, the final coat of mortar, after being levelled and allowed to stiffen, is scraped with a steel straight edge, or a board studded with nails or any other form of tool convenient for this purpose, so as to give a rough surface.

***(d) Textured finishes*:** With the aid of suitable tools ribbed stucco or fan textures can be created in the final coat.

***(e) Smooth finish*:** This type of finish has got a level and a smooth surface. The wooden float is normally used as the steel float gives surface much more liable to craze.

***(f) Machine applied finishes*:** A variety of finishes in which the final coat is applied by machine which throws or spatters the material on the wall. The machines may be manually operated or mechanically operated and a sort of gun is used to throw the material.

As far as the aesthetic values are concerned the choice of the type or colour will depend upon the environment within which the building is to fit in. Smooth finishes have a defect that the surface becomes crazy and has patchy appearance on account of variation in the surface structure due to floating. Lime dissolved by the mixing water either from cement or lime in the mix gets deposited to some extent on the outer surface as the water dries out, and forms calcium carbonate which tends to cover the true colour of the finish. This deposit may be more for some areas than others and thus gives a patchy appearance. Such effect is marked under slow drying conditions than finishes applied in warm dry weather. Pebble dash and rough cast are least liable to change in appearance over long periods. Deposits of dirt from atmospheric pollution are more visible on white and pale colours than on natural grey colour. Dirt also appears badly on heavily textured finishes than on white textured and scraped finish.

An external rendering, unless it is painted, should not require any maintenance over a long period of time. If a light-coloured or white-coloured finish is required it may be necessary to provide painting periodically. A rendered finish is often required to give protection against rain penetration through a wall. For this reason cracks in the rendering must be avoided as much as possible. Pebble dash or rough cast finishes are most suitable for such conditions. If the undercoat is resistant, other types of finishes may be used. Eaves protection greatly reduces the amount of rain falling on the wall. Similarly sills made with effective drips may throw away water clearly from the wall and thereby weaker types of rendering can be adopted.

The type of background on which the rendering is to be applied depends on the choice of the material to be used. For dense and smooth background mechanical keying is essential. For these surfaces spatter dash treatment can be given or a light mesh be secured to the walls for additional adhesion. In moderately strong materials like bricks, the joint should be irregular unless a spatter-dash treatment is to be used. Cement, lime and sand mix are most suitable in rendering coats on these materials. Strong and dense type of mixtures such as 1 : 3 cement sand mix may not exclude rain water from the wall although it may be very impermeable as it is liable to crack. A crack in a dense rendering once formed may allow more water than the one in a porous rendering since rain water runs down impervious surface into the crack. High alumina cement mix may sometimes be useful for greater resistance to the action of the sulphates.

22.8.2 Materials for External Finishes

The grading of the aggregates may be same as for the ordinary plasters except that whenever rough cast plaster is to be used, fine grades may be added. For pebble dash finishes pea shingle, crushed hard limestone or marble chippings or other crushed stone of suitable appearance in a 5 mm to 10 mm grading may be used. Texture finishes require fine sand than the rough finishes. Fine sand is also advantageous for pebble dash as it holds the dry pebble better when thrown. Binders may be lime or cement containing suitable pigments in suitable proportions. The total quantity of the binding agent in the rendered finishes is governed partly by the requirement that the amount of such material shall be just sufficient to fill in the voids between the coarse sand particles. For normal sand the amount of fine material required to give the maximum density is about 30 to 40 per cent. A normal mix contains one part of cement binder to $2\frac{1}{2}$ to 3 parts of sand or other aggregates. If it is made richer then there is a risk of cracking. The more leaner the mix the more difficult it is to apply. The function of the cement is to provide strength and durability. A mix of 1 : 3 cement sand is more than necessary. The use of this mix particularly on rigid backing of lower strength is likely to give rise to cracking. The use of this mix should therefore be restricted to conditions where the background is strong and good adhesion can be obtained and also where a very dense and impermeable mix is essential. In pebble dash also where plasticity for holding the pebbles first thrown and retaining them is very essential, a 1 : 3 mix is suitable. For all other work mixes containing lower cement content are satisfactory and addition of lime may improve the workability. Mixes containing portland cement, lime and sand in the proportions 1 : 1 : 5 to 6 by volume are suitable.

22.8.3 Application of External Plasters

Scaffolding may be needed for applying the external rendering because of the greater height involved. Raking of the mortar joints in all cases is essential unless a spatter dash treatment is to be given. Similarly mechanical keying may have to be given for various surfaces. Old-brick-work or poured concrete should be thoroughly wire brushed and washed down, before any rendering is applied.

Spatter dash treatments are applied by throwing, either from a trowel or a scoop coarse materials on to the surface. The mixes used should be made of suitable consistency and the workability should be varied according to the degree of the suction of the backing material. The surfaces should be covered with a thin layer and no attempt is to be made to level or smoothen the coating.

Undercoats other than spatter dash may be applied by throwing the mix or by laying it over the trowel. The thickness of these courses varies on account of the uneveness of the wall but preferably should not be greater than 15 mm or less than 3 mm. When the undercoat has started to harden, it should be combed or scratched to a depth of 3 mm to 5 mm except when finishing treatment is done with machines. The undercoat should be allowed to dry as long as possible before any further coat is applied.

Finishing treatments may be laid depending on the texture, *e.g.*, pebble dash, scraped and textured finishes. The final coat may be laid or thrown while for machine finishes, automatic plastering guns may be used. The thicknesses of these coats vary between 5 mm to 10 mm. Some machine finishes can be as thin as 3 mm. The finishing on the outside should be carried out continuously without any break. Suitable curing must be carried out to prevent rapid hardening of the binder and cracking.

POINTING

Pointing should be avoided as much as possible as this operation involves raking out of the joints good mortar. A weaker mortar replaces it as the latter cannot be worked into the joints properly. The main purpose of pointing is to give a good appearance to the brickwork or stone masonry. It also serves the additional purpose of preventing the ingress of moisture into the wall if the pointing is done carefully. For these reasons all types of faceworks are pointed.

Pointing should preferably be done when the mortar in the joints is comparatively "green" and fresh. Before pointing any brickwork, all the joints should be raked out to a depth of about 10 mm. For good work the joints should be struck after each day's work. The dust of mortar and other dirty matter is cleaned, the surface is well-wetted and washed with water. This washing should progress with the pointing.

The joints of the pointed work must be clearly defined, the lines being regular and uniform in breadth. It is not desirable to cover the brick edges with mortar so as to hide defects in brickwork.

The surface of the pointed work may be kept flush with the brickwork. The mortar will be filled and pressed into the joints with a pointing trowel and finished off level with the edges of the bricks to give a smooth surface.

For deep pointing also called struck pointing, the mortar shall be filled in the joints flush with the masonry or brickwork with a pointing trowel and then pressed in with proper pointing tools. The pointing tool for horizontal joints may have a suitable shape so as to give weathered or struck joints as the case may be. For vertical joints the pointing tool may be triangular in shape so as to form "Vee" joints. While using the tools, care should be taken to press the mortar and not to cut it away.

For ruled pointing, the mortar is filled, and pressed into the joints with a pointing trowel and finished off level with the edge of the bricks. It is then ruled along the centre of all joints with a half round tool, 10 mm wide.

The other variety of pointing is the "tuck" type. In this case the joint between the bricks is filled with mortar, may be of a different colour. A narrow ridge with raised parallel edges of about 4 mm in thickness is made in white colour. It fits into the backing joint, as a recess is made in that to receive the former.

For lime pointing, slaked lime is mixed with two parts of surkhi and kept under water for a period of 12 hours. This is then passed through a screen of 6 meshes to 10 mm. In exceptional cases, colouring matter may have to be added to match the colour of the bricks.

Whenever pointing is liable to get affected by dampness or salt action, cement pointing is used. A mix of 1 : 3 cement sand is used.

After pointing, the face of the wall should be cleared of all surplus matter adhering to it. No washing is done till the pointing has set. Lime pointed work is kept wet for 5 days and cement pointed work for 10 days, after the completion of the job. It should be protected during this period from the extreme changes in the weather.

QUESTIONS

1. What are the objectives of plastering? Define the following items:
   Background, blistering, cracking, crazing, dubbing out, finishing coat, flaking, float, gauging, grinning, raking, peeling, undercoat.
2. Describe with the help of sketches the tools used for plaster work.
3. What are the materials used in plastering? Discuss their suitability.
4. Discuss, in detail, the design considerations for plaster work.
5. Describe, briefly, the following:
   (*a*) Lime plastering.
   (*b*) Cement plastering.
   (*c*) Plastering on lathing.
6. State the different types of plasters adopted for internal and external finishes of buildings. What do you understand by water-proof plaster?
   What types of plaster would you adopt for
   (*a*) an ordinary residential building of a middle class family,
   (*b*) a big insurance building, and
   (*c*) a modern cinema house?
7. (*a*) Why is plastering done is more than one coat?
   (*b*) What do you understand by the term 'pointing' and how is it done?
8. What factors will you keep in mind while selecting a suitable external finish?
   Describe the various types of finishes and the materials used for them.
9. What are the various defects in plastering? How will you rectify them?

REFERENCES

1. **Sawyer, J.T.**: *Plastering.*
2. **N.B.O.**: *Plastering of buildings.*
3. **H.M.S.O.**: *External rendered finishes.*
4. **H.M.S.O.**: *Plastering on Metal lathing.—Building Research Station Digest.*
5. **British Standard Institute**: *Code of Practice—Internal plastering.*
6. **H.M.S.O.**: *Principles of modern buildings.*
7. **H.M.S.O.**: *Defects in internal plastering—Building Research Digest.*
8. **Punjab P.W.D.**: *Specifications.*

Painting, Distempering and White-washing

PAINTING

Besides protecting various building units from the weathering action of the atmosphere, the main object of using a paint is to provide a decorative finish to obtain a clean, colourful and pleasing surface. Such surfaces are also hygienically good and present healthy surroundings to live in.

An ordinary paint consists, essentially, of two ingredients, *i.e.*, a base which is a solid material and forms the main body for covering the surfaces and a liquid called vehicle which carries this solid matter and distributes it evenly on the surface. The vehicle also acts as a binder for the solid matter and enables it to adhere to the surface. Additional components like pigments needed for giving different types of colours, thinning agents for greater covering capacities and driers which give the property of drying at a quick rate, are added to the paints.

The base may consist of white lead, red lead, zinc white, iron oxide or graphite, out of which the white lead is mostly used. This has got an additional advantage of being easily applied and has a good characteristic of covering the surface well but this is liable to discolour on account of the fumes of hydrogen sulphide present in the atmosphere. The covering coats in such a type of atmosphere have to be given with some other materials. Zinc oxide, on the other hand, is non-poisonous and is not affected by sulphur fumes. Lithophone is also extensively used for interior work. This is a mixture of zinc sulphide and barium sulphate prepared by the process of mutual precipitation. This base is good and cheap and it is liable to change colour on account of exposure to daylight. Red lead is used for painting iron surfaces to which it adheres well and protects the metal.

The liquid vehicle generally consists of raw linseed oil, refined linseed oil, pale boiled linseed oil or stand oil. Raw linseed oil is thin but takes a long time to dry. It is, therefore, boiled and during this process a drier such as litharge is added. Then, this become thicker and darker. For delicate work either raw linseed oil which is thin is used along with driers or poppy or nut oils. Pale boiled or double boiled linseed oil is more suitable for painting plastered surfaces. Stand oil, made from linseed oil by allowing it to stand in the sun for getting thicker by heat treatment, forms a durable film with a clear finish.

Pigments consist of natural earth colours such as ochres, siennas and iron oxide and calcined colours such as lime black, Indian red, carbon black and red lead and precipitates such as Pursian blue, chrome yellow, etc. These are liable to fade away by bleaching action and are also subjected to a change in colour under the influence of sulphurated hydrogen, moisture and heat.

Liquid and paste driers are used for hardening of paints. Liquid driers are finely divided compounds of metals such as manganese, lead, cobalt, etc. and dissolved in a volatile liquid. Paste driers consist of compounds of these metals mixed with larger percentages of filling material and ground in linseed oil. Litharge, zinc sulphate and red lead are also used as driers. The main function of a drier is to absorb oxygen from the air and transfer it to linseed oil which consequently gets hardened.

The thinning agent commonly used is terpentine oil. This material makes the paint thinner and facilitates it to spread out. It helps the penetration on porous surfaces such as wood and plaster but at the same time reduces the gloss of the linseed oil. It should, therefore, be used more carefully particularly in the finishing coat.

23.1 TYPES OF PAINTS

In the following paragraphs, the various types of paints are briefly described:

23.1.1 Oil Paint

The basic composition of oil paint is a base dispersed in raw or refined linseed oil plus driers and thinners. Additions of pure linseed oil, stand oil, dehydrated castor oil are also made. The characteristics of these paints are that they are easy to apply, have good opacity, a moderate flow, low gloss and are cheap. Such paints are self-priming and do not need any special primer or under-coat. For general decoration, it is a common practice to apply, on unpainted surfaces, at least three coats of varying compositions, *i.e.*, primers, undercoats and finishing coats.

23.1.2 Cement Paints

The term cement paint is applied to paints which are based on Portland lime with a pigment, filler and accelerator. They are normally mixed with water before use. Titanium dioxide is used to improve the opacity. Calcium or aluminium stearates or other compounds having similar properties are added as water repellent agents for exterior use. Calcium chloride or other accellators are added for adequate setting and hardening before it dries. Pigments are incorporated to give the required tint. Cement paint is most suitable for application to concrete and cement surfaces and also to brick work. This is unsuitable for surfaces which are liable to be subjected to dampness. It can, however, be used for preventing rain penetration into the walls.

23.1.3 Synthetic Rubber Paints

Synthetic rubber paints are better than the previous types. They offer complete resistance to lime and are suitable for application over fresh concrete or other masonry surfaces. These surfaces need not be completely dry. These paints are moderate in cost and are easy to apply. Weather and sunlight have very slow action on them. They have excellent chemical resisting property. They offer good resistance to water ahd stand up to heavy rains. The strong binding properties result in a low rate of mechanical attrition by rain, sun, wind or cold. Quick drying and minimum penetration of the resins contribute towards a good covering and non mottling qualities. They are fast drying and have excellent colour uniformity.

23.1.4 Emulsion Paints

A simple mixture of a pigment and a binding material is not usually workable and hence thinning agents are essential. Alternatively, the pigment can be finally dispersed in water to form an emulsion; and suitable materials like glue, casein or modified celluloses are added to keep the paint in good condition. These types of paints are called emulsion paints. The various types of emulsions are briefly enumerated below:

Alkyl emulsion paint. The binding medium in this paint is a certain synthetic resin. They give a flat finish, are porous and are not inherently resistant to alkalies. They are available in a range of varieties. Those containing low binder content are similar to best oil-bound distempers and those having high binder content are similar to oil paints.

Polyvinyl acetate paint. The binding material is polyvinyl acetate which is a synthetic resin. The type of finish can range from a matt to oil-gloss. These paints are normally permeable to water vapours and allow drying to continue through them, if they are used on damp walls. However, those paints which have got a fair gloss are not permeable to water and may blister if used on damp walls. These paints increase corrosion on steel and should not be applied directly, *i.e.*, without a primer.

Styrene based emulsion paint. Styrene which is a common type of synthetic is used in three different forms in emulsion paints, the most important type being the oil-modified styrene. This form gives a good finish, the film is permeable to water vapour and fairly resistant to alkalies.

High gloss emulsion paint. The binder is similar to that of a styrene based paint. The finish is equally glossy but is not permeable to water vapour and is also non-resistant to alkalies. This type of paint can be used on interior surface directly if the surface is not porous or needs a suitable primer if the surface is porous. It is not suitable for damp surfaces.

Oil bound distempers. These consist of an emulsion of drying oil or varnish with glue or casein and usually lithophone as pigment. There is considerable variation in the quality but some attain a fairly washable quality after a few months. These are permeable to water vapour and fairly resistant to alkalies. They are not suitable for use in conditions where condensation is likely to occur, *i.e.*, in kitchens and bath-rooms.

Emulsion paints based on tar or bitumen. These are used for different purposes. They are often porous and usually resistant to alkalies. They are restricted in the range of colours if bitumen content is high.

Advantages of emulsion paints are that they are easy to apply, the spreading of these finishes is so good that the time required to complete a given job is reduced by about 50 per cent and no marks or brushes are visible. They are fast drying paints and two coats of these paints can be applied on the same day. They have got an excellent adhesion property and form a tight resistant film even when applied over a wide range of temperature and on different surfaces. They have got good durability and can be applied over masonry surfaces or over previously painted surfaces. The colour retention is good and has pleasing effects. They have a higher life and it is seen that they can stand for a period of more than five years. Whenever the finished coats are washed, a look like a fresh paint appears. They offer resistance to alkali action.

23.1.5 Multi-Coloured Finishes

During the last few years, multi-coloured finishes are being produced rapidly. These multi-coloured paints are a suspension of lacquer particles in a water phase which contains a collective colloid. The pigment phase is poured into water during agitation and due to the incompatibility of the two systems, smaller globules are formed. They need a stabiliser to maintain the individuality of the particles. They have a high decorative value. They have got added advantage that they can be applied on highly porous surfaces without any shrinkage. They can also be applied to damp surfaces without any detrimental effect. Films of 0.15 mm and more can be spread in one application without sucking or wrinkling. These can be used on a new plaster without fear of alkali effect. These are used for decoration purposes in houses, basements and industrial buildings.

23.2 PROCESS OF PAINTING

The process of painting involves the application of a priming coat followed by an undercoat and one or two finishing coats. All these coats are important for giving a desired protection.

Primer coat. The primer cot has a double purpose, *i.e.*, it forms a layer which on one hand provides the adhesion of the paint to the given surface and on the other hand contains some reactive pigments which serve to protect the surface from the corrosive effect of weather. The selection of any primer for a given purpose depends upon its characteristic and the surface over which it has to be applied. For an absorbent surface, a primer should be of such a quality that a part of it should remain on the surface after absorption. This portion provides a key for adhesion. The proportions of oily and resinous materials are adjusted to suit the suction. Wood priming requires special consideration as it contains trapped moisture. In the case of steel, besides corrosion, another important factor which is to be considered is the expansion and contraction due to changes in the atmosphere and thus it has to stand this variation without any bad effect. For plastered surfaces, the presence of moisture must be taken into account while selecting a primer.

Undercoats. The main function of the undercoat relates to the filling properties, opacity, colour and support for the finishing coat. These are highly pigmented and thereby cover most of the irregularities. Heavy undercoats are generally avoided and hence suitable thinning agents must be mixed with a paint so that the painted surface is quite thin and there are no excessive marks left after the application. In the case of old work, care has to be taken that there is no loss of adhesion. A good undercoat should be such that after its drying, no clogging marks should be produced when it is rubbed with a sand paper. As this rubbing is essential before the application of any finishing coat, the tearing away of the film will result in a damage which is difficult to rectify.

Finishing coat. The choice of the finishing material depends upon the position where this is to be applied, *i.e.*, whether exterior or interior surface; the type of finish required *i.e.* glossy or matty; atmospheric conditions, etc. For interior surfaces, a paint based on natural resin varnish is good. For cold places, the paint should not become too brittle owing to the climatic conditions. In coastal areas or extremely wet places, growth of fungi presents problems in addition to corrosion.

The application of the paint on various surfaces is briefly described below:

23.2.1 Painting Woodwork

Selection of the right type of paint and the correct method of applying it, keeping in view the surface to which the paint is to be applied, are important for good painting. The characteristics of wood which have to be considered in this connection are its cellulose or porous nature, its liability to contain oil substances, its liability to contain water soluble substances, its moisture content and its capacity of absorbing water thereby causing swelling or shrinkage. The internal structures of hard woods differ markedly from those of soft woods. In the case of hard woods, there are smaller cellular units but they contain pores which are wider and deeper than soft woods and thereby present a difficulty in painting. Ordinary paints applied by brush or spray will neither fill nor cover these pores. The pores in the soft woods are smaller and, in general, may be filled by painting brushes. But in the case of soft woods, the presence of cellular structures leads to the absorption of the paint and hence a proper proportion of oily content and the pigments has to be maintained.

Hard woods do not contain resin and in general are free from objectionable oils. Teak and some other kinds of wood contain oils which tend to interfere with the drying, hardening and proper adhesion of a paint. They can, however, be painted successfully by the use of a special primer. In soft wood, sometimes there are excessive amounts of resins present, particularly near knots, which lead to unsatisfactory painting.

With well seasoned timber, the effect of water soluble substances is relatively unimportant but some priming paints may be of slow drying type and may not harden on hard woods rich in tannins (various organic compounds) such as wall nuts, etc. In other circumstances, similar troubles arise with soft woods. Water soluble substances may cause discoloration of paints.

The timber which is to be primed should contain about 10 to 20 per cent of moisture. The moisture content of dry seasoned timber may rise considerably high merely by absorbing water vapours and may reach 25 to 30 per cent in places where condensation occurs. The moisture in this case is not held in the wood cells but within the fibrous cell walls which consequently are in a swelling state.

Wood in contact with water would absorb more than 30 per cent of moisture easily till the cavities are filled. The adverse effect of moisture present within the walls is due to shrinkage as the moisture is absorbed or dries out. The paints do not prevent entirely such drying or absorption. The changes in wood fibres will, therefore, weaken the adhesion of the paint film and thereby may peal off the wood surface.

Woodwork should generally be given four coats of paints, *i.e.*, one coat of primer, two undercoats and one finishing coat. For most of the soft woods of good varieties, white lead in linseed oil or similar material containing 10 per cent of red lead may be used for priming coat. Higher percentage of red lead would produce a harder film which may need extra rubbing to ensure adequate adhesion of the next coat. This is more true if the primer has been applied a month or so before the application of the next coat. For lower grades of soft wood a slightly water resistant varnish is sometimes used. For general interior woodwork, linseed oil primers are sometimes preferred to lead primers on account of their non-poisonous nature. They are also quite satisfactory. They should be made from lithophone or titanium white with or without a proportion of pigment mixed with linseed oil. Quick drying primers should not be used on soft wood joinery.

Undercoats for use on wood should contain a more penetrative medium than the undercoats used on metals although the finishing coats may be the same in both the cases. This penetrative medium seeps into the primer coat and helps to restore any undue loss of oil brought about by absorption in wood. Good adhesion between the wood and the primer and between the primer and the undercoat is maintained, thereby preventing or delaying cracking, flaking and peeling. Proprietory undercoats usually contain a proper mixture of pigment, oil, varnish, thinner etc., which enables it to be satisfactorily used for both coats. Undercoats of lead if properly constituted with respect to the particular finishing coat are quite satisfactory for external use.

The proprietory paint used as finishing coat on woodwork is described as hard gloss paints, synthetic gloss paints and enamel paints. These paints are intended to give a hard film having a smooth and clear surface. Finishing paints used for an exterior work usually differ from those on the interior woodwork.

Painting process on new woodwork. If possible, the moisture content in timber should not be more than 15 per cent at the time of painting. Before applying the priming coat, the new woodwork should be made ready for painting by removing or sealing off certain defects which are sources of trouble. The timber should be dried a little to reduce the moisture content. Large knots should be removed by cutting or otherwise replaced by sound wood. Projections should be smoothened with gloss paper. Knots should be given one or two coats of shellac knotting or aluminium paints. It should be ensured that the primer should work well into the nail holes, etc. The primer should be applied firmly into the wood and sufficient time allowed for it to penetrate. Some of the primers may be good in this respect but should ensure proper adhesion with wood.

Subsequent painting is done at site after erection. If the primer has been damaged, it should be touched. When the primer is dry and before applying the undercoat, nails holes, etc., should be filled in by a stiff putty. This may consist of white lead and linseed oil or glazing putty with or without some red lead. When it gets dry, it should be rubbed with a water-proof abrasive paper. Sometimes fillers are used to fill in the pores or to get smooth finished surface. They usually consist of a mixture of finely powdered slate and size. When dry, the filler is rubbed smooth with water-proof abrasive paper. Oil bound water paint is also sometimes used as filler; but this is not so suitable as the other type mentioned earlier.

The subsequent coats should be applied carefully. Condition of the primer has to be observed before applying the first undercoat. The primer should be clean and not too hard as otherwise the undercoat will not stick well. Subsequent coats should be applied after about 24 hours' interval between each coat.

Repainting. The old paint should be completely removed if it is soft, or has cracked, blistered, peeled and shown loss of adhesion. Similarly it should be removed when it is highly discoloured or damaged by fungal growth. Small, loose patches of paint may be removed by a scraper. Interior paints should be rubbed with wet abrasive paper so as to provide a key and slightly rough surface to which the subsequent paint will adhere well. Exposed doors and windows should be touched with a primer and an undercoat as usual. Whenever it is desirable to remove old paint completely, blow lamp or other heating devices may be used. Alkaline removers should not be used on wood to be repainted. Whenever the use of a blow lamp is liable to cause damage to the paint, a paint remover of organic solvent type may be used. When the removal of the paint is complete, it should then be rubbed with wet abrasive paper and subsequent painting should be done in a manner similar to the one adopted for new woodwork.

Wood treated with tar oil as a preservative should be painted with dark coloured paints, or otherwise with light coloured paints if one or two sealing coats of good aluminium paints are given. A suitable type of paint should be used if the wood has been treated with other types of preservatives.

23.2.2 Painting of Iron and Steel

After the steel has been rolled, it has an oily scale on the surface. It is sometimes considered that if a continuous layer of oily scale could be maintained, the surface of steel would get an effective protective layer. But this is not possible as once the break occurs, progressive loosening takes place and uncovers the steel. Similarly, the oils and greases which may be present on the surface will not permit the paint to stick.

Mill scale can be removed in many ways. If the steel is stretched, the scale will flake away. This method is not easy to apply. Sand blasting may also be effective. Immersion in hot solutions of acids will remove the scale. Films of metallic phosphates can be created which are more resistant to rusting and form a good surface for painting. Thorough drying is essential before painting. Scrapers and grinding machines are used in big factories. Oxy-acetylene flame may be used along with the mechanical methods, but this method is not suitable for sections smaller than 5 mm in thickness.

Primers for painting steel should always contain pigments which resist corrosion. Red lead and chromates of lead and zinc are commonly used. Red lead in linseed oil is the oldest primer still abundantly used, but this is unsuitable for dipping and spraying. Chromate based primers are more suitable for spraying. For dipping primers red oxide primers containing a proportion of zinc chromate and metallic lead primer are useful.

Red oxide of iron and micaceous iron ore paints are cheap and satisfactory for use as undercoats. Graphite paints and aluminium paints have good resistance to moisture. Finishing coats of bitumen based aluminium paints can be applied advantageously. Paints based on silicones or silicone-alkyds are fairly suitable for resisting higher temperatures.

23.2.3 Painting Other Metals

The metal surface should provide an adequate key for the paint film and should be clean, dry, free from dirt, grease, etc. Paints of suitable type should be used with each type of metal. For aluminium, the priming paint should consist of pigments like zinc chromate. These pigments should constitute about 20 per cent of the dry weight of the film. Priming paints on aluminium should be free from pigments containing graphite or lead. For zinc a primer based on zinc oxide is considered to be very satisfactory. For stopping corrosion on galvanised iron, the surface can be treated with a red lead primer. Special care should be taken where two dissimilar metals touch. Contacts of copper, nickel, and their alloys with aluminium and zinc should be avoided. All metallic surfaces should be easily accessible for repainting. Extra paint must be applied at corrugations, etc., where the paint tries to recede just after application.

23.2.4 Painting Plasters, Brickwork, Floors and Other Concrete Surfaces, Etc.

These surfaces present difficulties on account of the dampness or the presence of soluble salts. Usually efflorescence occurs on the surface of these materials which decrease the adhesion of the paint considerably and may lead to discoloration. When oil paints are to be used on alkaline surfaces, an alkali resisting primer must be applied.

For floor surfaces, enamel coatings are used. These surfaces have to resist alkali action, moisture, and abrasion. The floor enamel must be tough to stand the impact of the foot traffic. Floor enamels should have a high gloss. They should be of quick drying type so as to use the floor in the least time. Oil modified phenolics are occasionally used as floor finishes.

Areas subjected to the action of water like bathrooms, etc. present special problems. Oil bound water paints do not have a longer life. They have a certain amount of porosity. While absorbing moisture they swell and contract when the moisture evaporates leading to disintegration. High-gloss paints are more resistant to moisture but since they have an impermeable film the effects of condensation are not much perceptible. Emulsion paints and synthetic rubber paints withstand washing and scrubbing and give a tough film which makes them more suitable.

Cement paint is more suitable for application on concrete surfaces. The suction at the surface should be reduced to such an extent that the moisture is not removed quickly from the cement paint. The surface may slightly be wetted for this reason, but it should not present a damp surface. The paint available in a powder form is diluted with water to a creamy consistency. The paint is used within 2 to 3 hours after mixing. It is usually applied with coarse distemper brushes. Two coats with an interval of 24 hours are given. A fine spray of water is applied to the painted surface at intervals to cure the surface.

Silicate paints having a binding medium of sodium or potassium silicates can also be successfully used. They are sulphate resistant but may tend to promote efflorescence.

DISTEMPERING

Distempers are mainly composed of whiting, glue or casein which acts as a binder and suitable proportions of fast coloured pigments. Generally they are available in powder form (dry distempers) and as a paste (for oil bound distempers).

Dry distempers consist of 90 to 95 per cent of whiting and 4 to 5 per cent of glue. All colouring pigments must not be effected by alkaline reactions of lime. This is one of the cheapest forms of decoration with the exception of lime wash. These distempers give a pleasing coloured appearance which will stand a certain amount of dry rubbing but can be readily removed with washing.

Oil bound distempers are prepared by grinding pigments with a medium composed of an emulsion of a drying oil in water. Linseed oil is generally used but tung oils are often employed. The emulsion is formed by agitating the oil and water together in the presence of an emulsifying agent which is generally glue or casein. To ensure full resistance to putrefaction and other types of growth, 0.3 per cent to 0.5 per cent of phenol is added. These are generally supplied in a paste or dry form which are readily thinned with water to brush consistency. On evaporation of water these dry to a porous film, the glue acting as a temporary binder during the drying of the oils. This film hardens to stand a moderate degree of washing and is more durable for internal decoration. These types of distempers are more valuable for early decoration of new plasters, brick-work, etc., as the porousness of the finish allows it to be applied to relatively damp surfaces without entrapping moisture. A good finish is retained for a longer time. Cracking and peeling is not prevalent. However, these are not well suited for decoration of bathrooms, kitchens and other places exposed to steam or heavy condensation as wetting and drying for prolonged times induces peeling and flaking.

Distempers should not be applied before one year after plastering work has been completed. This is due to certain chemical reactions which take place in the plaster and which are likely to affect the distempered surface. The walls may be temporarity finished with a white wash or colour wash made of whiting mixed with water and sized and tinted as required. This is better than lime wash as it has to be removed before a new application of distemper is to be made.

New plaster for distempering should have a fine polished surface. All irregularities must be removed. The plaster must be dry at the time of application of distemper. Before starting work, the plaster must be sized with a coat of equal parts of size and alum dissolved in hot water. The size should not be made too concentrated nor too thick. Decomposed size should not be used. Sometimes a special priming coat may have to be given. All old colour wash or white wash must be removed completely with sand papers.

For application on new plasters or on which some sort of lime wash has been applied, distempers made with lime proof pigments shall be employed. Enough distemper needed for one room, is mixed at a time so as to attain uniformity in the tint. Distemper is applied quickly leaving no dry edges. The brush is dipped and strokes are applied crosswise on the wall followed immediately by up and down strokes. Two men usually work, one working from ceiling downwards as far as he can reach and the second following him applying the distemper below. Patchy overlaps should be avoided. The brushes must be cleaned at the end of each day's work and old brushes caked with dry distempers should never be used.

WHITE WASHING

White washing is made from pure fat lime which may be slaked at site. Slaking is carried out in a tub until the mixture has a creamy consistency. This is allowed to rest for about 24 to 48 hours. This mixture is then strained through coarse cloth. To each cubic metre of this mixture, 3 kg of gum boiled with 10 kg of rice are added.

The white wash is applied with a brush to the specified number of coats. Each consists of four strokes of brush, one in each direction. Every coat of white wash is to be allowed to dry before applying the next coat. When each coat of white wash dries, wash should not show any sign of cracking nor should it come off readily on fingers when rubbed.

The white wash, when completed, shall form an opaque coat of white colour through which the old wash cannot be seen. It should also present a smooth regular surface free from powdery matter.

In the case of re-white washing, the surface shall be cleaned and freed from any foreign matter and old loose white wash. If old white wash is discoloured by smoke, a wash of wood ashes and water is applied before the coat of white wash. White wash should not be applied to a very smooth trowelled plaster as it will not adhere well. The plaster should also be dry before white wash is applied.

23.5 COLOUR WASHING

This is made from pure slaked fat lime and mixed with necessary pigment to give the desirable shade. The pigment should be of such a type so as not to get affected by the presence of lime. The colouring matter is added to the prepared white wash and the mixture is thoroughly strained through clean and fine cloth. This mixture is kept constantly stirred with a stick whilst being applied. Sufficient colour wash is kept ready before a room is taken in hand so as to get a uniform tint. New or scrapped surfaces are given one coat of white wash and one or two coats of colour wash. Old surfaces having satisfactory white or colour wash should be given one coat of colour wash. If a lighter colour is to be given the older coloured surface should be scrapped off and a coat of white wash applied before the new colour wash is given. Each coat of colour wash is allowed to dry before a new one is applied. When the surface is completed, the walls should present a uniform colour free from blots, lines or cuts, and must have a regular surface. Colour should not crack or come off readily on the fingers when rubbed.

QUESTIONS

1. What are the essential constituents of a paint? What functions are performed by these constituents?
2. What are the materials commonly used for the various constituents of paints? What are their functions?
3. Describe briefly the various types of paints commonly employed.
4. What are the various processes involved in painting? How would you proceed to paint
   (*a*) a new wooden panelled door, and
   (*b*) a new steel roof truss of a workshop?
5. Describe in detail the method of removing old paint from woodwork and repainting it.
6. Describe briefly how you will apply paint on:
   (*a*) Plastered work. (*b*) Brickwork.
   (*c*) Floors, and (*d*) Concrete surfaces.
7. What is meant by distemper? What are the various types of distempers? Describe briefly the process of distempering.
8. Write short notes on:
   (*a*) White washing, and
   (*b*) Colour washing.

REFERENCES

1. **H. M. S. O., Building Research Station Digest No. 30**: *Painting woodwork.*
2. **H. M. S. O., Building Research Station Digest No. 50**: *Emulsion paints.*
3. **H. M. S. O., Building Research Station Digest No. 64**: *Painting metals in buildings.*
4. **Chadda, U. R.**: *Painting and Our Problems—A paper read at the Research Workers Conference at Roorkee in April 1960.*
5. **Swamy, Y. S.**: *Protective and decorative paints used in the modern buildings.*
6. **H. M. S. O., Building Research Station Digest No. 17**: *Colour washes on external walls.*
7. **H. M. S. O., Building Research Station Digest No. 28**: *Painting Asbestos cement.*
8. **Chaudhary, J. R.**: *Building paints and coatings.*
9. **Bureau of Indian Standards**: *Various standards on paints.*

Thermal Insulation

HEAT TRANSFERENCE

A difference of temperature between the inside and outside or between different parts of a building will result in a transfer of heat from the warmer to the cooler areas. This transfer of heat may take place by any one or more of the three methods, *viz.*, conduction, convection and radiation. The transfer of heat through solid building materials takes place mainly by conduction, the amount of heat transferred depending on the temperature difference between the two surfaces, the thickness of the intermediate media, the area of the exposed materials, the time through which the heat flow takes place and the rate of heat transfer or the conductivity of the intervening materials.

The characteristic of the material which determines the rate of heat transfer is called the thermal conductivity and is expressed in terms of the number of thermal units which will pass through one square metre of a material, 10 mm thick, in one hour, for one degree Centigrade temperature difference. The conductivity of a material depends on the type of the material which may be a good conductor or a bad conductor of heat. For the same material the rate of flow depends on the density. Certain types of materials have an optimum density for which the heat flow is the least. Any variation from this value will increase its conductivity. If the material contains air spaces within it, the conductivity will get decreased. Any substance containing moisture in it will transfer heat quickly as the water present in it is a better conductor of heat than air. The conductivity also increases with the mean of the inside and outside temperatures although the difference in temperature may remain the same.

Convection and radiation also play an important role in the transfer of heat through building materials. Large air surfaces exist within walls or internal structures of a building. These may not contain air in a dead state and therefore the heat gets transferred by the process of convection, that is, currents of air transmit from one side to the other. Similarly the presence of polished surfaces on the exposed side of a building may retard the inflow of heat as a large quantity would get reflected.

Briefly speaking, the following factors are most important in the transfer of heat from the exterior into the interior:

(*i*) Thermal insulation value of the exterior shell of the building, that is, the top floor, external walls and the roof,

(*ii*) The difference in temperature between the outside and the inside; for a greater difference there will be greater loss of heat, (*iii*) The area of the external part of the building; for greater areas the heat transfer is more,

(*iv*) The areas of windows and doors. The transfer of heat through a glazed window is about three times more than a typical wall construction. Hence the windows should be put in a correct orientation with respect to the movement of the sun,

(*v*) The rate of air movement through any cavities in the enclosing structure. Air spaces through the roof or below a suspended floor may cause considerable heat transference unless special precautions are taken.

For determining the heat loss from a building to the outside in winter months the following additional factors have to be taken into account:

(*i*) The rate of air change. This is the rate at which the heated air inside the building is displaced by cold air from outside. Air flows will occur through the open doors and windows or through ventilators, etc.

(*ii*) Next is the position of flues. The structure enclosing the flue is heated by the gases which pass through it. If the flue is built in an external wall, a greater proportion of heat will be lost through this wall directly by conduction.

(*iii*) The type of heating appliance and the design of the flue throat. In the heating appliance of flue, appreciable heat loss can occur if an excessive heated air passes up the flue. The restrictions to the air flow provided by the appliance and by the cross-sectional area of the flue throat are important.

The heat transmission losses are calculated by multiplying the area A by the proper co-efficient U for a particular type of construction or material and by the temperature difference between the two surfaces. This can be stated by the equation,

$$H = A \times U \times (T - T_o)$$

where H = Heat transmitted through the materials of walls, roofs, etc.

A = Area of the surface

T = Room temperature

T_o = Outside temperature,

U = Overall co-efficient of transmission or the amount of heat transmitted per unit area of surface for a temperature difference of one degree between the air of inside and the outside.

The heat transmitted through each surface is calculated separately and the total transmission loss for each room or space is the sum of individual transmission losses through each surface of the room. Due allowance is also given for the infiltration losses which occur on account of leakage of air through the crevices, etc. present in the walls or doors.

To achieve insulation effects, it is essential to either use insulating materials or devise special construction methods or alternatively adopt both. For a material to be considered as a good insulating substance it must have a low heat conductivity, that is, a high degree of heat resistance per unit of thickness and must be installed in an adequate thickness. Generally it should not have a greater thermal conductivity than 1260 Calories. The thickness must be sufficient so as to provide adequate heat resistance of the material.

24.1 TYPES OF INSULATING MATERIALS

The types in which these substances are fabricated are (*i*) loose fills, (*ii*) blanket insulation, (*iii*) bats, (*iv*) insulating boards, (*v*) slab or block insulations, (*vi*) reflective sheet materials, (*vii*) light weight aggregates, etc.

Loose fill insulations consist of fibrous materials like rock wool, slag wool, glass wool, cellulose or wood fibre wool. They also consist of granular loose materials of mineral or vegetable nature. Mineral wool is a fibrous material and is made of three materials—rock, slag and glass. These materials are conveyed to a large melting pot and the rock is melted under a high temperature. As it leaves the furnace, it is acted upon by a blast of steam which carries the beads of wool into an annealing chamber.

An inert light weight granular insulating material known as vermiculite is manufactured by exploding an aluminium magnesium silicate material. This treatment is done under the action of high temperature which causes the flakes of the mineral to get converted into cellular granules.

Blanket insulations are flexible fibrous materials supplied in rolls or otherwise. They are made principally from mineral wool, processed wood fibre, cotton and animal hair. They vary in thickness from 10 to 80 mm.

Bat insulating materials are similar to the above type but are smaller in size and greater in thickness, common thicknesses being 50 mm, 70 mm and 90 mm. They are available in smaller sizes suitable for framing units. Structural insulating board is made by reducing wood cane or other materials to a pulp and then reassembling fibres into boards. Adhesives are used to keep the fibres in correct position. Various types of insulating boards are available in different sizes and thicknesses.

Slab insulations are small rigid units of about 25 mm in thickness and range in size up to 600 mm × 1200 mm or more. Slab insulations are also known as blocks or boards and consist of cork board slabs, mineral wood slabs, vermiculite slabs, cellular glass slabs, cellular rubber slabs or wood fibres bound together with cement.

Reflective insulations depend entirely on their surface characteristics for their heat resistant properties. They are generally used along with the air spaces so that the reflective insulation surface is exposed. Bright metallic surfaces are more efficient. Sheet or blanket aluminium reflective materials, aluminium foils, surfaced gypsum boards, steel sheet reflective insulations or reflective coating applied to paper or other surfaces are the various types of reflective insulations commonly adopted.

Cement and concrete products have lower heat resistances but with the use of light weight aggregates such as blast furnace slag, burnt clay aggregate, vermiculite etc., the insulating resistance of concrete materials gets increased.

24.1.1 Methods of Application

Fill insulations may be poured in between the vertical studs after the outside sheathing is placed and as the wall to be insulated are receiving the plaster base or interior finish. This insulation should be poured in heights not greater than 1000 mm at a time and must be packed to a proper density between the covering sides. If a vapour barrier is to be provided it should be kept on the warmer side of the wall. These types of insulating materials may be poured from bags into the spaces between the ceiling joists to the desired depth. Whenever intricate inner spaces are present these materials may be placed with the aid of compressed air.

The blanket insulations of widths greater than the distance between the vertical studs are used. This enables the edges to be folded and nailed or stapled into the wood framing members. Special nailing or stapling flanges are provided in certain cases to facilitate the application of the insulation. Whenever possible the blanket insulation may be installed in a manner so as to leave sufficient air gaps on both sides of it. While fixing this type of insulating material to a masonry wall special furring strips are used to hold it and the interior finish.

The bat insulations are installed between the exterior finish and the interior covering material and air space may be left on one side.

Structural insulating boards are used as materials for walls or roof sheathing, as a plaster base or other decorative finishes. They are nailed to the woodwork supporting them. They may be covered with different finishing materials like wooden planks, concrete blocks, plaster, shingles, etc.

24.2 INSULATION OF ROOFS

In drier tropics where there is a marked day-night temperature range, it is desirable for the buildings used during the day to have a type of construction which will warm up slowly. The underside of the roof slab in such a case will be the warmest when the general temperature is falling. The extent of this time-lag depends on the mass of the roof, the heavier the roof the longer the lag. With a 100 mm solid concrete slab it is about $2\frac{1}{2}$ hours, whereas with a 200 mm solid slab it will be about 6 hours. The heavier the slab the smaller the variation in temperature on the underside. In more humid tropics where day-night temperature range is less, the mass of the roof is less important provided the temperature of the underside is prevented from rising above the shade temperature. To ensure this a white surface treatment or some form of sun-shade is essential.

For rooms which are used only at night the roof should be capable of cooling down as rapidly as possible and a light structure is more appropriate.

According to the recommended practice the thermal transmittance value for roofs of houses and flats should be about 0.2 to 0.40. Generally it is better to put a thermal insulation above the slab but it may sometimes be advantageous to place it underneath for other reasons.

Thermal insulation may be provided by the slab itself, a water-proof layer at the top being given reflective treatment sufficient to keep it reasonably hot. A sound absorbent ceiling may be added which would contribute to thermal insulation. Another typical construction would consist of the concrete portion acting as a screed placed on wood wool slabs carried on inverted T-sections. This type of roof is light, has a high thermal insulation and is free from condensation troubles. Suspended ceilings, along with a constant depth of water being maintained over the slab, may be useful for industrial structures.

24.3 INSULATION BY AIR SPACES OR CAVITIES

Air spaces or cavities form an integral part of many modern constructions. The dimension of the cavity, the emissivity of the surfaces facing the cavity and the ventilation rate are important as the heat transfer takes place mainly by convection and radiation. It is now accepted that for a vertical air space unventilated and enclosed by surfaces of high emissivity, the resistance to heat flow across it will be the greatest when its width is at least 20 mm. In a horizontal cavity such as that in a flat roof or floor the convection currents tend to cross the width of the cavity when the direction of the heat flow is upwards; with downward heat flow they try to get eliminated as the air is warmer at the upper part of the cavity. A value of 1.10 is adopted for the resistance of cavity when the cavity is unventilated, is at least 20 mm wide, is horizontal, vertical, or inclined and has enclosing surfaces of ordinary building materials having an emissivity of 0.9. If a similar cavity is faced with material of low emissivity such as an aluminium foil (having emissivity of 0.1 or less) a value of 2.0 is considered to be a suitable resistance for this cavity. Ventilation by cooler air outside will reduce the thermal resistance of a cavity.

24.4 ECONOMICS OF INSULATION

The following factors are taken into account while evaluating the economic feasibility of a particular type of insulation:

(1) Cost of providing and fixing the insulation.

(2) The area of exposed surfaces and windows.

(3) The cost of heating or cooling.

(4) The standard of insulation maintained.

The insulation material may also be used as a load bearing member so that it could economise in the constructional cost of the building itself. Such use can only be made of rigid type of insulating materials. It should also be ensured that the insulating material has got an adequate resistance to fire, fungi and insect attack. The insulating material should also not absorb excessive quantities of moisture. It should also not be liable to change in its size due to expansion or contraction.

24.5 CONDENSATION

Condensation on the interior surfaces of factories in which higher percentages of humidities are maintained is usually common. Moisture usually deposits on the underside of the ceilings. In residences, wall and ceiling condensation is not so important as the window condensation. The water vapour, instead of getting deposited at the inner surface of the ceiling, may pass through the roof or get deposited within the wall.

Whenever warm humid air comes into contact with surfaces which are below the dew point temperature, condensation of water vapour will take place. To prevent condensation on any surface, it is necessary to maintain the temperature of that surface above the dew point. The probable dew point temperatures can be estimated by knowing the relative humidity and the temperature conditions which are existing or going to exist in a building. After the dew point temperature has been established to the next step is to determine as to how much insulation must be added to the wall or roof structure to maintain the interior surface above the dew point temperature at all times. The thickness of insulation must be sufficient to prevent surface condensation in the coolest possible weather. Whenever extremely severe conditions are encountered and excessively thick insulations are needed, additional precautions are taken. If the moisture is an outcome of manufacturing process, the humidity may be reduced by passing air through a dehumidifier. If high humidity is essential for the product under manufacture, ceiling condensation may be prevented by installing a vapour-tight ceiling under the roof deck and passing warm dry air through the space between roof and ceiling.

The water vapour within the living space gets first diffused within the building which consequently increases the relative humidity and vapour pressure. This vapour will then be removed by either leakage or through permeable spaces in walls or ceiling as most of the building materials are permeable. The rate of movement of vapour through a material is a function of the vapour pressure through two sides. This permeability is based on the area, pressure difference, thickness, time and the property of the material called permeance. The unit of permeance is called perm and this is the vapour transmission in grains per sq. m per hour per cm of mercury vapour pressure difference. For asphalt roofing felt, the value of permeance is about 0.25; for card board it is about 0.1 while for wood it ranges from 0.18 to 0.8 depending on thickness, etc.

The inside vapour pressure is usually higher in winter than outside and hence the direction of travel is usually from inside to outside through the walls or the ceilings and condensation occurs wherever the temperature goes below the dew point. Materials which restrict the movement of water vapour to one perm are used as vapour barriers. Vapour barriers are of two types, namely, the paint or liquid type and the sheet or membrane type. In general a continuous heavy coat of any water resisting material such as wax, varnish or bitumen is likely to have adequate vapour resistance. The vapour barrier should retain its vapour resisting qualities for the life of the building or if a paint is used until it is renewed. In the liquid type oil bases, rubber bases, bitumen and aluminium paints are included. The other type of vapour barriers consist of two sheets of heavy kraft paper with bitumen between them, aluminium foil mounted on one or both sides of paper and bituminous coated papers having bright finish.

Vapour barriers should always be installed on the warmer side of the wall and at a place where it will be above the dew point of the air-vapour mixture of the room. When insulation is applied to the top ceiling and the attic space is unheated the underside of the roof and the attic walls may become cool and increase the possibility of condensation if water vapour is permitted to pass through the ceiling to the attic space. To prevent such a condition attic spaces should be suitably ventilated. Definite areas of ventilation have been maintained for different types of constructions by the various building authorities.

24.6 THERMAL TREATMENTS IN TROPICAL AND SUB-TROPICAL REGIONS

Indian climate can be broadly classified into three types, *viz.*, hot arid, hot humid and warm humid types. There are cold regions also but considerations of thermal insulations are more important in the first two types named above; and hence these are dealt with briefly in the succeeding paragraphs.

The hot arid climate is characterised by high air temperatures, dry air and dry ground. Under these conditions there will be little clouds or moisture in the air to screen off the sun's rays so that the resulting high intensity of direct solar radiation heats the ground further. The dryness of the ground reduces the plant life considerably and as such the soil reflects the solar radiation to a great extent. The evaporation can take place easily and hence a lot of heat can be radiated to the outer space.

One of the most important protective functions of the house is to screen the occupant from the incoming radiation. This may lead to difficulties as the enclosure of the house would get hot and thereby pass some heat into the interior. This would further deteriorate the conditions if the space within is not freely ventilated. It is better that this enclosure is exposed as little as possible and also it passes less heat into the interior. Roof is the surface which is exposed to the greatest radiation in the zone from 30° north to 30° south latitude. The walls facing the equator are subjected to a moderate heating effect during the day. East and west walls are subjected to somewhat higher heating effects for a part of the day. Internal barriers such as trees may be useful for screening the building from the radiation effects. Bushes, shrubs or artificial screens set on the eastern and western sides of the house are more effective for protecting these walls as the low altitude of sun in the morning and evening gives the main heating load. The wall facing the equator does not gain advantage by this effect as that precedes high angle sunshine. Houses in the east-west alignment may help each other unless the ventilation problems are not affected. Advantage can be taken by placing walls of store rooms on western or eastern sides of the building. Grass or other covered vegetation may be grown in front of the building so as to reduce the reflection of heat. Projections of the sloping roof can protect the walls as they will cast shadows on the wall surfaces. Either one projection or different at different elevations can be used. Such projections are more effective with a high altitude sun and can be ideally used on the wall facing the equator or on both north and south walls for a building situated near the equator. Such projections are advantageously placed over door and window openings. Vertical projections can be used to exclude morning or evening sun from window openings on walls facing the equator.

Spraying roof or wall structures which are exposed to sun can be effective in reducing the heat load on the structure. Spraying should be intermittent to permit effective evaporation from the surface.

The smaller the surface presented by the house to the solar radiation, the lesser will be the heat gained. Least proportional heat load is experienced in hot seasons by the tall building with its long axis running east-west. A tall building of square floor design is also suitable irrespective of the orientation but it is not of great advantage during cold season. A single-storeyed building has the highest heat load under hot conditions but is fair to retain the same under cold conditions when it would be more desirable. In tropical regions where the sun is at a fairly high altitude during most part of the day the slope of the roof has little effect on the total heat load. This is because of the fact that there is no effect on solar projection of the house by having a pitch which is less than the altitude of the sun.

Surfaces exposed to strong radiation will be heated by that part of radiation which is not reflected but absorbed. Movement of air over the surface will remove some of the heat and thus reduce the quantity of heat conducted into the interior. Selection of the site and construction of the building in a manner that it does not affect the freedom of wind movements is very important.

Insulation against the transference of heat may also be adopted. In hot climates heating is not constant; it rises during the day and decreases by night. Under these circumstances any insulating material which will store a large quantity of heat during heating period and give it back when heat is removed is most suitable. This property of a material depends on its thermal conductivity, specific heat and also its density. When heat is continuously applied to one side of an insulating material, time is required to reach equilibrium and there will be a temperature gradient across each part of the insulator and the property which determines the rate of heat transfer between certain distances at fixed temperatures is the thermal conductivity. This situation gets changed when the heat is applied to the outer surface. It is used up in warming particles of that surface and only after they have been warmed heat passes on to the next layer. Here the heat gets delayed again till it passes through the next layer. The rate at which the heat front advances through the material will be affected by the amount of heat needed in warming each layer. The rate at which heat directly applied to surface is diffused is called thermal diffusivity and it involves thermal conductivity, density and specific heat of the material. Materials like wood and clay are good as they have low diffusivity and are therefore usually employed as a building material. The storing capacity of heat of a material is more important in the case of roofing material. The quantity of heat transmitted should be small and hence such a type of construction should be adopted that the maximum temperature of the internal surface comes about twelve hours after the maximum temperature of the external surface has occurred. The heat in such a case will arrive at the interior surface when it is not objectionable for the comfort of the people.

Ventilation of the rooms must be suitably controlled to ensure comfortable living. The entry of external hot air should be reduced during the day-time in summer. Small doors or windows fitted with movable shutters can be helpful in this connection. These windows can be placed higher up in the walls to reduce the chances of radiation falling directly on the occupants of the room. If any air spaces are kept within the roof they should be suitably ventilated.

If the air is drawn through the ground at greater depth say about 4500 mm from the ground level, cooling can be obtained in summer and heating in winter with its help. Basements in buildings utilise this principle but greater comfort can be derived if the air is passed through a fairly long tunnel before it is taken into the building.

Air passed over water gets cooled and its water content is increased. Under hot dry conditions use of wet towelling materials or screens which are kept damp and placed in such a position that air can pass through them before coming in contact with the occupants are adopted as a simple means of giving comfort. This screen should be relatively thin as movement of air is done naturally through it. A box cooler containing absorbent materials like cinders or wood wool over which water drips and air being pulled through the device by a fan is effective in cooling rooms.

The internal heat developed within the room must be suitably controlled. The size of combustion must be provided with openings so that air is removed quickly within the combusting area to prevent transference of heat from this chamber. Reflecting materials can be used to diverge the heat towards the interior.

24.6.1 Warm Humid Climates

These areas have got moderately high temperature but are accompanied by moist air and damp ground. The presence of moisture in the air reduces certain amount of radiation. The moisture combined with moderate heat is favourable for growth of vegetation which further reduces the reflectivity of radiation. The amount of heat that is to be removed from a human body is mainly by evaporation which is usually difficult. Air movement which would facilitate evaporation gets reduced on account of plantation.

The problem in such climate is more important as people cannot have more comfort in sticky conditions and difficulties are created on account of prevalence of insects, dampness, etc. Special attention has to be paid to the flooring of a building. In an humid area an earthen floor would be more difficult. Impervious material, resting on earth, like concrete will keep a constant temperature and it will be relatively dry unless its temperature falls below the dew point of the atmosphere of the house. Although the average intensity of direct solar radiation is less it is still considerable on roof. Walls on the other hand, are subjected to much less reflected radiation from the ground. As the trees are usually abundant they can be used to shade the roof and the walls. Mutual shading of houses is not very desirable as the measures would restrict the air movement between and through the houses which is very essential. Full advantage should be taken of the vegetation in preventing the reflection from ground. Use of attached shading elements such as sun breakers is helpful. The heat absorbed by these shading elements is returned and not added to the house itself. The solar projection of the roof should be minimised with a high altitude of sun. This is proportional to the horizontal area of the roof. The shape of the roof is important only when its slope exceeds the solar altitude. The roof of a multi-storeyed house would be better than that of a single-storeyed house. However, on account of greater wall heights it will become difficult to give proper shading effects.

Use of high reflectivity materials is helpful in reflecting major portion of the heat radiation but the superiority of such reflecting surfaces is not felt fully in warm humid conditions.

It is desirable to have air movement as much as possible on outer surfaces especially those which are of a damp nature. In dry periods spraying may be used on roof to supplement this effect. Similarly higher buildings are liable to have greater movements.

Insulation is less effective in this case than in hot dry climates. For walls which are suitable there will be no requirement for insulation against radiation.

Since the transference of heat by conduction is of lesser importance in warm humid climates, basements and other devices for heat exchange with the ground are not generally effective and may lead to difficulty due to higher water table.

The vapour pressure inside the house may become excessive due to washing, cooling, etc. and it is desirable that the external air replaces the internal. Natural air currents may be effective in this connection or else some type of fans may be used. The velocity of incoming air can be kept within limits with the aid of partitions, etc.

For other types of climates the factors enumerated above may be kept in view so as to conform to the particular prevailing conditions.

24.7 TREATMENT OF DIFFERENT UNITS IN A BUILDING

Special considerations are necessary for some of the units in a building from the comfort point of view. In addition to the previously stated principles, the following considerations should be kept in view:

(a) Sleeping areas: In hot dry areas full advantage should be taken at night of relatively cool sky as well as of cold air. This would mean that sleeping in the open air would be beneficial. Rain, early morning sun, etc., may be minor difficulties in this connection. In warm humid areas or when sleeping outside is not possible, quarters on the north-eastern sides with good opportunity for cool air are better. Openings should preferably start from the floor level. Where air conditioning or fan is available, its use may be justified in sleeping areas so as to remove the disturbing degree of heat.

Verandah which is simply an attached shading device is more functional in warm humid than in hot dry weather. Its disadvantage may be that it converts main rooms of the house into darker areas.

(*b*) ***Kitchen*:** In hot climates it is easier to separate the kitchen from the house, so that the heat and the smell will not fill in the living area. The shady side of the building is a better position for the kitchen within the house and attention should be paid to direction of breeze. In hot dry climates special ventilation may be restricted to the carriage of heat from cooking areas. In warm climates free ventilation directed outwards from the cooking range is desirable.

QUESTIONS

1. What are the factors responsible for heat transference? How will you calculate the transmission losses?
2. What are the various materials used for insulation? Describe briefly their methods of application.
3. Write short notes on:
   (*i*) Insulation of roofs.
   (*ii*) Insulation by cavities.
   (*iii*) Economics of insulation.
   (*iv*) Condensation.
4. What are the various considerations of thermal treatment in tropical and sub-tropical regions?

REFERENCES

1. **U.S.A. Housing and Home Finance Agency**: *Physiological objectives in hot weather housing.*
2. **H.M.S.O.**: *Thermal Insulation of buildings.*
3. **Commonwealth Expt. Building Stations, Australia**: *Designing houses for Australian climates.*
4. **Ray-Chaudhury, B.C.**: *Thermal behaviour of buildings.*
5. **I.S.: 3792–1978**: *Guide for heat insulation of non-industrial buildings.*

Ventilation

25.1 PURPOSES OF VENTILATION

Ventilation may be defined as a process of supplying fresh or outside air by natural or mechanical means into or the removal of an inside air from an enclosed space.

Ventilation of buildings is required for the following purposes:

(*i*) to supply fresh air to prevent depletion of the oxygen content of the air and prevent undue accumulation of carbon dioxide and moisture.

(*ii*) to dilute inside air to prevent undue rise in concentration of body odours and other contaminants such as tobacoo smoke.

(*iii*) to remove any products of combustion or other contaminants in air such as dust, fumes, gases, mists, vapours, etc. present in air which are likely to be injurious or offensive to the occupants.

(*iv*) to provide such thermal environments which will assist in the maintenance of heat balance of the body in order to prevent discomfort and injury to health of the occupants. This is done by removing body heat and heat dissipated by electrical or mechanical equipment or solar heat gains.

(*v*) to control the concentration of bacteria within limits, by oxidising colonies of bacteria and fungus to prevent their proliferation.

(*vi*) to keep humidity within limits.

Fresh air or outside air should meet the criteria given in Table 25.1. In addition it should be such that the concentration of any contaminant in the air should be limited to within one-tenth the threshold limit value (*TLV*) of that contaminant. *TLV* refers to air borne concentration of contaminants accepted by the American Conference of Governmental Industrial Hygienists and represents conditions under which it is believed that nearly all occupants may be repeatedly exposed, day after day, without adverse effect.

25.2 ESSENTIAL FACTORS NECESSITATING VENTILATION

25.2.1 *Maintenance of Carbon Dioxide Concentration of Air within Safe Limits and to Provide Sufficient Oxygen Content in Air for Respiration*

Even in the worst ventilated rooms the content of carbon dioxide in air rarely exceeds 0.5 to 1 per cent and is, therefore, incapable of producing any ill effect. The amount of air required to keep the carbon dioxide concentration down to 1 per cent is very small. The change in oxygen content too is also small under normal conditions to have any ill effects, the oxygen content may vary quite appreciably without noticeable effect, if the carbon dioxide concentration is unchanged. The concentration of carbon dioxide or reduction in oxygen content is thus not sufficiently critical to provide a basis for fixing rates of ventilation for residential buildings.

Table 25.1: Maximum allowable contaminant concentrations for ventilation air contaminants' Annual Average (Arithmetic Mean)

Contaminant	*Annual average (Arithmetic mean)* $\mu g/m^3$	*Short term level (Not to exceed more than once in a year)* $\mu g/m^3$	*Averaging period* r
Suspended particulates	60	150	24
Sulphur oxides	80	400	24
Carbon monoxide	20000	30000	8
Photo-chemical oxidant	100	500	1
Hydro-carbons (not including methanes)	1800	4000	3
Nitrogen oxide	200	500	24
Odour-Essentially unobjectionable			

25.2.2 *Control of Odours*

Odours are disturbing and when present they cause headache and loss of appetite. It is, therefore, desirable that rate of ventilation is estimated on the basis of removal of noticeable body odour and other odours such as from tobacco smoke, cooking, etc.

Where no products of combustion or odour contaminants are to be removed from air, the amount of fresh air required for dilution of inside air to prevent vitiation of air by body odours, depends on the air space available per person and the degree of physical activity. The amount of air decreases as the air space available per person increases and it may vary from 20 m^3 to 30 m^3 per person per hour. In rooms occupied by only a small number of persons such an air change will automatically be attained in cool weather by normal leakage around windows and other openings and this may easily be secured in warm weather by keeping the openings open.

No standards have been laid down under the *Factories Act* 1948 as regards the amount of fresh air required per worker or the number of air changes per hour. Over-crowding requires that at least 14 m^3 to 16 m^3 of space shall be provided for every worker and for the purpose of that section no account shall be taken of any space in a work room which is more than 4.25 m above the floor level.

Vitiation of the atmosphere can also occur in factories by odours given off due to contaminants of the product itself, say for example, from tobacco processing in a '*Bidi*' factory. Here the ventilation will have to be augmented to keep odours within unobjectionable levels.

25.2.2.1 Recommended values for air changes

The following standards of general ventilation are recommended (Table 25.2) based on maintenance of required oxygen, carbon dioxide and other air quality levels and for the control of body odours when no products of combustion or other contaminants are present in the air:

Table 25.2: Air changes schedule

Sl. No.	*Application*	*Air changes/hour*
1.	Assembly rooms	4–8
2.	Bakeries	20–30
3.	Banks/Building Societies	4–8
4.	Bathrooms	6–10
5.	Bedrooms	2–4
6.	Billiard rooms	6–8
7.	Boiler rooms	15–30
8.	Capes and Coffee bars	10–12
9.	Canteens	8–12
10.	Cellars	3–10
11.	Churches	1–3
12.	Cinemas and theatres	10–15
13.	Club rooms	12, Min.
14.	Compressor rooms	10–12
15.	Conference rooms	8–12
16.	Dairies	8–12
17.	Dance halls	12, Min.
18.	Dye works	20–30
19.	Electroplating shops	10–12
20.	Engine rooms	15–30
21.	Entrance halls	3–5
22.	Factories and workshops	8–10
23.	Foundries	15–30
24.	Garages	6–8
25.	Glass houses	25–0
26.	Gymnasium	6, Min.
27.	Hair dressing saloon	10–15
28.	Hospitals-sterlizing	15–25
29.	Hospital wards	6–8
30.	Hospital domestic	15–20
31.	Laboratories	6–15
32.	Launderettes	10–15
33.	Laundries	10–30
34.	Lavatories	6–15
35.	Lecture theatres	5–8
36.	Libraries	3–5
37.	Living rooms	3–6
38.	Mushroom houses	6–10
39.	Offices	6–10
40.	Paint shops (not cellulose)	10–20
41.	Photo and X-Ray dark room	10–15
42.	Public houses bars	12, Min.
43.	Recording control rooms	15–25
44.	Recording studios	10–12
45.	Restaurants	8–12
46.	School rooms	5–7
47.	Shops and Supermarkets	8–15
48.	Shower baths	15–20
49.	Stores and warehouses	3–6
50.	Squash Courts	4, Min.

(*Contd.*)

Sl. No.	Application	Air changes/hour
51.	Swimming baths	10–15
52.	Toilets	6–10
53.	Utility rooms	15–20
54.	Welding shops	10–30

Note: The ventilation rates may be increased by 50 per cent where heavy smoking occurs or if the room is below ground.

25.2.3 Removal of Products of Combustion

Products of combustion discharged from *chullahs*, stoves, gas applicances, etc. used in a kitchen are likely to accumulate there and may also permeate into other rooms. Similarly *angithis* used for heating rooms in certain colder parts of the country result in the production of carbon monoxide and other gases. Natural ventilation can play here significant role in controlling concentration of these products of combustion.

25.2.4 Maintenance of Satisfactory Thermal Environment in a Room

Environmental factors like air temperature, humidity and air movement together with some other factors, such as clothing, level of activity, food, etc. have a direct influence upon bodily processes. Maintenance of thermal equilibrium of the body is very essential for securing thermal comfort and for avoiding heat stress. Heat transfer between human body and the environment occurs through conduction, convection, radiation and evaporation, the relative magnitude of each process varying with changes in ambient conditions. However under hot environments, evaporation is most important process of heat loss from the human body for securing thermal comfort. As the room air or especially, the air around body becomes more nearly saturated due to humidity, it becomes more difficult to evaporate perspiration and a sense of discomfort is felt. A combination of high humidity and high air temperature proves very oppressive. In such circumstances even a slight movement of air near the body gives relief.

Specially in hot weather, when thermal environment inside the room is worsened by heat given off by machinery, occupants and other sources, the prime need for ventilation is to provide such thermal environment as will assist in the maintenance of heat balance of the body in order to prevent discomfort and injury to health. Excess of heat either from increased metabolism due to physical activity of persons or gains from a hot environment has to be offset to maintain normal body temperature (37°C). Heat exchange of the human body with respect to the surroundings is determined by the temperature and humidity gradient between the skin and the surroundings and other factors, such as age of persons, clothing, etc. and the latter depends on air temperature (dry bulb temperature), relative humidity, radiation from the solid surroundings and rate of air movement. The volume of outside air to be circulated through the room is, therefore, governed by the physical considerations of controlling the temperature, air distribution or air movement. Air movement and air distribution may, however, be achieved by recirculation of the inside air rather than bringing in all outside air. However, fresh air supply or the circulated air will reduce heat stress by dissipating heat from body by evaporation of the sweat, particularly when the relative humidity is high and the air temperature is near body temperature.

25.2.4.1 Limits of comfort and heat tolerance

Thermal comfort is that condition of thermal environment under which a person can maintain a bodily heat balance at normal body temperature and without perceptible sweating. Limits of comfort vary considerably according to studies carried out in India and abroad. In terms of effective temperature, the upper limit of comfort may be 27.5°C for every day work in industry. This is also the temperature for most efficient production. Air movement is necessary in hot and humid weather for body cooling. A certain minimum desirable wind speed is needed for achieving thermal comfort at different temperatures and relative humidities. Such wind speeds are given in Table 25.3. These are applicable to sedentary work in offices and other places having no noticeable sources of heat gain. Where somewhat warmer conditions are prevalent, such as in godowns and machine shops and work is of lighter intensity, and higher temperatures can be tolerated without such discomfort, minimum wind speeds for just acceptable warm conditions are given in Table 25.4. For obtaining values of indoor wind speed above 2.0 m/s, mechanical means of ventilation may have to adopted.

Table 25.3: Desirable wind speeds (m/s) for thermal comfort conditions

Dry bulb Temperature °C	*Relative humidity (Percentage)*						
	30	40	50	60	70	80	90
28	*	*	*	*	*	*	*
29	*	*	*	*	*	0.06	0.19
30	*	*	*	0.06	0.24	0.53	0.85
31	*	0.06	0.24	0.53	1.04	1.47	2.10
32	0.20	0.46	0.94	1.59	2.26	3.04	†
33	0.77	1.36	2.12	3.00	†	†	†
34	1.85	2.72	†	†	†	†	†
35	3.20	†	†	†	†	†	†

* None

† Higher than those acceptable in practice.

Table 25.4: Minimum wind speeds (m/s) for just acceptable warm conditions

Dry bulb Temperature °C	*Relative humidity (Percentage)*						
	30	40	50	60	70	80	90
28	*	*	*	*	*	*	*
29	*	*	*	*	*	*	*
30	*	*	*	*	*	*	*
31	*	*	*	*	*	0.06	0.23
32	*	*	*	0.09	0.29	0.60	0.94
33	*	0.04	0.24	0.60	1.04	1.85	2.10
34	0.15	0.46	0.94	1.60	2.26	3.05	†
35	0.68	1.36	2.10	3.05	†	†	†
36	1.72	2.70	†	†	†	†	†

* Note

† Higher than those acceptable in practice.

There will be a limit of heat tolerance when air temperatures are excessive and the degree of physical activity is high. The limit is determined when the bodily heat balance is upset, that is, when the bodily heat gain due to conduction, convection and radiation from the surroundings exceeds the bodily heat loss, which is mostly by evaporation of sweat from the surface of the body. The limits of heat tolerance for Indian workers are based on the study conducted by the Chief Adviser Factories, Government of India, Ministry of Labour and are given in his report on Thermal Stress in Textile Industry. According to this report, where workers in industrial buildings wearing light clothing are expected to do work of moderate severity with the energy expenditure in the range 273 to 284 W (watts) the maximum wet bulb temperature shall not exceed 29°C and adequate air movement subject to a minimum air velocity of 30 m/min shall be provided, and in relation to the dry bulb temperature, the wet bulb temperature of air in the work room, as far as practicable, shall not exceed that given in Table 25.5.

Table 25.5: Maximum permissible wet bulb temperatures for given dry bulb temperatures

Dry Bulb Temperature	*Maximum Wet-Bulb Temperature*
°C	°C
30	29.0
35	28.5
40	28.0
45	27.5
50	27.0

Notes 1: These are limits beyond which the industry should not allow the thermal conditions to go for more than 1h continuously.

2: Figures given in this table are not intended to convey that human efficiency at 50°C will remain the same as at 30°C, provided appropriate wet bulb temperatures are maintained. Efficiency decreases with rise in the dry bulb temperature for a given wet bulb temperature attained and efforts should be made to bring down the dry bulb temperature as well as much as possible. Long exposures to temperature of 50°C dry bulb / 27°C wet bulb may prove dangerous.

3: Refrigeration or some other method of cooling is recommended in all cases where conditions would be worse than those shown in this table.

25.3 METHODS OF VENTILATION

General ventilation involves providing a building with relatively large quantities of outside air in order to improve general environment of the building. This may be achieved in one of the following ways:

(*a*) natural supply and natural exhaust of air;

(*b*) natural supply and mechanical exhaust of air;

(*c*) mechanical supply and natural exhaust of air; and

(*d*) mechanical supply and mechanical exhaust of air.

25.3.1 Control of Heat

Although it is recognized that general ventilation is one of the most effective methods of improving thermal environmental conditions in factories, in many situations, the application of ventilation should be preceded by and considered along with some of the following other methods of control. This would facilitate better design of buildings for general ventilation, either natural or mechanical or both, and also reduce their cost.

25.3.1.1 Isolation

Sometimes it is possible to locate heat producing equipment, such as furnaces in such a position as would expose only a small number of workers to hot environment. As far as practicable, such sources of heat in factories should be isolated.

In situations where relatively few people are exposed to severe heat stress and their activities are confined to limited areas as in the case of rolling mill operators and crane operators, it may be possible to enclose the work areas and supply conditioned air to such enclosures.

25.3.1.2 Insulation

A considerable portion of heat in many factories is due to the solar radiation falling on the roof surfaces, which in turn, radiate heat inside the building. In such situations, insulation of the roof or providing a false ceiling or double roofing would be very effective in controlling heat. Some reduction can also be achieved by painting the roof in heat reflective shades.

Hot surfaces of equipment, such as pipes, vessels, etc. in the building should also be insulated to reduce their surface temperature.

25.3.1.3 Substitution

Sometimes, it is possible to substitute a hot process by a method that involves application of localized or more efficiently controlled method of heating. Examples include induction hardening instead of conventional heat treatment, cold rivetting or spot welding instead of hot rivetting, etc.

25.3.1.4 Radiant shielding

Hot surfaces, such as layers of molten metal emanate radiant heat, which can best be controlled by placing a shield having a highly reflecting surface between the source of heat and the worker, so that a major portion of the heat falling on the shield is reflected back to the source. Surfaces of tin and aluminium have been used as materials for shields. The efficiency of the shield does not depend on its thickness, but on the reflectivity and emissivity of its surface. Care should be taken to see that the shield is not heated up by conduction and for this purpose adequate provision should be made for the free flow upwards of the heated air between the hot surface and the shield by leaving the necessary air space and providing opening at the top and the bottom of the sides.

25.3.2 Volume of Air Required

The volume of air required shall be calculated by using both the sensible heat or latent heat gain as the basis. The larger of the two figures obtained should be used in actual practice.

25.3.2.1 Volume of air required for removing sensible heat

When the amount of sensible heat given off by different sources, namely, the sun, the manufacturing processes, machinery, occupants and other sources, is known and a suitable value for the allowable temperature rise is assumed, the volume of outside air to be provided for removing the sensible heat may be calculated from:

$$Q = \frac{2.9768\,K_s}{t} \qquad \text{...(1)}$$

where

Q = Quantity of air in m^3/hn,

K_s = sensible heat gained in W and

t = allowable temperature rise in °C

5.3.2.2 Allowable temperature rise

Temperature rise refers mainly to the difference between the air temperatures at the outlet (roof exit) and at the inlet openings for outside air. As very little data exist on allowable temperature, rise values for supply of outside air in summer months, the values given in Table 25.6 related to industrial building may be used for general guidance.

Table 25.6: Allowable temperature rise values

Height of outlet opening m	*Temperature rise* °C
6	3 to 4.5
9	4.5 to 6.5
12	6.5 to 11

Notes 1: The conditions are limited to light or medium heavy manufacturing processes, freedom from radiant heat and inlet openings not more than 3 to 4.5 m above floor level.

2: At the working zone between floor level and 1.5 m above floor level, the recommended maximum allowable temperature rise for air is 2 to 3°C above the air temperature at the inlet openings.

25.3.2.3 Volume of air required for removing latent heat

If the latent heat gained from the manufacturing processes and occupants is also known and a suitable value for the allowable rise in the vapour pressure is assumed:

$$Q_2 = \frac{4127.26 \times K_1}{h} \qquad ...(2)$$

where

Q_2 = quantity of air in m^3/hn

K_1 = latent heat gained in W, and

h = allowable vapour pressure difference in mm of mercury.

In majority of the cases, the sensible heat gain will far exceed the latent heat gain, so that the amount of outside air to be drawn by ventilating equipment can be calculated in most cases on the basis of the equation (1).

Ventilation is also expressed as m^3/hn per m^2 of floor area. This relationship fails to evaluate the actual heat relief provided by a ventilation system, but it does give a relationship which is independent of building height. This is a more rational approach, because, with the same internal load, the same amount of ventilation air, properly applied to the work zone with adequate velocity, will provide the desired heat relief quite independently of the ceiling height of the space, with few exceptions. Ventilation rates of 30 to 60 m^3/hn per m^2 have been found to give good results in many plants.

25.4 NATURAL VENTILATION

The rate of ventilation by natural means through windows or other openings depends on:

(*a*) direction and velocity of wind outside and sizes and disposition of openings (wind action), and

(*b*) convection effects arising from temperature of vapour pressure difference (or both) between inside and outside the room and the difference of height between the outlet and inlet openings (stack effect).

25.4.1 Ventilation of Non-Industrial Buildings

Ventilation in non-industrial buildings due to stack effect, unless, there is a significant internal load, could be neglected, except in cold regions, and wind action may be assumed to be predominant.

In hot arid regions, the main problem in summer is to provide during day protection from sun's heat so as to keep the indoor temperature lower than those outside under the sun and for this purpose windows and other openings are generally kept closed and only minimum ventilation is provided for the control of odours or for removal of products of combustion.

In hot humid and warm humid regions, the problem, in the design of non-industrial buildings is to provide free passage of air to keep the indoor temperature's as near to those outside in the shade as possible and for this purpose the buildings are oriented to face the direction of prevailing winds and windows and other openings are kept open on both windward and leeward sides.

Adequate number of circulating fans should be installed to serve all interior working areas during summer months in the hot arid and hot/warm humid regions to provide necessary air movement at times when ventilation due to wind action alone does not afford sufficient relief.

In winter months in cold regions, the windows and other openings are generally kept shut, particularly during night; and ventilation necessary for the control of odours and for the removal of products of combustion can be achieved either by stack action or by some infiltration of outside air due to wind action.

25.4.2 Ventilation of Industrial Buildings

In providing natural ventilation of all industrial buildings having significant internal heat loads due to manufacturing process, proper consideration should be given to the size and distribution of windows and other inlet openings in relation to outlet openings so as to give, with due regard to orientation, prevailing winds, size and configuration of the building and manufacturing processes carried on maximum possible control of thermal environment.

In the case of industrial buildings wider than 30 m, the ventilation through windows may be augmented by roof ventilation.

25.4.3 Design Guidelines for Natural Ventilation

25.4.3.1 By wind action

(*a*) Inlet openings in the building should be well distributed and should be located on the windward side at a low level and outlet openings should be located on the leeward side near the top, so that incoming air stream is passed over the occupants. Inlet and outlet openings at high levels may only clear the top air without producing air movement at the level of occupancy. When outlets serve also as inlets, they shall be located at the same level.

Maximum air movement at a particular plane is achieved by keeping the sill height of the opening at 85 per cent of the height of the plane. The following levels of occupancy are recommended:

(1) For sitting on chair = 0.75 m

(2) For sitting on bed = 0.60 m,

(3) For sitting on floor = 0.40 m.

(*b*) Inlet opening should not, as far as possible, be obstructed by adjoining buildings, trees, sign boards or other obstructions or by partitions inside the path of air flow.

(*c*) Greatest flow per unit area of openings is obtained by using inlet and outlet openings of nearly equal areas.

For a total area of openings (inlet and outlet) of 20 to 30 per cent of floor area, the average indoor wind velocity is around 30 per cent of outdoor velocity. Further increase in window size increases the available velocity but not in the same proportion. In fact, even under most favourable conditions the maximum average indoor wind velocity does not exceed 40 per cent of the outdoor velocity.

(*d*) Where the stream of wind is quite constant and dependable, the openings may be readily arranged to take full advantage of the wind. Where the wind direction is quite variable, the openings shall be so arranged that as far as possible there is approximately equal areas on all sides and the openings shall be located at the same levels. Thus, no matter what the wind direction is, there are always some openings directly exposed to wind pressure and others to air suction and effective movement through building is assured.

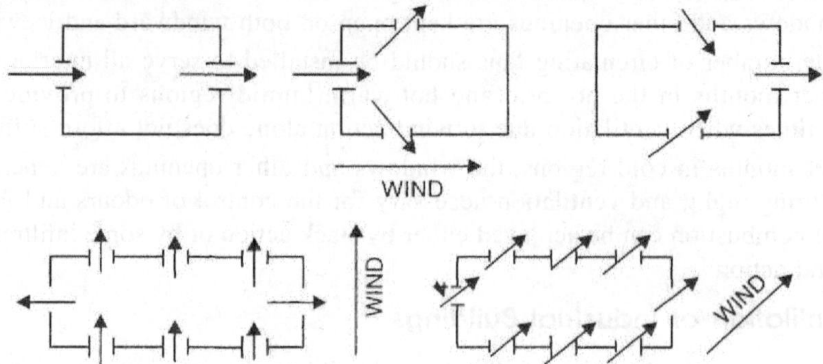

Figs. 25.1 to 25.5: Movement of wind through buildings.

25.4.3.2 *By stack effect*

Natural ventilation by stack effect occurs when air inside a building is at a different temperature than air outside. Thus in heated buildings or in buildings wherein hot processes are carried on and in ordinary buildings during summer nights and during pre-monsoon periods, the inside temperature is higher than that of outside, cool outside air will tend to enter through openings

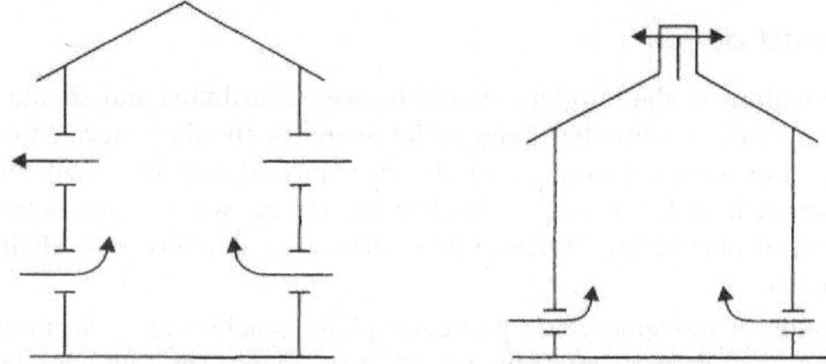

Figs: 25.6 to 25.7: Ventilation due to stack effect.

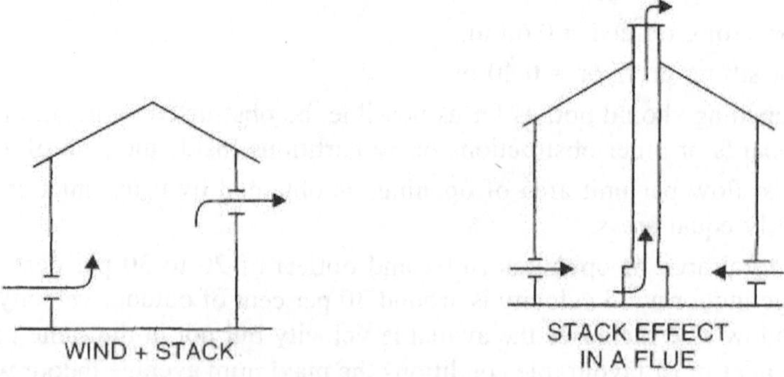

Figs. 25.8 to 25.9: Action of wind and stock effect

at low level and warm air will tend to leave through openings at high level. It would, therefore, be advantageous to provide ventilators as close to ceilings as possible. Ventilators can also be provided in roofs, as for example, cowl, ventpipe, covered roof and ridge vent.

25.5 MECHANICAL VENTILATION

25.5.1 General

Where adequate air changes specified or for providing thermal environment within the limits specified in Table 25.6 whichever is higher, cannot be obtained by natural ventilation, mechanical ventilation either by exhaust of air or by positive ventilation or a combination of the two shall be provided; and in case of positive ventilation where necessary, air before being brought into the area of occupancy may be cooled by evaporative cooling or by air-conditioning.

Fans and other equipment for mechanical ventilation may be located in convenient positions having regard to the intake of fresh air, accessibility for maintenance and noise control.

25.5.2 Exhaust of Air

Exhaust fans are provided in walls on one side of the building or in the attic and roof to draw large volumes or air through the building. These fans are usually of propeller type, since they operate against little or no resistance. It is important that windows and other openings near the fans are kept closed, as otherwise the fans would draw outside air from these openings and cause what is termed as "short circuiting". Adequate inlet openings shall be provided on opposite side of the building so as to limit inlet velocities.

When fans are centrally located on an attic or other unused space and arranged to draw proportionately from several areas of occupation or from exhaust appliances with duct work, these are predominantly centrifugal type so as to overcome the resistance from duct work.

25.5.3 Positive Ventilation

Positive ventilation is provided by centrally located supply fans which are usually of the centrifugal type or sometimes axial flow types, since this application requires duct work with a wide range of satisfactory and quiet operation against high pressures. Considerable advantage may be achieved by incorporating the ducts and risers with air discharge outlets into the building structure and by having the interior surfaces carefully finished to render them smooth and airtight and treated to prevent the possibility of dust being scoured from the walls by the passing air.

Unit ventilators may be provided for individual rooms and may be placed against outside wall near the central line of the room.

Both central system and unit ventilators could be equipped to provide, besides the function of ventilating, cooling by evaporative cooling or by cooling coils. Typical installations are equipped with a system of controls that permits ventilating and cooling effect to be varied, while the fans are operating continuously, in accordance with the room requirements.

25.5.4 Combined Systems

The combined systems with positive ventilation and with exhaust of air have the advantage of providing better control conditions and better distribution of air over the entire area of occupancy, particularly in wider buildings. By supplying sufficient volumes of air in proportion to heat load generated in the respective areas at suitable velocities at the required areas through duct work and by extracting the air in the return ducts in proportion to the supply air quantities and recirculating the air or a part of it after properly mixing it with cool fresh air, completely satisfactory ventilation is obtained. In a combined system, it is preferable to provide slight excess of exhaust if there are adjoining occupied spaces and a slight excess of supply if there are no such spaces. Unit exhausters can also be used to match unit ventilators' exteriors and located along the outside wall.

25.5.5 Evaporative Cooling

In regions where high day-time temperatures prevail with reasonably low humidities, evaporative cooling may be employed effectively to lower the temperature of the air to near the wet bulb temperature and produce an air supply cool enough to take care of the indoor sensible heat loads without exceeding the upper safe limits given in Table 25.5 or temperature rise values at area of occupancy given in Table 25.6. By positive ventilation, this air may be supplied to produce cooler environments with lower air volume than would be required as per equation (1) or (2) as greater temperature rise than given under Table 25.6 may be tolerated. Although the relative humidity of supply air will be increased, due to the large sensible heat loads, the resultant relative humidity of the air will be sufficiently lowered after mixing with the inside air to produce body cooling.

Evaporative cooling with positive ventilation using a central system consisting of a water spray chamber and a fan to supply outside air into the area of occupancy through a distribution duct is preferable to spray head system which will only humidify the air within space. Where a spray head system only humidifies the air, the cooling capacity of the air will be improved very little; and none of the air which absorbs the heat given off by men and machinery and other sources is removed from the building under these conditions.

25.6 AIR-CONDITIONING

Where the desired temperatures and humidities cannot be obtained by mere ventilation, air-conditioning may be resorted to.

25.6.1 Ventilation for Contaminants Control

When contaminants are given off during the manufacturing process, efficient local exhaust ventilation and or dilution ventilation to reduce their concentration below the threshold limit value (TLV) shall be provided.

Recommended capture velocities for some of the manufacturing processes which are likely to give rise to the contaminants depending upon their condition of dispersion, are given in Table 25.7 as a guide.

25.6.2 Make-up Air

Sufficient make-up air shall be brought into the work room by natural filtration or by positive ventilation at suitable points in relation to the exhaust points to replace the air exhausted by local exhaust ventilation or by dilution ventilation, and the air may be efficiently filtered or treated, when necessary.

Table 25.7: Range of recommended capture velocities

S. No.	*Condition of dispersion of contaminant*	*Examples*	*Capture velocity m/s*
(*i*)	Released at practically no velocity into quiet air	Evaporation from tanks degreasing, etc.	0.25-0.5
(*ii*)	Released at low velocity into moderately still air	Spray booths; intermittent container filling low speed conveyor transfers; welding, plating; pickling	0.5-1.0
(*iii*)	Active generation into zone of rapid air motion	Spray painting in shallow booths; barrel filling; conveyor loading, crushers	1.0-2.5
(*iv*)	Released at high initial velocity into zone of very rapid air motion	Grinding, abrasive blasting, tumbling	2.5-10

25.7 DETERMINING RATE OF VENTILATION

25.7.1 Natural Ventilation

This is difficult to measure as it varies from time to time. The amount of outside air through windows and other openings depends on the direction and velocity of wind outside (wind action) and/or convection effects arising from temperature or vapour pressure differences (or both) between inside and outside of the building (stack effect).

25.7.2 Wind Action

For determining the rate of ventilation based on wind action the wind may be assumed to come from any direction within 45° of the direction of prevailing wind. Ventilation due to external wind is given by the following formula:

$$Q = KAV \tag{3}$$

where

Q = the rate of air flow in m^3/hr;

K = coefficient of effectiveness, which may be taken as 0.6 for wind perpendicular to openings and 0.3 for wind at an angle less than 45° to the openings;

A = free area of inlet openings in m^2; and

V = wind speed in m/hr.

The value of the coefficient of effectiveness K depends on the direction of the wind relative to the opening and on the ratio between the areas of two openings. Figure 25.10 gives the increase in values of K by the percentage of unequal areas expressed as ratios of the two openings.

25.7.3 Stack Effect

Ventilation due to convection effects arising from temperature difference between inside and outside is given by:

$$Q = 7.0\, A\, \sqrt{h(t_r - t_o)}$$

where

Q = the rate of air flow in m^3/hr;

A = free area of inlet openings in m^2;

h = vertical distance between inlets and outlets in m;

t_r = average temperature of indoor air at height h, in °C; and

t_o = temperature of outdoor air in °C.

The equation is based on 0.65 effectiveness of openings. This should be reduced to 0.50 if conditions are not favourable.

When areas of inlet and outlet openings are unequal, 'A' given in equation (3) will be the smaller area and the volume of air will be increased according to the percentage given in Fig. 25.10.

When both forces (wind and thermal) act together in the same direction, even without interference, the resulting air flow is not equal to the two flows estimated separately. Flow through any opening is proportional to the square root of the sum of the two heads acting on that opening.

Wind velocity and direction, outdoor temperature, and indoor distribution can not be predicted with certainty, and refinement in calculation is not justified. A simple method is to calculate the sum of the flows produced by each force separately. Then using the ratio of the flow produced by thermal forces to the aforementioned sum, the actual flow due to the combined forces can be approximated from Fig. 25.11. When the two flows are equal, the actual flow is about 30 per cent greater than the flow caused by either force acting independently (see Fig. 25.11).

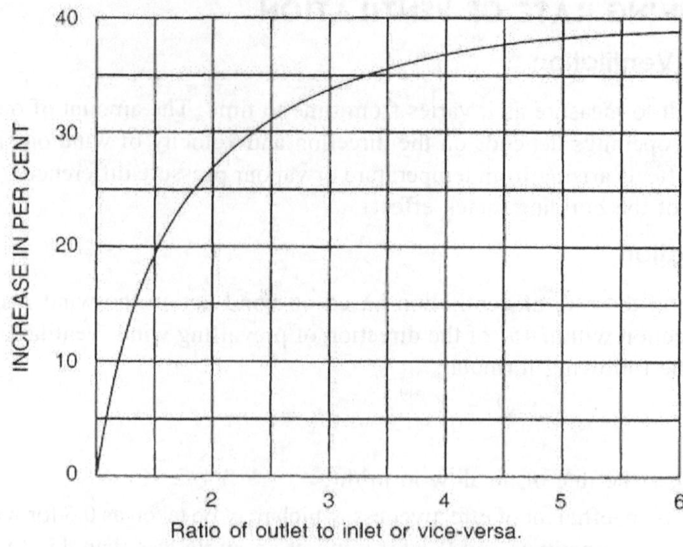

Fig. 25.10: Increase in flow caused by excess of one opening over another.

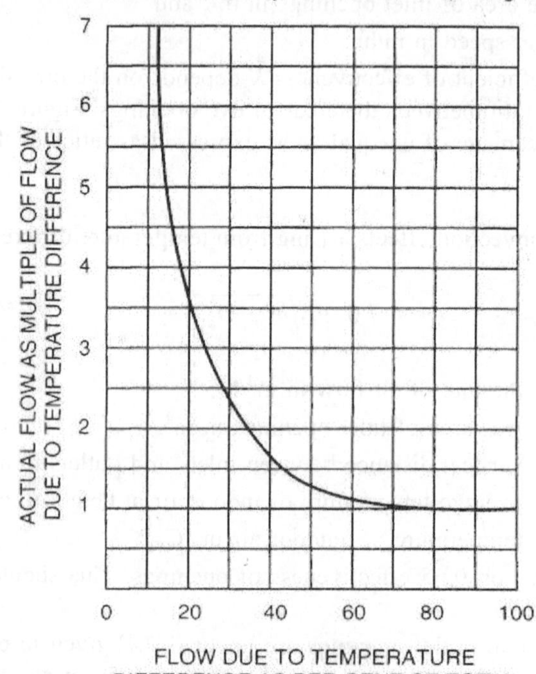

Fig. 25.11: Determination of flow caused by combined forces of wind and temperature difference.

Judgement is necessary for proper location of openings in a building specially in the roof, where heat, smoke and fumes are to be removed. Usually windward monitor openings should be closed, but if wind is so slight that temperature head can overcome it, all openings may be opened.

25.7.4 Mechanical Ventilation

The volume of outside air by positive ventilation shall be measured using proper instruments, such as properly calibrated anemometer, velocity meter and pitot tube.

To measure the average velocity of air flow, it is necessary to make a traverse of the instrument over the cross-sectional area of the inlet openings or ducts and obtain the average velocity from these results. The volume of air is given by:

$$Q = AV$$

where

Q = volume of air in m^3/hr,

A = free area of inlet openings or ducts in m^2, and

V = average velocity of air in m/hr.

When ventilation is achieved only by exhaust of air, the volume of exhaust air should be measured in the same manner as in the case of positive ventilation by measurement of air velocity and area of exhaust ducts or openings and multiplying one with the other.

25.6.5 Combined Effect of Different Methods of Ventilation

When combination of two or more methods of general ventilation is used, the total rate of ventilation shall be reckoned as the highest of the following three, and this rule shall be followed until an exact formula is established by research:

(*a*) 1.25 times the rate of natural ventilation

(*b*) Rate of positive ventilation and

(*c*) Rate of exhaust of air.

25.6.6 Air Movement

The rate of air movement of turbulent type at the working zone shall be measured either with a Kata thermometer (dry silvered type) or heated thermometer or properly calibrated thermocouple anemometer. Whereas anemometer gives the air velocity directly, the Kata thermometer and heated thermometer give cooling power of air and the rate of air movement is found by reference to a suitable nomogram using the ambient temperature.

QUESTIONS

1. Define ventilation. What functions are served by it?
2. Discuss the essential factors necessitating ventilation.
3. How are the minimum standards of ventilation fixed? Discuss the minimum standards based on control of body odours.
4. Briefly describe the different methods of ventilation.
5. How will you determine the volume of air required?
6. What factors affect natural ventilation? Give the general rules and design guidelines.
7. Describe the different methods of mechanical ventilation.
8. When is air-conditioning resorted to? Give the design principles.

REFERENCES

1. **Bureau of Indian Standards**: *National Building Code of India 2005, Part 8 section 1: Lighting and Ventilation.*
2. **IS: 3362-1977**: *Code of practice for natural ventilation of residential buildings.*

Air Conditioning of Buildings

Air conditioning may be defined as the process of treating air so as to control simultaneously its temperature, humidity, purity and distribution best suited to the physiological needs of occupants or to the requirements of industry. It may be employed to preserve and maintain health, comfort and convenience of people in their homes. Industrial air conditioning is concerned with materials and may be used for the preparation of materials such as artificial silk or preservation of fruits. Commercially air conditioning is employed for theatres, restaurants, shops, offices, etc. for maintaining comfort of the occupants within these premises.

Air conditioning consists of refining air in order that it may be used to the best advantage within an enclosure for the better health, comfort and convenience of people. The outside area which is unconditioned has got a very high or low temperature, unsuitable humidity and velocity, and may be full of dust, etc. Conditioned air would mean an atmosphere having controlled temperature, humidity and velocity conditions with dust cleanliness as much as possible.

In summer the air is hot and for conditioning the processes needed are to cool, dehumidify, clean and circulate the air properly. For winter months the air is to be heated, humidified, cleaned and circulated properly.

A feeling of comfort is a good indication of healthier atmospheric conditions. The control of temperature, air motion and relative humidity, is absolutely necessary in an air conditioning system. It is evident that a human being will feel comfortable in a certain temperature, air velocity and humidity. But since people move from one room to another or from one place to another and perform different duties, hence such conditions are different for different people. The feeling of comfort will also vary with outside weather conditions. Research has been carried out in different countries to develop the values of temperature, humidity as well as air velocity which give comfortable conditions for different seasons of the year. The investigators have made considerable tests to determine the temperature range which is suitable to majority of people; and this is called the comfortable zone. There is a difference between the comfortable zone in winter and in summer. This is due to the clothing worn in the two seasons and also due to the changes in the body.

For comfort air-conditioning, dry bulb and wet-bulb temperatures may be adopted as given in Table 26.1 for summer and in Table 26.2 for winter.

Table 26.1: Inside design conditions for summer

Sl. No.	*Optimum Conditions*		*Maximum Conditions*	
	Dry Bulb Temperature °C	*Wet Bulb Temperature* °C	*Dry Bulb Temperature* °C	*Wet Bulb Temperature* °C
(*i*)	23.3	19.4	25.9	21.8
(*ii*)	23.9	18.4	26.1	21.6
(*iii*)	24.4	17.6	26.7	20.9
(*iv*)	25.0	16.8	27.2	20.1
(*v*)	25.6	16.0	27.8	19.4
(*vi*)	26.1	15.2	28.3	18.8
(*vii*)	—	—	28.9	18.1
(*viii*)	—	—	29.4	17.5

Table 26.2: Inside design conditions for winter

Sl. No.	*Optimum Conditions*		*Maximum Conditions*	
	Dry Bulb Temperature °C	*Wet Bulb Temperature* °C	*Dry Bulb Temperature* °C	*Wet Bulb Temperature* °C
(*i*)	21.4	17.8	18.3	15.0
(*ii*)	21.7	17.3	18.9	13.4
(*iii*)	22.2	16.4	19.4	12.0
(*iv*)	22.8	15.3	19.7	10.8
(*v*)	23.3	14.4	—	—
(*vi*)	23.6	13.4	—	—

For practical reasons the limits, which have been set up, have to be modified to keep in view the working conditions of the people. In hot weather when a person comes to a cold room at 22°C a feeling of extreme cold or shock is created. Similarly, when he leaves this cold room and comes in contact with the outside temperature he feels a sense of discomfort because the room is cold and he is subjected to a higher temperature instantaneously. Hence large temperature difference between indoor and outdoor air must be removed, as they are harmful to the body. In summer the human body adapts itself to higher temperature than in winter. When conditioned spaces are used by people only for a short interval of time a thermal shock of more than 11°C between outside and inside air should be avoided. Whenever people have to sit for longer periods within an air-conditioned enclosure, *e.g.*, those working in offices, etc., higher dry bulb temperature can be maintained. A temperature of 21°C to 22.5°C is required in such cases for comfort regardless of outside temperature.

Regulation of air velocity is very important. The increased air velocity adds to the effect of decreasing the effective temperatures below outside temperatures. Unequal air movement upsets the regulation of conditioned air. Adequate movement of air should always be provided in an air-conditioned enclosure, but velocities in excess of 0.5 m/s in the zone between floor level and 1500 mm level should generally be avoided in the case of comfort air-conditioning. 0.25 m/s should be the permissible limit of air movement in this zone. Air velocity in the vicinity of a supply or return air-grille will be higher than this value but since these inlets/outlets are at greater heights, a feeling of discomfort is not created.

Humidity control is equally important with conditioning air. If air is very dry a great strain is put on the membranes of nasal passages which may become dry and irritated. For practical reasons relative humidity is kept between 35 to 65% whereas desirable limit between 40 to 60% may be adopted. During the hot season 40 to 50% relative humidity is most comfortable and for cold weather 50 to 60% is best. But these values for winter season in cold areas may create condensation on window glasses.

In India not much research has been done on the comfort of people at different temperatures and relative humidities. But by experience it has been seen that the comfort zone for this country extends from 23.5°C with 60% relative humidity to about 30°C with 45% relative humidity. These values assume that velocity of air does not exceed 10 metre per minute. It should be economically possible to achieve these values and hence it is always desirable to have a temperature difference of 11°C between interior and exterior temperatures as stated earlier. In the plains of our country the climate is dry and hence if correct relative humidity is maintained along with a suitable velocity of air a comfortable condition can be maintained. For health reasons, air conditioned spaces where people leave or enter frequently, it is desirable to have entrance lobbies which have a temperature midway between interior and exterior enclosures so that a person may not get sudden shock while leaving or entering. For efficient and economical air conditioning it is desirable to keep the inside atmosphere as free as possible from direct heat transmission, humidity or dust from exterior. For this reason suitable solid barriers in the form of doors, windows, roofs, etc. are created to offer resistance to these elements. The object of selecting a particular barrier is to prevent the transmission as much as feasible; and hence for prevention of heat transmission, materials having low conductivity should be adopted. Doors and windows may be made of double glazed type. A tight construction using impermeable materials helps in reducing their air leakage. Creation of exterior barriers like sun shades, sun breakers, etc. prevents the direct effect of sun's rays into the room. A combination of various barriers would be more helpful.

26.1 SYSTEMS OF AIR-CONDITIONING

Basically four systems of air conditioning are adopted:

(1) ***Central system*:** A central system uses conditioned air for all the requirements and for one or more seasons at one focal or central point. The conditioned air is then distributed to the various rooms or enclosures which are being treated.

(2) ***Self-contained system*:** The self-contained system (*e.g.* a room unit) conditions air within the unit itself.

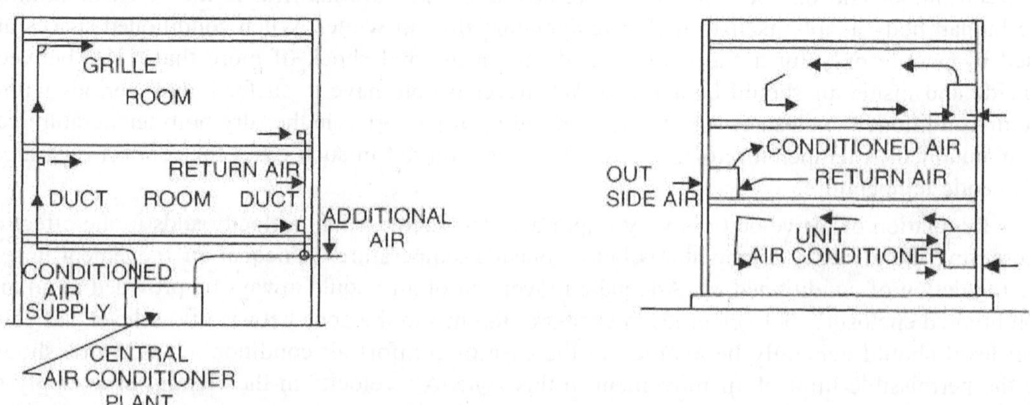

Fig. 26.1: A central system of air conditioning. **Fig. 26.2:** Self-contained (unit system) of air conditioning.

(3) *Semi-conditioned system*: This usually conditions air for all the common and one or both the seasonal requirements within unit but the heating and cooling mediums are generated at other point and delivered to the individual units.

(4) *Combined system*: A combined system may consist of the combination of (*a*) central and self-conditioned system, (*b*) central and semi-contained system, and (*c*) self and semi-contained system.

26.2 AIR-CIRCULATION

Circulation of air is an essential requirement in an air conditioned system. The total minimum outside fresh air introduced into an enclosure by an air-conditioning plant or unit is related to the number of occupants in the enclosure at any time, whether they are smokers or non-smokers and to the cubic contents of the enclosed space in the manner specified in table 26.3.

Table 26.3: Minimum fresh air requirements

Sl. No.	*Application*	*Smoking*	*Air requirement in m^3/min per person*		
			Recommended	*Minimum*	*per m^3 of Floor Area*
(1)	Apartments	Some	0.56	0.28	—
(2)	Banking space	Occasional	0.28	0.21	—
(3)	Board rooms	Very heavy	1.40	0.56	—
(4)	Departmental stores	None	0.21	0.14	0.015
(5)	Directors' rooms	Very heavy	1.40	0.84	—
(6)	Drug stores*	Considerable	0.28	0.21	—
(7)	Factories	None	0.28	0.21	0.03
(8)	Garages	—	—	—	0.30
(9)	Hospitals:				
	(*a*) Operating rooms (all fresh air)	None	—	—	0.60
	(*b*) Private rooms	None	0.84	0.70	0.10
	(*c*) Wards	None	0.56	0.28	—
(10)	Hotel rooms	Heavy	0.84	0.70	0.10
(11)	Kitchens:				
	(*a*) Restaurant	—	—	—	1.20
	(*b*) Residence	—	—	—	0.60
(12)	Laboratories	Some	0.56	0.42	—
(13)	Meeting rooms	Very heavy	1.40	0.84	0.38
(14)	Offices:				
	(*a*) General	Some	0.42	0.28	—
	(b) Private	None	0.70	0.42	0.08
		Considerable	0.84	0.70	0.08
(15)	Restaurants:				
	(*a*) Cafeteria	Considerable	0.34	0.28	—
	(*b*) Dining room	Considerable	0.42	0.34	—
(16)	Retail shop	None	0.28	0.21	—
(17)	Theatre	None	0.21	0.14	—
		Some	0.42	0.28	—
(18)	Toilets (exhaust)	—	—	—	0.60

* In case exhaust air required is more than fresh air specified, then fresh air requirements will take exhaust considerations into account.

In hospital operation theatres, the extent and quality of outside air depend upon the type of anaesthesia used. Depending on the emergency, 100 per cent outside air should normally be used and no recirculation of air should be allowed, since anaesthetic gases are explosive. The refrigerant used should be such as not to form an explosive mixture with the anaesthesia used.

Basically air circulation can be achieved by the following apparatus:

(1) Air pumps.

(2) Air delivery system consisting of supply and return ducts.

(3) Air distribution system consisting of inlets and outlets.

Air pumps used in air-conditioning are of the following two types:

(*a*) Flow, propeller or fan type

(*b*) Radial flow, centrifugal or blower type.

The propeller type is used to produce air at relatively lower pressures. They generally have two to eight blades of various grades, thicknesses and shapes. The static pressure against which fan will deliver air depends on its design. Fan operating against relatively high static pressure will deliver air from outer corners of the blades or propellers owing to the high linear speed of these sections. These fans have advantage of lower initial cost and economic operation. However, they create excessive noise.

Multi-blade or centrifugal fans are generally of straight blade, forward curved blade, and backward curved blade types. The forward curved multi-blade fan is very suitable for air conditioning works because of its low speed, quiet operation and large capacity. The backward curved blade fans are also suitable for air conditioning work. These have also the advantage that they operate more quietly and their higher speeds permit direct connection to motors. Their power requirements also do not increase considerably as the static pressure increases. While selecting a particular type of fan, its efficiency, cubic contents of the air which it gives, static pressure required by the system, type of available motive power, permissible noise level, etc., must be kept in view. Air conditioning fan must give to the air the needed velocity and pressures to overcome the resistance to the overflow set up by the ducts and also must maintain the desired velocity of the air at the exit from the supply grilles.

The air delivery system consists of the supply ducts, the return ducts, dampers and duct insulation. Ducts should be made of sheet metal to the required size and shape. The supply ducts are carefully calculated, shaped and designed for volume velocity and pressure so that proper distribution with the aid of the supply outlet may be obtained in each room or enclosure. Similarly, care should be taken for return ducts. Dampers manually or automatically operated are placed in the ducts to control the direction, velocity and volume of circulating air. Supply ducts from the air conditioned unit to the supply outlets should be properly insulated to save loss of heat, etc.

Outlets or inlets should be in the form of grilles or registers. The grille is an outlet or inlet having no damper or air control device for control. The position of the outlet and the inlet and their relation to one another are of great importance. The outlet for the supply of the conditioned air to a room is best placed in the ceiling or one side of a wall high enough from the floor to prevent direct discharge of air upon the occupants. In the case of rooms with high walls the outlets should be placed lower down to prevent distribution within the upperside of unoccupied spaces. As a rule, for wall outlets, the height is not less than 2 metres from the floor and not less than 450 mm from the ceiling. Supply outlets are usually placed on interior or cross-partitions. The object of the supply outlet should be to distribute conditioned air into the room without any draught. Inlets for the return of fowl air from individual rooms are placed at the beginning or the receiving end of the return air duct. They may be either grilles or registers depending upon the control desired and economical limitations.

26.3 CLEANING OF AIR

Atmospheric dust consists of minute particles of sand, ash, soot, chemicals, bacteria and other micro-organisms. Clean country air usually contains small concentration of dust ranging from 0.2 to 0.5 grains per 100 cu. m. of air and hence it is not harmful for inhabitants. In large steel and manufacturing concerns the dust content may be as high as 1.5 grains per 100 cu. m. which is harmful to human body. The purpose of obtaining clean air into a room is to reduce the concentration of the dust to a small fraction so that it becomes harmless. To accomplish this it is necessary to provide air filters or cleaners. Their object is to remove harmful particles from the air efficiently. They should be workable under a wide range of velocities. They should also be capable of holding a moderate amount of dust without decreasing their efficiency further and should be able to be cleaned automatically or manually. The air that passes out of them should not get contaminated with the moisture or chemicals used in the cleaner itself. Various types of air filters used are detailed below:

26.3.1 Panel or Unit Filters

Use of these is normally confined to the smaller systems up to a maximum air flow of about 5 m^3/s. They are constructed in various size ranges, such as 450 mm × 450 mm, 500 mm × 500 mm and 600 × 600 mm, and operate at face velocities between 1.0 and 2.5 m/s.

The characteristics of this type of filters are set out as follows:

(a) Viscous impingement filter—The efficiency range of these filters is 80 to 90 per cent. These are serviced by washing and reoiling, and are generally used for industrial or engine air intakes.

(b) Permanent dry type filter—The efficiency of these filters is up to 95 per cent. They may be of fabric, processed from plastics or brush type. They normally operate with initial resistance of 25 to 55 Pa rising to 50 to 125 Pa with use. Their efficiency increases with dust load, but volume flow is reduced. They are serviced by washing, or by compressed air or vacuum cleaning.

(c) Dry replaceable media type—The efficiency range of these filters is 95 per cent. They consist of fabric or wadding media on metal frames. They are serviced by replacing filter media upon a rise of 1.5 to 2 times the initial resistance which is normally between 35 and 100 Pa. They are commonly used as main filters or as prefilters to high efficiency filters, and

(d) Throw-away type—The efficiency of these filters is upto 95 per cent. They are generally of panel construction with filter media of fabric, plastic, glass or metal fibre mounted in a rigid wood, metal or cardboard frame. They are used in smaller plants where servicing requirement of permanent type would be disadvantageous.

26.3.2 Automatic Air Filters

These are primarily used on larger systems, where they usually give economic advantage over panel filters. These filters are obtainable down to a rating of 0.5 m^2/s. Air flow velocity is usually between 2.00 and 2.75 m/s. Velocities higher than this can give a reduction in overall size, but leads to a penalty in increased resistance and possibility of carry over of the filter fibres or the wetting oil.

(a) Automatic viscous type—The efficiency of these filters is up to 90 per cent. Operation is normally time-switch controlled and the filters maintain a constant operating resistance of 50 to 125 Pa according to type.

Servicing of these filters requires removal by hand of the dust deposited as a sludge from the viscous fluid tank and mechanical maintenance is required for timer mechanism, motor bearing, etc; and

(b) Automatic dry type—These can be obtained with various grades of filter media giving efficiencies between 95 and 98 per cent depending on media. Air flow pressure drop for the media are 100 to 125 Pa standard and 125 to 150 Pa increased density. Servicing consists mainly of replacement of media.

26.3.3 Electrostatic Cleaners

The efficiencies of these devices are 95 to 99 per cent. These are efficient in removal of large quantities of dust, especially in smaller particle range. Electrostatic cleaner is a two stage unit, the first stage being a series of fine wires that give the dust particles a positive electrical charge and the second stage, a series of parallel plates generally coated with a water soluble viscous solution charged negative or negative and positive alternately.

Servicing is by switch off and washing down and recoating as necessary. Cells require regular inspection for burnt plates, as also inspection and replacement of ionizing wires, rectifiers, etc., as necessary. This equipment operates on high voltage and suitable interlocks are required to ensure that access to live parts is possible only when power is off.

26.3.4 High Efficiency Particulate Air (*Hepa*) or Absolute Filters

The efficiencies of these filters is 99 to 99.995 per cent. They are invariably panel filters, most commonly 600 mm × 600 mm × 300 mm rated for 0.50 m^3/s at initial resistance of 125 to 300 Pa. Standard absolute filters use treated or glass paper media in honeycomb formation in mild steel or wooden case. Other filters are constructed of materials giving suitability for operation in high humidities, high temperatures and chemically corrosive atmosphere and are generally used in conjunction with a pre-filter. The life of these filters can vary from a few months to several years, depending on location and efficiency of the pre-filter.

26.3.5 Carbon Pack Filter

This is used to remove odours, fumes, vapours, gases, etc. from air. Carbon may be activated or catalytic and is, to some extent, selective. Air flow face velocity is comparatively low, usually of the order of 1.75 m/s. These filters quickly become clogged with atmospheric dust and use of 'throwaway' pattern or those suitable for regeneration by the manufacturer is recommended.

26.3.6 Selection Factors for Filter

When selecting a filter, the following particulars should be considered:

(*a*) the air flow rate for which filter is designed;

(*b*) the face velocity, which is the average velocity of air (m/s) entering the effective face area of the filter;

(*c*) resistance, that is, the difference between static pressure upstream and downstream specified for clean and dirty conditions;

(*d*) efficiency, that is, measure of the ability of the filter to remove dust from the air;

(*e*) dust holding capacity; which is the mass of dust that a filter can retain air flow during a rise in pressure drop from its initial clean resistance to some arbitrary maximum value, usually twice the value of pressure drop when clean;

(*f*) economic considerations should take into account the initial capital cost, cost of replacement of materials during cleaning operations and of the labour involved; and

(*g*) possible fire hazards.

26.3.7 Location and Installation

(1) Filters are normally placed upstream of the main supply fan between the pre-heat coil (if fitted) and the cooling coil. Besides producing clean room air, they protect cooling coils and other apparatus from deposition of dust. The system should be arranged to provide an even air velocity distribution across the filter face.

(2) Where a high degree of filtration is required, high efficiency filters, normally used in conjunction with pre-filters, are placed down-stream of the fan and should be the last item of equipment before the discharge point. This ensures that any air leakage is outward and the contaminated air is not drawn into the system. Another advantage is also that any contamination from the air handling equipment is captured by the final filter.

(3) Fresh air intakes should be as remote as possible from the concentration of surface of roof dirt and positioned to avoid intake of fumes and odours. Weather louvers with a wire mesh bird screen should be fitted. In some cases, the louvers may need acoustic treatment to reduce noise from or into the system.

(4) Adequate access to facilitate servicing of the filters should be provided and doors, ladders, electric lighting, etc. should be included where necessary. A manometer indicating differential pressures across the filter bank should be fitted to determine the need for filter change.

(5) All ducts should be clean and free from dust before filters are installed.

(6) The frame holding the filter media when in position should form an effective seal, so that no air bypasses the filter.

(7) Doors and latches giving access to live high voltage conductors of electrostatic precipitators should be equipped with locks.

26.4 METHODS OF REMOVING EXCESSIVE HEAT FROM AIR

The excessive heat which finds its way into the air-conditioned space or gets created must be removed by cooling and dehumidifying equipment. The latent heat is first changed to sensible heat which is achieved by condensing the excessive water vapour in the air. The exact amount of heat that is required to evaporate the moisture in air gets reduced when water vapour is condensed. This heat transfers into sensible heat in the condensing medium and results into an increase in temperature of medium. The latent and sensible heat in a building raises the temperature of the moisture content of air inside the building. This air is circulated through the heat removing equipment where the excess moisture and sensible heat are removed. After the excess heat has been removed the air together with additional amount brought from outside for ventilation purposes, is also passed through cooling equipment and then circulated through the building.

The methods of cooling generally adopted are **(1)** surface cooling, **(2)** spray cooling, and **(3)** evaporative cooling.

In surface type cooling the air is passed over coils or fins in which cold water or refrigerants are circulated. The temperature of these cooling media must be below the desired temperature of the air. Where air passes over the coils the dry bulb and wet bulb temperatures get reduced to the dew temperature of the air. When air is cooled below this point the moisture is removed. The amount of coil surface needed to cool and dehumidify the air depends on the average temperature difference between the air entering the coil and leaving the coil. This difference is 6°C to 12°C when water or brine is used. Where refrigerant is itself evaporated in the coils, there is little difference between entering and leaving temperatures. In this process of cooling and dehumidifying air, the air leaves at about 80% to 85% saturation, although the coil temperature may be at the dew point; the air leaving the coil has got a temperature a few degrees higher.

For a spray type the air passes through the spray of water which has been cooled to a temperature below the desired dew point of air and then passed through the eliminator plates which have vertical baffles to take out any water that is present in the air in the form of fog or small drops. When air enters the fine water spray a sufficient amount of water is evaporated and the heat required to evaporate this water is supplied by the sensible heat of the entering air. As this heat is taken from the air, the dry bulb temperature gets reduced.

In evaporative cooling the air to be cooled passes through a spray of air, the air by dehumidification gets cooled to the wet bulb temperature until the dry and wet bulb temperatures are the same. Refrigerants can also be used for cooling the air. Various methods for accomplishing this are used. In one of the methods the air is passed directly over the medium which absorbs the latent heat from the air and cools it. Alternatively water is passed over ice and cold water is circulated through the cooling tubes. Volatile refrigerants of inflammable or non-inflammable character may be used. Carbon sulphide, methyl chloride, chloroform, ether are the substances of the former category, while carbon dioxide, sulphur dioxide, etc. belong to the latter category. The vapours of the volatile refrigerant enter the cylinder on the compression side and get compressed. It is then passed through a condenser which is supplied with cooling water at a suitable temperature so that high vapour pressure is changed into a still higher vapour pressure, but the temperature being reduced. This liquid after being collected in a vessel is allowed to pass through an expansion valve into a chamber or set of tubes. At this place the pressure gets suddenly reduced and the liquid gets transformed into vapour. While doing this it absorbs large amount of heat from the material surrounding the tubes, which may be air, to be used in air-conditioning.

26.5 DEHUMIDIFYING AIR

Dehumidification may be accomplished by condensation or desiccation. Condensation can be done with the aid of surface cooling or spray cooling as described earlier. The desiccation is the removal of moisture from the substance and is carried out by two methods, namely:

Adsorption and Absorption: Materials which hold moisture on their surfaces are called Adsorbers. The moisture does not get absorbed into these materials but mainly remains on their surfaces. Usually activated alumina and silica jels are used as adsorbents. The air is passed through the beds of small particles of these adsorbents and dries in this manner. After the adsorption of a certain amount of moisture, the adsorbent is reactivated by heating it to a higher temperature.

Absorbents usually are solutions of ammonia, calcium or other salts, and these remove moisture from air. These salts are mixed with water and kept at certain concentration and temperature. The salt solution after being cooled is sprayed over glass wool or other material. The air to be dehumidified passes over the material and the excess moisture in air is absorbed by strong salt solution. This type of working removes all excess latent heat but does not remove sensible heat which can be removed by cooling further.

26.6 HUMIDIFICATION

In winter the outside air, because of its lower temperature, has very low humidity.

When air is humidified the heat evaporation must be added. This heat may be added to the spray water before actual humidification or it may be taken as sensible heat from the air and transformed to latent heat of evaporation for humidification. Humidification may be accomplished by (*i*) injecting a direct spray of water into room, (*ii*) introducing moistured air, (*iii*) a combined process. Washers may be considered as an indirect class of humidifying agents. The humidification of air with washer may be affected by the use of recirculating spray of water without prior treatment of entering air. Otherwise air may be pre-heated; or alternatively a heated spray of water may be used while injecting fine water sprays into the air inside any enclosure. It should be ensured that the sprays are very fine so that water does not spread on any object within the enclosure.

26.7 TYPES OF AIR-CONDITIONERS

26.7.1 Unitary Air Conditioner

These are self-contained air conditioning units comprising a compressor and evaporator with fans for evaporator and air-cooled condenser. Unitary air conditioners are generally installed in windows and, therefore, they are also known as window air conditioners. It is designed to provide free delivery of conditioned air to an enclosed space, room or zone. It includes a prime source of refrigeration for cooling and dehumidification and means for circulation and filtration of air. It may also include provision to exhaust room air as also induce fresh air for ventilation in the room. In addition to basic cooling unit, there are several other optional features available, such as:

(*a*) Means for heating during winter months.

(*b*) Reciprocating or rotary compressor.

(*c*) Swing louvers for better distribution of air in the room.

(*d*) In addition to normal, dust filters, indoor air quality filters, such as bactericidal enzyme filters for killing bacteria, low temperature catalyst filter for removal of unpleasant odours, electrostatic filters to trap particles of smoke as well as suspended matters present in the air.

(*e*) Digital LCD remote control which also indicates room temperature.

26.7.1.1 Capacity

Most of the manufacturers supply unitary air conditioners in capacities of 3 500 W (1 TR), 5 250 W (1.5 TR) and 7 000 W (2 TR). However, some of them may be able to supply window air conditioners of 1 750 W (0.5 TR) and upto 10 500 W (3 TR) alongwith intermediate range. The capacity of windows air conditioners is rated at outside dry bulb temperature of 35°C and wet bulb temperature of 30°C and they are suitable for 230 V, single phase 50 Hz power supply. Nominal capacity of all the window air conditioners has to be de-rated due to high ambient temperatures in summer months in most of Indian cities. Also, generally a voltage stabilizer has to be installed to ensure that window air conditioner gets stabilized rated voltage.

26.7.1.2 Suitability

Unitary air conditioners are suitable for bedrooms, office cabins, general office area, hotel rooms and similar applications where normal comfort conditions are required uto a distance of 6 m from unitary air conditioner.

26.7.1.3 Noise level

Noise level of window air conditioner inside the conditioned room should be as low as possible. However it should not exceed 65 dBA for 5 250 W (1.5 TR) or smaller capacity window air conditioners. Air conditioners with rotary compressors will have lower noise level as compared to those provided with reciprocating compressors.

26.7.1.4 Location

Unitary air conditioners should be mounted preferably at the window sill level on an external wall where hot air from air-cooled condenser may be discharged without causing nuisance. There should not be any obstruction to the inlet and discharge air of the condenser. Also while deciding location of the window air conditioners, care should be taken to ensure that the condensate water dripping does not cause nuisance. The opening for the air conditioner is generally made a part of window or wall construction at the planning stage.

26.7.1.5 Limitations

Room air conditioners are not generally recommended in the following situations:

(*a*) The width of the area exceeds 6 m.

(*b*) Area requiring close control of temperature and relative humidity.

(*c*) Internal zones where no exposed wall is available for the installation of room air conditioners.

(*d*) Sound recording rooms where criteria for acoustics are stringent.

(*e*) Special applications like sterile rooms for hospitals and clean room applications where high filteration efficiency is desired.

(*f*) Operation theatres where 100 per cent fresh air is needed and fire hazard exists depending on the type of anaethesia being used.

(*g*) Where required to comply with the recommended fresh air requirement for ventilation.

26.7.2 Split Air Conditioner

Split air conditioner has an indoor unit and an outdoor unit interconnected with refrigerant piping and power and control wiring. Indoor unit comprises of a filter, evaporator and evaporator fan for circulation of air in the conditioned space. Outdoor unit has a compressor, air-cooled condenser with condenser fan housed in a suitable cabinet for outdoor installation. Split air conditioner includes primary source of refrigeration for cooling and dehumidification and means for circulation and cleaning of air, with or without external air distribution ducting.

Split air conditioners may be provided with either reciprocating compressor or scroll compressor. Scroll compressor generally consumes about 10 to 12 per cent the power compared to reciprocating compressor.

Various split air conditioners available may be categorised as under:

(*a*) Exposed indoor unit, which is either a high wall unit or a floor-mounted unit.

(*b*) Furred-in units (ceiling suspended unit), which is mounted in the ceiling and provided with a duct collar and grille.

(*c*) Ducted indoor unit, which requires ducting for air distribution.

26.7.2.1 Suitability

Split air conditioners are suitable for wide range of applications including residences, small offices, clubs, restaurants, showrooms, departmental stores, etc.

26.7.2.2 Capacity

Split air conditioners are available in following capacities:

(*a*) Indoor exposed units, 3 500 W (1 TR), 5 250 W (1.5 TR), 7 000 W (2 TR) or two indoor units of 3 500 W (1 TR) or 5 250 W (1.5 TR), connected with one outdoor unit of 7 000 W (2 TR) or 10 500 W (3 TR) capacity. These units are available with corded and cordless remote control.

(*b*) Furred-in Units are available in capacities of 3 500 W (1 TR) and 5 250 W (1.5 TR) and may be provided with one outdoor unit or two outdoor units with two furred-in indoor units. These units are available with corded and cordless remote control.

(*c*) Ducted split air conditioners (ceiling suspended ducted units) are available in capacities of 10 500 W (3 TR), 17 500 W (5 TR), 26 250 W (7.5 TR) and 52 500 W (15 TR). Ducted split air conditioners with scroll compressors are available in spacities of 19 250 W (5.5 TR) and 29 750 W (8.5 TR).

26.7.2.3 Location

Split air conditioner indoor unit is mounted within the air conditioned space or above the false ceiling from where the air distribution duct is taken to the conditioned space to distribute the air. When the indoor unit is mounted in the false ceiling, inspection panel must be kept in the false ceiling to attend to the indoor unit including periodic cleaning of air filter. Outdoor unit is mounted at the nearest open area where unobstructed flow of outside air is available for air cooled condenser.

26.7.2.4 Installation

Ceiling suspended indoor units are provided with rubber grommet to reduce vibration. Outdoor units are mounted on a steel frame in an open area so that the fan of the air cooled condenser can discharge hot air to the atmosphere without any obstruction. Care should be taken to ensure that free intake of air is available to the outdoor air cooled condenser. Also precaution should be taken that hot air from any other outdoor unit does not mix with the intake of the other outdoor air cooled condenser.

26.7.2.5 Limitations

Split air conditioners are generally not recommended for:

(*a*) For areas where fresh air is required for ventilation.

(*b*) Where distance between indoor exposed unit or furred-in unit exceeds 5 m from the outdoor unit for units up to 7 000 W (2 TR) capacity. For larger ducted split air conditioners, the vertical distance between the indoor unit and the outdoor unit should not exceed about 6 m for units with reciprocating compressors. The horizontal distance between the indoor unit and outdoor unit should not exceed about 10 m for reciprocating compressors.

(*c*) Area requiring close control of temperature and relative humidity.

(*d*) Sound recording rooms where criteria for acoustics are stringent.

(*e*) Special applications like sterile rooms for hospitals and clean room applications where high filtration efficiency is desired.

(*f*) Large multi-storey building where multiplicity of the compressors may entail subsequent maintenance problems.

(*g*) Where the length of air distribution ducting may exceed about 20 m.

26.7.3 Packaged Air Conditioner

Packaged air conditioner is a self-contained unit primarily for floor mounting, designed to provide conditione air to the space to be conditioned. It includes prime source of refrigeration for cooling and dehumidification and means for circulation and cleaning of air, with or without external air distribution ducting. It may also include means for heating, humidifying and ventilating air.

The unit comprises a compressor, condenser and evaporator, which are interconnected with copper refrigerant piping and refrigerant controls. It also includes fan for circulation of air and filter. The unit is provided with compressor and fan motor starter and factory-wired safety controls.

Compressor is a device, which compresses low-pressure low temperature refrigerant gas to high-pressure high temperature super heated refrigerant gas. Compressors may be reciprocating type or scroll type for packaging unit applications.

Condenser condenses high pressure high temperature refrigerant gas to liquid refrigerant at approximately the same temperature and pressure by removal of sensible heat of refrigerant by external means of water cooling or air cooling.

The packaged units are also available with microprocessor-based controller installed in the unit for digital display of faults as also several other functions. The packaged unit can also be provided with winter heating package or humidification package. The packaged unit may be provided with either water-cooled condenser or a remote air cooled condenser with interconnected copper refrigerant piping. The units are available with reciprocating compressor as also scroll compressor, which consume about 10 to 12 percent lesser power. In a water-cooled condenser unit, condenser-cooling water is circulated through the cooling tower with necessary piping and pumpsets.

The water cooled condenser packaged unit gives higher capacity at lower power consumption as compared to an air cooled condenser packaged unit which gets considerably de-rated in capacity and also consumes more power in peak summer months in most of the cities of our country due to high ambient temperature.

Packaged units are generally available with vertical air discharge or horizontal air discharge.

26.7.3.1 Suitability

Packaged units are suitable for wide range of applications including offices, clubs and restaurants, showrooms and departmental stores, and computer rooms, etc.

26.7.3.2 Capacity

Normally the packaged air conditioners are manufactured in sizes of 17 500 W (5 TR), 26 250 W (7.5 TR), 35 000 W (10 TR) and 52 500 W (15 TR). Packaged units with scroll compressors are also available in capacity up to 58 100 W (16.6 TR).

26.7.3.3 Location

The packaged unit can be mounted within the air conditioned space with discharge air plenum or in a separate room from where the air distribution duct is taken to the conditioned space. While detecting location for the packaged unit, provision must be kept for proper servicing of the unit.

26.7.3.4 Installation

The packaged units are normally mounted on a resilient pad which prevents vibration of the unit from being transmitted to the building.

26.7.3.5 Limitations

Packaged air conditioner are not generally recommended for:

(*a*) Large multi-storey buildings where multiplicity of the compressors may entail subsequent maintenance problems.

(*b*) Where the length of air distribution ducting may exceed approx. 20 m.

(*c*) Where the vertical distance of air-cooled condenser from the packaged unit exceeds about 10 m. The sum of horizontal and vertical distances should be generally kept within 15 m.

(*d*) Special applications like sterile rooms for hospitals and clean room applications where high filtration efficiency is desired.

(*e*) Operation theatres where 100 percent fresh air is needed and fire hazard exists depending on the type of anesthesia being used.

26.8 HEATING

The installation for the air-conditioning system may be used advantageously for the central heating system with such additions as a hot water or boiler and hot water coils or strip heater banks.

For supplying heat to a building generally the steps involved are (1) fuel is burnt, (2) the generated heat by the combustion of fuel is used to heat the air or water, and (3) the heated air, water or steam is used to supply the heat lost from the building.

The heating equipments as described below are generally used:

26.8.1 Hot Water Heated Coils

Central heating systems using hot water usually required not more than one or two rows of tubes in the direction of air flow, in order to produce the desired heating capacity. To achieve high efficiency without excessive water pressure drop through the coil, various circuit arrangements are used.

Generally, the resistance to the hot water flow through the heater should not exceed 4 kPa in low pressure hot water heating installations. In high pressure hot water installations, the resistance to the water flow will probably be determined by other factors, for example, the need to balance circuits.

The heaters should be served from hot water flow and return mains with sufficient connections to each row or bank of tubes or sections to give uniform distribution of the heating medium.

The flow connections to the heater should generally be arranged at the lowest point of the heater, and the return connections at the highest, to aid venting. The expansion of the tubes when the heater is in operation should be considered and the necessary arrangements made to accommodate expansion and contraction.

Thermometer wells should be fitted in the pipes near the inlets and outlets of all air-heating coils so that the temperature drop through the heater can be readily observed.

26.8.2 Electric Air Heater

The air velocity through the heater should be sufficient to permit the absorption of the rated output of the finned tube heaters within its range of safe temperatures and the exact velocity determined in conjunction with the manufacturers of the heater. Electrical load should be balanced across the three-phase of the electrical supply.

Where automatic temperature control is required the heater should be divided into a number of sections dependent upon the degree of control to be effected.

Each section of heater elements, which may be two rows of elements should have its own busbars and connection and be capable of withdrawal from the casing, thus enabling the elements to be cleaned or repaired whilst the remainder is in operation. Each section should be capable of being isolated electrically before being withdrawn from the casing.

All heaters should be electrically interlocked with the fan motors, so that the electric heater will be switched off when the fan is stopped or when the air velocity is reduced to a level below that for which the heater has been designed.

The air velocity over the face of the heater is of particular importance in the design of electric air heaters, and the manufacturers should be given details of the maximum and minimum air velocities likely to occurs.

With all electric air heaters, care should be taken to preclude the risk of fire under abnormal conditions of operation, by the use of a suitably positioned temperature sensitive trip of the manual reset type to cut off the electric supply.

QUESTIONS

1. What is the object of air conditioning? What do you understand by the term 'comfort zone'?
2. What are the various factors which have to be taken into account while designing an air conditioning system?
3. Briefly describe the various systems of air conditioning. How is the correct circulation of air within air-conditioned building ensured?

4. Write short notes on:
   (*i*) Cleaning of air for an air-conditioning system.
   (*ii*) Humidification of air.
   (*iii*) Cooling of air for an air-conditioning plant.
   (*iv*) Dehumidifying air.
5. Describe the various types of air-filters used for cleaning of air.
6. Give the general features, capacity, power consumption, noise level and limitations of the following types of air conditioners:
   (*a*) Unitary air conditioner
   (*b*) Split air conditioner; and
   (*c*) Packaged in air conditioner.

REFERENCES

1. **Rummel A.T. Wand Vogelsang L.O.**: *Practical air-conditioning.*
2. **H.M.O.S.**: *Principles of modern building.*
3. **Merrit**: *Building construction handbook.*
4. **Hull W**: *Air conditioning primer.*
5. **Deshpande R.S.**: *Sanitary engineering.*
6. **H.M.S.O.**: *Principles of natural ventilation—A building research station digest.*
7. **Bureau of Indian Standards' New Delhi**: *National building code of India 2005--Part 8, Section 3: Air-conditioning, Heating and Mechanical Ventilation.*
8. **IS 1391 Part 1 : 1992**: *Unitary air conditioners (second revision).*
9. **IS 1391 Part 2 : 1992**: *Split air conditioners (second revision).*
10. **IS 8148 : 2003**: *Specification for packaged air conditioners (first revision).*

CHAPTER

Acoustics, Sound Insulation and Noise Control

GENERAL PRINCIPLES

Acoustics is a science which deals with the design and construction of different units of buildings to get proper acoustical conditions and also with the correction of the corresponding defects in existing rooms. It also includes the sound absorption or dissipation of exterior noise.

Sound is transmitted in the form of waves which are a series of compressions and rarefactions created in the medium through which it travels. The average sound travels at ordinary temperature with a speed of 340 m per second. This velocity depends on the medium through which it travels. The velocity of sound through water is 1310 m per second and through brick and steel is 3600 and 4900 m per second respectively.

Reflection of sound is one of the important considerations in any acoustical design problem and this leads to most of the difficulties. Repeated reflection between parallel walls sets up flutter of sound which is undesirable. In smaller rooms this sets up disturbing resonance. In the case of curved walls which are convex, the reflected sounds get diverged. Walls of focusing nature are objectionable as they concentrate the reflected sound at definite places.

If the reflected sound reaches the audience 1/15th of a second after the direct sound from the speaker, an echo is heard. If the time lap becomes less than 0.05 second, a beneficial effect called "reinforcement" results. If the compressions of direct sound combine with the compressions of the reflected sound, the result would be the creation of louder sound than either of the two alone. A similar case occurs when rarefactions get combined. If, however, the compressions of direct sound combine with the rarefactions of the reflected sound, the resulting sound will be weaker. If a whole number of half waves get set up between parallel walls, reinforcement of sound or resonance is created. If the reflected sound reaches the audience later by 0.05 second than the direct sound, then blurring of speech gets created. If a series of sound of slightly different frequencies combine at about the same time, an effect of loudness called beats is created, as at certain intervals the compressions of one wave get combined with the compressions of the other wave.

Sound in a building is generally classified as speech, music and noise. Three characteristics, *i.e.*, frequency, intensity, and tone structure, are important for common types of sound. Frequency is defined as the number of vibrations (cycles) per second and the unit is Hertz (Hz). Range of frequency is greater for music than for speech. Intensity of sound is defined as flow of sound energy per second through unit area. This is expressed in terms of bel which is the fundamental division of a logarithmic scale used to express the ratio of two mean square values of specified quantities *i.e.* sound pressure, sound power or sound intensity. The number of bels denoting such a ratio being the logarithm to the base 10 of this ratio. In bel units range of intensities that can be perceived by the ear is enormous ranging from one unit for a faint sound to 1200 units for a very loud sound. For convenient expression of these units a term decibel (*dB*) is used which means one-tenth of a bel. Some values for various types of sounds are given in Table 27.1.

Table 27.1: Intensity of sound

Feeling	*Intensity in bels*	*Decibels*	*Type of Sound*
Deafening	1200 1100	120 110	Thunder, artillery Nearly riveter Boiler factory
	1000	100	
Very loud	900	90	Loud street noise Noisy factory Police whistle
	800	80	
Loud	700	70	Noisy office Average street noise Average radio Average factory
	600	60	
Moderate	500	50	Noisy home Average office Average conversation Quiet radio
	400	40	
Faint	300	30	Quiet home or private office Average auditorium Quiet conversation
	200	20	
Very faint	100	10	Rustle of leaves Whisper Soundproof room Threshold of audibility
	1	0	

The tone characteristic is that particular property by which sounds of the same frequency and intensity can be distinguished from each other. To achieve good results, it is necessary to note the limitation of hearing which a human person can have. The range of frequencies which can be heard by human ear ranges from 20 Hz to 20000 Hz and the intensity varies by about 120 decibels in the region of 1,000 cycles.

When a speaker addresses an audience, the sound he utters proceeds outward in the form of spherical waves until they strike the boundary of the room where they can be reflected, transmitted and absorbed in various amounts depending on the character of the walls.

Though the reflection of sound produces some beneficial effect in the loudness, it unfortunately introduces possibilities for defective acoustics. When the walls of the room are hard and smooth, very little energy is lost at each impact and many reflections take place before the sound dies down. This prolongation or reverberation as it is called is the main difficulty in the auditoriums.

If sounds persist too long, then successive words of a speech will overlap and produce confusion. Special materials acting as absorbers can be introduced which work by reducing the time taken for the dying down of a sound. Reverberation is less objectionable for music than for speech since the prolongation of musical tones is usually desired but the mixing of words of a speech is highly disadvantageous. While constructing halls used both for music and speaking, it is common to choose an average time of reverberation that is somewhat too long for the best conditions for speech and somewhat too short for music, yet it is somewhat satisfactory for both. Reflection of sound may create echoes, interference, diffraction, blurring and resonance. Perfect acoustical conditions in an auditorium may be expected when an average sound has a suitable loudness for all its audience with no echo for distortion of the original sound and when it dies out quickly so as not to interfere with the succeeding sounds.

Reverberation as stated earlier is the most usual difficulty in acoustics. Large experimental work on corrections of reverberation in auditoriums was done by Prof. W. C. Sabine. He devised a formula $T = \dfrac{0.16V}{\sum Ai}$, where T is time of reverberation or the standard time taken for sound in a room to decay to one millionth of its initial value, V is the volume of the room in cubic metre, (m^3) Ai is equivalent sound absorbing area in the room in square metres (m^2). Each A_i is determined by multiplying the area of the surface in m^2 by the absorption coefficient, α. The surface of walls, ceiling, floor and each significant feature is to be considered. When making correction in the reverberation, the sum of absorption value of each material in the room is taken into account.

27.1 SOUND INSULATING MATERIALS

All materials absorb sound but to a varying extent. The sound energy has to be converted into the heat energy ultimately. On striking the solid material the sound waves experience greater resistance than while travelling in air. If sound waves strike a resilient and porous surface, considerable energy will be dissipated as heat in passing through its pores which are interconnected through a series of small channels. The resultant absorption is relatively very high. An open window does not interfere with the free passage of sound and it is considered as 100% absorbent, since it transmits the entire sound. The absorbing capacity of other materials is compared with this open window unit as a standard. The capacity of different materials in absorbing sound depends on the frequency of the sound. Sound insulation values of various types of materials and construction are given in table 27.2.

Table 27.2: Sound insulation values for various types of materials and construction

Examples of sound reduction values

Sl. No.	*Material or Type of Construction*	*Average Weight kg/m²*	*Average Sound Reduction (dB)*
	Single Panel Continuous Construction		
1.	19.0 mm particle board	12.0	26
2.	3-mm sheet glass	9.5	25
3.	10 mm gypsum plasterboard	10.0	25
4.	6.5 mm plate glass	17.0	28
5.	6.0 mm asbestos cement sheet	12.25	26
6.	50 mm wood-wool cement slab plastered 12.5 mm both sides	70.0	35
7.	200 mm hollow dense concrete block plastered 12.5 mm both sides	245	45

(Contd.)

8.	Half brick (115 mm) brickwork, plastered 12.5 mm both sides	270	45
9.	One brick (230 mm) brickwork, plastered 12.5 mm both sides	490	50
10.	Two brick (460 mm) brickwork, plastered 12.5 mm both sides	930	55
	Semi-discontinuous Constructions		
1.	20 mm boarding on timber joists, with 12.5 mm gypsum plasterboard ceiling with skim-coat plaster		30–35
2.	20 mm boarding on timber joists, with expanded metal lath and 20 mm gypsum sanded plastering ceiling		35–40
3.	As in (2), plus pugging on trays at 80 kg/m^2, the pugging to be airtight (for example, sand and mortar)		45–48
4.	As in (2), plus floating floor boarding on battens on resilient quilt		43–45
5.	Timber stud partition with metal lath and 20 mm plaster both sides (80 kg/m^2)		35
6.	Double partition of half (115 mm) brick each with 50 mm cavity and thin wire ties and 12.5 mm plaster both sides		50–55
7.	Cupboards used as partitions		25–35
	Discontinuous Constructions		
1.	Two leaves of 60 mm concrete plastered and separated by a 50 mm air-space associated with suspended ceiling and floating floors:		
	(*a*) Measured across the walls		50–60
	(*b*) Measured through the floor		65–75
2.	As in (1), but with a third leaf or 50 mm concrete between the two discontinuous shells; each air-space 50 mm, measured across the walls or through the floor		66–75
3.	As in (1), but with walls of half (115 mm) brick, plastered:		
	(*a*) Measured across the walls		60–70
	(*b*) Measured through the floor		65–75
	The values given for these three items are for walls without wall ties. Wire ties effect a slight reduction (about 5 dB) and strip metal tie a considerable reduction in these values. To achieve sound insulation values above 55 dB, considerable attention need be paid to the question of flanking transmission. The construction must be fully discontinuous to achieve the above values.		

Note: Variations in sound insulation due to the use of different plasters do not appear to be significant.

Sound reduction values of typical windows

Sl. No.	*Description of Window*	*Average Sound Reduction dB*
1.	Open, jn average furnished domestic room or in class-rooms occupied by pupils	5
2.	Open, but with opening reduced to about 0.5 m^2	10
3.	Closed, single 3.0 mm glass (9.5 kg/m^2)	25
4.	Closed, double 3.0 mm (9.5 kg/m^2) glass, 100 mm apart	35
5.	Closed, double 3.0 mm (9.5 kg/m^2) glass, 150 mm apart	40
6.	Closed, single 5.5 mm (17 kg/m^2) plate glass	25
7.	Closed, double 5.5 mm plate glass, 15 cm apart	45

Impact sound insulation of typical floors other than timber floors

Sl. No.	*Treatment*	*Improvement Over Bare Concrete Floors (dB)*
	Resilient Surface Material on Concrete	
1.	Bare concrete	0
2.	Linoleum	5
3.	Wood blocks or thin carpet or rubber	5–10
4.	Carpet or underfelt	10
	Floating Floor	
	(*a*) 50 mm *Dense Concrete Screed on*:	
1.	Mineral wool/glass wool quilt 25 mm nominal thickness (density 80 – 120 kg/m^3)	15–20
2.	Mineral wool batt (density 80 kg/m^3), two nominal 25 mm layers	25
	(*b*) *Boarding on battens on*:	
1.	Clips, unlined	5–10
2.	Asbestos or felt pads, 12.5 mm	5–10
3.	Fibreboard pads, 12.5 mm	10
4.	Felt pads, 25 mm or rubber pads, 12.5 mm	10–15
5.	Mineral wool batts, 25 mm nominal thickness	15–20
6.	Glass wool batts, 25 mm nominal thickness or rubber pads 25 mm	15–20
	Suspended Ceiling (see Notes 1 and 2)	
1.	10 mm (single coat or 12.5 mm (two coats) plasters on 12.5 mm fibreboard on 50 mm × 50 mm battens in clips	5–10
2.	5 mm (single coat) or 12.5 mm (two coats) plaster on 10 mm plasterboard, on battens in felt-lined clips	10–15

Sound insulation of timber floors

Sl. No.	*Treatment*	*Improvement Over Floor Described in Sl. No. 1*
1.	Boarding on joist with ceiling of lath and plaster, or plaster on plasterboard	0
2.	As in (1), plus carpet or underfelt	5–10
3.	As in (1), plus floating floor of boarding on battens on 12.5 mm fibreboard	5
4.	As in (1), plus floating floor of boarding on battens on 25 mm mineral wool battens	10–15
5.	As in (1), plus pugging of sand or ashes 80 kg/m^2 or mineral wool 10 kg/m^2	8–10
6.	As in (1), plus pugging of sand or ashes 100 kg/m^2	10
7.	As in (1), but with floor replaced by boarding on cross battens on 25 mm mineral wool batts	5–10
8.	As in (7), plus pugging of sand 50 kg/m^2	20
9.	Boarding on joists with separate joists for ceiling	5
10.	Carpet on underfelt with boarding on joists for ceiling as above	10–15

(Contd.)

11.	Boarding on joists, pugging of sand or ashes 10 kg/m^2 with ceiling as above	15
12.	Floating floor, boarding on battens on 25 mm mineral wool batts on sub-boarding, with joists and ceiling as above	15–20

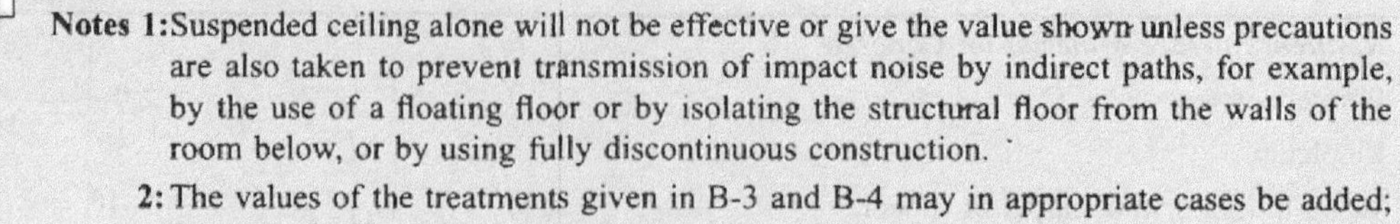

Notes 1: Suspended ceiling alone will not be effective or give the value shown unless precautions are also taken to prevent transmission of impact noise by indirect paths, for example, by the use of a floating floor or by isolating the structural floor from the walls of the room below, or by using fully discontinuous construction.

2: The values of the treatments given in B-3 and B-4 may in appropriate cases be added; for example, the value of pugging may be added to the value of a floating floor, giving a total noise reduction of 20 dB.

Absorption materials may be classified into four groups: The first group includes soft materials like hair felt which are very good absorbers as they have large pores with inter-connected channels. They are now replaced by rock wool, asbestos, etc. Semi-hard materials in the form of porous fibre boards which are stiff serve as sound absorbent materials as well as building panels. This forms the second group of these materials. In the third group are included the porous tiles of masonry and other products which are installed on the wall surfaces. In the fourth group are the classified acoustical plasters which have the advantage that they resemble the ordinary plasters and can be used where the tiles do not fit in the architectural design. However, they need extra care while fixing and may have to be specially treated. The main properties of good absorbing material are its sound absorbing efficiency, its initial cost, fixing cost, its durability, appearance, fire and vermin proof qualities, weight and reflection of light.

27.2 OPTIMUM TIME OF REVERBERATION

The time taken for the sound of a room to decay is the main factor for getting good acoustical conditions. Too long a time results in an overlap of speech. The reverberation time of greater than 5 seconds is considered to be very bad, between 5 and 3 seconds as bad, between 3 and 2 seconds as fairly good, between 2 and $1\frac{1}{2}$ seconds as good and between $1\frac{1}{2}$ and $\frac{1}{2}$ seconds as very good. This shows that the shorter times of reverberations are better. It has been seen that the presence of people in an auditorium reduces the time of reverberation. It would be desirable to make the rooms independent of the volume of the listeners. While considering the optimum time of reverberation the frequency of sound is to be taken into account. Short optimum reverberation is needed for reproduced sound than for original sound as some reverberation is already introduced at the time of recording. The values of optimum reverberation time which are corrected for the presence of audience are given in Table 27.3.

Table 27.3: Reverberation time and audience factor

Type of building	*Optimum reverberation by Sabine formula*	*Audience factor for design purposes*
Law Courts, Conference and Committee Rooms	1 to 1.5 sec	one-third.
Parliament House, Council Chambers	1.3 to 1.5 sec	Quorum.
Public Lecture Hall	1.5 to 2 sec	one-third.
Music Concert Hall	1.6 to 2 sec	full.
Cinema Theatres	1.3 sec	two-thirds.
Churches	1.8 to 3 sec	two-thirds.
Very Large Halls	2 to 3 sec	two-thirds.

27.3 ACOUSTICAL CORRECTION

The important elements of acoustics of an auditorium are the elimination of echoes and focal concentrations that set up unequal distribution of sound in a room and a reduction in the time of reverberation. Uniformity in the distribution of sound is brought about by adjusting the shape of the auditorium and its interior surfaces so as to control the reflection of sound. The reduction in time of reverberation is usually accomplished by installing sound absorbing materials. The formula for the optimum time as stated earlier is $T = 0.16\ V/\Sigma Ai$. This means that if the volume can be reduced considerably, there can be a reduction in the optimum reverberation. This solution is not practicable as the size of the room is always to be fixed at a certain value.

The reduction of the factor 0.16 does not seem to be possible. The value of 0.16 as a factor means the reduction of a sound level of 60 decibels to about 15 decibels which is hardly heard.

The usual practice is to increase the factor ΣAi and this is possible because it needs only the increase of sound absorbing materials which can be added. The procedure for the acoustical correction would be as below:

1. To calculate the total volume of the vacant spaces within the enclosure which has to be corrected.
2. To measure the surface areas of the different portions of the enclosures and note down their corresponding absorption coefficients.
3. To calculate the total absorption units available in the room by multiplying the surface of different materials by corresponding absorption coefficients and get total absorption quantity available in the room.
4. Compare the optimum time which should be obtained from Sabine formula with the data collected above.
5. If the absorption as provided is insufficient, then make up the deficiency by adding absorption units to obtain optimum reverberation. The additional units of absorption needed may be found by adding surface areas of materials with greater absorption coefficients.

PLANNING AND DESIGN AGAINST OUTDOOR NOISE

27.4 GENERAL

Planning against noise should be an integral part of town and country planning proposals, ranging from regional proposals to detailed zoning, and three-dimensional layouts and road design within built-up areas. Noise nuisance should be fully recognized in zoning regulations.

Noise is either generated by traffic (road, rail and underground railway) or it arises from zones and buildings within built-up areas (industry, commerce, offices and public buildings). For planning, the noise survey should examine all the possible causes of noise and consider the various factors causing actual nuisance.

Noise by night, causing disturbance of sleep, is more of nuisance than noise by day. For this reason, housing colonies that adjoin areas with heavy traffic movement during the night are liable to cause serious complaints. Also, the factories that work by night are liable to cause serious complaints if housing estates adjoin them. While planning, care should be taken that housing colonies are adequately set back from busy airports, state and national highways, factories, main railway lines and marshalling yards.

There are two aspects of defence by planning. The first is to plan so as to keep the noise at a distance. Under this aspect comes the separation of housing from traffic noise by interposing buffer zones, and the protection of schools and hospitals by green belts, public gardens, etc. The second is the principle of shading or screening. This consists of deliberately interposing a less vulnerable building to screen a more vulnerable one or by providing a solid barrier, such as a wall, between the source and the location to be protected.

27.5 TRAFFIC NOISE LEVELS

27.5.1 For Air Traffic

For guidance, approximate noise levels due to various types of aircrafts, measured on ground, when the aircrafts fly overhead at a height of 450 m, are given in Table 27.4.

Table 27.4: Typical noise level of some aircraft types

Sl. No.	*Type of aircrafts*	*Flyover noise levels at 450 m with take-off thrust (EPN dB)*
(*i*)	Boeing 737	107
(*ii*)	Boeing 747-200	103
(*iii*)	Airbus A 300	101
(*iv*)	Concorde SST	114

27.5.2 For Rail Traffic

Noise levels of some typical railway traffic are given in Table 27.5.

Table 27.5: Typical noise levels of railway trains

Sl. No.	*Type of train*	*Noise level at 30 m, measured on the side or in the direction of train, dB (A)*
(*i*)	Steam train, 60 kmph	85
(*ii*)	Diesel train, 60 kmph	83
(*iii*)	Electric train, 60 kmph	77

27.5.3 For Road Traffic

The level of noise generated by road traffic depends upon such factors as the number of vehicles passing per hour, the type of traffic, the preponderance of heavy vehicles, average speed, gradient and smoothness of traffic flow. The smoothness of traffic flow also affects variability of the noise and is governed by such things as roundabouts and traffic lights, and the volume of traffic and pedestrian movement with their effects on stopping, starting and overtaking. The level of traffic noise fluctuates continuously and the way it does has a considerable effect on the nuisance caused. For assessing traffic noise, noise is measured in dB (A). Because of the fluctuating nature of traffic noise, a unit known as L_{A10} or L_{eq} may be used. Typical noise levels due to different volumes of traffic flow with a varying mix of vehicles are given in Table 27.6.

Table 27.6: Typical noise levels due to free-flowing road traffic

Sl. No.	*Type of traffic*	*L_{A10} 30 m from edge of road, dB (A)*
(*i*)	5000 vehicles per 18-hour day (10 per cent heavy vehicles), 50 kmph	65
(*ii*)	10,000 vehicles per 18-hour day (20 per cent heavy vehicles), 60 kmph	70
(*iii*)	10,000 vehicles per 18-hour day (40 per cent heavy vehicles), 80 kmph	75
(*iv*)	20,000 vehicles per 18-hour day (40 per cent heavy vehicles), 80 kmph	77

Note: The values are applicable to free-flowing traffic without honking.

27.6 PLANNING AND DESIGN

27.6.1 Air Traffic

Near airports two sources of aircraft noise should be considered.

(a) Flyover noise

Flyover noise is that which occurs under flight paths close to airports and is the most serious and common problem. As the aircraft passes overhead, the noise level at any particular location rises to a peak and then decreases.

(b) Ground noise

The noise emitted by an aircraft during ground operations is less variable in direction than flyover noise, but is usually of a longer duration.

Aircraft noise may disturb sleep, rest and communication, and as such may be considered potentially harmful to health. It is important that no new development is carried out within areas where the expected noise levels will cause mental and physical fatigue or permanent loss of hearing. In case development in such areas is essential, adequate sound insulation shall be provided for the building.

As the problems caused by aircraft noise have become more acute, a number of methods have been devised for evaluating noise exposure in the vicinity of airports. They all combine many factors into a single number evaluation. A commonly used criterion is the *noise exposure forecast* (NEF). The NEF is used primarily to develop noise contours for areas around airports. It has been accepted generally that noise exposure forecast levels greater than NEF 40 are unacceptable to people while levels less than NEF 25 are normally acceptable.

While it is theoretically possible to provide sufficient insulation to achieve an acceptable indoor noise environment in the area of very high outdoor noise, there is a level above which aircraft noise seriously affects living conditions no matter how much sound insulation has been applied to the dwelling unit. For this reason it is recommended that no residential development be allowed beyond the NEF 35 level.

During summer months, the windows are normally kept open for adequate ventilation. In view of this, no matter how much sound insulation is provided for the building structure, the noise level inside the room can never be less than 10 dB below the outdoor noise level. For very critical buildings, such as buildings necessary for maintaining and supplementing the airport services, and for commercial development such as hotels, it is possible to provide sealed windows and to centrally aircondition the entire building. However, it is not feasible for most of the residential developments in the country. In such cases proper zoning regulations and siting of vulnerable buildings away from aircraft noise are of vital importance.

27.6.2 Rail Traffic

This is a very serious source of noise in built-up areas, both by day and by night. Railway cuttings reduce the spread of noise, whereas embankments extend it. The elevated railway on viaducts or embankments is very common in built-up areas. The elevation increases exposure to noise but in addition the construction of the viaduct may affect the propagation of noise. In this respect solid embankments are preferable to built-up arches, which tend to act as sound boxes. Worst of all are the steel bridges, which greatly magnify the noise due to vibration. Uphill gradients are another feature tending to increase noise, especially of heavy goods trains.

Wherever possible, no residential or public building zone should abut the railway lines, especially on the marshalling yards which are particularly objectionable because of the shrill, clanging and intermittent noise they generate, often at night. The appropriate zones alongside railway lines are industrial and commercial buildings other than office buildings. Where these precautions are not practicable and housing has to abut the railway lines, every attempt may be made to house as few people as possible in the vicinity of the railway lines.

The underground transportation system is being introduced in India. Experience with subway trains in the western countries has indicated that they can be both noisy and uncomfortable to passengers. Also wayside vibration can be a major cause of disturbance for the neighbouring community. Very high noise levels are propagated to long distances by the underground high speed railway, as a result of wheel rail interaction. Both air-borne noise and ground or structure-borne vibrations are potential sources of complaints. Noise control measures, therefore, need to be considered for the following:

(*a*) In stations, where high noise levels are produced at the arrival and departure of trains:

(*b*) In tunnels, during high speed train movement;

(*c*) Where an underground rail transit system passes close to existing structures or high-rise buildings adequate attention should also be paid to the problem of ground vibration transmitted to the building, and proper isolation should be provided for critical areas; and

(*d*) In transit cars, where sound insulation is of vital importance to provide comfortable conditions for the commutators.

27.6.3 Road Traffic

Convoys of long-distance heavy trucks at night moving past through built-up areas cause serious noise complaints. On busy roads, the noise of continuous traffic may be a worse nuisance than that of railways. At least the same precautions may, therefore, be taken in the planning of dwellings in relation to arterial and trunk roads as with railways. Care may be taken that local housing roads do not provide short-cuts for heavy traffic through residential areas. Hilly roads present the additional noise of gear changing. Trees with heavy foliage planted on both sides of carriageway help slightly to muffle the noise, provided the foliage extends for a considerable distance (30 m or more).

Road traffic may give rise to serious nuisance particularly on busy thoroughfares, between continuous high buildings in main streets, at the traffic lights, near bus stops, on steep slopes and in parking spaces and enclosed yards.

For zoning and planning new buildings in urban areas it is recommended that external L_{A10} is limited to a maximum of 70 dB(A) when the dwellings are proposed to have sealed windows and 60 dB(A) when the dwellings are proposed to have open windows. Indeed it is desirable to confine major new residential development to locations subject to L_{A10} levels substantially lower than those given above.

It is recognised, however, that within the large urban areas, the use of sites where the external L_{A10} is greater than 60-70 dB(A) cannot always be avoided. In that case it is suggested to utilize such design solutions as barrier blocks in order to reduce external L_{A10} noise levels to at least 60-70 dB(A) at any point 1.0 m from any inward looking facade. When the orientation of site and the density of development are such that this cannot be fully achieved, some form of dwelling insulation will have to be provided. It should be appreciated that where open windows are a must, the occupants would have to put up with discomfort if the above conditions are not met.

Certain other methods can often be utilized to provide economical and effective protection from noise:

(*a*) Methods may be adopted to improve the smoothness of flow and reduce number of stopping and starting. This leads to an improvement even if it leads to increased flows. Flow linking of traffic lights, for example, may reduce noise nuisance.

(*b*) Use of roads passing through residential areas may be prohibited to heavy commercial vehicles. An alternative would be to limit use by commercial vehicles to certain times of the day.

(*c*) Use of honking may be prohibited near sensitive buildings, such as hospitals and the like.

(*d*) Barriers may be provided to shield sites from noise.

27.7 ZONING

The zoning of the different cities shall be done by the town planning authorities, taking into account, besides other aspects, the noise levels from different occupancies.

27.8 GREEN BELTS AND LANDSCAPING

Where relief from noise is to be provided by means of green belts, these may be of considerable width and be landscaped. (In case of railway tracks, a minimum distance of 50 m to 70 m may be provided between the buildings and the tracks.) The extent of relief that may be derived from the above may be estimated only after considering other environmental factors. Only thick belts of planting (greater than 30 m) are of real value. Strong leafy trees may be planted to act as noise baffles. Shrubs or creepers may also be planted for additional protection between tree trunks; artificial mounds and banks should be formed where practicable. As little hard paving and as much grass as possible may be used. The creation of green belt is particularly advisable on the perimeter of aerodromes, along railway lines and arterial roads, through or past built-up areas and adjoining noisy industrial zones.

27.9 HIGHWAY NOISE BARRIERS

Barriers are often the most effective means of reducing traffic noise around residential areas. They have the great advantage that they generally protect most or all of the site. In nearly all situations a well-designed barrier of even a modest height (say 3 m) can at least ensure that all areas of open space are free from excessive noise levels.

There are two types of barriers that can be built to protect sites; one which are built solely for the purpose of reducing noise and two, which form part of the building complex (barrier blocks). Free standing walls and artificial mounds are typical examples of the first type while single and multi-storeyed dwellings and/or garages are the most common form of the second.

Of the two types, barrier blocks are more widely used because they are cheaper and also tend to form a more effective barrier overall because of their greater height and width. Barrier walls or mounds are more limited in their effect than barrier blocks for they protect little more than the area of the site close to ground level essentially because of the lack of height, as continuous walls much higher than 3 m are often difficult to construct.

PLANNING AND DESIGN AGAINST INDOOR NOISE

27.10 ACCEPTABLE INDOOR NOISE LEVELS IN BUILDINGS

The generally acceptable noise levels inside buildings from the point of view of comfort, economy and practical consideration under the conditions prevailing in this country may be taken as given in Table 27.7.

Table 27.7: Acceptable indoor noise levels for various buildings.

Sl. No.	*Location*	*Noise Level dB(A)*
(*i*)	Auditoria and concert halls	20–25
(*ii*)	Radio and TV studios	20–25
(*iii*)	Cinemas	25–30
(*iv*)	Music rooms	25–30
(*v*)	Hospitals and cinema theatres	35–40
(*vi*)	Apartments, hotels and homes	35–40
(*vii*)	Conference rooms, small offices and libraries	35–40
(*viii*)	Court-rooms and class-rooms	40–45
(*ix*)	Large public offices, banks and stores	45–50
(*x*)	Restaurants	50–55

27.11 VULNERABLE BUILDINGS

Some buildings or parts of buildings are specially vulnerable to noise, for example, recording and radio studios, hospitals and research laboratories. These should not be sited near loud noise sources. Most vulnerable buildings contain some areas which are themselves noisy and in such buildings the less vulnerable elements should be planned to act as noise buffers. Most noisy buildings also contain quiet accommodation, which equally may be planned to act as a buffer between the noisy part of the building and adjoining vulnerable buildings.

The details of site and internal planning and insulation requirements are covered under individual occupancies (5 to 12) as applicable to the respective character and sources of noise in different buildings.

27.12 SOURCES OF NOISE NUISANCE IN RESIDENTIAL BUILDINGS

Outdoor Noise

The main sources of outdoor noise in residential areas are traffic (aeroplane, railways, roadways), children playing, hawkers, services deliveries, road repairs, blaring loud-speakers and various types of moving machinery in the neighbourhood and building operations.

Indoor Noise

As far as indoor noises are concerned, conversation of the occupants, footsteps, banging of doors, shifting of the furniture, operation of the cistern and water-closets, playing of radios, television, music system, cooling and ventilating machinery, etc. contribute most of the noise emanating from an adjacent room or an adjacent building. Noise conditions vary from time to time and noise which may not be objectionable during the day may assume annoying proportions in the silence of the night when quiet conditions are essential.

In the case of flats the main sources of noise are from other flats and from stairs, lifts and access balconies. Plumbing noise is another cause. In semi-detached buildings, outdoor noises from streets are noticed more than indoor noises from neighbours.

27.13 RECOMMENDATIONS FOR NOISE CONTROL IN RESIDENTIAL BUILDINGS

27.13.1 Site Planning

The most desirable method is to locate the residential buildings in a quiet area away from the noisy sources like the industrial areas, rail tracks, aerodromes, roads carrying heavy traffic, etc.

To minimize ground reflection, the dwellings should be surrounded by the maximum amount of planting and grassed areas and the minimum amount of hard surfacing. This applies particularly to high density areas. Where for maintenance reasons a large amount of hard paving is necessary, it should be broken up by areas of planting and grassing. Narrow hard paved courts should be avoided between adjacent tall buildings.

Roads within a residential area should be kept to a minimum both in width and length, and should be designed to discourage speeding. Area-wise planning, with zones from which vehicular traffic is altogether excluded, will greatly help to reduce noise. Through traffic roads should be excluded from residential areas, but where sites have to be developed adjacent to existing major roads, the same principles should be observed in the siting of blocks as with railway lines.

Play areas for older children should be sited as far away from dwellings as possible. Special care should be taken with old people's dwellings. They should not be placed immediately adjacent to service entries, play spaces, or to any entrances where children may tend to congregate.

27.13.2 Internal Planning

The orientation of buildings in a locality should be planned in such a way as to reduce the noise disturbance from neighbourhood areas. The non-critical areas, such as corridors, kitchens, bathrooms, elevators and service spaces, may be located on the noisy side and the critical areas, such as bedrooms and living space, on the quiet side.

(i) Windows and Doors—Windows and doors should be kept away from the noisy side of the building as given below, wherever possible:

(*a*) When windows of a building, particularly those of bedrooms in apartments or flats, face roads carrying heavy traffic or other noises where the external noise is of the order of 80 to 90 dB(A), the building should be located at a distance of about 30 m from the road, but a distance of 45 m or more, where possible, should be aimed at for greater relief from noise;

(*b*) When the windows are at right angles to the direction of the above type of noise, the distance from the road should be arranged to be about 15 to 25 m; and

(*c*) In case another building, boundary wall or trees and plantations intervene between the road traffic and the house/flat, further noise reduction is achieved and in such cases the above distances may be reduced suitably.

(ii) Layout plans—It is desirable that rooms adjoining party walls and above/below party floors should be of similar use. By this means, bedrooms are not exposed to noise from adjoining living rooms, and there is less risk of disturbance of sleep.

In semi-detached houses, the staircase hall and kitchen should adjoin each other on each side of the party wall, thus providing a sound baffle between rooms requiring quiet conditions.

Open fireplaces on party walls should be avoided as far as possible; bedrooms should not be planned alongside access balconies, and preferably not underneath them. Where the approach is by an internal corridor, a sound baffle may usefully be provided by arranging internal passages and bathrooms between the corridor and the living room or bedrooms.

Water-closets should not be planned over living rooms and bedrooms, whether within the same dwelling or over other dwellings. Soil pipes should not be carried in ducts which adjoin living rooms or bedrooms unless the side of the duct next to these rooms is a solid wall containing no inspection openings. Refusè chutes should not be planned next to living rooms or bedrooms.

27.13.3 Sound Insulation

(i) Reduction of Air-borne Noise. The weighted sound reduction index, R_w, of partitions between individual rooms or apartments of a building unit shall be as given in Table 27.8. These values may, however, be suitably increased, where required, for critical areas.

Table 27.8: Sound insulation between individual rooms (air-borne)

Sl. No.	Situation	Rw, dB
(i)	Between the living room in one house or flat and the living room and bedrooms in another	50
(ii)	Elsewhere between houses or flats	45
(iii)	Between one room and another in the same house or flat	35

Notes 1: Where communicating doors are provided, all doors should be so designed as to provide recommended insulation between the rooms.

2: There are cases when a set of houses or flats have to be built for the people who work at night and sleep during the day. It is desirable to consider the design of at least one such room in each of the houses or flats which will provide an insulation of about 45 dB in that room.

3: The insulation values referred to are applicable with doors and windows shut.

(ii) Suppression of noise at the source itself. All items of equipment that are potentially noisy should be selected with care. Water-closet cisterns should not be fixed on partitions next to bedrooms or living rooms. Plumbing pipes should be isolated from the structures. Lift motors should be mounted on resilient supports. Access doors from machine rooms to internal staircases should be well fitting and of solid construction.

(iii) Reduction of air-borne noise transmitted through the structure. Reduction of air-borne noise requires the use of rigid and massive walls without any openings. Openings are the major cause of penetration of noise through a barrier. While designing it should be borne in mind that all components should provide a sound transmission compatible with that of the rest of the barrier so that an equivalent amount of sound energy is transmitted through each portion of the barrier.

Ventilating ducts or air transfer openings where provided should be designed to minimize transmission of noise. For this purpose, some sound attenuating devices may be installed in these openings.

All partitions should be sealed effectively where they abut against rest of the structure. All doors and windows should be properly gasketed where a high degree of sound insulation is desired.

(iv) Reduction of structure-borne noise. This requires the use of discontinuous or non-homogeneous materials in the construction of the structure.

(v) Reduction of impact noise. The floor of a room immediately above the bedroom or living room shall provide 1 impact sound pressure level not greater than 60 dB. For example, 150 mm thick concrete floor with thick carpet (12 mm) covering would satisfy this requirement.

(vi) Main staircases in blocks of flats are often highly reverberant. Some of the surfaces at least (for example, the soffits of stairs and landings) should be finished with sound absorbent materials wherever required.

ACOUSTICAL DESIGN OF IMPORTANT BUILDINGS

27.14 LAW COURTS AND COUNCIL CHAMBERS

It is important that law courts and council chambers be protected from the intrusion of outdoor noise and from indoor noise arising both from ancillary offices and circulation spaces. The general recommendations on site planning apply to law courts and muncipal buildings, but in the larger buildings at least, further protection against outdoor noise can be obtained by planning offices and other rooms around the court rooms or chambers, and separating the offices from the central rooms by means of corridors. This arrangement is usually convenient to the function of the buildings.

The wall between the corridors and the central rooms should have a sound reduction index, R_w of not less than 50 dB (for example 230 mm brick) to insulate against air-borne noise in the corridors. Entrances from halls or corridors into court rooms or council chambers should be through baffle lobbies with two sets of quiet action doors. Sound absorbing treatment on ceilings and upper parts or walls or entrance lobbies is recommended.

The whole of the floor of the court room or chamber including steps and seating areas set aside for the public should have a resilient floor finish to reduce the noise of footsteps and shuffling of feet. Any tip-up seats should be quiet in action.

Sound absorbing treatment applied for acoustic purposes serves also to reduce the build-up of noise within the room and, part of the treatment should be applied in a band to the perimeter of the ceiling to absorb intruding outdoor noise. It is often desirable to keep the centre part of the ceiling free of absorbent material for acoustic reasons.

27.15 LIBRARIES, MUSEUMS AND ART GALLERIES

Quiet conditions for reading and study are essential in these types of buildings and, since their occupancy is not noise producing, intruding noise is more noticeable and distracting. Every opportunity therefore should be taken to plan for noise defence, both in respect of siting of the building and internal planning. When possible, stack rooms, store rooms and administrative offices should be planned to screen reading rooms, print rooms and lecture rooms from noise sources. In public libraries, the reference library and lecture rooms should receive first consideration; the lending library, newspaper and periodical rooms have a higher background noise and are secondary in importance.

In large libraries, museums and art galleries echoes from lofty, large domed or concave ceilings are often a nuisance. Small noises such as footsteps, coughs, chair scraping and closing of books are reinforced by reverberation, and concave surfaces even when treated with a sound absorbent may focus these noises. Treated flat ceilings, if not too high, obviate these troubles. Books on shelves in libraries constitute a valuable wall absorbent.

Floor finishes are important. The impact noise of footsteps on marble, terrazzo or wood block flooring, and especially on hardwood strip and batten flooring, can be disturbing both within the room in which the noise is generated and the rooms below. On solid floors, resilient floor finishes, such as rubber, cork and linoleum on an underlay, are highly desirable. In the children's sections of libraries and museums they are essential. In existing buildings, rubber linoleum or vinyl asbestos tiles laid over the floor in the traffic areas are often a solution to the problem.

Reference libraries in universities, research establishments, office buildings and science buildings having machines and testing benches, should be planned in a quiet part of the building. Walls enclosing the library should normally have a sound reduction index, R_w of not less than 50 dB (for example 230 mm brick) and baffle lobbies should be planned between the library and halls and corridors. Walls facing on to corridors or other noisy areas should not have fanlights unless they are double glazed and non-operable.

27.16 AUDITORIA AND THEATRES

The sources of noise that have to be considered in concert halls, opera house, theatres and similar auditorium buildings are as follows:

(*a*) Outdoor noise entering through walls, roofs, doors, windows or ventilation openings;

(*b*) Noise from any other hall in the same building, especially if let out separately for revenue;

(*c*) Noise from foyers, service rooms and other ancillary rooms, particularly rehearsal rooms; and

(*d*) Noise from air conditioning plant, etc, and the cross-transmission of other internal noises *via* ventilating duct system.

Because of greatly increased outdoor noise, all auditorium buildings now need more care in siting than formerly. For listening to speech or music, a very low background noise level is desirable; in concert halls especially the quietest possible conditions should be provided because the pauses and moments of silence which are an essential element of music cannot otherwise be given full value. Therefore, sites at cross-roads or close to steel railway bridges, religious places or near churches where bell ringing is practiced, should be avoided unless very high standards of structural sound insulation are contemplated. Sites adjoining underground railways may also prove unsatisfactory at basement levels owing to low-pitched noise or rumble transmitted through the ground; special isolation measure need to be adopted for isolating large buildings from ground vibration of this sort.

Whenever possible, for concert halls and theatres on city sites a noise survey of the site should be made; a suitable sound reduction value for the structure of the building can then be chosen so as to keep down to certain maximum noise levels within the auditorium. The maximum octave-band sound pressure levels (SPL) recommended are given in Table 27.9.

Table 27.9: Maximum Sound Pressure Levels Due to External and Mechanical Equipment Noise in Auditoria (dB)

Type of Auditorium	*Centre Frequency (Hz)*							
	63	125	250	500	1 000	2 000	4 000	8 000
Concert Halls [dB(A)-25]	51	39	31	24	20	17	14	13
Drama Theatres and Cinemas [dB(A)-30]	55	44	35	29	25	22	20	18

The minimum standard of sound reduction index, R_w likely to be required for the envelope of an auditorium in a city to protect it against external noise is of the order of 65 dB for a concert hall or 55-60 dB for a theatre. This reduction should be provided on all sides, but it would be reasonable to make the R_w for the roof 5 to 10 dB less provided the building is not unduly exposed to noise from aircraft in flight. Surrounding the auditorium with ancillary rooms and foyers is an obvious and invaluable planning method of obtaining the required insulation against outdoor noise.

Ventilation intakes and returns are vulnerable features in the defence against external noise. They should be positioned so as to avoid exposure to noise, and in addition sufficient length of both inlet and outlet ducts should be provided with carefully designed silencers. The ventilation system should also be designed to avoid transmitting or adding to internal noise.

The most serious internal noise problem arises when there are two halls meant for separate use in the same building, especially if one of them is a concert hall. The latter is a very loud potential source of noise and requires a high standard of protection against extraneous noise. In these circumstances it is doubtful whether a 'single' wall can be adequate for insulating the two halls unless it is designed with a wide unbridged cavity. Separation by planning is preferable.

Other sources of internal noise are rehearsal rooms, scenery bays and workshops, stages of other halls where rehearsals or, erection of stage sets might be in progress and foyers and bars where loud conversation might occur. The insulation of the internal walls should be adequate to protect the auditorium from these noise sources and the insulation should not be by-passed by openings, doorways, etc. The general noise due to banging of doors also needs to be taken care of; soft sealing materials should be provided for all doors to ensure quiet closing.

27.17 CINEMAS

The main objective of the design should be to control noise from adjacent screens, the projection area, the foyer, and outside the cinema. The first of these, controlling noise from adjacent screens, is likely to be the most difficult with modern digital sound systems. As most cinemas are air conditioned, there will be some noise from services. To ensure reasonable listening conditions, this should be limited to 30 dBA. This will provide some masking of the noise from adjacent screens, but a high performance partition will still be essential. Masonry or lightweight construction may be used, and a typical performance specification for a lightweight wall separating two screens is given in Table 27.10. Cinema design, however, normally requires specialist acoustic advice.

Table 27.10: Typical Sound Insulation Specification for Wall Separating Two Cinema Screens

Octave Band *Hz*	*Sound Reduction Index* *R dB*
63	38
125	44
250	50
500	61
1 000	57
2 000	58
4 000	57
8 000	55

SOUND INSULATION OF NON-INDUSTRIAL BUILDINGS BY CONSTRUCTIONAL MEASURES

The desirability of insulating residences, educational buildings, office buildings and hospitals where disturbing sounds are of higher magnitude is important. Both the outdoor noise and indoor noise must not be transmitted either inside or outside.

The noise which is present in a building comes mainly through the air while a part may be transmitted through the structure itself. The latter part may be due to the noise of the machinery in contact with structure or due to impact of the people travelling on the floor spaces, etc. Broadly speaking, the sound energy may be transmitted into an adjoining room by three ways, namely:—

(1) By passing through the air passage in the openings of ventilators, windows, skylights, etc.

(2) By setting the separating wall in vibration so as to create sound waves on the other side.

(3) An elastic wave motion in which compressions and rarefactions of the sound are transmitted from particle to particle in the wall and the same are communicated to the air on the other side.

Mainly the first type of transmission is important. The second type becomes prominent only when the enclosing walls are of thin construction and the third type is prevalent wherever mechanical vibration is present.

The sound insulation of non-industrial buildings can basically be achieved by the following three methods:

(*a*) Use of specially non-porous and porous rigid materials so as to reduce the vibration.

(*b*) Adopting hollow and composite wall construction.

(*c*) Sound insulation of floors and ceilings by construction measures.

These are discussed below:

27.18 NON-POROUS AND POROUS RIGID MATERIALS

27.18.1 Sound Insulation Values of Non-porous Rigid Partitions

The sound insulation of non-porous, homogeneous, rigid constructions, such as plastered solid brick masonry walls, varies as the logarithm of the weight per unit area and would thus increase the thickness of wall. There is, however, a limit beyond which it requires an excessive increase in thickness to produce only a small increase in sound insulation (*see* Table 27.9). The relation between the weight of a rigid partition and its sound insulation is approximately such that every time the weight is doubled there is an increase in sound insulation of about 6 dB. Sound insulation values in relation to weight per square metre of wall are given in Table 27.11.

Table 27.11: Sound insulation values according to weight of material per square metre

Surface density of wall area in kg/m²	*Transmission loss (sound insulation value average for 128 and 2048 Hz) dB*	*Surface density of wall area in kg/m²*	*Transmission loss (sound insulation value average for 128 and 2048 Hz) dB*
5	22.8	300	49.1
25	33.2	350	50.0
50	37.6	400	50.9
100	42.0	450	51.6
150	44.7	500	52.3
200	46.4	550	52.9
250	47.9	600	53.6

27.18.2 Sound Insulation of Porous Materials

Porous rigid materials. The relation between weight per square metre and the sound insulation values given in Table 27.9 does not apply in the case of porous rigid materials, like porous concrete masonry, cinder concrete, etc., because of their sound absorptive properties which provide about 10 per cent higher insulation than the non-porous variety of the same weight. In order to secure the best insulation from such porous materials, it is recommended that the porous partition should be plastered at least on one side and if possible on both the sides.

Non-rigid or flexible porous materials. These are materials, such as felt, mineral wool and quilt, and they by themselves provide low sound insulation as compared to rigid materials and, therefore, they are not generally used for the purpose of noise reduction and isolation. However, a composite construction employing a combination of rigid materials and porous absorbers may be adopted where weight is an important factor. Such a construction produces better insulation per unit weight of the partition.

27.19 HOLLOW AND COMPOSITE WALL CONSTRUCTION

27.19.1 Heavy Weight Construction

Where a high degree of sound insulation is required, neither a non-porous rigid material nor a porous material by itself can provide the desired insulation in a reasonable thickness and weight. For example, about 1000 mm thick brick wall (1,950 kg/m^2) would be required to secure an isolation of about 60 dB. A double wall construction with each wall about 250 mm thick with an intervening air space of 100 mm would, however, provide about the same insulation at half the weight. When cavity walls are provided, the provision of sound absorption inside the cavity also helps in improving the sound insulation of the wall provided the two walls are acoustically decoupled from each other all along the perimeter.

27.19.2 Light Weight Construction

For light weight construction, a variety of materials, such as metal lath and plaster, fibreboards, plywood and plasterboards fixed on studs with an intervening space may be employed. For equal weight these provide greater insulation than solid masonry. Tests conducted on such partitions have shown that sound is mainly transmitted through the studs and very little through air space. Structural cross-connections or ties between the two partitions should, therefore, be avoided, but where they cannot be avoided, only the minimum number required should be used. These ties where provided should be made of wire or, even more preferably, of a suitable flexible material. Construction with staggered studs (*see* Fig. 27.1) which makes the two partitions independent, is recommended.

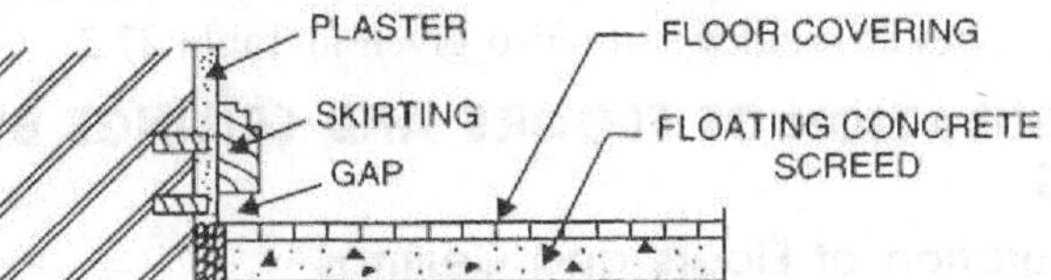

Fig. 27.1: Typical sectional plan of a sound-proof partition.

27.19.3 Filling of Cavity in Partitions

The filling of the cavity in a partition with acoustic materials like mineral wool or glass wool does not always produce satisfactory results. If the acoustic fill is rigid, it may form a bridge across the two partitions and thereby considerably reduce the insulation. In the case of hollow light weight partitions, in which the two panels are not structurally separated, the acoustic fill helps to increase the insulation only slightly. However, when the two panels are decoupled by means of staggered studs, resilient supports, etc. the acoustic fill can effect considerable increase in the insulation value, and should be provided.

27.19.4 Composite Wall

Composite walls of laminated construction may be designed to obtain any sound insulation desired, the maximum value is obtained when all the layers of alternate porous and rigid materials are separated by an air space in which case the value approaches the sum of the value of individual units, provided the air space is more than the wave-length of the sound at the lowest frequency and the layers of rigid material are isolated from each other. Such a construction, however, occupies more space and needs great care in installation in order to obtain the anticipated efficiency. Such composite wall may be used in special cases where high sound insulation is necessary.

The desired sound insulation may be achieved in any one or more of the following three ways:

(*a*) By adopting a continuous construction by using homogeneous materials where the extent of sound insulation is controlled by the weight per square metre. Where the requirements are not relatively stringent and are of the order of 50 dB, solid construction of (One to one and a half brick) 230 mm to 345 mm is considered suitable.

(*b*) By adopting semi-discontinuous construction where air cavities are employed with cross-connections between the two partitions, the cross-connections being kept as few as possible and made preferable of a flexible material, provided adequate discontinuity is maintained between the floor, the wall and the ceiling.

(*c*) By adopting a fully discontinuous construction where complete discontinuity in the structure is introduced by means of an air cavity or an elastic acoustic material, and the air cavity is filled with an acoustic material. This method would meet the requirements of the highest insulation within reasonable weight per square metre.

27.19.5 Overall Insulation

In providing sound insulation in any room the question of overall sound insulation should be considered. Some areas, like sound-proof doors, windows, etc., may provide insulation lower than that of the other surfaces and thus would adversely affect the overall insulation. All the areas should, therefore, be designed keeping in view the sound energy transmitted through the individual components so as to attain the desired overall sound insulation.

27.19.6 Examples of Sound Reduction Values

Values of sound reduction for types of (*a*) continuous construction, (*b*) semi-discontinuous construction, and (c) discontinuous construction are given in Table 27.2. Sound reduction values obtainable with various types of windows are also given in Table 27.2.

27.20 SOUND INSULATION OF FLOORS AND CEILINGS BY CONSTRUCTION MEASURES

27.20.1 Sound Insulation of Floors and Ceilings

Both air-borne and structure-borne sounds may be transmitted through floors and ceilings. In most of the cases where the ceilings and floor construction is of the solid type like cement concrete, these have sufficient weight and rigidity to provide adequate insulation for air-borne sounds, but offer poor insulation for structure-borne or impact sounds. Insulation against impact sounds may be done by the following methods.

27.20.1.1 By using a resilient surface material on floors

This helps to dampen the impact noises, but has no appreciable effect on air-borne sound. Linoleum, insulation board, cork, bituminous mastic and carpet are some of the materials usually employed, the softer the materials used, greater is the insulation effect. An improvement of 5 to 10 dB over a bare concrete floor is obtained with such materials.

27.20.1.2 By providing a floating floor construction

(*a*) ***Concrete floors.*** This is an additional floor constructed and isolated or floated from the existing concreted floor by means of a resilient material and, therefore, does not let the impact and consequent vibration to be transmitted to the room below. It also provides useful improvement in the insulation of air-borne sounds. A typical construction is shown in Fig. 27.2. The cement concrete used may be about 50 mm thick which is poured over a resilient material like quilted mineral or glass wool. It is important that a waterproof paper be used in between, and both the quilt and paper lapped so as to prevent concrete from getting through.

(*b*) ***Wooden floors.*** In the case of floors constructed on wooden joists, the problem of sound insulation becomes more difficult particularly in the presence of heavy mechanical impact sounds. Sketches in Fig. 27.3 to 27.5 show methods of insulating such floors employing mineral or glass wool quilt for isolation purposes. Resilient mountings may be used to obtain even more satisfactory results. A further improvement in the insulation of such floors is achieved by employing a pugging or a 'deadening' material in the air space between the wood joists (*see* Figs. 27.3 to 27.7). Either sound absorbent type materials like mineral wool or other materials like sand or ashes may be used, the latter being more effective because of the fact that the efficiency of the 'pugging' depends on the weight of the material used. In order to obtain useful improvement, at least 70 kg/m^2 of sand 'pugging' is usually employed. Mineral wool pugging (at least 15 kg/m^2) is used mainly in conjunction with thin walls of 100 mm thickness or less.

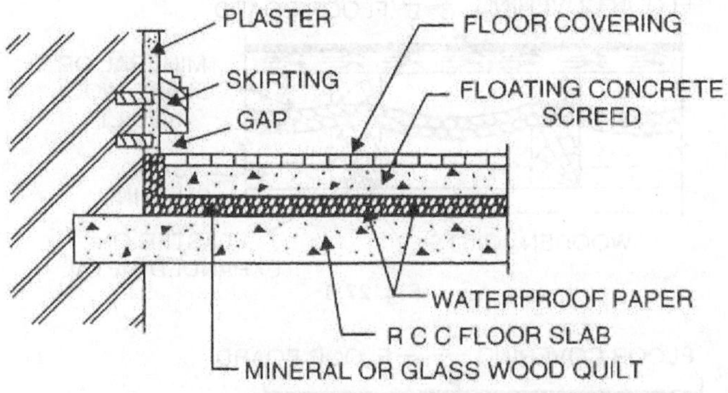

Fig. 27.2: Concrete floating floor construction.

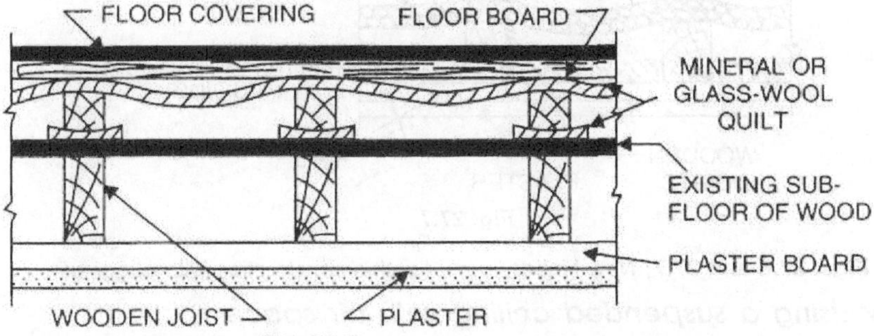

Fig. 27.3

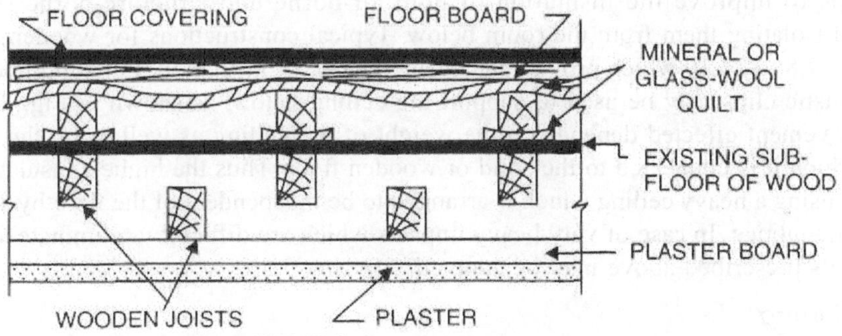

Fig. 27.4

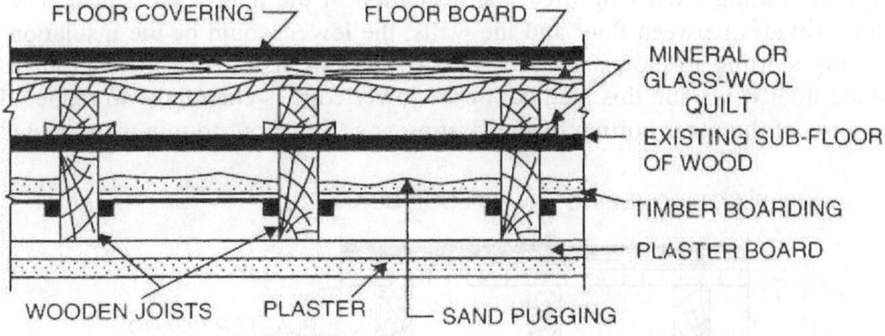

Fig. 27.5

Figs. 27.3 to 27.5: Existing timber floors, floating construction.

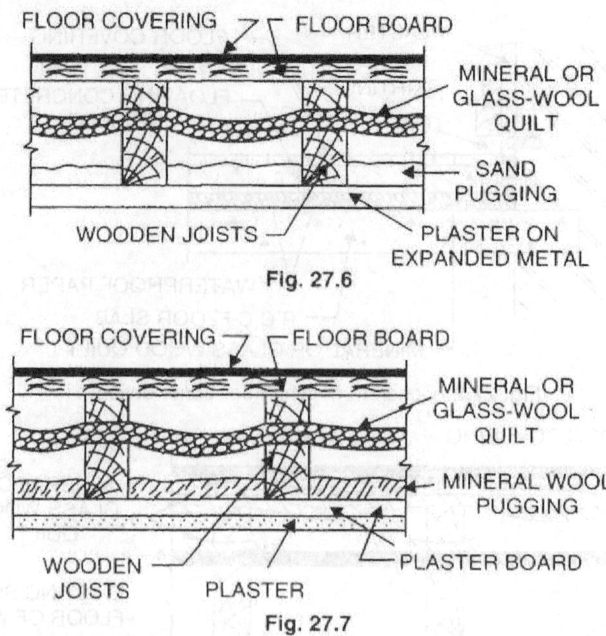

Fig. 27.6

Fig. 27.7

Figs. 27.6 and 27.7: New timber floors, floating construction with pugging.

27.20.1.3 By using a suspended ceiling with air space

This helps to improve the insulation of both air-borne and structure-borne sounds by attenuating and isolating them from the room below. Typical constructions for wooden floors are shown in Figs. 27.8 to 27.10 which provide increasing degree of insulation. For solid floors, metal hangers of acoustic clips may be used to support the ceiling below, as shown in Fig. 27.11. The extent of improvement effected depends on the weight of the ceiling as well as on the structural rigidity with which it is connected to the solid or wooden floor. Thus the highest insulation could be achieved by using a heavy ceiling which is arranged to be independent of the floor by supporting it on resilient mountings. In case of very heavy impacts which are difficult to eliminate altogether, both the methods prescribed above may be adopted.

27.20.1.4 Skirting

The type of skirting fixed will affect the insulation of the floor a great deal. The larger the contact area it provides between floor and the walls, the lower would be the insulation. A typical method of fixing skirting is shown in Fig. 27.2. An air gap or a resilient material between the skirting and the floor is used in this method, or the lower edge is chamfered to reduce the contact area. Again, most of the inner portion of the skirting is scooped out to minimize contact with the walls.

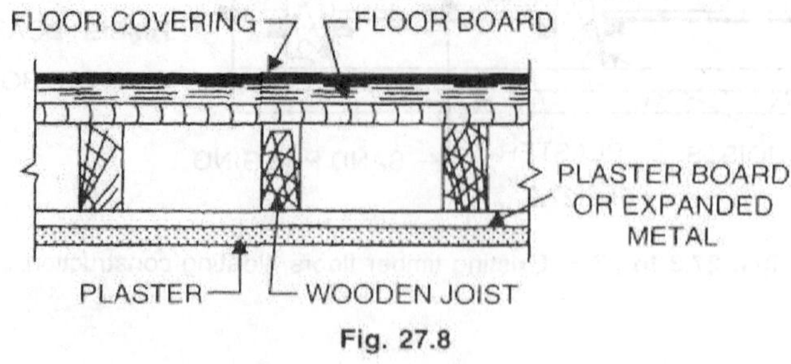

Fig. 27.8

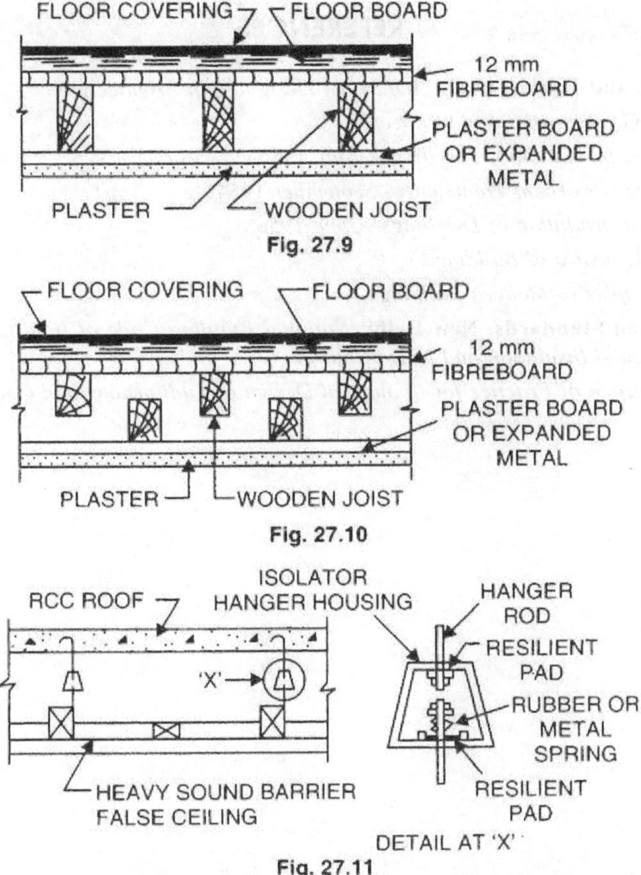

Fig. 27.9

Fig. 27.10

Fig. 27.11

Figs. 27.8 to 27.11: Timber floors, floating construction with suspended ceiling.

QUESTIONS

1. What is meant by the term 'acoustic'? Briefly describe as to how effects are created in a room from acoustical point of view.
2. How is the acoustics corrected in an auditorium? What is the optimum time of reverberation for a public lecture hall?
3. What is meant by 'sound insulation'? How is the sound transmitted from a room?
4. Describe the various types of floor constructions which are sound-proof. What steps are taken to make the wood joist floor sound insulated?
5. Write short notes on:
   (*i*) Acoustical treatment of a classroom.
   (*ii*) Sound absorbing materials.
6. Give the recommendations for sound insulation in a residential building.
7. How will you carry out the planning and design of a building against noise from (*i*) air traffic, (*ii*) rail traffic and (*iii*) road traffic?
8. Describe the methods adopted for sound insulation of a residential building against indoor noise.
9. Give details of the constructional measures taken for sound insulation of non-industrial buildings.

REFERENCES

1. **Kndusan, V. O. and Harris, C M.**: *Acoustical Designing in Architecture.*
2. **Richardson, E.G.**: *Acoustics for Architects.*
3. **H.M.S.O.**: *Sound Insulation of Dwellings—May* 1956.
4. **H.M.S.O.**: *Sound Absorbent Treatments—September* 1954.
5. **H.M.S.O.**: *Sound Insulation of Dwellings—June* 1956.
6. **Watson, F.R.**: *Acoustics of Buildings.*
7. **H.M.S.O.**: *Principles of Modern Buildings.*
8. **Bureau of Indian Standards, New Delhi**: *National Building Code of India* 2005—*Part 8, Section 4 : Acoustics, Sound Insulation and Noise Control.*
9. **IS: 2526-1963**: *Code of Practice for Acoustical Design of Auditoriums and Conference Halls.*

Shoring, Underpinning and Scaffolding

SHORING

Shoring is used to prevent a structure from failure and collapse which has become damaged due to foundation settlements or other causes. It is also used to give temporary supports to a structure which is undergoing alterations or where alterations of adjacent foundations are being carried out. This type of support may be given externally or internally and in certain cases shores may be placed from both sides of the wall to produce additional stability.

Temporary shores are generally made of timber but for heavier loads steel beams or suitably braced sections are adopted. A shore may also be made of concrete or masonry. Timber shores suffer from the usual defect of shrinkage on account of the changes in weather. They are, therefore, to be attended to regularly.

Walls when supported by shores may be considered to be acted upon by the vertical and horizontal forces and also inclined thrust of floors, roofs, etc. To maintain equilibrium the overturning forces must be resisted by the supporting shores. Therefore it is essential that the lines of action of the overturning forces in floors and roofs, the forces in walls and the reaction of the shores must meet at one point.

The type of shores based on their supporting characteristics are usually classified as raking shores, horizontal or flying shores and dead or vertical shores.

28.1 RAKING SHORES

These consist of timber pieces, placed in an inclined position with one end resting against a defective wall and the other supported on the ground, maybe on suitable sole plates so as to distribute the pressure on the soil. An inclined timber shore can be used for considerable height of wall and usually shores of 9 to 12 m in length are used. Far greater heights additional supports are given at lower elevation for more restraint. Braced timber shoring is used when heavier loads are to be supported, the bracing preventing the timbers from buckling.

Walls which are likely to bulge or crack, vertical timber plates and a series of two or more shores may be required in such cases. Similarly, for additional strength along the length of the wall, horizontal timber members are placed at the head of the shores. Steel sections may be used alternately if greater pressures exist. When a large area of wall is cracked or likely to crack both

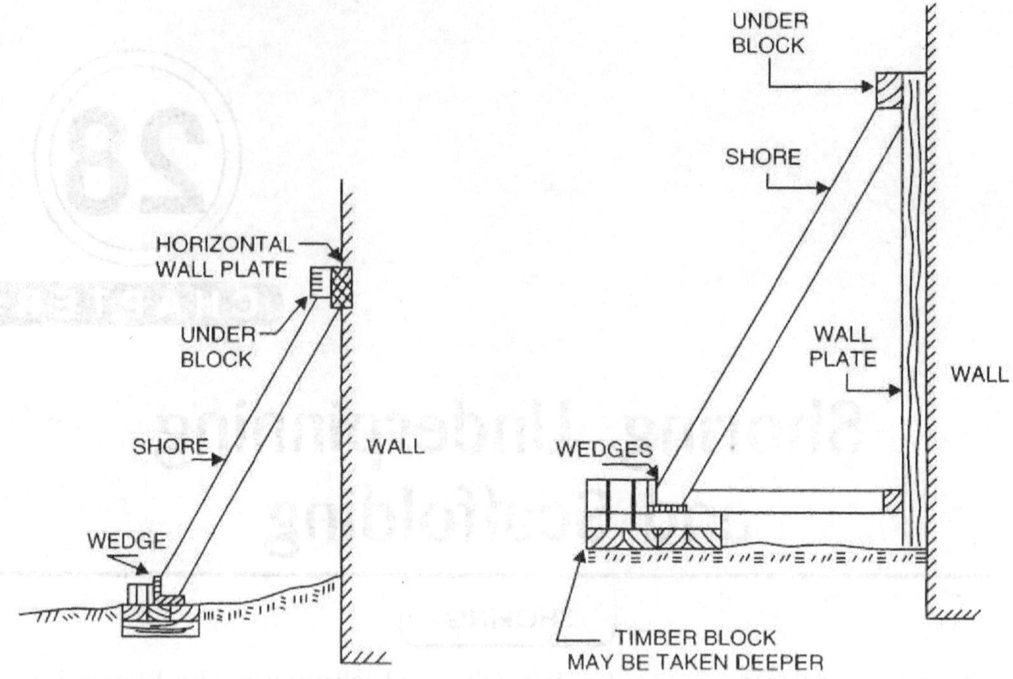

Fig. 28.1: A simple raking shore. **Fig. 28.2:** Raking shore for a defective wall.

vertically and horizontally, the shoring system should be provided with both horizontal and vertical wall plates. These consist of pieces of timber inclined at suitable angles. For practical purposes these angles vary from 60° to 75°. These shores are fixed in systems having one or more timbers placed in the same vertical plane but at different angles to support the entire height of the wall. The horizontal distance is not generally greater than 2.5 m but they may be placed at the piers which support the wall. A wall plate consisting of a 50 mm to 70 mm plank is placed against the wall to receive the ends of the shores. This is fixed with wrought iron wall hooks driven into the joints of the masonry at the back. One piece wall plate should be preferred. Wooden needles with cleats housed into the wall plate are used as a support for the top end of the raking wooden shore. This end of the raking shore should normally be placed where there is some resisting member such as a floor or a roof at the back of the wall. The shore should be notched so as to prevent its lateral movement. The foot of the shore rests on a sole plate embedded into the ground in an inclined position. This sole plate consists of thick wooden

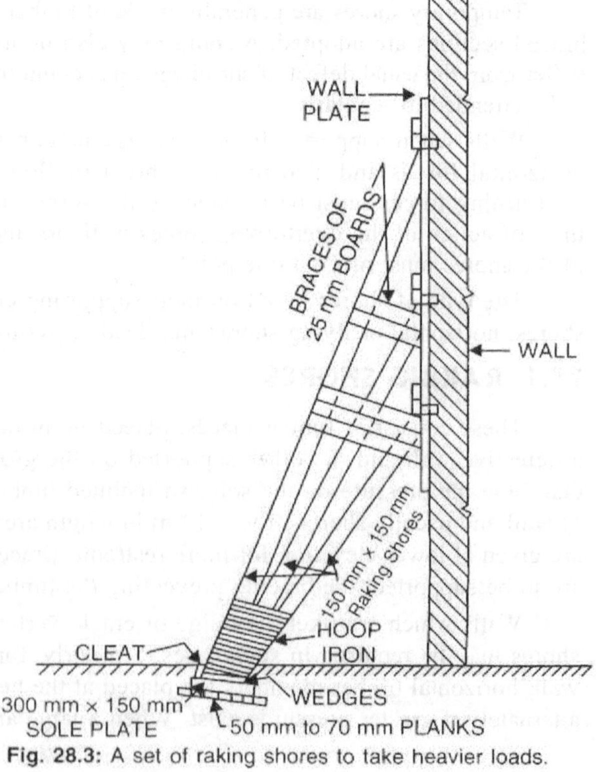

Fig. 28.3: A set of raking shores to take heavier loads.

planks and is not inclined exactly at right angles to the shores but is fixed at about 1 in 24 out of the perpendicularity to enable easy tightening with a crowbar. On soft ground the sole plate should be of such an area as to distribute the pressures evenly. The shores are tightened with a crowbar inserted in the foot of the shore. Wedging should not be done at this place as the vibrations caused thereby might be detrimental to the building. Once the shore is placed in position it is fixed firmly with iron dogs.

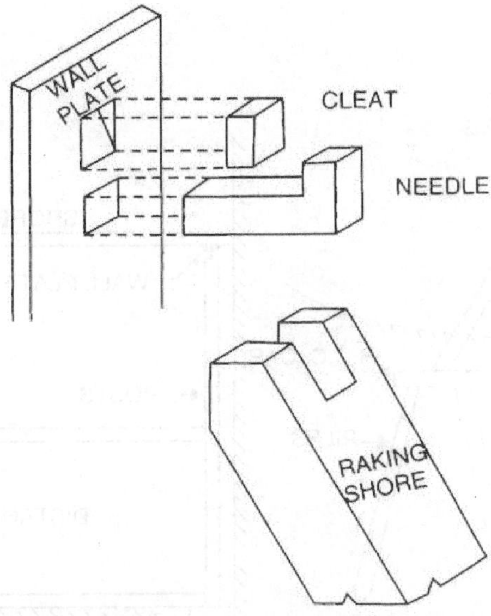

Figs. 28.4 and 28.5: Details at the head of a raking shore.

When more shores are used in one plane, wooden boards are nailed to the sides of the shores at intervals. They may also be nailed to the wall plate. The bottom portion of the shores may be connected with hoop iron. Where excessive settlements are anticipated, permanent cleats may be left so as to permit jacking operations.

Occasionally balanced shores may be used for supporting walls from both sides. For example, in case of a retaining wall which is unstable on account of the earth backing being removed, a permanent and reinforced concrete shore to a retaining wall may consist of an RCC frame made in the shape of a right-angled triangle. It has got a tie which prevents the shore from bending back.

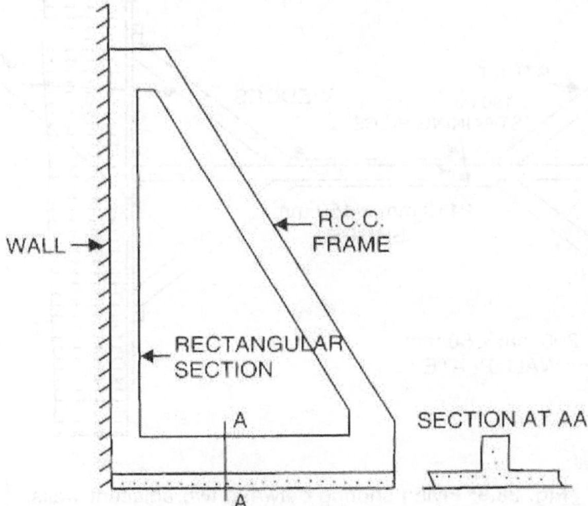

Fig. 28.6: R.C.C. framed shore.

For adequate support to a wall in the case of permanent shores, effective measures have to be taken to ensure a proper distribution of load. When the bearing pressure is reasonably high, mass or reinforced concrete foundations can be used. For low bearing capacities cast-in situ piles are used.

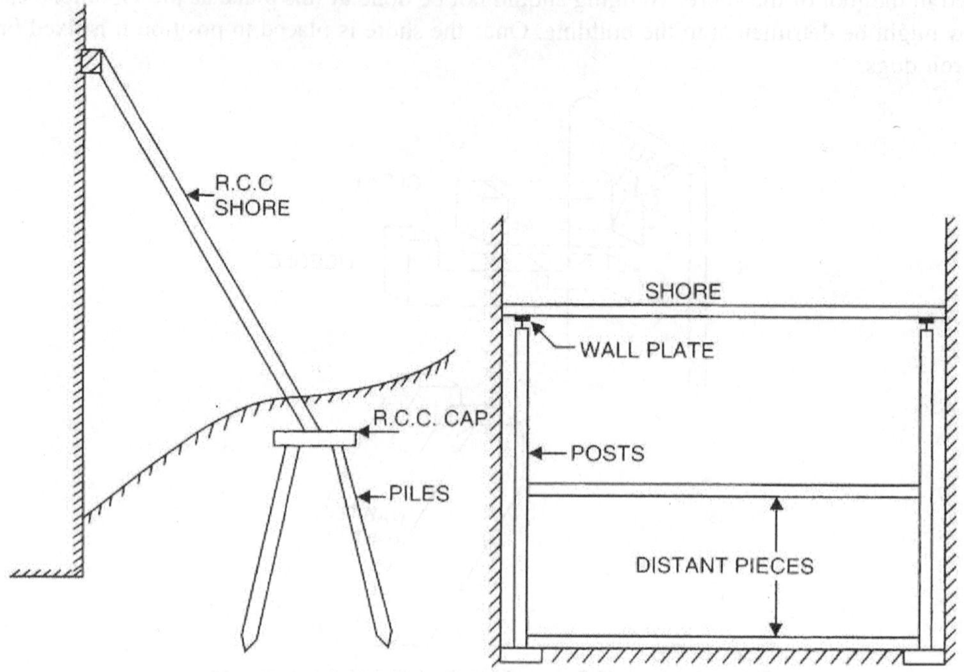

Fig. 28.7: R.C.C. shore with piles. **Fig. 28.8:** A simple type of horizontal shore.

28.2 HORIZONTAL SHORES

These are also called flying shores and are used to temporarily support two parallel walls. Generally, the maximum distance which can be supported with the aid of flying shores is considered to be 10 m. Such type of shoring work is used in residential areas where out of a number of houses, one house has to be removed.

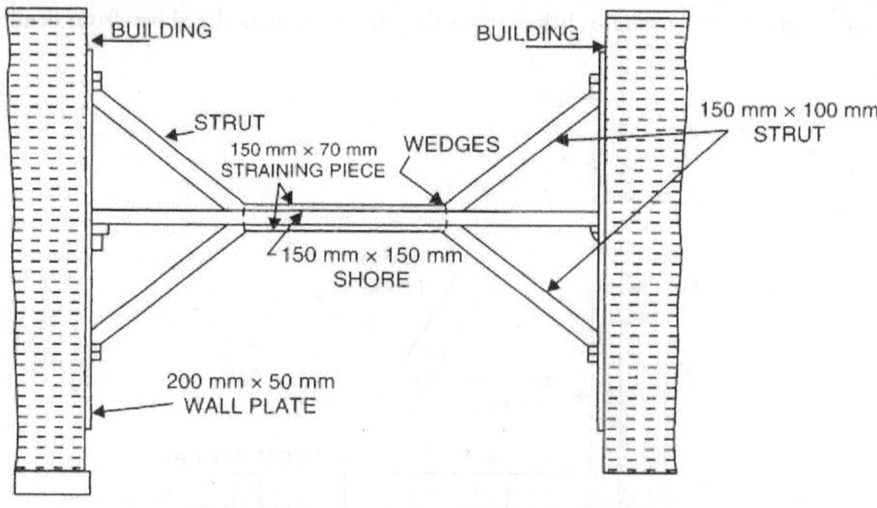

Fig. 28.9: Flying shoring between two adjacent walls.

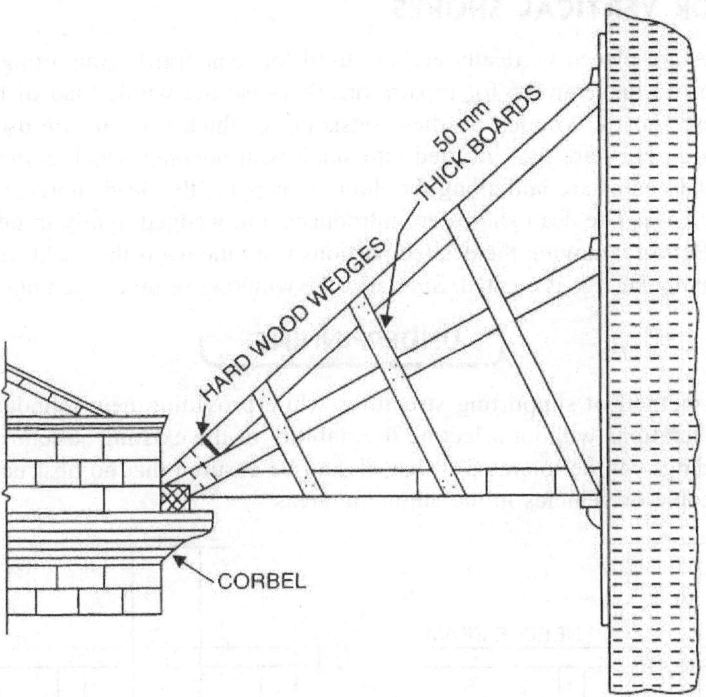

Fig. 28.10: Flying shoring of a building wall from an adjacent old building corbel.

Horizontal timbers are placed or steel joints are placed running over horizontal members which are duly supported on vertical columns. A horizontal beam is placed at the footing of these columns to keep the distance between the columns constant.

Whenever the head room below the shores has to be kept clear, special methods have to be adopted. These consist in fixing vertical wall plates on the two plates in a manner similar to raking shores. Horizontal timber shore is placed on needles which are fixed into the wall plates. This horizontal shore is tightened with wedges on one side. Inclined struts are added for additional strength. The foot of these struts is stopped against a horizontal straining piece, the other foot being fixed to the wall plates. For taller building two or more flying shores may be used which are suitably braced with each other.

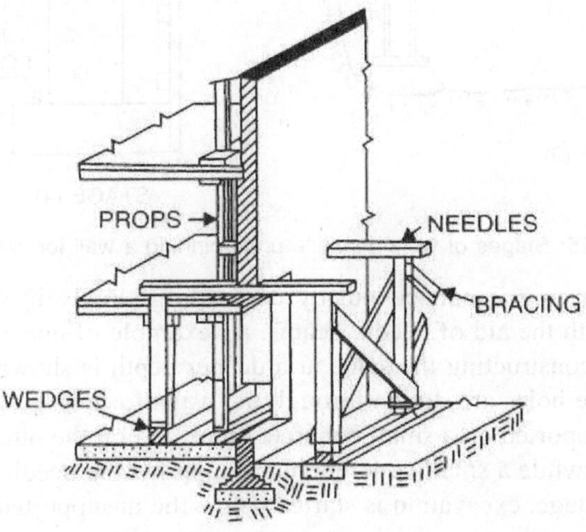

Fig. 28.11: Use of dead shores for creating an opening in a wall.

28.3 DEAD OR VERTICAL SHORES

These shores are placed vertically and are used for temporarily supporting walls, the lower parts of which are to be removed for repairs or otherwise the whole load of the roofs, etc., is supported by these shores. Wooden needles consisting of thick sections are used to transfer the load from the walls. They are first inserted into small wall opening which is made just sufficient for size. Horizontal beams are laid along the floors to support the dead shores, the load is evenly distributed on the base. The dead shores are introduced and wedged tightly in-between the beams and the needles. Before removing the desired portions from the walls the usual walls are supported with props, so that no damage is created. Similarly, the windows or other openings are duly strutted.

UNDERPINNING

This is the method of supporting structures while providing new foundations or carrying out repairs and alterations without affecting the stability of the existing structures. No damage to surrounding buildings can be tolerated. It has also to be ensured that no obstruction is created to the passage of people or vehicles in the adjoining areas.

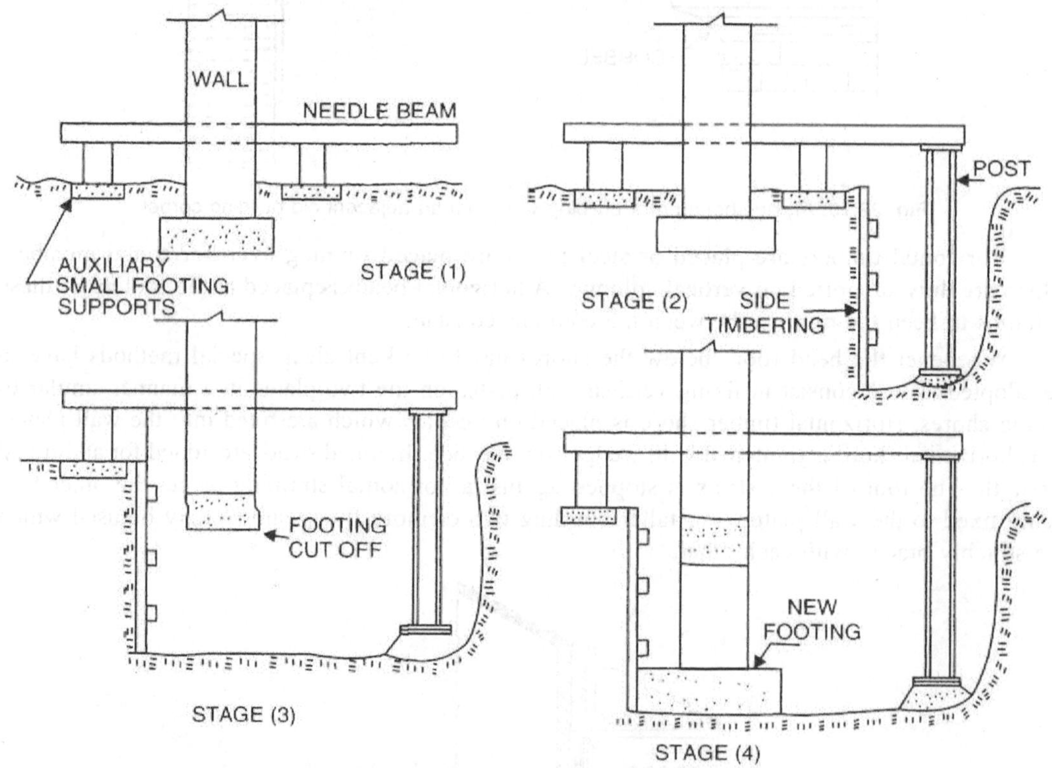

Figs. 28.12 to 28.15: Stages of working while underpinning a wall for casting a new footing.

Generally underpinning is carried out by excavating slowly in steps and supporting the structure thus opened with the aid of needle beams. As example of underpinning for replacing the footing of the wall and constructing the same at a deeper depth is shown in Figs. 28.12 to 28.15. In the first place, suitable holes are driven through the walls for engaging needle beams. One end of the needle beam is supported on a small concrete block and on the other side a major portion of it is left without support while a small concrete block supports the needle pin just near the wall on that side. In the second stage, excavation is started below the unsupported end of the needle beam.

Once the excavation is suitably supported with the aid of side timbering and carried down to the required depth, the unsupported end of the needle beam is supported on the vertical post supported on a concrete base. In this manner, the concrete block gets relieved of any load. The excavation trench is widened further and the soil below the wall footing is also removed. The additional offsets of the old footing are cut and thereafter the new footing is constructed at the required depth. This process is repeated in small stretches so as to complete the desired length of the wall.

Another method of underpinning is depicted in Figs. 28.16 and 28.17. This method is used where enough space is not available for the repeated use of the needle beams due to some obstruction on the other side. The longer arms of the needle beams are placed on the inside of the building, the other end supporting the wall.

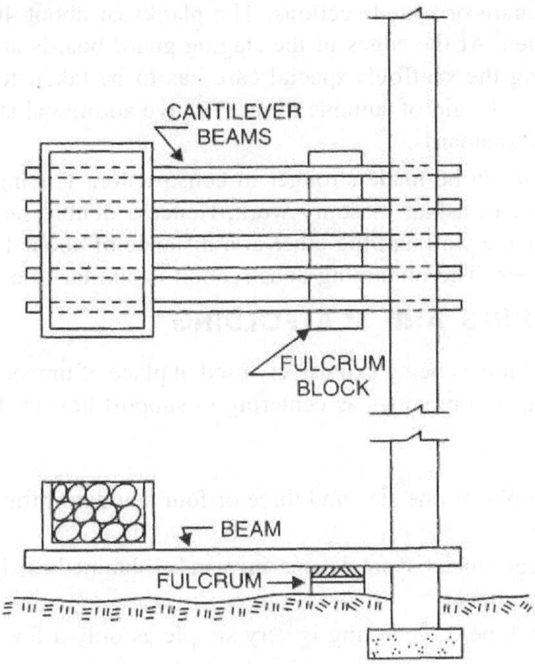

Figs. 28.16 and 28.17: Supporting a wall with cantilevered needle beams.

Heavier weights are placed on the unsupported ends of the needle beams and a fulcrum block is placed near the wall so that these needle beams act as cantilevers in supporting the walls. After these beams are placed in this manner, the footings are removed and repaired.

There are many other methods of underpinning depending on the type of structure to be supported and the circumstances within which it lies. Underpinning of columns may be carried out by erecting four small grillage footings on its sides and spanning the beams over these footings. The load of the main column which is to be underpinned is transferred to these beams through strong jacks and the column footing can be repaired.

SCAFFOLDING

28.4 TIMBER SCAFFOLDING

These are temporary erections constructed to support a number of platforms at different heights raised for the convenience of workers so as to enable them to work easily and raise the needed materials. Generally scaffolds are classified into two types, namely, bricklayer's and mason's. They are constructed of timber either sawn or in circular sections. In India Sal "*Ballies*" (round poles of 100 mm to 150 mm diameter) are abundantly used. For important and large works steel tubular scaffolds are used.

The bricklayer's scaffold consists of a number of uprights called standards and usually placed at not more than 2.5 m centres. They may be circular in section of about 120 mm to 150 mm average diameter. These rest on hardwood planks with wedges, on stone slabs laid on rammed ground, or in a drum containing rammed earth. For adequate stability they may be driven a few metres into the ground. Horizontally placed poles called ledgers are lashed to these standards and spaced at a vertical distance of 1.5 to 2 m apart. Such frames of standards and ledgers are placed at about 1.5 m away from the building face. They are connected to the building with putlogs, which bear on the wall, and are laid as the latter is built at one end and the ledgers at the other end. They may be of square or round sections. The planks of about 40 mm thickness are used as a platform for workmen. At the edges of the staging guard boards are fixed duly nailed to the standards. While building the scaffolds special care has to be taken for connecting the various units with each other with the aid of suitable ropes. To give additional strength, a pair of diagonal braces are fixed with the standards.

Mason's scaffolds are to be made stronger in construction. Placing of putlogs into the wall is not possible in the case of ashlar masonry work. Hence a double set of standards and ledgers is erected one being near the wall and the other at a distance of about 1.5 m from it. The putlogs rest both ends on these sets. The remaining construction is similar to a bricklayer's scaffold.

28.5 STEEL CENTERING AND SCAFFOLDING

Tubular steel scaffolding is being extensively used in place of timber scaffolding for temporary staging and also for posts and bracings as centering to support heavier loads.

Advantages

(1) ***Adaptability***: Tubes of one size and three or four standard fittings can be used to support vertical loads of all types.

(2) ***Availability***: Large stocks of steel centering can be obtained easily while it may be difficult to procure timber in large stocks.

(3) ***Simplicity***: This type of centering is very simple as only a few shapes and sizes have to be dealt with.

(4) ***Interchangeability***: Scaffold tubes can both be used for making a scaffolding or formwork.

(5) ***Ease of erection***: The methods of erection are standardized and easier erection is possible as no tedious cutting or lengthening is necessary as is the case with timber.

(6) ***Scrap value***: Their scrap value is high as they do not get damaged while fixing or dismantling.

(7) ***Strength***: Their strength is constant while that of timber varies on account of flaws, etc.

(8) ***Resistance to fire***: These are not liable to damage by fire..

Disadvantages

(1) ***Initial cost***: These need a lot of investment.

(2) ***Fittings***: The fittings may get lost as they are small in size and they are difficult to be replaced later on.

(3) ***Connections with timber***: It is difficult to attach timber members with steel tubes economically and hence they are left supported on the centering without any rigid connection.

(4) ***Skilled labour***: To fix a tubular centering, it is essential to employ very skilled labour which is always not possible.

(5) ***Painting***: These tubes should be painted periodically to keep them free from corrosion.

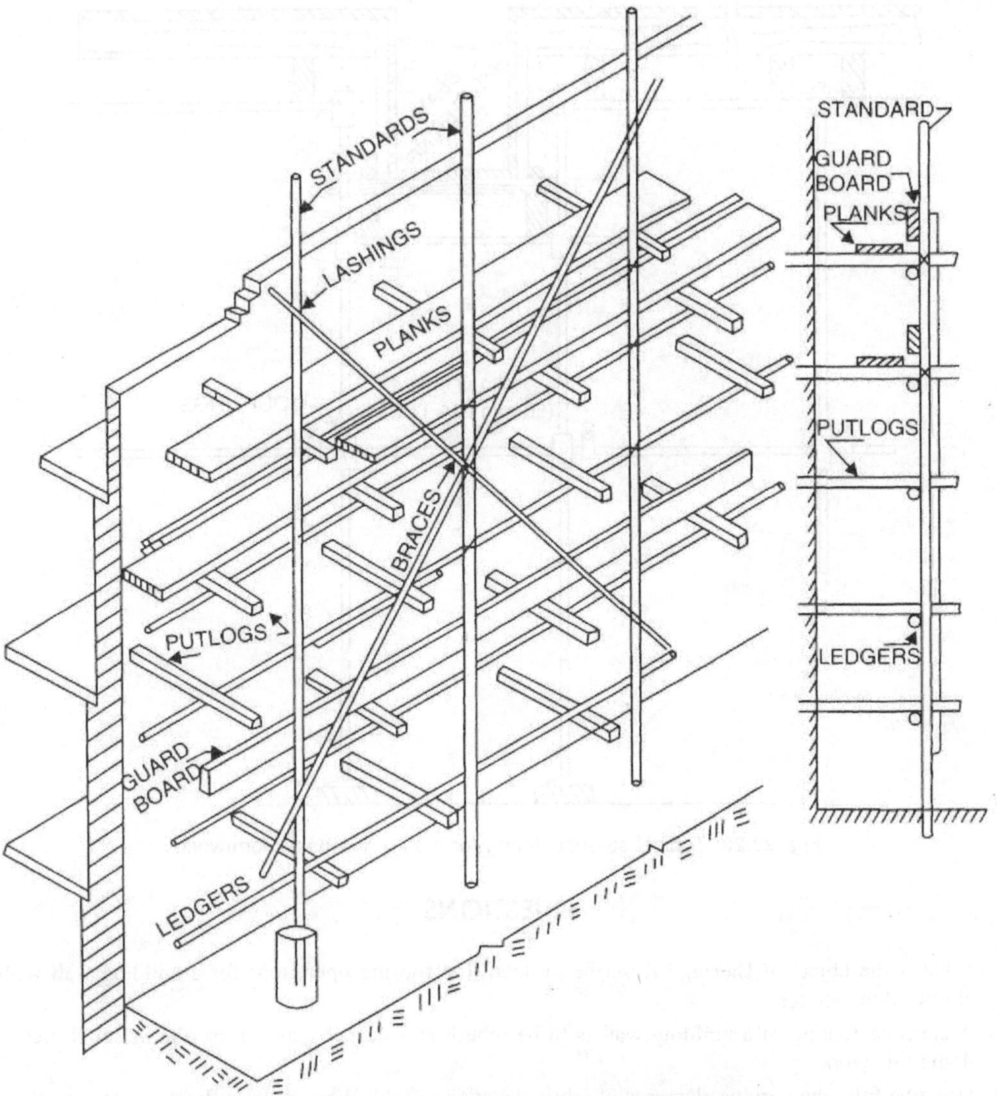

Figs. 28.18 and 28.19: Details of bricklayer's scaffolding.

The usual methods of constructing timber forms should be modified when tubular supports are to be used since the loads must be carried near the couplers to avoid bending and deflection in the horizontal tubes as the uprights are at greater distances apart and do not take load concentrically. A typical type of centering for supporting beam and slab formwork is shown in Fig. 28.20. There are several other arrangements possible.

Some methods of adjusting form heights are always required as it is not easy to wedge up the timber formwork resting on the tubular centering. Suitable screwed adjusting couplings are used and these works on the principle of jacks.

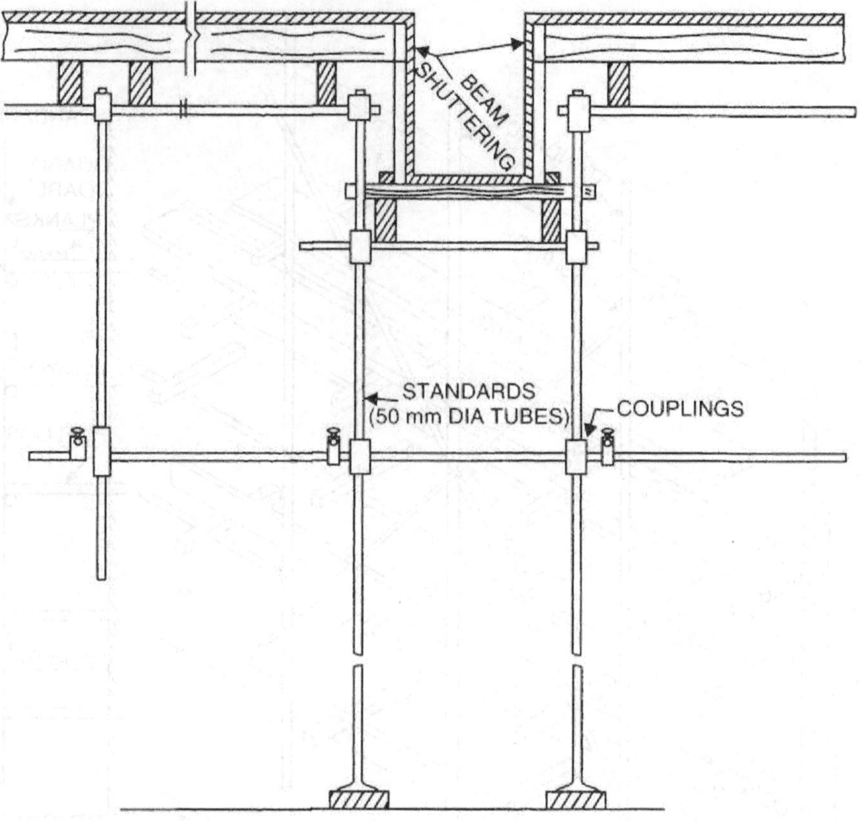

Fig. 28.20: Typical steel centering for a slab and beam formwork.

QUESTIONS

1. What is the object of shoring? Describe in detail the shoring operations for a building wall with the aid of raking shores.
2. A concrete footing of a building wall is to be rebuilt at a deeper depth. Describe in detail method of doing this work.
3. Describe fully the various elements of a bricklayer's scaffold. What is the difference between this type of scaffolding and the one used by the masons?
4. Write notes on:
   (*a*) Horizontal shores.
   (*b*) Vertical shores.
   (*c*) Steel centering for a slab and beam formwork.

REFERENCES

1. **Henry, F.D.C.**: *Design and Construction of Engineering Foundation.*
2. **Hunter, L.E.**: *Strengthening and Underpinning of Structures.*
3. **Wynn**: *Formwork—Concrete Publication Ltd.*

Fire and its Protection

FIRE

Protection of a building against fire can be attained by the use of the materials and construction techniques which aim at giving adequate resistance to the spread of fire within the building through its external walls. The building components should be structurally stable for a reasonable amount of time so that the occupants can leave within a reasonable time. In the planning of the building care should also be taken that sufficient appliances are available to the occupants so that they can leave the building safely and quickly.

The characteristics of a typical fire-resisting material would be that it should not get disintegrated under greater heat; the expansion of this material should not be excessive so as to damage the structure; and its contraction on account of sudden cooling from a hot state should not be so rapid as to break it into pieces. None of the materials has got all these qualities but many of them can resist the action of fire for a considerable amount of time. Stone is a bad material in this respect as it cannot resist sudden cooling. Once it becomes hot and is cooled, it breaks into pieces. Granite when subjected to excessive heat crumbles to sand or cracks and falls to pieces with a series of small explosions. Limestones get calcined and turn into quick lime. Sandstones resist fire better but also disintegrate after some time. Bricks are practically fire-proof and fire-bricks from the refractory are the best. Terracotta is better for fire resisting qualities than bricks but is very costly. This material is however used in the construction of fire-resisting floors. Cast iron breaks into fragments if contracted suddenly and hence is not a suitable material from fire-resisting considerations. Wrought iron and steel try to get twisted due to softening of material and the consequent reduction in resistance to the effects of tension and compression. At about 600 degrees Centigrade yield stress is reduced to one-third than at normal temperature. It also expands considerably and hence allowance should be made for these movements.

Timber is a combustible material but on exposure to fire it gets charred. This charring gives a protective coating to inner portions of the timber and prevents it from rapid combustion. At higher temperatures volatile gases get created which finally catch fire. Timber can be made fire-resistant by using thicker sections instead of employing a number of smaller sections. A number of corners and the area of exposed surface should be reduced as much as possible. Timber should not be treated with oil paints which are liable to catch fire.

The fire protection which concrete can give depends largely on the type of aggregates used. Natural stone aggregates expand on heating and thereby try to crack the concrete. Crushed brick aggregates or slags are more resistant but a concrete with these aggregates may fail on account of dehydration of cement. Pumice foamed slag, blast furnace slag, crushed brick, crushed limestone form better type of aggregates. Siliceous aggregates, flint, gravel, granite, etc., are inferior types as far as the resistance to fire is concerned. The cracks generally start from the reinforcement and as such adequate cover should be given to the steel.

Asbestos cement materials although are not combustible but crack explosively when they get heated in a building fire, hence such materials have no adequate resistance to fire. Plaster boards and some boards incorporating asbestos fibre can be used in assisting the required degree of fire resistance for steel work or improving the fire resistance of timber floors or other materials. Materials used for lining of walls and ceilings can be classified by the "surface spread of flame" test. This test measures the rate of spread of flame and classifies the materials into various classes.

29.1 DEVELOPMENT OF FIRE

Once some material gets ignited, the fire tries to spread and if there are openings in the walls and floors the fire can spread to other parts of the building. Even though there may be no opening, the fire can increase the temperature of the structure itself and transmit sufficient heat so as to ignite the material on the other side. A fire in a building creates a strong draught. Staircases and lift shafts try to act as flues which increase the spread of fire. Before the fire gets transmitted to another enclosure poisonous gases may pass ahead of it and thereby endanger the lives of the occupants there.

The more the material present in an enclosure for burning, the longer and hotter will be the fire. The amount of material is called fire-load and is more clearly defined as the number of Kilo-calories that could be liberated per square metre of floor area by the combustion of any building or any of its part. It is determined by multiplying the weights of all combustible materials by their calorific values and dividing by the floor area under consideration. For some of the materials the calorific values are given below:

Table 29.1: Calorific values

Material	*Calorific value Kilo-calories/kg*
Timber	45500
Paper	38900
Straw	33400
Bitumen	89500
Oil	94500 to 111100
Rubber	94500

29.2 TYPES OF CONSTRUCTION FOR FIRE RESISTANCE

The design of any building and the type of materials used in its construction are important factors in making the building resistant to a complete burn-out and in preventing the rapid spread of fire, smoke or fumes, which may otherwise contribute to the loss of lives and property.

The fire resistance of a building or its structural elements is expressed in hours against a specified test load which is expressed in kcal/m, and against a certain intensity of fire. The types of construction according to fire resistance can be classified into four categories, namely, Type 1 Construction, Type 2 Construction, Type 3 Construction and Type 4 Construction. The fire resistance ratings for various types of construction for structural and non-structural members should be as given in Table 29.2.

For buildings above 15 m in height non-combustible materials shall be used for construction.

Table 29.2: Fire resistance ratings of structural elements (in hours)

Structural element		*Types of construction*			
		Type 1	*Type 2*	*Type 3*	*Type 4*
1. Exterior walls					
(*a*) Fire separation less than 3.7 m	(*a*) Bearing	4	2	2	1
	(*b*) Non-Bearing	2	$1\frac{1}{2}$	1	1
(*b*) Fire separation of 3.7 m or more but less than 9 m	(*a*) Bearing	4	2	2	1
	(*b*) Non-Bearing	$1\frac{1}{2}$	1	1	1
(*c*) Fire separation of 9 m or more	(*a*) Bearing	4	2	2	1
	(*b*) Non-Bearing	1	1	1	1
2. Fire resisting walls		4	2	2	2
3. Fire separation assemblies (like fire check doors)		4	2	2	2
4. Fire enclosures of exitways, hallways and stairways		2	2	2	2
5. Shaft other than exitways, elevator; and hoistways		2	2	2	2
6. Exitway access corridors		1	1	1	1
7. Vertical separation of tenant spaces		1	1	1	1
8. Dwelling unit separations		1	1	1	1
Non-load bearing partitions		⟵ At least half an hour ⟶			
9. Interior bearing walls, bearing partitions, columns, girders, trusses (other than roof trusses) and framing	(*a*) Supporting more than one floor	4	2	2	2
	(*b*) Supporting one floor only	3	$1\frac{1}{2}$	1	1
	(*c*) Supporting a roof only	3	$1\frac{1}{2}$	1	1
10. Structural members supporting wall		3	$1\frac{1}{2}$	1	1
11. Floor construction, including beams		3	$1\frac{1}{2}$	1	1
12. Roof construction	(*a*) 5 m or less in height to lowest member	2	$1\frac{1}{2}$	1	1
	(*b*) More than 5 m but less than 6.7 m in height to lowest member	1	1	1	1
	(*c*) 6.7 m or more in height to lowest member	0	0	0	0

29.3 FIRE ZONES

29.3.1 Demarcation

The city or area under the jurisdiction of the Authority can be demarcated into distinct zones, based on fire hazard inherent in the buildings and structures according to occupancy which shall be called the 'Fire Zones'.

29.3.2 Number and Designation of Fire Zones

The number of fire zones in a city or area under the jurisdiction of the Authority depends upon the existing layout, types of building construction, classification of existing buildings based on occupancy and the expected future development of the city or area. In large cities or areas, three fire zones may be necessary, while in smaller ones, one or two may be adequate.

The fire zones in land use development plan shall be designated as follows:

(*a*) Fire Zone No. 1,

(*b*) Fire Zone No. 2, and

(*c*) Fire Zone No. 3.

29.3.2.1 Fire Zone No. 1

This comprises areas having residential (Group A), educational (Group B), institutional (Group C) and assembly (Group D), small business (subdivision E-1) and retail mercantile (Group F) buildings, or areas which are under development for such occupancies.

29.3.2.2 Fire Zone No. 2

This comprises business (Subdivisions E-2 to E-35) and industrial buildings (Subdivisions G-1 and G-2), except high hazard industrial buildings (Subdivision G-3) or areas which are under development for such occupancies.

29.3.2.3 Fire Zone No. 3

This comprises areas having high hazard industrial buildings (Subdivision G-3), storage buildings (Group H) and buildings for hazardous uses (Group J) or areas which are under development for such occupancies.

GENERAL REQUIREMENTS OF ALL INDIVIDUAL OCCUPANCIES

29.4 GENERAL

All buildings should satisfy certain requirements which contribute, individually and collectively, to the safety of life from fire, smoke, fumes and panic arising from these or similar causes. There are, however, certain general principles and common requirements which are applicable to all or most of the occupancies. They are given below:

29.4.1 Occupation of Building Under Construction

A building or portion of the building may be occupied during construction, repairs, alterations or additions only if all means of exit and fire protection measures are in place and continuously maintained for the occupied part of the building.

A high rise building during construction shall be provided with the following fire protection measures, which shall be maintained in good working condition at all the times:

(*a*) Dry rise of minimum 100 mm diameter pipe with hydrant outlets on the floors constructed with a fire service inlet to boost the water in the dry riser and maintenance should be as per the laid down requirements (*b*) Drums filled with water of 2 000 litres capacity with two fire buckets on each floor; and

(*c*) A water storage tank of minimum 20 000 litres capacity, which may be used for other construction purposes also.

29.4.2 Maximum Height

Every building shall be restricted in its height above the ground level and the number of storeys, depending upon its occupancy and the type of construction. The maximum permissible height for any combination of occupancy and types of construction should necessarily be related to the width of street fronting the building, or floor area ratios and the local fire fighting facilities available.

29.4.3 Floor Area Ratio

The comparative floor area ratios for different occupancies and types of construction are given in Table 29.3.

Table 29.3: Comparative floor area ratios for occupancies facing one public street of at least 9 m width

Occupancy Classification	*Type of Construction*			
	Type 1	*Type 2*	*Type 3*	*Type 4*
Residential	UL	2.0	1.4	1.0
Educational	UL	2.0	1.4	1.0
Institutional	UL	1.5	1.0	0.8
Assembly	UL	1.0	0.7	0.5
Business	UL	2.9	2.3	1.6
Mercantile	8.0	1.8	1.4	1.0
Industrial	7.5	1.9	1.6	1.3
Storage (*see* Note 4)	6.0	1.5	1.3	1.0
Hazardous (*see* Note 5)	2.8	1.1	0.9	NP

UL—Unlimited

NP—Not permitted

Notes **1:** The FAR values given in this table are subject to overall restrictions on the heights of buildings in the case of educational, institutional, assembly, storage and hazardous occupancies.

2: This table has been prepared, taking into account the combustible content in the different occupancies as well as the fire resistance offered by the type of construction.

3: This table should be modified by the Authority, taking into account the other aspects as given below:

(*a*) Density in terms of dwelling units per hectare:

(*b*) Traffic considerations;

(*c*) Parking spaces;

(*d*) Local fire-fighting facilities; and

(*e*) Water supply, drainage and sanitation requirements.

4: The FAR values specified in this table may be increased by 20 per cent for the following services:

(*a*) A basement or cellar; space under a building constructed on stilts and used as a parking space and air-conditioning plant room used as accessory to the principal use;

(*b*) Watchman's booth, pumphouse, garbage shaft, electric cabin or substation and other utility structures meant for the services of the building under consideration;

(*c*) Projections and accessory buildings as specifically exempted under the Code; and

(*d*) Staircase room and lift rooms above the topmost storey: architectural features; and chimneys and elevated tanks of dimensions as permissible under the Code; the area of the lift shaft shall be taken only on one floor.

5: In so far as single storey storage and hazardous occupancies are concerned, they would be further governed by volume to plot area ratio (VPR) to be decided by the Authority.

Each portion of a building, which is separated by one or more continuous fire-resisting walls, having a fire resistance of not less than 4 hr, extending from the foundation to 1 m above the roof at all points, may be considered to be a separate building for the calculation of maximum permissible height and floor area, provided openings, if any, in the separating wall are also protected by fire assemblies of 4 hr.

29.4.4 Open Spaces

Every room intended for human habitation shall abutt on an interior or exterior open space or on an open *verandah* open to such interior or exterior open space.

The open spaces inside and around a building have essentially to cater for the lighting and ventilation requirements of the rooms abutting such open spaces, and in the case of buildings abutting on streets in the front, rear or sides, the open spaces provided shall be sufficient for the future widening of such streets.

Open spaces separate for each building or wing. The open spaces shall be separate or distinct for each building and where a building has two or more wings, each wing shall have separate or distinct open spaces for the purposes of lighting and ventilation of the wings.

However, separation between accessory and main buildings more than 7 m in height shall not be less than 1.5 m. For buildings up to 7 m in height no such separation shall be required.

For high-rise buildings the following additional provisions of means of access to the building shall be ensured:

(*a*) The width of the main street on which the building abutts shall not be less than 12 m and one end of this street shall join another street not less than 12 m in width;

(*b*) The road shall not end in a dead end; except in the case of residential building, up to a height of 30 m.

(*c*) The compulsory open spaces around the building shall not be used for parking; and

(*d*) Adequate passageway and clearances required for fire-fighting vehicles to enter the premises shall be provided at the main entrance; the width of such entrance shall be not less than 4.5 m. If an arch or covered gate is constructed, it shall have a clear head room of not less than 5 m.

29.4.5 Mixed occupancy

When any building is used for more than one type of occupancy, it shall conform to the requirements for the most hazardous of the occupancies. If mixed occupancies are separated by a separating wall of 4 hr fire rating, the occupancies shall be treated individually.

29.4.6 Openings in Separating Walls and Floors

At the time of designing openings in separating walls and floors, particular attention shall be paid to all such factors as will limit fire spread through these openings and maintain fire rating of the structural member.

For Types 1 to 3 construction a door-way or opening in a separating wall on any floor shall be limited to 5.6 m^2 in area with a maximum height of 2.75 m and a maximum width of 2.1 m. The wall openings shall be protected with fire-resisting doors or steel rolling shutters. All openings in the floors shall be protected by vertical enclosures extending above and below such openings, the walls of such enclosures having a fire resistance of not less than 2 h and all openings therein being protected with a fire-resisting assembly having a fire-resistance rating as specified in (8).

Type 4 construction, openings in the separating walls or floors shall be fitted with 2 h fire-resisting assemblies.

Openings in walls/floors which are necessary to be provided to allow cables/electrical wirings/ telephone cables/plumbing and fixtures, etc. shall be protected by enclosures in the form of ducts/ shafts having a fire resistance not less than 2 hr. Medium and low voltage wiring running in shafts/ ducts shall either be armoured type or run through metal conduits. Further the space between the conduits and the walls shall be filled in by a filler material, compressed asbestos fibre mixed with cement or any other material having fire resistance rating of not less than one hour.

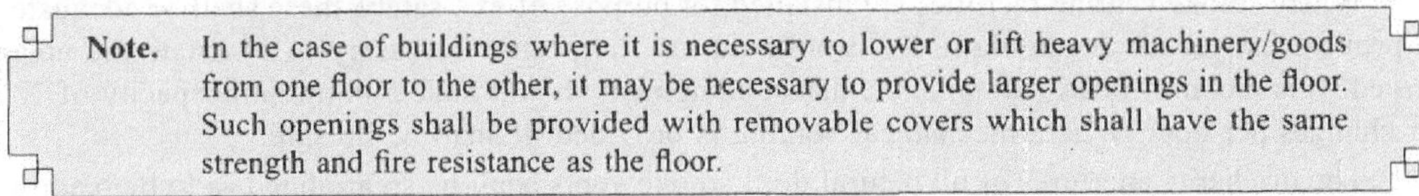

Note. In the case of buildings where it is necessary to lower or lift heavy machinery/goods from one floor to the other, it may be necessary to provide larger openings in the floor. Such openings shall be provided with removable covers which shall have the same strength and fire resistance as the floor.

Every vertical opening between the floors of a building should be enclosed or protected to provide the following:

(*a*) Reasonable safety to the occupants while using the means of egress by preventing spread of fire, smoke or fumes through vertical openings from floor to floor. Further it should be ensured to provide a clear height of 2.1 m in the passage/escape path of the occupants.

(*b*) Limitation of damage to the building and its content.

29.4.7 Fire Stop or Enclosure of Openings

When openings are permitted, these should not exceed three-fourths the area of the wall in the case of an external wall and should be protected with fire resisting assemblies or enclosures having a fire resistance equal to that of wall or floor. Such assemblies or enclosures should be capable of preventing the spread of smoke or fumes through the openings so as to facilitate the safe evacuation of occupants in case of fire.

29.4.8 Electrical Installation

Adequate provision should be made for requirements regarding electrical installations from the point of view of fire safety.

29.4.9 Air-conditioning and Ventilation

Air-conditioning and ventilating systems shall be so installed and maintained as to minimize the danger of spread of fire, smoke or fumes thereby from one floor or fire area of another, or from outside into any occupied building or structure.

Air-conditioning and ventilating systems circulating air to more than one floor or fire area shall be provided with dampers designed to close automatically in case of fire and thereby prevent spread of fire or smoke. Such a system shall also be provided with automatic controls to stop fans in case of fire, unless arranged to remove smoke from a fire, in which case these shall be designed to remain in operation.

Air-conditioning system serving large places of assembly (over 1,000 persons), large departmental stores or hotels with over 100 rooms in a single block shall be provided with effective means for preventing circulation of smoke through the system in the case of a fire in air filters or from other sources drawn into the system and shall have smoke sensitive devices for actuation. From fire safety point of view, separate air handling units for the various floors shall be provided so as to avoid the hazards arising from spread of fire and smoke through the air-conditioning ducts.

29.4.10 Smoke Venting

Smoke venting facilities, for safe use of exits in windowless buildings, underground structures, large area factories, hotels and assembly buildings (including cinema halls) shall be automatic in action, with manual controls in addition.

Natural draft smoke venting shall utilize roof vents or vents in walls at or near the ceiling level; such vents shall be normally open, or, if closed, shall be designed for automatic opening in case of fire, by release of smoke sensitive devices.

Where smoke venting facilities are installed for purpose of exit safety, these shall be adequate to prevent dangerous accumulation of smoke during the period necessary to evacuate the area served. It is recommended that smoke exhaust equipment should have a minimum capacity of 12 air changes per hour. Where mechanical venting is provided, it shall be firesafe.

The discharge apertures of all natural draft smoke vents shall be so arranged as to be readily susceptible to opening by fire service personnel.

Power operated smoke exhausting systems may be substituted for natural draft vents only by specific permission of the Authority.

29.4.11 Heating

Installation of chimney and heating apparatus should be provided.

29.4.12 Additional Precautions

In addition to the factors covered above, there are certain aspects, applicable to particular occupancies only, which may affect the spread of fire, smoke or fumes and thus the safe evacuation of the building occupants in case of fire. Some such aspects are:

(*a*) interior finish and decoration;

(*b*) seating, aisles, railings and turnstiles in places of assembly;

(*c*) service equipment and storage facilities in buildings other than storage buildings; and

(*d*) hazards on stage, in waiting spaces, projection booths, etc. in theatres and cinemas.

29.4.13 Surface Interior Finishes

The use of combustible surface finishes on walls (including external facade of the building) and ceilings affects the safety of the occupants of a building. Such finishes tend to spread the fire and even though the structural elements may be adequately fire resistant, serious danger to life may result. It is, therefore, essential to have adequate precautions to minimize spread of flame on wall, facade of building and ceiling surfaces.

The finishing materials used for various surfaces and decor shall be such that it shall not generate toxic smoke/fumes.

The susceptibility to fire of various types of wall surfaces is determined in terms of the rate of spread of flame. Based upon the rate of spread of flame, surfacing materials shall be considered as divided into four classes as follows:

***Class 1** Surfaces of very low flame spread.*

***Class 2** Surfaces of low flame spread.*

Class 3 Surfaces of medium flame spread.

Class 4 Surfaces of rapid flame spread.

The situation under which materials falling into various classes shall be used in building construction is given below:

Class 1	*Class 2*	*Class 3*
May be used in any situation	May be used in any suitation, except on walls, facade of the building and ceilings of staircases and corridors	Should be used only in living rooms and bedrooms (but not in rooms on the roof) and only as a lining to solid walls and partitions. Not on staircases or corridors or facade of the building

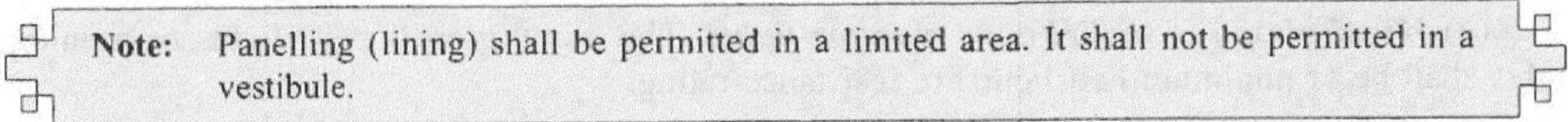

Note: Panelling (lining) shall be permitted in a limited area. It shall not be permitted in a vestibule.

Materials of Class 4 which include untreated wood fibreboards may be used with due fire retardant treatment as ceiling lining, provided the ceiling is at least 2.4 m from the top surface of the floor below, and the wall surface conform to requirements of relevant class. Class 4 materials shall not be used in kitchens, corridors and staircases. Some types contain bitumen and, in addition to risk from spread of fire, emit dense smoke on burning; such materials shall be excluded from use under these conditions and shall also not be used for construction of ceiling where the plenum is used for return air in air-conditioned buildings.

When frames, walls, partitions or floors are lined with combustible materials, the surfaces on both sides of the materials shall conform to the appropriate class, because there is considerable danger from fire starting and rapidly spreading within the concealed cavity unknown to the occupants whose escape may be hampered thereby.

29.4.14 Glazing

Building of types 1 to 4 construction shall employ one of the two types of glazing described below except that Type 4 construction may have the alternative of hardwood sashes or frames or both.

Wired glass shall comply with the following requirements:

(*a*) ***Wired glass***: The wired glass shall be of minimum half hour fire resistance rating.

(*b*) ***Sashes and frames.*** The sashes or frames or both shall be entirely of iron or other suitable metal such as stainless steel securely bolted or keyed into the wall, except in the case of panels in internal doors.

(*c*) ***Setting of glass.*** The panels of glass shall be set in rebates or grooves not less than 6.5 mm in width or depth, with due allowance for expansion, and shall be secured by hard metal fastenings to the sashes or frames independent of any lead, cement or putty used for weather-proofing purposes.

Electro-copper glazing shall comply with the following requirements:

(*a*) ***Electro-copper glazing***: The electro-copper glazing shall be of minimum half hour fire resistance rating.

(*b*) ***Sashes and frames.*** The sashes or frames or both shall be entirely of iron or other hard metal, securely bolted or keyed into the wall, except when in panels in internal doors.

(*c*) ***Fixing of sectional lights.*** The sectional lights shall be set in rebate or grooves not less than 6.5 mm in width or depth with due allowance for expansion and shall be secured by hard metal fastenings to the sashes or frames independent of any lead, cement or putty used for weather-proofing purposes.

Maximum permissible area shall be 5 m^2 for efficient protection by wired glass or electro-copper glazing.

Casement

Hard metal casements, not exceeding 0.8 m^2 fitted with wired glass or electro-copper glazing secured to the frames by hard metal hinges not more than 600 mm apart and by fastenings at top, centre and bottom shall be permissible.

29.4.15 Skylights

Wired glass for skylights or monitor lights shall comply with the following requirements:

(*a*) ***Wired glass for skylights or monitor lights.*** The wired glass for skylights or monitor lights shall be of minimum half hour fire resistance rating.

(*b*) ***Frames and glazing.*** The frame shall be continuous and divided by bars spaced at not more than 700 mm centres. The frame and bars shall be of iron or other hard metal, and supported on a curb either of metal or of wood covered with sheet metal. The toughened glass shall be secured by hard metal fastenings to the frame and bars independent of any lead, cement or putty used for weather-proofing purposes.

29.4.16 Louvers

Louvers wherever provided shall be of minimum half hour fire resistance rating.

29.4.17 Glass of Facade for High Rise Buildings

Glass of facade for high rise buildings, etc. shall be minimum one hour fire resistance rating.

LIFE SAFETY OR EXIT REQUIREMENTS

29.5 TYPES OF EXITS

An exit may be a doorway, corridor, passageway(s) to an internal staircase, or external staircase, or to a *verandah* or terrace(s), which have access to the street, or to the roof of a building or a refuge area. An exit may also include a horizontal exit leading to an adjoining building at the same level.

Lifts and escalators shall not be considered as exits.

29.5.1 General Exit Requirements

Every building meant for human occupancy shall be provided with exits sufficient to permit safe escape of occupants, in case of fire or other emergency.

All exits shall be free of obstructions.

No building shall be so altered as to reduce the number, width or protection of exits to less than that required.

Exits shall be clearly visible and the routes to reach the exit shall be clearly marked and sign posted to guide the population of the floor concerned. Signs shall be illuminated and wired to an independent electrical circuit or an alternative source of supply. The colour of the exit sign shall be green.

Note: This provision shall not apply to A-2 and A-4 occupancies up to 15 m in height.

Where necessary, adequate and reliable illumination shall be provided for exits.

Fire doors with 2 hour fire resistance shall be provided at appropriate places along the escape routes particularly at the entrance of lift lobby and stair well where a 'funnel or flue effect' may be created, inducing an upward spread of fire and smoke.

All exits shall provide continuous means of egress to the exterior of a building or to an exterior open space leading to a street.

Exits shall be so arranged that they may be reached without passing through another occupied unit.

29.5.2 Occupant Load

For determining the exits required, the number of persons within any floor area or the occupant load shall be based on the actual number of occupants, but in no case less than that specified in Table 29.5.

Table 29.5: Occupant load

Sl. No.	*Group of occupancy*	*Occupant load floor area in m^2/person*
(*i*)	Residential	12.5
(*ii*)	Educational	4
(*iii*)	Institutional	15*
(*iv*)	Assembly:	
	(*a*) with fixed or loose seats and dance floors	0.6*
	(*b*) without seating facilities including dining rooms	1.5†
(*v*)	Mercantile:	
	(*a*) street floor and sales basement	3
	(*b*) upper sale floors	6
(*vi*)	Business and industrial	10
(*vii*)	Storage	30
(*viii*)	Hazardous	10

* Occupant load in dormitory portions of homes for the aged, orphanages insane, asylums, etc. where sleeping accommodation is provided, shall be calculated at not less than 7.5 m^2 gross floor area/person.

† The gross floor area shall include, in addition to the main assembly room or space, any occupied connecting room or space in the same storey or in the storeys above or below, where entrance is common to such rooms and spaces and they are available for use by the occupants of the assembly place. No deductions shall be made in the gross area for corridors, closets or other subdivisions; the area shall include all space serving the particular assembly occupancy.

Mezzanine. The occupant load of a mezzanine floor discharging to a floor below shall be added to that floor occupancy and the capacity of the exits shall be designed for the total occupancy load thus established.

29.5.3 Capacities of Exits

The unit of exit width used to measure the capacity of any exit shall be 500 mm. A clear width of 250 mm shall be counted as an additional half unit. Clear widths less than 250 mm shall not be counted for exit width.

Occupants per unit exit width shall be in accordance with Table 29.6.

Horizontal exit allowance. When horizontal exit is provided in buildings of mercantile, storage, industrial business and assembly occupancies, the capacity per storey per unit width of exit of stairways in Table 29.6 may be increased by 50 per cent; and in buildings of institutional occupancy it may be increased by 100 per cent.

Table 29.6: Occupants per unit exit width

Sl. No.	*Group of occupancy*	*Number of occupants*		
		Stairways	*Ramps*	*Doors*
(*i*)	Residential (*A*)	25	50	75
(*ii*)	Educational (*B*)	25	50	75
(*iii*)	Institutional (*C*)	25	50	75
(*iv*)	Assembly (*D*)	40	50	60
(*v*)	Business (*E*)	50	60	75
(*vi*)	Mercantile (*F*)	50	60	75
(*vii*)	Industrial (*G*)	50	60	75
(*viii*)	Storage (*H*)	50	60	75
(*ix*)	Hazardous (*J*)	25	30	40

29.5.4 Arrangement of Exits

Exits shall be so located that the travel distance on the floor shall not exceed the distance given in Table 29.7.

The travel distance to an exit from the dead end of a corridor shall not exceed half the distance specified in Table 29.7, except in educational, assembly and institutional occupancies in which case it shall not exceed 6 m.

Whenever more than one exit is required for any room space or floor of a building, exits shall be placed as remote from each other as possible and shall be arranged to provide direct access in separate directions from any point in the area served.

Table 29.7: Travel distance for occupancy and type of construction

Sl. No.	Group of occupancy	Construction	
		Types 1 and 2	Types 3 and 4
		m	m
(*i*)	Residential (*A*)	30.0	22.5
(*ii*)	Educational (*B*)	30.0	22.5
(*iii*)	Institutional (*C*)	30.0	22.5
(*iv*)	Assembly (*D*)	30.0	30.0
(*v*)	Business (*E*)	30.0	30.0
(*vi*)	Mercantile (*F*)	30.0	30.0
(*vii*)	Industrial (*G*)	45.0	*
(*viii*)	Storage (*H*)	30.0	*
(*ix*)	Hazardous (*I*)	22.5	*

Note 1: For fully sprinklered building, the travel distance may be increased by 50 per cent of the values specified.

Note 2: Ramps shall be protected with automatic sprinkler system and shall be counted as one of the means of escape.

* Construction of type 3 or 4 is not permitted.

29.5.5 Number of Exits

All buildings which are 15 m in height or above and all buildings used as educational, assembly, institutional, industrial, storage and hazardous occupancies and mixed occupancies with any of the aforesaid occupancies, having area more than 500 m^2 on each floor shall have a minimum of two staircases. They shall be of enclosed type, at least one of them shall be on external walls of buildings and shall open directly to the exterior, interior open space or to an open place of safety. Further, the provision or otherwise of alternative staircase shall be subject to the requirements of travel distance being complied with.

29.5.6 Doorways

Every exit doorway shall open into an enclosed stairway, or a horizontal exit of a corridor or passageway providing continuous and protected means of egress.

No exit doorway shall be less than 1000 mm in width except assembly buildings where door width shall not be less than 2000 mm. Doorways shall be not less than 2000 mm in height.

Exit doorways shall open outwards, that is, away from the room, but shall not obstruct the travel along any exit. No door, when opened, shall reduce the required width of stairway or landing to less than 900 mm; overhead or sliding doors shall not be installed.

In the case of buildings where there is a central corridor, the doors of rooms shall open inwards to permit smooth flow of traffic in the corridor.

Exit door shall not open immediately upon a flight of stairs; a landing equal to at least the width of the door shall be provided in the stairway at each doorway; the level of landing shall be the same as that of the floor which it serves.

Exit doorways shall be openable from the side which they serve without the use of a key.

Mirrors shall not be placed in exit ways or exit doors to avoid confusion regarding the direction of exit.

29.5.7 Revolving Doors

Revolving doors shall not be used as required exits, except in residential, business and mercantile occupancies, but shall not constitute more than half the total required door width. In mercantile buildings where there is a possibility of congregation of a large number of people (more than 200), revolving doors shall not be permitted.

Where the revolving doors provided are completely made of glass (plate glass), a red circle or emblem or logo at 1.5 m level above sill shall be painted on the glass. The thickness of the glass shall be not less than 12 mm.

29.5.8 Corridors and Passageways

Exit corridors and passageways shall be of the width not less than the aggregate required width of exit doorways leading from them in the direction of travel to the exterior.

Where stairways discharge through corridors and passageways, the height of corridors and passageways shall be not less than 2.4 m.

All means of exit including staircases, lifts, lobbies and corridors shall be adequately ventilated.

29.5.9 Internal Staircases

Interior stairs shall be constructed of non-combustible materials throughout.

Interior staircase shall be constructed as a self-contained unit with an external wall constituting at least one of its sides and shall be completely enclosed.

A staircase shall not be arranged round a lift shaft.

Hollow combustible construction shall not be permitted.

No gas piping shall be laid in the stairway. Ducting in stairway may be permitted if it is of one hour fire resistance rating.

Notwithstanding the detailed provision, the following minimum width shall be provided for staircases:

(*a*) Residential buildings (dwellings)	1.0 m
(*b*) Residential hotel buildings	1.5 m
(*c*) Assembly buildings like auditorium, theatres and cinemas	2.0 m
(*d*) Educational buildings, upto 30 m in height	1.5 m
(*e*) Institutional buildings like hospitals.	2.0 m
(*f*) All other buildings	1.5 m

The minimum width of tread without nosing shall be 250 mm for internal staircase of residential buildings. This shall be 300 mm for assembly, educational, institutional, business and other buildings. The treads shall be constructed and maintained in a manner to prevent slipping.

The maximum height of riser shall be 190 mm for residential buildings and 150 mm for other buildings and the number shall be limited to 15 per flight.

Handrails shall be provided at a minimum height of 1000 mm to be measured from the base of the middle of the treads to the top of the handrails. Balusters/railing shall be provided such that the width of staircase does not reduce (Fig 29.1).

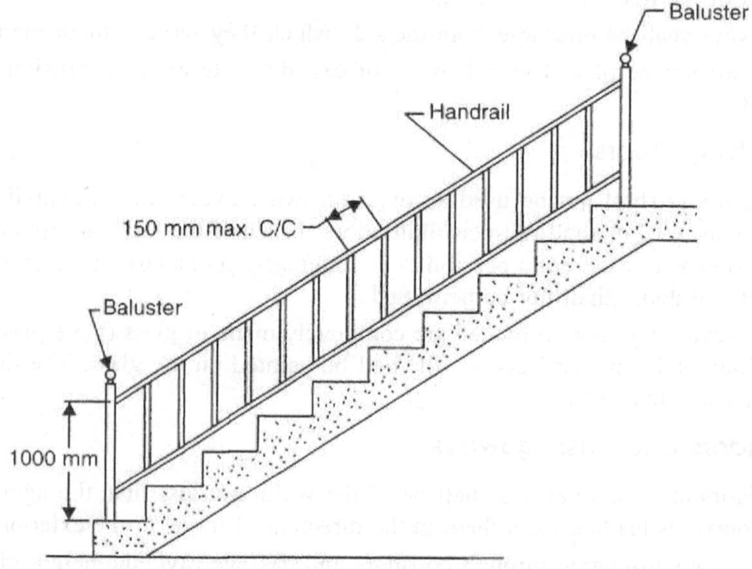

Fig. 29.1: Typical detail for handrail/blusters of a staircase

The number of people in between floor landings in staircase shall not be less than the population on each floor for the purpose of design of staircase.

The design of staircase should take into account the following:

(*i*) The minimum headroom in a passage under the landing of a staircase and under the staircase should be 2.2 m.

(*ii*) For buildings 15 m in height or more, access to main staircase shall be through a fire/smoke check door of a minimum 2 hour fire resistance rating. Fire resisting rating may be reduced to one hour for residential buildings.

(*iii*) No living space, store or other fire risk shall open directly into the staircase.

(*iv*) External exit door of staircase enclosure at ground level shall open directly to the open space or through a large lobby.

(*v*) The main and external staircases shall be continuous from ground floor to terrace level.

(*vi*) No electrical shafts/A.C. ducts or gas pipes, etc. shall pass through or open in the staircases. Lifts shall not open in staircase.

(*vii*) No combustible material shall be used for decoration/wall paneling in the staircase.

(*viii*) Beams/columns and other building features shall not reduce the headroom/width of the staircase.

(*ix*) The exit sign with arrow indicating the way to the escape route shall be provided at a suitable height from the floor level on the wall and shall be illuminated by electric light connected to corridor circuits. All exit way marking signs should be flush with the wall and so designed that no mechanical damage shall occur to them due to moving of furniture or other heavy equipments.

Floor indication board shall be placed on the wall immediately facing the flight of stairs and nearest to the landing. It shall be of size not less than 0.5 m × 0.5 m.

(*x*) Individual floors shall be prominently indicated on the wall facing the staircases.

(*xi*) In case of single staircase it shall terminate at the ground floor level and the access to the basement shall be by a separate staircase.

29.5.10 Pressurization of Staircases (Protected Escape Routes)

Though in normal building design, compartmentation plays a vital part in limiting the spread of fire, smoke will readily spread to adjacent spaces through the various leakage openings in the compartment enclosure, such as cracks, openings around pipes ducts, airflow grills and doors, as perfect sealing of all these openings is not possible. It is smoke and toxic gases, rather than flame, that will initially obstruct the free movement of occupants of the building through the means of escape (escape routes). Hence the exclusion of smoke and toxic gases from the protected routes is of great importance.

Pressurization is a method adopted for protected escape routes against ingress of smoke, especially in high-rise buildings. In pressurization, air is injected into the staircases, lobbies or corridors, to raise their pressure slightly above the pressure in adjacent parts of the building. As a result, ingress of smoke or toxic gases into the escape routes will be prevented. the pressurization of staircases shall be adopted for high rise buildings and building having mixed occupancy/ multiplexes having covered area more than 500 m^2.

The pressure difference for staircases shall be as under:

Building Height	***Pressure Difference***	
	Reduced Operation (Stage 1 of a 2-Stage System)	Emergency Operation (Stage 2 of a 2-Stage System or Single Stage System)
	(Pa)	(Pa)
Less than 15 m	8	50
15 m or above	15	50

If possible, the same levels shall be used for lobbies and corridors, but levels slightly lower may be used for these spaces if desired. The difference in pressurization levels between staircase and lobbies (or corridors) shall not be greater than 5 Pa.

Pressurization system may be of two types:

(*a*) Single-stage, designed for operation only in the event of an emergency, and

(*b*) Two-stage, where normally a level of pressurization is maintained in the protected escape routes and an increased level of pressurization can be brought into operation in an emergency.

The normal air-conditioning system and the pressurization system shall be treated as an integral one, especially for a two-stage system. When the emergency pressurization is brought into action, the following changes in the normal air-conditioning system shall be effected.

(*a*) Any re-circulation of air shall be stopped and all exhaust air vented to atmosphere:

(*b*) Any air supply to the spaces/areas other than escape routes shall be stopped;

(*c*) The exhaust system may be continued provided:

(1) the positions of the extraction grills permit a general air flow away from the protected escape route entry;

(2) the construction of the duct work and fans is such that, it will not be rendered inoperable by hot gases and smoke; and

(3) there is no danger of spread of smoke to other floors by the path of the extraction system which can be ensured by keeping the extraction fans running.

29.5.11 External Stairs Cases

An external staircase is desirable to be provided in a high rise building and should comply with the following:

All external stairs shall be directly connected to the ground.

Entrance to the external stairs shall be separate and remote from the internal staircase.

Care shall be taken to ensure that no wall opening or window opens on to or close to an external stairs.

The route to the fire escape shall be free of obstructions at all times.

The external stairs shall be constructed of non-combustible materials, and any doorway leading to it shall have the required fire resistance.

No external staircase, used as a fire escape, shall be inclined at an angle greater than 45° from the horizontal.

External stairs shall have straight flight not less than 1250 mm wide with 250 mm treads and risers not more than 190 mm. The number of risers shall be limited to 15 per flight.

Hand rails shall be of a height not less than 1000 mm and not exceeding 1200 mm.

Balusters shall be provided with a maximum gap of 150 mm.

Spiral fire escape. The use of spiral staircase shall be limited to low occupant load and to a building not exceeding 9 m in height.

A spiral fire escape shall be not less than 1500 mm in diameter and shall be designed to give adequate headroom.

Unprotected steel frame staircase will not be accepted as a means of escape. However, steel, staircase in an enclosed fire rated compartment of 2 hours will be accepted as a means of escape.

29.5.12 Horizontal Exits

The width of horizontal exit shall be same as for the exit doorways. A horizontal exit shall be equipped with at least one fire smoke door of minimum 1 hour fire resistance of self-closing type.

Further, it is required to have direct delivery to the fire escape staircase for evacuation.

For buildings more than 24 m in height, refuge area of 1.5 m^2 or an area equivalent to 0.3 m^2 per person to accommodate the occupants of two consecutive floors, whichever is higher, shall be provided as under:

Before area shall be provided on the periphery of lies or preferably on a cantilever projection and to the air at least on one side protected with suitable stages.

* ***For floor above 24 m and Up to 39 m***—One refuge area on the floor immediately above 24 m.
* ***For floor above 39 m***—One refuge area on the floor immediately above 39 m and so on after every 15 m. Refuge area provided in excess of the requirements shall be counted towards FAR.

Note: Residential flats in multi-storeyed building with balcony, need not be provided with refuge area, however flats without balcony shall provide refuge area as given above.

Where there is a difference in level between connected areas for horizontal exits, ramps, not more than 1 to 10 in slope shall be provided, steps shall not be used.

Doors in horizontal exits shall be openable at all times from both sides.

29.5.13 Ramps

Ramps shall comply with all the requirements applicable for stairways regarding enclosure, capacity and limiting dimensions except where specified for special uses and occupancies.

The slope of ramp shall not exceed 1 in 10. In certain case steeper slopes may be permitted but in no case greater than 1 in 8.

For all slopes exceeding 1 in 10 and wherever the use is such as to involve danger of slipping, the ramp shall be surfaced with approved non-slipping material.

29.5.14 Fire Tower

Fire towers are the preferred types of escape routes for storeyed buildings and their application shall be considered as the safest route for escape. Their number, location and size shall depend on the building concerned, and its associated escape routes.

In high rise buildings with buildings, over 8 storeys or 24 m in height, at least one required means of egress shall be a fire tower.

The fire towers shall be constructed of walls with a 2 hr fire-resistance rating without openings other than the exit doorways, with platforms, landings and balconies having the same fire-resistance rating.

29.5.15 Fire Lifts

Where applicable, fire lifts shall be provided with a minimum capacity for 8 passengers and fully automated with emergency switch on ground level. In general, buildings 15 m in height or above shall be provided with fire lifts.

In case of fire, only fireman shall operate the fire lift. In normal course, it may be used by other persons.

Each fire lift shall be equipped with suitable inter-communication equipment for communicating with the control room on the ground floor of the building.

The number and location of fire lifts in a building shall be decided after taking into consideration various factors like building population, floor area, compartmentation, etc.

29.5.16 Emergency and Escape Lighting

Emergency lighting shall be powered from a source independent of that supplying the normal lighting.

Escape lighting shall be capable of:

(*a*) Indicating clearly and unambiguously the escape routes.

(*b*) Providing adequate illumination along such routes to allow safe movement of persons towards and through the exits,

(*c*) Ensuring that fire alarm call points and fire-fighting equipments provided along the escape routes can be readily located.

Escape lighting luminaries should be sited to cover the following locations:

(*a*) Near each intersection of corridors,

(*b*) at each exit door,

(*c*) Near each change of direction in the escape route,

(*d*) Near each staircase so that each flight of stairs receives direct light,

(*e*) Near any other change of floor level,

(*f*) Outside each final exit and close to it,

(*g*) Near each fire alarm call point,

(*h*) Near fire-fighting equipment, and

(*i*) To illuminate exit and safety signs as required by the enforcing authority.

29.5.17 Illumination of Means of Exit

Staircase and corridor lights shall conform to the following:

(*a*) The staircase and corridor lighting shall be on separate circuits and shall be independently connected so that it could be operated by one switch installation on the ground floor easily accessible to fire fighting staff at any time irrespective of the position of the individual control of the light points, if any. It should be of miniature circuit breaker type of switch so as to avoid replacement of fuse in case of crisis;

(*b*) Staircase and corridor lighting shall be connected to alternative supply. The alternative source of supply may be provided by battery continuously trickle charged from the electric mains; and

(*c*) Suitable arrangements shall be made by installing double throw switches to ensure that the lighting installed in the staircase and that corridor does not get connected to two sources of supply simultaneously. Double throw switch shall be installed in the service room for terminating the stand-by supply.

29.5.18 Fire Detection and Warning

In buildings of such size, arrangement or occupancy where a fire may not itself provide adequate warning to occupants, automatic fire detection and alarm facilities shall be provided, where necessary, to warm occupants early to the existence of fire, so that they may escape, and to facilitate the orderly conduct of the fire exit drills.

FIRE PROTECTION

In buildings of such size, arrangement or occupancy that a fire may not itself provide adequate warning to occupants, automatic fire detecting and alarm facilities shall be provided, where necessary, to warn occupants of the existence of fire, so that they may escape, or to facilitate the orderly conduct of fire exit drills.

Various types of fire detectors are available depending upon the response to the generation of heat, smoke or flame. These are known as heat detectors, smoke detectors and flame detectors.

29.6 FIRE EXTINGUISHERS FIXED FIRE-FIGHTING INSTALLATIONS

All buildings depending upon the occupancy, use and height shall be protected by fire-extinguishers wet riser, downcomer, automatic sprinkler installation, high/medium pressure water spray, foam, gaseous or dry powder system, etc. in accordance with the following provisions. The minimum requirements for fire-fighting installations shall be as specified in Table 23 of National Building Code of India 2005-Part 4; Fire and Life Safety to which reference may be made. The typical arrangements of downcomer and wet riser installations are shown in Figs. 29.2 and 29.3. The wet riser shall be designed for zonal distribution ensuring that unduly high pressures are not developed in risers and hose pipes.

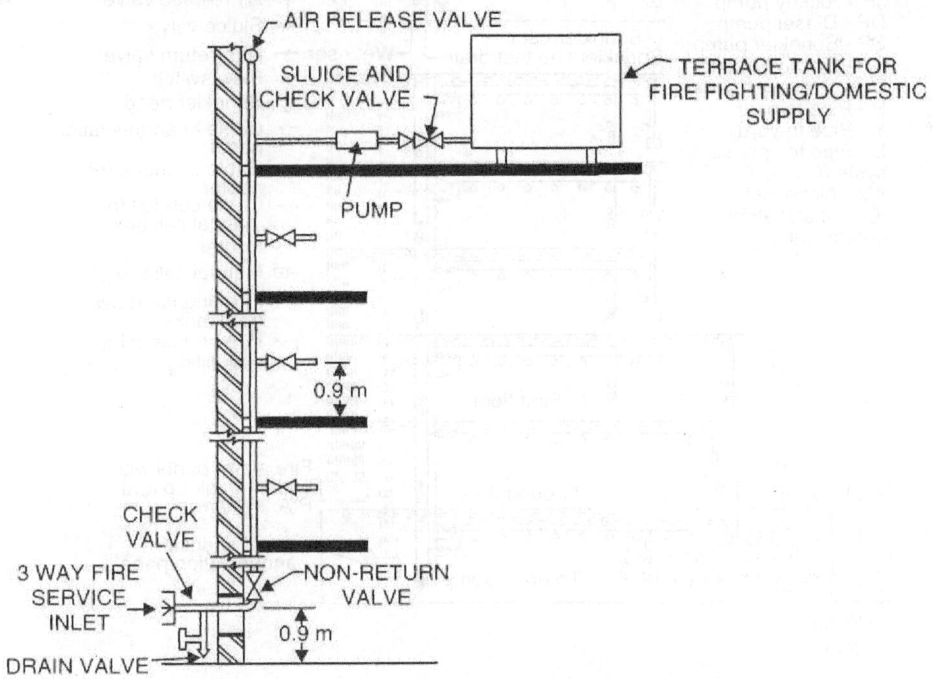

Fig. 29.2: Typical arrangement of downcomer for buildings above 15 m but not exceeding 30 m in height.

First-aid fire-fighting appliances should be provided and installed.

In addition to wet riser or downcomer, first-aid hose reels shall be installed on all the floors of buildings 15 m in height or more. The first-aid hose reel shall be connected directly to the riser/down-comer main and diameter of the hose reel shall not be less than 19 mm.

29.6.1 Static Water Storage Tanks

A satisfactory supply of water for the purpose of fire-fighting shall always be available in the form of underground terrace level static storage tank with capacity specified for each building with arrangements of replenishment by mains of alternative source of supply at the rate of 1,000 litres per minute for underground static tank. When this is not practicable, the capacity of static storage tank(s) shall be increased proportionately in consultation with the local fire brigade.

The static storage water supply required for the above-mentioned purpose shall entirely be accessible to the fire engines of the local fire service. Provision of suitable number of manholes shall be made available for inspection, repairs and insertion of suction hose, etc. The covering slab shall be able to withstand the vehicular load of 45 tonnes equally divided as a four point load when slab forms a part of pathway/driveway.

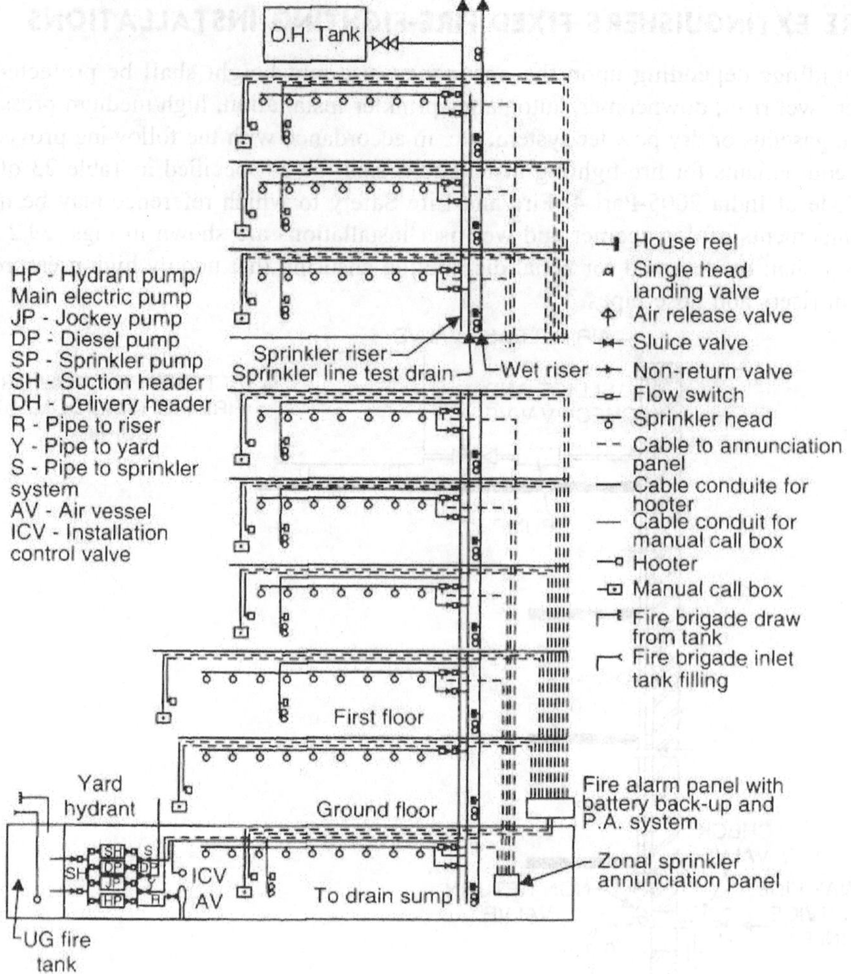

Fig. 29.3: Typical arrangement of wet riser and total sprinkler system of building other than apartment exceeding 30 m in height.

The domestic suction tank connected to the static water storage tank shall have an overflow capable of discharging 2,250 litres per minute to a visible drain point from which by a separate conduit the overflow shall be conveyed to a storm water drain.

(*a*) To prevent stagnation of water in the static water storage tank, the suction tank of the domestic water supply shall be fed only through an overflow arrangement to maintain the level therein at the minimum specified capacity (*see* Fig. 29.6).

(*b*) The static water storage tank shall be provided with a fire brigade collecting head with 4 number 63 mm dia (2 number 63 mm dia for pump with capacity 1400 litres/minute) instantaneous male inlets arranged in a valve box at a suitable point at street level and connected to the static tank by a suitable fixed pipe not less than 150 mm in dia to discharge water into the tank when required at the rate of 2,250 litres per minute, if tank is in the basement or not approachable for the fire-engines.

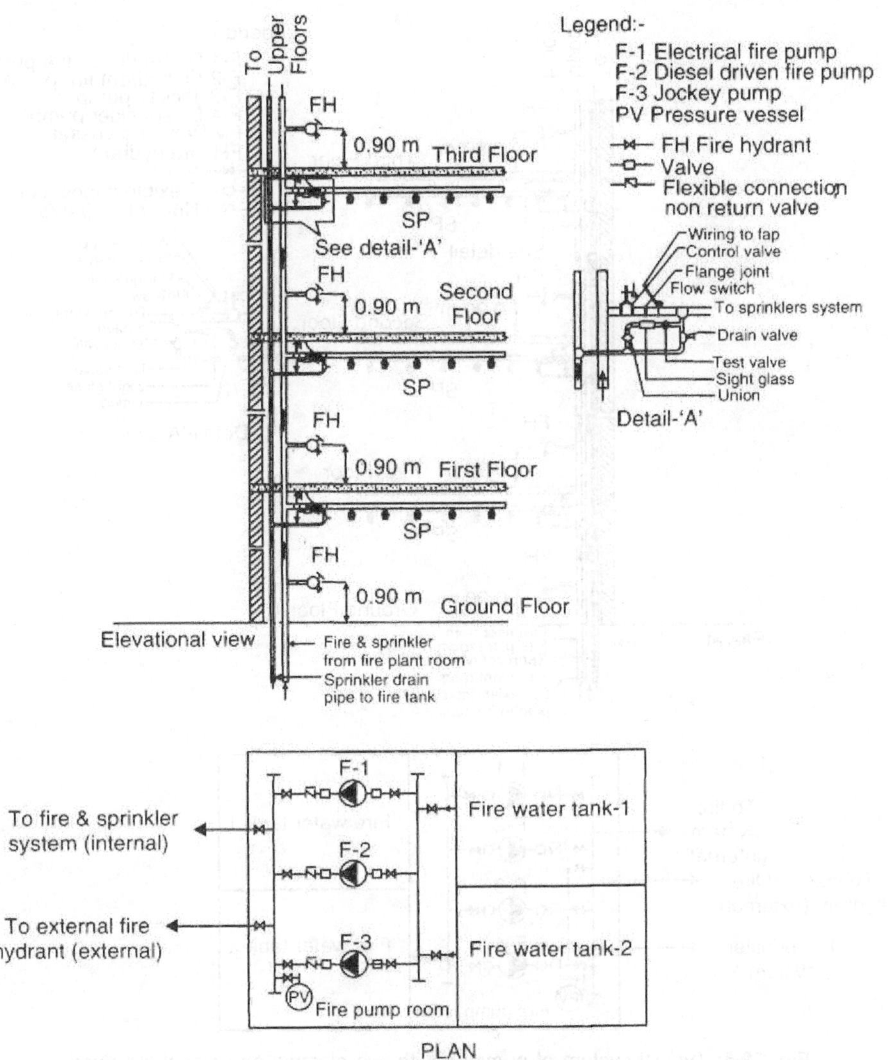

Fig. 29.4: Typical system of pumping with one electric and one diesel fire pump.

29.6.2 Automatic Sprinklers

Automatic sprinklers shall be installed in:

(*a*) basements used as car parks or for storage occupancy, if the area exceeds 200 m^2.

(*b*) multi-level basements, covered upper floors used as car parks and for housing essential services ancillary to a particular occupancy or for storage occupancy, excluding any area to be used for sub-station, A.C. plant and DG set.

(*c*) any room or other compartment of a building exceeding 1125 m^2 except as in (*g*).

(*d*) departmental stores or shops, if the aggregate floor area exceeds 500 m^2.

(*e*) all non-domestic floors of mixed occupancy which constitute a hazard and are not provided with staircases independent of the remainder of the buildings.

(*f*) godowns and warehouses as considered necessary.

(*g*) on all floors of the buildings other than residential and educational buildings, if the height of the building exceeds 15 m (45 m in case of group housing and apartments).

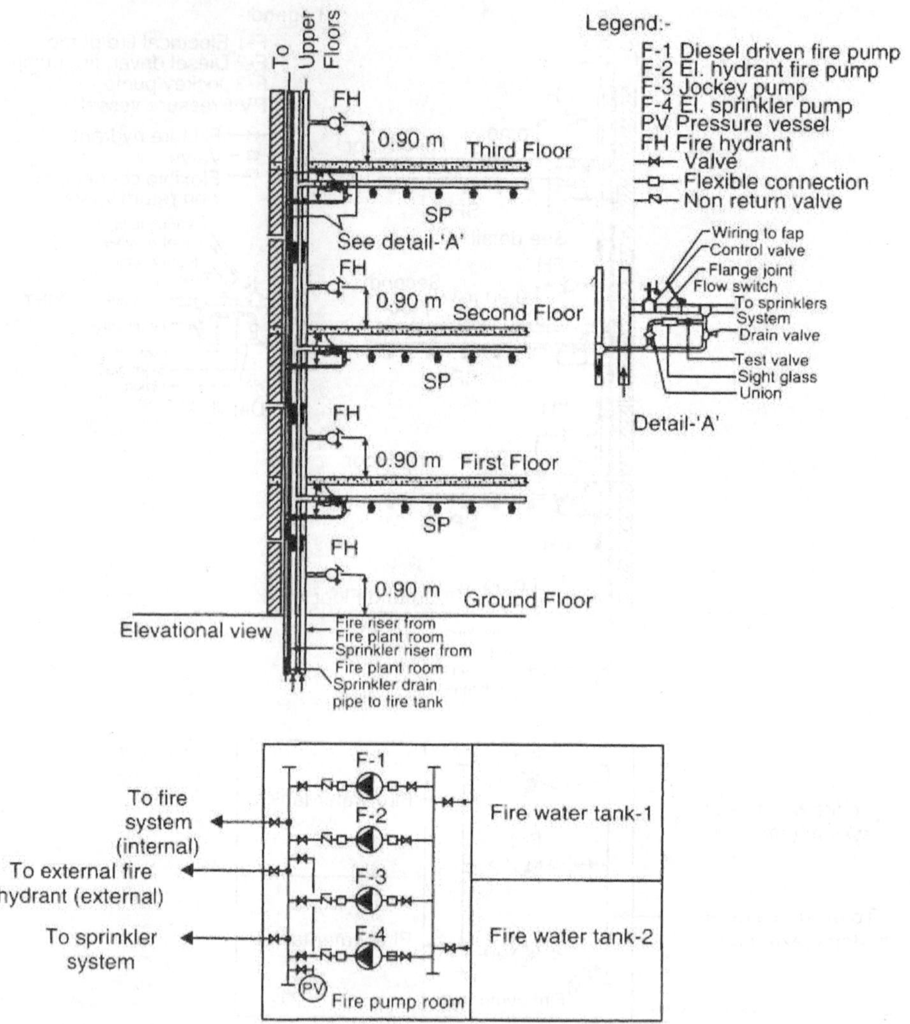

Fig. 29.5: Typical system of pumping with two electric; one diesel fire pump

(*h*) dressing rooms, scenery docks, stages and stage basements of theatres.

(*i*) in hotels, hospitals, industries, low and moderate hazard mercantile buildings of height 15 m or above.

(*j*) in hotels below 15 m, if covered area at each floor is more than 1000 m^2

(*k*) false ceiling voids exceeding 800 m in height.

(*l*) canteen provided in upper floors of D_1 and D_2 occupancies shall be sprinklered.

It is desirable that all high-rise buildings should be fully sprinkled irrespective of their height and occupancy. If selective sprinklering is adopted, there is a real danger of a fire starting on one of the lower unsprinkled floors gathering momentum spreading upwards from floor to floor through the unsprinkled floor and reaching the first sprinklered floor as a fully developed fire. In such an event, the sprinklers can be rendered useless/ineffective.

29.6.3 Automatic High Velocity Water Spray or Emulsifying System

Automatic high velocity water spray or emulsifying system shall be provided for protection of indoor oil-cooled transformers.

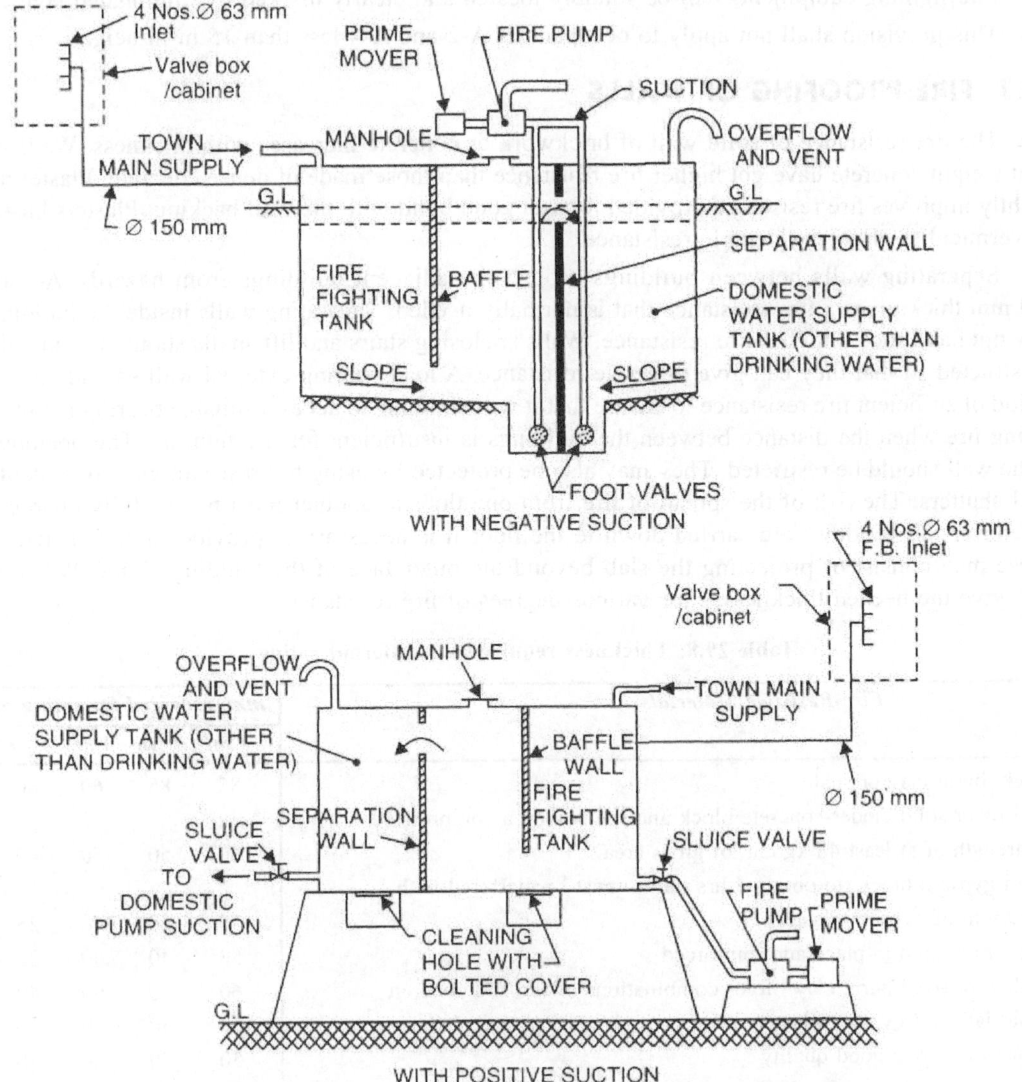

Figs. 29.6 and 29.7: Typical arrangement for providing combined fire fighting and domestic water storage tank

29.6.4 Fixed Foam Installation

Fixed foam generating system shall be provided for protection of boiler rooms with its ancillary storage of furnace oils in basement. Such installations can be low, medium or high expansion type.

29.6.5 Carbon Dioxide Fire Extinguishing System

Fixed carbon dioxide fire extinguishing installation shall be provided on premises where water or foam cannot be used for fire extinguishing because of the special nature of the contents of the buildings/areas to be protected. For some special fire risk, carbon dioxide may not be suitable and it may be necessary to provide BCF (Bromochloro-difluoromethane)—Halon 1211 or BTM (Bromochloro-trifluoromethane)—Halon 1301. The use of these types of installations should, however, be discouraged as halons are ozone depleting substances (ODS) and their use is being phased out throughout the world.

Fire-fighting equipment shall be suitably located and clearly marked by illuminated signs. This provision shall not apply to occupancies A-2 and A-4 less than 15 m in height.

29.7 FIRE-PROOFING OF WALLS

The fire resistance of solid wall of brickwork or concrete increases with thickness. Walls of light weight concrete have got higher fire resistance than those made of dense concrete. Plastering slightly improves fire resistance provided it has a good bond with the wall backing. Plasters based on vermiculite give considerable resistance.

Separating walls between buildings protect the adjacent buildings from hazards. A wall 230 mm thick gives a fire resistance that is normally needed. Separating walls inside the building may not have such a higher fire resistance. Walls enclosing stairs and lift shafts should be suitably constructed so that they can give effective resistance. A load bearing external wall should have a period of sufficient fire resistance to ensure that it will continue to act as a suitable bearing member during fire when the distance between the buildings is insufficient for fire fighting. The openings in the wall should be restricted. They may also be protected by using fire resistant glass or suitable steel shutters. The risk of the spread of fire from one floor to another must be carefully guarded. Whenever the windows are carried down to the floor it is necessary to provide suitable barriers. These may consist of projecting the slab beyond the outer face of the building. Table 29.8 and 29.9 give the needed thicknesses for various degrees of fire resistance.

Table 29.8: Thickness required for different rating

Fire-Resistive Materials	*mm, required for rating of*			
	4 hr.	*3 hr.*	*2 hr.*	*1 hr.*
Brick, burnt clay or shale ...	85	85	60	60
Hollow or solid cinder-concrete block and tile having a compressive strength of at least 48 kg/cm^2 of gross area ...	60	50	50	40
Solid gypsum block (to obtain 4 hrs rating must be plastered with 10 mm of gypsum plaster) ...	50	50	40	25
Gypsum poured in place and reinforced ...	50	40	40	25
Hollow or solid burnt clay tile or combinations of tile and concrete ...	60	50	50	40
Metal lath and gypsum plaster ...	60	50	40	20
Cement concrete good quality ...	50	50	40	25
Cement concrete poor quality ...	100	70	50	40
Cement concrete, poor quality with wire mesh ...	70	50	50	40
Hollow gypsum block (to obtain 4 hr rating must be plastered with 10 mm of gypsum plaster) ...	70	70	70	70

29.8 FIRE-PROOFING OF STRUCTURAL STEEL

For fire-proof construction the minimum resistance needed for columns and girders is of four hours' duration whereas for beams it may be about three hours. In semi-fireproof constructions these values are three hours and two hours respectively.

The modern steel frame building is fire-proof with concrete. The exterior columns may be fire-proofed with concrete and then encased in masonry. The columns may be protected by wrapping by a mesh or expanded metal mesh and then covering by two coats of cement plaster. A combination of terracotta and concrete for fire proofing the steel columns is quite suitable. The terracotta blocks surround the columns whereas the projections, etc., in column section are filled with concrete. It may be preferable to use either terracotta or concrete as otherwise the difference in thermal expansion in the two materials may lead to cracking.

Table 29.9: Thickness for different types of construction

Construction and materials	***Minimum thickness in millimetres (excluding plaster) for period of***				
	6 hrs.	**4 hrs.**	**2 hrs.**	**1 hr.**	**$\frac{1}{2}$ hr.**
Bricks and solid block construction:					
Bricks of clay, concrete or sand-lime built as a solid wall:					
No plaster ...	200	200	200	100	100
Built as a cavity wall ...	230	—	—	—	—
Concrete Blocks:					
Built as a cavity wall 100 mm outer leaf: inner leaf of solid or hollow concrete blocks ...	100	70	—	—	—
Reinforced concrete:					
With minimum concrete cover of 25 mm to reinforcement ...	200	170	100	70	70
Hollow block construction:					
Clay blocks:					
Plastered at least 10 mm thick on each side and shells not less than 5 mm thick:					
1 cell in each block and each block not less than 50 per cent solid ...	—	—	—	100	70
1 cell in each block and each block not less than 30 per cent solid ...	—	—	—	150	—
2 cells in each block and each block not less than 50 per cent solid ...	—	—	200	100	—
2 cells in each block and each block not less than 30 per cent solid ...	—	—	—	150	—
Concrete blocks:					
Plastered at least 10 mm thick on each side and 1 cell in wall thickness:					
Good aggregates ...	—	210	110	70	60
Poor aggregates ...	—	—	—	210	70
Hollow Partitions, steel or timber studding:					
Plaster or metal lathing:					
Portland cement plaster, Portland cement lime plaster or gypsum plaster ...	—	—	—	15	10
Plasterboard with or without gypsum plaster:					
15 mm thick plasterboard on each side ...	—	—	—	—	4 (neat single coat)
10 mm thick plasterboard on each side ...	—	—	—	9	nil.*
25 mm woodwool slab on each side plastered ...	—	—	—	1	1

*Thickness of plaster only.

29.9 FIRE-RESISTANCE OF WOODEN JOIST FLOORS

Under fire, such types of floors should not collapse or deflect more than 1/30 of the span within thirty minutes. The flames should not penetrate through the upper surface within 15 minutes. The average temperature rise should not increase more than 120°C within 15 minutes. The type of ceiling is very important as it protects the joist and also prevents the penetration of flame in-between the wooden boards if they are butt-jointed as there would be excessive gaps present. The fire resistance of a floor depends much on the detailed design of the various elements.

29.10 GENERAL

A building can never be made completely fire-proof. It should have structural elements which can withstand the effects of fire for suitable interval so that adequate protection to the occupants is afforded. Limitations to the height and the floor areas of buildings which are not fire-proof are imposed by the various building codes. Adequate means of escape must be provided if a fire breaks out. The location and design of lifts and stairs should be well worked out. All doors leading to them should be fire-proofed.

QUESTIONS

1. What is meant by fire protection construction? How is the development of fire harmful to the occupants and how is it measured?
2. Give the classification of buildings according to its fire resistance. How will you divide the area into fire zones.
3. Discuss the general requirements of all individual occupancies against fire.
4. Specify the typical fire-fighting installations for the following types of occupancies:
   (*a*) Residential building–lodging and rooming houses above 24 m in height but not exceeding 35 m.
   (*b*) One or two family private dwellings and flats.
5. Write short notes on
   (*a*) Fixed fire-fighting installations
   (*b*) Automatic sprinklers
   (*c*) Smoke venting facilities.
6. What are the different types of exits? Give their general requirements.

REFERENCES

1. **H.M.S.O.**: *Principles of Modern Building.*
2. **Bureau of Indian Standards, New Delhi**: *National Building Code of India 2005—Part 4, Fire and Life Safety.*
3. **Mitchell, G.A.**: *Building Construction, Vol. II.*
4. **Voss, F.**: *Fire-proof Construction.*
5. **Huntington, W.C.**: *Building Construction.*
6. **H.M.S.O.**: *Fire Resistance of Board and Joist Floors for Small Houses and Flats.*
7. **I.S. 1649-1989**: *Code of Practice for Safety of Buildings (General). Details of Construction (first revision).*
8. **Deshpande, R.S. and Vassak G.V.**: *A Treatise on Building Construction.*

Earthquake Protection

30.1 EARTHQUAKE ACCELERATIONS

Earthquakes consist of a set of waves originating from the epicentre *i.e.* the centre of disturbances and causing ground movements or vibrations in all directions. The movements are complicated due to superimposed vibrations. The horizontal vibrations are greater in magnitude than the vertical ones and these are mainly considered in the constructional aspects of buildings. Properly designed and constructed buildings can resist the strongest recorded earthquake shocks.

The horizontal acceleration is expressed as a fraction of the gravitational acceleration and is denoted in terms of Rossi-Forrel scale. Acceleration greater than 0.025 g creates damaging forces. The most destructive forces are caused by horizontal earth motion. When the underneath surface of a building gets moved slightly, the building tends to remain in its original position because of its inertia. The acceleration of the horizontal movement varies and its maximum value is only adopted for measuring the intensity of earthquake forces. If the acceleration of the horizontal earth movement is one-tenth of the acceleration due to gravity, it is assumed that the stresses in the structure caused by the earthquake are similar to those produced by the horizontal static forces of the same magnitude acting on the building.

The earthquake may occur in any direction but the movements can generally be split up into two components, one acting along the main axis of the building and the other perpendicular to it.

As stated earlier, a building is to withstand horizontal seismic acceleration at any plane. In order to obtain the horizontal acceleration the mass or dead weight above the plane under consideration is taken into account and also the effective imposed load above that plane is considered. Every building has its natural period of vibration. When vibrations are introduced in a building by an earthquake having the same period of vibration as the building itself, vibrations get intensified in magnitude and the building is liable to get damaged. The vibration period of earthquakes is about 1 to 2.5 seconds. The self-vibration period of R.C.C. buildings varies from 0.3 to 0.5 seconds for buildings up to 30 m in height; for greater heights the vibration may be similar to those of earthquakes. This means that resonance is likely to occur in buildings which are higher than 30 m and greater heights should, therefore, be avoided.

Table 30.1: Intensities of acceleration due to earthquakes

Rossi-Forrel scale	*Horizontal acceleration*	*Remarks*
1.	0.0005 *g*	Almost imperceptible, being felt only by an experienced observer.
2.	0.0001 *g*	Feeble, being felt by a small number of persons at rest.
3.	0.0025 *g*	Very slight, being felt by several persons at rest.
4.	0.005 *g*	Slight, being felt by several persons in motion and resulting distance of movable objects, doors, windows, etc.
5.	0.01 *g*	Weak, being felt generally by everyone and resulting in disturbance of furniture, etc.
6.	0.0125 *g*	Moderate, resulting in general awakening of those asleep and disturbance of trees and shrubs.
7.	0.025 *g*	Strong, resulting in fall of plaster, no serious damage to buildings but general panic.
8.	0.05 *g*	Very strong, resulting in cracks in walls of buildings, fall of chimneys.
9.	0.1 *g*	Severe, resulting in some buildings being partially or wholly destroyed.
10.	0.3 *g*	Violent, destructive resulting in disasters, landslides, cracks in ground, etc.

India has been divided into four zones: Zone I comprises those areas in which minor damages are caused to the structures in the event of an earthquake; zone III comprises those areas in which moderate but appreciable damages are caused; and zone IV comprises those areas in which servere damages are caused and zone V in which very severe damages are caused. For the purpose of design the acceleration in the four zones is given below:

Seismic zone	*Rossi-Forrel scale*	*Horizontal acceleration*
II	1 to 5	upto 0.01 *g*
III	6 to 7	(0.01 to 0.025) *g*
IV	8 to 9	(0.025 to 0.1) *g*
V	10	(0.1 to 0.3) *g*

The magnitude of the horizontal force is obtained by multiplying the mass of the equivalent dead load by the seismic acceleration enumerated above for the different zones.

The horizontal force at a plane is thus determined by calculating load on the building above the plane. Application of the force is similar to that of horizontal wind loads. For multi-storeyed building the coefficient of seismic acceleration may be reduced by the factor $\frac{4.5}{m + 4.5}$ to allow for increase in flexibility with the increase in the number of storeys. (m is the number of storeys above the one under consideration.)

For parapets and cantilever projections near the exterior walls the seismic coefficient may be taken as 40% *g*. For chimneys projecting above the roof its value may be taken as 20 % *g*. Tall structures standing side by side but not monolithic with each other should have some distance between them. Structures such as retaining walls, which depend for the stability on their weights, should be designed for the same amount of vertical forces as horizontal forces.

The overtaking moment of building should be kept within 50 per cent of its moment of stability.

30.2 GENERAL PRINCIPLES FOR EARTHQUAKE RESISTANT CONSTRUCTION OF BUILDINGS

The general principles, given below, should be observed in the construction of earthquake resistant buildings.

30.2.1 Lightness

Since the earthquake force is a function of mass, the building should be as light as possible consistent with structural safety and functional requirements. Roofs and upper storeys of buildings in particular should be designed as light as possible.

30.2.2 Continuity of Construction

As far as possible, the parts of the building should be tied together in such a manner that the building acts as one unit. Floor slabs over expansion joints should be continuous as far as possible. Concrete slabs should be rigidly connected or integrally cast with the support beams.

Through planes of weakness should be avoided by staggering openings in load bearing walls where possible. Otherwise reinforced concrete framing or R.C. brickwork of adequate design should be provided around the opening.

30.2.3 Projecting and Suspended Parts

Projecting parts should be avoided as far as possible. If they have to be provided, these should be properly reinforced and firmly tied to the main structure.

Ceiling plaster should be avoided. When it is unavoidable, the plaster should be as thin as possible. Suspended ceiling should be avoided as far as possible. Where provided, they should be light and adequately framed and secured.

30.2.4 Shape of Building

In order to minimise torsion, the building should have a simple rectangular plan and be symmetrical both with respect to mass and rigidity so that their centres coincide with each other. For asymmetrical sections in plan, elevation or mass, provision should be made for torsional and other effects due to earthquake forces in the structural design or the parts of different rigidities should be separated through separation or crumple sections. The length of such building between separation sections should not preferably exceed three times the width.

Buildings having plans with shapes like L, T, E and Y should preferably be separated into rectangular parts by providing separation sections, as shown in Figs. 30.1 to 30.5.

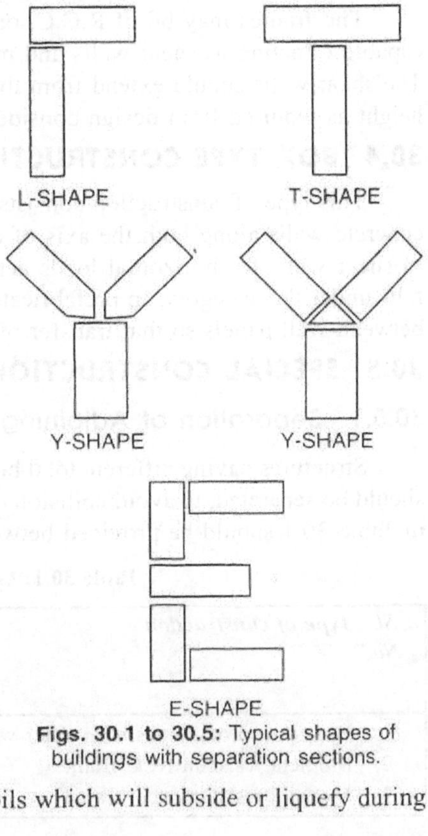

L-SHAPE T-SHAPE Y-SHAPE Y-SHAPE E-SHAPE

Figs. 30.1 to 30.5: Typical shapes of buildings with separation sections.

30.2.5 Strength in Various Directions

The structure should be designed to have adequate strength against earthquake effects along both the horizontal axis. The design should also be safe considering the reversible nature of earthquake forces.

30.2.6 Foundations

The structure should not be founded on such loose soils which will subside or liquefy during an earthquake resulting in large differential settlement.

30.2.7 Ductility

The main structural elements and their connections should be designed to have a ductile failure. This will enable the structure to absorb energy during earthquakes to avoid sudden collapse of the structure.

30.2.8 Damage to Non-structural Parts

The non-structural parts should be connected to the structural framing in such a manner that the deformation of the structural frame leads to a minimum damage of the non-structural elements.

30.2.9 Fire-safety

Fire frequently follows an earthquake and therefore, buildings should be constructed to make them fire resistant.

TYPES OF CONSTRUCTION

The types of construction usually adopted in buildings are as follows:

(*a*) Framed construction, and

(*b*) Box type construction.

30.3 FRAMED CONSTRUCTION

This type of construction is suitable for multi-storeyed and industrial buildings and consists of frames with flexible (hinged) joints and bracing members. Steel multi-storeyed building or industrial frames and timber construction usually are of this type. Such buildings should be adequately strengthened against lateral forces by shear walls and/or other bracing systems in plan, elevation and sections such that earthquake forces shall be resisted by them in any direction.

The frames may be of R.C.C. or steel with semi-rigid or rigid joints. The walls are rigid, capable of acting as shear walls and may be of R.C.C. or brickwork, reinforced or unreinforced. The shear walls should extend from the foundation either to the top of the building or to a lesser height as required from design considerations.

30.4 BOX TYPE CONSTRUCTION

This type of construction consists of prefabricated or *in situ* masonry, concrete or reinforced concrete walls along both the axis of the building. The walls support vertical loads and also act as shear walls for horizontal loads acting in any direction. All traditional masonry construction falls under this category. In prefabricated construction attention should be paid to the connections between wall panels so that transfer of shear between them is ensured.

30.5 SPECIAL CONSTRUCTION FEATURES

30.5.1 Separation of Adjoining Structures

Structures having different total heights or storey heights and different dynamic characteristics should be separated, to avoid collision during an earthquake. The minimum gaps of width specified in Table 30.1 should be provided between them.

Table 30.1: Gap width for adjoining structures

Sl. No.	*Type of construction*	*Gap width/storey in mm*	
		Design seismic coefficient less than 0.08	*Design seismic coefficient 0.08 or more*
1.	Box system or frames with shear walls	7.5	15.0
2.	Moment resistant RCC frame	10.0	20.0
3.	Moment resistant steel frame	15.0	30.0

Note. Minimum total gap shall be 25 mm.

30.5.2 Separation or Crumple Section

In framed construction members should be duplicated on either side of the separation or crumple section. As an alternative, in certain cases such duplication may not be provided, if the portions on either side can act as cantilevers to take the weight of the building and other relevant loads.

Typical details of separation and crumple sections are shown in Figs. 30.6 to 30.12.

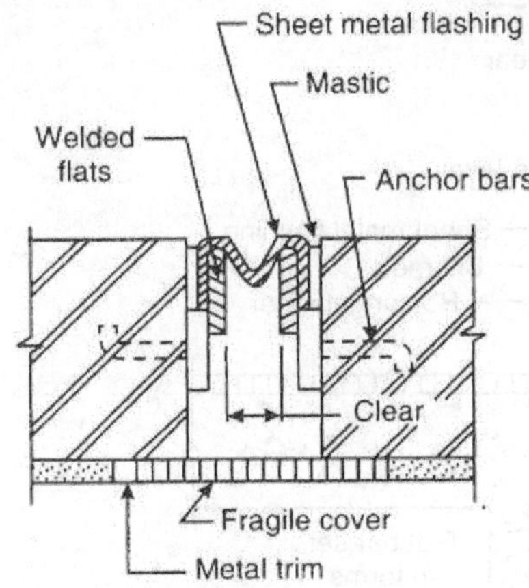

Fig. 30.6: Brick walls.

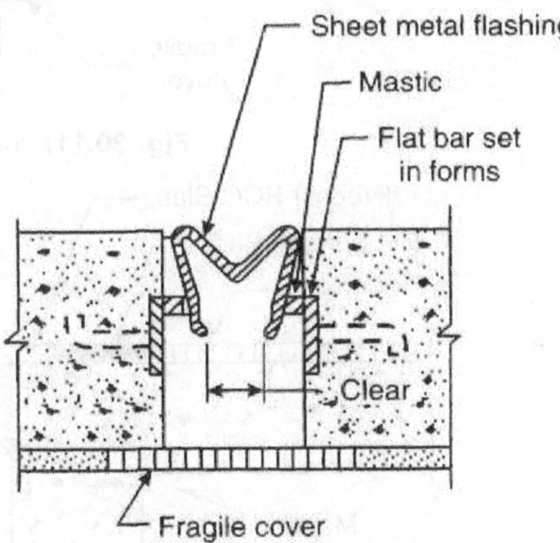

Fig. 30.7: Concrete walls.

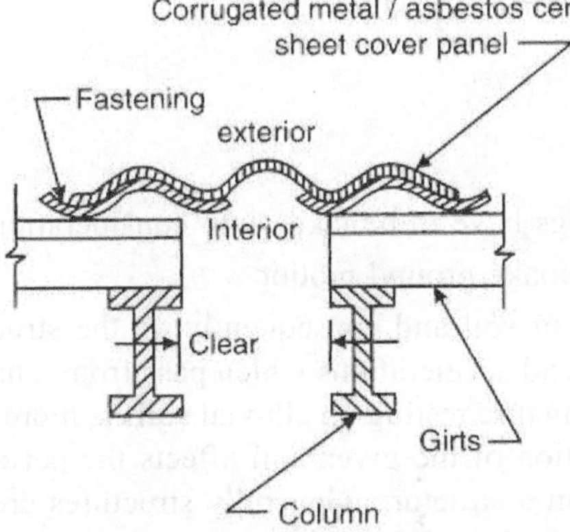

Fig. 30.8: Metal sliding industrial work.

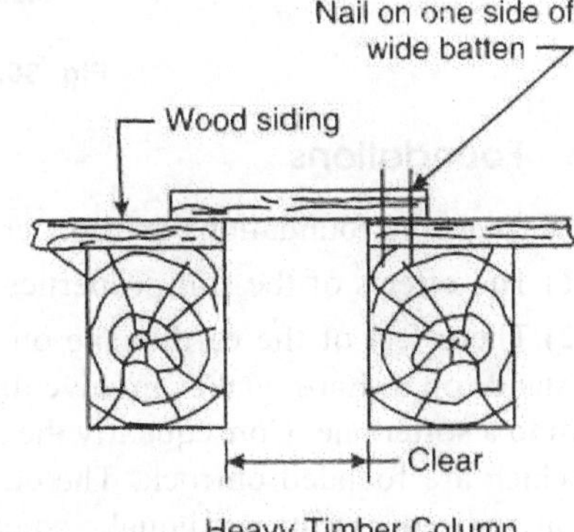

Fig. 30.9: Wood sheathing industrial work.

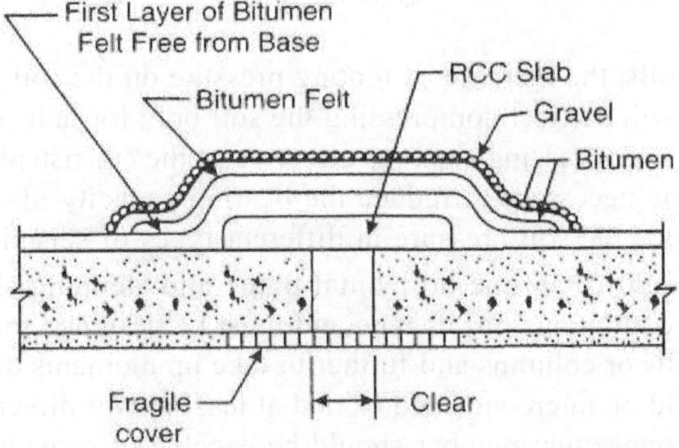

Fig. 30.10: RCC slab on roof surface.

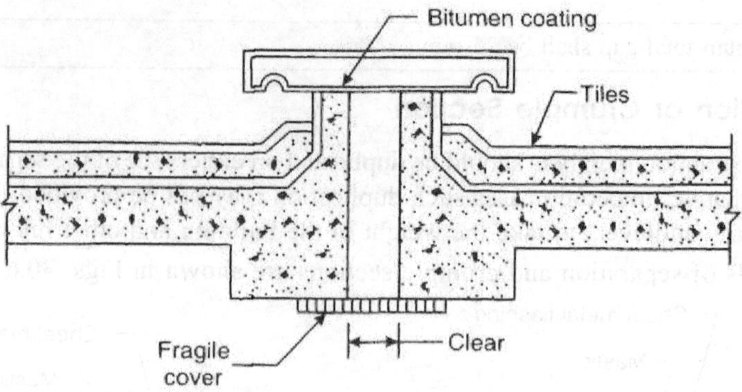

Fig. 30.11: Joint at terrace level.

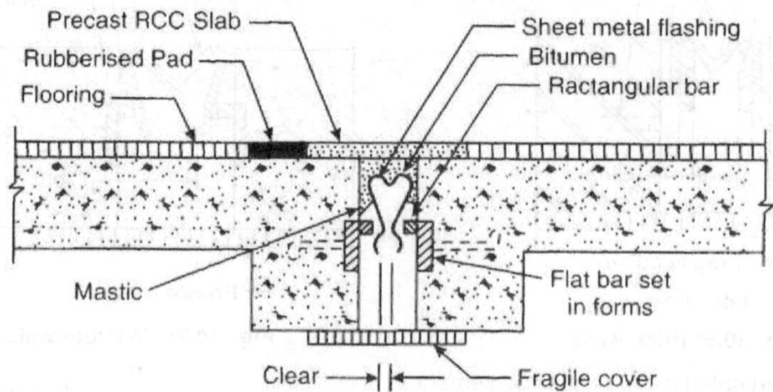

Fig. 30.12: Joint at floor level.

30.5.3 Foundations

As far as the foundations are concerned, two things have to be taken into consideration:

(1) The effects of the soil properties on the earthquake ground motion.

(2) The effect of the earthquake on the structure of soil and consequently on the structure which stands on it. Earthquakes involve displacements and accelerations which pass from a harder medium to a softer one. Consequently the shaking of structure resting on alluvial soils is more than those which are founded on rock. The elastic deformation of the given soil affects the period of vibration and reduces the earthquake stresses induced in a structure. Generally structures erected on alluvial soil need special attention. Effects of earthquake on firm soils are negligible. Loose soil deposits or soils which are heavily charged with water may be consolidated by the vibrations set up by an earthquake.

In the case of firm soils, the increase of footing pressure on the soil lasts momentarily as the vibration of the structure will be over compressing the soil only for half of the period of vibration at a time. This temporary over-pressing is not as effective as the consistent over-stress due to static loads. Hence it may not be necessary to reduce the bearing capacity of soil in all cases. But the general practice is to reduce the soil pressure in different types of seismic areas.

The foundations should be in one horizontal plane and stepping should not be permitted. They may be of isolated, continuous or raft type and must be designed to resist in every direction the moments set up in walls or columns and further to take up moments due to earthquake effects. Isolated foundations should be interconnected or tied at least in two directions to form a complete foundation frame. Each connecting member should be capable of resisting direct compression or tension. Solid concrete raft, where provided, should be cast integrally with ribs and beams.

30.5.4 Roofs and Floors

Flat roof or floor should not preferably be made of terrace or ordinary bricks supported on steel, timber or reinforced concrete joints nor they should be of a type which is likely to be loosened and parts or all of which may fall, in the event of an earthquake.

For pitched roofs, corrugated iron or asbestos sheets should be used in preference to country, Allahabad or Mangalore tiles or other loose roofing units. All roofing materials should be properly tied to the supporting members. Heavy roofing materials should generally be avoided.

All roof trusses should be supported on RCC or reinforced brick work. The holding down bolts should have adequate length.

Jack arched roofs or floors, where used, should be provided with mild steel ties in the end spans. Where the number of spans is large, such ties should also be provided in every fourth span.

30.5.5 Staircases

The interconnection of the stairs with the adjacent floors should be appropriately treated by either providing sliding joints at the stairs to eliminate their bracing effect on the floors or the design and construction should be such as to afford adequate strength to the stairs to transmit shear between the adjacent floors. Large stair halls should preferably be separated from the rest of the building by means of separation or crumple sections.

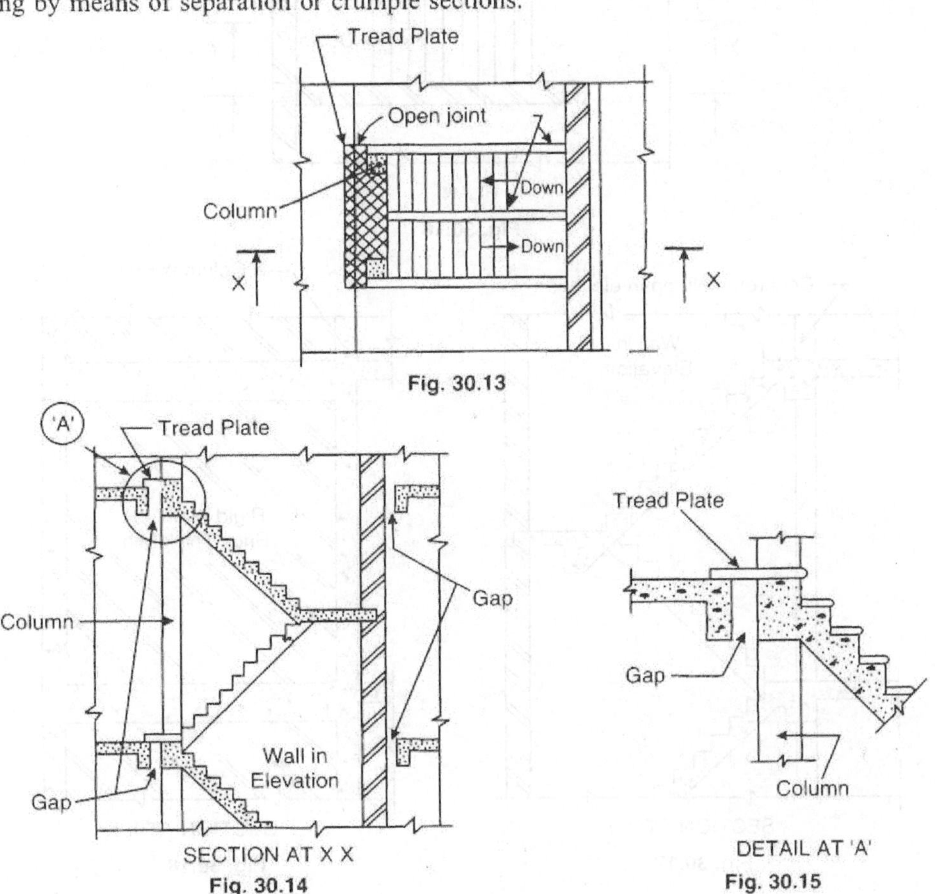

Fig. 30.13

Fig. 30.14

Fig. 30.15

Figs. 30.13 to 3.15: Separated staircase.

Three types of stair construction can be adopted; namely:

(*a*) ***Separated staircases***: One end of the staircase rests on a wall and the other end is carried by columns and beams which have no connection with the floors. A typical example is shown in Figs. 30.13 to 30.15.

(*b*) ***Built-in staircases***: When stairs are built monolithically with floors, they can be protected against damage by providing rigid walls at the stair opening. An arrangement in which the staircase is enclosed by two walls is given in Figs. 30.16 to 30.18.

(*c*) ***Staircases with sliding joints***: In case it is not possible to provide rigid walls around stair openings for built-in staircase or to adopt the separated staircases, the staircases should have sliding joints so that they will not act as diagonal bracing.

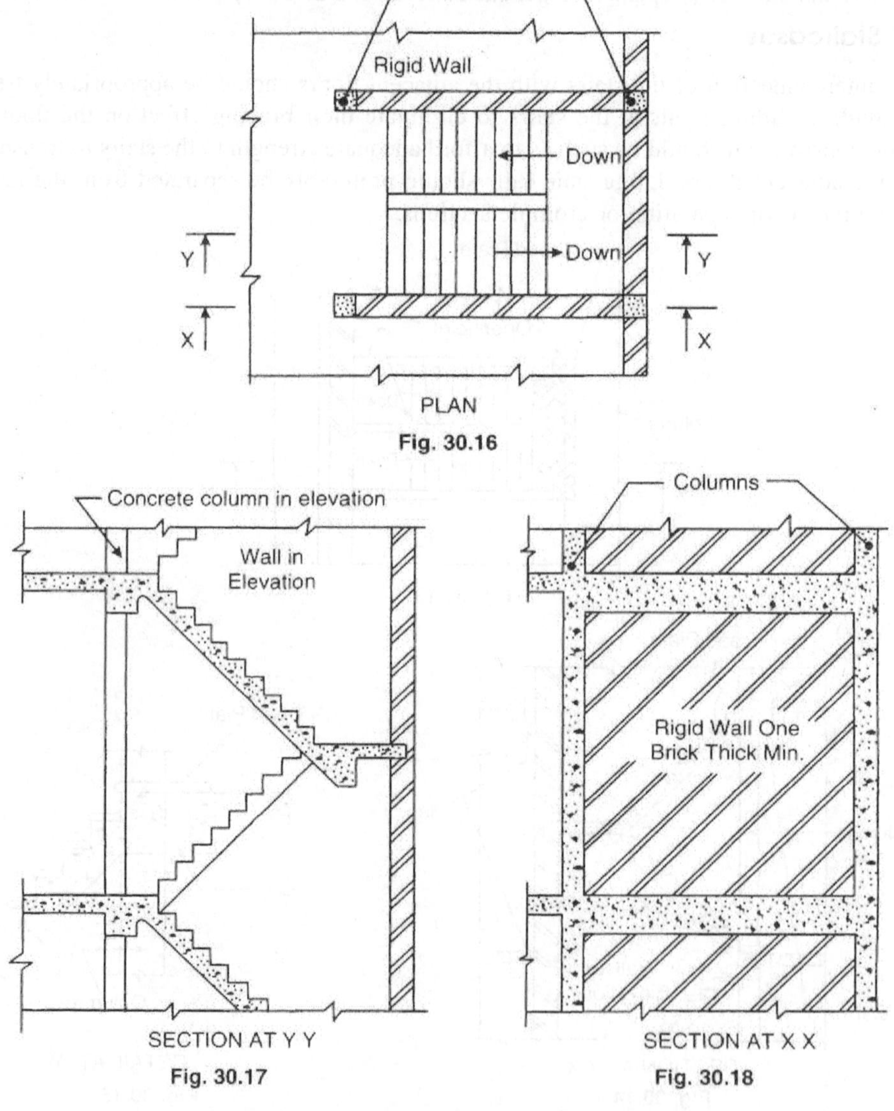

Fig. 30.16

Fig. 30.17

Fig. 30.18

Figs. 30.16 to 30.18: Rigidly built-in staircase.

QUESTIONS

1. How does earthquake affect the safety of buildings? How will you take into consideration the earthquake forces?
2. Discuss the general principles involved in the earthquake resistant design of buildings.
3. What are the types of construction normally adopted for earthquake resistant buildings?
4. What special construction features will you recommend for:
   (*a*) Separation of adjoining structures;
   (*b*) Foundations;
   (*c*) Roofs and floors; and
   (*d*) Staircases.

REFERENCES

1. **Bureau of Indian Standards IS: 4326–1976**: *Code of Practice for Earthquake Resistant Design and Construction of Buildings.*
2. ***Concrete Association of India***: *Earthquake Resisting Buildings.*
3. **Jaikrishna, Dr.**: *A Seismic Design of Structures in India.*

Index